D0849721

Also by Lawrence H. Aller

ASTROPHYSICS

Nuclear Transformations, Stellar Interiors, and Nebulae

ASTROPHYSICS

The Atmospheres of the Sun and Stars

By

LAWRENCE H. ALLER

UNIVERSITY OF CALIFORNIA, LOS ANGELES

SECOND EDITION

THE RONALD PRESS COMPANY · NEW YORK

3
VR-VR

Library of Congress Catalog Card Number 63-11836
PRINTED IN THE UNITED STATES OF AMERICA

To My Students and Colleagues here and abroad, whose help and encouragement have made this book possible

PREFACE

Despite a growing number of excellent monographs, reviews, and compendia covering various branches of astronomy and astrophysics, research workers as well as students and teachers of astronomy and physics have need for a book covering, not only fundamentals necessary for an understanding of the field, but also modern developments. This volume is intended for those who wish to learn something of methods employed for studies of the atmospheres of the sun and stars, spectroscopic properties of incandescent gases, and the influence of the sun upon the earth and its environment.

After a brief astronomical introduction, essentials of physics necessary for a study of stellar atmospheres and other branches of astrophysics are given in some detail. Rapid advances in astrophysics and allied sciences have made necessary a complete revision of the First Edition. Particularly noteworthy has been the progress in radio astronomy, plasma physics, and space astronomy. In particular, the articles on basic physics, solar phenomena, and solar-terrestrial relationships have been generously expanded.

Plasma physics utilizes a number of well-established astrophysical techniques to obtain the temperature and density of an incandescent gas from its spectrum. Workers in this field may find the material on atomic and molecular spectra, the gaseous state, radiation emission and absorption, and strengths and breadths of spectral lines especially useful. The final two chapters in the book are intended primarily for those interested in the outer layers of the sun and their influence on the environment of the earth.

Throughout this book I have tried to stress not merely the results but also the methods by which they are obtained. The reader is shown how each principle or important formula is applied to some definite numerical problems concerned with the interpretation of stellar atmospheres or radiating plasmas. For example, I have shown in detail how the energy flux of a star may be calculated and compared with observations, how the chemical composition of a stellar atmosphere may be deduced from its spectral lines by the curve of growth technique or by line profile analysis. In the largely descriptive chapters on solar phenomena, use is made of numerous excellent photographs obtained in France, Australia, Japan, and the United States. Considerable original material is included.

Special emphasis has been placed on current problems and the important role the stratoscope, rockets, satellites, and other tools of space research can play in solving them. It is essential to realize that astrophysics is very much of a science "in the making."

LAWRENCE H. ALLER

Los Angeles, California
June, 1963

ACKNOWLEDGMENTS

It would have been impossible to produce a book of this scope without the enthusiastic cooperation of many astronomers, students, and former students who have offered helpful comments, suggestions, illustrations, and material in advance of publication. Valuable advice and illustrations for the Second Edition have been supplied by H. Arp, R. G. Athay, H. W. Babcock, A. H. Barrett, D. E. Billings, K. H. Böhm, E. Böhm-Vitense, P. Boyce, R. J. Bray, D. Chalonge, J. W. Chamberlain, S. Chapman, A. Cowley, C. Cowley, J. P. Cox, W. C. DeMarcus, H. R. Dickel, R. B. Dunn, F. Edmonds, G. Elwert, H. Friedmann, R. G. Giovanelli, L. Goldberg, J. L. Greenstein, H. Griem, G. Herzberg, R. Hobbs, H. S. Hogg, H. G. Horak, R. Howard, J. Jugaku, C. C. Kiess, I. King, R. B. King, H. von Kluber, A. C. Kolb, R. Kraft, J. D. Kraus, G. Kron, M. Krook, M. Kundu, W. Liller, R. E. Loughhead, D. B. McLaughlin, D. H. McNamara, S. Matsushima, R. Michard, S. N. Milford, J. Milligan, P. Millman, W. Mitchell, D. Mugglestone, P. Mutschlecner, G. Newkirk, E. P. Ney, E. N. Parker, N. R. Parsons, J. Pawsey, D. M. Popper, M. Schwarzschild, J. A. Simpson, H. J. Smith, E. Spiegel, P. Stanger, T. Stecher, J. Strong, Z. Suemoto, R. G. Teske, R. N. Thomas, R. Tousey, A. B. Underhill, E. Upton, H. C. van de Hulst, M. S. Vardya, G. Wallerstein, C. Warwick, J. Warwick, D. Wentzel, B. Westerlund, P. Wild, R. Williamson, B. Bell, and K. O. Wright.

Special thanks are due to my Michigan colleagues, Orren Mohler, Helen Dodson Prince, William Howard, and F. T. Haddock, for critical comments on the manuscript and for illustrations, and to Mrs. Dorothy Mayer for her careful typing of the manuscript. The material on statistical mechanics has been adapted from notes on the subject prepared by Professor Menzel for his course in astrophysics at Harvard University.

The Observatory
University of Michigan
August 31, 1962

CONTENTS

LIST OF PHYSICAL CONSTANTS

Velocity of light	$c = 2.99773 \times 10^{10}$ cm sec^{-1}
Planck's constant	$h = 6.6252 \times 10^{-27}$ erg sec
Electron mass	$m = 9.1083 \times 10^{-28}$ gram
Electronic charge	$\varepsilon = 4.8029 \times 10^{-10}$ esu
Boltzmann's constant	$k = 1.38044 \times 10^{-16}$ erg deg^{-1}
Volume of a mole	22.4146×10^{3} cm^{3}
Gas constant per mole	$R_0 = 8.31436 \times 10^{7}$ erg mol^{-1} deg^{-1}
Avogadro's number (number of atoms or molecules per mole)	$N_0 = 6.0251 \times 10^{23}$
Loschmidt's number	$n_0 = 2.68731 \times 10^{19}$ cm^{-3}
Density of oxygen gas (0°C)	1.429×10^{-3} gram cm^{-3}
Radius of first Bohr orbit	$a_0 = 0.529172 \times 10^{-8}$ cm
Stefan-Boltzmann constant	$\sigma = 5.6687 \times 10^{-5}$ erg cm^{-2} deg^{-4} sec^{-1}
Second radiation constant	$hc/k = 1.4388$ cm deg
Wien displacement law constant	$\lambda_m T = 0.289782$ cm deg
Bohr magneton	$\mu_0 = \varepsilon h/4\pi mc = 0.92731 \times 10^{-20}$ erg gauss^{-1}
Ratio: mass proton/mass electron	$M_p/m = 1836.12$
Mass of proton	$M_p = 1.67239 \times 10^{-24}$ grams
Mass of hydrogen atom	$M_H = 1.6734 \times 10^{-24}$ grams
Wavelength associated with 1 ev (electron volt)	$\lambda_0 = 12397.7 \times 10^{-8}$ cm
Frequency associated with 1 ev	$\nu = 2.41814 \times 10^{14}$ sec^{-1}
Energy associated with 1 ev	$E = 1.60206 \times 10^{-12}$ erg
Conversion factor atomic mass units to Mev	1 amu = 931.04 Mev
Energy equivalent of electron mass	$mc^2 = 0.51097$ Mev
Constant of gravitation	$G = 6.670 \times 10^{-8}$ dynes cm^2 gram^{-2}
Standard atmosphere (pressure)	1,013,246 dynes cm^{-2} atmosphere^{-1}
Melting point of ice	273.16°K
Mechanical equivalent of heat	4.185 joules calorie^{-1}
Acceleration of gravity	$g_0 = 980.665$ cm sec^{-2}

The numerical values are taken from the data of R. T. Birge, *Phys. Rev.* **60,** 176, 1941; from J. W. M. DuMond and E. R. Cohen, *Rev. Mod. Phys.* **20,** 82, 1948; and from E. U. Condon and H. Odishaw (eds.), *Handbook of Physics* (New York: McGraw-Hill Book Co., Inc., 1958), Sec. 7, 143.

ASTROPHYSICS

The Atmospheres of the Sun and Stars

CHAPTER 1

A Brief Survey of Basic Data

1–1. Scope and Nature of Astrophysical Problems. Astronomy differs from other physical sciences in that most of the phenomena with which it deals cannot be handled by direct experimental methods. Stars, nebulae, planets, and even the moon (at the present epoch) are known only by means of the radiation they emit, absorb, or reflect. Unfortunately, much of this radiation cannot be observed at the surface of the earth.

Fig. 1–1A shows the electromagnetic spectrum from the domain of x-rays to the long-wave radio broadcast band. Wavelengths are usually expressed in angstroms (1 angstrom unit = 10^{-8} cm). Observations by an earth-bound astronomer must be made through two "windows," one in the optical region, the other in the radio-frequency region, neither of which is entirely clear of atmospheric obstruction.

Virtually all radiation shortward of 2.9×10^{-5} cm (2900 angstroms, written as 2900A or λ2900), that is, frequencies greater than 1.03×10^{15} cps, are extinguished in the earth's atmosphere. Absorptions by molecules of O_2 and N_2 and by atoms of O and N (which are produced by photodissociation in the higher atmosphere) takes place at shorter wavelengths (< 2400A). At heights below 100 km, ozone is formed in appreciable quantities from a combination of the free oxygen atoms with the O_2 molecules. Ozone has a few bands longward of (i.e., wavelengths greater than) 3000A, but the extinction is nearly total below 2900A. The opacity is a maximum at 2550A and then falls off to a small value near 2200A, but before this wavelength is reached other atmospheric gases produce a total extinction. By firing rockets above the ozone layer, it is possible to observe the ultraviolet spectrum of the sun (see Chapter 9).

In the infrared occur strong absorptions by O_3, CO_2, N_2O and particularly water vapor. Fig. 1–1B shows the optical window on an enlarged scale. Although the bands are depicted in the figure as continuous structures, they actually consist of numerous, discrete, more or less closely packed lines, between which it is occasionally possible to see radiations of extra-terrestrial origin. Notice that the bulk of the solar radiation falls in a relatively clear optical window, whereas the heat radiated by a planet at the temperature of the earth would be largely obstructed by water-vapor absorption. For an account of the earth's atmosphere and absorptions it produces, see, e.g., Goldberg (1954).

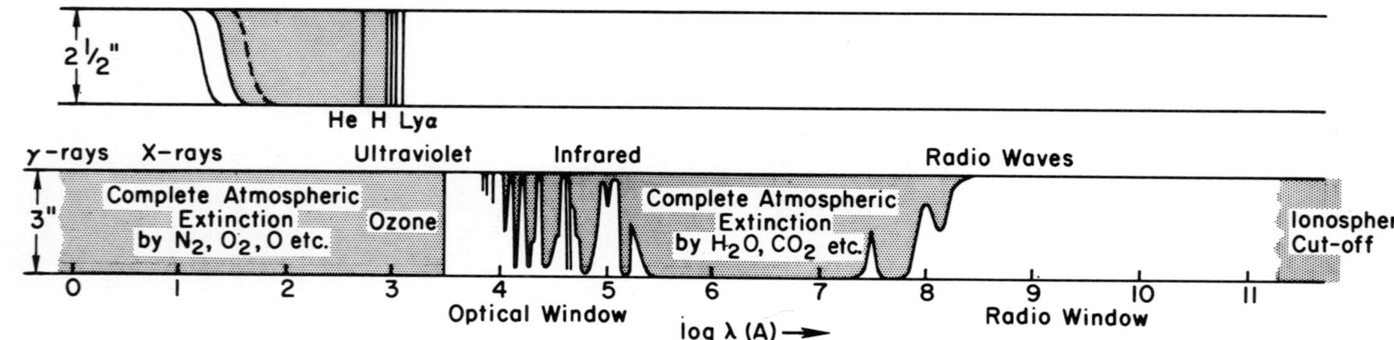

FIG. 1–1A. THE ELECTROMAGNETIC SPECTRUM

The electromagnetic spectrum is exhibited on a logarithmic scale from γ-rays and x-rays to the radio spectrum. The abscissa is log λ (angstroms). In the lower strip the positions of the atmospheric windows are indicated. The top strip depicts the range accessible to detectors flown in a satellite above the earth's atmosphere. The radiation from a distant star is absorbed by the hydrogen Lyman lines and continuum and by the helium absorption produced by the interstellar medium. No such blocking is produced for radiation from the sun.

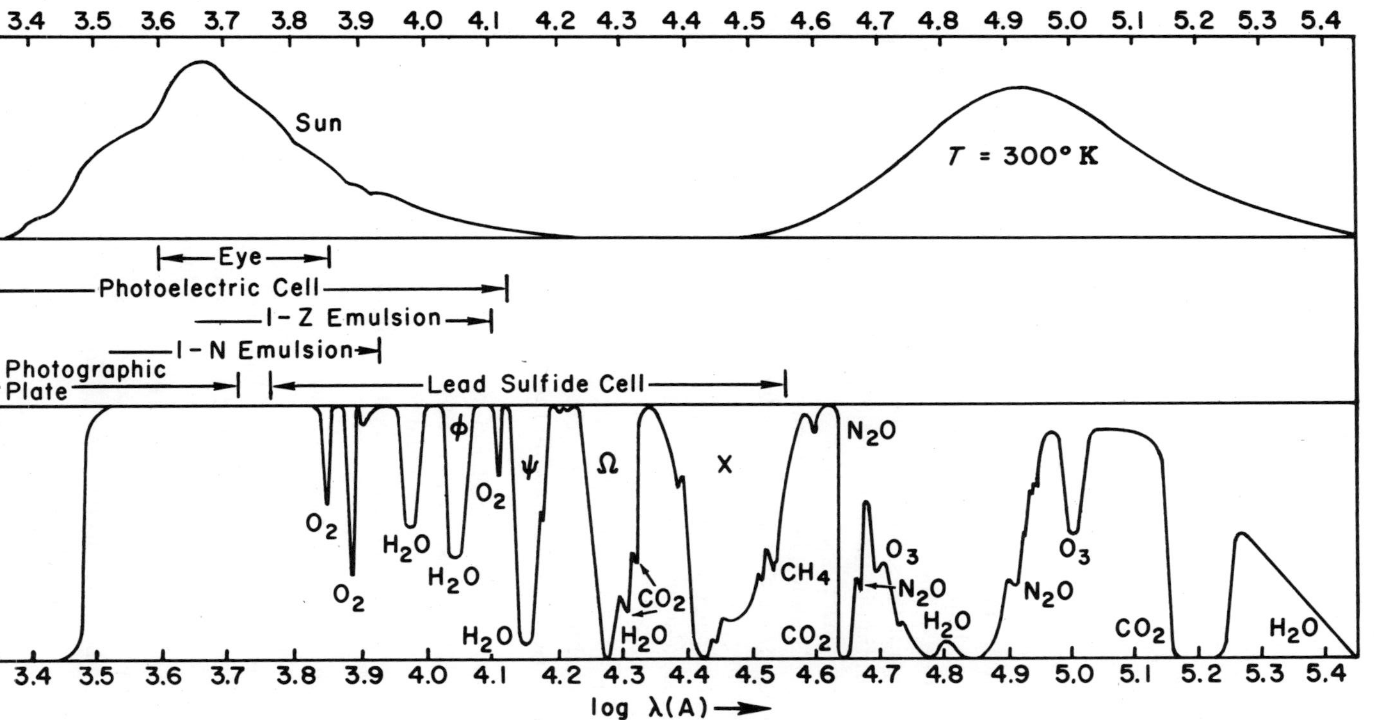

Fig. 1–1b. The Optical Window in the Earth's Atmosphere

The region from 0.2μ to 25μ is shown in greater detail. The infrared-band absorption due to water vapor and other atmospheric constituents, here depicted as continuous, actually consists of discrete molecular lines. In some spectral regions, these lines overlap so badly as to produce a complete extinction of the outgoing radiation. The approximate spectral ranges of certain radiation detectors is exhibited (see also Fig. 1–4). Thermocouples, bolometers, and Golay detectors are sensitive throughout the entire region.

The top strip shows the *relative* energy distributions in the solar spectrum and in a black body at 300°K (which is in the range of planetary temperatures).

Whereas astronomy and astrophysics developed with the aid of only the optical window, the exploitation of the radio window, opened by K. G. Jansky in 1933, has already yielded enormous dividends. The radio window extends over more octaves than does the optical window. Its high-frequency end is limited by water-vapor absorption, the low-frequency end by the ionosphere. The latter cut-off varies in frequency and sometimes permits observations to frequencies as low as 5 mc.

The top strip in Fig. 1–1A illustrates one reason why the astrophysicist is eager to obtain observations from satellites and space stations. Although an important region between about 20A–912A in the radiation of distant stars and nebulae is still obstructed by continuous hydrogen and helium absorption produced by an interstellar medium, we should be able to detect x-rays and the rest of the electromagnetic spectrum. Fortunately, the radiation of the sun can be studied throughout its entire spectral range, although if the satellite flies too low there may be trouble with residual line absorption in the 100A–1000A range produced by gases high in the earth's atmosphere.

Nor is the earth's atmosphere the only obstacle; we can observe only the outermost layers of a star. The interior lies concealed from view; conditions there may be inferred only with the aid of universal physical laws. Stars and nebulae tell us what they please; skilled interpretation of more or less slender clues is required to obtain additional knowledge. The problem is to apply known physical laws to interpret observational data. Thus we may learn the structure, temperature, and composition of stellar atmospheres, the state of stellar interiors, and the conditions prevailing in the gaseous nebulae and in the interstellar and interplanetary media.

Let us describe briefly the character of our observational data: positions of stars and nebulae may be measured upon the celestial sphere with a high order of precision, those of radio-frequency sources with a lesser precision, such that it is sometimes difficult to identify them with optically recognizable objects.

Apparent brightnesses of stars and nebulae can be measured with various light-detecting devices such as the eye, photographic plate, or photoconductive cell or photoelectric cell. The fluxes from certain nebulae and stellar systems can also be measured in the radio-frequency region. To be able to convert apparent luminosity to true luminosity, one must know the distance of the star or nebula. Visual double stars of established distances and orbital periods provide most of our knowledge of stellar masses. Stellar radii and densities may be found for the components of eclipsing binaries.

The physical nature and chemical compositions of stellar atmospheres may be deduced from their continuous and dark-line spectra.

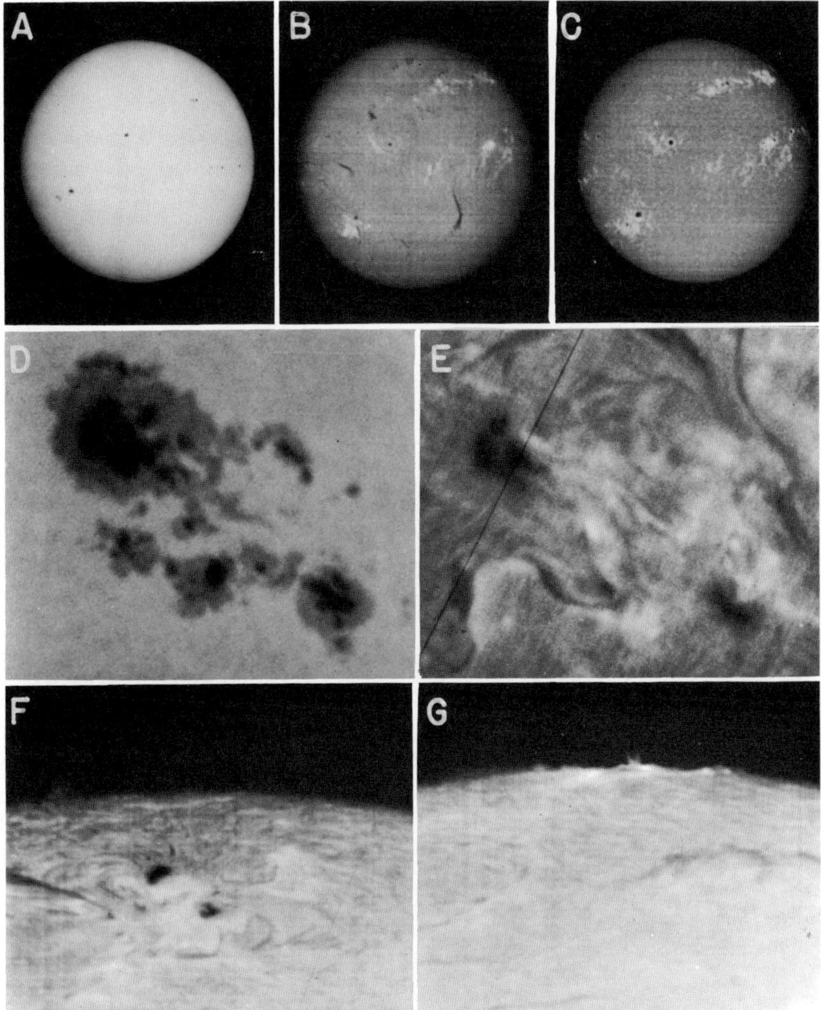

FIG. 1–2. PHOTOGRAPHS OF SOLAR DISK PHENOMENA

A. Direct or white light photograph of solar disk (August 18, 1939).

B. Photograph of disk in the light of the red line of hydrogen. (Hα spectroheliogram.)

C. Photograph of disk in the light of ionized calcium (λ3933 of Ca II). (Ca II or "*K*" line spectroheliogram.)

Notice the dark filaments upon the disk in the Hα image and the bright areas (plages) in the vicinity of spots on both the Hα and the Ca II photographs.

D. Direct photograph of sunspot group of September 1, 1939.

E. Hα spectroheliogram of the same sunspot group.

F. Hα spectroheliograms of a sunspot group approaching the limb of the sun, September 14, 1939.

G. The same sunspot group observed on September 16, 1939.

(McMath-Hulbert Observatory, University of Michigan.)

Much more detailed information may be obtained for the sun. Direct photographs show granules, spots, and faculae. Spectroheliograms show plages, flares, and prominences (see Fig. 1–2). The structure of the inner corona may be studied with the coronagraph. Active areas near or above sunspots also may be studied in the radio-frequency region.

Gaseous and diffuse nebulae and the interstellar medium present special problems of observation and interpretation both in the radio-frequency and optical regions. One may measure the shapes and surface brightnesses of nebulae in both the continuous and discrete line radiations that they emit. Hot ionized hydrogen clouds are studied in the optical region; cold neutral hydrogen clouds are observed with the 21 cm line in the r-f region. The entire galaxy seems to be enveloped in a vast, low density halo of extremely high energy particles that produces a strong r-f emission, although it remains invisible to optical observations.

Among questions which astrophysics seeks to answer are: What are the densities, temperatures, and compositions of stellar atmospheres? How are they constructed? Are the atmospheric strata in balance with gravity like those of the earth and planets, or do atmospheres consist of jets and filaments hurled from the depths of the star, the whole gaseous envelope resembling a vast fountain of heated gases? The sun poses many additional problems, the most fundamental of which is the origin of the solar cycle of 11.5 years and its attendant phenomena, sunspots, flares, and associated prominences. How are sunspots produced? What makes the spectacular flares which cause particles to be accelerated to high energies and hurled from the sun to the accompaniment of high energy radiation such as x-rays, as well as enhanced r-f emission? What is the nature of the interaction between this enhanced solar radiation and resultant streams of particles with the upper atmosphere of the earth or the belt of high-energy particles that surrounds our planet?

The gaseous nebulae, the interstellar medium, and the galactic halo pose numerous riddles. Although the densities, temperatures, masses, and compositions of nebulae that derive their energy from imbedded stars are found in a relatively straightforward way, objects such as the Crab Nebula apparently draw on sources of power not yet identified.

Are the interstellar obscuring clouds composed of dust, droplets, or large chunks of matter—of ice, metals, or silicates? What is the role of the gaseous component? How is the material collected together to form stars? What mechanism produces the high-energy particles that make up the r-f radiation from the halo of this and other galaxies?

Why do stars differ in their chemical compositions; e.g., certain of them contain technetium, an unstable element whose lifetime is very much shorter than that of the earth?

These are but a few of the questions that might be asked. We shall see that some of them can be answered in a fairly satisfactory manner—to others we can supply but the crudest conjectures. The interpretation of the observational data of modern astrophysics requires knowledge of almost all branches of physics, from aerodynamics to nuclear transformations. At the top of our list, we place the theory of the absorption and emission of radiation. We must become familiar with definitions of flux, intensity, and energy density, as well as such topics as Kirchhoff's, Planck's, Wien's, and the Stefan-Boltzmann laws. Furthermore, since we are frequently concerned with the properties of heated gases, we need some acquaintance with the gas kinetic theory.

In some respects, stellar atmospheres are close to thermal equilibrium—the state of affairs that would prevail in a box whose walls were kept at constant temperature. This situation may be studied with the aid of the Boltzmann, ionization, and dissociation equations that are derived from statistical mechanics. Because a stellar atmosphere cannot be in strict thermodynamic equilibrium, these relations should be labeled "handle with care," and we must be prepared to take into account deviations that may occur.

Since we study the behavior of atoms and molecules by means of their spectra, an understanding of atomic and molecular structure is necessary. The arrangement of the energy levels in the atom determines the character of the spectrum. The intensity of a spectral line will depend on (among other factors) the number of atoms capable of absorbing it, and the probability coefficient for the transition. Except at very high temperatures, most of the atoms in any stage of ionization are in the ground level and thus can absorb lines arising from this level (resonance lines). Thus the calcium *H* and *K* lines attain great strengths in the sun. On the other hand, the lines of the more abundant oxygen and carbon are weak in the sun, simply because the observable lines arise from high-energy levels; at the relatively low temperature of the sun, only a few of these atoms are in these levels. The role of transition probabilities is exhibited by the λ3302 and the *D* pairs of sodium; both are resonance lines, but the former are inconspicuous, whereas the latter are strong. This happens because the absorption probability is much larger for the *D* pair. Atomic transition probabilities are usually found experimentally, although some of them can be calculated by quantum mechanics.

The production of the so-called forbidden lines in the spectra of the solar corona, novae, and nebulae, however, depends on the cross-section for collisional excitation of certain metastable levels, whence the atom can return to the ground level with the emission of forbidden lines. These target areas can be found only by difficult quantum mechanical calculations.

The development of radio astronomy has brought out the importance of non-thermal radiation, more specifically, emission produced by the directed motion of charged particles. Examples include the r-f radiation from solar flare bursts and the synchrotron emission from the Crab Nebula.

Magnetic fields play a much more important role in astrophysical phenomena than previously suspected, from the van Allen radiation belt surrounding the earth to the spiral arms of the galaxy. They are found not only in the sunspots where Hale measured fields of thousands of gauss more than half a century ago, but also in the plages (bright floculli areas) and at other places on the solar surface. They appear to control the motions of many prominences, certain coronal streamers, the structures of certain peculiar stars, the shapes of gaseous nebulae, and the motions of the high-energy particles in the galactic halo.

Shock waves, presumably arising from acoustical disturbances moving through gases of varying density, may account for the high temperature of the corona and peculiarities of the atmospheres of supergiant stars. Aerodynamics has been employed to study the interaction of clouds in the interstellar medium.

The solid-state theory has become important to the astrophysicist in connection with the accretion of gas molecules by small crystals in interstellar space, in the analysis of comets, and of the structure of solid cores of planets.

Nuclear physics is involved in connection with processes of energy generation and element building in stars.

1–2. Stellar Distances. Much of the work of the astronomer is devoted to the measurement of stellar masses, radii, and luminosities. In order to make these measurements, it is usually necessary to know the distance.

For the nearer stars, we may measure the distance by the trigonometric method. The distance of a star is related to its *parallax* (the angular radius of the earth's orbit about the sun as seen from the star) by

$$r = \frac{1}{p} \tag{1–1}$$

If p is measured in seconds of arc, r is the distance in *parsecs*. (One parsec is 206,265 times the distance of the earth from the sun, i.e., 206,265 *astronomical units* or 3.085×10^{18} cm.) The trigonometric method fails for the more remote stars, and indirect procedures of a statistical nature must be employed. Among these are the following: (1) analysis of the components of motion (with respect to the sun) in the line of sight (radial velocity expressed in kilometers per second) and perpendicular thereto (proper motion expressed in seconds of arc per year); (2) the association of a star with a nearby moving cluster; (3) membership of the star in a star cluster; (4) dynamical parallaxes (these may be found only for visual binary stars and depend on the existence of a correlation between mass and luminosity); (5) galactic rotation; and (6) membership of the star

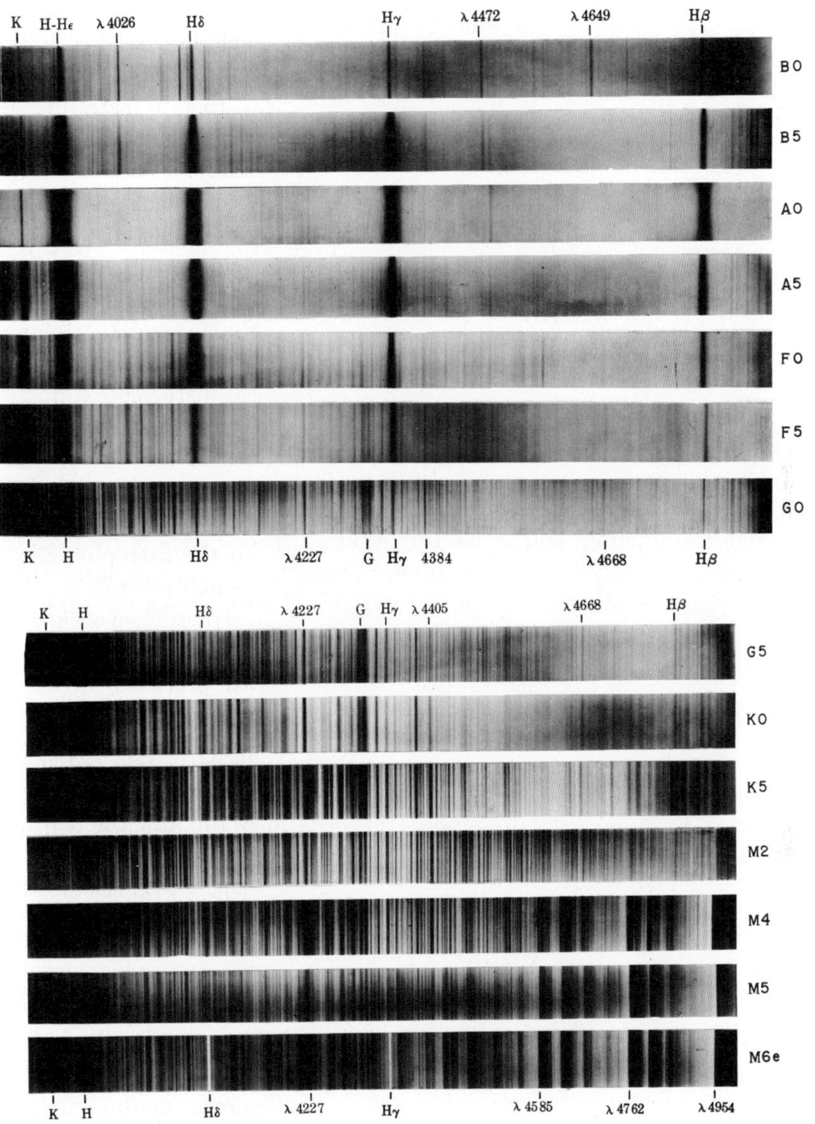

Fig. 1–3. The Spectral Sequence from *B*0 to *M*6

Notice the decline in the intensity of the He I, λ4472 and λ4026 lines from *B*0 to *A*0, the great increase in the strength of the hydrogen lines to a maximum at *A*0 and subsequent decline, and the steady increase in the intensities of the metallic lines; *H* and *K* of ionized calcium (Ca II), λ4227 of neutral calcium (Ca I), and λ4384, λ4668, and λ4405 of neutral iron (Fe I), toward the lower temperatures. The bands of compounds become increasingly conspicuous toward the lower temperatures. The bands of titanium oxide, whose heads (see Chapter 2) fall at λ4585, λ4762, and λ4954, are strong in the *M*-type stars. The last strip, *M*6*e*, is the spectrum of Mira Ceti. (Photographed by R. H. Curtiss and W. C. Rufus of the University of Michigan Observatory.)

in a system in which variable stars or other bright stars, clusters, or nebulae of known luminosity can be recognized.

These methods are described in various astronomical texts (see, e.g., Russell-Dugan-Stewart (1938), Van de Kamp (1952), and McLaughlin (1961) for methods 1, 2, and 4). Method 3 is described in Art. 1–5, and the following qualitative remarks may be made concerning methods 5 and 6.

Our entire stellar system (or galaxy) is in rotation in such a way that stars at different distances from the center move about it with different velocities. Hence differential rotation effects will occur, depending on the distance of the star from the sun and the angle between it and the galactic center. The method is applicable only to stars that move in very nearly circular orbits about the center.

Certain stellar systems, such as the galactic bulge or external galaxies, contain variable stars, particularly those of the RR Lyrae type and "Classical" Cepheids whose intrinsic luminosities are known. Hence by a comparison of the apparent and actual brightnesses the distances of the systems can be found, provided that absorption of light by intervening interstellar particles can be evaluated. Both methods 3 and 5 which can be employed for remote stellar systems require accurate photometry.

Finally, we may mention that the spectrum of a star (see Fig. 1–3) is determined not only by its temperature and chemical composition, but also by its intrinsic luminosity. Hence criteria may be set up for the establishment of spectroscopic parallaxes which must be calibrated, however, by trigonometric and statistical methods (see Chapter 8). This calibration is accurate for dwarf stars like the sun, but becomes difficult for stars of intrinsically high luminosity.

1–3. Stellar Brightnesses, Colors, and Luminosities. The observed brightness of a star depends not only on its distance, total power output, and energy distribution, but also on the sensitivity and color response of the detector (eye, photographic plate, or photoelectric cell plus filter). The effects of atmospheric extinction, etc., can be evaluated by a well-defined set of observations (see Chapter 6). Thus the apparent brightness of a star may be expressed as

$$b = c_0 \int f(\lambda)\, T(\lambda)\, R(\lambda)\, d\lambda \tag{1–2}$$

where $f(\lambda)\, d\lambda$ is the flux of energy in ergs sec^{-1} cm^{-2} per unit wavelength interval outside the earth's atmosphere, $T(\lambda)$ is the transmission of the optical system of the telescope, and $R(\lambda)$ is the response function for the detector. Here c_0 is a constant that depends on the aperture of the telescope and the unit chosen for the particular system of stellar brightnesses.

Stellar brightnesses are expressed in terms of a quantity called a *magnitude.*[1] Magnitude scale is logarithmic, a difference of five magnitudes corresponding to an apparent luminosity ratio of 100. A first magnitude star is $\sqrt[5]{100}$ or 2.512 times brighter than one of the second magnitude which in turn is 2.512 times brighter than a third magnitude star. The magnitude difference between two stars is related to their relative apparent brightnesses b_1 and b_2 by

$$\frac{b_1}{b_2} = (2.512)^{m_2 - m_1} \quad \text{or} \quad m_2 - m_1 = 2.5 \log \frac{b_1}{b_2} \qquad (1\text{–}3)$$

Notice that Eq. 1–3 defines merely the magnitude scale, i.e., the relation between a brightness *ratio* and a magnitude *difference.* We still have to define the zero point of the scale, i.e., just how bright is a star that is called magnitude 1.0. The zero point is defined in terms of a group of stars near the north pole of the sky (north polar sequence) whose zero point was in turn established with respect to the magnitudes in yet older catalogues. Ultimately, the definition of the zero point may be traced back to the catalogues of classical antiquity. Negative as well as positive magnitudes may be defined. On the above-defined (Pogson) scale the apparent photovisual [2] magnitude of the sun is -26.73 ± 0.03 (Kron and Stebbins, 1957).

The magnitude of a star depends on its color or spectral energy distribution and the wavelength sensitivity of the device used to observe it.[3] Light-measuring devices employed by astronomers differ in their response to light of different colors. The eye is most sensitive to the green and relatively insensitive to the violet and deep red (see Fig. 1–4). Ordinary photographic emulsions which are most sensitive from the blue down to the ultraviolet limit of transmission of the atmosphere have been used to determine *photographic* magnitudes.[4] Orthochromatic plates used in conjunction with yellow filters have a color sensitivity roughly similar to that of the eye. Magnitudes determined with such a combination are designated as *photovisual* (*pv*) magnitudes. Red- or infrared-sensitive plates can be employed with appropriate filters to determine *red* and *infrared* magnitudes.

Earlier, photographic plates were widely used to set up magnitude scales for photometric work. The zero point of the two systems was so

[1] Because of the physiological character of the eye, its response to light is roughly logarithmic. A series of lamps arranged by brightness in a geometrical progression will be appreciated by the eye as being arranged in an arithmetical progression of luminosity.

[2] Actually, the photovisual magnitude of the sun is nearly equivalent to the visual magnitude.

[3] For example, Edmondson's red variable, BN Monocerotis, has a *photographic* magnitude of 14, a *photovisual* magnitude of around 9, and a *photored* magnitude of about 7.5.

[4] Such magnitudes were measured originally with reflectors with silvered mirrors or with refractors. With aluminized surfaces, it is necessary to use a filter such as GG1 to cut out the ultraviolet.

adjusted that the mean photographic magnitudes of certain white stars of spectral class $A0$ (see Art. 1–4 and Fig. 1–3) between 5.5 and 6.5 would equal the mean magnitudes, as measured on the visual photometric system

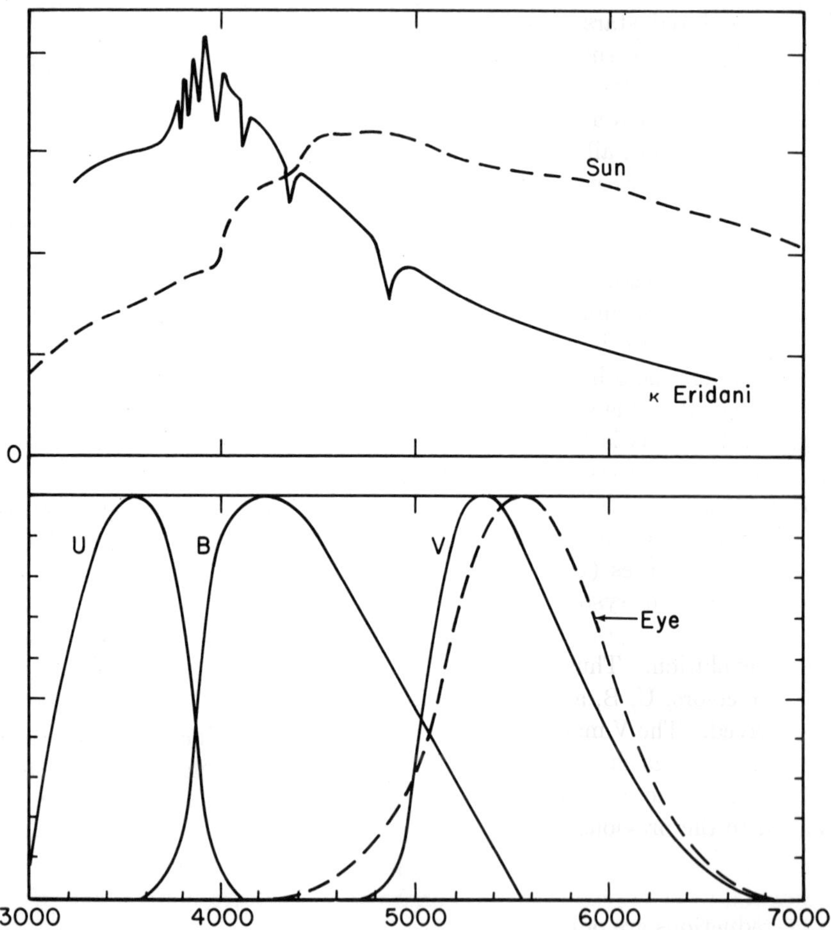

Fig. 1–4. Sensitivity Curves of the Eye and of the UBV Photometric Sytem

Lower: The curves labeled U, B, V give the sensitivity function for the combination of 1P21 photomultiplier tube and appropriate filters as employed by Johnson and Morgan to define their three-color system. For comparison, the sensitivity curve of the eye is also shown.

Upper: Energy distributions in the sun (T ∼ 5800°K) and in a hotter star (T ∼ 10,000°K) are shown as representatives of typical stars. Notice the abrupt change in the energy distribution of κ Eridani near the limit of the Balmer series of hydrogen at λ3650. One advantage of the Johnson-Morgan three-color system is shown in that the U color refers to the region shortward of the Balmer jump, while the B color refers to the region immediately longward of this jump. Some of the older "photographic" magnitudes often straddled the discontinuity, others did not, thus yielding discordances that varied in an erratic manner from one star to another.

established by Pickering at Harvard. Then the difference between the photographic and photovisual magnitude of a star is its international *color index*, viz.:

$$\mathrm{CI} = m_{\mathrm{ptg}} - m_{\mathrm{pv}}$$

Yellow and red stars are fainter photographically than visually. The color index of the sun is +0.53. BN Monocerotis has a color index of about 5 magnitudes. The color indices of blue stars are negative but never by more than a few tenths of a magnitude.

In modern work all magnitude and color standards are set up by photoelectric photometry which can achieve a much greater accuracy. On the other hand, only one star can be observed at a time, whereas a photographic plate can record the images of many stars simultaneously. Hence current photometric procedures employ both techniques, the photocell establishes colors and magnitudes of a selected group of stars, and a photographic plate is used as an interpolation device to get magnitudes and colors for the others.

Through the blue and visual spectral regions, one of the most sensitive photoelectric cells is the RCA type 1P21 multiplier phototube that employs an antimony-caesium surface. It is nearly always used in conjunction with appropriate filters. The sensitivity curves for the three most widely used filter-cell combinations are shown in Fig. 1-4. These are the response curves (U = ultraviolet, B = blue, and V = visual) for the Johnson-Morgan system (1953) of three-color photometry which has proved extremely useful for work on problems of galactic structure and stellar evolution. Thus each observation gives the magnitude of the star in three colors, U, B, and V, from which two color indices, U-B and B-V are derived. The V magnitudes are very close to the old visual magnitudes and may be regarded as essentially equivalent to them except, of course, the photoelectric data have a much higher accuracy. The B-V colors are related to the previously defined color index roughly by [5]

$$\mathrm{CI} = -0.18 + 1.09(\mathrm{B\text{-}V})$$

Such reductions are only approximate; actually the transformation is nonlinear and multivalued. See Chapter 6.

The Johnson-Morgan system has a number of important advantages in that it reduces difficulties caused by the Balmer jump in the older magnitudes (see Fig. 1-4), includes stars of all luminosity and spectral classes well distributed over the sky, and permits one to assess the effects of space reddening.

A star may be red because it is actually cool, or it may be red because of interstellar absorption. A single color measurement cannot determine

[5] If the blue magnitude is measured with a 103a0 plate and GG1 filter, Arp finds that $\mathrm{CI} = -0.29 + 1.18(\mathrm{B\text{-}V})$.

which of these causes is operative, but, as was demonstrated by Becker by the use of three colors, one can evaluate the amount of space reddening.

The zero point of the colors is fixed so that for a selected group of white stars (spectral class $A0$V, see Art. 1–4), which are unaffected by space absorption, the mean U-B and B-V color indices are zero. A star of photoelectric-visual magnitude zero (i.e., V = 0) and zero color index (B-V = 0) has a flux just outside the earth's atmosphere of 3.8×10^{-9} erg cm^{-2} sec^{-1} A^{-1} at λ5560, according to the calibration by Arthur Code (1960).

Among photoelectric magnitude systems, mention must also be made of the six-color photometry of Stebbins and Whitford (1943, 1945) who used a caesium-oxide cell enclosed in a metal-covered refrigerated glass tube that yielded data over the wavelength range λ3300–λ12,500. They used filters to secure observations at effective wavelengths of λ3530, λ4220, λ4880, λ5700, λ7190, and λ10,300. This very broad base line has many advantages; the chief disadvantage is that it is expensive of telescope time. Stellar color measurements amount to evaluations of their energy distributions with broad band-pass filters. It is also possible to work with narrow band-pass filters or to make direct energy distribution measurements (see Chapter 6).

In the infrared, where photoelectric cells and photographic plates become insensitive, photoconductive lead sulfide or lead telluride cells, bolometers, and thermocouples may be employed. In the region from 1 to 3.5μ, the photoconductive cells are about a hundred times more sensitive than the best thermocouple or bolometer. The latter detectors, however, have the advantage that they are sensitive to all radiation in the optical region; their operation depends upon the actual heating of the sensitive element by the stellar radiation. Magnitudes measured by thermocouples are called *radiometric magnitudes.*

For many problems we need a number which characterizes the total amount of energy received from the star at a point just outside the earth's atmosphere. We call this quantity the apparent *bolometric magnitude.* Therefore, one bolometric magnitude corresponds to a step of 4 db in power output. The magnitude measured by the eye, photographic plate, photoelectric cell, photoconductive cell, or thermocouple must be corrected for the fact that the detector may not be sensitive to all wavelengths, and for the absorption in the earth's atmosphere. These corrections are called *bolometric corrections.* (See Chapter 6.)

The apparent brightness of the star as measured by an observer can be converted to its intrinsic luminosity only when the distance of the star is known. We define as its *absolute magnitude,* M, the magnitude the star would have if placed at a distance of 10 parsecs. Let m be the apparent magnitude, M the absolute magnitude, and r the distance of the star in parsecs. Let l refer to the amount of light received by a particular radiation detector from the star at its true distance, and L to that which would be received at the standard distance r_0 of 10 parsecs. Then

$$\frac{l}{L} = \left(\frac{r_0}{r}\right)^2$$

and

$$M - m = 2.5 \log l - 2.5 \log L = 5 \log r_0 - 5 \log r \qquad (1\text{–}4)$$

Since $\log r_0 = 1$, we have

$$M = m + 5 - 5 \log r \qquad (1\text{–}5)$$

The apparent magnitudes of distant stars are often affected by the absorption of light in the interstellar medium. If this absorption amounts to A magnitudes then the star is actually brighter than would be judged from its apparent magnitude and distance. Therefore,

$$M = m + 5 - 5 \log r - A \qquad (1\text{–}6)$$

The absolute magnitude will refer to the same type of magnitude as the system in which the apparent magnitude was measured, visual, photographic, U, B, V, or radiometric. The absolute bolometric magnitude is a measure of the true luminosity of the star.

The absolute photovisual magnitude of the sun is $+4.84$ (Stebbins and Kron, 1957), which means it would be nicely visible to the eye on a clear night at a distance of 10 parsecs. The intrinsically faintest known star, the companion to BD $+ 4° 4048$, discovered by van Biesbroeck, has an absolute magnitude of $+19$; it is a million times fainter than the sun. The brightest stars in our galaxy or the Magellanic clouds may be a million times as bright as the sun. Most of the stars in a volume of space in our part of the galaxy are fainter than the sun, while most of those visible to the eye are actually brighter, since brighter stars are visible at greater distances than fainter ones.

1–4. Stellar Spectra. The surface temperatures, chemical compositions, radial velocities, and even the luminosities of stars may be deduced from their spectra. Most stars show a continuous spectrum upon which are superposed dark absorption lines. This dark-line spectrum is commonly called the *Fraunhofer spectrum.*

Most stars in the solar neighborhood may be grouped into a small number of spectral classes, which form a continuous sequence (see Fig. 1–3). The order of the Harvard spectral sequence is

$$\begin{array}{l} \qquad\qquad\qquad\qquad\quad R - N \\ O - B - A - F - G - K - M \\ \qquad\qquad\qquad\qquad\qquad\quad\; S \end{array}$$

The spectral sequence O to M is continuous, whereas the carbon R and N

stars form one side branch and heavy-metal *S* stars another. Decimal subdivisions are also employed. Thus *F*5 denotes a spectrum about halfway between *F* and *G* in appearance, whereas *B*8 indicates a spectrum closer to *A*0 than to *B*0. On this system the sun is near *G*2 or *G*3 (subdivisions not employed in the Henry Draper Catalogue of stellar spectra but required for refined work).

In classes *O* and *B* both emission-line and absorption-line objects are observed.[6] Prominent in class *O* are the lines of hydrogen, H I; neutral helium, He I; ionized helium, He II, as well as doubly ionized oxygen, nitrogen, and carbon, O III, N III, and C III. In class *B*, the helium, He I, and hydrogen lines strengthen, while ionized helium disappears. Singly ionized carbon, nitrogen, and oxygen are important and the lines of doubly ionized iron, Fe III, and ionized magnesium, Mg II, put in their appearance. In the later *B*'s the helium lines become very weak and lines of singly ionized metals such as Fe II, Cr II, and Ti II appear. The hydrogen lines attain their greatest strength near *A*2, the lines of He I disappear, and the metals gain in prominence. Throughout classes *A* and *F* the hydrogen lines weaken and those of metals strengthen, until near *G*0, the most prominent features are the *H* and *K* lines of ionized calcium, Ca II, the *G-band*, which is a mixture of atomic lines and molecular bands, and lines of common metals, Fe, Ti, Mg, Cr, etc. In class *K* low temperature lines of the metals strengthen, and finally in class *M* the titanium-oxide bands dominate the spectrum.

An important fact about the spectral sequence is that it is also a color sequence. The *O* and *B* stars are intrinsically blue, the *A* and *F* stars are white, the *K* stars orange, and the *M* stars red. In other words, the *O* and *B* stars must be the hottest and the *M* stars the coolest in the spectral sequence. As Saha pointed out in 1922, the spectral sequence from *O* through *G*, *K*, and *M* is essentially a temperature sequence and spectral variations arise from differences in temperature rather than from differences in composition. (For a temperature scale, see Table 6–5.) On the other hand, the *R* and *N* branch comprises stars with strong bands of the carbon compounds; class *S* features bands of ZrO rather than bands of TiO. Classes *R* and *N* branch off between *G* and *K*; class *S* branches off between *K* and *M*. In the later *K* stars, the TiO bands appear and steadily increase in intensity with diminishing temperature through class *M*. Other stars, however, which also merge into the spectral sequence near class *K* show ZrO bands or the bands of the carbon compounds. These spectroscopic differences arise from actual differences in chemical composition; they cannot be accounted for by changes in atmospheric tempera-

[6] A numerically small but spectroscopically interesting group of hot, broad-lined emission-line stars called *Wolf-Rayet stars* are denoted as *W*. *WC* designates these objects with strong carbon and oxygen lines, *WN* those with strong nitrogen lines.

tures and pressures. Furthermore, certain metal deficient stars (often called *subdwarfs*) in which the H/metal ratio is often a hundred times as great as in the sun, cannot be fitted into this spectral classification, while special dichotomies have to be provided for certain peculiar A stars that show abnormally strong lines of various metals, including the rare earths.

1–5. Correlation Between Spectrum and Luminosity. If the absolute magnitudes of the stars in the neighborhood of the sun are plotted against their spectral classes, the points do not scatter all over the diagram as one might expect, but fall in a definite pattern. The resultant plot is called the *Russell diagram* or the *Russell-Hertzsprung diagram,* often abbreviated in the literature as the HR or RH diagram (see Fig. 1–5). Such a plot was first prepared by H. N. Russell in 1913. Most stars cluster along a curve running diagonally from highly luminous blue stars at one corner to dim red ones at the other, color and luminosity changing progressively. This group of stars is called the *main sequence* (or *dwarf sequence*). The sun is a main sequence star. Its absolute magnitude is about 4.8 and its spectral class is $G2$.

The rest of the diagram is more sparsely populated. The *giants* form a group around absolute magnitude 0 to +1.0 extending from about $G0$ to the later (i.e., redder) spectral classes. At the top of the diagram is a thin sprinkling of the highly luminous *supergiants* ($M = -3$ to -8) in which all spectral classes are represented. These stars are rare; their enormous candlepower enables them to be seen at great distances with the result that among naked-eye stars there is a large proportion of supergiants. The group includes the *Classical Cepheids* which have played so important a role in distance determinations for external galaxies.

Beneath the giants and above the dwarfs in spectral classes F to K is the group of *subgiants,* whereas far below the main sequence are the famous "*white dwarfs*" whose sizes are comparable with those of planets but whose masses are similar to that of the sun. Hence their densities exceed that of the sun by many thousandfold. Yet in spite of this fact, or more accurately because of it, the internal structures of these stars appear to be better understood than those of giants or supergiants.

The terms giant, supergiant, and dwarf, originally invoked to express the relative luminosities of the stars, apply to their sizes as well. The emissivity of matter per square centimeter depends primarily on its temperature. Hence a supergiant ten thousand times as luminous as the sun, but of the same surface temperature, must have very close to ten thousand times the surface area or a radius a hundred times as great. The giant Capella has a radius sixteen times that of the sun, while the red supergiant α Herculis, which has ten thousand times the sun's luminosity but a much lower surface temperature, has a diameter 800 times that of the sun. Even

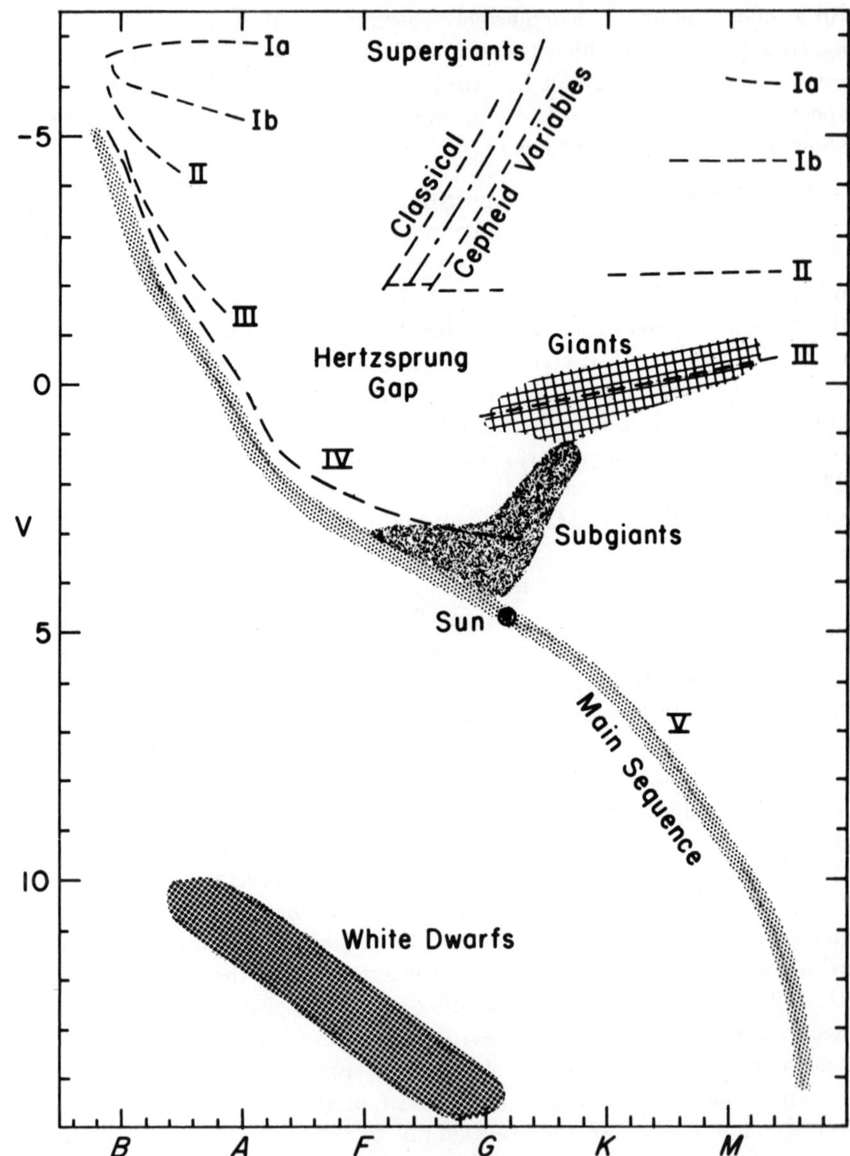

Fig. 1–5. Spectrum-Luminosity Correlation

Absolute visual magnitude V is plotted as ordinate against spectral class as abscissa. The dotted lines labeled I, II, III, IV denote the Morgan-Keenan (MK) luminosity classes. Luminosity class V is the main sequence. The calibration of the luminosity classes in terms of absolute magnitudes is taken from discussions by Morgan and by Johnson and Iriarte. The position of the classical Cepheid variables is that suggested by Sandage.

this star is outdone by VV Cephei whose vast atmosphere would envelop Saturn if its center were placed in the position of the sun.

On the other hand, the most luminous main-sequence stars of whose sizes we have any precise knowledge, the components of the eclipsing system Y Cygni, which are ten thousand times as bright as the sun, have radii only six times that of the sun. Krueger 60A, a typical red dwarf, has a radius about half that of the sun, while its luminosity is 0.017 that of the sun.

Spectra of dwarfs, giants, and supergiants of the same spectral class differ in small but noticeable details. Hence it is possible to assign luminosity classes on the basis of the appearance of the spectrum or from spectrophotometric measurements. The Yerkes system of classification, as set forth in the *Atlas of Stellar Spectra* by Morgan and Keenan (1943), adds a suffix I, II, III, IV, V to the Draper class to denote intrinsic luminosity. Class I is now subdivided into classes Ia and Ib to denote highly luminous and ordinary supergiants, respectively, whereas II represents bright giants. In the earlier spectral classes, III and IV denote objects falling above the main sequence, whereas in later spectral classes, III denotes giants and IV subgiants. Class V always refers to main-sequence stars.

The calibration of the luminosity classes has been attempted by a number of investigators, e.g., Morgan and Keenan (see Hynek, 1951), Johnson and Iriarte (1958), and Petrie (1956). The distances of dwarf stars may be measured directly by the trigonometric method and are comparatively accurate. The distances of giants and supergiants, however, must be established by statistical methods that utilize their motion, membership in clusters, and galactic rotation. These calibrations are difficult, and unsatisfactory discrepancies exist between the results by Petrie and by Morgan.

From a star's intrinsic luminosity (expressed by its absolute magnitude M) and its apparent magnitude m, one can compute its distance modulus

$$y = m - M \tag{1–7}$$

Then, if the amount of space absorption can be estimated from three-color photometry, its parallax may be found from Eq. 1–6. A stellar parallax derived in this way is called a *spectroscopic parallax.* Once these fundamental calibrations have been worked out, spectroscopic parallaxes may be determined more rapidly and, except for the nearest stars, more accurately than the trigonometric parallaxes. One may also obtain a photometric parallax if M is estimated from $(B\text{-}V)_0$, i.e. (B-V) corrected for space absorption.

When absolute magnitude is plotted against spectral type for stars whose trigonometric parallaxes, apparent magnitudes, and spectra are determined, the main sequence shows an appreciable scatter. Some of this

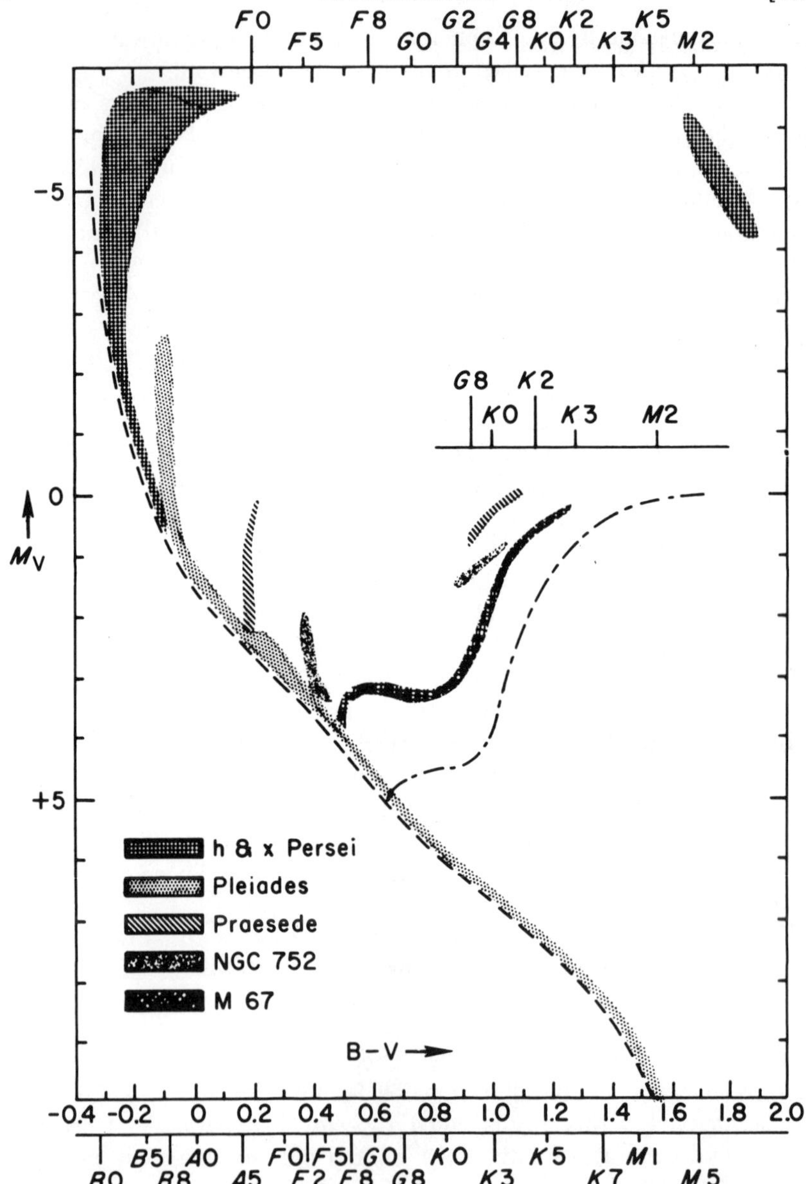

Fig. 1–6. Color Luminosity Relation for Certain Galactic Clusters

Absolute visual magnitude is plotted against (B-V) color index for five representative clusters. The dotted line at the left represents the zero-age main sequence. The dot-dash curve emerging from the main sequence near +5 and curving upward to the right represents a boundary beyond which no stars have been identified. Spectral class equivalents of color on the Morgan-Johnson system are given at the bottom for the main sequence, near the center for giants, and at the top for super giants. (Adapted from studies by A. Sandage, H. L. Johnson, and O. C. Wilson. The intrinsic colors of the supergiants are taken from Kraft and Hiltner, *Ap. J.* **134,** 850, 1961.)

scatter probably arises from observational error but most of it may be intrinsic, due mainly to different ages and chemical compositions of the stars. It is to be noted that when accurate colors and magnitudes of stars in galactic clusters, such as the Pleiades or Hyades, are determined (by photoelectric photometry), the color-magnitude plot gives a well-defined curve with little scatter. Colors may be measured more accurately than spectral types may be estimated.

Following Sandage (1957), let us now compare the color-magnitude arrays for a number of galactic clusters (see Fig. 1–6). The lower parts of main sequences of all galactic clusters are found to coincide very closely with one another, but the upper regions differ. The main sequence in the bright galactic double cluster h and χ Persei reaches above $M = -5$, that of the Pleiades reaches to -2.15, but the portion brighter than $M = 0$

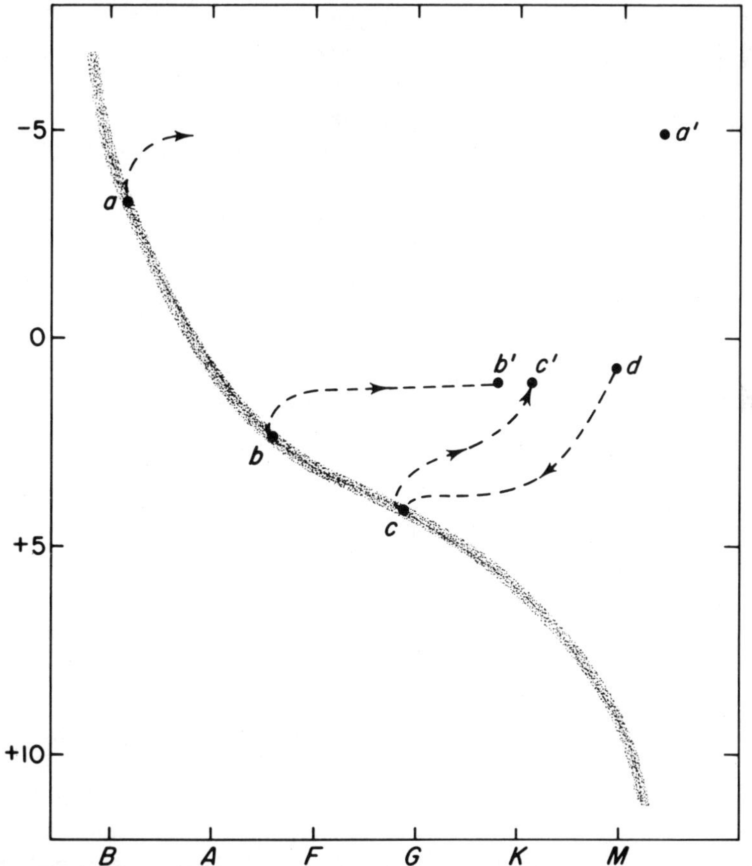

Fig. 1–7. Schematic Evolutionary Tracks of Stars from the Main Sequence

Hypothetical tracks are shown for stars of three different masses. (See Art. 1–5.) The dotted lines are schematic only and do not represent the actual evolutionary track of any star.

deviates from the main sequence defined by h and χ Persei. NGC 752 and Praesepe possess main sequences that deviate at even fainter magnitudes, while the sequence in M67 turns off without a break through the subgiant and into the giant regions. NGC 752 and Praesepe have small giant branches; h and χ Persei has supergiant stars.

The interpretation of Fig. 1–6 is to be sought in terms of stellar evolution. The main sequence represents stars that are shining by converting hydrogen into helium in their cores. When the core hydrogen has all been burned into helium, the star departs from the main sequence to the right and becomes a subgiant, giant, or supergiant, depending on its initial mass. In this stage the hydrogen burning takes place in a shell surrounding the inert core. Fig. 1–7 illustrates the evolutionary argument. A star is formed from the material of the interstellar medium and shines by contraction until it reaches the main sequence (line dc). Then it shines by converting hydrogen into helium in its core. As it does so, it departs from the main sequence, slowly at first, and then more rapidly, until it reaches the giant region at c'. The main sequence star at b passes from the neighborhood of the main sequence across the Hertzsprung gap to the giant region at b', while the much more massive star a evolves quickly to the supergiant phase at a'. The luminosity of a star depends critically upon its mass (see Art. 1–7), brighter stars burning out much more quickly than fainter ones. Thus, h and χ Persei is much younger than the Pleiades and Praesepe is younger than NGC 752, whereas M67 is older than any of the others. Ages are estimated from turning-off points at which the color-luminosity plot for a given cluster leaves the main sequence; e.g., the age of h and χ Persei is a few million years, whereas that of M67 is greater than 7×10^9 years. Accurate age assignments depend on the calculation of good stellar models and evolutionary tracks.

The dotted line in Fig. 1–6 represents the zero age main sequence adopted from Sandage (1957) and Johnson and Iriarte (1958). This represents the locus of all stars that have just started to burn hydrogen into helium after they have reached the main sequence. The dot-dash line represents the boundary of the subgiant region for the solar neighborhood; all such stars lie to the left of this line (Wilson, 1959). The only exception would be stars in the process of formation.

1–6. Color-Luminosity Diagrams for Globular Clusters, Stellar Population Types. The advent of photoelectric photometry made possible the construction of precise color-magnitude arrays or diagrams for globular clusters. See Fig. 1–8. These show a main sequence that does not extend to magnitudes brighter than about $M_v = +4$, an extensive well-populated giant branch that lies considerably above the giant branch of M67, and a horizontal branch around $M_v = +0.5$, the RR Lyrae variables and certain very blue stars of absolute magnitude near zero. Noticeable dif-

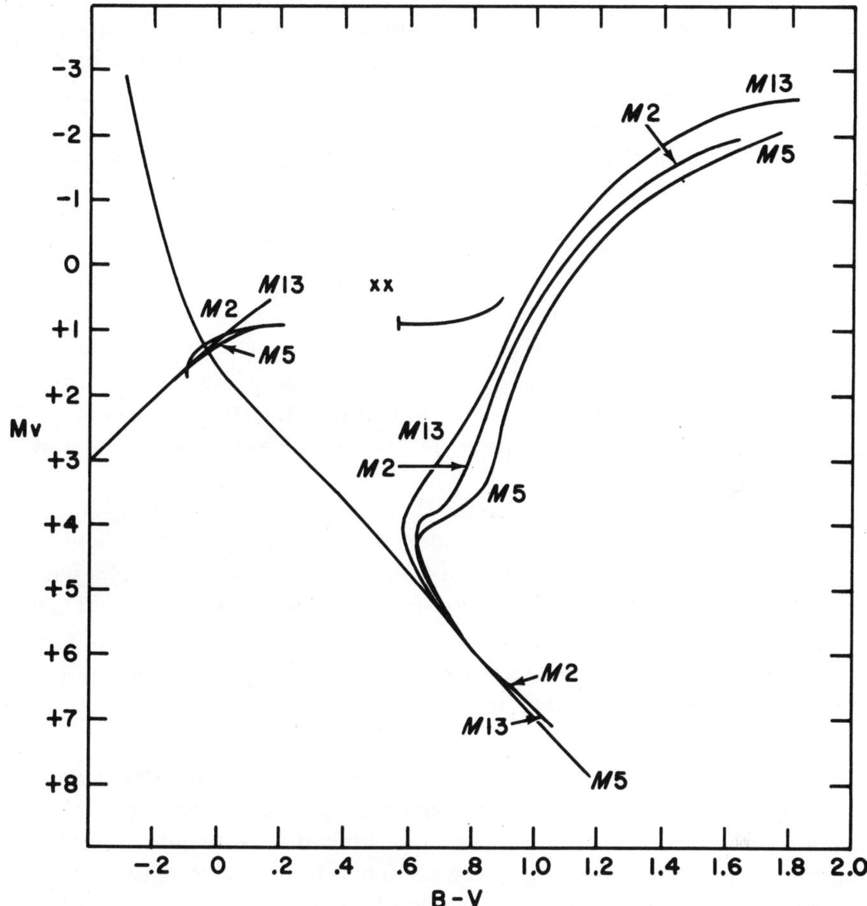

FIG. 1–8. SCHEMATIC COLOR MAGNITUDE DIAGRAM FOR THREE GLOBULAR CLUSTERS

These schematic color-magnitude diagrams for *M*2, *M*5, and *M*13 have been fitted together on the assumption that their main sequences fit the fainter portion of the Hyades main sequence. Some correction has been made for the effects of spectrum lines upon the colors. The composite diagram illustrates the age differences in the clusters and the difference in the RR Lyrae absolute magnitudes in *M*13 relative to *M*2 and *M*5. (Courtesy Halton Arp, *Astron. Jour.* **64**, 445, 1959.)

ferences exist from one globular cluster to another; for example, the giant branches in *M*3 and *M*92 are displaced in color index with respect to one another; some clusters have many variables or blue stars; others have very few. All globular clusters must be certainly very old, since the main sequence does not extend far above the sun's position. On the other hand, *M*67 is comparable, yet it shows a different color-luminosity diagram. The most essential difference between the globular and galactic clusters is to be traced to the differences in their chemical compositions. The galactic clusters have a metal to hydrogen ratio much larger than do the globular clusters and seem to contain greater quantities of elements like C, N, O, Si, S as well. The chemical composition difference affects not only the color and appearance of the spectrum, but the very structure of the star itself and the course of its evolution.

The globular clusters seem to be made up of these ancient metal-deficient stars, which Baade (1944) called population type II. The galactic clusters and stars associated with spiral arms (comprising stars of a vast age range) are assigned to population I. Various intermediate groups have been assigned by sundry workers who have used criteria of velocity and spatial distribution as well as chemical composition (see Chapter 8). There is some advantage in adopting the nomenclature of Arp (1958) and assigning to type II only the metal-deficient stars such as occur in globular clusters and similar systems. Probably elliptical galaxies, the central bulge of our galaxy, and the Andromeda spiral all contain high-metal-content stars thus resembling $M67$.

In summary, we must emphasize that *the UBV magnitudes of a star depend not only on its intrinsic luminosity and surface temperature, but also on its chemical composition.* The main sequence of a family of stars with a metal/H ratio a hundred times lower than that of the sun would not coincide on an M_v − (B-V) diagram with the standard main sequence. Hence, in comparing color-magnitude arrays for clusters or even in interpreting data for a single star, the role of chemical composition must be kept in mind.

1–7. Stellar Masses and Dimensions, Mass-Luminosity Correlation. If stars occurred only singly or in huge clusters, our knowledge of their masses and dimensions would be severely limited. Fortunately, stars of a great range of mass and luminosity are found in binary systems. Visual binaries are our most important source of data on stellar masses, while double stars that occur as eclipsing binary systems provide a wealth of additional facts. Two stars move about each other in elliptical orbits whose orientation with respect to the observer is such that an eclipse will occur when one passes in front of the other. The inclination of the orbit and the radii of the stars determine whether the eclipse will be partial, annular, or total.

The observational data for an adequately observed system consist of a light curve, which shows the variation of the brightness of the system as a function of time, and measurements of the velocity of one or both of the components in the line of sight. From the light curve it is possible to derive the relative radii of the stars in terms of the size of the orbit and inclination i of the orbit to plane of the sky. The spectroscopic measurements give the radial velocity curve from which one may obtain $a_1 \sin i$, the semimajor axis a_1 of the orbit of the brighter star multiplied by the sine of the inclination. If both spectra are visible, we find $(a_1 + a_2) \sin i$.

A combination of the photometric and spectroscopic data therefore yields a. With the size of the orbit given in kilometers, the radii of the stars can be computed immediately.[7] Furthermore, from a and the period P, we can find the sum of the masses from Kepler's third law, which may be written in the form

$$\mathfrak{M}_1 + \mathfrak{M}_2 = \frac{a^3}{P^2} \tag{1-8}$$

where a is measured in astronomical units, P in years, and $\mathfrak{M}$ in terms of the sun's mass. The individual masses follow from

$$\frac{\mathfrak{M}_1}{\mathfrak{M}_2} = \frac{a_2}{a_1}$$

The relative depths of the primary and secondary eclipses give the ratio of the surface brightnesses of the two stars. If the parallax of the system is known, and the sizes of both components are calculated with the aid of spectroscopic observations, one can compute the absolute surface brightnesses of the stars in ergs sec^{-1} cm^{-2} per unit wavelength interval. Since this quantity defines the brightness temperature of the star (see Chapter 6), we have a check on the system of stellar temperatures. Such temperature calibrations have been attempted for the $B3$ star, μ' Scorpii, and the $M0$ star, YY Geminorum.

In addition to radii and masses, much other useful information can be found from eclipsing binary systems. Often the stars tidally perturb each other; the amount of distortion depends on their separation and concentration of mass towards the center. Because of these distortions, the stars do not attract each other as point masses, and as a consequence, the whole elliptical orbit slowly rotates in space, a phenomenon called *motion of the line of apsides.* From the rates of this motion, much has been learned concerning density distributions in stellar interiors. Theoretically, the deformation of the figure of the star by tidal action could be determined from the shape of the light curve. However, the interpretation of the latter outside of eclipse is complicated by reflection of the light of one star from the surface of the other and by the phenomenon of *gravity darkening.* This effect causes the amount of energy radiated per unit area to be proportional to the acceleration of gravity at that point. The pole of a rapidly rotating star is therefore hotter than its equator. Usually rotations of eclipsing stars are synchronized exactly with their orbital revolution. The stars keep the same sides pointed toward each other, just as the moon

[7] Further information on stellar diameters comes from certain red supergiants whose angular diameters have been measured directly with the Michelson stellar *interferometer* and from bright blue stars whose angular diameters can be measured with the photon-correlation interferometer (Brown and Twiss, 1954, 1956). If the parallax of the star is known, the true diameter can be found immediately from the angular diameter.

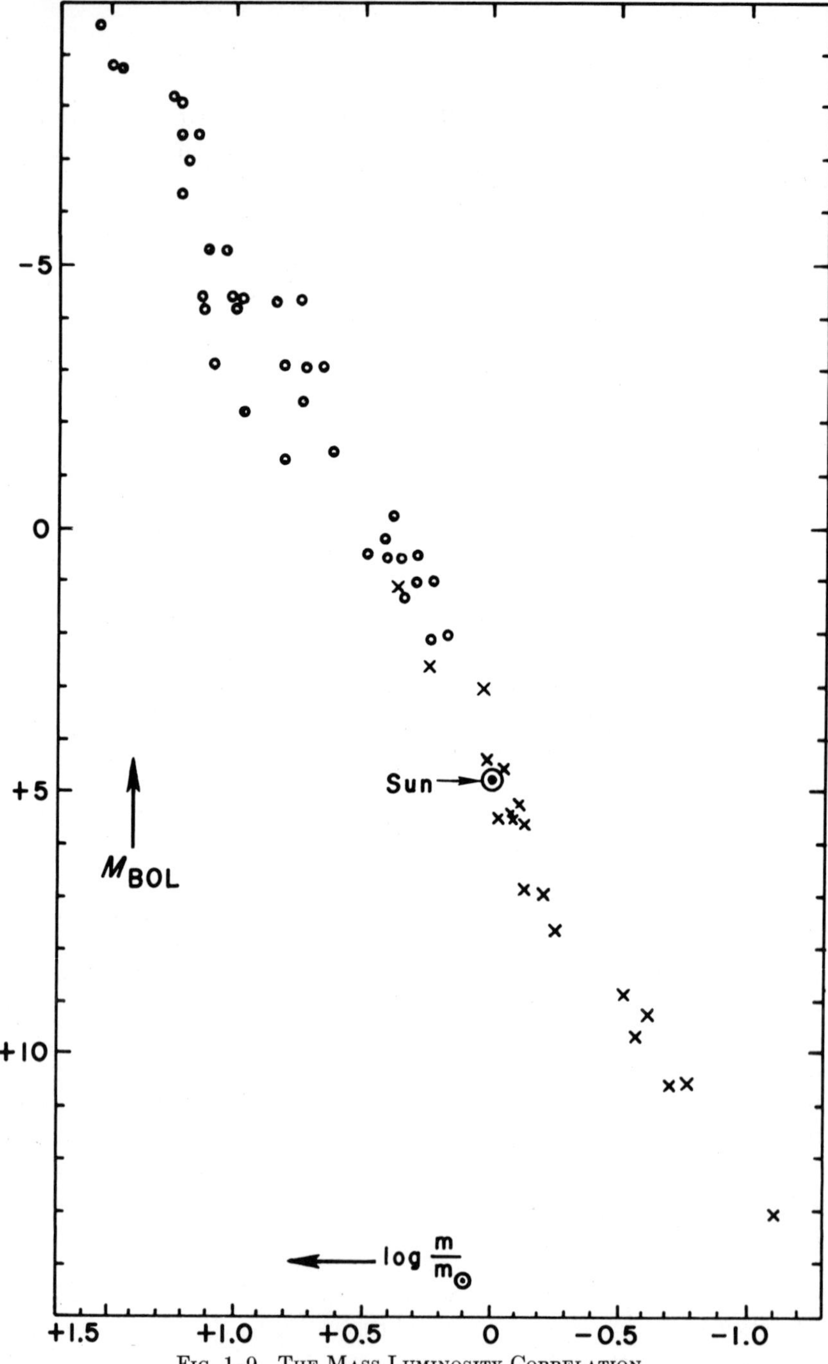

Fig. 1–9. The Mass Luminosity Correlation

Bolometric absolute magnitude is plotted against the logarithm of the mass. The circles denote data from eclipsing binaries; the crosses, data from visual binaries. The large spread among the eclipsing stars may result in part from the circumstance that in close binaries the evolution of each star is affected by the presence of its neighbor.

keeps the same face pointed toward the earth. In a few systems, the smaller, heavier component of the pair spins more rapidly than it revolves, but in all instances the direction of rotation is in the same sense as the orbital revolution.

The *darkening to the limb*, a phenomenon observed in the sun, depends on the wavelength, surface temperature, gradient, and source of opacity in the stellar atmosphere. Its dependence on wavelength can be inferred from good light curves in different colors, but its absolute value may be established only with difficulty.

The most abundant eclipsing stars are the W Ursae Majoris variables — dwarf stars closely comparable to the sun. They revolve around each other almost in contact and are highly distorted. The information obtainable from eclipsing binaries is so extensive that Henry Norris Russell referred to the "Royal Road of Eclipses" (1948). Unfortunately, it is not a road without booby traps. One cannot apply results and correlations obtained for components of eclipsing stars to single stars in an uncritical fashion. The evolution of a pair of stars in an eclipsing binary may differ profoundly from the evolution of two separate stars of the same mass and chemical composition. Mass exchange between the components may occur, and much of the original mass may be lost. Eclipsing binaries often contain main sequence stars paired with subgiants.

The components of eclipsing binaries are usually separated by only a few million kilometers, whereas those of visual binaries (which are seen as double stars when viewed in the telescope) are separated by an astronomical unit (1.494×10^8 km) or more. The orbit of the binary gives the period P and semimajor axis a'' in seconds of arc. If the parallax p'' is known, we may find $a = a''/p''$, the semimajor axis in astronomical units. Then from Kepler's third law, Eq. 1–8, it is possible to solve for the sum of the masses. Hence a good mass determination requires an accurate parallax. The individual masses, $\mathfrak{M}_1$ and $\mathfrak{M}_2$, of a visual double can be found only if the orbit of each component can be measured in relation to the center of gravity of the pair. That is, the absolute motions of the stars upon the background of the sky must be determined.

Fig. 1–9 shows the observed correlation between mass and luminosity. Consider, first, the visual binaries. The masses are taken from compilations by various investigators; the bolometric corrections to the absolute visual magnitudes are from Chapter 6. For the fainter stars, the relation between bolometric magnitude and mass is fairly well represented by

$$M_{\text{bol}} = 5.3 - 7 \log \mathfrak{M} \qquad M_{\text{bol}} > 7.0 \tag{1–9}$$

whereas for the brighter stars,

$$M_{\text{bol}} = 4.70 - 9.5 \log \mathfrak{M} \qquad M_{\text{bol}} < 7.0 \tag{1–10}$$

or

$$\begin{aligned} L &\sim \mathfrak{M}^{3.8} \text{ for the brighter stars} \\ L &\sim 0.60\mathfrak{M}^{2.8} \text{ for the fainter stars} \end{aligned} \tag{1–11}$$

Here $\mathfrak{M}$ is the mass of the star in terms of the solar mass.

The eclipsing binary data (Kopal, 1952), which are included for comparison, show a deplorable scatter. Part of this is produced by the uncertainties in the luminosities which are taken from the spectral class on the assumption the star falls on the main sequence and from uncertainties in the bolometric corrections. Stars with subgiant companions are excluded, but many of the remaining objects may have been affected by evolutionary effects. Quite generally, the brighter components of eclipsing binaries tend to agree with the normal mass-luminosity relationship, although the faint components are often discordant. If two members of an eclipsing system are main-sequence stars, the fainter star will be smaller, cooler, and less massive than the brighter one. Therefore, it cannot totally eclipse the more luminous one. Observational selection tends to favor deeper eclipses, that is, to pick objects in which a normal main-sequence star is accompanied by a larger, less dense, cooler companion. This latter star tends to be a subgiant, whose mass, luminosity, and radius may be abnormal. The masses of the tenuous companions turn out to be smaller than one would estimate from their luminosities, the more tenuous the star, the greater the deviation.[8]

We must emphasize that the mass-luminosity correlation holds only for main-sequence stars. It fails for giants, supergiants, subgiants and white dwarfs. Consider, for example, the giant stars b' and c' in Fig. 1–7 Both now have the same luminosities but have evolved from stars of different masses on the main sequence. Further, the mass-luminosity relation appears to depend on the chemical composition of the star (Eggen, 1963).

The surface gravity, g, of a star depends on its mass and radius. If $\mathfrak{M}$ and R are measured in terms of the solar mass and radius

$$g = g_0\mathfrak{M}/R^2 = 2.74 \times 10^4\mathfrak{M}/R^2 \text{ cm sec}^{-2} \tag{1–12}$$

From the data of eclipsing variables, Russell has found that the surface gravities of main-sequence stars can be well represented by the empirical formula

$$\log g = 3.79 + 0.64\theta \tag{1–13}$$

where $\theta = 5040/T$. For a crude reconnaissance,

$$\log g = 5.45 - 2.70\theta \tag{1–14}$$

and

$$\log g = 3.42 - 2.06\theta \tag{1–15}$$

[8] Mass luminosity correlations have been prepared by many workers, among whom we mention Kuiper (1937), Russell and Moore (1940), Parenago and Massevich (1950), Petrie (1950), Plaut (1953), van de Kamp (1954), Strand (1954, 1957), and Eggen (1956).

for giants and supergiants, respectively. The surface gravity g and temperature T, together, determine the spectrum of a star of normal composition. When we distinguish giants and dwarfs by spectroscopic criteria, we separate them in terms of their surface gravities. It must be remarked that the effective surface gravities of supergiant stars are often smaller than the values computed from their radii and masses. Perhaps the atmospheres of such stars are not in ordinary mechanical equilibrium. (See Chapters 7 and 8.)

With this brief sketch of some of the basic astronomical data, we turn next to an account of the physical background required for an interpretation of stellar atmospheres. Chapter 2 summarizes salient facts on spectroscopic nomenclature, and atomic and molecular structure. Chapter 3 treats of the gas laws, the distribution of molecular velocities, ionization, excitation, and dissociation, while following chapters deal with the interaction of radiation with matter, the interpretation of stellar spectra, the surface phenomena of the sun, and solar terrestrial relationships. We shall pay special attention to the sun, whose interpretation is one of the most fundamental tasks of astrophysics. Most of our discussion is concerned directly with solar or closely allied problems.

REFERENCES

Goldberg, L. 1954. "The Absorption Spectrum of the Atmosphere," *The Earth as a Planet*, G. P. Kuiper (ed.). University of Chicago Press, Chicago, p. 434.

The classical reference on fundamentals of general stellar astronomy, even though much of the material is out of date, is:

Russell, H. N., Dugan, R. S., and Stewart, J. Q. 1938. *Astronomy*. Ginn & Co., Boston, vol. 2.

A concise, up-to-date treatment of the general aspects of astronomy is given by:

Struve, O., Pilians, H., and Lynds, B. 1958. *Elementary Astronomy*. Oxford University Press, Fair Lawn, N. J.

McLaughlin, D. B. 1961. *Introduction to Astronomy*, Houghton Mifflin Co., Boston.

Abell, G. 1963. *Exploration of the Universe*. Holt, Rinehart, and Winston, New York.

See also:

Hynek, J. A. (ed.). 1951. *Astrophysics, A Topical Symposium*. McGraw-Hill Book Co., Inc., New York, chaps. 1, 9, 10, 11.

van de Kamp, P. 1952. *Basic Astronomy*. Random House, New York, chaps. 8, 9, 13, 14, 17, 22.

A comprehensive, modern, but more advanced treatment of stars and stellar systems is contained in:

Pecker, J. C., and Schatzman, E. 1959. *Astrophysique Générale*. Masson & Cie, Paris.

For a discussion of the older work on stellar magnitudes and photometry see:

Weaver, H. F. 1946. *Pop. Astr.* **54**, 211, 287, 339, 451, 504.

The UBV photometric system is described in:

JOHNSON, H. L., and MORGAN, W. W. 1951. *Ap. J.* **114,** 522.
———. 1953. *Ap. J.* **117,** 313.

The six-color photometry is discussed by:

STEBBINS, J., and WHITFORD, A. E. 1943. *Ap. J.* **98,** 20.
———. 1945. *Ap. J.* **102,** 318.

Spectral classification is discussed in the introduction to the Henry Draper Catalogue, Harvard College Observatory. The Yerkes system is given in:

MORGAN, W. W., KEENAN, P. C., and KELLMAN, E. 1943. *An Atlas of Stellar Spectra.* University of Chicago Press, Chicago.

See, particularly, the excellent review article by:

FEHRENBACH, C. 1958. *Encyclopedia of Physics.* S. Flügge (ed.). Springer-Verlag, Heidelberg, Germany, vol. 50, pp. 1–92.

For an account of the spectrum-luminosity relation see:

ARP, H. C. 1958. *Encyclopedia of Physics.* S. Flügge (ed.). Springer-Verlag, Heidelberg, Germany, vol. 51, pp. 75–133.

The stellar evolutionary picture is summarized by:

BURBIDGE, E. M., and BURBIDGE, G., in the same volume, pp. 135–295.

For visual, spectroscopic, and eclipsing binaries see:

BINNENDIJK, L. 1960. *Properties of Double Stars.* University of Pennsylvania Press, Philadelphia.

For eclipsing binary stars see:

KOPAL, Z. 1946. *Introduction to Study of Eclipsing Variables.* Harvard Observatory Monograph No. 6. Cambridge University Press, London.
———. 1952. *Transactions of the International Astronomical Union.*
———. 1955. *Ibid.*
GAPOSCHKIN, S. 1958. *Encyclopedia of Physics.* S. Flügge (ed.). Springer-Verlag, Heidelberg, Germany, vol. 50, p. 225.
PIERCE, N. L. 1951. *Astrophysics, A Topical Symposium.* J. A. Hynek (ed.). McGraw-Hill Book Co., Inc., New York.

For visual binaries see:

VAN DE KAMP, P. 1958. *Encyclopedia of Physics.* S. Flügge (ed.). Springer-Verlag, Heidelberg, Germany, vol. 50, p. 187.

CHAPTER 2

Atomic and Molecular Spectra

2–1. Introduction. Before the modern quantum theory was developed, the interpretation of spectra had progressed scarcely beyond a rudimentary level. The chemical origin of most strong lines in the spectra of the sun and similar stars had been established. *Kirchhoff's three laws of spectral analysis* gave the relation between the general nature of the source and the kind of spectrum emitted, viz., (1) an incandescent solid, liquid, or a gas under sufficient pressure gives a continuous spectrum, (2) an incandescent gas under low pressure emits a spectrum of discrete bright lines, and (3) an incandescent gas placed in front of a hotter source of continuous radiation will produce a dark-line or absorption spectrum superposed upon a continuous background. These absorption lines will fall at precisely the same wavelengths that the gas regularly emits. In addition to bright- and dark-lined atomic spectra, complicated "band" spectra of molecules were also recognized and studied.

Thus the *Orion nebula,* which emits a bright-line spectrum, was correctly interpreted as an incandescent gas, whereas the dark-line spectra of the sun and stars could be explained in terms of Kirchhoff's third law. It was generally believed that the bright surface or "photosphere" of the sun emitted a continuous spectrum, and relatively cooler gas in the overlying atmosphere, then called the "*reversing layer,*" absorbed certain wavelengths characteristic of the elements present. C. A. Young described a striking demonstration of the second and third laws obtained by observing the 1870 eclipse with a spectroscope. Before totality, the dark-line or *Fraunhofer spectrum* was evident; the instant the photosphere of the sun was covered by the moon, the dark lines disappeared and were replaced by bright lines in exactly the same positions. Before totality, the light reaching the observer passed through the cooler gases of the sun's atmosphere and produced a dark-line spectrum in the usual way. When the photosphere is blocked from view, the observer receives radiation from only the upper atmosphere and thus obtains the characteristic bright-line spectrum of an incandescent gas. We now know this picture to be somewhat oversimplified. The Fraunhofer and continuous spectra of the sun and stars actually are both produced in the same layers. Only the tiniest fraction of the atoms in the solar atmosphere, namely those in the attenuated chromosphere, produce the bright lines of the flash spectrum observed by Young and others.

The source of the solar continuous spectrum was a puzzle for many years. Kirchhoff's first law would suggest that it was produced by a hot gas under pressure (since the temperature of the solar atmosphere is too high to permit solids or even liquids to exist). Later studies by J. Q. Stewart and others showed that the pressure at the bottom of the solar atmosphere was much too low to admit this explanation. Atoms and molecules are known to emit and absorb continuous as well as discrete line spectra, but the continuous absorption in the solar atmosphere is produced primarily by the negative hydrogen ion.

The astronomers and physicists of the late nineteenth century could make a qualitative chemical analysis of stars and nebulae. Kirchhoff, who pioneered in these investigations, was able to identify many of the lines in the solar spectrum with such familiar elements as hydrogen, iron, nickel, calcium, chromium, titanium, sodium, magnesium by the simple technique of matching side by side the solar spectrum and spectra of selected elements produced in a flame or arc. Similar studies of stellar spectra by Huggins, Lockyer, and others, established the essential identity of the chemical compositions of the stars and the earth. Elements abundant in the earth were usually well represented in the sun and stars, and elements rare on earth were often missing. The recognition of the sameness of matter throughout the universe was the first great discovery of astrophysics. Nevertheless, some abundant elements are poorly represented in spectra of stars like the sun.

Atomic carbon and oxygen—very abundant constituents of the earth—are represented in the sun by moderately conspicuous lines in the infrared. The strongest lines of these and many other elements lie in the astronomically inaccessible ultraviolet below $\lambda 2900$. No lines of abundant neon appear in the normally observable solar spectrum at all. The appearance of molecules of carbon and nitrogen, CN, oxygen and hydrogen, OH, and the like show these elements to be abundant in the sun.

Gaseous nebulae presented other difficulties. Their spectra showed familiar emissions of hydrogen and helium, but the strongest nebular radiations had never been produced on earth. For many years astronomers spoke of a hypothetical and mysterious nebulium until advances in physics showed there was no room in the periodic table for such an element. Ultimately, these radiations were identified with atoms of oxygen, neon, nitrogen, argon, sulfur, and others, stripped of several electrons.

Any quantitative analysis of the sun and stars presents formidable difficulties. How is the intensity of a spectral line related to the abundance of the relevant element? Can the temperature of a stellar atmosphere be determined from its spectrum? These questions could be answered only when the structure of the atom was understood and when processes of emission and absorption of energy could be expressed in quantitative terms.

Thus the relationship between the structure and spectra of atoms proves of crucial importance to astrophysics, and our survey of basic principles can well begin with a brief excursion into this problem.

The fundamental building block of all complicated substances of nature, the *atom*, consists of a positively charged core or *nucleus* which contains nearly all its mass, surrounded by one or more negatively charged, constantly moving *electrons*. Each electron carries a *negative charge* of 1.60×10^{-19} coulombs or 4.80×10^{-10} esu. In the complete (neutral) atom the positive charge on the nucleus equals the sum of the negative charges on the electrons. For example, eight electrons surround the nucleus of an oxygen atom which carries a positive charge of $8 \times 4.80 \times 10^{-10} = 3.84 \times 10^{-9}$ esu.

Simplest of all atoms is hydrogen, which consists of a positively charged particle called a *proton* and a single negatively charged electron. The charge on the proton is the same as that on the electron, but the mass of the proton, 1.67239×10^{-24} gm, exceeds that of the electron, 9.1083×10^{-28} gm, by a factor of 1836.12.

The hydrogen nucleus, or proton, is a fundamental particle of nature like the electron. All other atomic nuclei are more complicated, containing both protons and particles of about the same mass but no charge, called *neutrons*. Thus ordinary carbon of atomic number 6 (determined by the number of protons) and atomic weight 12 has 6 protons and 6 neutrons. The carbon *isotope* of atomic weight 13 has 7 neutrons and 6 protons.

2–2. Spectral Series. The outer or electronic structure of the atom determines the spectral lines that will be radiated. A basic postulate of the quantum theory is that an atom can exist in certain definite energy states; when it jumps from one of energy W_2 to another of lesser energy W_1 it emits radiation in accordance with the law:

$$W_2 - W_1 = h\nu \tag{2–1}$$

or if $W_2 - W_1$ be denoted as E,

$$E = h\nu \tag{2–2}$$

This represents the basic relation between energy and frequency ν. Here h is a fundamental quantity of nature called *Planck's constant;* $h = 6.6252 \times 10^{-27}$ erg sec.

Hydrogen, simplest of all atoms, shows the simplest of all spectra. In 1885 Balmer showed that the wave numbers $\tilde{\nu}$ of all visible hydrogen lines could be expressed by

$$\tilde{\nu} = R\left(\frac{1}{2^2} - \frac{1}{n^2}\right) \tag{2–3a}$$

where $n = 3, 4, 5, \ldots$, and R is a constant the same for all values of n, i.e., for all lines of a series. Subsequently, *series* in the ultraviolet and infrared were found that could be represented by the following formulas

$$\text{Lyman series} \qquad \tilde{\nu} = R\left(\frac{1}{1^2} - \frac{1}{n^2}\right); \quad n = 2, 3, 4, \ldots \tag{2–3b}$$

$$\text{Paschen series} \qquad \tilde{\nu} = R\left(\frac{1}{3^2} - \frac{1}{n^2}\right); \quad n = 4, 5, 6, \ldots \tag{2–3c}$$

Brackett and Pfund found other series yet farther in the infrared. The same constant R was involved in all of these formulas. In this chapter R is expressed in *wave number units*.

Alkali metals show somewhat similar series that may be represented by formulas of the type

$$\tilde{\nu} = T' - R/(n + \delta)^2 \tag{2–4}$$

where T' corresponds to the wave number of the series limit and δ is nearly constant for a given series. Its numerical value varies from one series to another (see Art. 2–5).

Quite generally the wave numbers of spectral lines can be written as

$$\tilde{\nu} = T' - T'' \tag{2–5}$$

where T' and T'' are *term values* which are usually expressed in wave number units and correspond to the atomic energy levels. No theory based on classical physics could explain formulas of the type in Eqs. 2–3, 2–4, or 2–5; an entirely new approach was required.

2–3. The Bohr Atom. In order to explain the hydrogen spectrum, Bohr postulated that the electron traveled in a circular orbit of radius a and angular velocity ω around the positively charged proton. During this time it does not radiate; it emits energy only when it jumps from one orbit to another. See Fig. 2–1. In any one of these stable orbits, the integral of the angular momentum $ma^2\omega$ around the path is postulated to be an integral multiple of Planck's constant h, viz.:

$$\int_0^{2\pi} p_\phi \, d\phi = 2\pi m a^2 \omega = nh \tag{2–6}$$

The acceleration of the electron in its orbit is $ma\omega^2$, while the force due to the coulombic attraction of the proton is $Z\varepsilon^2/a^2$. Hence

$$ma\omega^2 = Z\,\varepsilon^2/a^2 \tag{2–7}$$

and therefore the orbits permitted to the electron have radii given by

$$a = \frac{h^2}{4\pi^2 m\varepsilon^2 Z} n^2 = 0.529446 \times 10^{-8} n^2 \text{ cm} \tag{2–8}$$

where m and ε denote the mass and charge on the electron, respectively, Z is the atomic number (1 for hydrogen) and the integer n is called the *principal quantum number.* The total energy of the electron in the orbit n is the sum of the kinetic energy

$$\frac{1}{2}mv^2 = \frac{1}{2}ma^2\omega^2 = \frac{Z\,\varepsilon^2}{2a} \tag{2–9}$$

and the potential energy of the electron at a distance a from the nucleus, viz., $-Z\varepsilon^2/a$, since potential energy is counted as zero when the proton and electron are at an infinite distance apart and at rest. Hence

$$\underset{\text{Total energy}}{W_n} = \underset{\text{Kinetic energy}}{\frac{Z\varepsilon^2}{2a}} - \underset{\text{Potential energy}}{\frac{Z\varepsilon^2}{a}} = -\frac{2\pi^2mZ^2\varepsilon^4}{h^2}\frac{1}{n^2} \tag{2–10}$$

is the energy in the nth orbit, it being understood that $W = 0$ corresponds to the complete detachment of the electron from the atom. Accordingly, W_n is the energy necessary to detach the elec-

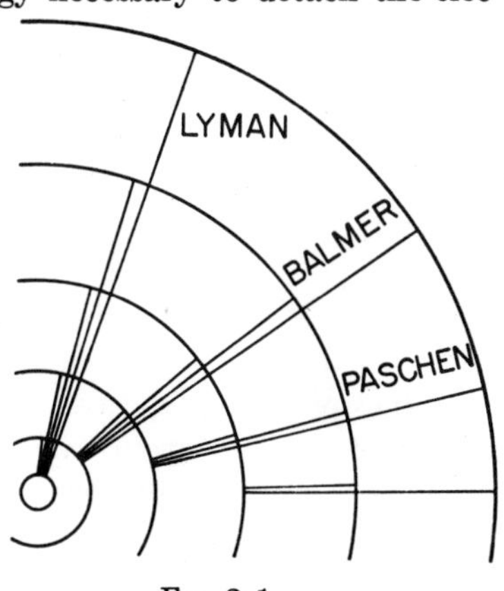

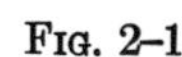

Fig. 2–1

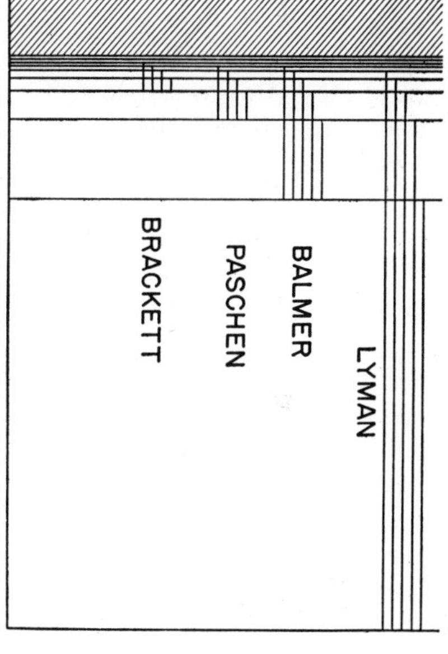

Fig. 2–2

Fig. 2–1. The Bohr Model of the Hydrogen Atom

Transitions corresponding to the Lyman, Balmer, and Paschen series are depicted.

Fig. 2–2. Energy Level Diagram for Hydrogen

The energy levels (each of which corresponds to a Bohr orbit) are depicted as horizontal lines. The transitions are shown as vertical lines. If the lowest level is taken as 0 ev, the first excited level lies at 10.2 ev and the ionization potential (the highest horizontal line) lies at 13.6 ev. The cross-hatched area represents the continuum which corresponds to a complete detachment of the electron from the atom.

tron from the nth level, i.e., ionize the atom from the nth level. In particular, the ionization energy from the lowest level is

$$|W_1| = \frac{2\pi^2 m Z^2 \varepsilon^4}{h^2} \tag{2–11}$$

and the energy radiated in the transition (n—n') is:

$$W_n - W_{n'} = \frac{2\pi^2 m Z^2 \varepsilon^4}{h^2}\left(\frac{1}{n'^2} - \frac{1}{n^2}\right) \tag{2–12}$$

The Bohr theory correctly predicted the frequencies of the lines of the hydrogen spectrum. Thus, Eqs. 2–3a, b, c are obtained from Eq. 2–12 by setting $n' = 2$, 1, and 3, respectively. We may represent energies of the Bohr orbits in the hydrogen atom by means of an *energy-level diagram.* In accordance with Eq. 2–10, the zero of energy corresponds to complete detachment of the electron from the atom. The energy of the lowest level is $-W_1$, that of the second level is $-W_1/4$, and so on. If we depict the transitions as vertical lines connecting possible energy states we get Fig. 2–2. Thus, lines ending on the first level represent radiation of the Lyman series, those ending on the second level radiation of the Balmer series, etc. Downward transitions represent emission of energy by the atom; upward transitions represent absorption of energy.

Although Eq. 2–10 gives the energy in ergs if cgs units are employed, term diagrams are often plotted with wave-number units or electron volts (ev). The *excitation potential* of a level in electron volts is the potential through which a bombarding electron must drop in order to acquire sufficient energy to excite an atom from the ground level to the level in question. One electron volt is equivalent to 1.6021×10^{-12} ergs. Similarly, energies expressed in electron volts are related to energies expressed in wave-number units by [1]

$$W\ (\text{ev}) = 1.2398 \times 10^{-4}\,\tilde{\nu} \tag{2–13}$$

The energy-level diagram illustrates why lines of various spectral series crowd together and ultimately coalesce into a *continuous spectrum.* The larger the orbit the weaker the proton-electron attraction until the addition of a minute amount of energy suffices to detach the electron completely, i.e., ionize the atom.

In transitions between two discrete levels, an atom absorbs or emits the precise amount of energy for a spectral line to be emitted or absorbed. In order to eject an electron completely, however, an atom may absorb any amount of energy equal to or greater than that required to go from its initial state to the one represented by $W = 0$. The excess energy imparts a velocity, v, to the free electron, viz.,

$$h\nu - |W_n| = \tfrac{1}{2}mv^2$$

[1] We may verify this result by noting that an electron acquires a velocity v by falling through a potential $V/300$ statvolts, so that $V\varepsilon/300 = \frac{1}{2}mv^2$.

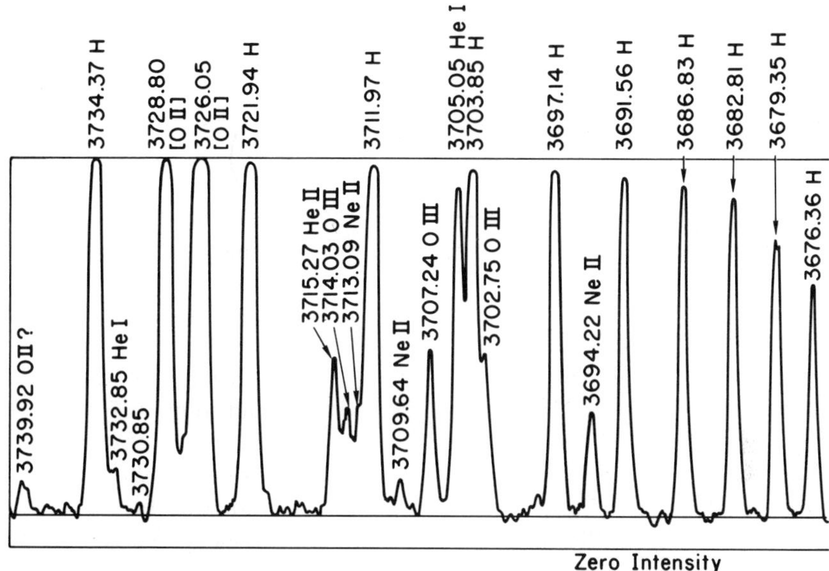

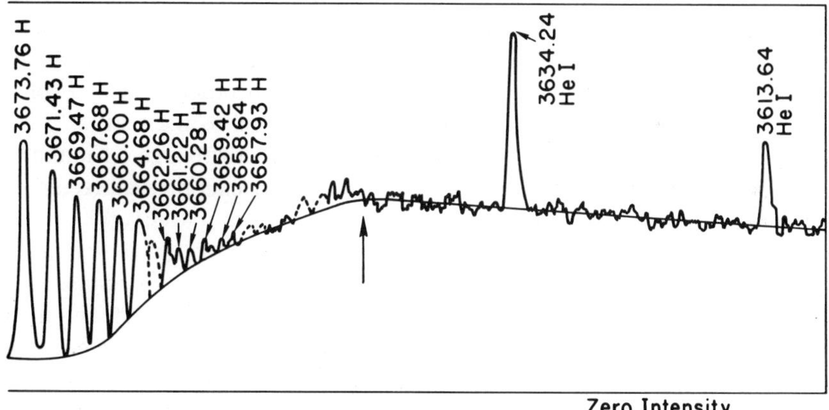

FIG. 2-3. CONFLUENCE OF HYDROGEN LINES AND BALMER CONTINUUM IN THE GASEOUS NEBULA NGC 7009

This is a tracing of a portion of a spectrogram secured August 31–September 1, 1961, with the coudé spectrograph at the Mount Wilson Observatory. Lines of He I, He II, O III, Ne II, and forbidden transitions of [O II] are also present. The dotted lines indicate portions of the spectrogram affected by mercury-vapor lights in Los Angeles. The arrow indicates the theoretical position of the Balmer limit.

In hot stars a strong continuous absorption is observed beyond the limit of the Balmer series (see Chapters 5 and 6). This continuum corresponds to the photo-ionization of hydrogen atoms from the second level. Conversely, the capture of free electrons of various velocities by protons in the second energy level gives a bright continuous spectrum observed beyond the Balmer limit in planetary nebulae (see Fig. 2–3).

Ionized helium He^+, whose spectrum is denoted as He II, displays a hydrogen-like spectrum. Its *ionization potential*, however, is four times greater than that of hydrogen, 54.40 ev instead of 13.595 ev. Since $Z = 2$, Eq. 2–11 shows that the wave numbers of the He II lines will be given by

$$\tilde{\nu} = 4R\left(\frac{1}{n'^2} - \frac{1}{n^2}\right)$$

The value of R for ionized helium, $R_{\text{He II}} = 109{,}722.264 \text{ cm}^{-1}$ is slightly different from the value $109{,}677.581 \text{ cm}^{-1}$, for hydrogen, because of the finite mass of the electron (see Art. 2–19). An interesting result is that alternate lines of the Brackett (4–n) series of He II almost coincide with the Balmer (hydrogen—H I) lines. This series was first discovered in the stars and called the *Pickering series.* Doubly ionized lithium shows a similar term structure, but this element is so rare that these lines are never observed in stellar spectra. The general structure of the spectrum of any singly ionized atom resembles that of the atom immediately preceding it in the periodic table.

2–4. The Wave Atom. As we shall see shortly, the planetary atom picture may be modified to produce a useful model for prediction of lines emitted and absorbed by more complex atoms. This refined Bohr model or *vector model* is valuable for an enumeration of the kinds of levels of complex atoms, but fails to predict positions of energy levels correctly.

The modern quantum theory does not assign electrons to planetary orbits. Instead it is concerned with finding for each particular energy level a wave amplitude, ψ, whose square expresses the probability of finding an electron at a given point at a given time. Not only the locations of energy levels, but also such quantities as the probabilities of transitions between them may be computed by these techniques. Although the idea of definite planetary orbits must be discarded, the concept of the energy-level diagram (which involves no assumption concerning the atom model) remains intact. In quantum mechanics the behavior of the electrons in an atom is completely described by this function $\psi\ (r, \theta, \phi, t)$ which is the solution of a partial differential equation called the *Schrödinger equation.* This equation is obtained by writing down a classical expression for the total energy of the system including all internal interactions. For simpler problems, such as the hydrogen atom, exact solutions can be found, but for many-body problems as are represented by complex atoms it is necessary to use perturbation methods.

2–5. Spectra of the Alkali Atoms. Next to that of hydrogen, the alkali atoms Li, Na, K, and so on, display the simplest spectra. In depicting the Bohr model of an alkali atom it is useful to refer to Sommerfeld's generalization of the Bohr theory to elliptical orbits. He showed that the allowed orbits were of such a shape that the ratio of minor to major axis was k/n, where k and n are integers, with circular orbits when $k = n$. In a coulomb field, such as exists in the hydrogen or ionized helium atom, the energy of the orbit depends only on n and not at all on k (except for a small effect due to the variable speed of the electron in its orbit). The angular momentum depends on k.

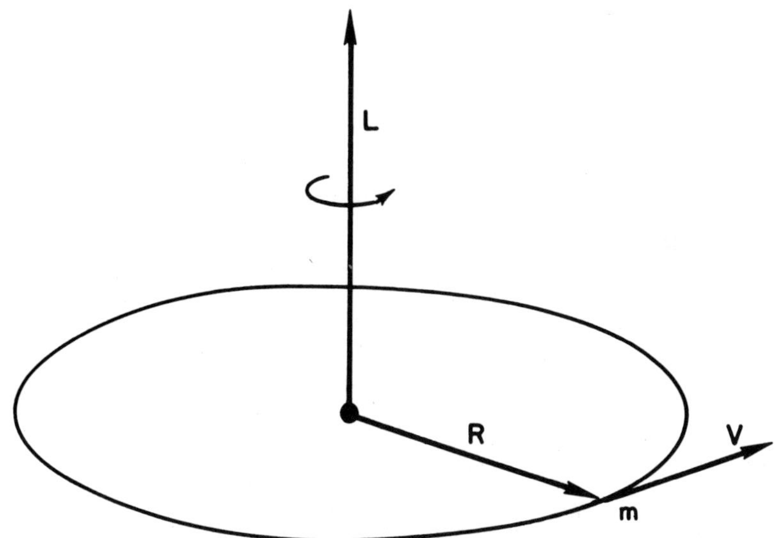

Fig. 2–4. Definition of Angular Momentum

We may represent the angular momentum of an electron in an orbit as a vector. Consider a mass m moving in a circular orbit of radius R with a speed V. The angular momentum mRV is represented as a vector perpendicular to the orbit in the sense shown in Fig. 2–4. In vector notation we write

$$\mathbf{L} = m\mathbf{R} \times \mathbf{V}$$

Thus, if the velocity of vector $\mathbf{V}$ makes an angle θ with respect to the radius $\mathbf{R}$, $L = mRV \sin \theta$. In the classical Sommerfeld orbit the angular momentum was $k(h/2\pi)$.

In the quantum picture, the electron retains an angular momentum even though the concept of an orbit is discarded. In the vector model we define an *orbital angular momentum quantum number* $l = k - 1$ such that the angular momentum is $l\ (h/2\pi)$. Atomic states with $l = 0$ are called s states; $l = 1$ corresponds to a p state; $l = 2$ to a d state; and $l = 3$ to an f state. Now s states correspond to orbits of increasingly greater eccentricity as n increases. In hydrogen the energy depends on n and hardly at all on l, but in other atoms the energy depends on both n and l. Notice that the $3s$ level in sodium falls far below the corresponding hydrogenic $n = 3$ level, the $3p$ level falls closer, and the $3d$ only slightly below the hydrogenic level.

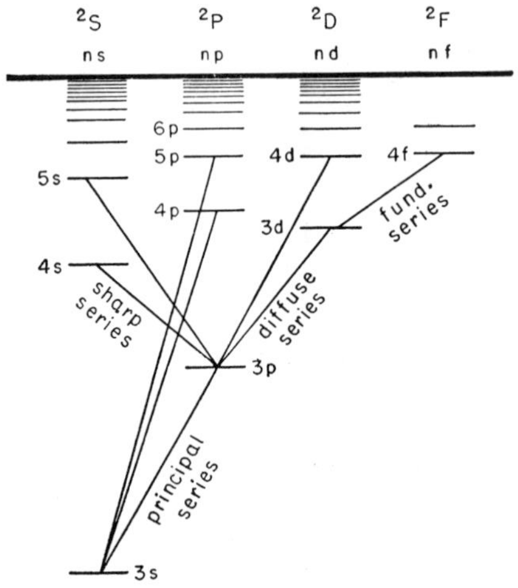

FIG. 2–5. THE ENERGY LEVELS OF SODIUM

The ten inner electrons of sodium are held in tightly bound shells, whereas the eleventh electron which is responsible for the optical spectrum is held only loosely. The highly eccentric $3s$ orbit penetrates into this inner electron cloud where the optical electron experiences a greater electrostatic force from the nucleus. That is, the screening produced by the inner electrons is not complete. Hence the electron is more tightly held in this $3s$ orbit than in the $3p$ or $3d$ configurations which correspond to more nearly circular orbits that penetrate the inner core slightly or not at all. See Fig. 2–5. Hence for them the shielding by the ten core electrons is more nearly complete. The observed spectral lines in sodium belong to the following series:

Sharp series	$= 3p - ns$	$n = 4, 5, 6$, etc.
Principal series	$= 3s - np$	$n = 3, 4, 5$, etc.
Diffuse series	$= 3p - nd$	$n = 3, 4, 5$, etc.
"Fundamental" series	$= 3d - nf$	$n = 4, 5, 6$, etc.

The wave numbers of the lines may be represented by Eq. 2–4 with the same value of T' for the sharp and diffuse series, but with different values of δ. Transitions always take place between levels in adjacent columns in the energy level diagram, i.e., between levels whose l-values differ by unity. In the alkalis different l-values correspond to substantially different energies. The strongest lines in sodium correspond to the $3s$—$3p$

transition, the first line of the *principal series.* This is also called a *resonance line* since it involves the ground level. Transitions between higher levels give what are called *subordinate lines.*

A conspicuous characteristic of alkali levels is their doubling which Uhlenbeck and Goudsmit suggested to be a consequence of the interaction between the magnetic moment of the orbital motion and that of the *spin* of the optical electron. All the p, d, f, etc., energy levels are doubled. The s-levels are not doubled because $l = 0$, and the orbital angular momentum and magnetic moment vanish. That is, the "fine structure" of atomic energy levels depends on the magnetic interactions within the atom itself. That such interactions occur should come as no surprise since the electrons moving in orbits constitute tiny electric currents. The spin of an electron produces yet another tiny current; these microscopic circuits then interact with one another through their magnetic effects, the general character of which can be understood in terms of a simple classical picture.

The magnetic moment of a bar magnet is simply ml, where m is the pole strength and l is the distance between the two poles. A current i, in electromagnetic units, moving in a plane loop of area A has a magnetic moment $\mu = Ai$ which may be regarded as a vector perpendicular to the area A. If the current is produced by an electron moving in a closed orbit with a period T, then the total magnetic moment of the system is

$$\mu = \varepsilon A/cT$$

if ε is measured in electrostatic units. Now the area of the orbit A is simply $p_\phi T/2m$ where p_ϕ is the (constant) angular momentum. Since $p_\phi = kh/2\pi$ in the Sommerfeld orbit, the magnetic moment is

$$\mu = k \frac{h}{2\pi} \frac{\varepsilon}{2mc} = k \frac{\varepsilon h}{4\pi mc} \tag{2–14}$$

where

$$\mu_0 = \frac{\varepsilon h}{4\pi mc} = 0.92731 \times 10^{-20} \text{ erg gauss}^{-1} \tag{2–15}$$

is called the *Bohr magneton.* The ratio of the magnetic moment to the angular momentum is

$$\frac{\mu}{p_\phi} = \frac{\varepsilon}{2mc} \tag{2–16}$$

for the orbital motion of the electron. This ratio for total magnetic and mechanical moments defines a factor g, the Landé splitting factor, discussed later on. From the splitting of lines in a magnetic field (Zeeman effect) it is found that

$$\frac{\mu_s}{p_s} = \frac{\varepsilon}{mc} \tag{2–17}$$

for electron spin, however (see Art. 2–8).

If a magnet of moment $\boldsymbol{\mu}$ is placed in a magnetic field $\mathbf{H}$ such that the angle between $\mathbf{H}$ and $\boldsymbol{\mu}$ is θ, its potential energy will be $\mu H \cos\theta$ if the potential energy is taken as zero when $\theta = \pi/2$. Note that since the electron carries a negative charge, $\boldsymbol{\mu}$ and $\mathbf{p}$ (regarded as vectors) are directed in opposite directions.

Suppose we apply an external magnetic field to an electron moving in an orbit, gradually increasing the field strength up to some value H. If H is applied perpendicular to the plane of the orbit, the effect will be that although the size and shape of the orbit is not changed, it will precess at an angular velocity

$$\omega_L = H \frac{\varepsilon}{2mc} \tag{2–18}$$

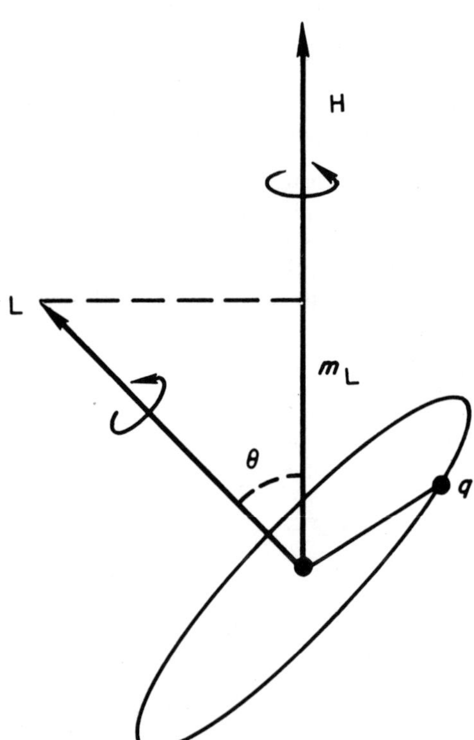

FIG. 2–6. PRECESSION OF AN ELECTRON ORBIT IN A MAGNETIC FIELD

Now suppose that the orbit is placed such that its normal makes an angle θ with respect to the direction of the field (see Fig. 2–6). The effect of the magnetic field will now be to produce a gyroscopic effect, the orbit precessing uniformly about the direction of $\mathbf{H}$. The velocity of this Larmor precession, $\omega = H\varepsilon/2mc$, does not depend on the size, eccentricity or orientation of the orbit. If p_H is the projection of the vector $\mathbf{p}$ on $\mathbf{H}$, the quantum theory suggests that p_H will be an integral multiple of $h/2\pi$, i.e., $p_H = m_l(h/2\pi)$, or $m_l/k = \cos\theta$.

That is, if there were no electron spin or if the spins of the electrons in the atom added together in such a way as to cancel out to zero (as can actually occur), the effect of an external magnetic field upon a Sommerfeld orbit of angular momentum $k(h/2\pi)$ would be to cause this orbit to take up an orientation in the field such that the projection of the angular momentum on the field would be $m_l(h/2\pi)$. The corresponding shift in energy would be

$$\delta E = E_m - E_0 = \frac{\varepsilon H h}{4\pi m c} m_l \qquad (2\text{–}19)$$

According to quantum mechanics the angular momentum is not $k(h/2\pi)$ but actually $\sqrt{l(l+1)}\,(h/2\pi)$ where $l = k - 1$, and m_l is restricted to the values

$$-l, -l+1, \ldots, l-1, l$$

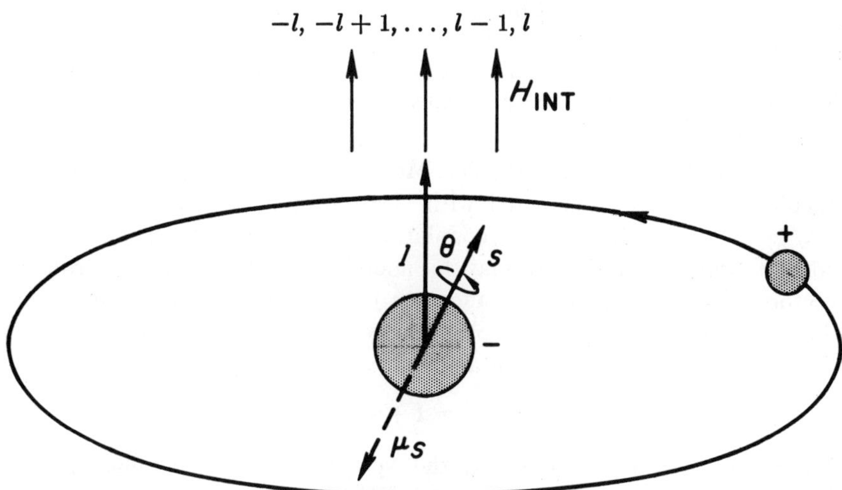

FIG. 2–7A. ORIENTATION OF A SPINNING ELECTRON IN THE MAGNETIC FIELD PRODUCED BY ITS ORBITAL MOTION

The spinning electron is depicted at rest with the proton in motion about it.

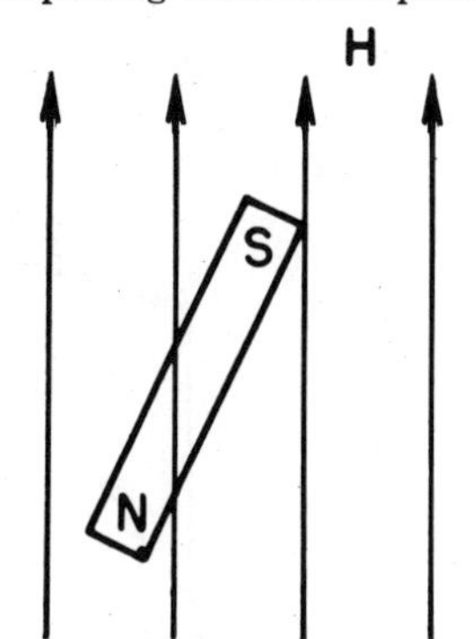

FIG. 2–7B. EQUIVALENT MAGNET FOR SPIN DIRECTED AS SHOWN IN FIG. 2–7A

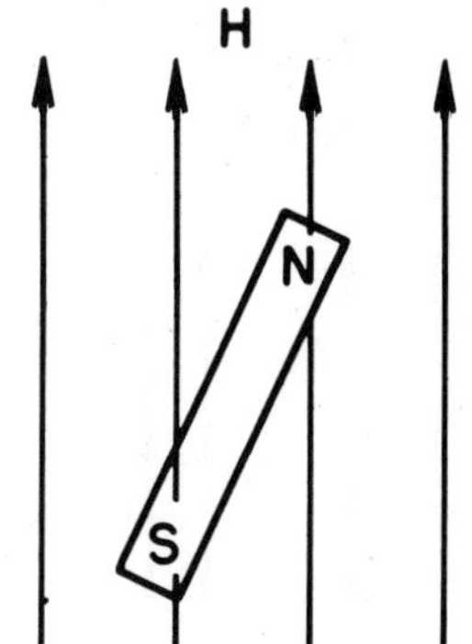

FIG. 2–7C. EQUIVALENT MAGNET FOR SPIN OPPOSITELY DIRECTED TO WHAT IS SHOWN IN FIG. 2–7A

Now consider the problem of magnetic interactions in the alkali atom in the absence of an external magnetic field. The spinning electron finds itself in a magnetic field produced by the electron's orbital motion. The magnetic moment of the spinning electron becomes oriented in this internal magnetic field, whose direction and magnitude is determined by the orbital angular momentum. Hence the magnitude of the spin-orbit interaction will be proportional to $\boldsymbol{l} \cdot \mathbf{s} = ls \cos \theta$ where θ is the angle between the vectors defined by the spin and the orbital angular momentum.

Fig. 2–7 shows a classical representation of the spin-orbit interaction. The spinning electron is depicted *at rest*, with the nucleus and core of the atom in motion with respect to it. This moving positive charge is equivalent to a current flowing in a wire and produces a magnetic field directed as shown. The spinning electron makes an angle θ with respect to the direction of the angular momentum vector $\boldsymbol{l}$, while $\boldsymbol{\mu}_S$ is directed oppositely to $\mathbf{s}$. The situation shown corresponds to the state of greater energy. (See Fig. 2–7B.) An opposite direction of spin corresponds to lesser energy since the equivalent magnet is then more nearly lined up with the field (see Fig. 2–7C).

2–6. Orbital and Spin Angular Moments. In the vector model the angular momentum associated with electron spin is always numerically $\frac{1}{2}(h/2\pi)$. It and the orbital angular momentum $l(h/2\pi)$ have the important property that they may be added and subtracted like ordinary vectors. The total angular momentum $\mathbf{j}$ of an alkali atom in which only the outermost electron plays a role in the production of the optical spectrum is the vector sum of the orbital and spin angular momenta, thus

$$\mathbf{j} = \boldsymbol{l} + \mathbf{s} \tag{2–20}$$

Numerically $j = l + s$ or $l - s$, the spin is lined up parallel or antiparallel to the orbital motion vector. For an s-level, $j = \frac{1}{2}$; for a p-level, $j = \frac{1}{2}$ and $\frac{3}{2}$, and so on. (See Fig. 2–8a.)

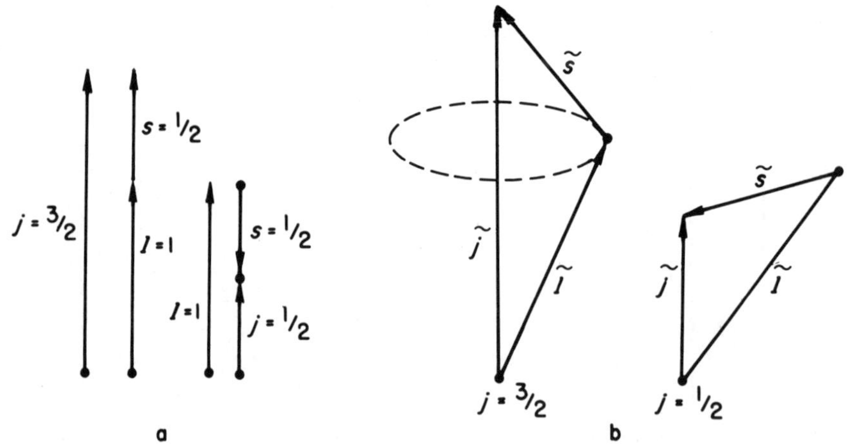

Fig. 2–8. Vector Model for an Alkali Atom

In each instance the angular momentum quantum number $l = 1$.

(a) Simple vector model representation.

(b) Vector model representation with $\sqrt{l(l+1)}$, $\sqrt{s(s+1)}$, and $\sqrt{j(j+1)}$.

Quantum mechanics shows the angular momenta actually to be $\sqrt{s(s+1)}\,(h/2\pi)$, $\sqrt{l(l+1)}\,(h/2\pi)$, and $\sqrt{j(j+1)}\,(h/2\pi)$, but the approximation $s(h/2\pi)$, $l(h/2\pi)$, and $j(h/2\pi)$ suffices to give the number and kind of energy levels (see Fig. 2–8b). Here we use the notation $\tilde{X} = \sqrt{x(x+1)}$, $\tilde{X} = \tilde{l}$, $\tilde{s}$, or $\tilde{j}$ to denote these quantum mechanical values. Hence, for our purposes it is sufficient to retain the simpler vector model procedures, although we shall occasionally use the $\tilde{\mathbf{L}}$, $\tilde{\mathbf{S}}$, and $\tilde{\mathbf{J}}$ vectors to remind ourselves of the connection between the predictions of the vector model and the exact theory.

We designate a level by the *notation*

$$n^2(L)_J$$

where n is the total quantum number, L (which only for one-electron spectra—hydrogen and the alkalis—is equal to l) is the *azimuthal quantum number*. The superscript 2 indicates that the levels are doubled, and the subscript J (here equal to j) denotes the angular momentum quantum number. The L-value symbol is chosen according to the scheme

L–value	0	1	2	3	4	5
Term symbol	S	P	D	F	G	H

The sodium ground level for which $l = L = 0$ and $j = l + \frac{1}{2} = 0 + \frac{1}{2}$ is written $3s^2\mathrm{S}_{1/2}$. Likewise $3p^2\mathrm{P}_{1/2}$ and $3p^2\mathrm{P}_{3/2}$ refer to the two lowest p-levels with $J = 1/2$ or $3/2$. The sodium D lines are represented by the transitions

$$3s^2\mathrm{S}_{1/2}\text{—}3p^2\mathrm{P}_{3/2}\ \lambda5889.953 \qquad 3s^2\mathrm{S}_{1/2}\text{—}3p^2\mathrm{P}_{1/2}\ \lambda5895.923$$

2–7. The Vector Model for Complex Atoms. For atoms with several electrons responsible for the optical spectrum, the positions, numbers, and kinds of energy levels can be computed by quantum mechanics. The vector model, nevertheless, predicts correctly the kind and number of energy levels in complex atoms; hence it is a useful device for remembering results of quantum mechanical calculations.

We suppose that to each electron we can assign n, l, and s values appropriate to the size and shape of the corresponding Bohr orbit and electron spin. The magnetic interactions attributed to the spins and orbits of the individual electrons are amenable to direct calculation by quantum mechanics. The *vector* $\boldsymbol{l}$'s and $\mathbf{s}$'s of the individual electrons are then added to get the total angular momentum $\mathbf{J}$ of the whole atom in a particular energy level. In order to perform this addition, we are guided by the results obtained from the experimental study of energy levels of individual atoms and ions.

It turns out that for light atoms the appropriate mode of vector combination is to add the $\mathbf{s}$'s of the individual electrons to form the total *spin* vector $\mathbf{S}$ and the individual $\boldsymbol{l}$'s to form the resultant angular momentum $\mathbf{L}$, viz.,

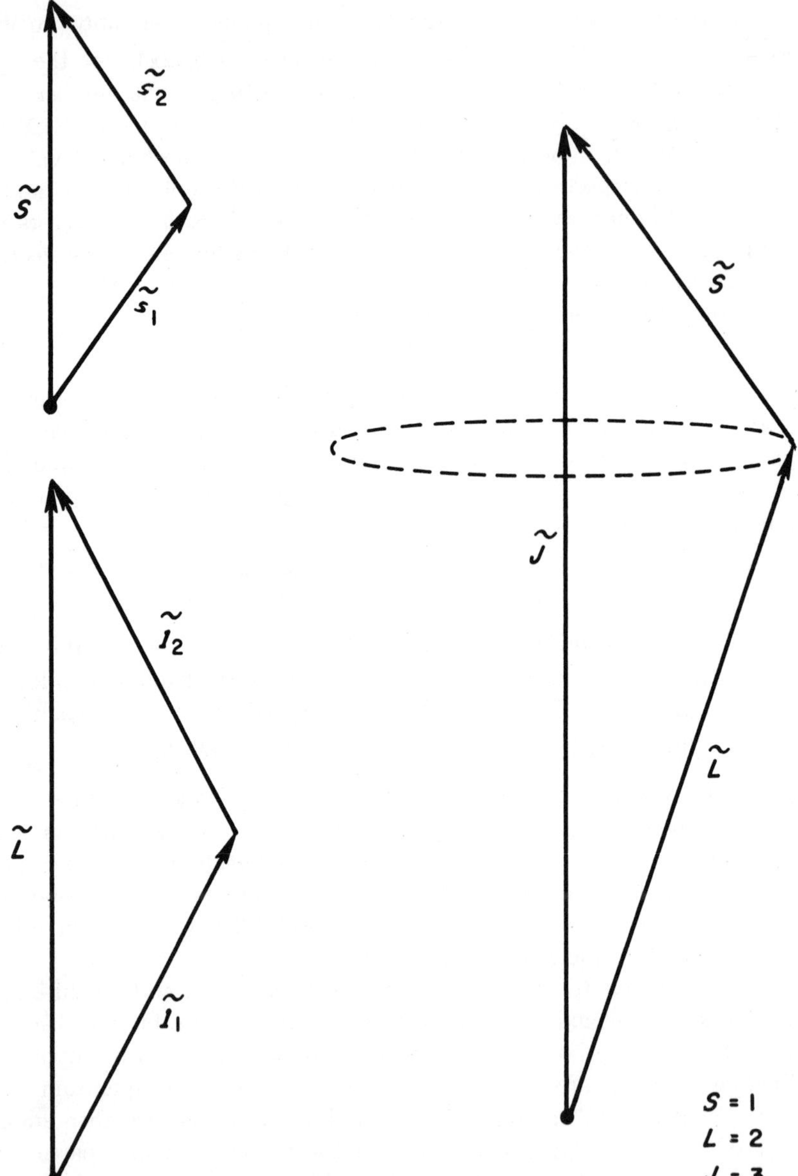

FIG. 2–9. VECTOR MODEL FOR AN ATOM WITH TWO OPTICAL ELECTRONS

We depict the combination of vectors to form a particular level corresponding to $S = 1$, $L = 2$, and $J = 3$. *Left:* The two individual orbital angular momentum vectors $\tilde{l}_1$ and $\tilde{l}_2$ combine to form the combined orbital angular momentum vector $\tilde{\mathbf{L}}$ corresponding to $L = 2$. The two spin angular momenta $\tilde{\mathbf{s}}_1$ and $\tilde{\mathbf{s}}_2$ are combined to form the vector $\tilde{\mathbf{S}}$ corresponding to $S = 1$. *Right:* The vectors $\tilde{\mathbf{L}}$ and $\tilde{\mathbf{S}}$ are combined to form the vector $\tilde{\mathbf{J}}$.

$$\mathbf{L} = \Sigma\, \mathbf{l}_i \qquad\qquad \mathbf{S} = \Sigma\, \mathbf{s}_i \tag{2–21}$$

The sum of these vectors

$$\mathbf{J} = \mathbf{L} + \mathbf{S} \tag{2–22}$$

gives the total *angular momentum* in a particular energy state; this mode of vector coupling is called *LS coupling*.[2] In Fig. 2–9 two orbital angular momentum vectors, each with $\mathbf{l} = 1$, are combined to form $\mathbf{L} = 2$, while the two spin orbital momentum vectors combine to form $\mathbf{S} = 1$. Then **L** and **S** combine to form $\mathbf{J} = 3$ as is shown on the right-hand side of the figure. The diagrams for $\mathbf{J} = 2$ and $\mathbf{J} = 1$ are not shown. The vectors **L** and **S** precess around **J**. We use here the notation $\tilde{J} = \sqrt{J(J+1)}$, etc.

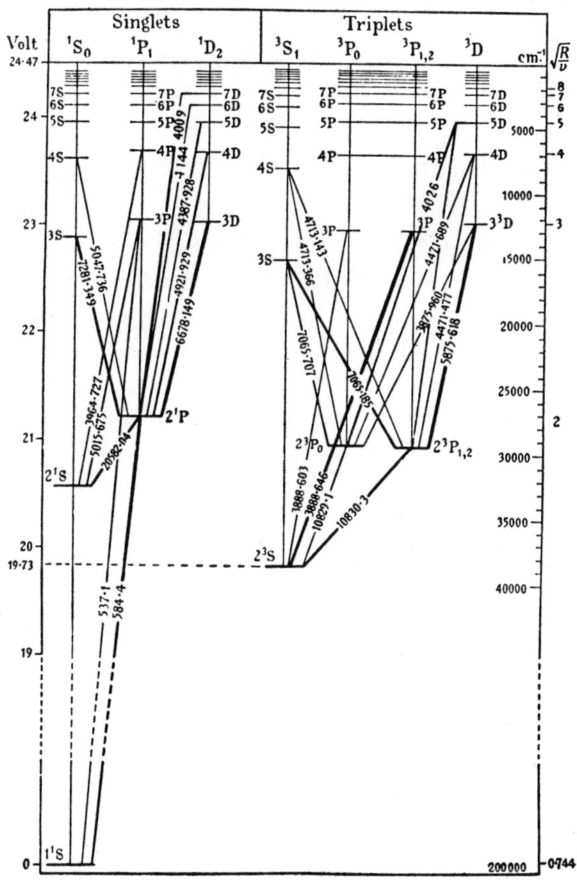

Fig. 2–10. Term Diagram for Helium

(Courtesy, O. Struve and K. Wurm, *Astrophysical Journal*, University of Chicago Press, **88**, 87, 1938.)

[2] The use of the letter *S* to denote both the term with $L = 0$ and the total spin vector is unfortunate. It arises from the fact that the notation S for the term $L = 0$ was firmly established before the **S** vector was invented.

Helium may serve as a simple example of a two-electron spectrum (see Fig. 2–10). The lowest level is represented by two 1s electrons, viz.:

$$1s^2 \qquad l_1 = 0 \quad l_2 = 0 \qquad\qquad L = l_1 + l_2 = 0$$

Pauli's exclusion principle (Art. 2–9) requires that the two s-electrons in the $n = 1$ have oppositely directed spins, i.e., $\mathbf{S} = \mathbf{s}_1 + \mathbf{s}_2 = \frac{1}{2} - \frac{1}{2} = 0$. Also, $\mathbf{J} = 0$. Levels for which $\mathbf{S} = 0$ are called singlet levels (denoted by the superscript 1). The ground level of helium is denoted as $^1\text{S}_0$, where the subscript 0 means $\mathbf{J} = 0$ and the S means $\mathbf{L} = 0$.

The first set of excited levels falls about 20 ev above the ground level. One electron remains in the 1s level; the other is excited to an $n = 2$ level, viz.:

Unexcited electron	1s	$n_1 = 1$	$l_1 = 0$ and	$s_1 = 1/2$
Excited electron	2s	$n_2 = 2$	$l_2 = 0$	$s_2 = 1/2$
or	2p	$n_2 = 2$	$l_2 = 1$	$s_2 = 1/2$

The 1s2s combination gives $L_1 = l_1 + l_2 = 0$ (S-term). The spins $\mathbf{s}_1$ and $\mathbf{s}_2$ may be added in two ways

$$\mathbf{S} = \mathbf{s}_1 + \mathbf{s}_2 = 1/2 + 1/2 = 0 \quad S = 0, L = 0, J = 0 \quad {}^1\text{S}_0$$
$$= 1 \quad S = 1, L = 0, J = 1 \quad {}^3\text{S}_1$$

Triplet and singlet terms of the same nl values do not coincide in energy (see Fig. 2–8). All terms in helium are singlets with $S = 0$ or triplets with $S = 1$.

The 1s2p combination results in a triplet $p(^3\text{P})$ term consisting of three levels of the same L and S but different J's and a singlet p (^{1}P) term which has but a single level. Now $L = l_1 + l_2 = 1$ and $S = 0$ or $S = 1$. Choosing $L = 1$, $S = 1$, vector addition gives $\mathbf{J} = \mathbf{1} + \mathbf{1} = 0$, 1, or 2, corresponding to $^3\text{P}_0$, $^3\text{P}_1$, and $^3\text{P}_2$ levels. With $\mathbf{L} = 1$, $\mathbf{S} = 0$, $\mathbf{J}$ has only one value, 1, and there results a $^1\text{P}_1$ level. The superscripts 1 and 3 refer to the *multiplicity of the term*, i.e., the number of levels when $L > S$. A specification of the n- and l-values of the individual electrons is said to give the *configuration*. In its ground state the configuration of helium is 1s^2. The excited configurations are ss, sp, sd, etc., the l-value of the second electron fixes the L-value of all terms.

The observed transitions in helium consist of jumps between triplet terms or between singlet terms, but not between singlet and triplet terms, i.e., there are ^{1}S—^{1}P, ^{3}S—^{3}P jumps but not ^{1}S—^{3}P transitions. The totality of transitions between 2 terms comprises a multiplet. In Fig. 2–10 notice that the ^{1}S term in helium lies lower than the ^{1}P term. The 1s2s^1S—1s^2 ^{1}S transition is strictly forbidden; the 2s^1S level is called a *metastable level*. In helium, which is in pure LS coupling, the ^{3}S level is also metastable, and atoms may escape from such levels only by going to a higher level or by giving up their energy to a passing electron (*superelastic collision*).

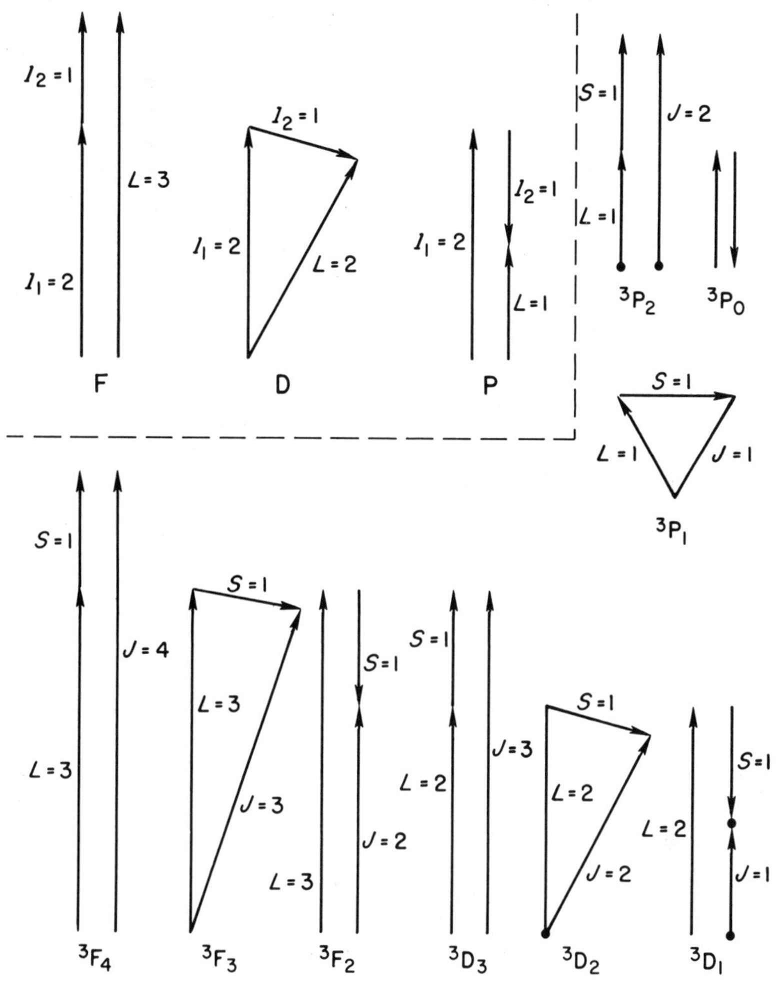

FIG. 2–11. ALLOWED LEVELS FOR A $1s^22s^22p3d$ CONFIGURATION

The $1s^22s^2$ electrons give only a closed shell with a $^1S_0$ term; the $2p3d$ electrons give singlet and triplet P, D, and F terms. The vector diagrams for the triplet levels only are depicted.

Now consider the problem of more complex atoms. Two electrons that have the same n- and l-values are called *equivalent electrons*. The normal state of carbon or doubly ionized oxygen is $1s^22s^22p^2$ which means there are two electrons in a closed shell with $n = 1$, $l = 0$, two in a shell with $n = 2$, $l = 0$, and two in a shell with $n = 2$, $l = 2$. The s electrons are all held in tightly bound shells for which $\mathbf{L} = 0$, $\mathbf{S} = 0$, $\mathbf{J} = 0$. There are three sets of equivalent electrons. Suppose now one of the $2p$ electrons becomes excited to a $3d$ orbit so the configuration of the atom is now $1s^22s^22p3d$. The p and d electrons are not equivalent; their l's may be combined in all possible ways to form **L** and the **s**'s similarly to form **S**. The p and d electrons determine the entire L, S, and J of the excited atom. The possible L's are

$$\mathbf{L} = \mathbf{1} + \mathbf{2} = 3(\text{F term}),\ 2(\text{D term}),\ 1(\text{P term})$$

See the upper left-hand box of Fig. 2–11. Since there are two electrons, $\mathbf{S} = 1$ (triplets) or $\mathbf{S} = 0$ (singlets), depending on whether the spins are parallel or opposed, and the terms are ^{3}P, ^{3}D, ^{3}F, or ^{1}P, ^{1}D, ^{1}F. For each term we must add the L and S values to get the J values of the individual levels. There results $^1P_1$, $^1D_2$, $^1F_3$, $^3P_{2,1,0}$, $^3D_{3,2,1}$, and $^3F_{4,3,2}$ (see Fig. 2–11). The vector schemes are shown for only the triplet levels since $S = 0$ for the singlets $L = J$.

The arithmetical sum of the l-values of the individual electrons defines the *parity* of the *configuration*. If the l-sum is odd, we use the superscript ^{0}. In the pd configuration $\Sigma l = 1 + 2 = 3$ so the levels are denoted as $^1P_1{}^0$, $^1D_2{}^0$, $^1F_3{}^0$, $^3P_2{}^0{}_{,1,0}$, $^3D_3{}^0{}_{,2,1}$, $^3F_4{}^0{}_{,3,2}$. In modern *spectroscopic notation* the designation of a given level is

$$^{(2S+1)}L_J{}^0 \text{ (when odd)}$$

First write the symbol for L according to the notation S, P, D, etc. The multiplicity, equal to $2S + 1$, is written in the upper left-hand corner and the J-value is added in a subscript in the lower right-hand corner. If the term is odd, a superscript ^{0} is placed in the upper right-hand corner. The complete spectroscopic designation of a level involves also the nl values of the electrons. For example, the ground level of carbon is $1s^22s^22p^2\ ^3P_0$, that of nitrogen is $1s^22s^22p^3\ ^4S^0{}_{3/2}$. We denote the spectrum of a neutral atom by a I following the chemical symbol; II, III, IV denote successive stages of ionization. Thus, O I means neutral oxygen, O II singly ionized oxygen, and O III doubly ionized oxygen.

The value of the spin vector **S** and hence the multiplicity, $2S + 1$, depends on the number of electrons engaged in the production of the outer spectrum. Two electrons give singlets and triplets. With three electrons we see that since the s vectors must be added parallel or antiparallel $S = 3/2$ or $1/2$. Four electrons will give $S = 2$, 1, or 0, i.e., quintets, triplets, and singlets. Oxygen supplies a useful illustration. The O I ground level is $1s^22s^22p^4\ ^3P_2$, but the excitation of one of the $2p$ electrons may yield $2p^33d$, etc., configurations. We omit the closed $1s^22s^2$ shells which contribute nothing. Four electrons are now involved, and there occur quintets, triplets, and singlets. Singly ionized oxygen contains three optical electrons, and there occur quartets and doublets, whereas doubly ionized oxygen with two optical electrons has only triplets and singlets. If three electrons are removed there will be doublets only as the $2p$ electron is excited. As the number of electrons active in the production of the optical spectrum changes, the multiplicity is alternately even (quartets, doublets, etc.) and odd (triplets, singlets, etc.). The arc spectrum of neutral manganese provides a classical example. It contains terms of high multiplicity, quartets, sextets, and octets, the ground level being $3d^5 4s^2\ {}^6S_{3/2}$. The Mn II energy level diagram is made up of singlets, triplets, and quintets.

The spectra of ions with the same number of outer electrons resemble one another in term structure except that with increasing atomic number the corresponding spectral lines become shifted to higher and higher frequencies. A series of ions with the same number of outer electrons is said to form an *isoelectronic sequence.* Thus the $1s^2 2s^2 2p^2$ isoelectronic sequence contains C I, N II, O III, F IV, and Ne V. Isoelectronic sequences have played a very important role in the identification of lines observed in the spectra of gaseous nebulae and of the solar corona.

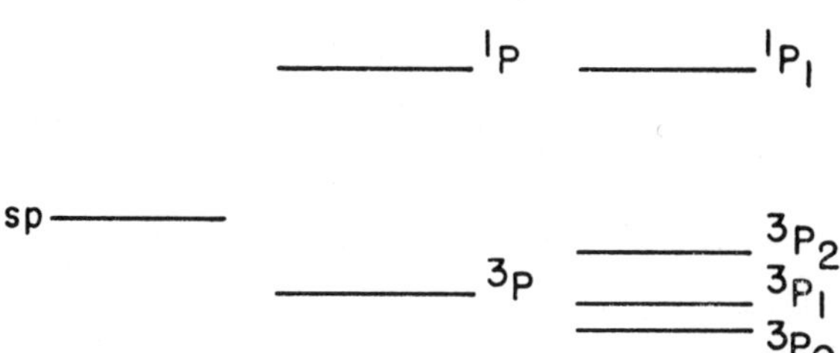

Fig. 2–12. Origin of Terms and Levels

Quantum mechanical calculations show that the separation of the terms of a given configuration arises from the electrostatic repulsion between the individual electrons. Thus the separation between the ${}^1P$ and ${}^3P$ terms in an *sp* configuration arises from these electrostatic effects. The fine structure of a particular term, e.g., the splitting of a ${}^3P$ term into ${}^3P_0$, ${}^3P_1$, and ${}^3P_2$ levels arises from the magnetic effects of *spin-orbit interaction.* (See Fig. 2–12.) The spin orbit interaction in complex atoms is always much larger than for hydrogen because the electrons move in a non-coulombic field. For a single electron the spin-orbit interaction energy will be given by

$$\gamma_i = a_i\, ls \cos(l, s) \qquad (2\text{–}23)$$

where cos (l, s) denotes the angle between the orbit and spin vectors. In *LS* coupling the **l** vectors combine to form **L,** the **s** vectors to form **S** and the total spin-orbit interaction energy becomes

$$\Gamma = \sum \gamma_i = \sum a_i\, \overline{ls \cos(l, s)} = ALS \cos(L, S) \qquad (2\text{–}24)$$

which in terms of the quantum mechanical results reduces to

$$\Gamma = \tfrac{1}{2}A\,[J(J+1) - L(L+1) - S(S+1)] \qquad (2\text{–}25)$$

Here A depends on the configuration and term but not on J. Now Γ represents the displacement of the level in question from the center of gravity of the term.

Consider a ^3P term for which $S = 1$, $L = 1$ and $J = 0, 1, 2$. We find that $\Gamma(^3P_2) = +A$, $\Gamma(^3P_1) = -A$, and $\Gamma(^3P_0) = -2A$. Eq. 2–25 embodies *Landé's interval rule:* in a normal multiplet, the separation between the levels is proportional to the J-value of the upper level. This rule is extremely useful in deciding whether or not a term is in good LS coupling. If A is negative instead of positive the term is inverted. Inverted terms occur when the electron shell is more than half-filled. For example, an np-shell may contain only six electrons (see Art. 2–9). The O I atom with a $1s^22s^22p^4$ configuration has a ground ^3P term in which the levels are inverted (see Fig. 2–24), i.e., $^3P_2$ lies lowest and $^3P_0$ is highest, whereas the opposite is true for C I or O III which belongs to the same isoelectronic sequence.

Isoelectronic sequences provide some interesting illustrations of term splittings. Let us consider the simplest example, i.e., the separation between the two doublet levels in a p configuration $^2P_{1/2}$ and $^2P_{3/2}$. In wave number units the spin-orbit interaction for non-penetrating orbits is

$$\Delta\tilde{\nu} = \frac{5.822\ (Z - \sigma)^4}{n^3l(l + 1)}\ \text{cm}^{-1} \tag{2–26}$$

Here Z is the atomic number and σ is very nearly constant, and for an isoelectronic sequence one finds that very nearly

$$\Delta\tilde{\nu}\ (^2P_{3/2} - {}^2P_{1/2}) = \text{const}\ (Z - \sigma)^4 \tag{2–27}$$

Edlén used this law in connection with the isoelectronic sequence $3s^23p$, Al I, Si II, P III, S IV, Cl V, etc., to identify the lines of Fe XIV and Ni XVI in the solar corona. The spectrum of the inner corona is a continuum upon which are superposed broad emission lines whose identification remained a mystery until 1939. One of the strongest of these lines is λ5302 due to thirteen-times ionized iron.

Table 2–1, which is abstracted from a table by Edlén, illustrates the method of identification. The first two columns give the atomic number and the ion; the third column gives the term separation, $^2P_{3/2} - {}^2P_{1/2} = \Delta\tilde{\nu}$, in the wave number units cm^{-1}, and the fourth column gives $\sqrt[4]{\zeta}$ where

$$\zeta = \tfrac{2}{3}\ \Delta\tilde{\nu} \tag{2–28}$$

Notice that the differences run very smoothly. If we assume that the green coronal line (whose wave number is 18,852 cm^{-1}) belongs to Fe XIV, the resultant mean value of the difference tabulated in the last column is quite consistent with the run of the table. The ions in parentheses are those whose term splittings had not yet been observed; c denotes an observed coronal line. Fe X and Ni XII can be identified in the same way from the $3s^23p^5$ isoelectronic sequence. The intensity ratios of the iron and nickel lines give further support to the suggested identifications. These transitions occur between terms of the ground configuration. Hence they differ fundamentally from ordinary transitions. In fact they are called *forbidden lines.* (See Art. 2–12.) Edlén identified the other coronal lines in a more complicated but equally accurate way.

A comparison of term separations with the distances between individual levels composing a term will indicate how close this particular term of this

TABLE 2–1

GROUND TERM SPLITTING IN ISOELECTRONIC SEQUENCE $3s^23p$*

Atomic No.	Ion	${}^2P_{3/2} - {}^2P_{1/2}$ Term Separation cm^{-1}	$\sqrt[4]{\zeta}$	Difference
13	Al I	112.04	2.939	
				0.781
14	Si II	287.3	3.720	
				0.675
15	P III	559.6	4.395	
				0.622
16	S IV	950.2	5.017	
				0.599
17	Cl V	1,492	5.616	
				0.579
18	A VI	2,210	6.195	
				0.564
19	K VII	3,131	6.759	
				0.560
20	Ca VIII	4,305	7.319	
				0.552
21	Sc IX	5,759	7.871	
22	(Ti X)			
23	(V XI)			
				0.543
24	(Cr XII)			
25	(Mn XIII)			
26	Fe XIV	18,852.5c	10.588	
				0.538
27	(Co XV)	——	——	
28	Ni XVI	27,762c	11.664	

* P. Swings, "Edlén's Identification of the Coronal Lines," *Ap. J.* **98,** 119, 1943.

atom comes to LS coupling. If, as in helium, the separation of terms of a given configuration much exceeds the distance between levels, an LS coupling approximation is a good one. Otherwise, l's of individual electrons no longer combine strictly with one another to form L, nor do the s's combine to form S. Instead the $\mathbf{l}$ vector of a given electron may interact with its own spin as well as with the $\mathbf{l}$ vectors of other electrons. This is the condition of *intermediate coupling.*

An extreme condition is that of *jj coupling,* wherein the $\mathbf{l}$ and $\mathbf{s}$ vectors of each electron combine to form individual $\mathbf{j}$'s and these in turn combine to form a resultant $\mathbf{J}$. Departures from LS coupling become important in the heavier elements and noble gases. For example, consider the successive atoms in the sequence $np(n + 1)s$ (see Fig. 2–12). Sometimes, as in carbon, a single atom displays a transition from LS to jj coupling; the lowest level is in good LS coupling, whereas the higher levels show pronounced departures.

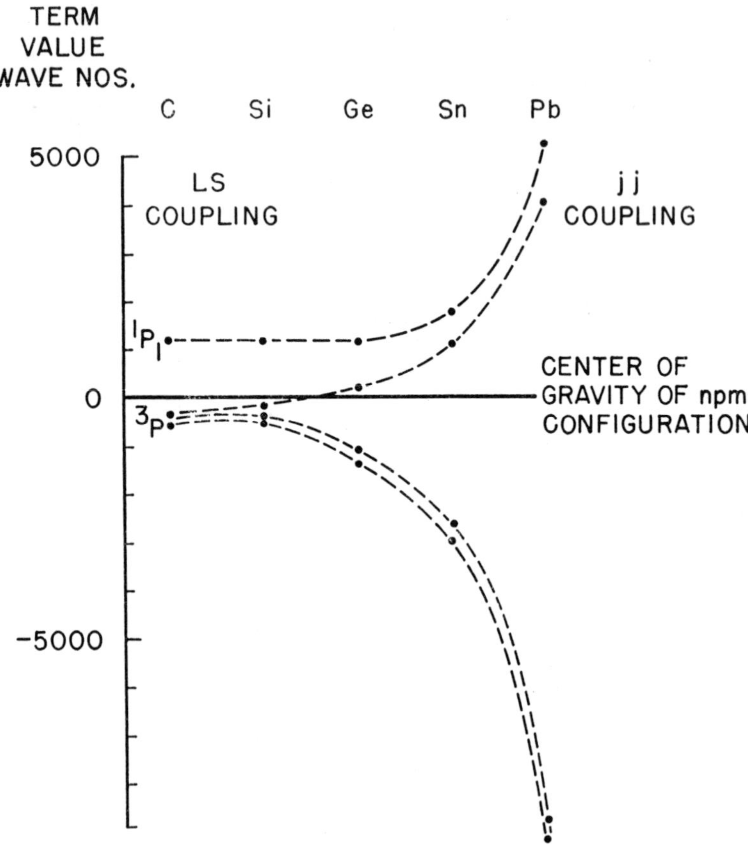

FIG. 2–13. DEPARTURES FROM *LS* COUPLING

The level separations of the first excited ^1P and ^3P terms in C, Si, Ge, Sn, and Pb, whose configurations are $2p3s$, $3p4s$, $4p5s$, $5p6s$, and $6p7s$, respectively, exhibit a transition from fair *LS* coupling to almost complete *jj* coupling. Notice that in carbon the triplet-singlet separation is much greater than is the splitting of the ^3P term, but that in the heavier atoms, $^3P_0$ and $^3P_1$ deviate in one direction and $^3P_2$ and $^3P_1$ in the other. The horizontal line represents the average energy of the configuration. Energies are plotted in wave number units.

Consider a $2pnp$ configuration in C I, N II, or O III. As n increases the l_1l_2 and s_1s_2 couplings weaken and as $n \rightarrow \infty$, $\boldsymbol{l}_1$ and $\mathbf{s}_1$ tend to couple together as do $\boldsymbol{l}_2$ and $\mathbf{s}_2$ to form $\mathbf{j}_1$ and $\mathbf{j}_2$ vectors which in turn couple to form a resultant $\mathbf{J}$. A $2pnp$ configuration yields ten levels, viz., $^1S_0$ $^1P_1$ $^1D_2$ and $^3S_1$ $^3P_{0,\,1,\,2}$ $^3D_{\,1,\,2,\,3}$, but when the atom is ionized the remaining $2p$ electrons gives rise to $^2P_{1/2}$ and $^2P_{3/2}$ levels. As n increases, one set of levels converges to the $^2P_{1/2}$, the other to $^2P_{3/2}$ in such a way that the l and s of the remaining electron tends to be preserved, and states of the same J do not cross in passing to the limit. (See Fig. 2–13.) Notice that the

^{2}P term splitting is comparable with that of the ^{3}P term. Germanium shows a good example of jj coupling (White, 1934).

The terms and levels of a pp (e.g., $2pnp$) configuration in the two types of coupling are:

LS coupling	^{3}S$_1$	^{3}P$_{0,\,1,\,2}$	^{3}D$_{1,\,2,\,3}$	^{1}S$_0$ ^{1}P$_1$ ^{1}D$_2$
jj coupling	$(3/2, 3/2)_{3,2,1,0}$	$(3/2, 1/2)_{2,1}$	$(1/2, 3/2)_{2,1}$	$(1/2, 1/2)_{1,0}$

We use the notation suggested by White $(j_1, j_2)_J$ to denote levels in jj coupling. Notice that the same number of levels with the same J-values are found in the two types of coupling.

For intermediate coupling the vector model is no longer applicable, but precise calculations can be made by quantum mechanical methods, and levels of the same J can be traced from the scheme of LS to that of jj coupling. That is, J always remains a "good" quantum number even though L and S may not be. Deviations from LS coupling are usually expressed in terms of a coupling parameter

$$\chi = \zeta/F \tag{2-29}$$

ζ measures the separation between levels of a given term and where F is an interaction term due to electrostatic repulsions between electrons, and measures the separation between the different terms of a configuration (e.g., between the ^{1}D and ^{3}P terms of a p^2 configuration). In pure LS coupling $\chi = 0$; as jj coupling is approached, χ increases without bound.

2–8. The Zeeman and Paschen-Back Effects. Under the influence of a magnetic field, a spectrum line is broken up into a number of components whose separation depends on the strength of the magnetic field and on the L, S, and J values of the levels involved in the transition. From a study

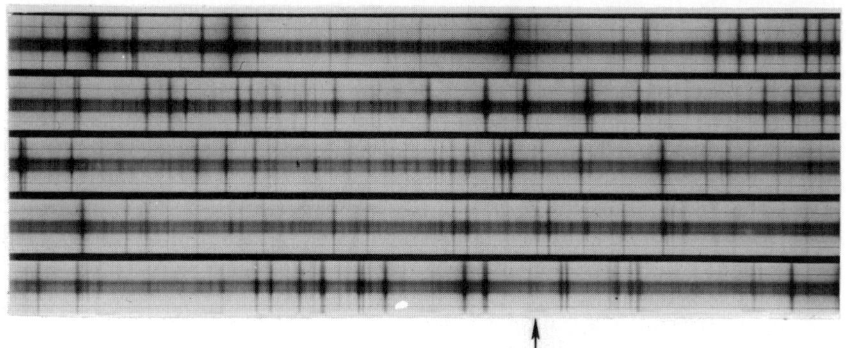

FIG. 2–14. ZEEMAN EFFECT IN SUN SPOTS

Observations of the wavelength range λ5180–λ5280 in a sunspot spectrum as observed at McMath-Hulbert Observatory, September 3, 1959, by O. C. Mohler. Notice the extreme effect on the λ5250 line (strip d). (McMath-Hulbert Observatory of the University of Michigan.)

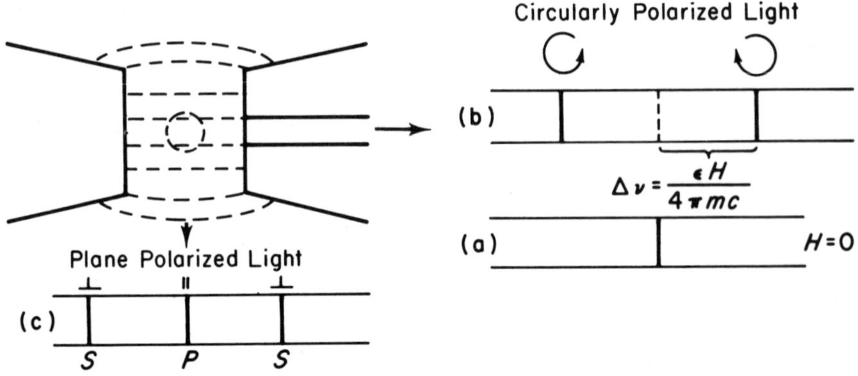

FIG. 2–15. SIMPLE ZEEMAN EFFECT

of this Zeeman effect a great deal can be learned about the nature of the energy levels involved, i.e., the type and multiplicity of the terms and deviations from LS coupling. Among astrophysical applications, use is made of the polarization and separation of the Zeeman components to assess the magnitude and direction of the fields in sunspots and plages (see Fig. 2–14), and to determine the strength of the fields in magnetic stars.

First, let us consider the simple (sometimes called "normal") Zeeman effect produced in singlets in a magnetic field (Fig. 2–15). Suppose the emitting gas is placed in a magnetic field and viewed along a direction perpendicular to the magnetic field. The singlet line a is observed to be split into three components c. The central component is undisplaced in frequency and is linearly polarized parallel p to the field direction. One component is displaced to a frequency $\nu_1 = \nu_0 - (H\varepsilon/4\pi mc)$ and the other to a frequency $\nu_2 = \nu_0 + (H\varepsilon/4\pi mc)$ where ε is given in electrostatic units, m and c in cgs units. These displaced s components are polarized perpendicular to the field.

If one views the emitting gas along a direction parallel to the field, e.g., through a hole bored through the pole piece, he finds that the central component is missing and the two outer components are circularly polarized in *opposite* directions b. H. A. Lorentz was able to explain this simple Zeeman effect on the basis of a purely classical model of the radiating atom. We give here, however, the interpretation based on the vector model. For singlets, $\mathbf{L} = \mathbf{J}$ and the projection M of the vector $\mathbf{J}$ on the direction of the magnetic field is $M = J, J - 1, \ldots, -(J - 1), -J$. There are $2J + 1$ values of M for each energy level M. Consider the $^1F_3 - {}^1D_2$ transition as an illustration (see Fig. 2–16A).

$^1F_3$ $M =$ 3 2 1 0 −1 −2 −3

$^1D_2$ $M =$ 2 1 0 −1 −2

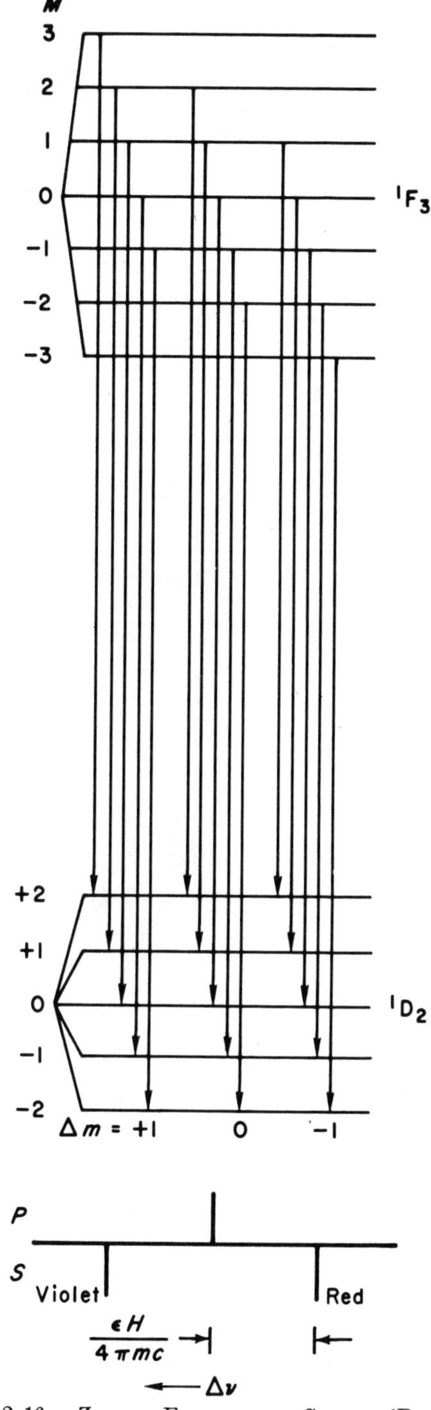

Fig. 2–16a. Zeeman Effect for a Singlet $^1D_2 — ^1F_3$

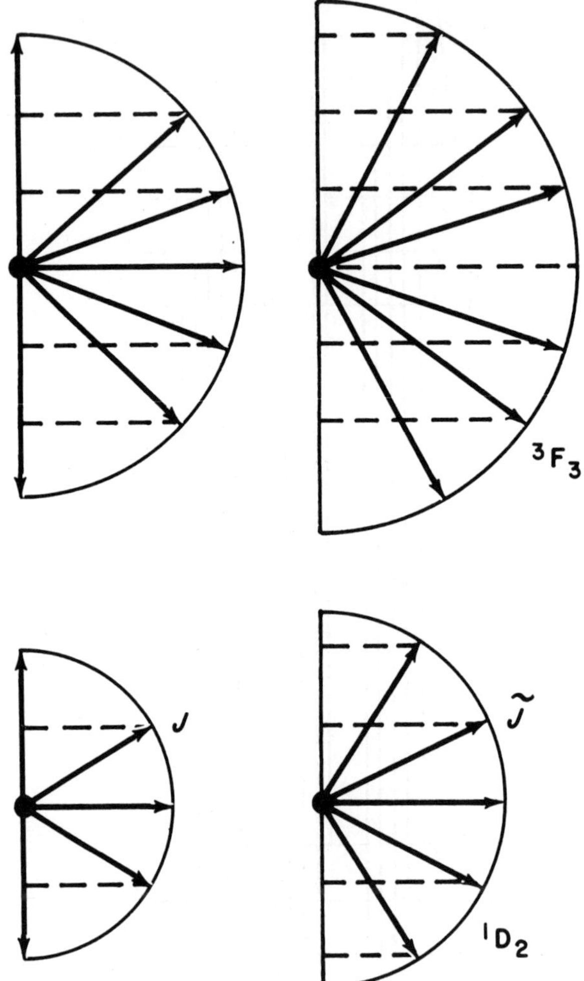

Fig. 2–16b. Projection of J Vector on a Magnetic Field for Singlets

Left: Vector model representation. *Right:* Representation according to the quantum mechanical picture $\tilde{J} = \sqrt{J(J+1)}$.

The selection rules on the transition are that $\Delta M = +1$ (giving the higher frequency component), $\Delta M = -1$ (giving the lower frequency component, and $\Delta M = 0$ (giving the undisplaced component). Notice that in spite of the large number of individual transitions, only three components are observed. The projection of the angular momentum on the field must always assume integral values. See Fig. 2–16b.

Transitions involving lines other than singlets show the "complex" or "anomalous" Zeeman effect. Instead of just three components, there are now several whose intensities and displacements form a symmetrical pattern with respect to the undisplaced position.

In order to evaluate the magnitude of the complex Zeeman effect, it is necessary to calculate for a particular atomic level the total magnetic moment, taking into account the differing ratios of the magnetic moment to angular momentum for electronic spin and orbital motion. Let us assume LS coupling. The total angular momentum $\mathbf{J}$ assumes an angle θ with the direction of the field such that the projection of $\mathbf{J}$ on $\mathbf{H}$ is an integer if J is an integer, a half-integer if J is a half-integer Now $\mathbf{J}$ precesses about the field $\mathbf{H}$, but $\mathbf{L}$ and $\mathbf{S}$, and therefore also $\boldsymbol{\mu}_S$, and $\boldsymbol{\mu}_J = \boldsymbol{\mu}_L + \boldsymbol{\mu}_S$ precess at a much greater rate about $\mathbf{J}$. Therefore, the quantity relevant to the total magnetic moment is the projection of $\boldsymbol{\mu}_J$ on the direction of the vector $\mathbf{J}$, i.e., μ_P. See Fig. 2–17. Hence, noting Eqs. 2–14, 2–15, 2–16, and 2–17, we take the projections of $\boldsymbol{\mu}_L$ and $\boldsymbol{\mu}_S$ on the direction J as follows:

$$\mu(L, J) = L \frac{h}{2\pi} \frac{\varepsilon}{2mc} \cos \alpha, \quad \mu(S, J) = S \frac{h}{2\pi} \frac{2\varepsilon}{2mc} \cos \beta \qquad (2\text{–}30)$$

whence

$$\mu_P = \left[L \cos \alpha + 2S \cos \beta \right] \frac{h}{2\pi} \frac{\varepsilon}{2mc} \qquad (2\text{–}31)$$

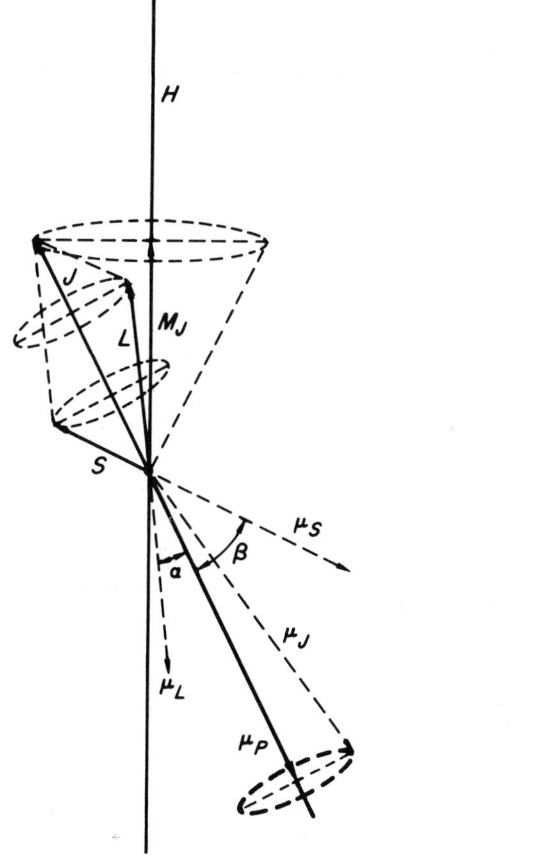

Fig. 2–17. Projection of J Vector on Magnetic Field for Complex Zeeman Effect

We now introduce the Landé g-factor by the definition

$$Jg = [L \cos \alpha + 2S \cos \beta] \tag{2–32}$$

By the cosine law

$$L \cos \alpha = \frac{J^2 + L^2 - S^2}{2J} \qquad S \cos \beta = \frac{J^2 - L^2 + S^2}{2J}$$

whence

$$g = 1 + \frac{J^2 + S^2 - L^2}{2J^2} \tag{2–33}$$

To obtain the correct quantum mechanical expression from this vector model result we must make the replacements

$$J^2 \rightarrow J(J+1) \qquad L^2 \rightarrow L(L+1) \qquad S^2 \rightarrow S(S+1) \tag{2–34}$$

whence

$$g = 1 + \frac{J(J+1) + S(S+1) - L(L+1)}{2J(J+1)} \tag{2–35}$$

and the ratio of magnetic and mechanical moments for the energy level defined by L, S, and J is

$$\frac{\mu(L, S, J)}{p\,(L, S, J)} = \frac{gJ \dfrac{h}{2\pi} \dfrac{\varepsilon}{2mc}}{J \dfrac{h}{2\pi}} = g \frac{\varepsilon}{2mc} = \frac{\omega_L}{H} \tag{2–36}$$

where ω_L is the rate of Larmor precession of the vector $\mathbf{J}$ in the field $\mathbf{H}$. If M denotes the projection of the vector $\mathbf{J}$ on the field, the energy shift of a given Zeeman state is given by

$$\Delta E = \frac{\varepsilon h}{4\pi mc} HgM \tag{2–37}$$

In general, g will differ from unity and will not be the same for both upper and lower levels.

As a simple example, we may consider the resonance $^2\mathrm{S} - {}^2\mathrm{P}$ lines of sodium or ionized calcium. For the $^2\mathrm{P}_{3/2}$ level

$$g = 1 + \frac{\frac{3}{2}\frac{5}{2} + \frac{1}{2}\frac{3}{2} - 1(2)}{2\,\frac{3}{2}\frac{5}{2}} = \frac{4}{3}$$

Similarly, one finds for the $^2\mathrm{P}_{1/2}$ and $^2\mathrm{S}_{1/2}$ levels $g = 2/3$ and $g = 2$, respectively. In general, S-levels give $g = 2$, since only electron spin determines the angular momentum. We calculate the patterns as follows:

$^2\mathrm{S}_{1/2} - {}^2\mathrm{P}_{3/2}$

M_J	$-3/2$	$-1/2$	$+1/2$	$+3/2$
$Mg(^2\mathrm{P}_{3/2})$	$-6/3$	$-2/3$	$+2/3$	$+6/3$
$Mg(^2\mathrm{S}_{1/2})$		$-3/3$	$+3/3$	
M_J		$-1/2$	$+1/2$	

$^2\mathrm{S}_{1/2} - {}^2\mathrm{P}_{1/2}$

M_J	$-1/2$	$+1/2$
$Mg(^2\mathrm{P}_{1/2})$	$-1/3$	$+1/3$
$Mg(^2\mathrm{S}_{1/2})$	$-3/3$	$+3/3$
M_J	$-1/2$	$+1/2$

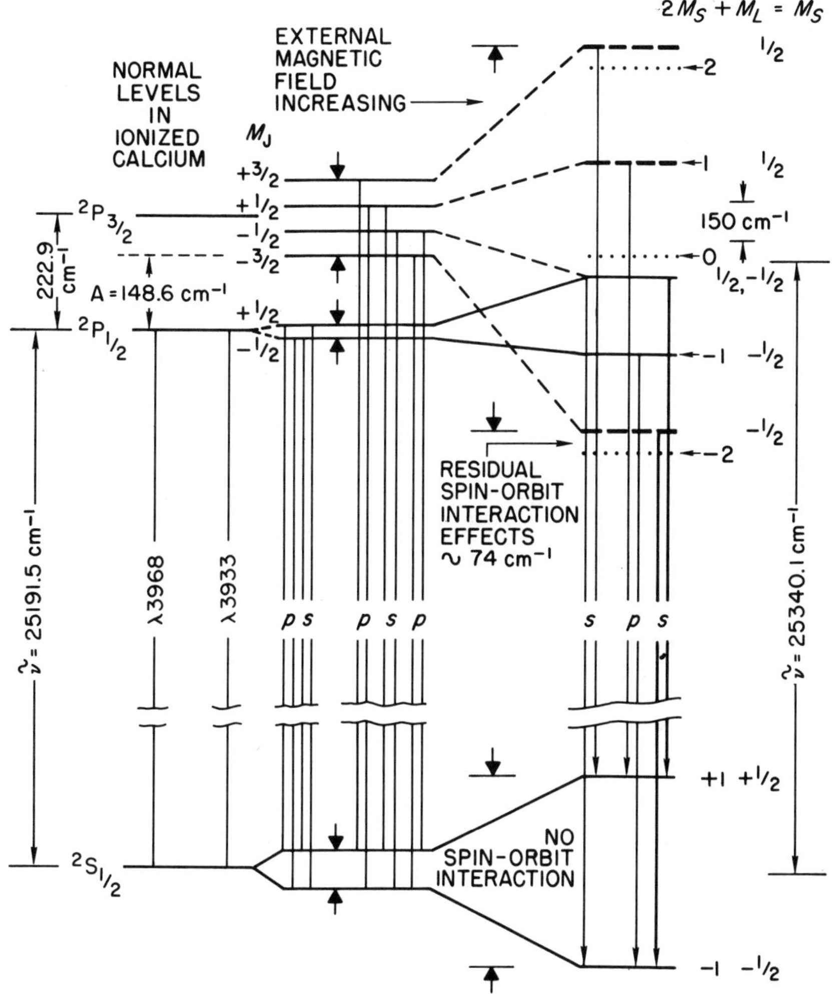

Fig. 2-18. Zeeman and Paschen-Back Effects for the H and K Lines of Ionized Calcium

The normal positions of the levels are shown at the left. Level separations are given in cm^{-1}. The Zeeman splitting is shown in the center sketch and the polarizations of the transitions are indicated. At the right the strong-field Paschen-Back pattern is shown; the scale of the splitting is shown reduced by a factor 2 compared with the left-hand side. The heavy arrows indicate the extent of the interaction between the atomic electrons and the external field; it is more than an order of magnitude greater than the residual spin-orbit interaction effects ~ 74 cm^{-1}.

The vertical arrows which correspond to $\Delta m = 0$ give the p components, the diagonals the s components. The pattern for the $^2S_{1/2} - {}^2P_{3/2}$ transition can be represented by $\Delta\nu = \pm 1/3$ (p) and $\Delta\nu = \pm 3/3$, $\pm 5/3$ (s), the displacements being

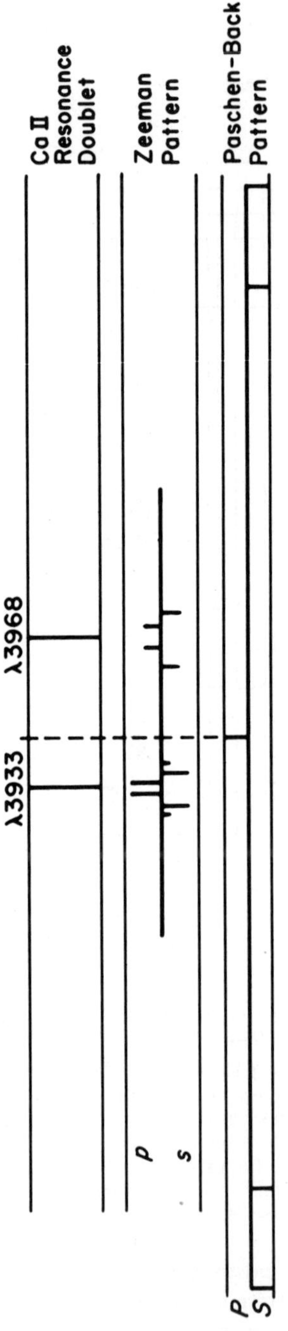

FIG. 2–19. EFFECT OF A MAGNETIC FIELD ON THE CA II RESONANCE LINES, λ3933 AND λ3868

The field free positions of the lines is depicted in the top strip. The lower two strips show the Zeeman and Paschen-Back effects respectively, the p and s components being depicted separately.

measured in units of $(\varepsilon H/4\pi mc)$. Similarly, the $^2S_{1/2}$ — $^2P_{1/2}$ transition gives $\pm 2/3$ (p) and $\pm 4/3$ (s) (see Fig. 2–18). The corresponding intensity pattern is depicted in the middle strip of Fig. 2–19.

In a transition like $^3P_1$—$^3D_2$, we find that $g = 7/6$ and $3/2$ for the D and P terms, respectively. Then the p components are found to be displaced by $\pm 2/6$, whereas the s components are shifted by $\pm 5/6$, $\pm 7/6$, and $\pm 9/6$, respectively. It is apparent that in transitions involving terms of such high multiplicity as occur in the rare earth, e.g., ^7H—^7G, quite elaborate Zeeman patterns can be obtained. Notice that the Zeeman patterns differ for different types of multiplets but are identical for successive members of a series.

In the Zeeman effect the coupling between the external magnetic field and **J** vector was much smaller than the coupling between **L** and **S** to form **J**. This statement implies that the precession of **L** and **S** about **J** is much faster than the precession of **J** about **H**. It further implies that the over-all spread of the Zeeman pattern is small compared with the separation $\Delta\nu \sim A$ of the lines in the multiplet.

We must now inquire what happens when the splitting produced by the external magnetic field exceeds the spin-orbit interaction. This is the *Paschen-Back effect*. The vectors **L** and **S** now precess almost independently of one another about the field, and the total energy of the atom in a particular state now consists of three contributions: (1) energy of precession of **L** about **H**, (2) energy of precession of **S** about **H**, and (3) the interaction between **L** and **S**. The change in energy of a particular energy state produced by the magnetic field depends on the projection of the corresponding magnetic moment on the magnetic field. Thus,

$$\Delta E_{L,H} = H\frac{\varepsilon}{2mc}M_L\frac{h}{2\pi}, \qquad \Delta E_{S,H} = H\frac{\varepsilon}{2mc}2M_S\frac{h}{2\pi}\ \text{ergs} \tag{2–38}$$

while the interaction between **L** and **S** depends on the circumstance that each precesses independently of the other about **H**, the interaction energy is $\sim AM_LM_S$. Hence the total energy of the state will be

$$\Delta E = (M_L + 2M_S)\frac{\varepsilon h}{4\pi mc}H + A\,M_LM_S \tag{2–39}$$

Notice now that the magnetic states of the entire term overlap. In Fig. 2–18, the Paschen-Back levels are shown at the extreme right; the states correlated with the $^2P_{3/2}$ level are indicated by dotted lines. The vertical scale for the Paschen-Back levels is chosen half that of the Zeeman and field free patterns on the left. Notice that the states belonging to $^2P_{3/2}$ spread over a huge range in energy, completely engulfing the states arising from $^2P_{1/2}$. A similar effect is shown in Fig. 2–20, where we depict the Paschen-Back effect on the ground term of O I, where A is negative. Notice that the number of Zeeman levels is $(2J + 1)$ for each level J, and the total number belonging to a given term is $\Sigma\,(2J + 1) = (2L + 1)\,(2S + 1)$ which is precisely the number of Paschen-Back levels. One may correlate the Zeeman and Paschen-Back levels by noting that $M = M_L + M_S$ does not change in going from a weak to a strong field and that when there is more than one level with the same M, no two levels having the same M will cross (for details, see White,

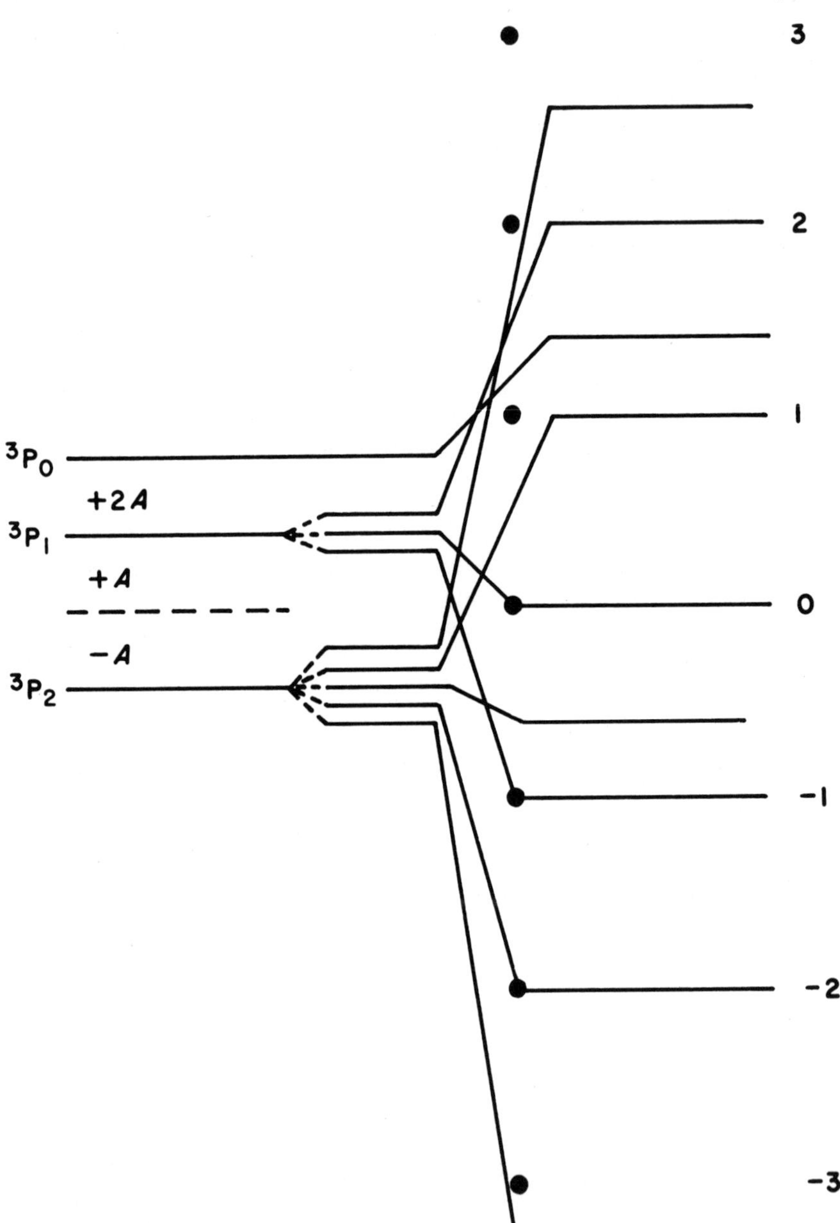

Fig. 2–20. Paschen-Back Effect on Ground Term of O I

1934). The selection rules for the transition between the individual Paschen-Back states are that $\Delta M_L = 0$ (p component), ± 1 (s component), but that $\Delta M_S = 0$ (for LS coupling), whereas for the Zeeman component $\Delta M_J = 0, \pm 1$. Some components that are permitted by the Zeeman selection rules no longer appear in the strong field. The calculation of intensities and positions of states for intermediate field strengths requires quantum mechanics and is rather difficult. One can also discuss the Zeeman and Paschen-Back effects for intermediate and jj coupling (see Condon and Shortley, 1935; White, 1934).

Fig. 2–19 depicts the Zeeman and Paschen-Back patterns for a doublet such as the resonance pair of Ca II. The Zeeman patterns of these lines can be observed, the p and s components being depicted in the conventional manner. The Paschen-Back effect is not observable for Ca II. Notice that the Paschen-Back pattern resembles the normal Zeeman triplet, but the s components instead of being single are doubled, the separation being comparable with the fine structure splitting. The Paschen-Back pattern in Fig. 2–20 is likewise purely theoretical.

In practice the Paschen-Back effect can be observed in the laboratory only for multiplets with very small spin-orbit splitting such as the resonance line of lithium, and even then huge fields are required. Hence, unlike the Zeeman effect, it is of little direct astrophysical importance; its principal value lies in the insight it gives in certain problems of atomic spectra and structure to which we now turn.

2–9. Pauli's Exclusion Principle and the Periodic Table. Our vector coupling model permits the calculation of terms for non-equivalent electrons. Thus a $2p3p$ configuration would give the terms ^{1}S, ^{1}P, ^{1}D, and ^{3}S, ^{3}P, ^{3}D. For a p^2 configuration, however, the only observed terms are

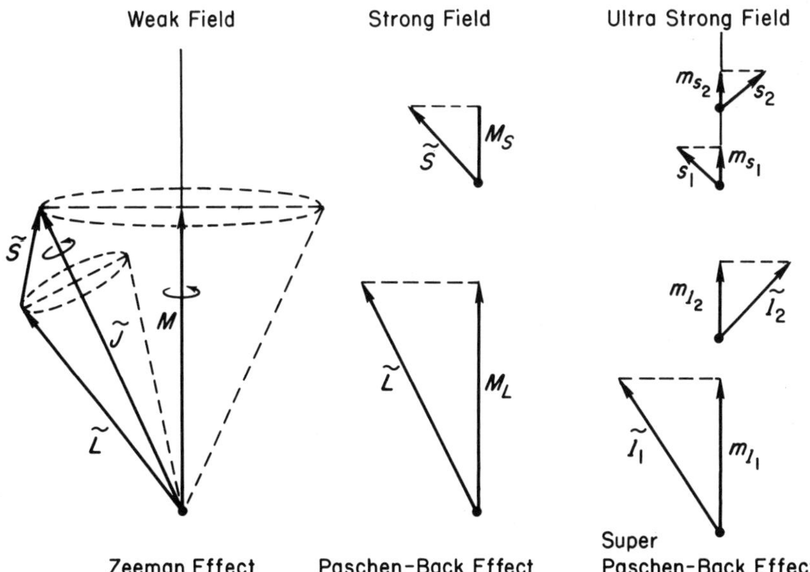

FIG. 2–21. VECTOR COUPLING SCHEMES FOR THE ZEEMAN EFFECT, THE PASCHEN-BACK EFFECT AND A HYPOTHETICAL SUPER PASCHEN-BACK EFFECT

^{1}S, ^{1}D, ^{3}P. The number of allowed terms is restricted by Pauli's exclusion principle. In order to explain this principle we consider what happens if the strength of the magnetic field increased to very large (experimentally unattainable) values. The situation is depicted in Fig. 2–21. When the field is weak **L** and **S** precess about **J** much faster than **J** precesses about the direction of the external field. As the field strength is increased, the coupling between L and S breaks down and we get the Paschen-Back effect; but if the field could be still further increased, the coupling between $\boldsymbol{l}_1$ and $\boldsymbol{l}_2$ to form **L** or between $\boldsymbol{l}_1$ and $\mathbf{s}_1$ to form **j**, etc., would be broken down and each vector would precess independently of the other with projections m_{l1}, m_{s1}, etc., on the field direction. This is the "super" Paschen-Back effect, and the energies of the states would now spread over a range in term value comparable with that of the energy spread of the entire configuration. That is, the magnetic splitting would have to be greater than the term separation, e.g., between the D, P, and S terms in a pp configuration.

It is important to note that the same number of states occur in a strong field as in a weak field situation. We may correlate "weak" and "strong" field states by noting that $M_L = \Sigma m_l$, $M_S = \Sigma m_s$, and $M_J = M_L + M_S$.

Pauli's principle states that *no two electrons in the same atom may have identical values of the four quantum numbers* n, l, m_l, m_s. For example, the 1s electrons in the ground level of helium have $n = 1$, $l = 0$, $m_l = 0$, and $m_s = \frac{1}{2}$ and $-\frac{1}{2}$, respectively. Hence $S = 0$ and there results a ^{1}S$_0$ level. No additional electron can be added in the $n = 1$ shell. In complex atoms the number of permitted terms involving equivalent electrons is greatly reduced. Pauli's exclusion principle explains the grouping of electrons in shells and hence the periodic table.

Consider a p electron for which $l = 1$, $s = \frac{1}{2}$. Now m_l can be $+1$, 0, or -1 and m_s can be $+\frac{1}{2}$ or $-\frac{1}{2}$. Using Condon and Shortley's notation, we write 1^+ for the combination $m_l = +1$, $m_s = +\frac{1}{2}$, -1^+ for $m_l = -1$, $m_s = +\frac{1}{2}$, and so forth. A p electron can have six possible combinations of m_l and m_s, viz., 1^+, 1^-, 0^+, 0^-, -1^+, -1^-, and hence six and only six p electrons can be placed in a shell. Similarly, there are ten d electrons in a closed shell since we have 2^+, 2^-, 1^+, 1^-, 0^+, 0^-, -1^+, -1^-, -2^+, -2^-. Similarly, a closed shell of f electrons contains fourteen electrons, and quite generally the maximum number of equivalent electrons will be $2l(l + 1)$.

To show how Pauli's exclusion principle can be applied to find the allowed states for a p^3 configuration, we proceed as follows. Combine the six electrons in all possible triads as indicated in Table 2–2 (top). Then the combinations are arranged as shown in the box with M_L and M_S as arguments. There are twenty zero-order states which can be identified with twenty Paschen-Back states; they are in turn identified with the Paschen-Back states of a ^{4}S, a ^{2}D, and a ^{2}P term.

TABLE 2-2

Calculation of Allowed Terms for a p^3 Configuration

	M_L	M_S		M_L	M_S		M_L	M_S		M_L	M_S
$1^+\ 1^-\ \ 0^+$	2	$+\frac{1}{2}$	$1^+\ 0^+\ -1^+$	0	$+\frac{3}{2}$	$1^-\ \ 0^+\ -1^+$	0	$+\frac{1}{2}$	$1^-\ -1^+\ -1^-$	-1	$-\frac{1}{2}$
$1^+\ 1^-\ \ 0^-$	2	$-\frac{1}{2}$	$1^+\ 0^+\ -1^-$	0	$+\frac{1}{2}$	$1^-\ \ 0^+\ -1^-$	0	$-\frac{1}{2}$	$0^+\ \ 0^-\ -1^+$	-1	$+\frac{1}{2}$
$1^+\ 1^-\ -1^+$	1	$+\frac{1}{2}$	$1^+\ 0^-\ -1^+$	0	$+\frac{1}{2}$	$1^+\ -1^+\ -1^-$	-1	$+\frac{1}{2}$	$0^+\ \ 0^-\ -1^-$	-1	$-\frac{1}{2}$
$1^+\ 1^-\ -1^-$	1	$-\frac{1}{2}$	$1^+\ 0^-\ -1^-$	0	$-\frac{1}{2}$	$1^-\ \ 0^-\ -1^+$	0	$-\frac{1}{2}$	$0^+\ -1^+\ -1^-$	-2	$+\frac{1}{2}$
$1^+\ 0^+\ \ 0^-$	1	$+\frac{1}{2}$	$1^-\ 0^+\ \ 0^-$	1	$-\frac{1}{2}$	$1^-\ \ 0^-\ -1^-$	0	$-\frac{3}{2}$	$0^-\ -1^+\ -1^-$	-2	$-\frac{1}{2}$

M_S \ M_L	2	1	0	-1	-2
$+3/2$			$1^+0^+-1^+$		
$+1/2$	$1^+1^-0^+$	$1^+1^--1^+$ $1^+0^+0^-$	$(1^+0^+-1^-)\ (1^+0^--1^+)$ $(1^-0^+-1^+)$	$1^+-1^+-1^-$ $0^+0^--1^+$	$0^+-1^+-1^-$
$-1/2$	$1^+1^-0^-$	$1^+1^--1^-$ $1^-0^+0^-$	$(1^+0^--1^-)\ (1^-0^+-1^-)$ $(1^-0^--1^+)$	$1^--1^+-1^-$ $0^+0^--1^-$	$0^--1^+-1^-$
$-3/2$			$1^-0^--1^-$		

$^4S_{3/2}$ $M_L = 0$ $M_S = -3/2, -1/2, +1/2, +3/2$

^2D $M_L = 2, 1, 0, -1, -2$ $M_S = -1/2, +1/2$

^2P $M_L = 1, 0, -1$ $M_S = -1/2, +1/2$

A similar analysis carried out for a p^2 configuration yields a ^3P, a ^1D, and a ^1S term.

The Pauli exclusion principle can also be applied to calculate allowed terms in jj coupling (see White, 1934.) Thus a d^2 configuration gives $^3F\,^3P$, $^1G\,^1D\,^1S$ terms in LS couplings and levels $(3/2, 3/2)_{2,0}$, $(5/2, 5/2)_{4,2,0}$, and $(3/2, 5/2)_{4,3,2,1}$ in jj coupling. Notice that the same number of levels with the same J-values is obtained in each instance.

We may now describe briefly how Pauli's exclusion principle is used to explain the periodic table. The successive completed shells of electrons are $1s^2$, $2s^22p^6$, $3s^23p^63d^{10}$, $4s^24p^24d^{10}4f^{14}$, etc., but the binding energies for s, p, d, and f electrons differ so greatly that the shells are not filled in the order of successive n's and l's when these quantities become large.

The $1s^2$ shell is filled with helium. For $Z = 3$ (lithium) the third electron has to be added in the $2s$ shell. The filling of the $2s$ and $2p$ shells gives us the second row of the periodic table, viz.:

$Z =$ 3	4	5	6	7	8	9	10
Li	Be	B	C	N	O	F	Ne
$2s$	$2s^2$	$2s^22p$	$2s^22p^2$	$2s^22p^3$	$2s^22p^4$	$2s^22p^5$	$2s^22p^6$

The theory of chemical valence is based on the principle that when a molecule is formed, it tries to attain a configuration resembling the inert gases. Thus lithium loses one electron to fluorine to form LiF, Ca $4s^2$ tends to lose two, aluminum $3s^23p$ tends to lose three or gain five. The third row of the periodic table parallels the second row very closely, viz.:

$Z =$	11	12	13	14	15	16	17	18
	Na	Mg	Al	Si	P	S	Cl	Ar
	$3s$	$3s^2$	$3s^23p$	$3s^23p^2$	$3s^23p^3$	$3s^23p^4$	$3s^23p^5$	$3s^23p^6$

One would expect a third electron to be added upon the completion of the $3p$ shell, but instead Ar acts as an inert gas and the next electrons in K and Ca are added in the $4s$ shell. With scandium $Z = 21$, the third shell starts to fill but irregularly and is completed with copper $Z = 29$, $4s3d^{10}$; with zinc the shells are filled up through $4s^2$. The $4p^2$ shell is filled from gallium ($Z = 31$) to krypton ($Z = 36$). With rubidium an electron is added to the deeply penetrating $5s$ orbit, while the $4d$ electrons are added from yttrium $Z = 39$ to cadmium $Z = 48$. The $5p$ shell is filled from In ($Z = 49$) to Xe ($Z = 54$) and a $6s$ electron is added with caesium. Before the $5d$ shell starts to fill the fourteen $4f$ electrons are added to make up the rare earth group. With lutecium ($Z = 72$) the $5d$ shell again begins to fill and by Hg ($Z = 80$) it is completed. The $6p$ shell is filled from Tl ($Z = 81$) to radon ($Z = 86$), the $7s$ shell is filled at radium and the $5f$ electrons are then added. The second group of rare earth elements includes thorium, protoactinium, uranium, and synthetic unstable nuclides from Np ($Z = 93$) to No ($Z = 102$).

Irregularities in the building up of the table are caused by the fact that the binding energy does not depend on n alone but on (n, l). Note that closed shells always give $^1S_0$ states for which $\Sigma m_s = M_S = 0$, $\Sigma m_l = M_L = 0$, etc. In x-ray spectra terminology, the $n = 1,2,3, \ldots$ shells are called K, L, M, etc.; $2s^2$ is called L_I, $2p^6$ L_{II}, and so on.

2–10. Complex Spectra; Parentage of Atomic Terms. If more than two electrons participate in the optical spectrum, attention must be paid to the *parentages* of the terms. Consider an atom of ionized sulfur. The terms of the $3s^23p^3$ ground configuration are $^4S(L = 0, S = \frac{3}{2})$, $^2D(L = 2, S = \frac{1}{2})$, $^2P(L = 1, S = \frac{1}{2})$. Let us add a $4p$ electron ($l = 1$, $s = \frac{1}{2}$) and calculate the resulting terms of $3s^23p^34p$ of S I. We add the l and s vectors of the $4p$ electron to the L and S values of the original terms. Note that $S = \frac{1}{2}$ added to doublet terms can give singlets and triplets; added to a quartet term it can give a triplet and a quintet. The addition of l to L gives $L^1 = L + l, L + l - 1, \cdots, L - l$. There results

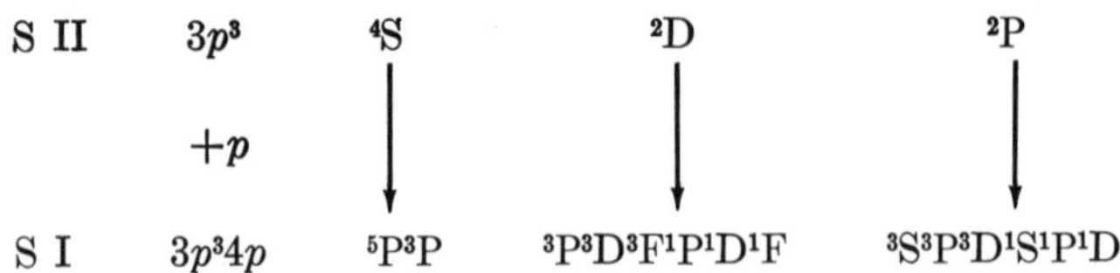

We denote the parentages as $(^4S)^5P$, $(^2D)^3F$, etc. Notice there are three distinct ^3P terms and two distinct ^3D's, ^1P's, and ^1D's based on different parents and possessing different energies. (See also White, 1934.)

Suppose, however, that the added p electron is equivalent to the others and a p^4 configuration results. We have an example of what Menzel and Goldberg called *fractional parentage*, the ^{3}P term of p^4 is based on all the terms of the p^3 configuration, ^{1}D is based on ^{2}P and ^{2}D, but ^{1}S is based only on ^{2}P, thus

$$^3\mathrm{P}: p^3(\tfrac{4}{3}\,{}^4\mathrm{S} + \tfrac{5}{3}\,{}^2\mathrm{D} + {}^2\mathrm{P}),\ {}^1\mathrm{D}: p^3(3^2\mathrm{D} + {}^2\mathrm{P}),\ {}^1\mathrm{S}: p^3(4^2\mathrm{P}).$$

The factor 4 enters because there are four equivalent p electrons.

2–11. Selection Rules for Radiation; Summary of Some Spectroscopic Definitions. Although all spectral lines represent transitions between distinct levels, all possible combinations of levels do not give observed lines. Certain selection rules must be obeyed. The most important of these rules for the ordinary type of radiation called dipole radiation (Chap. 4) are *Laporte's parity rule* and the restrictions of J, viz.,

1. The parity must always change. Transitions occur between different configurations such that $\Delta l = \pm 1$. Levels belonging to $2p^3$ may combine with levels in $2p^3d$ or $2p3s$, but not with levels in $2p3f$ or $2p3p$.
2. The change in J may be ± 1 or 0, except that the transition $J = 0$ to $J = 0$ never occurs.

Under the conditions of strict LS coupling:

3. L may change to $L \pm 1$ or not at all.
4. S must not change.
5. For jj coupling, the j value for only one of the electrons, say j_1, changes, i.e., $\Delta j_1 = 0, \pm 1$.

Selection rules (3) and (4) hold reasonably well for the low-lying levels of light atoms such as oxygen, but they fail in heavier atoms such as iron or titanium, where L and S are no longer "good" quantum numbers, and deviations occur in the direction of jj or "intermediate" coupling. When violations of rules (1) and (2) occur, the radiation is spoken of as "forbidden" (Art. 2-12). Sometimes, especially in heavy atoms such as gold, "double electron jumps" may occur, i.e., two optical electrons participate in a transition. Evidently what is at fault here is failure of configuration assignment; we cannot assign a given level to a unique configuration.

This is a good point to summarize some important spectroscopic definitions which will be useful in later applications. The quantities $nlSLJM$ (where nl is given for all electrons and M is the magnetic quantum number, i.e., the projection of J on the magnetic field) define a *Zeeman state* of an atom.

The quantities $nlLSJ$ define a *level* which includes $2J + 1$ states, e.g., $2p^2\ {}^3\mathrm{P}_2$.

The quantities $nlSL$, define an *atomic term*, the set of $(2S+1)\times(2L+1)$ states characterized by given values of L and S, e.g., $2p^2\ {}^3\mathrm{P}$.

A *polyad* is a set of terms of the same multiplicity based on the same parent term by the addition of one electron to an atom, e.g., $p^3d(^2D)$ plus a d electron gives $^1S^1P^1D^1F^1G$ (one polyad) and $^3S^3P^3D^3F^3G$ (another polyad).

1. Transitions between the individual states into which a given pair of levels is resolved by a magnetic field are called *Zeeman components.*
2. A transition between two levels, e.g., $^3P_2$—$^3D_3$, is called a *spectral line.*
3. The totality of transitions between two terms, e.g., ^3P—^3D, gives a *multiplet.*
4. A *supermultiplet* comprises the totality of transitions between two polyads.
5. All the jumps between two configurations constitute a *transition array.*

2–12. Identifications of Spectral Lines; Term Diagram; Forbidden Lines. The analysis of the spectrum of a given atom or ion with assignments of energy levels is often difficult. Various clues are employed: the changing appearance of the spectrum with temperature is helpful since at low temperatures only lines associated with the lower levels are strong. The *Zeeman effect* enables the kind of term to be identified, since the Zeeman pattern depends in a known fashion on the L, S, and J values of the terms involved. It also indicates if there are important deviations from LS coupling. The multiplicity of the spectrum is, of course, known. Individual multiplet patterns have to be isolated and identified (see, e.g., Herzberg, 1937). *Intercombination lines* which arise from transitions between terms of different S values serve to locate terms of differing multiplicity, e.g., the positions of singlets and triplets.

Analyses have been carried out for the spectra of all important elements in their neutral and first ionized stages, although the higher-level lines of some metallic ions have not been assigned. For some atoms many energy levels are known even for the higher stages of ionization. Few permitted lines of elements heavier than silicon and in ionization stages higher than the second are observed in absorption in stellar spectra. A thorough compilation of presently known data on atomic energy levels, as derived from analyses of optical spectra, has been given by Charlotte Moore (1949, 1952, 1958). For each ion these tables give for each energy level the configuration, the term, the j-value, and the height of the level above the ground term expressed in wave number units, as well as references to the original literature. Much work remains to be done in the analyses of rare earths and other heavier elements.

The indispensable aid for identification work is Charlotte Moore's "Multiplet Table of Astrophysical Interest."[3] The lines are grouped according to ion and multiplet, with laboratory intensities, excitation potentials, and J-values in successive columns. For the more abundant elements such as iron, the table lists the predicted as well as the observed lines. It also gives the predicted positions of many forbidden lines. A new edition of this important table is now in preparation.

[3] *Princeton Observatory Contribution No. 20,* 1945.

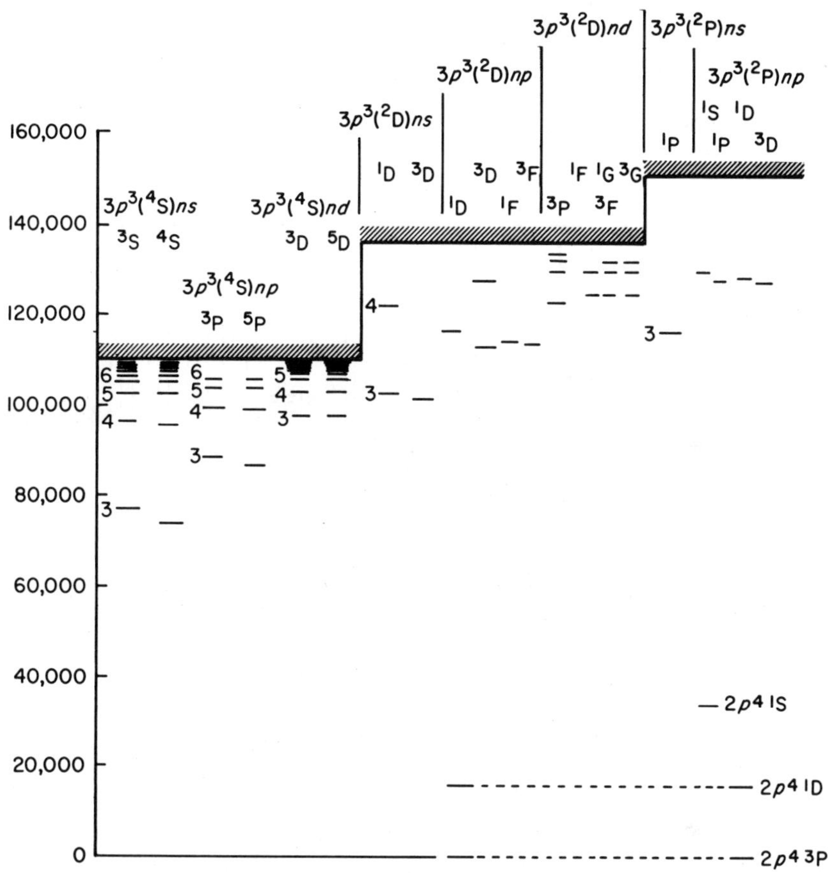

FIG. 2–22. TERM DIAGRAM FOR O I

In Fig. 2–22 we give a partial term diagram for O I as compiled from Moore's (1949) tables (see also White, 1934). The analysis is based mostly on the work of Edlén. Four optical electrons are involved. Notice that there are three groups of terms based on the three parent terms of the p^3 configuration, viz., ^4S, ^2D, and ^2P. The ground level partakes of all three parents while the ^1D term is based on ^2P and ^2D. We indicate this by representing the ^3P and ^2D terms as though they were 3 and 2 distinct terms, respectively, in separate parentage groups.

The normal ionization potential of the atom 13.614 ev corresponds to a removal of the outermost electron from a $3p^3(^4S)nl$ configuration. The O II atom then finds itself in its ground $^4S_{3/2}$ level. The terms based on the ^2D and ^2P parents have ionization potentials 1.96 and 4.17 ev higher. Intercombination lines connect the singlet, triplet, and quintet systems, as well as systems based on different parents.

In each group the term of highest multiplicity lies lowest; the P terms lie higher than the S terms, while D terms are yet higher. The resonance lines fall in the far ultraviolet and are not observable in the solar spectrum except from rockets and satellites. The strongest transitions in the observable solar spectrum are the $3s^5S—3p^5P$ quintet near λ7774 and the $3s^3S—3p^3P$ triplet near λ8446.

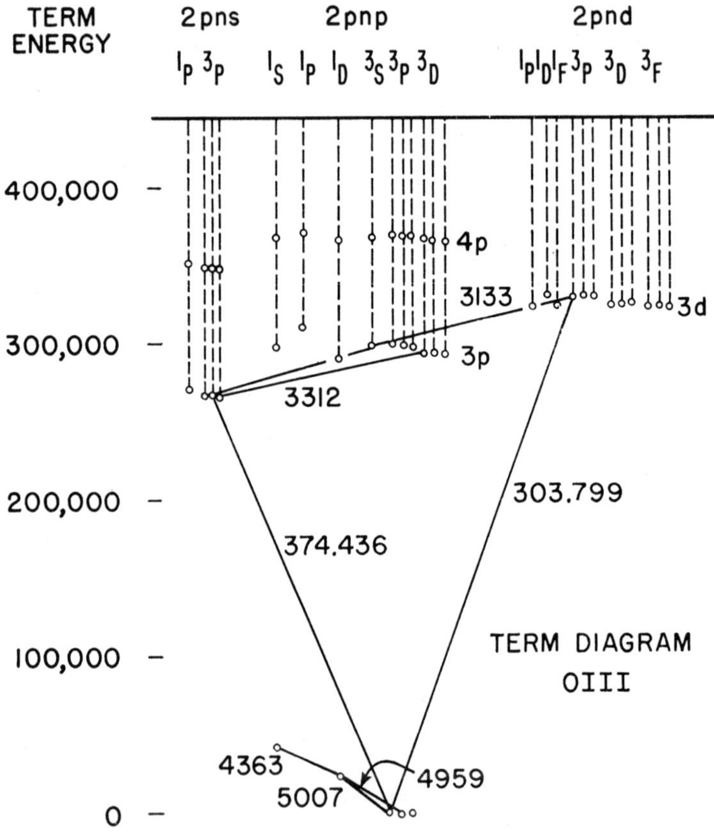

FIG. 2–23. TERM DIAGRAM FOR O III

The levels are plotted on a wave number scale. Notice that the ^3P, ^1D, and ^1S terms of the ground $2p^2$ configuration lie very much lower than the first excited levels. As with the alkalis, although less strikingly, the $3s$ term lies below the $3p$ and $3d$ terms. Ordinary (dipole-radiation) transitions can take place only between terms in adjacent configurations. The "forbidden" λ5007, λ4959, and λ4363 lines are strong in the spectra of gaseous nebulae.

Fig. 2–23 gives a portion of a similar diagram for O III. The top horizontal line represents the ionization potential of 54.886 ev. Transitions from $3d$ to $3p$ and $3s$ to $3p$ represent lines observable in O stars where the temperature is high enough to excite the O III lines. Similar term diagrams for elements like iron or manganese can become quite complicated (see White, 1934; Herzberg, 1944). In working with individual elements in various stellar and nebular spectra it is often helpful to construct partial energy level diagrams isolating features of particular interest in the problem at hand (see Merrill, 1957).

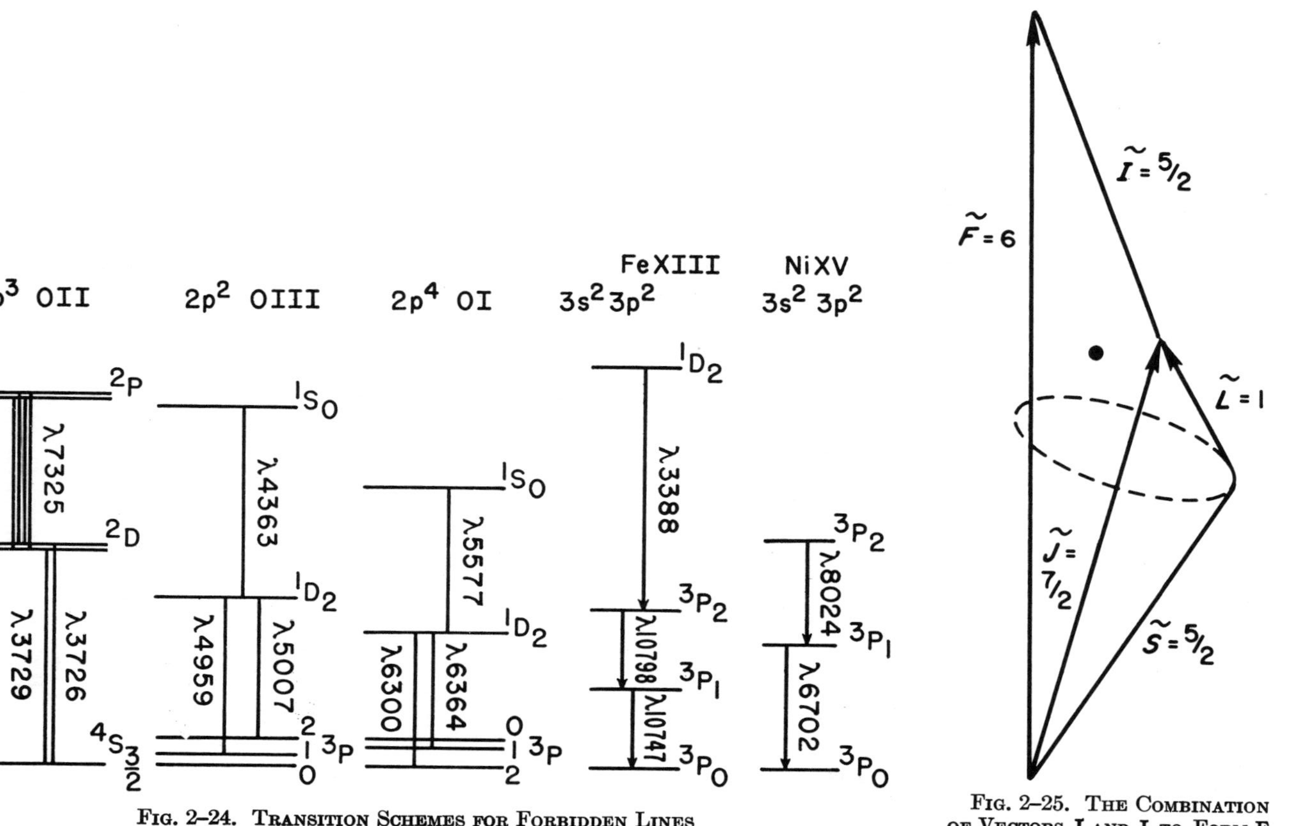

Fig. 2–24. Transition Schemes for Forbidden Lines

Fig. 2–25. The Combination of Vectors J and I to Form F

Notice that the ground configuration $2p^4$ of O I or $(2p^2)$ of O III which contains ^{3}P, ^{1}D, and ^{1}S terms lies far below the first excited $3s$ or $3p$ configurations.[4] Transitions from ^{1}S to ^{1}D or from ^{1}D to ^{3}P in the $2p^2$ or $2p^4$ configurations violate Laporte's rule, $\Delta l = \pm 1$, and are called *forbidden lines.* The ^{1}D and ^{1}S levels, which in O III lie 2.48 and 5.3 v, respectively, above the ground level, are called *metastable levels.* An atom in an ordinary high level of O III like $2p3s^3P_2$ will cascade downwards in a time of the order of 10^{-8} sec. If it finds itself in one of the low-lying metastable levels, it may remain there for a time of the order of 1 sec before it escapes to a lower level with emission of radiation.

Menzel, Payne, and Boyce proposed the following terminology for forbidden lines: In a p^2 or p^4 configuration, jumps between ^{1}S and ^{1}D are called *auroral transitions,* since the strong λ5577 line of the permanent aurora of the night sky is of this type. The strongest lines in the gaseous nebulae arise from ^{1}D — ^{3}P transitions; hence they are called *nebular transitions.* A *transauroral* line denotes one of the type ^{1}S — ^{3}P and is usually weak. In a p^3 configuration, e.g., O II, the auroral, nebular, and transauroral transitions are analogously ^{2}P — ^{2}D, ^{2}D — ^{4}S, and ^{2}P — ^{4}S. The lines identified by Edlén in the solar corona mostly represent transitions between levels of the ground terms. (See Fig. 2–24.) The symbol [ ] is used to denote forbidden lines.

Although forbidden lines usually appear in emission, I. S. Bowen utilized the faint [O I] λ5577, λ6300, and λ6363 absorption lines which appear in the solar spectrum to estimate the abundance of oxygen in the sun.

Relatively few forbidden lines can be produced experimentally. In general, they can be predicted only after an analysis of the spectrum has located the low-lying metastable levels. If the analysis is not sufficiently complete to locate the metastable levels, the extrapolation of the energy levels in an isoelectronic sequence may be employed. (See Art. 2–7.)

2–13. Hyperfine Structure. Up to this point we have treated the nucleus as a fixed point, ignoring any influence it may have on atomic spectra. Spectral lines of certain atoms were long ago discovered to possess an extremely narrow structure, which came to be called *hyperfine structure.* Whereas the scale of the ordinary fine structure (represented, e.g., by the D lines of sodium) may extend over several angstroms, this hyperfine structure is built on a scale that taxes the resolution of the best optical equipment. For example, the hyperfine patterns of the Mn λ6013, λ6016, and λ6021 lines are spread over a range of only 0.1A. The Doppler broad-

[4] *Hund's rule* states that in a ground configuration the term of highest multiplicity will lie lowest. If there are several terms of the same multiplicity, the term with the highest L value will be lowest. If the shell is less than half-filled, e.g., p or p^2, d, d^2, d^3, or d^4, the level of smallest J value will be lowest; otherwise the term will be inverted. Thus in O III, the $^3P_0$ level of the ground $2p^2$ configuration lies lowest, while in O I, F II, or Ne III $(2p^4)$ the $^3P_2$ level is lowest and the term is inverted (see Fig. 2–24).

ening of lines by the gas kinetic motions of the atoms in an ordinary arc is of this order of magnitude. Pauli suggested that hyperfine structure could be explained if it is supposed that the nucleus has an angular momentum $\hbar\mathbf{I}$. The maximum value of $\mathbf{I}$ in any spatial direction is called the *nuclear spin;* it assumes half-integer values for nuclei of odd mass number A and integral values for nuclei of even mass number.

Associated with the nuclear spin is a magnetic moment

$$\mu_{\mathrm{I}} = g_I\, \mu_N\, I \tag{2–40}$$

where the nuclear g-factor g_I is peculiar to a particular nucleus and can be positive or negative. The nuclear magneton

$$\mu_N = \frac{\varepsilon\hbar}{2Mc} = 5.05038 \pm 0.00036 \times 10^{-24} \text{ ergs gauss}^{-1} \tag{2–41}$$

is 1840 times smaller than the Bohr magneton (see Eq. 2–15). The total atomic angular momentum is $\mathbf{F}$, the vector sum of $\mathbf{I}$ and $\mathbf{J}$, thus

$$\mathbf{F} = \mathbf{I} + \mathbf{J} \tag{2–42}$$

The vectors $\mathbf{L}$ and $\mathbf{S}$ are added to form $\mathbf{J}$ which in turn combines with $\mathbf{I}$ to form $\mathbf{F}$. Now $\mathbf{J}$ precesses about $\mathbf{F}$, but the rate of precession is 1840 times smaller than the rate of precession of $\mathbf{L}$ and $\mathbf{S}$ about $\mathbf{J}$. The number of hyperfine levels will be

$$\mathbf{J} + \mathbf{I},\ \mathbf{J} + \mathbf{I} - 1, \ldots, |\,\mathbf{J} - \mathbf{I}\,|$$

Since $\mathbf{I}$ depends only on the nucleus and is not influenced by the actions of the atomic electrons, it is the same for all levels of an atom or ion. (Of course, I can be different for different isotopes of an element.) From levels for which $J > I$ we can get the nuclear spin from the number of hyperfine components of the spectral line. See Fig. 2–25.

The separation of the hyperfine levels depends on the magnetic interaction energy between the nuclear magnetic moment and the magnetic field produced by the orbital motion and spin of the electron. For a hydrogen-like atom this hyperfine magnetic interaction energy may be shown to be (e.g., Barrett, 1958)

$$E_{hfs} = g_I \left(\frac{m}{M}\right) \frac{\alpha^2 hcRZ^3}{n^3} \left[\frac{F(F+1) - I(I+1) - J(J+1)}{J(J+1)\,(2L+1)}\right] \tag{2–43}$$

whereas the ordinary fine structure arising from the interaction between $\mathbf{L}$ and $\mathbf{S}$ similarly is given by

$$E_{fs} = \frac{\alpha^2 hcRZ^4}{n^3} \left[\frac{J(J+1) - L(L+1) - S(S+1)}{L(L+1)\,(2L+1)}\right] \tag{2–44}$$

Here $\alpha = 2\pi\varepsilon^2/hc = 7.2973 \times 10^{-3}$ is the *fine structure constant,* R is the Rydberg constant, Z is the ionic charge and n is the effective quantum number of the hydrogen-like level. Since $g_I \sim I$, and the bracketed terms are comparable, the hyperfine structure will be of the order of $1/1840Z$ times the fine structure. The scale of the splitting in actual practice depends on both n and l, being greater for an s electron than for a p electron.

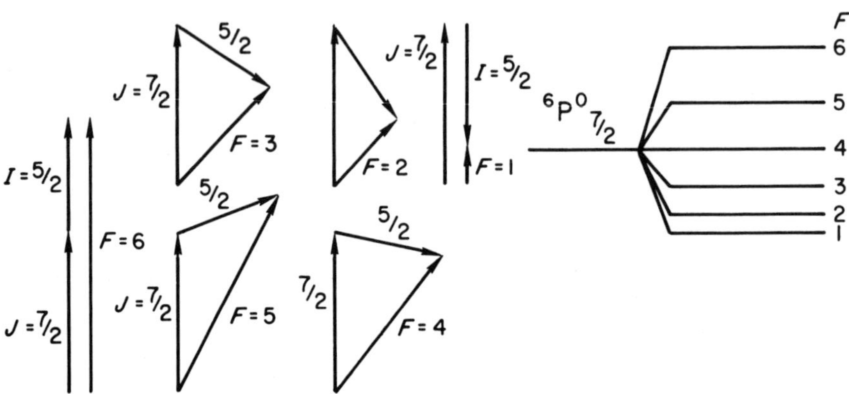

FIG. 2–26. VECTOR DIAGRAM AND HYPERFINE SPLITTING OF THE MN $^6P^0_{7/2}$ TERM

Fig. 2–26 shows the vector diagram for the hyperfine levels of a $^6P^0_{7/2}$ term, F = 1, 2, 3, 4, 5, 6 and their energy spacing which obeys an analogue of the Landé interval rule. The hyperfine structure of the Mn λ6021, $^6P^0_{7/2}$—$^6S_{5/2}$ transition is shown in Fig. 2–27. Except for a component of negligible intensity the pattern is confined to an interval of 0.075A. Transitions between hyperfine levels obey the selection rule $\Delta F = 0, \pm 1$ with jumps $F = 0$ to $F = 0$ forbidden. The Zeeman and Paschen-Back effects have also been observed.

Since the hyperfine splitting is generally smaller than the thermal Doppler broadening in the sun, one might expect that it could be ignored. Actually, Arthur Abt (1955) found that hyperfine structure can contribute appreciably to the widths of certain lines in the solar spectrum, while Hindmarsh (1955) compared manganese lines λ6016, λ6013, λ6027, which are affected by *hfs,* with the iron line λ6027 which is not so influenced. With the reasonable assumption that the two lines were formed in the same layers, he showed that the pronounced differences in shape (the Mn lines had a higher central intensity) could be explained by the effect of hyperfine structure.

For many problems in astronomy, particularly those pertaining to galactic structure and the interstellar medium, the most important spectrum line is the 21-cm transition of hydrogen, which fortunately falls in the radio-frequency window. The existence of this state cannot be demonstrated by optical-region experiments, but was established by the molecular-beam resonance technique.

Now $J = \frac{1}{2}$ and $I = \frac{1}{2}$ for the ground $^2S_{1/2}$ level of hydrogen. Here there can exist two hyperfine levels corresponding to $F = 1$ and $F = 0$. In the $F = 1$ level the proton and electron magnetic moments are parallel; in the $F = 0$ level they are anti-parallel. The magnetic interaction between the two spinning particles is analogous to that between two bar magnets. Since the proton is a positively charged particle its spin and magnetic moment vectors lie parallel to one another. Hence the electron spin flips from a position anti-parallel to the nuclear spin in the $F = 1$ level to a

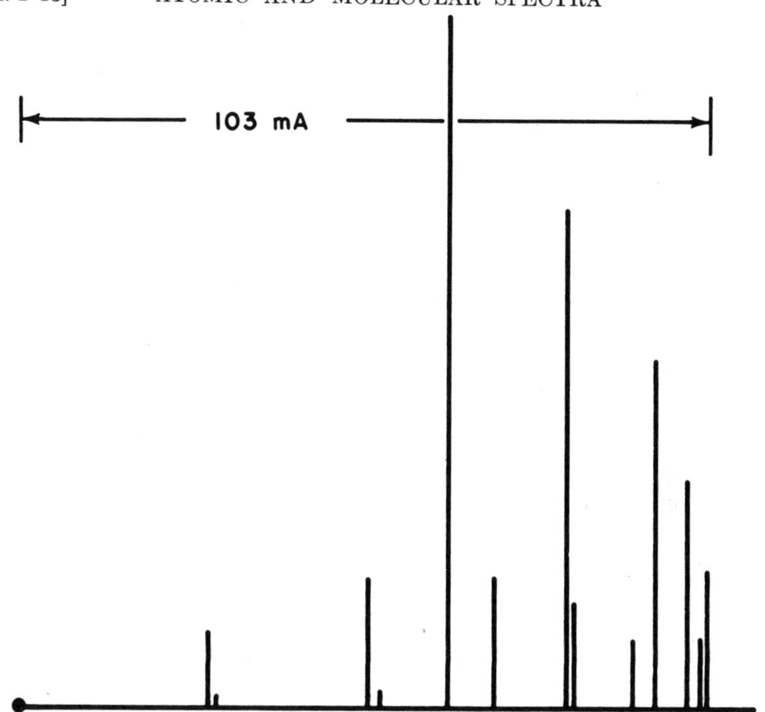

FIG. 2–27. HYPERFINE PATTERN OF THE MN λ6021 $^6P^0_{7/2}$ — $^6S_{5/2}$ TRANSITION (After W. R. Hindmarsh.)

parallel position in the $F = 0$ level. The probability of a spin flop with the emission of the 21-cm line is extremely low. A H atom in free space in the $F = 1$ level, where no collisions could disturb it, would remain there for about eleven million years!

This 21-cm (1420.405 mc) radiation has proved an important tool for galactic research. It is emitted by cold clouds of atomic hydrogen, and from the displacements and intensities of the line the velocity, temperature, and density of the emitting gas can often be inferred. The $F = 1$ level is split into three Zeeman states by a magnetic field; hence the 21-cm line is resolved into three components. Davies, Verschuur, and Wild (1962) found evidence for a field of 2.5×10^{-5} gauss in a cloud in the direction of the radio source Taurus A; they suggest a general magnetic field of the order of 5×10^{-6} gauss.

Heavy hydrogen (deuterium) has a radio-frequency line at 327.4 mc (91.6 cm) ($F = \frac{3}{2} \rightarrow \frac{1}{2}$), which has not been observed in the interstellar medium. Barrett has listed several other r-f transitions involving ground levels of relatively abundant atoms, viz., He^3 II, N^{14}, and Na^{23}.

In addition to hyperfine structure, mention may be made of another type of fine structure splitting exemplified by the "Lamb shift" in hydrogen. Fig. 2–28

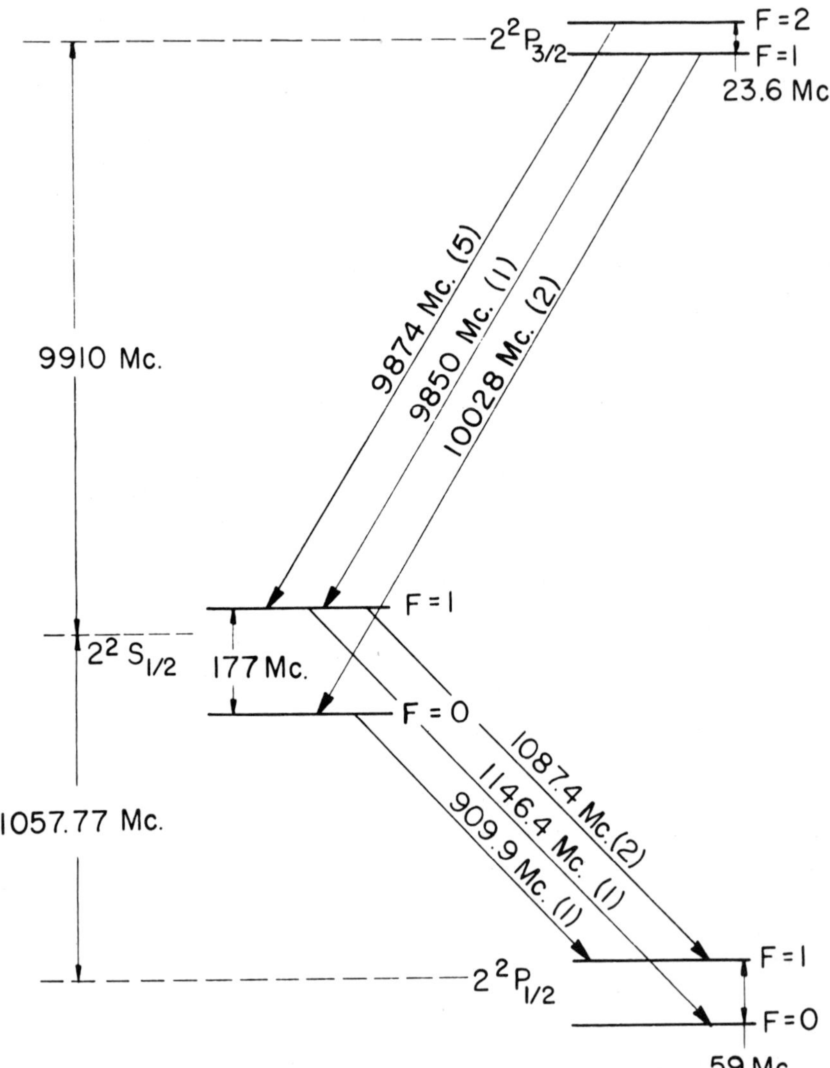

Fig. 2–28. Fine and Hyperfine Structure of Hydrogen ($n = 2$) Levels (Courtesy, Alan H. Barrett.)

shows the energy levels for the $n = 2$ term in H. Notice the hyperfine structure of the individual terms and also the fine structure (i.e., separation between ^2S and ^2P). Unlike the fine structure described previously, this phenomenon cannot be handled by any classical or semiclassical argument, although it is explained both qualitatively and quantitatively by modern quantum electrodynamics. The likelihood of a spontaneous jump from $^2S_{1/2}$ to $^2P_{1/2}$ or from $2^2P_{3/2}$ to $2^2S_{1/2}$ is very small; the $2^2S_{1/2}$ level can be regarded as metastable with a lifetime of about 0.14 sec. Transition probabilities are discussed in Chapters 4 and 7.

Spectroscopy has occupied an important position in the development of astrophysics; further details and applications are given in later chapters. We defer to Chapter 7 a discussion of the Stark effect, i.e., the broadening or splitting of a spectral line by an electric field. No large-scale electric field appears to exist in stellar atmospheres, but rapid motions of charged particles may produce intense ephemeral microscopic electric fields near a radiating atom. Hydrogen and helium lines are widened by these fields (Arts. 7–7, 7–8, 8–16 and 8–17).

2–14. Structure of Diatomic Molecules. In stars like the sun the predominant molecules are of the diatomic type such as OH, CH, and CN. Despite their simple structure from the chemical point of view, they show complicated spectra consisting of extensive bands composed of numerous closely packed lines.

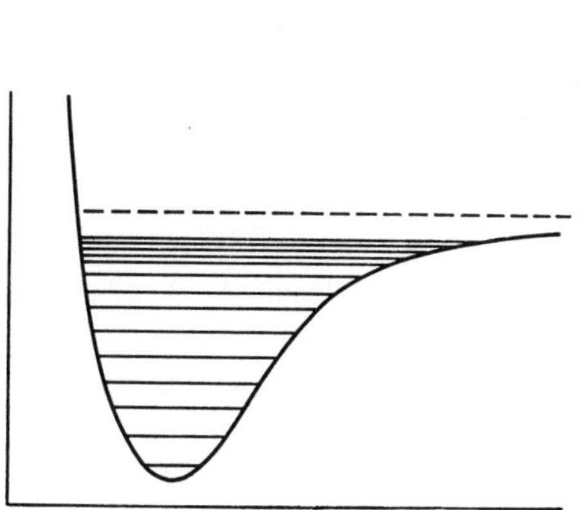

FIG. 2–29. VIBRATIONAL ENERGY LEVELS OF A DIATOMIC MOLECULE

The potential energy curve is indicated by a heavy solid curve. Thin solid horizontal lines indicate vibrational levels and the dotted line the energy corresponding to dissociation. The abscissa is the distance r between the component atoms. The minimum of the potential curve lies at the equilibrium distance r_0. The ordinate is the energy.

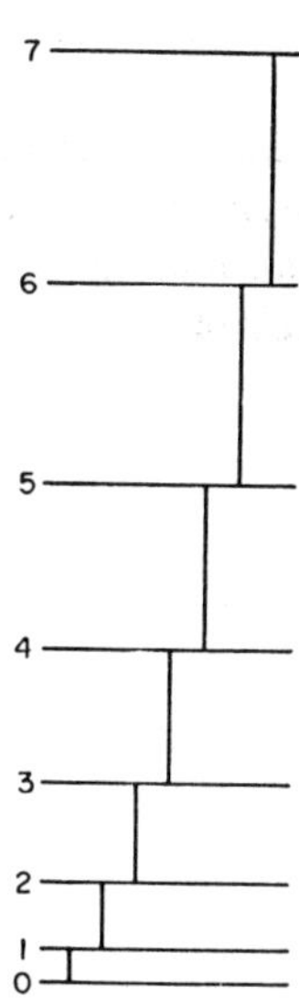

FIG. 2–30. PURE ROTATIONAL TRANSITIONS IN A DIATOMIC MOLECULE

Transitions are restricted to $\Delta K = 1$.

We may think of a molecule AB, e.g., CN, as composed of two ions A^+ and B^+, surrounded by a cloud of electrons and held together by electrostatic forces. The potential energy curve of the system (Fig. 2–29) shows that there exists a distance r_0 that corresponds to a minimum potential energy. At greater distances the potential curve gradually rises and approaches the horizontal dotted line which corresponds to the dissociation energy of the molecule. At distances smaller than r_0, the repulsive forces of the two atomic nuclei become increasingly important and prevents a close approach of the two nuclei. The horizontal solid lines represent the quantized vibrational energy states of the molecule. Notice that they fall closer together with increasing energy. Energies above the dotted line represent states wherein the molecule is dissociated and the free atoms are flying about. These states are unquantized.

Such a molecule may take up energy in several ways. First, it may rotate about an axis perpendicular to the line joining the two atoms. Second, the two component atoms may vibrate back and forth along the line joining them. Third, the molecule may be excited to definite electronic energy states analogous to atomic energy levels.

The rotational modes (not shown in Fig. 2–29) usually involve the smallest amounts of energy. Radiations corresponding to transitions from one rotational level to another are in the far infrared and microwave regions. The vibrational energies are greater and transitions between the corresponding levels produce bands in the relatively near infrared. The excitation of electronic states requires frequencies falling in the visible and ultraviolet part of the spectrum. Each type of energy; rotational, vibrational, and electronic, is quantized according to definite rules. The total energy of the molecule is:

$$E = E_{\text{rotational}} + E_{\text{vibrational}} + E_{\text{electronic}} \tag{2–45}$$

2–15. Rotational Energies of Diatomic Molecules. Quantum mechanics shows that in the first approximation the rotational energy of the molecule is

$$E_r = K(K+1)\frac{h^2}{8\pi^2 I} \tag{2–46}$$

Here

$$I = \frac{m_A m_B}{(m_A + m_B)} r^2 = \mu r^2 \tag{2–47}$$

is the moment of inertia of the molecule, μ is called the reduced mass, and the integer K, called the rotational quantum number, follows the rule

$$\Delta K = \pm 1 \tag{2–48}$$

in transitions from one rotational level to another. See Fig. 2–30. The wave number of the transition between rotational levels K' and K'' is

$$\tilde{\nu} = \frac{1}{hc}(E_{r'} - E_{r''}) = 2BK' \tag{2–49}$$

where

$$B = \frac{h}{8\pi^2 Ic} \tag{2–50}$$

A *pure rotation spectrum* would consist of a series of equidistant lines were it not for an increase in the size of the molecule due to the centrifugal force of rotation. This effect leads to a small correction term of the form $CK^2(K + 1)^2$ in Eq. 2–46. Pure rotation bands do not exist if the two atoms are identical.

The pure rotation bands of the HCl molecule may be represented by an equation of the form

$$\tilde{\nu} = 20.793m - 0.0016m^3 \tag{2–51}$$

neglecting the term in m^3, we put $m = K + 1$, so $2B = 20.793$, whence from Eq. 2–50, $I = 2.71 \times 10^{-40}$ gm cm² and $r = 1.29 \times 10^{-8}$ cm. The pure rotation spectra of astrophysically interesting diatomic molecules, such as CO, NO, or CS, fall in the microwave region extinguished by the earth's atmosphere. Pure rotational lines of more complex molecules may be observed eventually in the r-f spectra of planetary atmospheres.

2–16. Rotation and Vibration Bands. If the component atoms are displaced slightly from their equilibrium separation, r_0, the restoring force is proportional to $(r - r_0)$. If the displacements are large, this is no longer true as the potential energy curve is asymmetrical. Consequently, the energy levels, E_{vib}, do not show an even spacing. They fall closer and closer together, in accordance with the empirical formula.

$$E_{\text{vib}} = hc\omega(v + \tfrac{1}{2}) - hc\omega x_\epsilon\,(v + \tfrac{1}{2})^2 + \text{etc.} \tag{2–52}$$

Here ω (expressed in wave number units) and x_2 are found from the analysis of the band spectrum. If higher terms can be neglected, the energy necessary to raise the molecule from the lowest vibrational level to the point of dissociation (heat of dissociation D_0) can be shown to be related to x_2 and ω by

$$D_0 = \frac{\omega^2}{4\omega x_e}\,(1 - x_e)^2 \tag{2–53}$$

Other spectroscopic methods are often capable of giving a better value of D_0. Sometimes the *absorption continuum* above the dissociation limit (analogous to ionization continua of atoms) can be observed and yields a good value of D_0.

We now consider the character of the spectrum emitted or absorbed when a molecule jumps from one rotational and vibrational state to another rotation-vibration state. For the time being, we shall suppose that there is no change in electronic energy. We may write, with the aid of Eqs. 2–46 and 2–52:

$$\tilde{\nu} = \tilde{\nu}_r + \tilde{\nu}_v = \frac{1}{hc}[E_{r'} + E_{v'} - E_{r''} - E_{v''}] = B'K'(K' + 1) -$$
$$B''K''(K'' + 1) + \omega'(v' + \tfrac{1}{2}) - x'_e\omega(v' + \tfrac{1}{2})^2 - \omega''(v'' + \tfrac{1}{2}) +$$
$$x''_e\,\omega''(v'' + \tfrac{1}{2})^2 \tag{2–54}$$

Here the *vibrational energy* E_v is the energy of the vibration of the component atoms about their equilibrium separation when the molecule does not rotate. The rotational energy, E_r, is that added by the rotation of the molecule. One must also include the small energy interaction between rotation and vibration and the effects of the expansion of the molecule by centrifugal force.

The totality of transitions from one vibrational state of the molecule v' to another vibrational state v'' constitutes a band. A transition from a particular $v'K'$ state to another $v''K''$ state corresponds to a single line of the $v'v''$ band.

If the molecular potential curve (Fig. 2–29) were a parabola, the molecule would constitute a harmonic oscillator and the selection rule for the transitions would be $v' - v'' = 1$. To the extent that the potential curve approximates a parabola we would expect transitions of the type $v' = 0$ to $v'' = 1$, $v' = 1$ to $v'' = 2$, etc., to be the strongest. The curve is most nearly parabolic at low energies. For the lowest levels, the bands corresponding to $\Delta v = 1$ are indeed the most intense, but among higher vibrational levels, transitions in which $\Delta v = 2$, 3, etc., can become important. A pure vibrational transition is denoted as (v', v''). Thus, (1, 0) means a jump from the vibrational level $v = 1$ to the vibrational level $v = 0$. (See Fig. 2–31.)

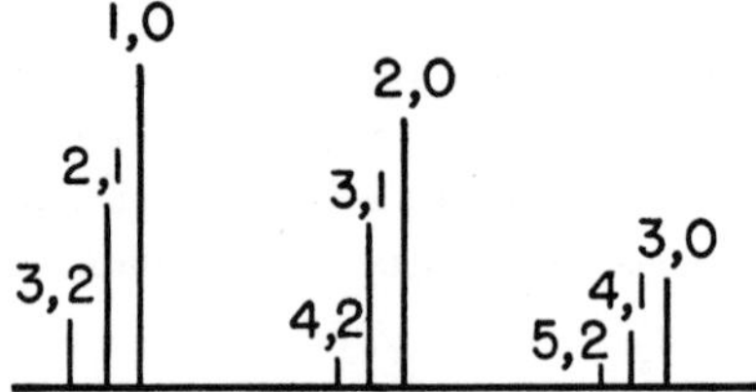

FIG. 2–31. VIBRATION BANDS FOR $\Delta v = 1, 2, 3$ (SCHEMATIC)

Now consider the rotational structure of the pure vibration-rotation bands. The moment of inertia of the molecule will depend upon the state of vibration. Because of the asymmetry of the potential curve, the average separation of the atoms in the molecule and therefore the moment of inertia is greater in the higher vibrational level. Hence $I_{v'} > I_{v''}$ and $B' > B''$. Neglecting the small rotational terms of higher order, the rotational structure of a particular vibration-rotation band will be given by

$$\tilde{\nu} = \tilde{\nu}_v + \tilde{\nu}_r = \tilde{\nu}_v + B'K'(K' + 1) - B''K''(K'' + 1) \qquad (2\text{–}55)$$

where $\tilde{\nu}_v$ is the wave number of the band. For a given change in v, $\tilde{\nu}_v$ is fixed, and since K can change by ± 1, there are two branches of lines depending on whether

$$K' = K'' + 1 \text{ (}R\text{-branch), or } K' = K'' - 1 \text{ (}P\text{-branch)} \qquad (2\text{–}56)$$

Then

$$\begin{aligned} \tilde{\nu} &= \tilde{\nu}_v + (B' + B'')(K + 1) + (B' - B'')(K + 1)^2 \quad &R\text{-branch} \\ \tilde{\nu} &= \tilde{\nu}_v - (B' + B'')K + (B' - B'')K^2 \quad &P\text{-branch} \end{aligned} \qquad (2\text{–}57)$$

See Fig. 2–32. Since $B' - B''$ is negative, the lines of the R-branch tend to crowd together; those of the P-branch to spread apart. Since the change in K cannot be zero, the line corresponding to the band origin does not appear. *Homonuclear molecules* such as C_2 or N_2 show no vibration-rotation bands.

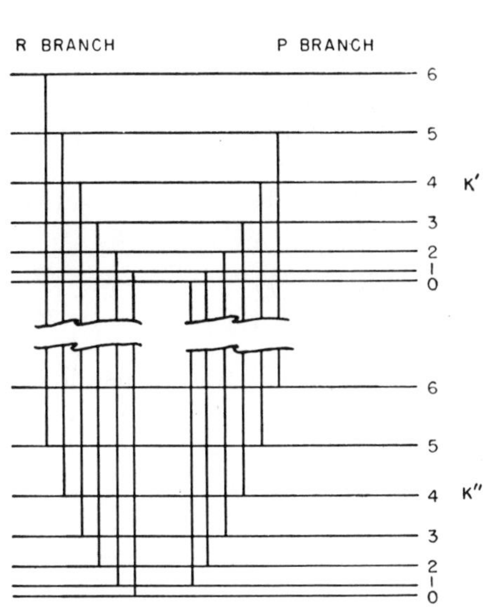

FIG. 2–32. *R* AND *P* BRANCHES IN PURE VIBRATIONAL TRANSITIONS

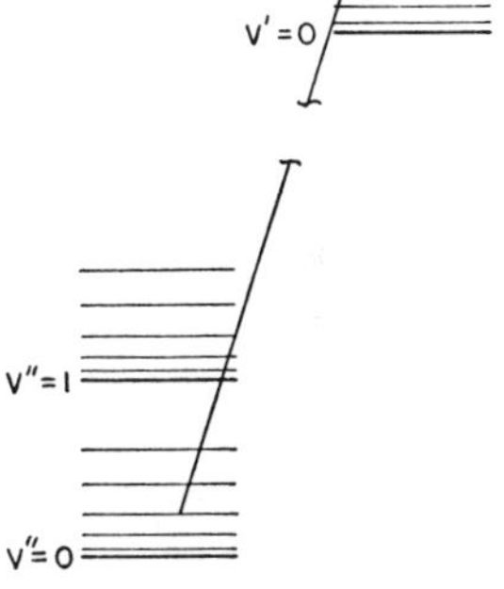

FIG. 2–33. ELECTRONIC TRANSITIONS

The particular transition illustrated herewith takes place from the 4th rotational state of the lowest vibrational level ($v'' = 0$) of the lower electronic state to the 5th rotational state of the vibrational level ($v' = 1$) of the upper electronic state. This transition corresponds to one molecular line of the entire band system.

Vibration-rotation bands of CO have been observed in the sun. Similar bands of polyatomic molecules, e.g., CO_2, CH_4, and N_2O are prominent in the infrared spectrum of the earth's atmosphere.[5]

2–17. Electronic Bands. The bands of greatest interest in stellar spectroscopy are the electronic bands in which E_e, E_v, and E_r all change. Since molecular electronic excitation potentials are comparable with atomic excitation potentials, the energies involved are much greater than the vibrational or rotational energies. That is (cf. Figs. 2–33, 2–34),

$$\tilde{\nu} = \tilde{\nu}_e + \tilde{\nu}_v + \tilde{\nu}_r$$

[5] The relative intensities of the bands, either in absorption or emission, will depend upon the temperature of the gas. At low temperatures most of the molecules are in the lowest vibrational level and the 0–1 band tends to be the strongest.

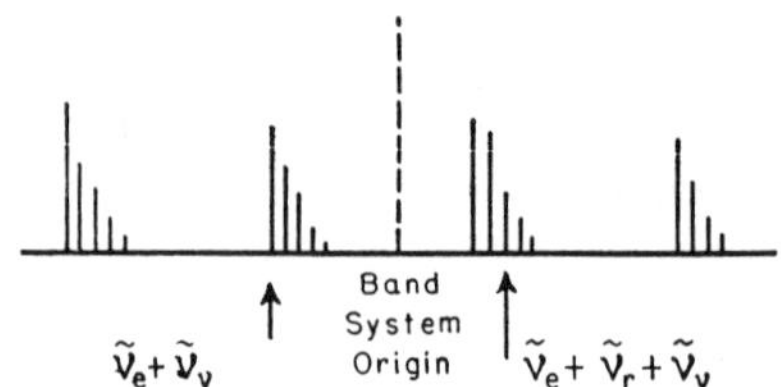

FIG. 2–34. SCHEMATIC RESOLUTION OF A BAND SYSTEM INTO INDIVIDUAL LINES AND BANDS

$$\tilde{\nu} = \frac{1}{hc}\{E_{e'} - E_{e''} + E_{v'} - E_{v''} + E_{r'} - E_{r''}\} \tag{2–58}$$

Here $\tilde{\nu}_e$ determines the position of the entire band system, $\tilde{\nu}_v + \tilde{\nu}_e$ locates the position of a particular band in the system, and $\tilde{\nu}_r + \tilde{\nu}_v + \tilde{\nu}_e$ fixes the position of a single line in the entire set of bands. Since the potential curve of an excited level usually differs from that of the ground level, I and therefore B will be different and $B' - B''$ may be positive, negative, or zero. Let

$$\begin{aligned}\tilde{\nu}_o = \tilde{\nu}_e + \tilde{\nu}_v = \tilde{\nu}_e &+ \omega'(v' + \tfrac{1}{2})[1 - x'(v' + \tfrac{1}{2})] \\ &- \omega''(v'' + \tfrac{1}{2})[1 - x''(v'' + \tfrac{1}{2})]\end{aligned} \tag{2–59}$$

denote the position of a band origin. In electronic transitions all values of Δv are permitted. All bands which have a common lower or a common upper level are said to form a *progression* or a series. Bands for which the change in v is constant are said to form a *sequence*. (See Fig. 2–35.)

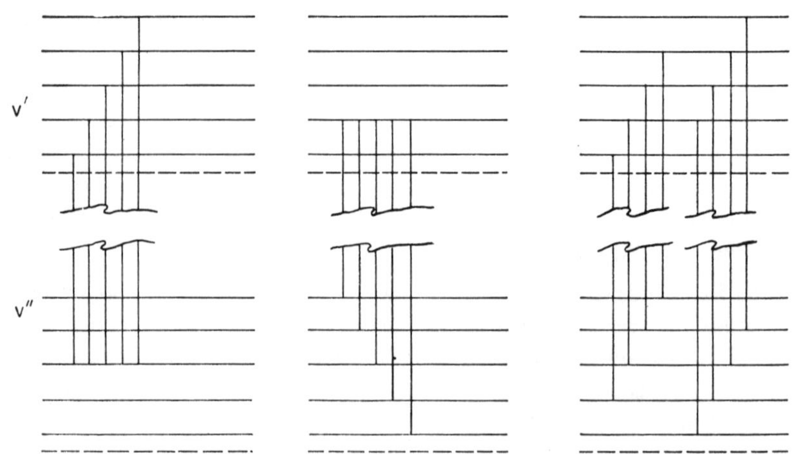

FIG. 2–35. PROGRESSIONS AND SEQUENCES

The bands at the left have a common lower vibrational level and are said to form a v' progression. The middle sketch depicts a v'' progression, while a sequence is shown at the right.

The relative intensities of the bands within a given band system depend on the potential curves of the molecules in the two electronic states and may be estimated with the aid of quantum mechanics in accordance with the *Franck-Condon principle.* If the bands are arranged in a (v', v'') array, it is found that the locus of the strongest bands is a parabola. If r'_O is nearly equal to r''_O, the strongest bands fall near the diagonal, $\Delta v = 0$, $\Delta v = -1$, $\Delta v = +1$, i.e., sequences are strongest. If r'_O and r''_O differ, the strongest bands belong to the progression, $v'' = 0, 1$ and $v' = 0, 1$.

The wave number of a particular line of an electronic band is

$$\tilde{\nu} = \tilde{\nu}_o + B'K'(K' + 1) - B''K''(K'' + 1) \tag{2–60}$$

As before, a change in K of $\pm$ 1 is possible, but $\Delta K = 0$ can also occur in some transitions. Hence in addition to the P- and R-branches, a Q-branch sometimes appears.

$$\begin{aligned}
\tilde{\nu} &= \tilde{\nu}_o + 2B' + (3B' - B'')K + (B' - B'')K^2 = R(K) \quad K' = K'' + 1\\
\tilde{\nu} &= \tilde{\nu}_o \qquad + (B' - B'')K + (B' - B'')K^2 = Q(K) \quad K' = K''\\
\tilde{\nu} &= \tilde{\nu}_o \qquad - (B' + B'')K + (B' - B'')K^2 = P(K) \quad K' = K'' - 1
\end{aligned} \tag{2–61}$$

For the vibration-rotation bands, $B' - B''$ is always negative and the lines tend to crowd together in the R branch. In the electronic transitions the bands can have differing structures depending on whether $B' - B''$ is positive, zero, or negative.

If $B' - B''$ is positive, the lines will crowd together, i.e., form a head in the P-branch and the individual lines will tend to spread out in the violet. We say that the band is degraded toward the violet. If $B' = B''$ the bands form no head at all, while if B' is less than B'', the head will be formed on the violet side of the band origin and the bands are degraded toward the red. Notice that we may represent the lines of the P- and R-branches by the same formula if we replace K by $(m - 1)$ in the P-branch and by m in the R-branch and suppose that the P-branch corresponds to negative m values. That is,

$$\tilde{\nu} = a + bm + cm^2$$

where a, b, and c are found from an analysis of the spectrum. If one plots m as ordinate and the wave number $\tilde{\nu}$ as abscissa, he obtains what is known as a *Fortrat parabola.* Each integral m value projected on the $\tilde{\nu}$ axis gives a line. The analysis of the spectrum by means of the Fortrat parabola shows how the band head arises as a consequence of the crowding together of the lines toward the vertex of the parabola. Although the band heads are the most conspicuous features of molecular spectra, their theoretical significance is less than that of the band origin, $\tilde{\nu}_o$, whose position can be found only after an analysis of the spectrum.

Electronic energy levels are labeled by a scheme somewhat similar to that employed for atoms, except that Greek letters are used in place of the Roman, viz.,

Σ	Π	Δ	Φ	(molecular)
S	P	D	F	(atomic)

The significant quantities are the projections of the orbital and spin angular moments, L and S, upon the line joining the nuclei of the two atoms, Δ and Σ. The sum of these projections is $\Omega = |\Lambda + \Sigma|$. As we would write $^3P_2$ for the atomic term for which $S = 1$, $L = 1$, and $J = 2$, we write $^3\Pi_2$ for the molecular term for which $S = 1$, $\Lambda = 1$, and $\Omega = 2$. In general we use the symbolism

$$^{2S+1}\Lambda_\Omega$$

The multiplicity is $2S + 1$, the kind of term is Λ, and the subscript denotes the value Ω may take. Multiplicity is indicated as in atomic spectra, e.g., $^1\Sigma$ $^1\Pi$ $^3\Sigma$ $^3\Pi$ $^2\Delta$, etc. A $^1\Sigma$ is analogous to a $^1S_0$ state, i.e., the resultant angular and spin momentum of the molecule is zero. The parity of a

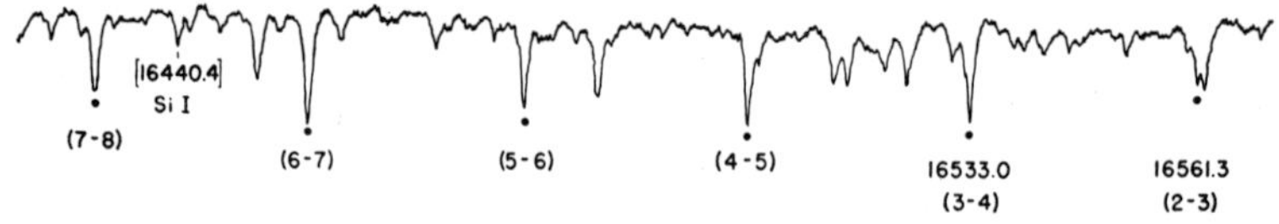

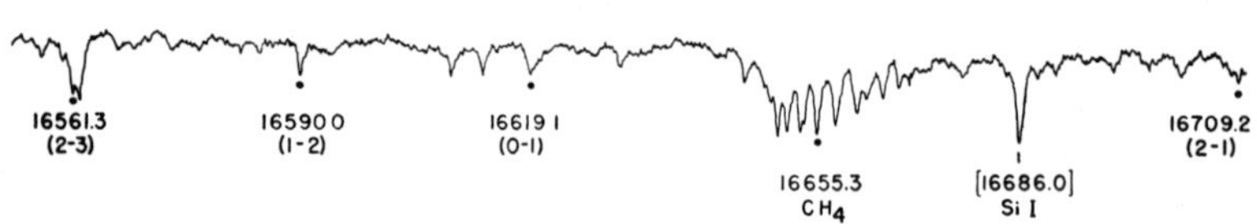

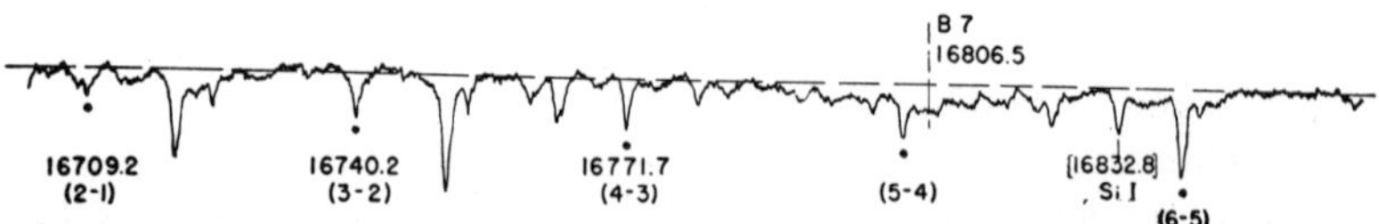

FIG. 2–36. A PORTION OF THE INFRARED SOLAR SPECTRUM AS OBSERVED WITH THE LEAD SULFIDE CELL

The tracing shows the diffuse λ16806.5 line of the Brackett series of hydrogen, whose computed position is indicated by the vertical broken line, and certain silicon lines. The [ ] indicate that the wavelengths were computed for a vacuum. The positive (0–1), (1–2), etc., and negative (3–2), (2–1), etc., branches of the band system of atmospheric methane CH_4 are indicated. The wavelengths are given for air and are indicated on the trace by dots. Most of the other lines on the trace are of atmospheric origin. (McMath-Hulbert Observatory, University of Michigan.)

molecular configuration is fixed by whether the sum of the l values of the individual electrons is even or odd, and is denoted by the letters g and u (gerade and ungerade).[6]

A few permitted band systems of astrophysical interest are:

$$^2\Sigma - {}^2\Pi,\ \mathrm{CH};\ {}^3\Pi^5 - {}^3\Pi,\ \mathrm{C_2},\ \mathrm{N_2},\ \mathrm{TiO},\ \mathrm{ZrO};\ {}^2\Delta - {}^2\Pi,\ \mathrm{OH}$$

The so-called "G band" in the solar spectrum contains the (O–O) sequence of $^2\Delta$—$^2\Pi$ of CH plus numerous atomic lines. Among forbidden electronic bands, the best known is the telluric[7] band of atmospheric oxygen. In spite of a low transition probability, this absorption band is strong because of the long air path. The type of molecular fine structure of greatest interest in radio astronomy is the so-called Λ doubling. In a molecule, the orbital angular momentum of the electrons is not constant, but its projection on the axis of the molecule is fixed. If $\Lambda \neq 0$, there exists a coupling between the orbital momentum $\mathbf{L}$ and the total angular momentum $\mathbf{J}$ that splits each rotational energy level into two levels, whose separation is in the radio range.

Identifications of molecules in spectra of the sun and stars are far from complete. In many instances, stars have not been observed with sufficient dispersion. For less abundant molecules and those that have few lines in accessible regions, blends can present serious problems.

One may make direct comparisons between individual rotational lines observed in the laboratory and in the sun and brighter stars. For faint stars it is necessary to rely on positions of band heads, an uncertain identification procedure except for strong bands. Unfortunately, laboratory studies are incomplete for many astrophysically important constants, and one must predict line positions from known molecular constants derived from an analysis of low excitation bands. Further, molecular bands observed in the sun and most stars are produced at a very much higher temperature than in the laboratory. Since the higher vibrational and rotational levels are occupied by much greater numbers of molecules, the emitted bands contain many more lines in the solar spectrum than in the laboratory spectrum.

In sunspots, atomic lines show a pronounced Zeeman effect; molecular lines do not. Most of the numerous weak sunspot lines have never been identified; probably most of them are of molecular origin.

[6] Symbols (+) and (−) are also used to denote spatial symmetry properties of the molecular electronic wave functions. See G. Herzberg, *Molecular Spectra and Molecular Structure* (Princeton, N. J.: D. Van Nostrand Co., Inc., 1950), p. 238.

[7] Telluric lines in the solar spectrum are lines which arise from absorption by molecules in the earth's atmosphere such as O_2 and H_2O. They can be identified by their marked increase in intensity as the sun gets lower in the sky and by the fact that they do not partake of the Doppler shift of solar rotation. See Fig. 2–36.

Some molecules such as H_2, known to be abundant in the sun, have no lines in the ordinarily accessible region. Molecular ions such as CH^+ are not observed in the stars since dissociation of a molecule usually occurs before ionization. On the other hand, Osawa has concluded that the H_2^+ continuous absorption is important in hotter stars.

2–18. Some Polyatomic Molecules of Astrophysical Interest. *Polyatomic molecules* abound in comets, in the coolest stars, and in planetary atmospheres. Bands of C_3 and SiC_2 have been identified in the cool carbon stars, and it has been suggested that a substantial contribution to the opacity may be made by the continuous absorption of these molecules.

As with diatomic molecules, there exist electronic states like $^1\Sigma$, $^1\Pi$, etc., and the vibrational states are split into rotational substates. The type and complexity of the spectrum depend on the spatial arrangement of the atoms. Carbon dioxide is a linear symmetrical molecule with three atoms arranged in a straight line, O—C—O. Water is neither linear nor symmetrical; it gives a complicated spectrum with overlapping rotational structure. In methane, CH_4, the four hydrogen atoms are arranged symmetrically in a tetrahedron with the carbon atom at the center.

The fundamental frequencies of vibration of the CO_2 molecule according to Ta You Wu are $\tilde{\nu}_1 = 1321.7$ cm^{-1}, $\tilde{\nu}_2 = 667.9$ cm^{-1}, and $\tilde{\nu}_3 = 2362.8$ cm^{-1}. These correspond to excitations of first one vibrational mode and then another. In the spectrum of Venus the higher overtones and combinations, $5\tilde{\nu}_3$, $5\tilde{\nu}_3 + \tilde{\nu}_1$, and $5\tilde{\nu}_3 + \tilde{\nu}_2$, are observed in appreciable strength. Methane has four fundamental frequencies: $\tilde{\nu}_1 = 2915$ cm^{-1}, $\tilde{\nu}_2 = 1520$ cm^{-1}, $\tilde{\nu}_3 = 3020$ cm^{-1}, and $\tilde{\nu}_4 = 1306$ cm^{-1}. Overtones of these bands are observed in the earth's atmosphere, but they are particularly strong in the spectra of the outer planets. In Neptune harmonics as high as $16\tilde{\nu}_4$ are observed. Since these high harmonics have never been observed in the laboratory, the amount of vapor in the light path must be considerable. There also appear bands in which the difference of two modes of excitation occurs, e.g., $5\tilde{\nu}_3 - \tilde{\nu}_2$, meaning that the lower state of vibration is excited. Ammonia bands appear in the spectra of Jupiter and Saturn although they are missing in the spectra of Uranus and Neptune where the substance must be frozen out.

Actual estimates of the amount of material in the line of sight are difficult because the band structures are sensitive to temperature, path-length, and density. Eventually, with the aid of appropriate laboratory studies, we may be able to interpret the strengths and structures of these bands in terms of the temperature and density of the planet's atmosphere.

Rotational bands of polyatomic molecules fall in the r-f region. Water vapor and oxygen bands limit the high-frequency end of the radio spectrum window.

2–19. Isotope Effects in Atomic and Molecular Spectra. Except for atoms of the lightest elements, different isotopes may be distinguished easily only with the aid of hyperfine structure data. For ionized helium and deuterium, one may utilize the fact that the Rydberg constant for an atom of atomic weight A is given by

$$R = \frac{R_0}{1 + \frac{1}{1837A}}$$

where $R_0 = 109{,}737.30$ is the Rydberg constant for an atom of infinite mass. One sees that the Pickering lines of He II do not quite coincide with the Balmer lines. For hydrogen of atomic weight 2 (deuterium), $R_D = 109{,}707.42$ as compared with $R_H = 109{,}677.68$. The corresponding wavelengths of the Hα line are:

$$\lambda(\mathrm{H}\alpha) = 6562.817 \qquad \lambda(\mathrm{D}\alpha) = 6561.032 \qquad \delta\lambda = 1.785\mathrm{A}$$

Menzel searched for deuterium in the solar spectrum, in the chromosphere, and in several stellar spectra, and concluded that within the stars, deuterium was less than 10^{-5} as abundant as ordinary hydrogen. Kinman (1956) found no evidence for deuterium in the sun, although he stated that an abundance ratio as low as 4×10^{-5} might go undetected. Deuterium has not been detected in the interstellar medium, although if it is present there with the terrestrial ratio, it should not be difficult to find by radio telescopes. In heavier atoms the above mentioned mass effect shift would be so small that the spectra of isotopes would not be separated from one another by detectable amounts. On the other hand, there is also an isotope shift that depends on the fact that a nucleus can be polarized by a sufficiently powerful electric field and that it occupies a finite volume. When an electron (e.g., in an s orbit) passes close to the nucleus it may distort the latter by its electrostatic effects. Since these effects will differ from one type of nucleus to another, the isotope shift may vary markedly in heavier elements. Hyperfine effects among different isotopes may be very large in the r-f range.

Isotope effects can be important in band spectra, since molecules composed of different isotopes will have different vibration frequencies. The binding forces are nearly the same in a $C^{12}C^{12}$ molecule as in a $C^{12}C^{13}$ molecule, for example, but the masses are different. Since the vibration frequency parameter ω (see Eq. 2–52) depends on the mass in the sense that the greater the mass the slower the vibration, the band origin or head of $C^{12}C^{12}$ will be shifted with respect to that of $C^{12}C^{13}$. Furthermore, the moment of inertia will be changed so that B will differ for molecules of different isotopes. Positions of corresponding lines will differ.

Diatomic molecules with two identical nuclei show alternating intensities of the lines in their band systems. The amount of the variation depends on the nuclear spin. When the two nuclei are isotopes of different mass, the alternation disappears.

Isotopic bands of carbon appear in *R*- and *N*-type stars where strong $C^{12}C^{12}$ bands are flanked by weaker bands of $C^{12}C^{13}$ and $C^{13}C^{13}$ (see Fig. 2–37). Evidently the $C^{12}C^{13}$ ratio, which is 90 on the earth, is much smaller in these stars. McKellar found a ratio of the order of 3, but the most recent work (Climenhaga, 1960) indicates that the ratio is closer to 4 or 5. The C^{13} isotope is responsible for some weak infrared telluric lines of CO_2.

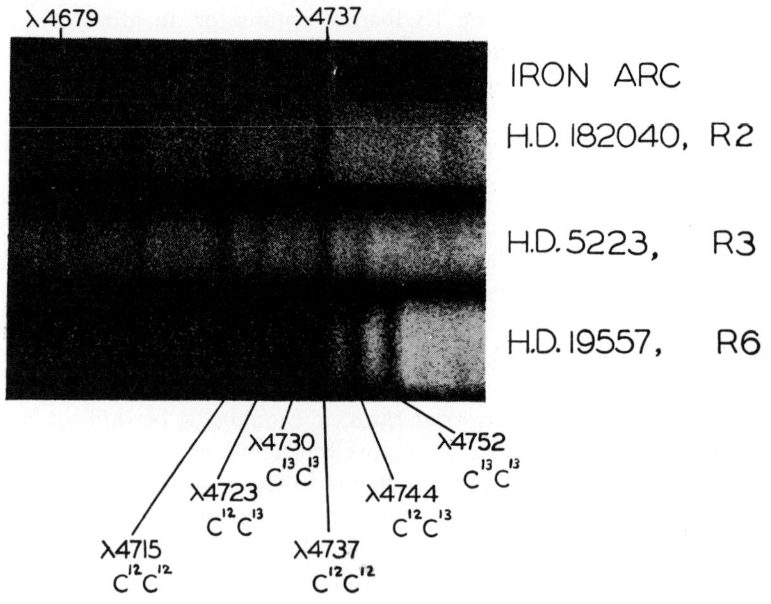

FIG. 2–37. THE ISOTOPIC BANDS OF CARBON

The λ4737 region of the spectra of 3 *R*-type stars is shown. In *HD* 182040 the $C^{12}C^{12}$ bands are present with moderate strength but no isotopic bands are seen with certainty. In *HD* 5223 the $C^{12}C^{13}$ bands are prominent, while in *HD* 19557, bands arising from all three molecular species are present with great strength. (Courtesy, Andrew McKellar, Dominion Astrophysical Observatory, Victoria, B. C.)

REFERENCES

General references on spectra:

WHITE, H. E. 1934. *Introduction to Atomic Spectra.* McGraw-Hill Book Co., Inc., New York.

HERZBERG, G. 1950. *Molecular Structure and Molecular Spectra.* I, "Diatomic Molecules." D. Van Nostrand Co., Inc., Princeton, N. J.

———. 1944. *Atomic Spectra and Atomic Structure.* Dover Publications, Inc., New York.

MOORE, CHARLOTTE. 1945. "A Multiplet Table of Astrophysical Interest," *Princeton Observatory Contribution No. 20.*

———. 1949, 1952, 1958. "Atomic Energy Levels," *Natl. Bureau of Standards Circ. 467.*

TOWNES, C. H., and SCHAWLOW, A. L. 1955. *Microwave Spectroscopy.* McGraw-Hill Book Co., Inc., New York.

Advanced theoretical treatment on spectra:

CONDON, E. U., and SHORTLEY, G. H. 1951. *Theory of Atomic Spectra.* Cambridge University Press, London.

SLATER, J. C. 1960. *Quantum Theory of Atomic Structure.* (2 vols.) McGraw-Hill Book Co. Inc., New York.

For a discussion of forbidden lines:

ALLER, L. H. 1956. *Gaseous Nebulae.* John Wiley & Sons, Inc., New York, chap. 5.

General reference on astronomical spectra:

MERRILL, PAUL W. 1956. *Lines of Chemical Elements in Astronomical Spectra.* Carnegie Institution of Washington, Publ. 610.

CHAPTER 3

THE GASEOUS STATE

The stars are gaseous throughout and in the layers accessible to observation, the perfect gas law is obeyed. Only in the deep interiors of certain stars do deviations from this law become important.

3–1. The Equation of State for a Perfect Gas. The fundamental equation of state for a perfect gas is

$$PV = RT \tag{3–1}$$

where P is the pressure, V is the volume, and T is the temperature in absolute degrees (°K). The constant R depends on the mass of gas involved. As our standard of quantity we usually take one gram-molecule or mole, which amounts to a volume of 22.4 liters at 0°C (273°K) and one atmosphere pressure. Then

$$PV = \mathcal{R}T \tag{3–2}$$

where $\mathcal{R} = 8.314 \times 10^7$ ergs deg^{-1} mole^{-1} if the pressure P is measured in dynes cm^{-2}. If P is measured in atmospheres (one atmosphere equals 1.013×10^6 dynes cm^{-2}) and the volume V is expressed in cm^3, $\mathcal{R}$ is 82.05 atmospheres deg^{-1} mole^{-1}. The number of molecules in a mole (Avogadro's number) must be established empirically; it is $N_0 = 6.025 \times 10^{23}$ molecules per mole. The Loschmidt number, the number of molecules per cm^3 under standard conditions, is 2.687×10^{19}.

It is sometimes convenient to define the gas constant per atom or molecule, instead of per mole. Thus, Boltzmann's constant is $k = \mathcal{R}/N_0 = 1.38044 \times 10^{-16}$ ergs deg^{-1}. The gas law then becomes

$$P = NkT \tag{3–3}$$

where N is the number of molecules per unit volume.

There is a third useful form of the perfect gas law. If the N_0 atoms or molecules in a mole, each of mass m, have a total mass of μ, the actual density in grams per cm^3 will be $\rho = Nm$, and since $N_0k = \mathcal{R}$,

$$P = \frac{\rho\mathcal{R}T}{\mu} = \rho\left(\frac{k}{\mu H}\right)T \tag{3–4}$$

where H is the mass of a particle of unit atomic weight.

An important corollary of the equation of state is Dalton's *law of partial pressure*. If a gas contains a number of non-reacting constituents each of which exerts a pressure of its own, the gas pressure is the sum of the pressures exerted by each constituent. Each kind of particle contributes its own partial pressure, $p_i = n_i kT$, so that the total gas pressure is

$$P_g = \Sigma p_i = kT\Sigma n_i = NkT \tag{3–5}$$

For example, a mixture of hydrogen and nitrogen exerting partial pressures p_H and p_N will exert a total pressure $P = p_H + p_N$.

The high temperatures that exist in the atmosphere of certain stars and in the interiors of all of them serve to break down the molecules into their constituent atoms, and finally the atoms themselves become stripped of their electrons. Let us illustrate by the history of a mass of hydrogen gas whose temperature is raised. Under normal conditions of temperature and pressure, hydrogen exists in a molecular form, each molecule consisting of two hydrogen atoms. If the temperature is raised to the order of that of the solar atmosphere, H_2 becomes dissociated into separate hydrogen atoms, and whereas formerly two grams contributed 6.03×10^{23} particles (hydrogen molecules), only one gram of the dissociated hydrogen gas now suffices to contribute the same number of particles (hydrogen atoms). At still higher temperatures the hydrogen atom itself becomes broken down into its constituent electron and proton, so that only a half gram of the completely ionized hydrogen contains 6.03×10^{23} particles (protons plus electrons), each capable of contributing as much to the pressure as the other. The molecular weight of H_2 is 2, that of atomic hydrogen is 1, and completely ionized hydrogen is $\frac{1}{2}$. Thus in the perfect gas law, Eq. 3–4, μ is itself a function of the temperature, and also of the pressure, since a high pressure tends to jam the electrons back into the atoms again.

Thus, it is clear that we will need ionization and dissociation equations so that we may calculate the numbers of ions, atoms, and molecules as a function of temperature and pressure for any given gas mixture. As soon as these numbers are known, μ can be calculated at once.

3–2. The Adiabatic Gas Law. Our gas law, Eq. 3–1, represents a relation between three variables, P, V, and T. If we keep the temperature constant, and vary the pressure or volume, we obtain Boyle's law. On the other hand, if the volume is maintained constant and the temperature is changed, the pressure will be proportional to the temperature, and we have Charles' law. If both P and T are changed, one cannot, in general, say anything about V. One special type of change merits our attention. Suppose a mass of gas is allowed to expand or contract and no heat is permitted to enter or leave it during the change. Such a change is called an *adiabatic process* and may be of importance in those regions of a star where the chief transport of energy is by convection currents rather than by radiation.

In a mass of a gas undergoing an adiabatic change, it may be shown that the pressure and volume are related by

$$PV^{\gamma} = \text{const} \tag{3-6}$$

where $\gamma = c_p/c_v$. Here c_p denotes the specific heat of a gas under conditions of constant pressure and c_v is the specific heat when the volume is kept constant. Now γ is always larger than 1, but approaches unity for complex atoms. It depends on the number of degrees of freedom of the atom or molecule, i.e., the minimum number of separate data we must have to describe the motion, e.g., (1) a point mass has three degrees of freedom, (2) a rigidly connected rotator has two degrees of rotation and three degrees of translation.[1] A complicated molecule may possess modes of vibration as well as rotation. For a monatomic gas $\gamma = \frac{5}{3}$. Furthermore, the specific heats c_p and c_v will depend upon modes in which the energy may be stored internally within the gas, by dissociation of molecules and by ionization of atoms, for example. A gas which is undergoing ionization at the temperature and density in question may have an effective γ different from that of a monatomic gas which is either wholly neutral or wholly ionized. This fact is of considerable interest in connection with theories of the origin of the solar convection zone and solar granules (see Chapter 9).

3–3. Relation Between Temperature and Mean Gas Kinetic Velocity. If we could look at individual molecules of a gas we would witness a hurly-burly of rapidly rushing particles, running hither and yon, colliding with one another and with the walls of the container. The impact of these particles with the walls and upon one another produces the gas pressure. Were we able to tag one of these molecules and follow it through the vicissitudes of its wanderings we would find it moving now fast, now slow, first in one direction and then, after a collision, in quite another. Or, if we took a couple of snapshots of the gas and measured the magnitude and direction of the motion of the molecules, we would find them to be moving in random directions and with different speeds. A few would be going with speeds considerably greater than the average, while others would be scarcely moving at all, but the majority would have speeds differing from the average by less than a factor of two.

The exact law of the distribution of velocities follows from the kinetic theory of gases or statistical mechanics. We derive it below in Art. 3–5, and discuss it in Art. 3–6, but first we shall establish the relation between the mean gas kinetic velocity and the temperature of the gas by means of a very elementary argument.

Let there be n molecules each of mass M in a cubical box of length L, whose sides are oriented parallel to the three coordinate axes x, y, z. A molecule moving in the x-direction with a velocity v_x strikes one of the

[1] The motion of a dumbbell whose center of gravity is fixed can be represented as the resultant of motion about two axes.

faces perpendicular to the x-axis and rebounds. In so doing it delivers a momentum $2Mv_x$ to the wall. The number of collisions per second with the same wall will be $v_x/2L$; hence the total momentum conveyed each second to the wall by all molecules is $\Sigma Mv_x^2/L$. The effects of the momentary short-range forces the molecules exert on one another during mutual collisions cancel out.

The momentum $\text{cm}^{-2}\,\text{sec}^{-1}$ is the pressure, thus

$$P = \frac{M}{L^3}\Sigma v_x^2 \qquad \text{gm cm}^{-1}\,\text{sec}^{-2} \tag{3–7}$$

We define a mean velocity such that

$$\bar{v}_x^2 = \frac{1}{n}\Sigma v_x^2 \tag{3–8}$$

Furthermore,

$$\bar{v}_x^2 = \bar{v}_y^2 = \bar{v}_z^2 = \tfrac{1}{3}\bar{v}^2_{\text{total}} \qquad v^2_{\text{total}} = v_x^2 + v_y^2 + v_z^2 \tag{3–9}$$

Hence

$$P = \frac{n}{3L^3}M\bar{v}^2 = \tfrac{1}{3}NM\bar{v}^2$$

where N is the number of molecules per unit volume but using Eq. 3–3 we see that the average energy per molecule is

$$\tfrac{1}{2}M\bar{v}^2 = \tfrac{3}{2}kT \tag{3–10}$$

Here $\bar{v}$ is the *root mean square* (rms) velocity since it is an average taken with respect to energy.

3–4. Thermodynamic Equilibrium. In order to have an adequate description of matter in situations of astrophysical interest, we not only need a P, N, T relationship of the type provided by the gas law, but also must know

1. Relative populations of the ground level and of excited states of atoms, ions or molecules.
2. Relative numbers of ions and neutral atoms or of ions in second and third stages of ionization, etc.
3. Relative numbers of atoms and molecules in a reacting assembly, e.g., N(O) N(H)/N(OH).
4. Velocity distribution of the atoms and electrons with respect, e.g., to the rms velocity defined in Eq. 3–10.
5. Intensity distribution with wavelength of the radiation present in the heated gas as a consequence of its temperature.

The nature of relations 1, 2, 3, and 5 under general conditions obtained in the outer envelope of a star, e.g., the corona or chromosphere, can be extremely complicated. We are concerned with properties of matter at temperatures ranging from 1500°K or 2000°K in atmospheres of red supergiants to several million degrees in highly excited regions of the solar corona.

In a first approximation, interest is centered in steady-state conditions. In a treatment of variable stars, the practice frequently has been to consider changes in terms of a series of separate stages, each of which can be treated as a steady state.

At the outset it is important to understand that a steady state *does not* imply thermal equilibrium. Consider radiation passing through a stratum of matter, e.g., within a stellar atmosphere. The state may be a steady one, i.e., not change with time, but it is certainly not an equilibrium one. A flow of energy from one side of a layer to the other implies a temperature gradient and under equilibrium conditions a temperature difference cannot exist. Nevertheless, the rates of certain atomic processes may be nearly the same as in thermal equilibrium and in a first reconnaissance of the problem, various relationships derived for strict thermodynamic equilibrium will prove useful. In thermodynamic equilibrium, (1) is given by Boltzmann's equation, (2) by the ionization equation, (3) by the dissociation equation, (4) by the Maxwellian velocity distribution, and (5) by Planck's law.

Let us picture the state of affairs in a gas in thermal equilibrium. Imagine a mass inclosed in a hypothetical box whose walls are maintained at a temperature T of the order of 5000–10,000°K. The atoms in the box move rapidly about, strike one another in more or less violent collisions, absorb and re-emit energy, and lose and recapture electrons. A condition obtains wherein each process is exactly balanced by its inverse. Every collision in which an electron gives up energy to an atom to excite it to a higher energy level is balanced by an encounter in which an excited atom unloads its excitation energy upon a passing electron (superelastic collision). Every ionization from a particular level is balanced by a recapture upon the same level.

We can express these ideas quantitatively by saying that quite generally the number of absorptions from n to n'' is equal to the number of transitions in which the atoms jump from n'' to n with the emission of radiation. The number of collisional excitations from n to n'' is equal to the number of inverse superelastic collisions, wherein atoms go from n'' to n and a passing particle carries off the excess energy. The number of ionizations from level n, in which the detached electron moves with a speed between v and $v + dv$, is equal to the number of recaptures of electrons from the same velocity range dv at v to the same level n. An assemblage of particles in which every process is balanced by its inverse, i.e., one in which *detailed balancing* occurs and in which the distribution of atoms among the various excited levels is given by the Boltzmann formula, is said to be in *strict thermodynamic equilibrium.*

Thermodynamics, which deals with macroscopic properties of matter, gives us much information about material in thermal equilibrium, although conventional thermodynamics is of relatively restricted use to the astronomer. A more powerful technique is that of statistical mechanics, which makes use of the fact that atoms possess known energy states. Application of formal procedures of statistical mechanics requires that thermal equilibrium exists and assumes that atoms or molecules interact with one another. It does not, however, insist on any knowledge of the details of how energy passes from one atom to another.

Statistical mechanics enable us to derive important relationships such as Maxwell's law of velocities, Planck's law (Chapter 5), Boltzmann's law, and the ionization and dissociation equations. In the following article we sketch the statistical mechanical arguments whereby these laws are established. It will be necessary to accept as axioms certain statements which are normally proven in texts on the subject, and whose validity requires extensive calculation. The following presentation is adopted from a derivation of the ionization and dissociation equations by Menzel.

3-5. Derivation of Boltzmann's Equation, the Maxwellian Distribution of Velocity and the Ionization Equation by Statistical Mechanics.[2] A single particle, be it classical oscillator, electron, atom, molecule, or photon is called a system and the aggregate of systems is said to constitute the assembly. If we could specify the position and momentum of each system at each instant of time we would have a complete description of the assembly, but with a host of interacting particles such detailed information is not possible, and the best we can do is to specify the chance of some particle occupying a given domain of momentum and energy.

Consider first a system such as a classical oscillator that has but one degree of freedom. Its "state" or "phase" can be represented as a point in a space of two dimensions with the momentum p and distance from origin q serving as coordinates. We call such a depiction a "phase diagram" and the plane is called "phase space." Consider a free particle with three degrees of freedom. At any instant it may be characterized by three space coordinates, usually denoted as q_1, q_2, and q_3 (e.g., x, y, z) and three velocity, or rather *momentum coordinates*, p_1, p_2, and p_3. All six of these numbers are needed to specify the position and momentum of each particle at any one time. If we know the forces acting upon the particles, classical mechanics asserts that theoretically it would be possible to tell what they would be doing at any later time: To indicate the state of a given particle, we could employ three separate graphs, one for each pair of momentum and space coordinates, or we could adopt a single point in a six-dimensional

[2] This article may be omitted without impairing an understanding of the rest of the presentation.

"phase space." The latter point of view is the more useful. Actually, the phase space is thought of as being divided into small rectangles of size $\Delta p \Delta q$ for each degree of freedom such that the location of the system within the rectangle supplies us with adequately accurate information.[3]

We introduce the concept of the *statistical weight* of the cell as being proportional to its area. The *a priori probability*, apart from any other restrictions that may be imposed on the assembly, that the representative point of a single system will fall in cell k is proportional to the cell's weight

$$g_k = K(dp_1 \ldots dq_s)_k \tag{3–11}$$

if the system has s degrees of freedom. Here K is a scaling constant. The a priori probability m systems will occupy cell j and n systems cell k is $g_j^m\, g_k^n$; in general, the weights will not be equal. Hence if n systems are distributed such that n_1 are in cell 1, n_2 in cell 2, etc., the *weight* of this particular distribution will be

$$g_d = g_1^{n_1} g_2^{n_2} \cdots = \Pi\, g_i^{n_i} \tag{3–12}$$

What counts are the *relative values* of the weights, not their absolute values. In the quantum theory the statistical weight of a level expresses the relative likelihood of an atom being found there, other things being equal. If two states have the same excitation potential, under equilibrium conditions their relative populations will be in proportion to their statistical weights. Since each Zeeman level is assigned unit weight, the statistical weight, $2J + 1$, equals the number of Zeeman states into which a level of inner quantum number J is resolved by a magnetic field. In the classical theory the statistical weight of a particle is $K\, dp\, dq$. It is frequently necessary to refer both classical and quantum statistical weights to the same unit, as, for example, in deriving the ionization equation.

Each state of a harmonic oscillator has unit weight, with momentum and coordinate variables obeying the relation

$$\oint p\,dq = (n + \tfrac{1}{2})h \tag{3–13}$$

In phase space the representative points trace out ellipses; the area between successive elliptical annuli is h, but for large n the classical and quantum systems asymptotically approach the same limiting value. In order to make the weightings consistent, we set $K = 1/h$ for one degree of freedom or $1/h^s$ for s degrees of freedom.

Consider a six-dimensional phase space corresponding to the three degrees of freedom of a free particle. If the particle is an electron, it will have two additional degrees of freedom as a consequence of the two spin directions. Hence, the statistical weight of a free electron will be

[3] One thinks immediately of the Heisenberg uncertainty principle, which suggests measuring the area of the rectangle $\Delta p \Delta q$ in terms of the quantity h. We shall see that this intuition is justified.

$$g = 2\,\frac{dp_1dp_2dp_3\,dq_1dq_2dq_3}{h^3} \tag{3–14}$$

One may show (Liouville's theorem) that the density of a given set of representative points in phase space does not change with the time. In other words, if the system is once in equilibrium, it will remain in equilibrium. It also justifies the practice of assigning weights proportional to the "volumes" $dp_1 \cdots dp_r\, dq_1 \cdots dq_r$ of the cells of phase space.

Let us now suppose that we have N identical particles which are to be placed in a series of boxes, labeled 0, 1, 2, $\cdots$. Suppose that N_0 are placed in the zeroth box, N_1 in the first box, etc.

$$N = N_0 + N_1 + N_2 + \cdots = \Sigma N_i \tag{3–15}$$

The total number of ways of doing this will be

$$\frac{N!}{N_0!N_1!N_2!\cdots} = \frac{N!}{\Pi N_i!} \tag{3–16}$$

since the total number of permutations of N things is $N!$ but the permutations of all particles in the rth box among themselves gives nothing new. In particular, notice that there is only one way in which all particles may be placed in the first box and zero in all others, N ways in which $(N - 1)$ particles can be placed in the zeroth box, and one in the second, etc. Maximum probabilities clearly occurs for the situation in which roughly equal numbers of particles appear in boxes of equal weight, provided that there are no further restraints on the problem. In order to get the relative probability of the distribution W, we multiply by g, the statistical weight, and find

$$W = \frac{N!}{\prod\limits_i N_i!}\prod_i g_i^N \tag{3–17}$$

which may be reduced to the true probability

$$P = W/C \text{ where } C = \Sigma W \tag{3–18}$$

The states of the assembly in which we are interested are those for which N_i is large. It can be shown[4] from the properties of the Γ function

$$\ln N! = N \ln N - N + \frac{1}{2}\ln(2\pi N) + \frac{1}{12N} - \left(\text{terms in } \frac{1}{N^2}\right) \tag{3–19}$$

so that for large N we can use the following approximation for Stirling's formula Eq. 3–19, viz.

$$N! \sim (N/e)^N \tag{3–20}$$

[4] For example, Margenau and Murphy, *Mathematics of Physics and Chemistry* (Princeton, N. J.: D. Van Nostrand & Co., 1943), p. 93; D. H. Menzel (ed.), *Fundamental Formulas of Physics* (New York: Dover Publications, 1960).

Using Eqs. 3–15, 3–17, 3–18, and 3–20 we can write

$$P = \frac{N^N}{C} \prod_i \left(\frac{g}{N_i}\right)^{N_i} \tag{3–21}$$

and solve for the maximum value of P for the situation in which the distribution is simply determined by the statistical weights and number of permutations. Set the derivative of ln P equal to zero, viz.

$$d \ln P = -d\, \Sigma\, N_i \ln \frac{N_i}{g_i} = -\sum_i \left(\ln \frac{N_i}{g_i} + 1\right) dN_i = 0 \tag{3–22}$$

Since N is a constant by Eq. 3–15, we must have as a second independent condition on the problem that

$$dN = \Sigma\, dN_i = 0 \tag{3–23}$$

We employ the method of Lagrange's undetermined multipliers, i.e., we multiply Eq. 3–23 by a constant a and add it to Eq. 3–22.

$$\sum_i (\ln N_i/g_i + 1 + a)\, dN_i = 0 \tag{3–24}$$

The variations, dN_i, are arbitrary; hence the term in brackets must vanish and

$$\ln \frac{N_i}{g_i} = -(1 + a) \quad \text{or} \quad N_i = g_i e^{-(1+a)} \tag{3–25}$$

From Eq. 3–15 we have that

$$\Sigma\, N_i = N = e^{-(1+a)}\, \Sigma\, g_i \tag{3–26}$$

or

$$N_i = \frac{g_i N}{\Sigma\, g_i} \tag{3–27}$$

which is the expression for the weighted mean, a result which checks with our intuition.

Now consider the situation in which an amount of energy, E, is to be distributed among N particles in the assembly, each particle taking a unit ε_0, ε_1, ε_2, etc., where $\varepsilon_{max} << E$. In addition to the constraints indicated by Eqs. 3–22 and 3–23 which are still valid, we now have that since

$$E = \varepsilon_0 N_0 + \varepsilon_1 N_1 + \cdots + \cdots = \sum_i \varepsilon_i N_i \tag{3–28}$$

$$dE = \sum_i \varepsilon_1 dN_i = 0 \tag{3–29}$$

Multiply Eq. 3–29 by an arbitrary constant, $-\ln \vartheta$, and Eq. 3–23 by another arbitrary constant denoted as

$$C_2 = \ln \frac{u}{N} - 1 \tag{3–30}$$

where the constants, u and ϑ will be identified later. Application of the method of Lagrangian multipliers yields

$$\sum_i \left(\ln \frac{N_i}{g_i} - \varepsilon_i \ln \vartheta + \ln \frac{u}{N}\right) dN_i = 0 \tag{3–31}$$

Since dN_i is arbitrary, Eq. 3–31 can be fulfilled only if the term in the parentheses vanishes for all values of i. Therefore,

$$\ln \frac{N_i}{g_i} = \ln \frac{N\vartheta^{\varepsilon_i}}{u} \tag{3–32}$$

or

$$N_i = \frac{N}{u} g_i \vartheta^{\varepsilon_i} \tag{3–33}$$

but since

$$\Sigma N_i = N = \frac{N}{u} \Sigma g_i \vartheta^{\varepsilon_i} \tag{3–34}$$

$$u = \Sigma g_i \vartheta^{\varepsilon_i} = \Sigma g_i e^{-\varepsilon_i/kT} \tag{3–35}$$

where we set

$$\vartheta = e^{-1/kT} \tag{3–36}$$

We must now prove that the T introduced in Eq. 3–36 is actually the temperature in the accepted meaning of the term. We establish this identity as follows:

Now the energy associated with N particles is

$$E = \Sigma N_i \varepsilon_i = \frac{N}{u} \Sigma g_i \varepsilon_i \vartheta^{\varepsilon_i} \tag{3–37}$$

Differentiate Eq. 3–35 with respect to ϑ and multiply by ϑ, viz.:

$$\vartheta \frac{du}{d\vartheta} = \Sigma g_i \varepsilon_i \vartheta^{\varepsilon_i} \tag{3–38}$$

whence by Eq. 3–37

$$E = N \frac{\vartheta \dfrac{du}{d\vartheta}}{u} = N\vartheta \frac{d}{d\vartheta} \ln u \tag{3–39}$$

so the *average* energy per system, i.e., particle, will be

$$\bar{\varepsilon} = \frac{E}{N} = \vartheta \frac{d}{d\vartheta} \ln u = \vartheta \frac{d}{dT} (\ln u) \frac{dT}{d\vartheta} = kT^2 \frac{d}{dT} \ln u \tag{3–40}$$

if Eq. 3–36 is valid.

Consider now classical gas particles, each one of which has three degrees of freedom; let there be n particles in a volume V. We suppose that there are no internal or body forces. Then

$$\varepsilon_i = \frac{M}{2} (v_x^2 + v_y^2 + v_z^2) = \frac{p_1^2 + p_2^2 + p_3^2}{2M} \tag{3–41}$$

$$g_i = \frac{dp_1 \cdots dq_3}{h^3} \tag{3–42}$$

Substitute Eq. 3–41 and Eq. 3–42 in Eq. 3–35 to obtain

$$u = \frac{V}{h^3} \iiint_{-\infty}^{+\infty} e^{-(p_1^2+p_2^2+p_3^2)/2MkT}\, dp_1\, dp_2\, dp_3 \tag{3–43}$$

since we replace the summation in Eq. 3–35 by an integration and note

$$V = \int dq_1\, dq_2\, dq_3 = \int dx\, dy\, dz \tag{3–44}$$

In Eq. 3–43 let us make the substitution

$$x^2 = \frac{p_i^2}{2MkT} \tag{3–45}$$

with $i = 1, 2,$ or 3, successively. Then

$$\sqrt{2MkT}\, dx = dp_i \tag{3–46}$$

and

$$\int_{-\infty}^{+\infty} \exp\left[-\frac{p^2}{2MkT}\right] dp = \sqrt{2MkT} \int_{-\infty}^{+\infty} e^{-x^2}\, dx = (2\pi MkT)^{1/2} \tag{3–47}$$

and

$$u(T) = \frac{(2\pi MkT)^{3/2} V}{h^3} \tag{3–48}$$

is the partition function for free particles. The number of systems with momentum components $dp_1\, dp_2\, dp_3$ and coordinate components falling in $dq_1\, dq_2\, dq_3 = dx\, dy\, dz$ will be (since $N = n/V$)

$$Nf(p, q)\, dp_1 \cdots dq_3 = \frac{Nh^3}{(2\pi MkT)^{3/2}} e^{-(p_1^2+p_2^2+p_3^2)/2MkT} \frac{dp_1 \cdots dq_3}{h^3} \tag{3–49}$$

Now replacing the momenta by Mv_x, Mv_y, Mv_z, the equation reduces to the form

$$Nf(v_x\, v_y\, v_z)\, dv_x\, dv_y\, dv_z = N\left(\frac{M}{2\pi kT}\right)^{3/2} e^{-M(v_x^2+v_y^2+v_z^2)/2kT}\, dv_x\, dv_y\, dv_z \tag{3–50}$$

which is the expression for Maxwell's distribution law of velocities. See Fig. 3–1. The mean kinetic energy $\bar{\varepsilon}$ of a molecule is found from Eq. 3–40.

$$\bar{\varepsilon} = \frac{\overline{E}}{N} = \frac{1}{2} M\, \bar{v}^2 = kT^2 \frac{d}{dT} \ln\left\{\frac{(2\pi MkT)^{3/2}}{h^3}\right\} = \frac{3}{2} kT \tag{3–51}$$

which is identical with our Eq. 3–10, establishing the validity of Eq. 3–36.

Note further that Eq. 3–33, with the aid of Eq. 3–36, now becomes

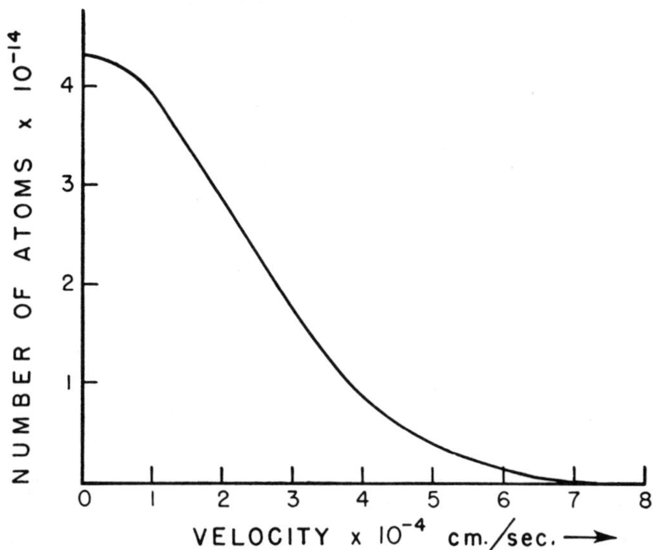

FIG. 3–1. MAXWELLIAN DISTRIBUTION OF VELOCITIES IN THE x-DIRECTION

We plot the number of argon atoms per cm³ with x-velocities between v and $v + dv$ for N.T.P. (273°K and 760 mm or one atmosphere pressure).

$$\frac{N_i}{N} = \frac{g_i}{u} e^{-\varepsilon_i/kT} \tag{3–52}$$

which is Boltzmann's equation, a relationship fundamental to statistical mechanics. Applications of Eqs. 3–50 and 3–52 to situations of astrophysical interest will be given in Arts. 3–6 and 3–7.

We may now apply statistical mechanics to the much more difficult problem of reacting systems in order to derive the equations for ionization and dissociations. Let the assembly contain X_M particles of type M and X_N particles of type N which combine to form particles of type MN, viz.:

$$M + N \rightleftarrows MN \tag{3–53}$$

For example, M could represent oxygen atoms, N hydrogen atoms, and MN the molecule OH. Alternately, M could denote OH radicals, N hydrogen atoms, and MN water molecules. In an ionization problem, M could represent calcium ions, N electrons, and MN neutral calcium atoms. In the following developments, we shall call M and N "atoms" or "particles" and MN "molecules."

In a given phase of the assembly, there will be M free particles of type M, N free particles of type N, and S molecules of type MN. The assembly is subject to the restraints:

$$X_M = M + S \qquad X_N = N + S \tag{3–54}$$

The values of M, N, and S determine the degree of dissociation of the assembly. We call each specification of M, N, and S a group noting that a given group may be formed in a variety of ways, each of which we designate as an *example.*

The number of permutations of the individual particles is $X_M!\ X_N!$ Permutations of free particles and molecules among themselves give no new examples. Hence the number of examples in a group is

$$\frac{X_M!\,X_N!}{M!\,N!\,S!} \tag{3–55}$$

For simplicity, we shall suppose that the particles M and N are not identical. For each given triad of numbers M, N, and S, there correspond a number of distributions. Consider first the distribution where the M free particles $M_0\ M_1\ M_2\cdots$ fall respectively in cells of weight $\mu_0\ \mu_1\cdots$ with energies $\varepsilon_0\ \varepsilon_1\cdots$. The N free particles $N_0\ N_1\ N_2\cdots$ similarly fall in cells of energies $\eta_0\ \eta_1\cdots$ and weights $\nu_0\ \nu_1\cdots$. The S molecules $S_0\ S_1\ S_2\cdots$ fall in cells whose energies and weights are $\zeta_0\ \zeta_1\cdots$ and $g_0\ g_1\cdots$, respectively. The number of permutations of the various particles is $M!\ N!\ S!$, but the permutations of the systems within the individual cells give no new complexions. Hence the number of complexions in the given distribution is

$$\frac{M!\,N!\,S!}{M_0!\,M_1!\cdots N_0!\,N_1!\cdots S_0!\,S_1!\cdots} \tag{3–56}$$

Corresponding to this distribution there will be a weight

$$\mu_0^{M_0}\mu_1^{M_1}\cdots\nu_0^{N_0}\nu_1^{N_1}\cdots g_0^{S_0}g_1^{S_1}\cdots \tag{3–57}$$

To obtain the total number of weighted complexions in the group specified by the numbers M, N, S, and $M_0\ M_1\cdots N_0\ N_1\cdots S_0\ S_1\cdots$ we multiply Eqs. 3–55 by 3–56 and 3–57.

$$W = \frac{X_M!\,X_N!}{M!N!S!}\,\frac{M!\,N!\,S!\,\mu_0^{M_0}\mu_1^{M_1}\cdots\nu_0^{N_0}\nu_1^{N_1}\cdots g_0^{S_0}g_1^{S_1}\cdots}{M_0!\,M_1!\cdots N_0!\,N_1!\cdots S_0!\,S_1!\cdots}$$

$$= \frac{X_M!\,X_N!\,\mu_0^{M_0}\mu_1^{M_1}\cdots\nu_0^{N_0}\nu_1^{N_1}\cdots g_0^{S_0}g_1^{S_1}\cdots}{M_0!\,M_1!\cdots N_0!\,N_1!\cdots S_0!\,S_1!\cdots} \tag{3–58}$$

where the relative probability W is related to the true probability by Eq. 3–18.

Now the total energy of the system, E, includes kinetic energies of translation, as well as internal energies and potential energy. Then

$$E = M_0\varepsilon_0 + M_1\varepsilon_1 + \cdots + N_0\eta_0 + N_1\eta_1 + \cdots + S_0\zeta_0 + S_1\zeta_1 + \cdots \tag{3–59}$$

$$= \sum_i M_i\varepsilon_i + \sum_j N_i\eta_i + \sum_k S_k\zeta_k \tag{3–60}$$

with the conditions

$$\sum_i M_i = M,\ \sum_j N_j = N,\ \sum_k S_k = S \tag{3–61}$$

Now, using Eqs. 3–18, 3–58, 3–54, and 3–20 we get

$$P = \frac{X_M{}^{X_M} X_N{}^{X_N}}{Ce^S} \prod_i \left(\frac{\mu_i}{M_i}\right)^{M_i} \prod_j \left(\frac{\nu_j}{N_j}\right)^{N_j} \prod_k \left(\frac{g_k}{S_k}\right)^{S_k} \tag{3–62}$$

so we rewrite Eq. 3–54 as

$$X_M = \sum_i M_i + \sum_k S_k \tag{3–63}$$

$$X_N = \sum_j N_j + \sum_k S_k \tag{3–64}$$

which we differentiate to obtain

$$dX_M = \sum_i dM_i + \sum_k dS_k = 0 \tag{3–65}$$

$$dX_N = \sum_j dN_j + \sum_k dS_k = 0 \tag{3–66}$$

since the total number of particles X_M and X_N is fixed. From Eq. 3–59 we obtain

$$dE = \sum_i \varepsilon_i dM_i + \sum_j \eta_j dN_j + \sum_k \zeta_k dS_k = 0 \tag{3–67}$$

By logarithmic differentiation of Eq. 3–62 we get

$$\begin{aligned} d\ln P = &- \sum_i \left(\ln\frac{M_i}{\mu_i} + 1\right) dM_i \\ &- \sum_j \left(\ln\frac{N_j}{\nu_j} + 1\right) dN_j - \sum_k \left(\ln\frac{S_k}{g_k} + 2\right) dS_k = 0 \end{aligned} \tag{3–68}$$

Now multiply Eqs. 3–65, 3–66 and 3–67 by the constants

$$\ln\frac{u_M}{M} - 1, \quad \ln\frac{u_N}{N} - 1, \quad \text{and } -\ln\vartheta \tag{3–69}$$

and add them to Eq. 3–68 to get

$$\begin{aligned} &\sum_i \left(\ln\frac{M_i}{\mu_i} + \ln\frac{u_M}{M} - \varepsilon_i \ln\vartheta\right) dM_i + \sum_j \left(\ln\frac{N_j}{\nu_j} + \ln\frac{u_N}{N} - \eta_j \ln\vartheta\right) dN_j \\ &\quad + \sum_k \left(\ln\frac{S_k}{g_k} + \ln\frac{u_M}{M} + \ln\frac{u_N}{N} - \zeta_k \ln\vartheta\right) dS_k = 0 \end{aligned} \tag{3–70}$$

whence there results

$$M_i = M\frac{\mu_i \vartheta^{\varepsilon_i}}{u_M} \qquad N_j = N\frac{\nu_j \vartheta^{\eta_j}}{u_N} \qquad S_k = MN\frac{g_k \vartheta^{\zeta_k}}{u_M u_N} \tag{3–71}$$

since the individual brackets must vanish in accordance with our previous arguments. In particular we have that

$$S = \sum S_k = \frac{MN}{u_M u_N} \sum_k \tilde{g}_k \vartheta^{\zeta_k} = MN\frac{u_S}{u_M u_N} \tag{3–72}$$

or

$$\frac{MN}{S} = \frac{u_M u_N}{u_S} \tag{3–73}$$

We may now use this result to obtain the ionization equation. Let the zero of energy correspond to the situation in which the gas is completely ionized and all particles are at rest. Let N refer to the ions, S to the atoms, and M to the electrons. The partition functions for the electrons, neutral atoms, and ions are, respectively,

$$u_e = \frac{g_e(2\pi mkT)^{3/2}}{h^3} V \qquad u_S = \tilde{u}_S \frac{(2\pi M_a kT)^{3/2}}{h^3} V$$
$$u_N = \tilde{u}_N \frac{(2\pi M_i kT)^{3/2}}{h^3} V \tag{3–74}$$

where

$$\tilde{u}_S = e^{\chi_0/kT} \sum g_{0,r}\, e^{-\chi_r/kT} \tag{3–75}$$

since χ_0 is the ionization energy, and the energy difference with respect to our above chosen zero point is

$$\chi = -\chi_0 + \chi_{0,r} \tag{3–76}$$

For the ions,

$$\tilde{u}_N = u_i(T) = \sum g_{1,r}\, e^{-\chi_{1,r}/kT} \tag{3–77}$$

where we use the notation $(0, r)$ and $(1, r)$ to denote energy levels in atoms and ions, respectively. Therefore, noting $M_a = M_i$, from Eq. 3–73,

$$\frac{N_1 N_e}{N_0 V} = \frac{(2\pi mkT)^{3/2}}{h^3} \frac{g_e u_1(T)}{u_0(T)} e^{-\chi_0/kT} \tag{3–78}$$

If we choose V as the unit volume, and note that $g_e = 2$, the ionization equation becomes

$$\frac{N_1 N_e}{N_0} = \left(\frac{2\pi mkT}{h^2}\right)^{3/2} \frac{2u_1(T)}{u_0(T)} e^{-\chi_0/kT} \tag{3–79}$$

3–6. The Law of Distribution of Velocities. Maxwell's velocity distribution law, Eq. 3–50, may be written down for the three velocity components v_x, v_y, v_z or in terms of molecular speeds. In particular, the number of molecules moving in the x-direction with velocities between v_x and $v_x + dv_x$ is

$$dN(v_x) = \frac{N}{\alpha\sqrt{\pi}} e^{-v_x^2/\alpha^2}\, dv_x \tag{3–80}$$

where α, the most probable speed, is given by

$$\tfrac{1}{2}M\alpha^2 = kT \tag{3–81}$$

and N is the number of molecules cm^{-3}. M is the mass of the molecule. That is, the velocities of the molecules in any specified direction follow a curve of the gaussian error type, whose dispersion is determined by the most likely speed of the molecule which in turn depends on temperature divided by molecular mass.

From Eq. 3–50 it follows that the number of molecules with velocities simultaneously in the range v_x to $v_x + dv_x$, v_y to $v_y + dv_y$, and v_z to $v_z + dv_z$ is the product of three factors of the type of Eq. 3–80

$$dN(v_xv_yv_z) = \frac{N}{\alpha^3\pi^{3/2}} e^{-v^2/\alpha^2} dv_xdv_ydv_z \qquad (3\text{–}82)$$

where

$$v^2 = v_x^2 + v_y^2 + v_z^2 \qquad (3\text{–}83)$$

Often we are less interested in the actual velocities in the x, y, z directions than in the total speeds of the molecules themselves. To derive Maxwell's law for speeds, we note that

$$dv_x\, dv_y\, dv_z = v^2 \sin\theta\, d\theta\, d\phi\, dv \qquad (3\text{–}84)$$

where (θ, ϕ, v) are polar coordinates replacing $v_x\ v_y\ v_z$. Here ϕ is the azimuthal angle and θ is the colatitude. If we substitute Eq. 3–84 in Eq. 3–82 or Eq. 3–50 and integrate from $\theta = 0$ to π and $\phi = 0$ to 2π we obtain Maxwell's law for speeds rather than velocities

$$Nf(v)\, dv = N(v)\, dv = 4\pi N\left(\frac{M}{2\pi kT}\right)^{3/2} v^2\, e^{-Mv^2/2kT}\, dv \qquad (3\text{–}85)$$

which gives the numbers of atoms $N(v)$ with *speeds* between v and $v + dv$ per unit volume. A plot of $N(v)$ against v shows a skew-shaped curve which rises rapidly from the origin to a maximum and then falls off less abruptly on the high energy side (see Fig. 3–2). The shape of the curve depends on the temperature. For low temperatures, the curve is steep and narrow, but as the temperature rises the curves flatten out since molecules then have a greater range of velocities. Few atoms have speeds greatly in excess of the average. From Eq. 3–85 one finds the following results:

Velocity	$N(v)/N(\alpha)$
1α	1.0000
2α	0.199
3α	0.0030
4α	0.000005
5α	0.0000000009

At the larger velocities, the exponential factor rapidly overpowers the v^2 factor.

At a given temperature, three kinds of speeds are of interest (see Fig. 3–2):

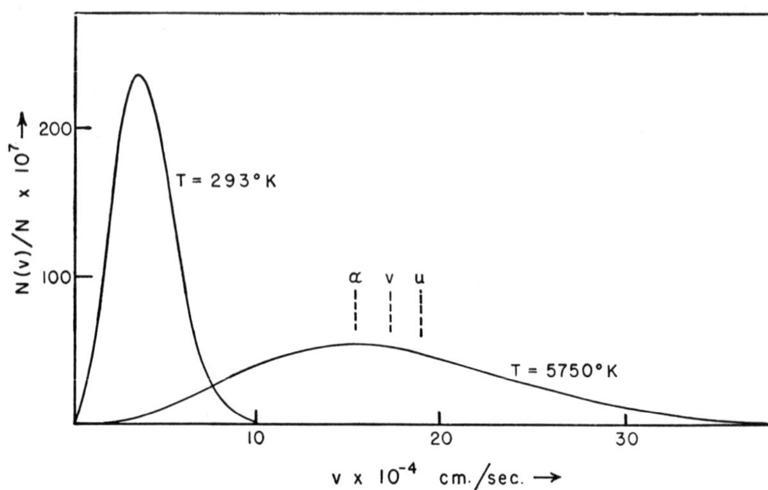

FIG. 3–2. MAXWELL'S LAW FOR SPEEDS

The velocity distribution for argon atoms at 293°K is compared with that for 5750°K. Ordinates are $N(v)/N$. The most probable, average, and root mean square velocities are denoted by α, v, and u, respectively.

1. α — the most probable speed (defined by Eq. 3–81) corresponds to the maximum of the $N(v)$ curve; it measures the dispersion of the molecular velocities along a fixed direction, e.g., x.
2. The average speed in the usual sense is

$$\bar{v} = \frac{1}{N}\int N(v)v\,dv = \frac{2\alpha}{\sqrt{\pi}} \tag{3–86}$$

3. The rms speed, u given by Eq. 3–10.

In astrophysical applications, α and u are of greater interest than $\bar{v}$.

As an elementary example, let us compute the values of α, $\bar{v}$, and u for nitrogen molecules at $T = 0°\text{C} = 273°\text{K}$. The mass of the nitrogen molecule is $28 \times 1.66 \times 10^{-24}$ gm since the molecular weight of nitrogen is 28 and 1.66×10^{-24} gm is the mass of an atom of unit atomic weight. Then α is computed from Eq. 3–81, viz.:

$$\alpha = \left[\frac{2 \times 1.380 \times 10^{-16} \times 273}{28 \times 1.66 \times 10^{-24}}\right]^{1/2} = 4.02 \times 10^4 \text{ cm sec}^{-1}$$

Similarly,

$$v = 2\alpha/\sqrt{\pi} = 1.1284\alpha = 4.54 \times 10^4 \text{ cm sec}^{-1}$$

$$u = \sqrt{\tfrac{3}{2}}\alpha = 1.2248\alpha = 4.93 \times 10^4 \text{ cm sec}^{-1}$$

The broadening of a spectral line, as a consequence of the kinetic motion of the radiating atoms, provides a useful illustration of Maxwell's law for the distribution of velocities. Let us suppose that each individual atom radiates monochromatically, i.e., we neglect all other effects that would make the atoms radiate split or fuzzy lines, e.g., density broadening (see

Chapter 7), disturbances produced by electric and magnetic fields acting upon the atom (see Chapters 2 and 7), and the phenomenon of "natural breadth" described in Chapter 4. Insofar as there are no large-scale motions of the gas, the forbidden lines radiated by the highly ionized atoms of nickel and iron in the solar corona provide a good example. Each atom radiates a sharp line of frequency ν_0 (or wavelength λ_0). If it is moving toward the observer with a velocity v_x, the observed frequency and wavelength of the emitted radiation will be changed by an amount given by

$$\frac{\Delta\nu}{\nu} = \frac{\Delta\lambda}{\lambda} = \frac{v_x}{c} \tag{3–87}$$

in accordance with Doppler's principle. Here c denotes the velocity of light. The number of atoms moving at any time with a velocity v_x toward the observer follows from Maxwell's velocity law, Eq. 3–80. If the intensity distribution within the spectral line is simply proportional to the number of atoms radiating at a frequency ν, then

$$I(\nu)\,d\nu = \frac{I_0}{\sqrt{\pi}}\,e^{-\frac{c^2}{\alpha^2}\left(\frac{\nu-\nu_0}{\nu}\right)^2}\,\frac{c\,d\nu}{\alpha\nu} \tag{3–88}$$

since

$$\left(\frac{\nu-\nu_0}{\nu}\right)^2 = \frac{v_x^2}{c^2} \quad \text{and} \quad dv_x = \frac{c\,d(\Delta\nu)}{\nu} = \frac{c\,d\nu}{\nu} \tag{3–89}$$

In wavelength units Eq. 3–88 is replaced by

$$I(\lambda)\,d\lambda = \frac{cI_0}{\lambda\alpha\sqrt{\pi}}\,e^{-\frac{c^2(\lambda-\lambda_0)^2}{\alpha^2\lambda_0^2}}\,d\lambda \tag{3–90}$$

Here I_0 denotes the total intensity integrated over the line; it is related to the central intensity, I_c, by

$$I_c = \frac{cI_0}{\lambda\alpha\sqrt{\pi}} \tag{3–91}$$

Notice that the line has a roughly bell-shaped profile; the top is rounded, and the intensity thereafter falls off abruptly with wavelength. If such a line is recorded upon a photographic plate of high contrast, there often results a broadened line image of a rather definite width. We define the half-width of the line as the width at which I has fallen to one half its maximum value. The wavelength at which $I = \frac{1}{2} I_c$ is given by

$$e^{-\frac{Mc^2(\lambda-\lambda_0)^2}{2\lambda_0^2kT}} = \tfrac{1}{2} \tag{3–92}$$

Thus the total half-width of the line is

$$\delta\lambda_a = 2\delta\lambda = 2(\lambda-\lambda_0) = 2\sqrt{\frac{2\lambda_0^2kT}{Mc^2}\ln 2} \tag{3–93}$$

whence

$$\delta\lambda_a = 7.16 \times 10^{-7}\,\lambda\,\sqrt{T/\mu} \tag{3–94}$$

where μ is the molecular weight, T is the temperature in absolute degrees, and λ is measured in angstrom units.

As an example consider the [Fe XIV] λ5303 line in the solar corona for which we take a gas kinetic temperature $T = 2{,}000{,}000°$K. Since $\mu = 56$, we obtain from Eq. 3–94, $\delta\lambda_a = 0.72$A. The quantity $\delta\lambda_a$ is not to be confused with the Doppler half-width parameter $(\Delta\lambda_D)$ that enters in the theories of line profiles and the curve of growth (Chapter 8).

3–7. Boltzmann's Law. The fundamental relationship, Boltzmann's law, Eq. 3–52, expresses the fraction of atoms in the rth level of an atom in terms of the total number N_i in the ith stage of ionization, viz.:

$$\frac{N_{i,r}}{N_i} = \frac{g_{i,r}}{u_i(T)}\, e^{-\chi_{i,r}/kT} \tag{3–95}$$

where $\chi_{i,r}$ is the excitation energy of the level r above the ground level, $g_{i,r} = 2J + 1$ is the statistical weight of the level (i.e., the number of Zeeman states into which the level is split by a magnetic field), T is the temperature and u_i is the *partition function*

$$u_i(T) = g_{i,1} + g_{i,2}\, e^{-\chi_{i,2}/kT} + \cdots = \sum_r g_{i,r}\, e^{-\chi_{i,r}/kT} \tag{3–96}$$

If χ is expressed in electron volts, Boltzmann's formula can be written as

$$\log \frac{N_{i,r}}{N_i} = -\frac{5040.4}{T}\,\chi_{i,r} + \log \frac{g_{i,r}}{u_i(T)} \tag{3–97}$$

or if we wish to compare the population of the rth level with that of the ground level,

$$\log \frac{N_{i,r}}{N_{i,1}} = -\theta\,\chi_{i,r} + \log \frac{g_{i,r}}{g_{i,1}} \tag{3–98}$$

where

$$\theta = \frac{5040}{T} \tag{3–99}$$

As an example, let us compare the relative populations of the ground and $n = 2$ levels for hydrogen, for $T = 6000°$K, 8000°K, 10,000°K, 15,000°K, and 20,000°K. The statistical weight of the ground level of hydrogen is two, that of level n is $2n^2$. The excitation potential of the second level is 10.15 ev.

$$\log \frac{N_2}{N_1} = -\frac{5040}{T} \times 10.15 + \log 4 = -\frac{51{,}160}{T} + 0.60$$

Temperature °K	$\theta = \frac{5040}{T}$	$\frac{N_2}{N_1}$
6,000	0.840	0.00000001
8,000	0.630	0.0000016
10,000	0.504	0.000031
15,000	0.336	0.00155
20,000	0.252	0.0110

Notice the rapid increase in relative numbers of atoms in the second level with an increase of temperature.

One difficulty in application of Boltzmann's formula is the calculation of the partition function. Clearly the summation in Eq. 3–96 must be truncated at some point. The most naïve approach is to compute the principal quantum number for the Bohr orbit whose radius equals the mean distance between atoms and ions. This procedure neglects the statistics of encounters, i.e., that the distance between particles fluctuates in a random manner and that an orbit may be so perturbed as to eject the electron before an actual interlocking of orbits occurs. In practice the summation will not be truncated at a well-defined point, but beyond some quantum number n there will be a steadily decreasing probability that the energy level will be occupied at the given density and temperature. Hence the expression for the partition function is written in the form

$$u = \sum_j p_j g_j \exp\left(-\frac{\chi_j}{kT}\right) \tag{3–100}$$

where p_j is the probability that the jth level is occupied at the density under consideration.

A perturbing ion of charge Z^1 approaching the considered atom may so disturb the potential field in which the bound electron moves that it escapes. Unsöld found that an electron moving in an orbit of quantum number n around an ion core of charge Z would be lost if the perturbing ion approaches to a distance r given by

$$\frac{RcZ^2}{n^2} = \frac{\varepsilon^2}{r}\left[Z + 2\sqrt{ZZ^1}\right] \tag{3–101}$$

Applications of these considerations show that one can write:

$$p_n = \exp\{-C(Z)P_\varepsilon n^6\theta\} \tag{3–102}$$

where $\theta = 5040/T$. In the hotter stars where the perturbing ions are mostly hydrogen, Elste and Jugaku (1957) found the following values for $C(Z)$ (dynes^{-1} cm^2 deg):

$C(Z)$ =	1.927×10^{-10}	0.1255×10^{-10}	0.02644×10^{-10}	0.00892×10^{-10}
Z =	1	2	3	4

These values of $C(Z)$ may be employed, however, also for cooler stars like the sun where singly charged ions produce the perturbations. Here $Z = 1$ for neutral atoms, 2 for singly ionized atoms, and so on.

3–8. The Ionization Equation. At any given temperature and density it is important to know not only the relative number of atoms in various excited levels but also the relative numbers of neutral and ionized atoms. In thermal equilibrium, atoms will lose electrons at a rate dependent upon

TABLE 3–1

IONIZATION POTENTIALS OF ATOMS OF ASTROPHYSICAL INTEREST

(*National Bureau of Standards Circular No. 467*, C. E. Moore)

			I	II	III	IV
1	H	hydrogen	13.60			
2	He	helium	24.58	54.40		
3	Li	lithium	5.39	75.62		
4	Be	beryllium	9.32	18.21		
5	B	boron	8.30	25.15		
6	C	carbon	11.26	24.38	47.87	64.48
7	N	nitrogen	14.53	29.59	47.43	77.45
8	O	oxygen	13.61	35.11	54.89	77.39
9	F	fluorine	17.42	34.98	62.65	87.14
10	Ne	neon	21.56	41.07	63.50	97.02
11	Na	sodium	5.14	47.29	71.65	98.88
12	Mg	magnesium	7.64	15.03	80.12	109.29
13	Al	aluminum	5.98	18.82	28.44	119.96
14	Si	silicon	8.15	16.34	33.46	45.13
15	P	phosphorus	10.48	19.72	30.16	51.35
16	S	sulphur	10.36	23.40	35.00	47.29
17	Cl	chlorine	13.01	23.80	39.90	53.50
18	Ar	argon	15.76	27.62	40.90	59.79
19	K	potassium	4.34	31.81	46.00	60.90
20	Ca	calcium	6.11	11.87	51.21	67.00
21	Sc	scandium	6.54	12.80	24.75	73.90
22	Ti	titanium	6.82	13.57	27.47	43.24
23	V	vanadium	6.74	14.65	29.31	48.00
24	Cr	chromium	6.76	16.49	30.95	
25	Mn	manganese	7.43	15.64	33.69	
26	Fe	iron	7.87	16.18	30.64	
27	Co	cobalt	7.86	17.05	33.49	
28	Ni	nickel	7.63	18.15	35.16	
29	Cu	copper	7.72	20.29	36.83	
30	Zn	zinc	9.39	17.96	39.70	
31	Ga	gallium	6.00	20.51	30.70	
32	Ge	germanium	7.88	15.93	34.21	
33	As	arsenic	9.81	18.63	28.34	
34	Se	selenium	9.75	21.50	32.00	
35	Br	bromine	11.84	21.60	35.90	
36	Kr	krypton	14.00	24.56	36.90	
37	Rb	rubidium	4.18	27.50		
38	Sr	strontium	5.69	11.03		
39	Y	yttrium	6.38	12.23		
40	Zr	zirconium	6.84	13.13		
41	Nb	niobium	6.88	14.32		
42	Mo	molybdenum	7.10	16.15		
43	Tc	technetium	7.28	15.26		
44	Ru	ruthenium	7.36	16.76		
45	Rh	rhodium	7.46	18.07		
46	Pd	palladium	8.33	19.42		
47	Ag	silver	7.57	21.48		
48	Cd	cadmium	8.99	16.90		
49	In	indium	5.78	18.86		
50	Sn	tin	7.34	14.63		

TABLE 3–1 (*Continued*)

			I	II	III	IV
51	Sb	antimony	8.64	16.50		
52	Te	tellurium	9.01	18.60		
53	I	iodine	10.45	19.09		
54	Xe	xenon	12.13	21.20		
55	Cs	caesium	3.89	25.10		
56	Ba	barium	5.21	10.00		
57	La	lanthanium	5.61	11.40		
72	Hf	hafnium	7.00	14.90		
73	Ta	tantalum	7.88	16.20		
74	W	tungsten	7.98	17.70		
75	Re	rhenium	7.87	16.60		
76	Os	osmium	8.70	17.00		
77	Ir	iridium	9.00			
78	Pt	platinum	9.00	18.56		
79	Au	gold	9.22	20.50		
80	Hg	mercury	10.43	18.75		
81	Tl	thallium	6.11	20.4		
82	Pb	lead	7.42	15.03		
83	Bi	bismuth	7.29	16.68		

TABLE 3–2A

PARTITION FUNCTIONS
(Temperatures in degrees Kelvin)

	3600	5700	8000	10,000		3600	5700	8000	10,000
Li I	2.0	2.4	3.0	3.6	Cu I	20.4	25.2	32.4	41
Be I	1.0	1.05	1.15	1.41	Cu II	1.0	1.05	1.29	1.74
C	9.1	9.3	9.5	9.8	Ga I	6.0	6.0	6.2	6.5
N	1.0	1.08	1.35	1.70	Ge I	7.6	8.5	9.8	10.3
O	8.5	8.9	9.1	9.4	Ge II	3.8	4.37	4.8	5.0
Na I	2.00	2.40	3.17	4.3	Rb I	2.2	2.4	8.3	
Mg I	1.0	1.05	1.20	1.48	Sr I	1.05	1.41	2.9	5.25
Mg II				2.0	Sr II	2.04	2.30	2.82	3.46
Al I	5.76	5.90	6.2	6.6	Y I	11.0	14.1	21.8	25.0
Al II				1.0	Y II	14.8	17.4	21.4	25.7
Si I	8.9	9.5	10.45	11.2	Y III			10.5	10.7
Si II	5.60	5.70	5.75	5.9	Zr I	23.4	40.8	67.5	98
P I	4.0	4.5	5.75		Zr II	41.7	51.2	67.5	89
P II	8.0	8.1	8.9	10	Zr III			25	27.6
S	8.5	8.5	8.7	9.0	Nb I	40	60	91	129
K I	3.5	4.1	7.8	15.5	Nb II	33	52.5	78	98
Ca I	1.05	1.32	2.62	4.7	Mo I	7.4	10.7	19	34
Ca II	2.04	2.30	2.82	3.55	Mo II	5.35	8.1	15.1	25
Sc I	11.0	13.8	21.0	30	Ru I	25.7	38	56	79
Sc II	20.9	24.0	29.5	33.2	Ru II	19.5	26	35.5	47
Ti I	23.0	34.0	54.0	81.2	Pd I	1.95	3.9	6.18	8.3
Ti II	48.0	56.2	72.5	85.0	Pd II	6.9	7.6	8.5	9.3
V I	38.0	54.0	76	107	Ag I	2.0	2.0	2.14	2.3
V II	35.5	49	58	69	Ag II	1.0	1.0	1.0	1.0
Cr I	8.35	12.0	18.6	25.8	Cd I	1.0	1.0	1.05	1.12
Cr II	6.2	8.1	11.7	17.4	In I	3.6	4.3	5.0	5.8

TABLE 3–2A (*Continued*)

	3600	5700	8000	10,000		3600	5700	8000	10,000
Mn I	6.0	7.1	9.3	13.2	In II	2.6	4.1	5.1	5.9
Mn II	7.2	8.3	9.5	10.8	Sn I	4.1	5.9	7.2	8.3
Fe I	24	31	44.6	60.1	Sn II	2.76	3.4	3.9	4.2
Fe II	37	47	56	67.5	Sb I	4.3	5.1	6.3	7.4
Fe III			25	29	Sb II	2.46	3.8	5.0	6.0
Co I	24.6	35	46.7	60	Ba I	1.52	3.55	8.1	14.1
Co II	22	31	40	47	Ba II	3.24	4.7	6.0	7.1
Co III			25.7	28.2	La I	19	22	31	
Ni I	28	32.3	37	41.7	La II	28	33	42	
Ni II	9	12.0	16.6	22	Pb I	1.20	1.82	2.63	3.56
Ni III			18	19	Pb II	2.0	2.14	2.34	2.50

TABLE 3–2B

Log $u(T)$

$\theta = 5040/T$

	Log P_e	0.1	0.14	0.18	0.23	0.3	0.4	0.5
H	2	2.77	2.20	1.78	0.92	0.42	0.31	0.30
	3	2.28	1.63	1.09	0.70	0.35	0.30	
	4	1.83	1.19	0.75	0.50	0.32	0.30	
He I	2	1.74	0.97	0.25	0.015	0.00	0.00	0.00
	3	1.50	0.57	0.10	0.007	0.00		
	4	1.04	0.28	0.04	0.003			
He II	2	0.34	0.30	0.30	0.30	0.30	0.30	0.30
C I	2					1.10	1.03	1.01
C II	2				0.85	0.80	0.78	0.78
	3	2.30	1.41	0.96	0.84	0.80	0.78	0.78
	4	1.88	1.16	0.91	0.84	0.80	0.78	0.78
C III	2				0.11	0.04	0.01	0.00
	3	.76	0.35	0.22				
	4	0.68	0.35					
N I	2					0.62	0.37	0.24
N II	2				1.05	1.02	0.99	0.98
	3	1.73	1.20	1.09	1.05			
	4	1.45	1.16	1.08				
N III	3	1.05	0.88	0.83	0.80	0.79	0.78	
O I	2					1.05	.99	0.97
O II	2			0.86	0.79	0.71	0.66	0.61
	3	1.43	0.97	0.86				
	4	1.22	0.95					
O III	3	1.15	1.08	1.05	1.02	1.04	1.01	
Ne I	2	2.25	1.33	0.54	0.07	0.01	0.00	0.00
	3	1.70	0.85	0.23	0.02	0.00		
	4	1.17	0.47	0.10	0.01	0.00		

TABLE 3–2B (*Continued*)

	Log P_ε	0.1	0.14	0.18	0.23	0.3	0.4	0.5
Ne II	2	1.28	0.80	0.78	0.78	0.78	0.78	0.78
	3	1.00	0.79					
	4	0.88	0.78					
Ne III	2	1.07	1.04	1.02	1.00			
Mg I	2					1.92	1.14	0.54
	3					1.47	0.77	0.33
Mg II	2	3.22	2.54	1.90	1.17	0.52	0.34	0.31
	3	2.70	2.04	1.43	0.81	0.42	0.33	0.31
	4	2.17	1.53	1.00	0.58	0.39	0.33	0.31
Mg III	2	0.00	0.00	0.00	0.00			
Al II	2	3.15	2.33	1.55	0.72	0.15	0.05	0.00
	3	2.66	1.86	1.13	0.48			
	4	2.19	1.43	0.81	0.37			
Al III	2	2.23	1.12	0.49	0.345	0.31	0.30	
	3	1.76	0.79	0.42	0.34			
	4	1.32	0.60	0.40	0.34			
Si I							1.07	1.05
Si II					0.84	0.78	0.78	0.77
Si III		0.65	0.38	0.23	0.11			
Si IV		0.47	0.37	0.33	0.31			
P II	2				1.17	1.07	1.04	
	3		1.77	1.34	1.12			
	4		1.54	1.23				
P III	2				0.81	0.79	0.78	
	3		1.14	0.86	0.81			
	4		1.00	0.55				
P IV	2		0.23	0.12	0.05	0.01	0.00	
S II	2				0.98	0.88	0.79	
	3		1.58	1.13	0.97			
	4		1.35	1.08				
S III	2		1.15	1.09	1.07	1.04	1.02	
Cl II	2				1.08	1.04	1.01	
	3		1.65	1.19	1.07			
	4		1.39	1.13				
Cl III	2		1.07	0.97	0.90	0.82	0.74	
A II	3		1.10	0.81	0.78	0.78	0.78	
A III	3		1.10	1.07	1.05	1.03	1.00	
A IV	2		1.00	0.92	0.86			

The data for H, N I, O I, C I, Si I, Si II are taken from de Jager and Neven, *Utrecht Publ.* **13,** 1957.

The data for He I and He II are from Elste and Jugaku (1957).

The data for Si III, Si IV, Mg II, Al II, Al III, Ne I, Ne II, Ne III are from Aller, Elste, and Jugaku (1957). The remaining numbers were calculated by J. Jugaku and A. Boury.

the temperature and the ionization potential of the atom. At a given temperature, calcium atoms (ionization potential = 6.11 ev) will lose electrons at a greater rate than will hydrogen atoms whose ionization potential is 13.60 ev. On the other hand, the rate at which ions can recapture electrons will depend on the electron density (or electron pressure since $P_\varepsilon = N_\varepsilon kT$).

Saha derived the ionization formula by thermodynamical considerations and in 1922 pointed out its importance for astrophysical problems. A more rigorous derivation was given by R. H. Fowler with the aid of statistical mechanics. In Art. 3–5 we presented a simpler derivation due to Menzel to obtain the Eq. 3–79.

$$\frac{N_1 N_\varepsilon}{N_0} = \frac{(2\pi mkT)^{3/2}}{h^3} \frac{2u_1(T)}{u_0(T)} e^{-\chi_0/kT}$$

We may easily show that this type of equation holds for ionization stages higher than the first. Actually, under most conditions in stellar atmospheres, only two stages of ionization prevail at any one time, say the qth and $(q + 1)$st. We can write,

$$\frac{N_{q+1} N_\varepsilon}{N_q} = \frac{(2\pi mkT)^{3/2}}{h^3} \frac{2u_{q+1}(T)}{u_q(T)} e^{-\chi_q/kT} \tag{3–103}$$

where N_q is the number of atoms in the qth stage of ionization, N_{q+1} is the number in the $(q + 1)$st stage of ionization, and χ_q is the energy necessary to ionize the atom from the qth stage of ionization to the $(q + 1)$-st stage.

For many problems, the electron pressure is a convenient parameter to employ in place of the electron density. Substituting $P_\varepsilon = N_\varepsilon kT$ in Eq. 3–79 we obtain

$$\frac{N_1 P_\varepsilon}{N_0} = \frac{(2\pi m)^{3/2} (kT)^{5/2}}{h^3} \frac{2u_1(T)}{u_0(T)} e^{-\chi_0/kT} \tag{3–104}$$

For purposes of numerical calculation, the logarithmic form is particularly useful, viz.,

$$\log \frac{N_1}{N_0} P_\varepsilon = -\frac{5040}{T} I + 2.5 \log T - 0.48 + \log 2u_1(T)/u_0(T) \tag{3–105}$$

where I here denotes the ionization potential in volts, P_ε is the electron pressure in dynes cm^{-2}, N_1 is the number of ionized atoms cm^{-3}, N_0 is the number of neutral atoms cm^{-3}, $u_1(T)$ is the partition function of the ionized atoms, and $u_0(T)$ is the partition function of the neutral atoms. The u's can be calculated with the aid of a term table for the atom or ion in question as a function of the temperature and electron pressure (see Art. 3–7).

Tables 3–1 and 3–2 give the ionization potentials and partition functions for a number of atoms and ions of astrophysical interest.

The lines of the permanent gases observed in the hotter stars and in the solar chromosphere arise from high levels much closer to the ionization limit than to the ground level. It is often useful to combine the Boltzmann and ionization equations in such a way as to refer the number of atoms, $N_{0,r}$, in the rth level of the neutral atoms, say, to the total number N_1 of ionized atoms. See Fig. 3–3. If we divide Eq. 3–104 by Boltzmann's Eq. 3–50, we obtain

$$\frac{N_1 P_\epsilon}{N_{0,r}} = \frac{(2\pi m)^{3/2}(kT)^{5/2}}{h^3}\frac{2u_1(T)}{g_{0,r}}e^{-(I-\chi_r)/kT} \tag{3–106}$$

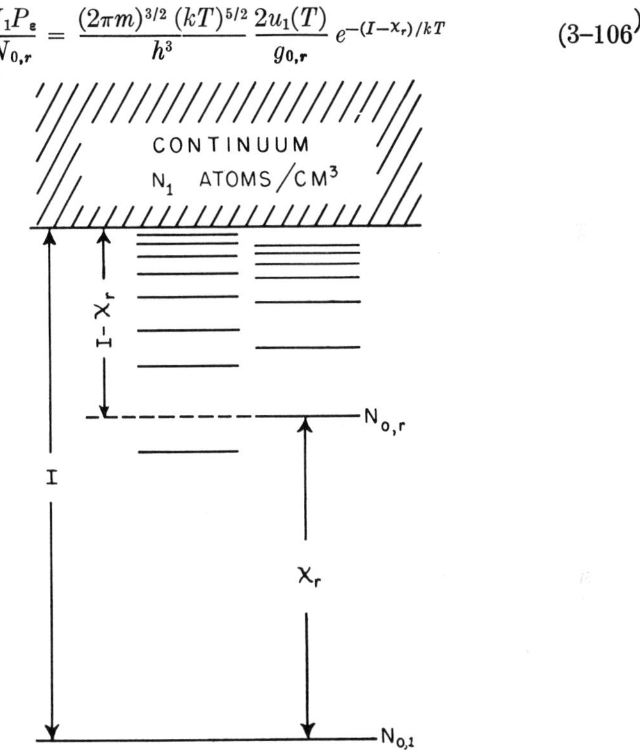

Fig. 3–3. Schematic Energy Levels

The ionization potential is I. The excitation potential of level r is χ_r. The number of atoms per cm³ in the ground level of the neutral atom is denoted as $N_{0,1}$, the number of atoms in level r is $N_{0,r}$, and the number of ionized atoms is N_1.

which relates the number of atoms in the rth level to the total number of singly ionized atoms. Here we denote the ionization potential by I and the excitation potential of the rth level simply by χ_r. We may generalize this expression for any stage of ionization. Since the numerical value of $I - \chi_r$ is much less than I or χ_r for the permanent gases, the Boltzmann correction is much smaller than it would be if we tried to relate the number in level r with the number in the ground level. Putting in numerical values as in Eq. 3–105, Eq. 3–106 reduces to the following

$$\log \frac{N_1 P_\varepsilon}{N_{0,r}} = -\frac{5040}{T}(I - \chi_r) + 2.5 \log T - 0.48 + \log \frac{2u_1(T)}{g_{0,r}} \qquad (3\text{–}107)$$

Examples: If the temperature of a stellar atmosphere is 5700°K and the electron pressure is 30 dynes cm^{-2}, what proportion of aluminum is neutral? The first ionization potential of aluminum is 5.98 ev, and $2u_1/u_0 = 0.34$. Let θ denote $5040/T$. Then,

$$\theta I = 5.27, \qquad 2.5 \log T = 9.39, \qquad \log P_\varepsilon = 1.48$$

Hence $\log N_1/N_0 = 1.68$, whence $N_1/N_0 = 47.7$ or $N_0/(N_1 + N_2) = 1/48.7 = 0.0205$, i.e., 2.1 per cent of all atoms are neutral and the rest are ionized. If $P_\varepsilon = 10$ dynes cm^{-2}, only 0.69 per cent of aluminum atoms are neutral. Since the second ionization potential of aluminum is 18.82 ev a negligible fraction of aluminum atoms have lost a second electron.

What fraction of calcium atoms are in the singly ionized condition in the atmosphere of Sirius if $T = 10{,}000$°K and $P_\varepsilon = 300$ dynes cm^{-2}? For calcium, $I = 6.11$ ev and $\log 2u_1(T)/u_0(T) = 0.18$, and

$$2.5 \log T = 10.00 \qquad \theta I = 3.08 \qquad \log P_\varepsilon = 2.48$$

$\log N_1/N_0 = 4.14$, i.e., there is practically no neutral calcium

We suspect that the calcium may be doubly ionized. Hence we should apply the ionization formula again. The second ionization potential is 11.87 ev. For Ca III, $u_0 = 1.0$ since the ground level is $^1S_0$ and all other terms lie so high as to make a negligible contribution. Hence $\log 2u_2/u_1 = -0.25$, $\theta I = 5.97$.

$$\log N_2/N_1 = 0.82 \text{ or } N_2/N_1 = 6.6$$

whence $N_1/(N_1 + N_2) = 0.132$ or 13 per cent remain singly ionized and the remainder are doubly ionized.

The strong Mg II doublet at λ4481 arises from transitions from the 3^2D term to the 4^2F term. Calculate, for $P_\varepsilon = 100$ dynes cm^{-2} and $T = 7200$°K, the fraction of magnesium atoms capable of absorbing λ4481.

The excitation potential of the lower 3^2D term of the transition is 8.83 ev, and the statistical weight of this term is, since $J = 3/2$ and $5/2$, $6 + 4 = 10$. The statistical weight of the ground $^2S_{1/2}$ term is 2. By Boltzmann's formula, the fraction of Mg II atoms in the ^{2}D term is given by

$$\log N(3^2\text{D})/N(3^2\text{S}) = -8.83\,\theta + \log 5 = -5.47, \text{ at } T = 8200\text{°K } (\theta = 0.7)$$

For magnesium the first ionization potential is 7.64 ev, the second is 15.03 ev. For Mg I, $\log 2u_1/u_0 = 0.43$, whence we find $\log N_1/N_0 = 2.24$, so we conclude that 99.6 per cent of the Mg is at least once ionized.

Is double ionization important? Here $\log 2u_2/u_1 = 0.00$ and we find $\log N_2/N_1 = -3.28$. Of all magnesium atoms, the total fraction in the excited 3^2D level and therefore capable of absorbing λ4481, will be

$$\frac{N(3^2\text{D})}{N(\text{total})} = 0.996 \times 10^{-5.47} = 3.4 \times 10^{-6}$$

Example: The following example will illustrate the use of Eq. 3–107. In the atmosphere of 10 Lacertae, it is found from the measurements of intensities of the

Balmer lines that the number of hydrogen atoms in the second level above each cm^2 of the visible surface of the star is given by log $N_{0,2} = 15.80$. With $T =$ 29,600°K and log $P_\varepsilon = 2.80$, we shall compute the corresponding number of hydrogen ions, N_1. With $I = 13.60$ ev, $\chi_2 = 10.20$ ev, $g_2 = 8$, and $u_1 = 1$, we find

$$\log \frac{N_1}{N_{0,2}} = 6.72, \text{ and } \log N_1 = 22.52$$

An application of the ionization equation will show that very few hydrogen atoms are neutral. Ionization equilibrium calculations of astrophysical interest have been given by Rouse (1961, 1962).

3–9. Tests of the Ionization Theory. The Spectral Sequence and Luminosity Effects. At Mount Wilson Observatory, A. S. King carried out qualitative laboratory tests of the ionization theory. Alkaline earth metals such as Ca or Mg have ionization potentials so low that even at temperatures attainable in an electric furnace an appreciable fraction of their atoms can be ionized. King found that the intensity ratio of λ3933 of ionized calcium to λ4226 of neutral calcium increased as the temperature increased, in harmony with theory. In another experiment the temperature was kept constant and small amounts of caesium (the most readily ionized of all the elements) were added. The ionization of caesium raised the electron pressure and decreased the ionization and hence the intensity of λ3933. King was unable to measure the electron pressure in these experiments; hence a quantitative check on this theory could not be made.

Astrophysical examples of ionization phenomena are numerous. The sunspots, which are refrigerated areas about 1200°K cooler than the surrounding bright surface of the sun (photosphere), provide a good illustration. The alkalis and alkaline earths become almost completely ionized in the region of the photosphere while their neutral lines are greatly strengthened in cooler spot areas in accordance with ionization theory.

The ionization theory's outstanding achievement was its interpretation of the "main line" of the spectral sequence from *O* through *G* and *K* to *M* as a temperature sequence. Proceeding from the cooler to hotter stars, lines of neutral elements gradually weaken and become replaced by those of ionized elements. Calcium provides a good illustration. In the coolest stars, it is mostly neutral and its resonance λ4227 line attains great strength (Fig. 3–8). In slightly hotter stars ionization begins to be appreciable, λ4227 weakens, while the resonance *H* and *K* lines of ionized calcium strengthen until in class *G* they dominate the spectrum. At still higher temperatures calcium becomes doubly ionized and the *H* and *K* lines fade away. When we deal with lines that arise from excited levels, i.e., subordinate transitions, we must apply both the ionization and Boltzmann formulas. With rising temperature, subordinate lines grow in intensity as the number of atoms capable of absorbing them increases, but then weaken as the

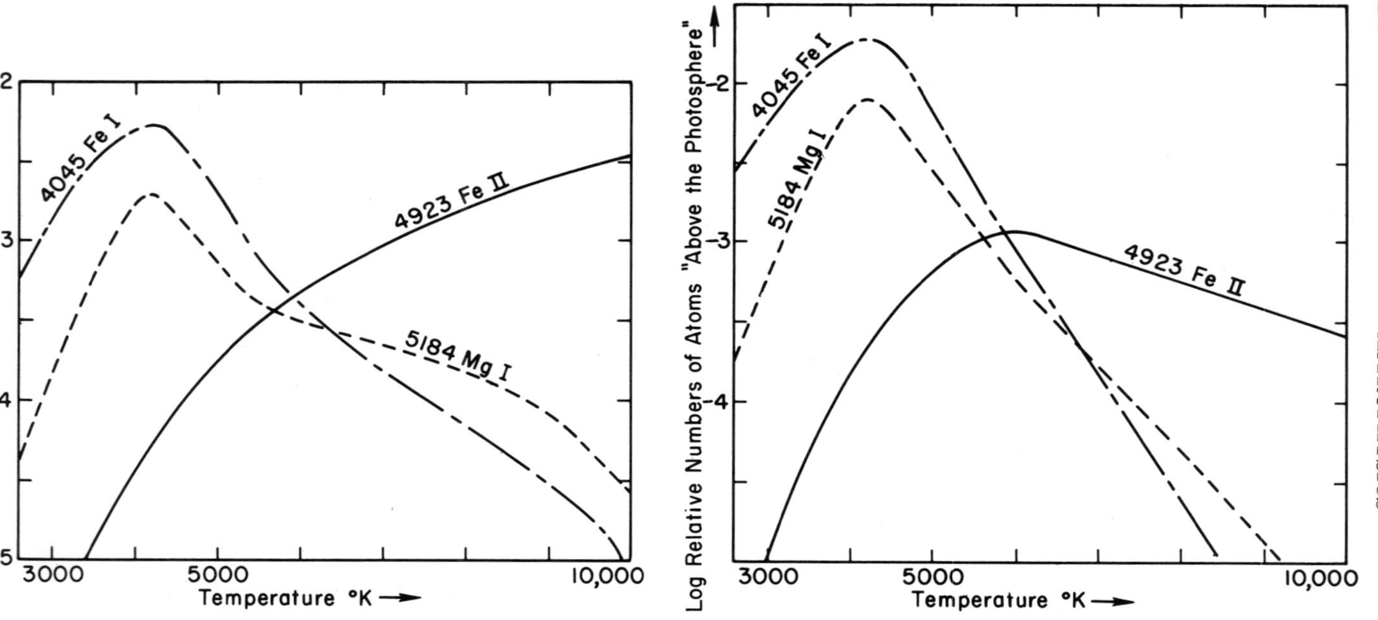

Fig. 3–4. Variations of Numbers of Absorbing Atoms Along the Main Sequence

Left: The logarithms of the relative number of atoms per unit mass of element concerned, capable of absorbing the line indicated is plotted against the temperature for several metallic lines. These are subordinate lines, arising from excited levels. Hence both the Boltzmann and Saha equations must be used. The variation of electron pressure and temperature along the main sequence is taken into account.

Right: Anticipating the results of Chapter 5, we now correct for effects of a variation of opacity of the stellar atmospheres with temperature. The hotter the star, the more opaque the atmosphere. Hence the number of atoms in the visible layers of the atmosphere, i.e., above the photosphere, steadily decreases with rising temperature.

atoms become ionized. Fig. 3–4 illustrates the predicted variation of the number of atoms capable of absorbing certain subordinate iron and magnesium lines as a function of temperature along the main sequence. For comparison, Fig. 3–5 shows the observed behavior of these same lines.

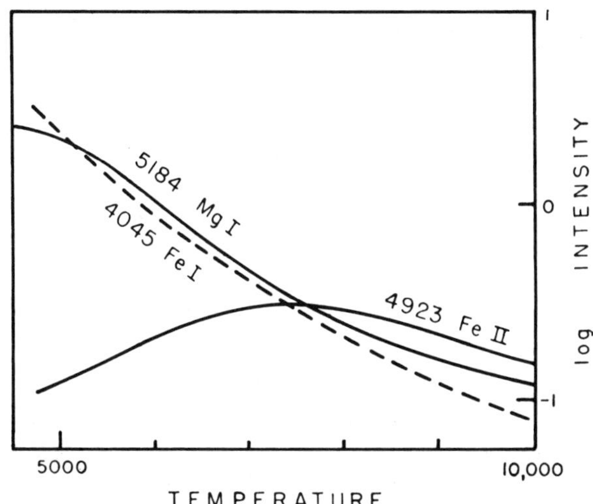

FIG. 3–5. OBSERVED VARIATIONS OF LINE INTENSITIES ALONG THE MAIN SEQUENCE
The logarithm of the observed intensities, expressed as equivalent widths (see Chapter 8), is plotted against the temperature. Compare with Fig. 3–4.

The behavior of the hydrogen lines is an excellent example. Since the Balmer series arises from transitions from the second to higher levels, a hydrogen atom must first be excited to the second level before it can absorb a Balmer line. Referring to our illustrative calculations in Art. 3–7, we notice that at the lower temperatures the number of hydrogen atoms in the second level is negligible, but that it increases rapidly with rising temperature. In harmony with this result, the hydrogen lines steadily increase in strength from *M* through *K*, *G*, and *F* to *A*, where these lines attain their maximum strength. Thereafter, with rising temperature, the intensity of the Balmer lines decreases as ionization makes serious inroads on the number of neutral hydrogen atoms. The Balmer lines are still present in the *O* stars, however, even though about one atom in a hundred thousand remains neutral. See Fig. 3–6.

Lines of neutral metals steadily weaken as the temperature rises. The effect is illustrated for sodium in Fig. 3–7. In stars hotter than *A*0, the metals become multiply ionized and their resonance lines fall in the unobservable ultraviolet. The subordinate lines involve very high levels and are very weak. No atom heavier than silicon is observed in absorption beyond the second stage of ionization.

In the *B* stars, the helium lines, together with those of hydrogen,

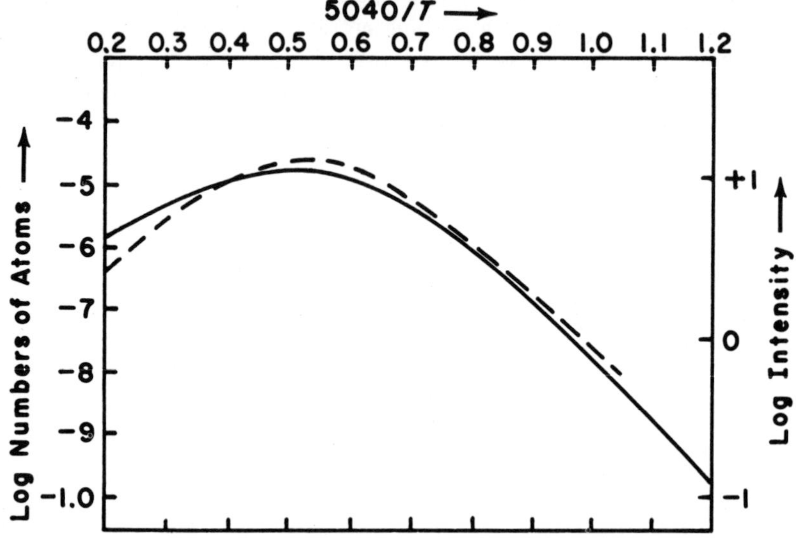

Fig. 3–6. Variation of Balmer Line Intensities Along the Main Sequence

The solid curve gives the relative numbers of hydrogen atoms in the second level and therefore capable of absorbing the Balmer lines. The dotted curve gives the logarithm of the equivalent width of Hγ. The vertical scale of the observed intensity curve is multiplied by a factor 5/2, since theory shows that the intensity (equivalent width) of strong hydrogen lines varies roughly as $N^{2/5}$. The changing opacity of the atmosphere is taken into account.

become the most conspicuous features of the spectrum. The *O* stars are so hot that hydrogen is nearly all ionized and even helium is doubly ionized. The light elements O, N, C, and Ne are prominent in various stages of ionization.

Among the *O* and *B* stars, bright lines often appear. Some stars show both bright and dark lines. Generally, the bright lines have been attributed to shells surrounding the star. The very coolest stars also occasionally show bright lines, especially the long period variables in class *M*, *N*, and *S*, certain dwarfs that appear to be subject to flarelike activity, and the *T Tauri variables* that are believed to be dwarf stars in the process of formation from the interstellar medium. These phenomena cannot be explained by the simple ionization theory and require invocation of appropriate non-thermal-equilibrium mechanisms.

A third interesting application of ionization theory may be found in luminosity effects in stellar spectra. Electron pressure, as well as temperature, affects ionization. The temperatures being equal, the star whose atmosphere has a lower density will exhibit a higher ionization, i.e., an earlier spectral class. When we differentiate stars according to the densities prevailing in their atmospheres, we are really separating them according

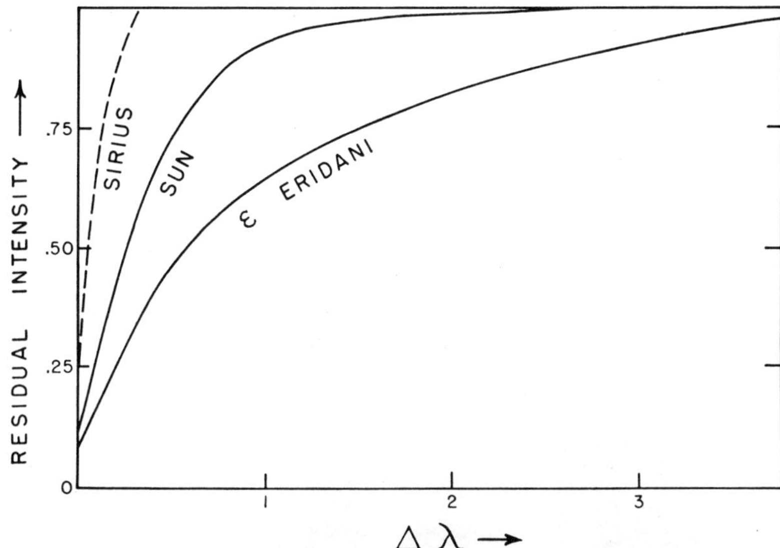

FIG. 3-7. THE VARIATION OF THE PROFILE OF λ5889 (SODIUM) ALONG THE MAIN SEQUENCE

The intensity at each point in the line is expressed in units of the continuum. Note the decrease of intensity from ε Eridani ($T = 4700°K$), through the sun (5700°K) to Sirius (10,000°K), because of the increasing ionization with rising temperature. The profile in Sirius is schematic, so adjusted as to give the correct equivalent width. The profile for the sun is taken from the Minnaert Atlas, whereas the ε Eridani profile is taken from a Mount Wilson coudé plate.

to their luminosities, since giant and supergiant stars have much lower densities than do dwarfs of the same temperature or spectral class. For example, a giant star of the same temperature as our dwarf sun would have an earlier spectral type, say $F8$, since ionization would be greater in the more tenuous atmosphere of the larger star. If we compare stars of the same spectral class, giants tend to be cooler than dwarfs. While the compensation of decreased temperature and electron pressure suffices to give two similar-looking spectra, lines of certain ions will not keep in step.

Neutral calcium, λ4227, provides one illustration. If we compare an $M4$ dwarf and giant, we find that they have about the same temperature, 3150°K. Elements such as iron are neutral in both stars. For the dwarf we assume $P_\epsilon = 0.5$ dynes cm^{-2}, and for the giant we assume $P_\epsilon = 0.008$ dynes cm^{-2}. The ionization potential of calcium is 6.11 ev, and we find that in the dwarf the calcium is 82 per cent neutral, while in the giant only 7 per cent is neutral. In stars of this spectral class, we would expect neutral calcium to be stronger in dwarfs than in giants, an effect which is substantiated by the observations (see Fig. 3-8).

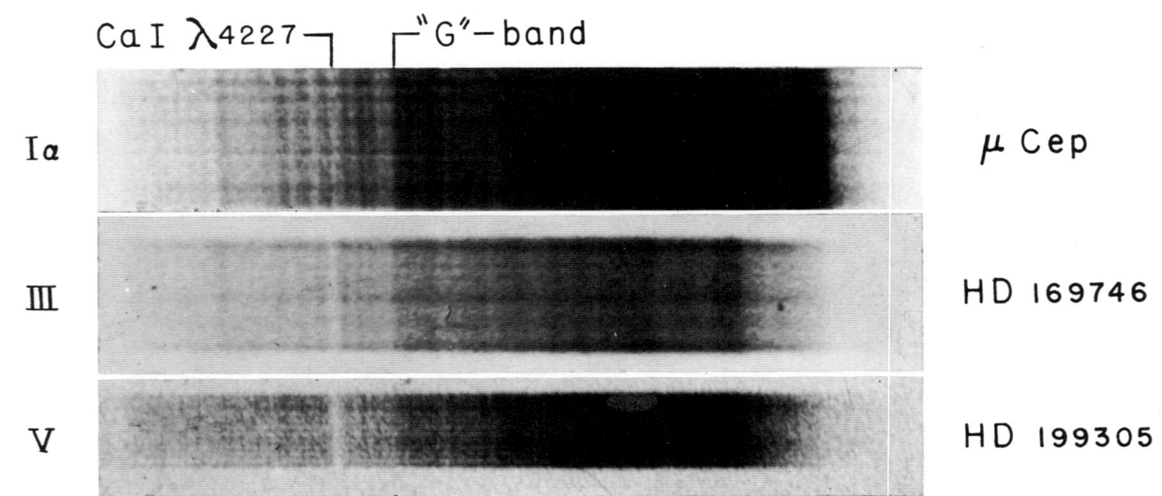

FIG. 3–8. COMPARISON OF THE CA I LINE, λ4227, IN DWARFS, GIANTS, AND SUPERGIANTS OF SPECTRAL CLASS *M*2

The symbols Ia, III, and V denote the luminosity class of the star in the classification scheme of the Morgan-Keenan-Kellman *Atlas of Stellar Spectra*. The supergiant μ Cephei is one of the brightest stars known, *HD* 169746 is a giant, and *HD* 199305 is a dwarf. Notice that the Ca I line, λ4227, the most conspicuous feature in the dwarf spectrum, steadily weakens in stars of increasing luminosity, until in μ Cephei it is weaker than the lines on either side. The red edge of the "*G*-band," λ4308, strengthens in the brighter stars. (Courtesy, P. C. Keenan and J. J. Nassau, *Ap. J.* **104,** 458, 1946, University of Chicago Press.)

Other examples could be cited, and we find that although the ionization theory gives a qualitative explanation of some absolute magnitude effects, it usually fails if quantitative comparisons are made. In addition, some striking absolute magnitude effects are not predicted by the ionization theory at all.

All absolute magnitude effects have to be established empirically. They will also depend on the dispersion of the spectograph employed; criteria that are useful at medium dispersion may be invisible with low dispersion. One uses certain spectral lines to establish the spectral class. Then one selects representative stars of different luminosities and compares their spectra to find line ratios that are sensitive to luminosity. For example, over a range from middle *F* to late *K*, the ratio 4077 (Sr II)/4045 (Fe I) is a very useful luminosity criterion, the ratio increases with increasing luminosity. Detailed classification criteria are given by Morgan and Keenan (1943) or in the review article by Fehrenbach (1958). The width and blackness of a spectral line will depend on many factors besides the fraction of atoms in the proper stage of excitation and ionization. These items include the quantitative relation between the number of absorbing atoms and the intensity of the line (curve of growth), which involves the degree of line broadening, the gas kinetic motions of the atoms and any large scale atmospheric motions (turbulence). The transparency and structure of the atmosphere are also important, while the presence of magnetic fields will further complicate the problem (see Chapters 8 and 9).

It is clear that even an application of ionization theory in a first approximation requires a knowledge of two things — the electron pressure and number of atoms acting in the layers producing the lines. If the continuous absorptivity is low, we may see to great depths in the star; if it is high, we may see only the thinnest portions of its outer layers. Although both line and continuous spectra are produced in the same layers, for many problems it is legitimate to think of the dark-line spectrum as originating in a layer which produces line absorption only, whereas a pure continuous spectrum is supplied by the photosphere. Thus the "number of atoms above the photosphere" will depend on the absorptivity of the outer layers, a quantity which will depend on the electron pressure and temperature. The temperature of the star may be estimated from its color, or more accurately from its spectral energy distributions interpreted with the aid of appropriate model atmospheres (see Chapters 5 and 6). We must estimate the electron pressure from the ionization equilibrium itself or from the broadening of certain ion-density sensitive lines such as those of hydrogen (see Chapters 7 and 8).

In Fig. 3–4 we show effects of varying atmospheric opacity along the main sequences on the number of active atoms capable of absorbing representative lines of Mg I, Fe I, and Fe II. Masses above the photosphere and mean electron pressures are taken from the "rough analysis" calculations of Chapter 5.

3-10. A Relation Between Gas Pressure, Electron Pressure, and Chemical Composition. For theoretical work on stellar atmospheres we shall need to know the relation between gas pressure and electron pressure. In the hottest stars all the atoms are ionized and every atom supplies at least one electron. Since hydrogen is overwhelmingly the most abundant constituent of the atmosphere and since it supplies one electron per atom, the electron pressure is very nearly half the gas pressure. On the other hand, in a star of "normal" composition, such as the sun, the bulk of the gas pressure is supplied by hydrogen, whereas the electron pressure comes solely from ionization of metals. In a subdwarf star the H/metal ratio may be a hundred times greater than in the sun, so that the domain in which hydrogen rather than metals supply the electrons extends to stars almost as cool as the sun. Hence gas pressure is related to electron pressure in a complicated way which depends on the assumed proportions of various abundant elements. The relative proportions of the metals among themselves is not critical in determining the P_g (P_ε, T) relationship, nor do elements like C, N, O, Ne, S, and A play a very significant role. Helium is important because over a large temperature range it contributes to the number of particles but not to the number of electrons. A *most crucial parameter* is the H/metal ratio.

We calculate the gas pressure P_g as a function of electron pressure P_ε and T as follows. Let N_a denote the number of atoms of all kinds (both neutral and ionized) per cm³, while N_ε denotes the number of electrons per cm³. Let x_E denote the fraction of the atoms of element E that have been ionized. We can neglect the second ionizations because of the overwhelming predominance of hydrogen. The less abundant elements can be grouped together and treated with a mean ionization potential. Thus

$$x_E = \frac{N_{1,E}}{N_{0,E} + N_{1,E}} \tag{3-108}$$

$$N_a = N_1 + N_2 + \cdots = \sum_E N_E \tag{3-109}$$

$$N_\varepsilon = N_1 x_1 + N_2 x_2 + \cdots = \sum N_E x_E \tag{3-110}$$

and

$$P_g = NkT = (N_a + N_\varepsilon)\, kT,\ P_\varepsilon = N_\varepsilon kT \tag{3-111}$$

Then

$$\frac{P_g}{P_\varepsilon} = \frac{N_a + N_\varepsilon}{N_\varepsilon} = \frac{1 + N_\varepsilon/N_a}{N_\varepsilon/N_a} \tag{3-112}$$

The problem reduces to one of computing N_ε/N_a as a function of T and P_ε for an assigned chemical composition. We choose the following relative abundances for the elements most significant to the problem at hand.

ABUNDANCES ADOPTED FOR P_g, P_ε, T RELATION
("Normal" Stars) log A = 4.13

Element	Abundance	Element	Abundance	Element	Abundance
H	1,000,000	Ne	2000	K	0.09
He	140,000	Na	2	Ca	2.4
C	250	Mg	19	Fe	6
N	410	Al	1.6	Ni	0.6
O	1300	Si	40	Misc. metals	2

The tabulation is based on solar abundances (Goldberg, Müller, Aller, 1960) and the results obtained for the B stars by various authors. The corresponding value of the ratio H/metals = 1.35×10^4 (by numbers).

The computational procedure is as follows: We first select a value of P_ε and $\theta = 5040/T$ and calculate x_E for each element. The entry in the table denoted as miscellaneous metals is treated as having an ionization potential like nickel with $2u/u_0 = 1.0$. With each x_E known, N_ε can be calculated and the ratio P_g/P_ε found. Then the quantity $P_g(P_\varepsilon, T)$ can be tabulated.

Fig. 3–9 and Table 3–3 show the results of the calculation. One notes, that at temperatures of 10,000°K and higher, $P_g \sim 2P_\varepsilon$, and that hydrogen (because

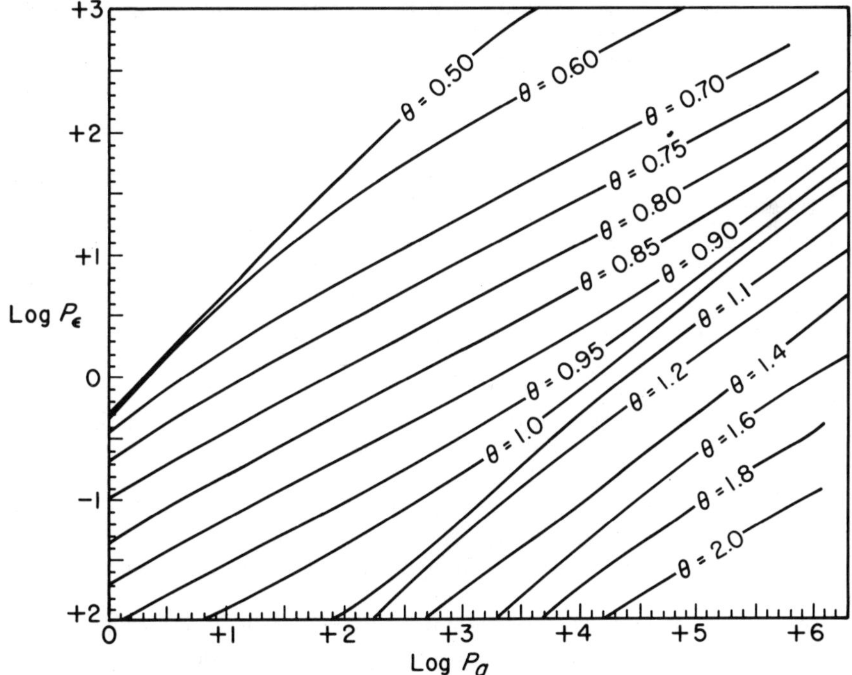

FIG. 3–9. THE RELATION BETWEEN GAS PRESSURE, ELECTRON PRESSURE, AND TEMPERATURE FOR NORMAL COMPOSITIONS, $P_g(P_\varepsilon, T)$
(Courtesy Anne and Charles Cowley.)

of its abundance) continues to supply the bulk of the electrons even when it is mostly neutral. At the lowest temperatures even the metals become mostly neutral and the electrons are supplied by such rare ones as potassium and sodium. In the intermediate regions ($\theta \sim 1.0$, log $P_\varepsilon \sim 1.0$) it is often sufficiently accurate for most purposes to divide the metals into two groups (Fe, Si, Mg, Ni, etc., and Al, Ca, Na, K), choose a mean ionization potential for each group, and calculate P_ε/P_g.

TABLE 3–3

RELATION BETWEEN GAS AND ELECTRON PRESSURE FOR A SOLAR-LIKE ABUNDANCE

(IBM 709 solution. Courtesy Anne and Charles Cowley.)

log P_ε	log P_g							
	$\theta = 0.50$	$\theta = 0.60$	$\theta = 0.70$	$\theta = 0.75$	$\theta = 0.80$	$\theta = 0.85$	$\theta = 0.90$	$\theta = 0.95$
−2.0							− .61	.07
−1.5						− .22	+ .36	1.07
−1.0				− .46	− .02	+ .63	1.34	2.04
− .5	− .17	− .16	− .05	+ .30	+. 91	1.62	2.33	2.98
0	+ .33	+ .34	+ .65	1.19	1.88	2.60	3.28	3.84
+ .5	+ .83	.87	1.48	2.15	2.87	3.57	4.19	4.58
1.0	1.34	1.45	2.41	3.12	3.84	4.51	4.99	5.26
1.5	1.84	2.14	3.38	4.11	4.81	5.38	5.72	5.94
2.0	2.37	2.96	4.36	5.08	5.72	6.16	6.43	6.66
2.5	2.94	3.89	5.35	6.03	6.56			
3.0	3.62	4.86	6.32					

log P_ε	log P_g						
	$\theta = 1.0$	$\theta = 1.1$	$\theta = 1.2$	$\theta = 1.4$	$\theta = 1.6$	$\theta = 1.8$	$\theta = 2.0$
−2.0	.78	1.92	2.22	2.65	3.27	3.64	4.20
−1.5	1.75	2.60	2.78	3.36	3.89	4.33	5.06
−1.0	2.65	3.20	3:39	4.08	4.53	5.09	5.95
− .5	3.46	3.79	4.06	4.76	5.22	5.95	6.88
0	4.14	4.42	4.76	5.41	5.97	6.85	7.85
+ .5	4.78	5.10	5.50	6.09	6.82		
1.0	5.44	5.83	6.22	6.83			
1.5	6.14	6.58					

The effect of an increase in the H/metal ratio on the $P_g(P_\varepsilon, T)$ relationship is shown in Fig. 3–10 where the same relative abundances of the metals and the C, N, O group is retained, as is the same H/He ratio, but the H abundance is now increased 31.6 times. That is, log $A = 5.63$, corresponding to one of the less extreme examples of population type II. At the lowest temperatures, the metals continue to supply the electrons, but hydrogen dominates throughout most of the ranges.

We apply these $P_g(P_\varepsilon, T)$ relationships in determinations of models for stellar atmospheres and in calculations of mean values of atmospheric parameters, $\bar{P}_g$, $\bar{P}_\varepsilon$, and mass above the photosphere (see Chapter 5).

3–11. Dissociation Equilibrium of Chemical Compounds. While the ionization theory gives a rational explanation of the spectral sequence for the hotter stars, it is satisfying that a closely analogous treatment of the

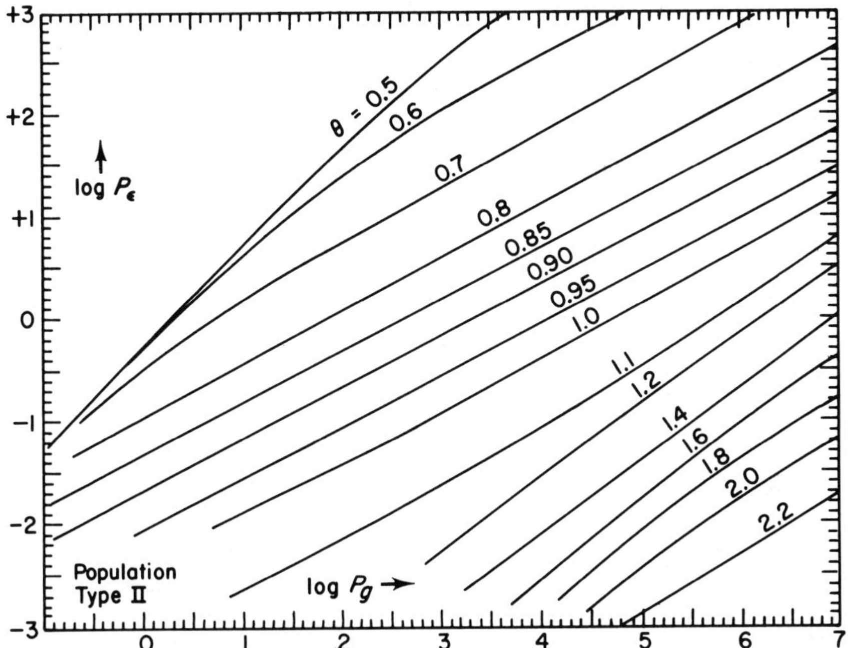

FIG. 3–10. THE RELATION BETWEEN GAS PRESSURE, ELECTRON PRESSURE, AND TEMPERATURE FOR METAL-DEFICIENT (SUBDWARF) STARS, $P_g(P_\varepsilon, T)$

formation and dissociation of molecules explains qualitatively, at least, the spectra of the cooler stars.

Suppose two atoms, A and B, combine to form the diatomic molecule AB according to the reversible reaction:

$$A + B \rightleftarrows AB \tag{3–113}$$

The numbers of combining atoms A and B and molecules AB are related by an expression similar to the ionization equation:

$$\frac{n(A)\,n(B)}{n(AB)} = K'\,(AB) \tag{3–114}$$

where $K'\,(AB)$ sometimes is called the dissociation "constant." It depends on the temperature and dissociation potential of the molecule. The explicit form of Eq. 3–114 may be obtained from Eq. 3–73 by inserting the appropriate partition functions for the rotation and vibration of the molecule together with those of the types we have already discussed in Art. 3–5. Hence the expression Eq. 3–114 may be rewritten in a form analogous to the ionization equation:

$$\frac{p_A p_B}{p_{AB}} = \frac{g_A g_B}{g_{AB}}\left[\frac{2\pi MkT}{h^2}\right]^{3/2}\frac{h^2}{8\pi^2 I}\left(1 - e^{-S}\right)e^{-D/kT} = K(AB) \tag{3–115}$$

in terms of the partial pressures, p_A, p_B, p_{AB}. Here g_A, g_B, g_{AB} represent the statistical weights of the ground levels of atoms A and B and the molecule AB. At the low temperatures at which molecules form, we can generally replace the partition functions involving the electronic states by their first terms. Here

$$M = \frac{M_A M_B}{M_A + M_B} \tag{3-116}$$

is the "reduced" mass. The term $h^2/8\pi^2 I$ comes from the rotational partition function, while the $(1 - e^{-S})$ term arises from the vibrational partition function. Here $S = hv/kT$ where v is the vibrational frequency of the molecule (see Chapter 2).

TABLE 3–4

LOGARITHM OF DISSOCIATION CONSTANTS, log K
(D_0 is expressed in electron volts)
(After P. Stanger, 1963)

	H_2	C_2	N_2	O_2	CH	CN
T	D_0 = 4.476	6.2	9.756	5.080	3.47	8.1
5000	7.682	6.068	3.041	7.796	8.248	4.57 :
4500	7.130	5.341	1.911	7.194	7.818	3.68 :
4000	6.442	4.434	0.501	6.445	7.284	2.525
3500	5.563	3.272	−1.309	5.485	6.600	1.017
3000	4.395	1.728	−3.716	4.211	5.692	−0.990
2500	2.768	−0.426	−7.080	2.434	4.436	−4.50
2000	0.337	−3.646	−12.112	−0.220	2.556	−7.980
1500	−3.695	−8.996	−20.483	−4.626	−0.558	−14.946

	CO	NH	NO	OH	CaH	MgH
T	D_0 = 11.2	3.8	6.48	4.35	1.70	2.49
5000	2.20	7.969	5.80	7.50	8.808	8.185
4500	0.92	7.499	5.00	7.00	8.567	7.874
4000	−0.67	6.915	4.112	6.356	8.313	7.488
3500	−2.736	6.169	2.896	5.506	7.965	6.995
3000	−5.475	5.178	1.260	4.378	6.506	6.344
2500	9.332	3.799	−0.971	2.805	6.870	5.439
2000	−15.088	1.741	−4.340	0.457	5.928	4.093
1500	−24.663	−1.672	−9.937	−3.438	4.375	1.867

	AlH	AlO	VO	TiO	ZrO
T	D_0 = 3.06	5.79	6.4	6.94	7.8
4000	7.931	5.180	4.221	3.670	2.403
3500	7.334	4.099	3.030	2.381	1.088
3000	6.543	2.662	1.446	0.669	−0.750
2500	5.443	0.658	−0.763	−1.723	−3.510
2000	3.804	−2.226	−4.066	−5.300	−7.516
1500	1.091	−7.310	−9.554	−11.239	−14.176

Table 3-4 gives the logarithm of the dissociation equilibrium "constants" for important compounds of H, C, N, and O as calculated by P. Stanger (1963). See also similar calculations by Pecker and Peuchot (1957), and by C. de Jager and L. Neven (1957). For some compounds the constants for the dissociation equilibria can be written in the form

$$\log K = A - \frac{B}{T} + \frac{1}{2}\log T \tag{3-117}$$

where, for example, $A = 10.69$ and $B = 35{,}100$ for TiO; the corresponding values for ZrO are 10.86 and 39,600, respectively.

Calculations of the dissociation equilibria have been made by a number of investigators among whom we may mention especially H. N. Russell (1934), and Neven and de Jager (1957). Russell postulated a series of model stellar atmospheres similar in composition to that of the sun but differing in temperature and pressure. Then he investigated the relative proportions of the various kinds of atoms and undissociated molecules, taking into account the variation of electron pressure among giant and dwarf stars, the variation in the thickness of atmospheric layers responsible for the formation of absorption lines, and other refinements as well. He made separate calculations for giant and dwarf stars of the same composition as the sun, and also considered mixtures in which carbon was more abundant than oxygen. Although the actual abundance of hydrogen is now known to be greater than Russell supposed, and the molecular constants he used have been superseded by more accurate values, the general qualitative picture he has given seems to be correct.

Neven and de Jager carried out calculations for different total pressures, temperatures, and abundances utilizing more recent data on molecular constants.

The calculations of dissociation equilibria for molecules may be carried out as follows: The total number of atoms of any given kind, e.g., carbon, equals the number of free atoms plus the number tied up in molecules, viz.:

$$n'(\mathrm{C}) = n(\mathrm{C}) + 2n(\mathrm{C_2}) + n(\mathrm{CN}) + n(\mathrm{CO}) + \cdots \tag{3-118}$$

Nitrogen atoms will satisfy a relation of the form

$$n'(\mathrm{N}) = n(\mathrm{N}) + 2n(\mathrm{N_2}) + n(\mathrm{CN}) + \cdots \tag{3-119}$$

and analogous expressions may be written for O, H, Ti, and so on. Now the relative values of $n'(\mathrm{C})$, $n'(\mathrm{O})$, $n'(\mathrm{H})$, etc., are known from the adopted abundances, and the quantities on the right-hand side of Eqs. 3-118 and 3-119 are to be computed.

Since the partial pressure is $p = nkT$, one can write a series of simultaneous quadratic equations of the form:

$$p'(\mathrm{C}) = p(\mathrm{C})\left\{1 + \frac{2p(\mathrm{C})}{K(\mathrm{C_2})} + \frac{p(\mathrm{N})}{K(\mathrm{CN})} + \cdots\right\} \qquad (3\text{–}120)$$

in which each p' and K is known. The solution of these simultaneous equations will give the quantities, p(C), p(N), p(H), and so forth. In stars of normal composition the problem is somewhat simplified by the fact that hydrogen is so overwhelmingly abundant and that other elements fall into groups that have little mutual influence and may usually be handled separately.[5] In the giant and supergiant stars particularly, ionization may deplete the numbers of metals available for molecular formation.

Adopting the hydrogen/metal ratio for "normal" stars, viz., 1.35×10^4, we may compare the concentrations of various compounds in giant and dwarf stars. In these calculations mean values of P_g, P_ε, T, and mass above the photosphere are adopted from Chapter 5 for the dwarf and giant sequences from $T = 6000°$K to 2500°K. Three composition groups are considered:

	C	N	O	Ti	Zr
(O) "standard"	250	125	800	0.05	0.0002
(CO) carbon = oxygen	800	125	800	0.05	0.05 and 0.10
(C) carbon > oxygen	1600	125	800	0.05	0.0002 and 0.05

The abundance of H is taken as 1,000,000. In the "standard" composition, oxygen is more abundant than carbon; in the second group they are of equal abundance; whereas in the last group, carbon is taken to be twice as abundant as oxygen. Many years ago R. H. Curtiss suggested that the splitting of the spectral sequence was due to differences in the chemical composition. The ordinary *M* stars represent objects in which oxygen exceeds carbon in abundance. Hence, such oxides as TiO, CO, and OH play important roles. In the carbon (*R*, *N*) stars, on the other hand, most of the oxygen gets tied up in the unobservable CO and the important bands are those due to carbon and its compounds. The *S* stars contain a larger proportion of metals of the second long row of the periodic table than do the sun and similar stars.

The results of the calculations are shown in Figs. 3–11, 3–12, 3–13, and 3–14, and are in as good agreement with those of Russell as the change in hydrogen abundance, atmospheric models, and molecular constants would permit. In the dwarf stars H_2 becomes the most abundant compound, and in all three groups the extremely stable compound CO is prominent. In the "oxygen" (*O*) stars essentially all of the carbon gets tied up in this compound. There remains sufficient oxygen for the oxides, OH, TiO,

[5] An iteration procedure works well for stars of normal composition (Stanger, 1963), but for atmospheres in which hydrogen is depleted, interlocking effects become very important. Formation of compounds involving H, C, N, and O, metallic compounds of Zr, Ti, etc. and ionization effects must all be handled simultaneously.

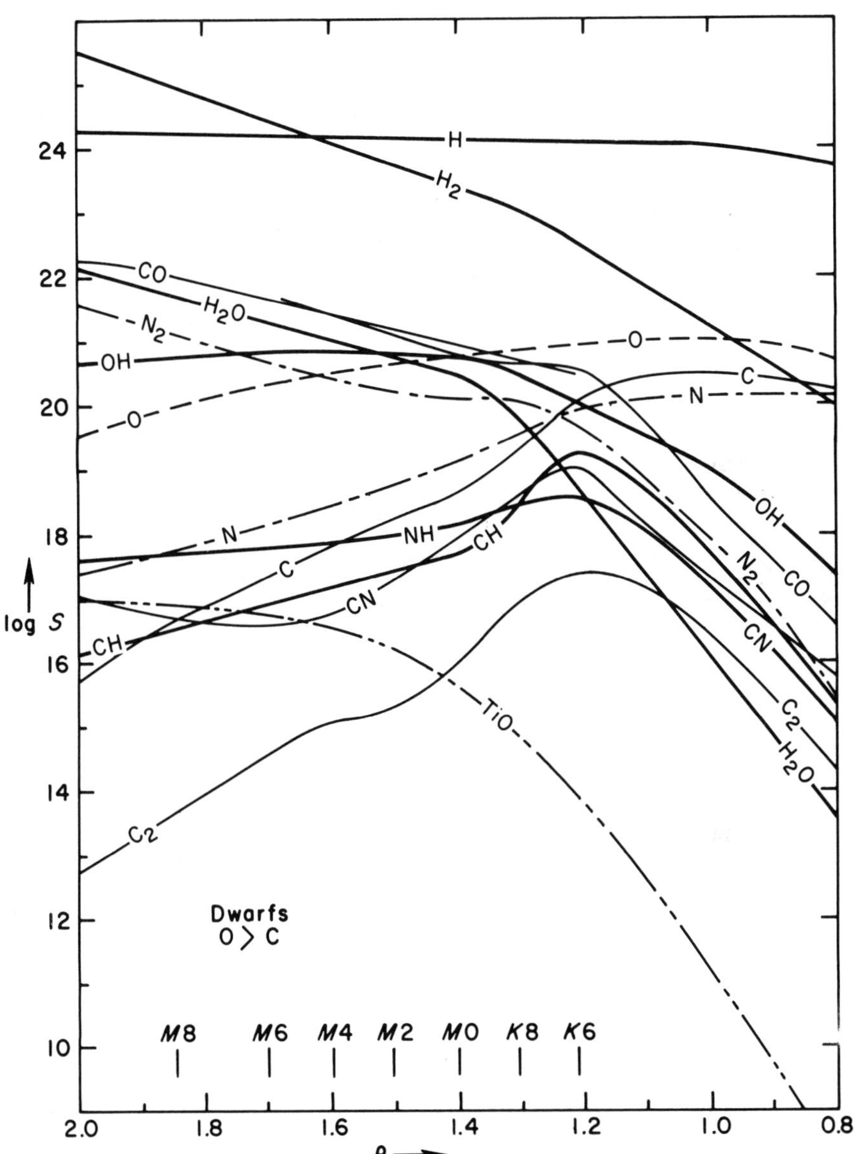

Fig. 3-11. The Behavior of Compounds in the Atmospheres of Main Sequence Stars

The ordinates are the logarithms of the numbers of molecules or atoms above each cm² of the photosphere, log S. Abscissas are $\theta = 5040/T$. The corresponding spectral classes are given also. The abundances assumed are those given in the first row of the unnumbered table of Art. 3-11.

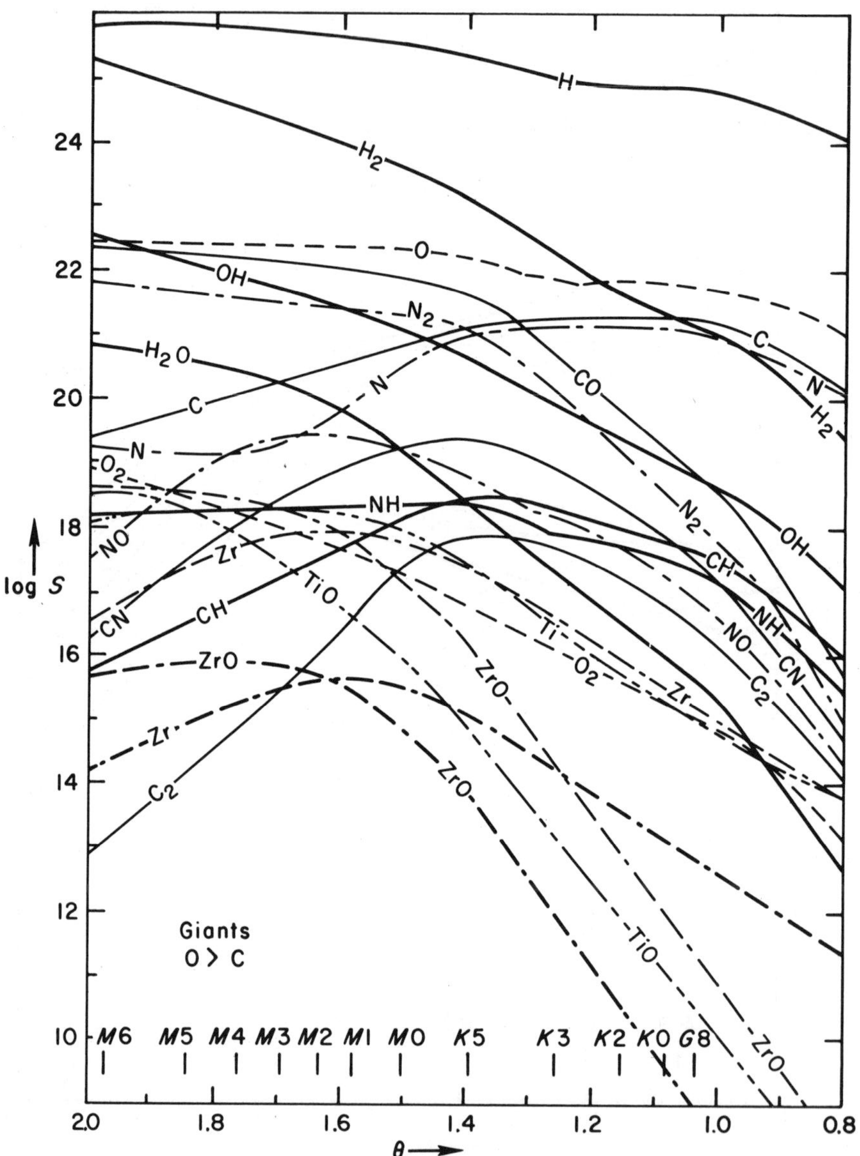

FIG. 3–12. THE BEHAVIOR OF COMPOUNDS IN THE ATMOSPHERES OF GIANT STARS (O > C)
Ordinates and abscissas are defined as for Fig. 3–11. Two values of the zirconium abundance are considered.

ZrO, etc., to form in appreciable quantities. Notice that at low temperatures large amounts of water are formed in the dwarf star.

In the second group (C, O) most of the oxygen and carbon are tied up

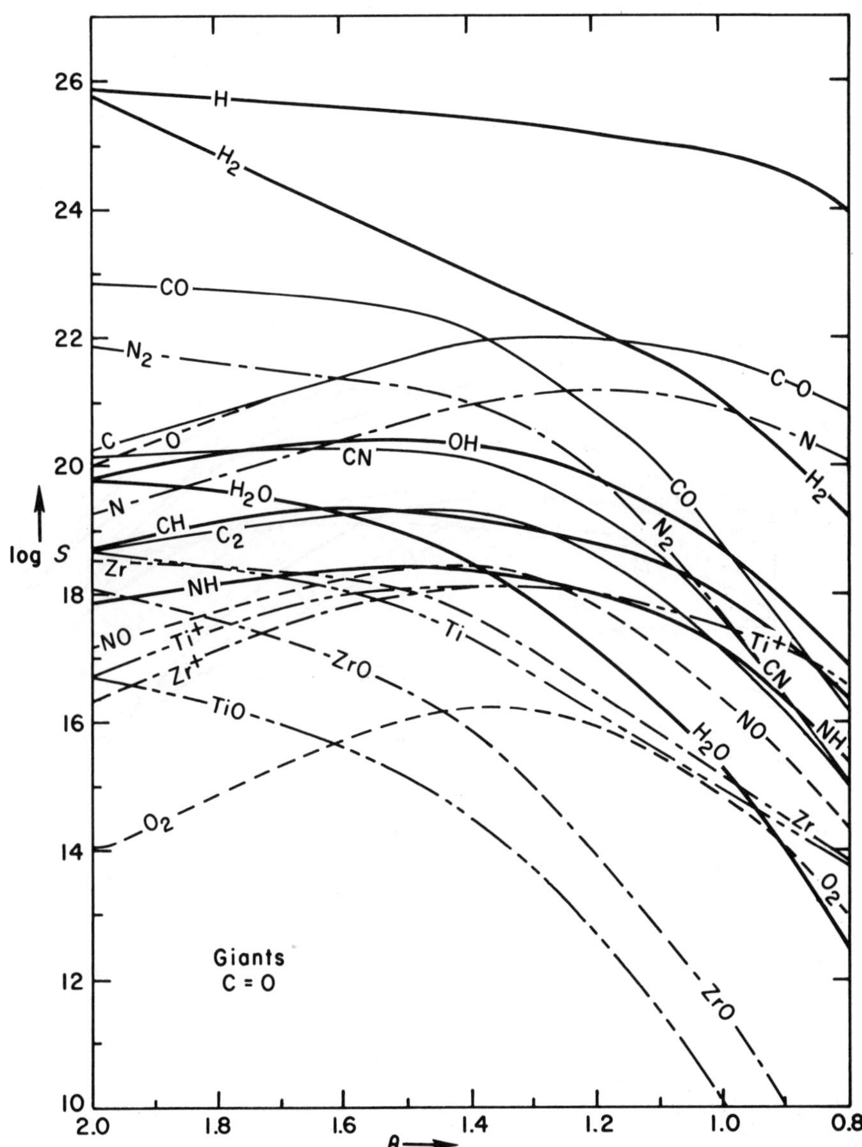

FIG. 3-13. THE BEHAVIOR OF COMPOUNDS IN THE ATMOSPHERES OF GIANT STARS (C=O)

The ordinates and abscissas are defined as for Fig. 3-11. The high zirconium abundance situation is illustrated (see text).

in carbon monoxide. Other oxides and water are greatly diminished in abundance, while the carbon compounds are not yet prominent. In the stars of the "carbon" (*C*) group virtually all of the oxygen gets locked up

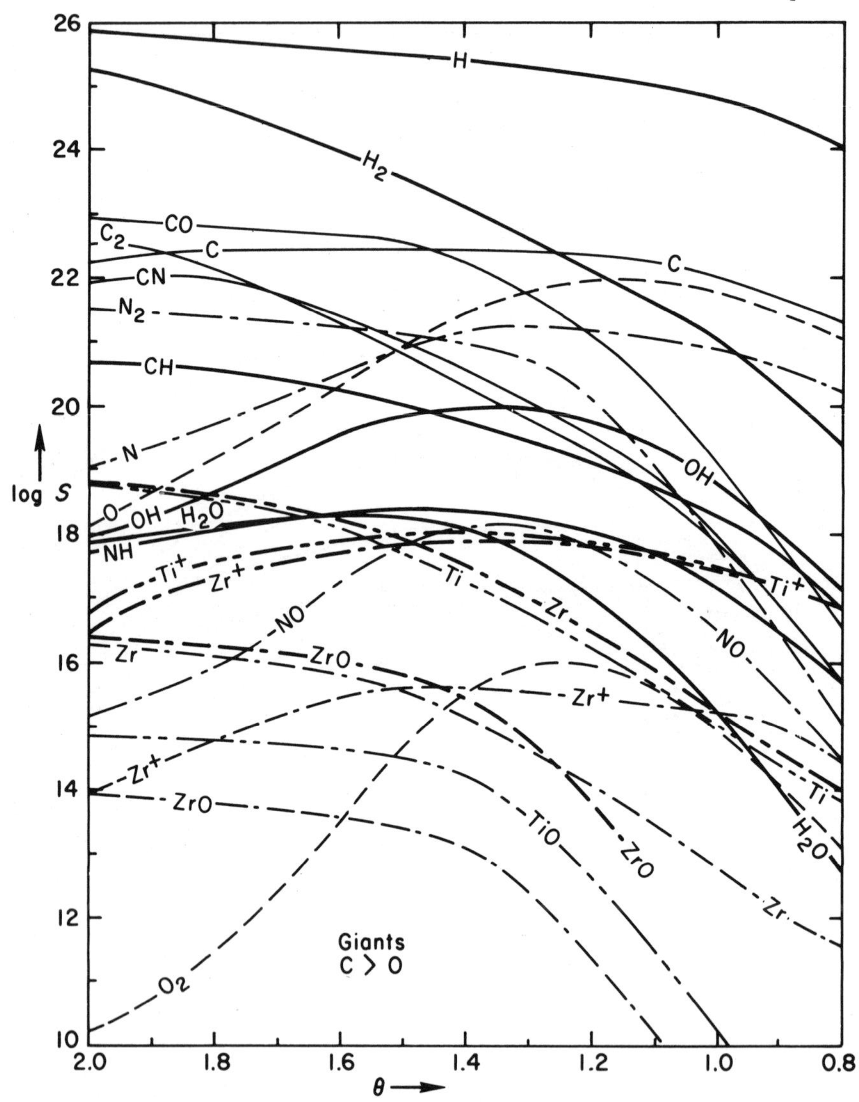

FIG. 3–14. THE BEHAVIOR OF COMPOUNDS IN THE ATMOSPHERES OF GIANT CARBON STARS (C > O)

Ordinates and abscissas are defined as for Fig. 3–11. Two values of the zirconium abundance are considered (see text).

in CO, the other oxygen compounds are suppressed, while CN, CH, and C_2 are plentiful.

Notice that the character of the equilibrium (and therefore of the emitted spectrum as well) is very sensitive to the C/O ratio. A change in carbon abundance by a factor of 2 from three-quarters of the oxygen abundance to three-halves of the oxygen abundance could completely alter the appearance of the spectrum of the star. Calculations with (C, O) and (C) compositions have been carried out only for giant and supergiant stars since these "abnormal" compositions are not found in dwarf stars. Likewise, the enhanced abundances of zirconium and similar elements are found only in giant or supergiant stars in advanced stages of their evolution. These increased quantities of carbon and heavy elements are believed to be manufactured in the stars themselves.

It is of interest to compare Figs. 3–11 and 3–12 showing results obtained for dwarfs and giants of "normal" composition. The TiO molecule achieves a greater concentration in the giants and supergiants than in dwarfs, whereas the formation of water is favored in the dwarfs. Stanger finds that formation of ZrO_2 and TiO_2 depletes ZrO and TiO by substantial amounts at temperatures below 2500°K in dwarfs and 2100°K in giants and supergiants.

Interpretations of the spectra of cool stars require, on the one hand, models of their atmospheres so that the changing molecular equilibrium with depth may be followed, and on the other hand, information on molecular line transition probabilities so that observed band intensities can be converted into numbers of absorbing molecules. Small differences in chemical composition can have a profound effect on the observed spectrum, as has been emphasized often.

The heavy metal stars present particularly interesting problems since these elements are presumably formed by nuclear processes occurring within the stars themselves. Anne Cowley (1962) finds evidence that in the S stars, not only is the abundance of zirconium probably enhanced, but also the C/O ratio is greater. The formation of ZrO is influenced not only by the abundance of Zr, but also by the C/O ratio.

PROBLEMS

3–1. For a pure gas one may derive the ionization equation from Boltzmann's equation. Note that the energy of a free electron will be, if χ_0 = ionization energy

$$E = \chi_0 + \frac{1}{2m}(p_1^2 + p_2^2 + p_3^3)$$

If $N_{0,1}$ is the number of neutral atoms in the ground level whose statistical weight is $g_{0,1}$ show that

$$\frac{dN_\varepsilon}{N_{0,1}} = \frac{2g_1'}{g_{0,1}} \frac{e^{-\chi_0/kT}}{h^3} e^{-p^2/2mkT}\, dp_1 \ldots dq_3$$

By integrating over the momenta and a space of volume V_0, verify that

$$\frac{N_\varepsilon}{N_{0,1}} = \frac{2(2\pi mkT)^{3/2}}{h^3} \frac{g_1'}{g_{0,1}} e^{-\chi_0/kT}\, V_0$$

where V_0 is of such a size as to contain only one ionized atom, i.e. $N_{1,1}V_0 = 1$. Then obtain Eq. 3–79.

3–2. Show that the relative numbers of atoms in the nth energy level in the qth stage of ionization are related to the number in the n'th energy level in the q'th state of ionization by

$$\frac{N_{nq}}{N_{n'q'}} = \left[\frac{2(2\pi mkT)^{3/2}}{h^3N_\varepsilon}\right]^{q-q'} \frac{g_{nq}}{g_{n'q'}} e^{-\chi_{nq,n'q'}/kT}$$

where $\chi_{nq,n'q'}$ denotes the total energy separation of the two levels.

3–3. Show that for hydrogen, the combined Boltzmann and ionization equation is

$$N_n = \frac{N_iN_\varepsilon}{T^{3/2}} \frac{h^3}{(2\pi mk)^{3/2}} n^2 e^{hRZ^2/n^2kT}$$

where N_i is the number of ions, N_n is the number of atoms in level n, and N_ε is the number of electrons cm^{-3}. R is the Rydberg constant in frequency units.

REFERENCES

Gas Kinetic Theory

Kennard, E. H. 1942. *Kinetic Theory of Gases.* McGraw-Hill Book Co., Inc., New York.

Jeans, J. H. 1925. *Dynamical Theory of Gases.* Cambridge University Press, London.

Chapman, S., and Cowling, T. G. 1952. *Mathematical Theory of Non-Uniform Gases, 2d ed.* Cambridge University Press, London.

Statistical Mechanics

Tolman, R. C. 1938. *Principles of Statistical Mechanics.* Oxford University Press, Fair Lawn, N. J.

Fowler, R. H. 1936. *Statistical Mechanics.* Cambridge University Press, London.

Mayer, J. E., and Mayer, M. G. 1940. *Statistical Mechanics.* John Wiley & Sons, Inc., New York.

Ter Haar, D. 1954. *Elements of Statistical Mechanics.* Holt, Rinehart & Winston, Inc., New York.

Relation Between Gas Pressure, Electron Pressure, and Chemical Composition

Extensive calculations were given by:

Strömgren, B. 1944. *Publ. Copenhagen Obs. No. 138.*

Tables giving the data plotted in Figs. 3–9 and 3–10 may be found in *Stars and Stellar Systems*, J. L. Greenstein (ed.), vol. 6, chap. 5. Chicago: University of Chicago Press, 1960.

Dissociation Equilibrium of Chemical Compounds in Stars

de Jager, C., and Neven, L. 1957. Liége Roy. Soc. Mem. **18,** 357.

Pecker, J. C., and Peuchot, M. 1957. Liége Roy. Soc. Mem. **18,** 352.

Stanger, P. 1963. In press.

CHAPTER 4

THE EMISSION AND ABSORPTION OF RADIATION

4–1. Fundamental Definitions. The most important astrophysical property of matter is that it absorbs and emits radiation. The wavelengths of the emitted radiations tell the kind of atoms or ions present in the source. Previous chapters have given a qualitative interpretation of stellar spectra. To make further progress we must consider quantity as well as quality of radiation, and for this purpose it will be necessary to define carefully such concepts as intensity, flux, and energy density.

a. *Specific Intensity.* In a volume of space through which radiation is passing, let us consider a surface S which may or may not coincide with an actual physical surface emitting radiation. We fix our attention on an element of area dA, centered at a point O, and on radiation passing per unit time through this element and confined within a truncated cone defined by dA and an elementary solid angle $d\omega$ (see Fig. 4–1). If the amount of this radiation in a frequency range ν to $\nu + d\nu$ is denoted by dE_ν, the specific intensity of the radiation I_ν in the interval $d\nu$ at ν is defined by

$$dE_\nu = I_\nu \cos\theta \, dA \, d\nu \, d\omega \qquad (4\text{–}1)$$

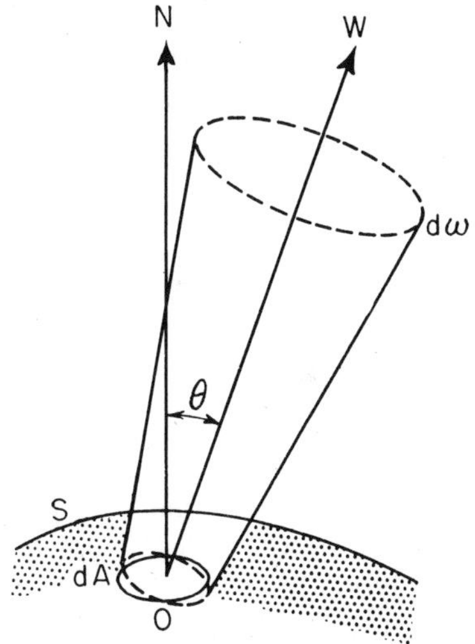

FIG. 4–1. DEFINITION OF SPECIFIC INTENSITY

where θ is the angle between the axis of the cone and the outward normal ON. In general, I_ν will depend both on the position of O in space, x, y, z, and on the direction of the pencil of rays OW specified by colatitude θ and azimuthal angle ϕ. Hence $I_\nu = I_\nu(x, y, z, \theta, \phi)$. At any point, the specific intensity of the radiation in a direction θ, ϕ, is defined as the amount of energy which flows per second in unit solid angle in unit frequency interval through a unit area placed perpendicular to the direction θ, ϕ. In particular the amount of energy flowing per second per cm^2 and in the frequency interval $d\nu$ in the cone $d\omega$ will be $I_\nu\,(\theta, \phi)\, d\nu\, d\omega$.

Thus, to obtain a quantity in energy units, we must multiply I_ν by $dA\, dt\, d\nu\, d\omega$, where dt is the element of time. An element of solid angle being

$$d\omega = \sin\theta \, d\theta \, d\phi$$

the mean value of the intensity averaged over all angles will be

$$J_\nu = \frac{1}{4\pi}\int I_\nu(\theta, \phi)\, d\omega = \frac{1}{4\pi}\int_{\theta=0}^{\pi}\int_{\phi=0}^{2\pi} I_\nu(\theta, \phi) \sin\theta \, d\theta \, d\phi \qquad (4\text{–}2)$$

We are sometimes interested in the total amount of radiation as integrated over all frequencies, viz.,

$$I = \int_0^\infty I_\nu \, d\nu \qquad (4\text{–}3)$$

b. *Radiation Flux.* A quantity that is often confused with intensity is the flux of radiation. To get the total amount of radiation passing through each cm² of a surface, per unit time, we integrate dE over all solid angles. Thus the flux $\mathfrak{F}$ is defined by

$$\mathfrak{F} = \int I \cos\theta \, d\omega = \int_0^\pi \int_0^{2\pi} I(\theta, \phi) \cos\theta \sin\theta \, d\theta \, d\phi = \pi F \quad (4\text{–}4)$$

The contribution of the backward flowing radiation occurs with a negative sign (because of the cosine factor). Hence it is customary to speak of an outward and an inward flux, viz.

$$\mathfrak{F}_{\text{out}} = \int_0^{\pi/2} \int_{\phi=0}^{2\pi} I \cos\theta \sin\theta \, d\theta \, d\phi;$$

$$\mathfrak{F}_{\text{in}} = -\int_{\pi/2}^{\pi} \int_{\phi=0}^{2\pi} I \cos\theta \sin\theta \, d\theta \, d\phi \qquad (4\text{–}5)$$

The net flux is

$$\mathfrak{F} = \mathfrak{F}_{\text{out}} - \mathfrak{F}_{\text{in}} \qquad (4\text{–}6)$$

For isotropic radiation I is independent of angle; hence the integral in Eq. 4–4 vanishes and F is 0, which means there is no preferential direction of radiation flow. The outward flux of radiation from a surface that emits uniformly in all directions, i.e., $I(\theta, \phi) = I_a$ is

$$\mathfrak{F} = 2\pi I_a \int_0^{\pi/2} \cos\theta \sin\theta \, d\theta = \pi I_a \qquad (4\text{–}7)$$

Here I_a is identical with the F introduced in Eq. 4–4.

Some illustrations may clarify the physical distinction between *flux* and *specific intensity*. If we neglect limb darkening in the sun, the specific intensity of solar radiation within solid angle defined by the solar disk is constant as long as the observer is close enough to the sun to see it as a disk. At half the sun's present distance, its surface brightness and hence specific intensity of its radiation would remain unchanged, but its angular area in the sky and the total amount of radiation received from it by each cm², i.e., *the flux,* would be quadrupled. By similar arguments it follows that an extended nebula will have the same surface brightness as long as the eye perceives it as an area.

On the other hand, for an emitter so distant as to appear only as a point, the quantity that counts is the energy actually reaching the observer, i.e., the flux. Thus the apparent brightness of a star, or nebula so distant as to be stellar in appearance, varies as the inverse square of the distance. Along any ray in empty space, the specific intensity I is constant, whereas the flux, under such circumstances, falls off according to the inverse square law.

No optical system of lenses or mirrors can increase the specific intensity of radiation from any object, even though the total amount of energy reaching the receiver may be greatly augmented. If a lens is used to concentrate the sun's rays, the solid angle from which radiation of a given specific intensity is received, is enlarged, i.e., the flux passing through the focus is increased.

In most astrophysical problems amenable to quantitative treatment, I does not depend on ϕ but only on θ. The pronounced solar limb darkening is an illustration in point. As Fig. 4–2 shows, this darkening is a consequence of a dependence upon θ of the specific intensity. It is greatest along the outward normal and gradually falls off as θ increases.

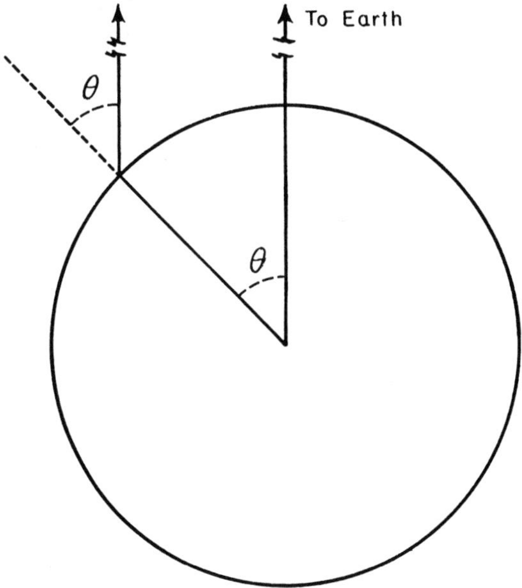

FIG. 4–2. EMERGENT RAY FROM THE SUN

As an illustration of some of these principles, consider the calibration of a surface brightness described as equivalent to sixty tenth-magnitude stars of the same color as the sun per square degree. The apparent photovisual magnitude of the sun is -26.73; its angular diameter is $32'$. Hence its surface brightness in visual magnitudes per square minute of arc is $-26.73 + 2.5 \log \pi(16)^2 = -19.47$. Measurements of the radiation received from the sun (Abbot, see Chapter 6) show that this corresponds to an emission rate of 6.25×10^{10} ergs cm^{-2} sec^{-1} (= antilog 10.796) by the solar surface. Sixty tenth-magnitude stars per square degree correspond to one sixtieth of a single tenth-magnitude star per square minute of arc or one star of 14.45 magnitude. Hence the surface brightness will be given by

$$\log S = (-19.47 - 14.45)\ 0.4 + 10.796 = -2.772$$

or

$$S = 0.169 \times 10^{-2} \text{ ergs cm}^{-2} \text{ sec}^{-1}$$

is the surface brightness in cgs units corresponding to sixty tenth-magnitude stars of solar color per square degree.

c. *Energy Density.* The energy density of radiation, u, is the amount of radiant energy per cubic centimeter which at any given instant happens to be passing through space at the point considered. It is related to the specific intensity I by the expression,

$$u = \frac{1}{c}\int I\, d\omega \tag{4–8}$$

For example, consider an enclosure with heated walls. Radiation will be flying about from one wall to the other and we can speak of the energy density as the amount of energy in ergs cm^{-3} in the enclosure. An elementary volume will be exposed to radiation from all directions and the amount passing in each second is obtained by summing over all angles, viz., $4\pi I$, i.e., it is simply the constant specific intensity multiplied by the number of solid radians in a sphere. The velocity with which the radiation travels is c; hence the amount of energy residing in a volume at any one time will be $4\pi I/c$ times the volume of the element. Thus the energy density of *isotropic* radiation is related to intensity by

$$u = \frac{4\pi}{c} I \tag{4–9}$$

Since the directly observed quantities are intensities or fluxes, it seems better to use I or F rather than u in theoretical discussions of stellar atmospheres or gaseous nebulae. As for stellar interiors where no direct observations can be made, there are some advantages in using u rather than I.

4–2. Black-Body Radiation. Often we are concerned not with the total radiation integrated over the entire spectrum, but rather with emission in particular frequencies. The precise character of the relation between emitted intensity and frequency is known for a perfect radiator or black body. A black body is an object that totally absorbs all radiation falling upon it. Such a surface is difficult to prepare experimentally, but if radiation is in equilibrium with its surroundings, as for example, in an enclosure whose walls are maintained at a fixed temperature T, each cm^2 of an internal wall will emit just as much radiation as it absorbs in every wavelength, and the condition of a "black" body is fulfilled. A small hole in the wall permits the emergent black-body radiation to be studied without disturbing its quantity or energy distribution. Then the intensity is given as a function of frequency and temperature by Planck's law,

$$B_\nu(T) = \frac{2h\nu^3}{c^2} \frac{1}{e^{h\nu/kT} - 1} \tag{4-10}$$

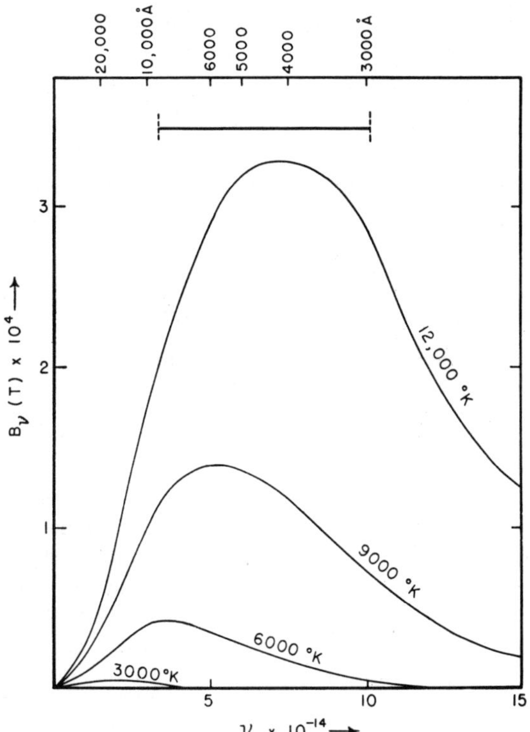

Fig. 4-3. The Energy Distribution from a Black Body

The ordinates are expressed in terms of ergs per unit frequency. The dotted lines indicate the limit reached by photographic plates ordinarily employed in astronomical work. Beyond λ2900 the earth's atmosphere is opaque.

where $h = 6.625 \times 10^{-27}$ erg sec (Planck's constant), c is 3×10^{10} cm sec^{-1} (velocity of light), and k is 1.3804×10^{-16} cgs units (Boltzmann constant). The energy density of radiation in the frequency range ν to $\nu + d\nu$ in ergs cm^{-3} may be obtained by multiplying Eq. 4-10 by $4\pi/c$, thus,

$$u(\nu, T)\, d\nu = \frac{8\pi h}{c^3} \frac{\nu^3}{e^{h\nu/kT} - 1}\, d\nu \tag{4-11}$$

We plot $B_\nu\,(T)$ for different temperatures in Fig. 4-3. Notice how the energy distribution changes as the temperature increases, the frequency of maximum energy shifts violetward, and the total amount of energy radiated rises rapidly. We find the total energy density by integrating Eq. 4–11 over all frequencies

$$u = \int_0^\infty u(\nu, T)\, d\nu = aT^4 \qquad (4\text{–}12)$$

where the radiation constant $a = 7.5634 \times 10^{-15}$ ergs cm^{-3} deg^{-4}. That is, the energy density of the black-body radiation depends on the fourth power of the temperature. This is the *Stefan-Boltzmann law* for the energy density of total radiation. It also may be derived by thermodynamics. Let us compute the rate of emission of energy $\mathfrak{F}_{BB}$ by a surface that radiates as a black body. If I does not depend on angle for all outward-flowing radiation, the outward flux by Eq. 4–7 is $\mathfrak{F}_{BB} = \pi I = \pi B$. Since $u = 4\pi B/c$ by Eq. 4–9, the amount of energy crossing each cm^2 sec^{-1} is $cu/4$ or

$$\mathfrak{F}_{BB} = \sigma_R T^4 \qquad (4\text{–}13)$$

where $\sigma_R = 5.6687 \times 10^{-5}$ ergs cm^{-2} deg^{-4} sec^{-1}.

Example: What is the amount of energy radiated per square centimeter by a star whose surface temperature is 5700°K? From Eq. 4–13

$$\mathfrak{F}_{BB} = 5.67 \times 10^{-5}\,(5700)^4 = 5.99 \times 10^{10} \text{ ergs cm}^{-2}\text{ sec}^{-1}$$

or about 60,000 kw m^{-2}.

We often speak of the black sphere temperature of interstellar space. This is the temperature that would be assumed by a perfect black body, or hypothetical black bulb thermometer that absorbed and emitted equally well in all frequencies. It is measured by the energy density u of the radiation field. A black sphere of radius r will intercept and absorb $\pi r^2 uc$ ergs sec^{-1}, and will emit $4\pi r^2 \sigma T^4$ ergs sec^{-1}. Since absorbed and emitted energies must be equal, $u = (4\sigma/c)T^4 = aT^4$, which defines an energy-density temperature of interstellar space. It does not necessarily correspond to the temperature of any real particle.

From the known distribution and luminosities of the stars, Eddington estimated the energy density at a typical point in interstellar space to be $u = 7.7 \times 10^{-13}$ ergs cm^{-3}. From $u = aT^4$ we find $T = 3.18$°K.

It is sometimes convenient to express Planck's law in wavelength rather than in frequency units. Since $\nu = c/\lambda$,

$$|\,d\nu\,| = c\,\frac{d\lambda}{\lambda^2}$$

and

$$B_\lambda\, d\lambda = \frac{2hc^2}{\lambda^5}\,\frac{1}{e^{hc/\lambda kT} - 1}\, d\lambda \qquad (4\text{–}14)$$

The wavelength at which the intensity of the emitted radiation is a maximum is found by differentiating B_λ with respect to λ and setting the derivative equal to zero. This calculation gives the *Wien displacement law:*

$$\lambda_{\max} T = 0.289782 \tag{4-15}$$

This law tells how the color of radiation changes as the temperatures of an emitting source is varied. As temperature is increased, the wavelength at which the maximum output of energy occurs decreases and the light becomes progressively bluer in quality.

As an example, let us compute the wavelength at which the radiation intensity is a maximum for a star whose temperature is 5000°K. From Eq. 4–15, we find $\lambda_{\max}$ to be 5.796×10^{-5} cm or λ5796. We emphasize that $\lambda_{\max}$ is the wavelength of maximum intensity when I is plotted on a wavelength scale. If we plot intensity I on a frequency scale so that I_ν represents the amount of energy per unit frequency interval at ν, the wavelength corresponding to the frequency at which I_ν is a maximum will be very different from $\lambda_{\max}$.

The fundamental law of radiation is Planck's law which gives the rate of emission of energy in ergs cm^{-2} sec^{-1} from a surface which radiates like a black body at a temperature T. If we determine the distribution of energy in the spectrum of the sun or a star, we can estimate its temperature from a comparison of the observed energy curve with the Planckian curve. Examples of this procedure will be found in Chapter 6.

The Stefan-Boltzmann law, Eq. 4–13, expresses the total rate of energy emission in terms of the temperature but says nothing about its quality, i.e., the frequency or wavelength distribution of the radiation. As we have seen, one can derive this law from Planck's law by integrating over all frequencies, but it can also be established by thermodynamic reasoning. Astrophysical applications of Stefan's law can be made to the sun and a few other stars whose dimensions and total radiation output are known. For example, various observers have measured the total amount of energy radiated by the sun. Since the solar surface area is known, the energy output per square centimeter can be found and its temperature computed.

In astronomical applications, certain asymptotic forms of Planck's law are often useful. If λ or T are so small that $hc/\lambda kT >> 1$ (as in the ultraviolet), the exponent $hc/\lambda kT$ will be much larger than unity and we can write

$$B_\lambda(T) \sim \frac{2hc^2}{\lambda^5} e^{-ch/\lambda kT} \tag{4-16}$$

the so-called Wien approximation to Planck's law. In frequency units

$$B_\nu(T) \sim \frac{2h\nu^3}{c^2} e^{-h\nu/kT} \tag{4-17}$$

In the far infrared and radio-frequency regions $h\nu/kT << 1$ so that Planck's law reduces to

$$B_\nu(T)\,d\nu = \frac{2kT}{c^2}\,\nu^2\,d\nu = \frac{2kT}{\lambda^2}\,d\nu \tag{4–18}$$

This is the *Rayleigh-Jeans formula.* It is particularly useful in discussions of thermally emitting sources in the r-f region, as we shall see in Art. 4–5.

4–3. Derivation of Planck's Law. Many years ago Lord Rayleigh and also Jeans suggested that one could calculate the number of degrees of freedom of electromagnetic waves in a box whose walls were maintained at a temperature T, on the assumption that with each degree of freedom was associated a kinetic energy $\frac{1}{2}kT$ and an equal potential energy. On this assumption Jeans was able to derive the correct asymptotic form of the radiation law for long wavelengths, but it remained for Planck to obtain a form of the law correct for all situations.

The radiation energy density is calculated on the assumption that for any given frequency it is equal to the number of degrees of freedom multiplied by the energy per degree of freedom per unit volume. Consider first the problem in one dimension where we may use the analogue of the transverse vibrations of a string. If l_0 is the length of the box, the string must vibrate in such a way that nodes occur at each end. Any possible wavelength λ is related to l_0 by the condition

$$n = 2\,l_0/\lambda \tag{4–19}$$

Further, the number of possible modes of vibration of the string for any wavelength λ greater than $\lambda_0 = 2l/n$ will be $2n$. The factor 2 arises from the fact that the string possesses two possible modes of vibration corresponding to two possible planes of polarization.

If n is very large, the number of modes between λ and $\lambda - d\lambda$ is related to the change dn by

$$dn = \frac{4l}{\lambda^2}\,d\lambda \tag{4–20}$$

In a three-dimensional box, the projections of the standing waves on each edge of the box must be an integer. For a cubical box of edge l_0 we have that

$$\frac{2l_0\cos\theta_1}{\lambda_0} = n_1 \qquad \frac{2l_0\cos\theta_2}{\lambda_0} = n_2 \qquad \frac{2l_0\cos\theta_3}{\lambda_0} = n_3 \tag{4–21}$$

Since the angle of incidence must equal the angle of reflection,

$$\cos^2\theta_1 + \cos^2\theta_2 + \cos^2\theta_3 = 1$$

whence

$$n_1^2 + n_2^2 + n_3^2 = \frac{4l_0^2}{\lambda_0^2}$$

for λ_0, while all wavelengths λ greater than λ_0 must satisfy

$$n_1'^2 + n_2'^2 + n_3'^2 < 4l_0^2/\lambda_0^2$$

Consequently, the number of permitted modes of vibration corresponding to $\lambda > \lambda_0$ will equal the total number of possible combinations of integers $n_1n_2n_3$ such that

$$n_1^2 + n_2^2 + n_3^2 \leq 4l^2/\lambda^2 \tag{4–22}$$

This constraint may be visualized in another manner. Consider a series of concentric spheres of successive radii $1, 2, 3, \cdots, 2l/\lambda$ and fix attention on the one octant for which $n_1 > 0$, $n_2 > 0$, $n_3 > 0$. Imagine this space is filled with unit cubes oriented parallel to the coordinate axes. Each corner of one of these cubes will fall on the surface of one of the successive spheres and the total number of points will equal the total number of unit cubes in this positive octant of a sphere of radius $2l_0/\lambda_0$. Since $2l_0/\lambda_0$ is assumed very large, the total number of permitted combinations is equal to the volume of the octant. Since each combination of n_1, n_2, and n_3 corresponds to two possible transverse vibrations, the total number of modes of vibration corresponding to $\lambda \geq \lambda_0$ is

$$n = 2\frac{1}{8}\frac{4}{3}\pi\left(\frac{2l_0}{\lambda_0}\right)^3 = \frac{8}{3}\pi\frac{l_0^3}{\lambda_0^3} \tag{4–23}$$

The number of permitted degrees of freedom between λ and $\lambda + d\lambda$ is then

$$dn = \frac{8\pi}{\lambda^4}\, l_0^3\, d\lambda = \frac{8\pi\nu^2}{c^3}\, l_0^3\, d\nu \tag{4–24}$$

since $d\nu = c/\lambda^2\, d\lambda$. To get the total density of ν radiation we must now multiply by the average energy per degree of freedom, and set the volume $l_0^3 = 1$. If we set the average energy per degree of freedom equal to kT per unit volume in accordance with the kinetic theory analogy as classical theory would suggest, we get

$$u_\nu d\nu = \frac{8\pi\nu^2 kT}{c^3} \tag{4–25}$$

which is the Rayleigh-Jeans law (see Eqs. 4–18 and 4–19).

Planck suggested that the oscillators comprising the radiation field could take up energy, not kT, but only in multiples of $h\nu$, viz.

$$\varepsilon_0 = 0 \qquad \varepsilon_1 = h\nu \qquad \varepsilon_2 = 2h\nu, \text{ etc.} \tag{4–26}$$

The partition function (Art. 3–5) is

$$u(T) = \sum g_i e^{-\varepsilon_i/kT} = 1 + e^{-\varepsilon/kT} + e^{-2\varepsilon/kT} + \cdots = \frac{1}{1 - e^{-\varepsilon/kT}} \tag{4–27}$$

since

$$\frac{1}{1-x} = 1 + x + x^2 + x^3 + \cdots \text{ for } x < 1 \tag{4–28}$$

and the statistical weight $g_i = 1$. The average energy per oscillator (see Eq. 3–40) is

$$\bar{\varepsilon} = kT^2 \frac{d}{dT} \ln u(T) = -kT^2 \frac{d}{dT} \ln (1 - e^{-\varepsilon/kT}) = \frac{\varepsilon}{e^{\varepsilon/kT} - 1} \tag{4-29}$$

where $\varepsilon = h\nu$. Hence

$$u_\nu \, d\nu = \frac{8\pi\nu^2}{c^3} \bar{\varepsilon} \, d\nu = \frac{8\pi h\nu^3}{c^3} \frac{1}{e^{h\nu/kT} - 1} \, d\nu \tag{4-30}$$

which is Planck's law.

4-4. The Coefficients of Absorption and Emission — Kirchhoff's Law. The flow of energy through the atmosphere of a star and the formation of absorption lines in stellar spectra or of emission lines in spectra of stars and gaseous nebulae all involve interaction of matter with radiation — absorption followed by subsequent re-emission. Consequently, it is essential to have a clear understanding of the exact meaning of such terms as emissivity, absorption coefficient, optical depth, and others.

Consider a small volume element dv of density ρ in a heated gas. The total amount of energy emitted by this elementary volume over all directions in the range ν to $\nu + d\nu$ will be

$$dE_\nu = j_\nu \rho \, dv \, d\nu \tag{4-31}$$

where j_ν, the *mass emission coefficient* of the material, is here defined as the total amount of energy emitted per second (per unit frequency interval) per gram of the radiating substance in all directions. (Some writers use j_ν as the emission per unit solid angle.)

The *absorption coefficient* can be defined as follows. Consider radiation of intensity $I_{\nu,0}$ falling perpendicularly upon a slab of thickness x (see Fig. 4-4). Some of this impinging energy will be absorbed and some will pass on through the layer. At any point within the slab, the loss of energy will be proportional to the density of the material and to the intensity of the radiation. That is, in a distance dx, the intensity of the beam will be cut down by an amount,

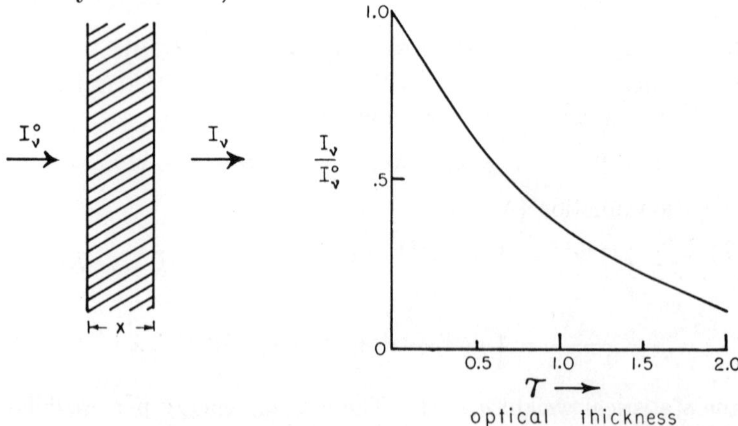

FIG. 4-4. THE EXTINCTION OF A BEAM IN A SLAB OF MATERIAL

$$dI_\nu = -k_\nu I_\nu \rho \, dx \tag{4–32}$$

where k_ν is defined as the mass absorption coefficient[1] for radiation of frequency ν. For strongly absorbing substances k will be large; for nearly transparent substances k will be small. We divide Eq. 4–32 by I_ν and integrate to obtain

$$\ln I_\nu = -k_\nu \rho x + \text{const}$$

$$I_\nu = I_\nu^0 \, e^{-k_\nu \rho x} \tag{4–33}$$

where I_ν^0 is the intensity of the incident beam. This means that when a beam of light passes through a slab of material its intensity falls off exponentially at a rate that depends upon the thickness of the slab, on its density, and on its mass absorption coefficient.

In general k_ν will depend on ρ and x. Hence

$$I_\nu = I_\nu^0 \, e^{-\tau_\nu} \tag{4–34}$$

where the quantity

$$\tau_\nu = \int_0^x k_\nu \rho \, dx \tag{4–35}$$

is called the *optical depth* or thickness of the slab. The larger the value of τ_ν, the greater will be the extinction in the slab (see Fig. 4–4).

Optical Thickness of the Slab	Ratio of Emergent to Incident Intensity
0.1	0.905
0.5	0.606
1.0	0.368
2.0	0.1353
3.0	0.050
5.0	0.007

As an example, in a filter of optical thickness 0.695 half the incident intensity will be lost, while for one of unit optical thickness, I will be cut down to 1/2.718, or 0.368 of its original value. Solar radiation supplies a second example. That emitted from the surface reaches us with undiminished specific intensity, whereas that emitted by a layer at optical depth τ is cut down to $e^{-\tau}$ of its original value.

Sometimes we are concerned with absorption by small solid particles such as grains in interplanetary space or in a comet. A small mass m exposed to radiation will absorb an amount

$$k_\nu m \, d\nu \int I_\nu (\theta, \phi) \, d\omega = 4\pi k_\nu m J_\nu \, d\nu \tag{4–36}$$

of radiation between ν and $\nu + d\nu$. The total radiation absorbed follows from an integration over all frequencies, viz.,

$$4\pi m \int k_\nu J_\nu \, d\nu$$

[1] Note that, in Chapter 4, k denotes the thermal absorption coefficient, whereas κ is employed for this purpose in Chapter 5 *et seq.* In Chapter 4, κ has already been used: see Eq. 4–71 and Eq. 4–128 in two entirely different contexts.

Kirchhoff showed that if emission by matter is controlled solely by its temperature as obtains, for example, for material in thermal equilibrium with its surroundings, an important relation exists between the coefficients of emission and absorption. The ratio j_ν/k_ν depends only on the temperature and not at all on the color, shape, or composition of the body, i.e.,

$$\frac{j_\nu}{k_\nu} = 4\pi B_\nu(T) \tag{4–37}$$

where $B_\nu(T)$ is the intensity of black-body radiation as given by Planck's law. In this form it is sometimes called the *Planck-Kirchhoff law* since the $B_\nu(T)$ function was not known in Kirchhoff's day. *Kirchhoff's law*, Eq. 4–37, expresses quantitatively the well-known fact that good absorbers are also good emitters, and that a heated substance emits more strongly at those frequencies where the absorption coefficient is high than at frequencies where absorption is low. In this connection we may recall the third law of spectrum analysis.

This relationship can be applied *only* for thermal emission. It is not valid, for example, for scattering of sunlight by air molecules where the particles are not in thermal equilibrium with the radiation falling upon them, nor does it apply to scattering of sunlight by free electrons in the corona. On the other hand, it appears to be valid for processes concerned with stellar continuous spectra and for the formation of certain absorption lines. (See Chapters 5, 7, and 8.)

4–5. Energy Measurements in the Radio-Frequency Region. The definitions developed in the preceding article are intended to apply throughout the entire electromagnetic spectrum. Special spectral regions, e.g., x-rays, and radio-frequencies, require somewhat different treatment than does the optical region. X-ray astronomy is discussed in Chapter 9. The chief characteristics of the radio-frequency region are the following: (a) The atmospheric window permits a broad band-pass in wavelength; (b) since $h\nu << kT$, we are always in the Rayleigh-Jeans region of any thermal source; and (c) the wavelengths are often comparable with the apertures of the antennas employed.

Very high spectral resolution $\Delta\lambda/\lambda$ is therefore frequently combined with very low angular resolution. In the optical region, atmospheric tremors or "bad seeing" limits the resolving power of the equipment, but in radio astronomy the angular resolving power usually does not approach that of the human eye and often is very much less. Hence there is abundant use of interferometer techniques and recourse to equipment of large physical size to gain resolution.

Suppose that there existed in the sky a single point source of radio-frequency emission. If we were to point a radio telescope at it, we would find that at its position (spherical coordinates θ_0, ϕ_0) the deflection of a recording potentiometer would be a maximum but at other positions near to it deflections also would be obtained. In fact, one would find that at some nearby point (θ, ϕ), $d(\theta - \theta_0, \phi - \phi_0) = d_0 \; f(\theta - \theta_0, \phi - \phi_0)$ where $f(\theta - \theta_0, \phi - \phi_0)$ is the antenna pattern. If the emitting source has an extension in space and the antenna is directed towards it, the power received is

$$P_T = \tfrac{1}{2}A_0 \iiint B(\theta, \phi, \nu) f_n(\theta, \phi, \nu) \, d\Omega \, \Delta\nu = kT_A\Delta\nu \tag{4–38}$$

which defines antenna temperature, T_A. The factor $\frac{1}{2}$ comes from the fact that only one plane of polarization can be observed at a time. A_0 is the maximum effective aperture of the antenna in square meters. We assume

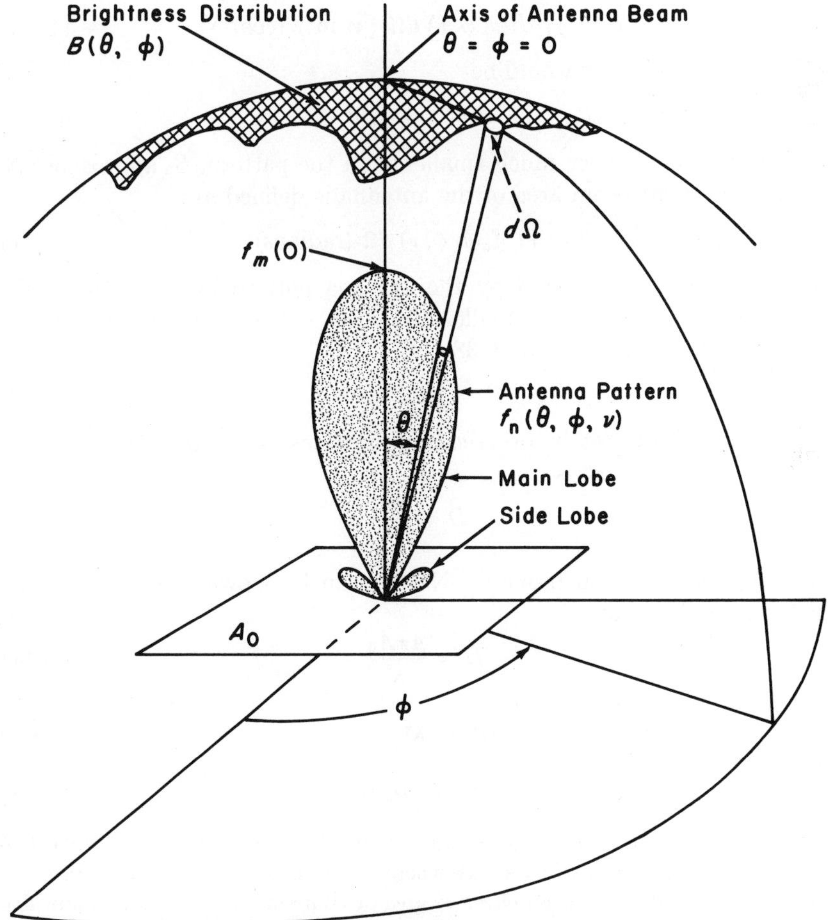

FIG. 4–5. ANTENNA PATTERN AND CELESTIAL SPHERE

(Adapted from Fig. 2, *Celestial Radio Radiation* by J. D. Kraus and H. C. Ko, 1957; RF Project 673, Scientific Report No. 1.)

that the antenna is directed at the brightest part of the source and that θ and ϕ are angular distances measured with respect to the axis of the antenna, i.e., $\theta = \phi = 0$ on the antenna axis (see Fig. 4–5). Now $f_n(\theta, \phi, \nu)$ is the antenna power pattern normalized to unity on the axis. It assumes a relatively simple form for a steerable paraboloid but is somewhat complicated for a fixed, asymmetrical array. Here $d\Omega = \sin\theta\, d\theta\, d\phi$ is the element of area.

The surface brightness of the radio source $B(\theta, \phi, \nu)$ is customarily expressed in mks units, i.e., as watts per square meter per frequency (in cps) per square degree, i.e., as w m^{-2} (cps)$^{-1}$ deg^{-2}. The brightness integrated over $d\Omega$ or the "flux density" is

$$S = \iint B(\theta, \phi, \nu)\, d\Omega \ [\text{w m}^{-2}\ (\text{cps})^{-1}] \qquad (4\text{–}39)$$

The "observed" flux would be

$$S_{\text{obs}} = \iint B(\theta, \phi, \nu)\, f_n(\theta, \phi, \nu)\, d\Omega \qquad (4\text{–}40)$$

and if the source is very much smaller than the pattern, S_0 approaches S.

The equivalent beam area of the antenna is defined as

$$\beta = \iint f_n(\theta, \phi, \nu)\, d\Omega \text{ (radians)} \qquad (4\text{–}41)$$

but the actual area covered by the antenna pattern is larger than this. If the antenna pattern is smaller than the extended source of surface brightness B_0, then from Eq. 4–38

$$P = \tfrac{1}{2} A_0 B_0 \beta \qquad (4\text{–}42)$$

A quantity of interest in practical applications is the directivity of the antenna. It is defined as

$$D = \frac{4\pi}{\beta} \qquad (4\text{–}43)$$

where β is expressed in radians. Now it can be shown (see, e.g., Kraus, 1950, pp. 24, 54) that

$$D = \frac{4\pi A_0}{\lambda^2} \qquad (4\text{–}44)$$

whence by Eq. 4–43

$$A_0\beta = \lambda^2 \qquad (4\text{–}45)$$

and Eq. 4–42 becomes

$$P = \tfrac{1}{2} B_0 \lambda^2 \qquad (4\text{–}46)$$

Example: At a certain frequency an antenna beam has an area $\beta = 80$ deg^2 and is directed at a point in the sky where the r-f brightness is 5×10^{-25} w m^{-2} (cps)$^{-1}$. If the antenna has an effective area of 20 m^2 and the receiver bandwidth is $\Delta\nu = 1$ mc what is the total power received if the source is much larger than the antenna pattern and S is constant over $\Delta\nu = 1$ mc?

$$P\Delta\nu = \tfrac{1}{2}\,(20)\,(80)\,(5 \times 10^{-25})\,(1 \times 10^{6}) = 4 \times 10^{-16} \text{ w} \qquad (4\text{–}47)$$

Suppose that a celestial source radiates as though it were at a uniform temperature T, then (see Eq. 4–18),

$$B = \frac{2kT}{\lambda^2} \tag{4–48}$$

Also suppose that the source were much larger than the antenna pattern.

$$P = \frac{1}{2} A_0 \iint \frac{2kT}{\lambda^2} f_n(\theta, \phi)\, d\Omega \tag{4–49}$$

but by Eq. 4–45

$$P = \frac{1}{2} A_0 \frac{2kT}{\lambda^2} \beta = kT \tag{4–50}$$

That is, the power received by the system would give a direct measurement of the equivalent black-body temperature of the source. If the source is a black body or if the optical thickness is large, the same value of T will be obtained at different frequencies, but if it is a non-thermal emitter, the equivalent black-body temperature T will vary markedly with ν. If the angular size of the body emitting at a temperature T is $\omega << \beta$, we can write, from Eqs. 4–38, 4–45, 4–49, and 4–50

$$P = kT_A = T\frac{\omega}{\beta} \tag{4–51}$$

Here T_A is called the *antenna temperature.* Physically, it can be regarded as *the equivalent temperature of the radiation resistance of the antenna.*

Often one measures a source against a background of cosmic noise and the significant quantity here is the rise in antenna temperature ΔT_a.

4–6. The Mechanical Force Exerted by Radiation — Radiation Pressure. That radiation exerts a force upon a surface intercepting it is predicted by Maxwell's electromagnetic theory of light, as well as by the quantum theory, which asserts that with each quantum of energy, $E = h\nu$, there is associated a momentum $h\nu/c$ in the same direction as the beam.

Radiation pressure can be computed as the net rate of transfer of momentum through a unit area of arbitrary orientation at the point considered. A similar definition holds for gas pressure, except that in this more familiar example the momentum is carried by the gas molecules themselves.

The amount of radiation passing per second through our arbitrarily chosen unit area at an angle θ with the normal and in a solid angle $d\omega$ will be $I \cos\theta\, d\omega$. The corresponding amount of momentum transported is $I \cos\theta\, d\omega/c$. Since the ray strikes the surface at angle θ, the component of momentum normal to the surface will be $(I \cos\theta\, d\omega/c) \cos\theta$.
Hence the total force will be

$$P_R = \frac{1}{c}\int I \cos^2\theta\, d\omega \tag{4–52}$$

Under the special circumstance where a beam of intensity I falls perpendicularly upon a totally absorbing surface, $\cos\theta = 1$, and $P_R = I_c$ but I/c in this example is simply the radiation density itself.

When radiation is isotropic,

$$P_R = \frac{4\pi I}{3c} = \frac{u}{3} \tag{4–53}$$

i.e., the radiation pressure is numerically equal to one-third the radiation density. If the radiation is compared with a gas, we may imagine the light quanta divided into three groups traveling in three mutually perpendicular directions. Thus at any one time, one-third of the photons or light quanta may be thought of as exerting a pressure in a given direction.

As an example, let us calculate the radiation pressure at a point in the solar atmosphere where the energy density corresponds to a temperature of 5700°K. From Eqs. 4–12 and 4–53 we find

$$P_R = 7.563 \times 10^{-15} \times \tfrac{1}{3}(5700)^4 = 2.65 \text{ dynes cm}^{-2}$$

as compared with a gas pressure (see Chapter 5) of about 80,000 dynes cm^{-2}. At the center of the sun, the temperature is about 14,700,000°K. At this point the radiation pressure is 1.18×10^{14} dynes cm^{-2} as compared with a total pressure of 2.25×10^{17} dynes cm^{-2}. Thus the radiation pressure is still only a small fraction of the total pressure. In a sufficiently luminous star, radiation pressure may contribute an appreciable fraction of the total pressure.

We must be careful to distinguish between the radiation pressure as just described and the net mechanical force exerted by radiation upon a mass of material.

A vane in quiet air is bombarded by molecules on both sides, but it remains motionless because the force on each side is the same. If there is a gust of wind there will be a net mean motion of the air molecules. The resultant mechanical force upon the vane will cause it to turn.

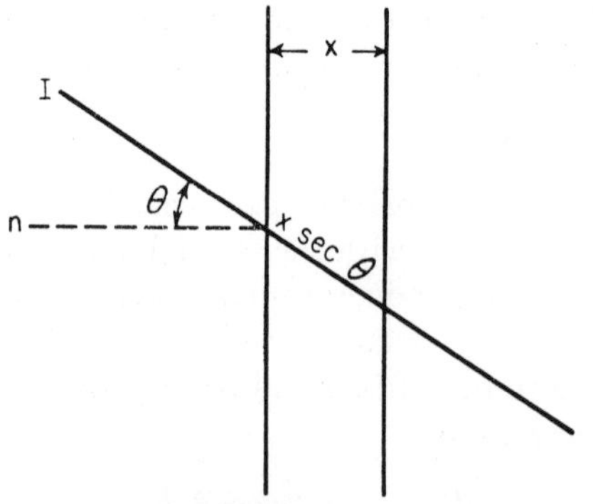

Fig. 4–6. The Oblique Passage of a Ray Through a Slab

Let a thin plane slab of thickness dx be illuminated by radiation of intensity $I_\nu(\theta)$. See Fig. 4–6. The amount of energy falling upon the slab per cm^2 sec^{-1} in the elementary cone $d\omega$ at an angle θ with the normal will be $I_\nu \cos\theta \, d\omega \, d\nu$. Since the path length of this ray through the slab is $dx \sec\theta$, the amount of energy absorbed per cm^2 sec^{-1} will be

$$de_\nu = I_\nu \cos\theta \, k_\nu \, \rho \sec\theta \, dx \, d\nu \, d\omega \tag{4–54}$$

As this radiation is absorbed, its momentum is conveyed to the material. The force supplied by de_ν normal to each unit area of the slab is $\cos\theta \, de_\nu/c$. The resultant mechanical force exerted by radiation of frequency between ν and $\nu + d\nu$ acting upon each cm^2 per sec is obtained by integrating $\cos\theta \, de_\nu/c$ over all angles of the incoming radiation.

$$\frac{1}{c}\int \cos\theta \, de_\nu = \frac{k_\nu\rho \, dx \, d\nu}{c}\int I_\nu \cos\theta \, d\omega$$

Thus

$$= \frac{k_\nu\rho \, dx \, d\nu}{c}\pi F_\nu \tag{4–55}$$

where πF_ν is the net flux in ergs cm^{-2} sec^{-1} (unit frequency interval)$^{-1}$.

The total mechanical force cm^{-2} sec^{-1} will be obtained finally by an integration over all frequencies. (The k employed here must be corrected for the effects of negative absorption. See Art. 4–11.) Thus the total amount of momentum abstracted from the radiation field by the slab will be

$$dp = \bar{k}\rho \frac{\pi F}{c} dx \quad \text{cm}^{-2}$$

where (4–56)

$$\bar{k} = \frac{1}{F}\int k_\nu F_\nu \, d\nu$$

is the *flux mean absorption coefficient* and πF is the total flux. See Fig. 4–6.

As an example, let us compute the mechanical force $f_g = dp/dx$ exerted by radiation upon a unit volume of gas in the atmosphere of an A-type star. With $T = 10{,}000°$K, $\pi F = \sigma T^4$, and $\bar{k} = 27$ (see Chapter 5) we get

$$f_R = 27\rho \times \frac{5.75 \times 10^{-5} \times 10^{16}}{3 \times 10^{10}} = 5.17 \times 10^2 \rho \text{ dynes}$$

as compared with a force of gravity upon the same volume element of $f_g = 2 \times 10^4 \, \rho$ dynes. In the atmosphere of a certain O star, $k = 3.0$, $T = 29{,}600°$K, and we find $f_R = 4.4 \times 10^3 \rho$ dynes, as compared with a force of gravity of $2.8 \times 10^4 \rho$ dynes. In the atmospheres of giant and supergiant stars, mechanical forces exerted by radiation may become comparable with or even exceed the ordinary force of gravity.

4–7. The Classical Picture of Radiation. We now seek quantitative information on the absorptivities and emissivities of atoms and molecules. For example, how may we obtain k_ν for the "D" lines of sodium? Within the scope of this book it will not be possible to give a complete discussion of this problem, but we shall cover developments essential to a quantitative interpretation of stellar and nebular spectra.

With the aid of the classical theory of absorption and emission of radiation it is possible to develop, with relative ease, certain expressions that can be transformed easily into the correct relationships, thus avoiding the complexities of a rigorous quantum mechanical treatment.

About a century ago Maxwell, from a thorough analysis of electrical and magnetic phenomena, showed that a system of electrical charges in motion would produce electric and magnetic waves in space. The velocity of these waves, 299,793 km sec^{-1}, which may be computed from electrical data alone, is precisely the velocity of light — a fundamental constant of nature. That is, waves of light are not waves of a material substance, but waves of electric and magnetic fields.

Visible light is but one form of *electromagnetic radiation* — akin to radio waves, heat radiation, x-rays, and γ-rays. These radiations differ chiefly by their wavelengths and exhibit similar phenomena of diffraction and polarization. A coarse wire grating produces interference with radio microwaves, a diffraction grating produces a spectrum with visible light, while crystals are used to examine x-ray spectra.

Electromagnetic waves can be polarized, proving that the displacements of the electric and magnetic fields occur perpendicular to the direction of propagation. In Fig. 4–7A we imagine that the wave comes perpendicularly out of the plane of the paper. The direction of vibration is then in the plane of the paper. Although the direction of vibration of ordinary light changes in a random fashion many times a second, it always remains in a plane perpendicular to the direction of propagation. If a piece of polaroid is placed in the beam, all vibrations save those in a single direction will be extinguished.

The transverse displacements consist of variations in the magnitude

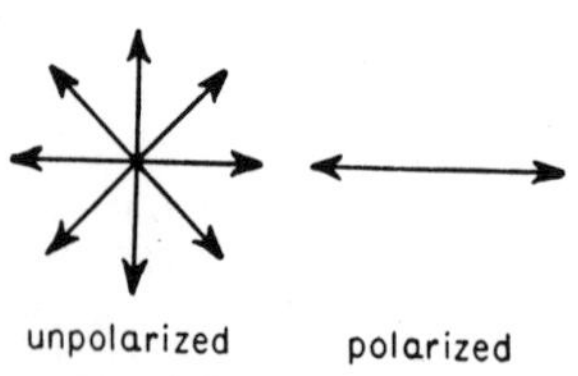

FIG. 4–7A. UNPOLARIZED AND PLANE POLARIZED LIGHT VIBRATIONS

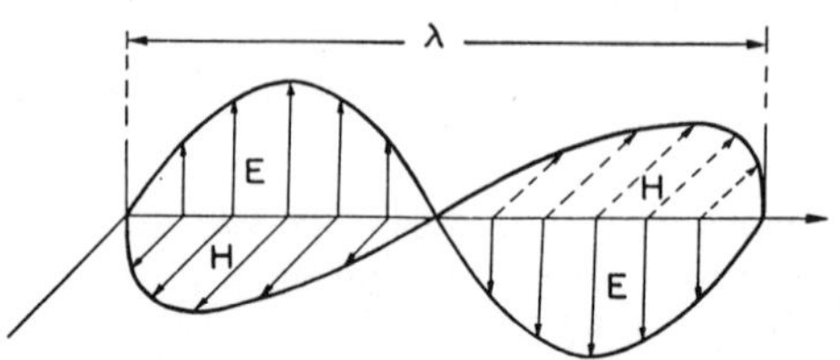

FIG. 4–7B. THE ELECTROMAGNETIC WAVE (Adapted from a diagram in *Sky and Telescope*, August, 1945, p. 8.)

of electric and magnetic fields that are directed perpendicular to one another and to the direction of propagation of the light wave (Fig. 4–7B). The flux of the radiation, i.e., the amount of energy traveling in the beam, is proportional to $E^2 + H^2$, viz.,

$$\mathfrak{F} = \frac{c}{8\pi}\left[E^2 + H^2\right] \tag{4–57}$$

where E and H are measured in cgs units and $\mathfrak{F}$ is in ergs cm^{-2} sec^{-1}. In a vacuum $E = H$ so $\mathfrak{F} = cE^2/4\pi$. In the mks system $\mathfrak{F} = EH$, but $H = E/120\pi$ ohms for a vacuum. Then $\mathfrak{F}$ is in watts per square meter.

Classical electromagnetic theory relates emission of energy with the motions of a system of charges. Let us suppose that a stationary charge of ε is placed at point A and a charge $-\varepsilon$ executes simple harmonic motion along the z-axis with an amplitude z_0. If z_0 is much smaller than the wavelength of the emitted light, an observer at O at a distance large compared with z_0 will receive an almost pure sine wave. As the charge moves up and down along the z-axis, it emits radiation which has a maximum intensity in the xy-plane and is zero along the z-direction. See Fig. 4–8.

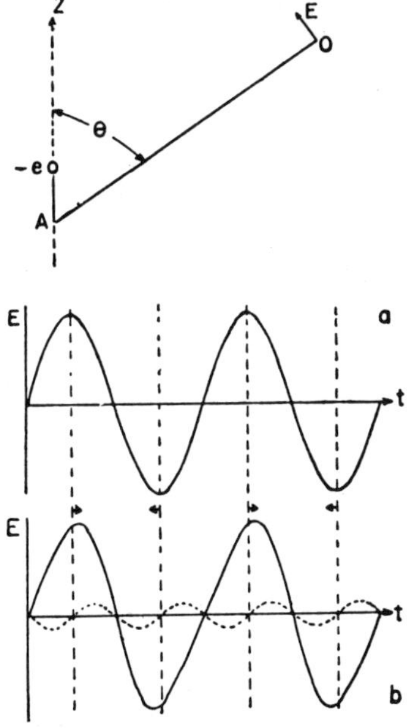

FIG. 4–8. THE RADIATION FROM A DIPOLE

The electric charge $-e$ vibrates around A along the direction Z. The electric vector **E** of the wave emitted is observed in O. If the amplitude of vibrations of the electric charge is negligibly small, the vibration of **E** is represented by the sine curve (a). For an amplitude which is not very small in comparison to the wavelength λ, a distortion occurs and the vibration of **E** can be decomposed into two vibrations with frequencies ν and 2ν (curve b), corresponding to electric dipole and electric quadrupole radiations. (Courtesy, S. Mrozowski, *Review Modern Physics* **16,** 164, 1944, Fig. 5.)

The distance of the charge $-\varepsilon$ from O varies as it executes its oscillations. The time interval between the maximum and minimum of the electric vector E will be smaller in the half of the cycle when the charge moves towards the observer than in the other half of the cycle. Hence there will be deviations from a pure sine pattern. It is found that the emitted wave can be represented as the sum of two vibrations,

$$E = a_d \sin 2\pi\nu t + a_q \sin 4\pi\nu t \tag{4-58}$$

The ratio a_q/a_d is proportional to z_0/λ. There are, in fact, also higher order terms but these fall off very rapidly in amplitude. We call the second term the electric quadrupole component of the radiation.

If z_0 is made smaller and smaller but the charges are made larger in such a way that the product

$$P_0 = z_0\varepsilon \tag{4-59}$$

remains finite, in the limit the vibrating system will radiate a pure sine wave or pure dipole radiation.

Now consider a system where a charge $+2q$ is placed at the origin and two charges $-q$ vibrate in opposite phases on either side. The dipole radiation from such a system will be zero and only quadrupole emission will appear. The intensity of emitted radiation is zero along the z-axis and also in the xy-plane, and will attain a maximum value at some intermediate angle. In general, if two charges oscillate along a line as in an antenna, the system emits as though it contained a dipole radiating a pure sine wave of frequency ν, and a quadrupole radiating at twice this frequency. When both are present and the amplitude of motion is much smaller than the wavelength of the emitted radiation, the quadrupole component is much smaller than the dipole component. See Fig. 4–9.

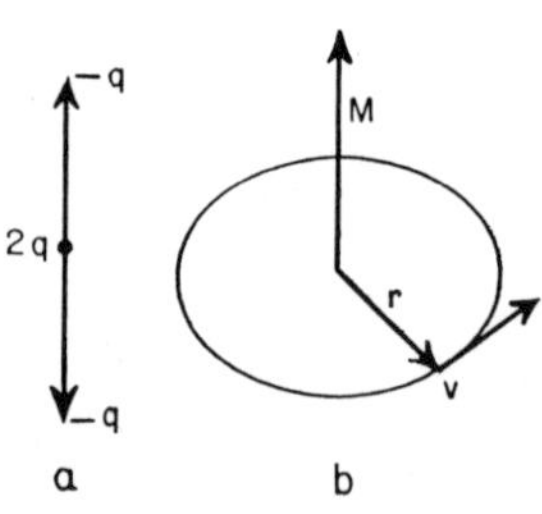

FIG. 4–9. (a) THE ELECTRIC QUADRUPOLE; (b) THE MAGNETIC DIPOLE

Suppose now that instead of oscillating along a straight line the charges move in a closed curved path, as in a loop antenna. Such a closed circuit has the properties of a magnetic dipole and if the charges surge first one way and then the other, the magnetic dipole will change sign and the system will radiate magnetic dipole radiation.

The concepts of electric dipole, electric quadrupole, and magnetic dipole radiation are useful for a theoretical treatment of the radiation of energy by atoms. Unless the electric dipole component of the radiation is missing, the others are of no importance.

On the classical theory of atoms radiating continuously, the energy emitted was simply proportional to the square of the amplitude of the emitted light wave. On the other hand, the quantum theory requires that atoms radiate energy in discrete pulses or *quanta* in which energy and frequency are related by

$$E = h\nu$$

where h is Planck's constant and the frequency ν of the emitted quantum determines its energy.

4–8. Absorption by a Classical Dipole. The simplest classical model of a radiating atom is a dipole wherein a charge ε is bound by an elastic restoring force, and is subject to a damping or dissipative force proportional to its velocity.

When such an atom is placed in a light beam, the atomic electron will try to follow the rapidly varying electric field and will bob up and down with the frequency of the light wave. If this frequency lies near the resonance or natural frequency of the atom, the oscillation of the charge may attain considerable amplitude. (In fact, its amplitude is limited only by the rate of dissipation of energy.)

We shall suppose that the electric field of the incident light wave has a sinusoidal variation that can be represented by

$$\mathbf{E} = \mathbf{E}_{0z}\, e^{i\omega t} \tag{4–60}$$

where the $\mathbf{E}$ vector is directed along the z-axis. That is: $\mathbf{E} = E_z$.

Let m denote the mass of the electron, ε its charge, and K the harmonic restoring force constant. There is also a damping force gz' which we assume to be proportional to the velocity. At any instant, the disturbing force acting upon the charge is εE_z. The equation of motion is therefore

$$\underset{\substack{\text{mass times}\\\text{acceleration}}}{mz''} = \underset{\substack{\text{dissipative}\\\text{force}}}{-gz'} - \underset{\substack{\text{restoring}\\\text{force}}}{Kz} + \underset{\substack{\text{external}\\\text{force}}}{\varepsilon E_{0z}\, e^{i\omega t}} \tag{4–61}$$

If $g/m = \gamma$, $K/m = \omega_0^2$, Eq. 4–61 reduces to

$$z'' + \gamma z' + \omega_0^2 z = \frac{\varepsilon E_{0z}}{m}\, e^{i\omega t} \tag{4–62}$$

To solve this equation we assume a particular solution,

$$z = z_0 e^{i\omega t} \tag{4–63}$$

where z_0 may be a complex number. The transient solution is of no interest here as it represents momentary effects when radiation first strikes the atom. When Eq. 4–63 is put in Eq. 4–62 and use is made of the relations

$$\omega = 2\pi\nu \quad \text{and} \quad \omega_0 = 2\pi\nu_0 \tag{4–64}$$

there results

$$z = \frac{\dfrac{\varepsilon}{4\pi^2 m} E_{0z}\, e^{2\pi i\nu t}}{\nu_0^2 - \nu^2 + i\gamma \dfrac{\nu}{2\pi}} \tag{4–65}$$

The complex form of z_0 merely means that the displacement z of the charged particle is not in phase with the incident wave, although it vibrates with the same frequency ν.

With the aid of certain formulas from classical electricity and magnetism, we shall now obtain expressions for the index of refraction and the absorption coefficient of the gas.

In any medium the velocity of light is

$$v = \frac{c}{n} = \frac{c}{\sqrt{\epsilon}} \tag{4-66}$$

where n is the index of refraction which may be complex, and ϵ is the dielectric constant. In free space $\epsilon = 1$.

The concept of a dielectric constant is usually introduced in connection with Coulomb's law. If two charges q_1 and q_2 are separated by a distance r in a homogeneous medium whose extent is very large compared with r, the electrostatic force between them will be

$$F = \frac{q_1 q_2}{\epsilon r^2}$$

If q_1 and q_2 are measured in esu and r in centimeters, F will be in dynes.

Now ϵ is related to the *polarizability* of the medium. Imagine a non-conducting slab placed between the plates of a condenser. The field between the plates tends to distort the atoms in such a way that the negative charges are pulled towards the positive plate and the positive charges towards the negative plate. An actual flow of electricity cannot take place in the medium; what happens is that the charges bound in the atoms are displaced from their normal positions. Although the material as a whole must remain neutral, the surfaces of the slab will become charged. The surface of the non-conducting slab near the positive condenser plate acquires an excess of negative charge while the other surface acquires a positive charge. These induced surface charges cancel some of the lines of force that originate on the condenser plate with the result that the field within the slab is weakened (see Fig. 4–10).

Under normal circumstances, the negative charges in an atom are symmetrically distributed about the nucleus, but if the atom is placed in an electric field, the electrons tend to be pushed one way and the positively charged nucleus the other. For small distortions the amount of this shift is proportional to the field. Let r denote the distance between the center of gravity of the positive charge and that of the negative charge. If two opposite charges, $+\varepsilon$ and $-\varepsilon$, are separated by a distance r, they are said to constitute a dipole of moment

$$\mathbf{p} = \varepsilon \mathbf{r} \tag{4-67}$$

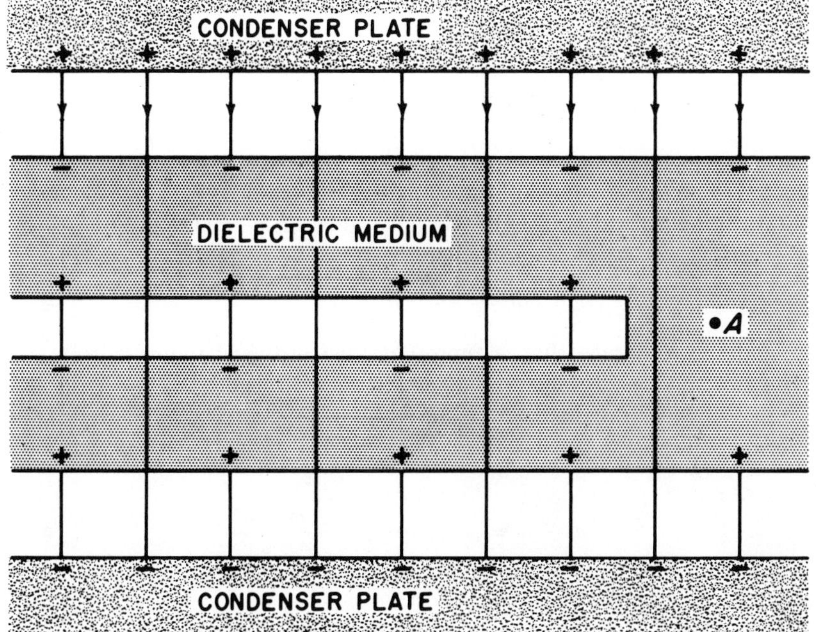

FIG. 4-10. ELECTRIC FIELD IN A DIELECTRIC

The positive direction of r is taken from plus to minus. Since $\mathbf{p}$ depends on the electric field we can write

$$\mathbf{p} = \alpha \mathbf{E}$$

where the constant of proportionality, α, is called the *susceptibility* of the medium. If there are N_ε electrons cm^{-3} the surface charge induced on the edge of the dielectric or on the walls of a slot cut parallel to the condenser plates will be $N\varepsilon r = Np$. At a point A in the dielectric, the electric field will be E, but within the slot it will be $D = \epsilon E$.

Since each charge ε contributes $4\pi\varepsilon$ lines of force, a surface charge $N\mathbf{p}$ will contribute $4\pi N\mathbf{p}$ lines of force. Hence the field within the slot will be $\mathbf{E} + 4\pi N\mathbf{p}$ and hence

$$\epsilon\mathbf{E} = \mathbf{E} + 4\pi N\mathbf{p} = \mathbf{E}(1 + 4\pi N\alpha), \qquad \epsilon = 1 + 4\pi N\alpha \tag{4-68}$$

where N is the number of electrons cm^{-3}. Since z is the net shift between the two charges comprising the dipole, $p = \alpha E_z = \varepsilon z$, and from Eqs. 4-60 and 4-65 we get

$$\alpha = \frac{\varepsilon z}{E} = \frac{\dfrac{\varepsilon^2}{4\pi^2 m}}{\nu_0^2 - \nu^2 + i\gamma\dfrac{\nu}{2\pi}} \tag{4-69}$$

From Eqs. 4–68 and 4–69 the dielectric constant is

$$\epsilon = 1 + \frac{\dfrac{N\varepsilon^2}{\pi m}}{\nu_0^2 - \nu^2 + i\dfrac{\nu}{2\pi}\gamma} \tag{4–70}$$

The physical significance of the complex dielectric constant is that the medium not only refracts light but also absorbs it. The complete index of refraction may be represented by

$$\tilde{n}^2 = \epsilon = (n - i\kappa)^2 \tag{4–71}$$

We shall demonstrate that n represents the ordinary index of refraction of the gas and κ represents the absorptivity. For gases, n is of the order of unity, so we can expand Eq. 4–70 by the binomial theorem and neglect all terms beyond the second,

$$n - i\kappa = 1 + \frac{N\dfrac{\varepsilon^2}{2\pi m}}{\nu_0^2 - \nu^2 + i\dfrac{\gamma}{2\pi}\nu} \tag{4–72}$$

We may reduce this formula to real and imaginary parts by multiplying the second term on the right by its complex conjugate. Then

$$n = 1 + \frac{N\varepsilon^2}{2\pi m}\frac{\nu_0^2 - \nu^2}{(\nu_0^2 - \nu^2)^2 + \left(\dfrac{\gamma}{2\pi}\right)^2\nu^2}$$

$$\kappa = \frac{N\varepsilon^2}{2\pi m}\frac{\nu\left(\dfrac{\gamma}{2\pi}\right)}{(\nu_0^2 - \nu^2)^2 + \nu^2\left(\dfrac{\gamma}{2\pi}\right)^2} \tag{4–73}$$

For many astrophysical applications we are interested in the behavior of the ordinary index of refraction n and the imaginary component κ near the resonance frequency ν_0. When $\nu \sim \nu_0$ we may put

$$\nu_0^2 - \nu^2 \sim 2\nu(\nu_0 - \nu)$$

and we obtain the following adequately accurate expressions

$$n = 1 + \frac{N\varepsilon^2}{4\pi m\nu}\frac{\nu_0 - \nu}{(\nu_0 - \nu)^2 + \left(\dfrac{\gamma}{4\pi}\right)^2} \tag{4–74}$$

$$\kappa = \frac{N\varepsilon^2}{8\pi m\nu}\frac{\dfrac{\gamma}{2\pi}}{(\nu_0 - \nu)^2 + \left(\dfrac{\gamma}{4\pi}\right)^2} \tag{4–75}$$

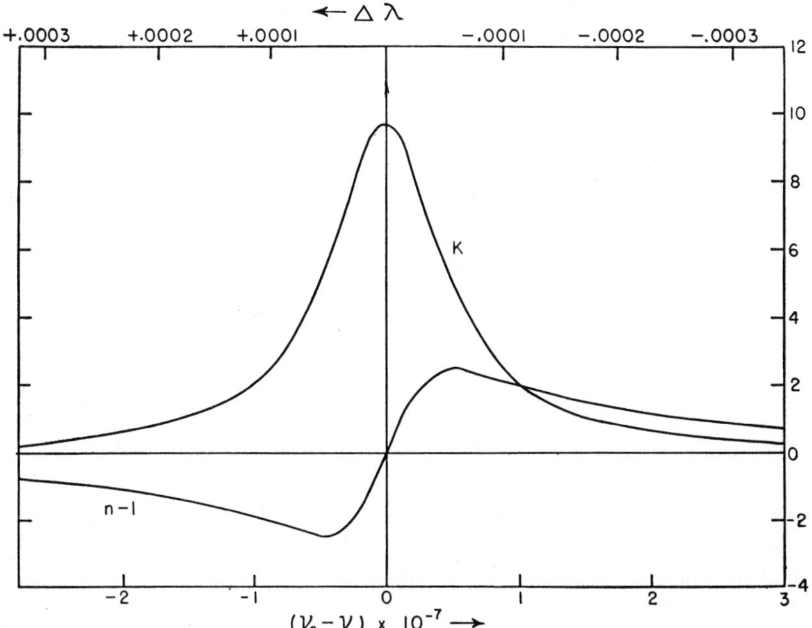

FIG. 4–11. THE VARIATION OF $(n - 1)$ AND κ WITH FREQUENCY NEAR RESONANCE

The curves are computed for sodium atoms absorbing at the λ5889 (D_1) line. Notice the anomalous behavior of the index of refraction near resonance. The ordinates are arbitrary.

The dependence of $n - 1$ and κ upon ν near the resonance frequency ν_0 is shown in Fig. 4–11. Notice that at ν_0, $n - 1$ changes sign while κ reaches a maximum. This behavior of n in the neighborhood of a resonance frequency is referred to as *anomalous dispersion*.

We shall now show how κ is related to the absorption coefficient, k, as defined in Eq. 4–32. Suppose a light beam of electric field $\mathbf{E}$, initial intensity I_0 and amplitude E_0 falls normally upon a slab of absorbing material. Let x denote the direction in which the beam is moving. Then E is a function of both x and t. It can be shown that in free space both E and H satisfy the wave equation. In particular, E satisfies

$$\frac{\partial^2 E}{\partial x^2} = \frac{1}{v^2}\frac{\partial^2 E}{\partial t^2} \tag{4–76}$$

where v is the *phase* velocity of light. A solution which satisfies the requirements (i.e., boundary conditions) of our problem is

$$E = E_0\, e^{2\pi i\nu\left(t - \frac{x}{v}\right)} \tag{4–77}$$

which is the equation of a wave traveling in the positive x-direction.

Since the velocity v is given by

$$v = \frac{c}{\tilde{n}}$$

where $\tilde{n}$ is a complex index of refraction, i.e., $\tilde{n} = n - i\kappa$, we have

$$E = E_0 \, e^{-2\pi\nu\kappa x/c} \, e^{2\pi i\nu \left(t - n\frac{x}{c}\right)} \tag{4–78}$$

which means that the amplitude dies away according to exp $(-\,\omega\kappa x/c)$. Since intensity varies as the square of the amplitude, we have

$$I = I_0 \, e^{-(4\pi\nu\kappa/c)x} = I_0 \, e^{-k\rho x} \tag{4–79}$$

where k is the *mass absorption coefficient* and ρ is the density.

4-9. Classical Atomic Absorption Coefficient. From Eqs. 4–75 and 4–79 we may now write down an expression for the absorption coefficient per unit volume, $k_\nu\rho$, viz.,

$$k_\nu\rho = \frac{4\pi\nu\kappa}{c} = \frac{N_0\varepsilon^2}{2mc}\,\frac{\gamma}{2\pi}\,\frac{1}{(\nu - \nu_0)^2 + (\gamma/4\pi)^2} \tag{4–80}$$

where N_0 is the number of bound electrons per unit volume with a natural frequency ν_0.

Calculations according to the quantum theory lead to this same formula with a somewhat modified meaning of γ and with N_0 replaced by $N_0 f$. Here N_0 is the number of atoms cm^{-3} capable of absorbing the frequency ν_0 and f is called the *oscillator strength*, the *Ladenburg f*, or *number of dispersion electrons of frequency ν_0 per atom.* For example, for the first few members of the Balmer series the f values are

$$\text{H}\alpha = 0.6408, \qquad \text{H}\beta = 0.1193, \qquad \text{H}\gamma = 0.0447, \qquad \text{H}\delta = 0.0221$$

according to data by Menzel and Pekeris (1936). Notice how f progressively decreases with increasing series number.[2]

From Eq. 4–80 we may compute the total amount of energy absorbed by the atom in a unit volume if I_ν does not vary appreciably with frequency near ν_0. If α_ν is the absorption coefficient per atom at frequency ν, then $k_\nu\rho = N_{\nu 0}\,\alpha_\nu$. The total amount of isotropic radiation absorbed in a unit solid angle per second in a range $d\nu$ will be $N_{\nu 0}\,\alpha_\nu I_\nu d\nu$. The total amount of energy absorbed per second over all solid angles per cm^3 per second will be

$$4\pi I_\nu N_{\nu_0} \int_0^\infty \alpha_\nu \, d\nu = 4\pi I_\nu \, \bar{k}\rho$$

We can write with sufficient accuracy

$$\int_0^\infty \alpha_\nu \, d\nu = \frac{\pi\varepsilon^2}{mc} \int_{-\infty}^{+\infty} \frac{(\gamma/4\pi^2)}{(\nu - \nu_0)^2 + (\gamma/4\pi)^2} \, d(\nu - \nu_0) = \frac{\pi\varepsilon^2}{mc} \tag{4–81}$$

[2] For the strongest doublets of alkali atoms f may be close to 1, e.g., for the two D lines of sodium, the sum of the f values is 0.98.

since the integral is of the form

$$\frac{1}{\pi}\int_{-\infty}^{+\infty} \frac{a}{x^2 + a^2}\, dx = \frac{1}{\pi} \tan^{-1} \frac{x}{a}\Big]_{-\infty}^{+\infty} = 1$$

If there are f oscillators, i.e., f electrons oscillating at a frequency ν_0, per atom, then

$$\int \alpha\, d\nu = \frac{\pi \varepsilon^2}{mc} f \qquad\qquad \bar{k} = \frac{\pi\varepsilon^2}{mc}\, n_0 f \tag{4–82}$$

where n_0 is the number of atoms with a natural frequency ν_0 per gram of material. Since an atom emits at the same frequencies as it absorbs, our formula for $\bar{k}$ tells us the frequency-intensity character of a spectral line emitted by a classical oscillator.

The qualitative difference between natural broadening as expressed in Eq. 4–80 and Doppler broadening (Chapter 3) will be apparent from the character of the formulas. At distances $\nu_0 - \nu$ from the line center much larger than $\gamma/4\pi$, the absorption coefficient in a naturally broadened line falls off as $(\nu_0 - \nu)^{-2}$ while that of an intrinsically narrow line broadened only by the thermal motions of the radiating atoms falls off as $\exp(-C(\nu - \nu_0)^2)$. (See Chapter 7.)

The shape of the line absorption coefficient for natural broadening depends on the value of the classical damping constant γ. The electromagnetic theory gives the rate at which the dipole emits energy and hence the speed with which the oscillations of a dipole of initial amplitude r_0 die down. This decay takes place exponentially and the rate of the decay will determine γ.

According to classical theory an accelerated electron of change ε will radiate energy at a rate [3]

$$\frac{dW}{dt} = -\frac{2}{3}\frac{\varepsilon^2 (z'')^2}{c^3} \tag{4–83}$$

where z'' is the acceleration. Although z'' is a vector, the scalar $(z'')^2$ is always positive unless $z'' = 0$. Hence, whatever the direction of the acceleration the moving charge will lose energy. One may show (see Problem 4–6) that the mean rate of radiation by a classical oscillator is

$$\overline{\frac{dW}{dt}} = -\frac{16\pi^4\nu^4}{3c^3} P^2 = -\frac{8\pi^2\nu^2\varepsilon^2}{3mc^3} W \tag{4–84}$$

where W is the energy, z_0 is the amplitude of displacement, and $P = \varepsilon z_0$. Therefore the energy of the oscillator will die away according to

$$W = W_0 e^{-\gamma t}$$

[3] See, for example, F. K. Richtmyer and E. H. Kennard, *Introduction to Modern Physics* (New York: McGraw-Hill Book Co., Inc., 1942), p. 73.

where

$$\gamma = \frac{8\pi^2\nu^2\varepsilon^2}{3mc^3} = \frac{0.2223}{\lambda^2} \text{ sec}^{-1} \ (\lambda \text{ in cm}) \tag{4–85}$$

is the classical damping constant. We may write $\gamma = 1/T$, where $T = 4.50\lambda^2$ is called the *mean lifetime* of the radiating system. It is the time required for the energy of the oscillating dipole to fall to $1/e$ of its maximum value. This is the γ in the equation

$$z'' + \gamma z' + \omega_0^2 z = 0 \tag{4–86}$$

which is obtained from Eq. 4–62 when the external force is zero. The solution is

$$z = z_0 e^{-\frac{1}{2}\gamma t} \cos \omega t + \text{const} \tag{4–87}$$

From Eq. 4–84 we may compute the law of scattering for particles whose radii a are much smaller than the wavelength of the incident light, λ. Such particles can be regarded as containing a multitude of oscillating electric dipoles all of which are in phase. Let P denote the electric moment per dipole. Then the total electric moment of a particle is proportional to its volume, i.e., the scattering or dissipation of energy will be proportional to the (volume)^2 or a^6. Then the scattering coefficient S is proportional to

$$\frac{a^6}{\lambda^4} \text{ or to } \pi a^2 \left(\frac{a}{\lambda}\right)^4 \tag{4–88}$$

As an example consider absorption by the two D lines of sodium. Since the mean wavelength of the two lines is $\lambda = 5.893 \times 10^{-5}$ cm it follows from Eq. 4–85 that:

$$\gamma \doteq 6.403 \times 10^7 \text{ sec}^{-1}$$

Now calculate the absorption coefficient per gram of sodium atoms at a distance of 2A from the center of λ5889 (D_1). Here Doppler broadening is not important. From Eq. 4–80, noting that

$$(\nu_0 - \nu) = 1.729 \times 10^{11} \text{ sec}^{-1}$$

and

$$N = \frac{1}{23 \times 1.660 \times 10^{-24}} = 2.619 \times 10^{22} \text{ atoms gm}^{-1}$$

we find that $k = 3.764 \times 10^4 f$. Since the f value for the D_1 line is $\frac{2}{3}$, the absorption coefficient $k(\Delta\lambda = 2\text{A}) = 2.51 \times 10^4$ per gram of neutral sodium.

We may use this result to make a rough estimate of the amount of neutral sodium above the photosphere of the sun. At a distance of 2A from the center of λ5889, the depression of the line below the continuum is about 1 per cent. If we make the naïve assumption that the light of the continuous spectrum is simply extinguished by the sodium atoms, the actual amount of material may be estimated with the aid of Eq. 4–33.

$$\frac{I}{I_0} = e^{-k\overline{\rho h}} \sim 1 - k\overline{\rho h} = 0.99, \text{ or } k\overline{\rho h} = 0.01$$

Here $\overline{\rho h}$ is the amount of neutral sodium above the photosphere, i.e.,

$$\overline{\rho h} = 4 \times 10^{-7} \text{ gm cm}^{-2}$$

Even if we allow for the fact that at a temperature 5700°K and $P_\varepsilon = 20$ dynes, 99.92 per cent of all sodium is ionized, the total amount of this metal above the photosphere of the sun would be of the order of 0.5 mg cm^{-2}.

4–10. Rayleigh and Thomson Scattering. We derived Eq. 4–73 from a consideration of light scattering. An electromagnetic wave of frequency ν_0 falling upon a charged particle sets it in motion with the same frequency. The vibrating particles then act as a new source of waves of the same frequency ν_0. The energy of these new waves is supplied at the expense of that of the incident wave. We say that the incident wave is *scattered.*

Scattering at frequencies considerably different from resonance frequencies is often important. Two limiting examples are of particular interest:

1. The frequency of the light is much lower than the resonance frequency, i.e., $\nu << \nu_0$ as in the scattering of sunlight by air molecules.
2. The frequency of the incident light is much greater than the resonance frequency, e.g., the scattering of x-rays by the outer bound electrons of an atom or scattering by free electrons.

If we put Eq. 4–85 in Eq. 4–73 (and make use of Eq. 4–79) we find for $(\nu - \nu_0) >> \gamma$

$$k\rho = \frac{4\pi\nu\kappa}{c} = \frac{8\pi N\varepsilon^4}{3m^2c^4} \frac{1}{\left[\left(\frac{\nu_0}{\nu}\right)^2 - 1\right]^2} \tag{4–89}$$

where N is the number of scattering electrons per cm^3 and $k\rho$ is thus the scattering coefficient per cm^3. When $\nu_0 >> \nu$, the scattering coefficient per cm^3 is

$$k\rho = \frac{8\pi N\varepsilon^4}{3m^2c^4} \frac{\nu^4}{\nu_0{}^4} = \left[\frac{8\pi N\varepsilon^4\lambda_0{}^4}{3m^2c^4}\right]\frac{1}{\lambda^4} \tag{4–90}$$

This is the Rayleigh scattering formula. When $\nu_0 << \nu$, Eq. 4–89 gives

$$k\rho = \frac{8\pi N\varepsilon^4}{3m^2c^4} = 0.6655 \times 10^{-24} N_\varepsilon \quad \text{(Thomson formula)} \tag{4–91}$$

which is the formula for scattering by free electrons. Note that electron scattering is independent of wavelength, all colors being treated impartially, whereas with Rayleigh scattering there is a strong color dependence. Both types of scattering are *non-isotropic.* Consequently, sunlight scattered by the earth's atmosphere or by the solar corona will be strongly polarized.

4–11. The Quantum Theory of Radiation. The quantum picture of emission and absorption of energy differs markedly from the classical picture for transitions involving bound levels. Instead of escaping as

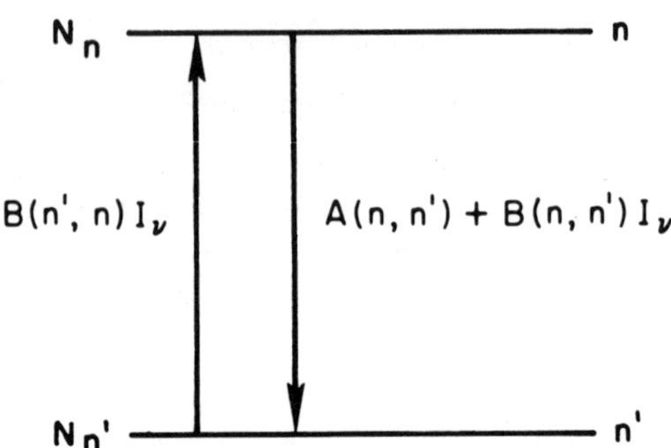

FIG. 4–12. TRANSITIONS FROM A LEVEL n
Absorption, spontaneous emission, and induced emission are indicated.

continuous emission from an accelerated charge, the energy springs from an atom in a discrete packet or quantum when the atom jumps from one energy level to another.

Emission of energy takes place when an atom passes from a level of higher energy to one of lower energy; absorption, when there is a transition from a lower level to a higher one. The quantitative discussion of the emission and absorption of energy may be approached most suitably by means of *Einstein's probability coefficients.* Suppose an atom is initially in an excited state n. See Fig. 4–12. Unless the transition is forbidden, there is a finite probability, $A(n, n')\,dt$, that in a time dt the atom will spontaneously (i.e., without any external influence) jump from level n to level n' with emission of energy, $h\nu = W_n - W_{n'}$. In other words, the Einstein coefficient $A(n, n')$ expresses the probability that, in a unit time, the atom will undergo the spontaneous downward transition from n to n' with emission of energy. The total number of downward n—n' transitions in a unit volume in a time dt will be $N_nA(n, n')\,dt$, where N_n is the number of atoms in level n per unit volume.

For example, if 10^8 atoms are maintained in level n by collisions, by absorption of radiation, by recapture from the continuum, etc., and if the A value for the n—n' transition is 10^6, the number of quantum emissions per second will be $10^8 \times 10^6 = 10^{14}$. That is, if the only permitted downward transition is n—n', a particular atom will remain in level n only 10^{-6} sec. We say that the level has a radiative lifetime of 10^{-6} sec.

In actual practice, the A values are of the order of 10^8 or 10^9 sec^{-1}; for the first member of the principal series and in higher members they may be as little as 10^3. For intercombination lines, the A values may be even smaller.

If the atom is exposed to radiation, two additional phenomena occur. For the moment, let us suppose the radiation to be isotropic and Planckian or "black body" in character. If we wish, we can think of the processes as taking place in an enclosure at some temperature T. The atoms will absorb energy. Transitions from a level n' to a higher level n will occur at a rate proportional to the number $N_{n'}$ in the level n' and to the intensity I_ν of radiation of frequency $\nu_{nn'} = (W_n - W_{n'})/h$. See Fig. 4–12. Thus the number of upward transitions per cm³ per sec will be

$$N_{n'}\, B(n', n)\, I_\nu(n'n)$$

where $I_\nu\,(n'n)$ is given by Planck's law and $B(n',n)$ is the *Einstein coefficient of absorption*. A second effect of the radiation upon the atom is not so obvious — we refer to the *induced emissions* (more correctly called (*negative absorptions*). If an atom is in an excited level n and radiation of frequency $\nu(n, n')$ corresponding to the permitted n—n' transition falls upon it, the likelihood of its cascading to level n' with emission of $\nu(n, n')$ is increased by an amount dependent upon the intensity of the incident light.

Hence we define a *coefficient of negative absorption* (induced emission) $B(n, n')$, such that the number of induced emissions of atoms from level n to n' per unit time and volume will be $B(n, n')I_\nu N_n$. The total number of atoms leaving n for n' in a time dt per unit volume will be

$$\text{(spontaneous plus induced emission)} = N_n[A(n, n') + B(n, n')\, I_\nu]\, dt \tag{4–92}$$

We shall avoid confusion if we call the induced emissions negative absorptions, since the induced radiation is returned in the beam in the same direction and phase as the incident radiation. It is not emitted in a random direction as are the spontaneously emitted quanta.

Under conditions of thermal equilibrium, the relative populations of the two levels n and n' is given by Boltzmann's formula (see Chapter 3).

$$\frac{N_n}{N_{n'}} = \frac{g_n}{g_{n'}} e^{-\chi_{nn'}/kT} \tag{4–93}$$

where $\chi_{nn'} = h\nu_{nn'} = W_n - W_{n'}$ is the energy difference between the two levels n and n', and g_n and $g_{n'}$ are the statistical weights of the two levels, $2J + 1, 2J' + 1$, where J and J' are the respective total angular momentum quantum numbers of level n and n'.

By the *principle of detailed balancing*, the number of downward transitions from level n to n' must equal the number of upward transitions from n' to n, viz.,

$$N_n[A(n, n') + B(n, n')\, I_\nu] = N_{n'}\, B(n'n)\, I_\nu \tag{4–94}$$

Let us henceforth write $\nu(nn')$ as $\nu_{nn'}$ or ν and $B(nn')$ as $B_{nn'}$, etc. If we now use Eq. 4–93 and recall that under conditions of thermal equilibrium I_ν is given by the Planck formula Eq. 4–10, we have

$$\frac{g_n}{g_{n'}} A_{nn'} = \frac{2h\nu^3}{c^2} B_{n'n} \frac{\left[e^{h\nu/kT} - \dfrac{g_n}{g_{n'}}\dfrac{B_{nn'}}{B_{n'n}}\right]}{e^{h\nu/kT} - 1} \tag{4–95}$$

In order for $A_{nn'}$, $B_{n'n}$ and $B_{nn'}$ to be independent of the temperature, i.e., constants of the atom only, the following relations must hold

$$g_n B_{nn'} = g_{n'} B_{n'n} \tag{4–96}$$

and

$$\frac{g_n}{g_{n'}} A_{nn'} = B_{n'n} \frac{2h\nu^3}{c^2} \tag{4–97}$$

We emphasize that the Einstein coefficients are atomic constants, which may be determined for a given transition by experiment or by quantum mechanical calculation. If one coefficient is known, the others may be found from Eqs. 4–96 and 4–97. Although we have considered conditions of thermal equilibrium in order to derive the relations between these atomic constants, it is important to realize that the same relations, Eqs. 4–96 and 4–97, will hold under all conditions.[4]

In the spectral region where $h\nu/kT >> 1$, spontaneous emissions are much more important than induced emissions, whereas the reverse is true when $h\nu/kT << 1$. In the ultraviolet spectral region, we may replace I_ν by the Wien approximation to Planck's law, Eq. 4–16. Then the number of induced emissions per cm³ per sec will be

$$B_{nn'} I_\nu = B_{nn'} \frac{2h\nu^3}{c^2} e^{-h\nu/kT} = A_{nn'} e^{-h\nu/kT} \ll A_{nn'} \tag{4–98}$$

since $h\nu/kT << 1$. Hence the induced emissions may be neglected in comparison with the spontaneous emissions.

In the far infrared or microwave region $h\nu/kT << 1$ and Planck's law may be replaced by the Rayleigh-Jeans approximation (Eq. 4–18):

$$B_{nn'} I_\nu = B_{nn'} \frac{2\nu^2 kT}{c^2} = \frac{c^2}{2h\nu^3} A_{nn'} \frac{2\nu^2 kT}{c^2} = A_{nn'} \frac{kT}{h\nu} \gg A_{nn'} \tag{4–99}$$

Hence the number of negative absorptions is much greater than the number of spontaneous emissions.

4–12. The Absorption Coefficient According to Quantum Theory. By quantum mechanics Weisskopf and Wigner derived the rigorous formulas for the shape of the line absorption coefficient. Their expression is identical with our classical formula if we replace N_{ν_0} by $N_{\nu_0} f$ and γ by a new constant Γ, viz.,

$$k_\nu \rho = \frac{\pi \varepsilon^2}{mc} N_{\nu_0} f \frac{\Gamma}{4\pi^2(\nu - \nu_0)^2 + \left(\frac{\Gamma}{2}\right)^2} \tag{4–100}$$

where the value of Γ must be computed on the basis of the following considerations. Consider an atom initially in a level n. Suppose it can go

[4] Many authors derive the relations for energy density and hence obtain a different form for Eq. 4–97. See Problem 4–2.

to a lower level n'. If the negative absorptions can be neglected, the rate at which atoms in this upper level descend to the lower ones is proportional to the sum of the A values of the corresponding transitions, thus

$$dN_n/dt = -N_n \sum A_{nn'} = -N_n\Gamma_n \tag{4–101}$$

$$\Gamma_n = \sum A_{nn'} \tag{4–102}$$

By integration we obtain

$$N_n = N_n{}^0 e^{-\Gamma_n t} \tag{4–103}$$

Now Γ_n is simply the reciprocal of the mean lifetime T_n of an atom in the level n, i.e.,

$$\Gamma_n = 1/T_n \tag{4–104}$$

where $N_n{}^0$ denotes the number of atoms per unit volume in level n at $t = 0$. If the lower level of the transition is the ground level or a metastable level of very long life we can write $\Gamma = \Gamma_n$. If the lower levels have short lifetimes $\Gamma_{n'} = 1/T_{n'}$ may be comparable with Γ_n itself. A rigorous treatment of the problem shows that the absorption coefficients for subordinate lines require a damping constant given by

$$\Gamma_{nn'} = 1/T_n + 1/T_{n'} = \Gamma_n + \Gamma_{n'} \tag{4–105}$$

or the sum of the reciprocal lifetimes of the upper and lower levels. Hence as soon as the A's are known, we can calculate the broadening factor for pure radiation damping.

At high temperatures, the radiation density becomes large and the atoms in a level n may escape in considerable numbers by negative absorptions, excitations to higher levels, and by ionization as well as by spontaneous emission. Hence these additional processes will play a role in fixing Γ_n and we must write:

$$\Gamma_n = \sum_{n'} A_{nn'} + \sum_{n'} B_{nn'} I(\nu_{nn'}) + \sum_{n''} B_{nn''} I(\nu_{nn''}) \tag{4–106}$$

where the first term refers to spontaneous transitions to lower levels, the second to negative absorptions to lower levels, and the third to ordinary absorption processes including photo-ionizations. In the above discussion we have neglected collisional processes. One may further remark that in the far infrared and microwave regions the lifetimes of a level may be determined by the negative absorptions rather than by the spontaneous emissions. Now Eq. 4–101 is replaced by

$$\frac{dN_n}{dt} = -N_n \left\{ A_{nn'} + B_{nn'} \frac{2kT\nu^2}{c^2} \right\} = -N_n A_{nn'} \left[1 + \frac{kT}{h\nu} \right] \tag{4–107}$$

where we have used Eqs. 4–18, 4–96 and 4–97. Since $kT/h\nu >> 1$, the lifetime (compare Eqs. 4–102 and 4–104) is now given by

$$T_n \sim \frac{h\nu}{kT} \frac{1}{A_{nn'}} \tag{4–108}$$

If the intensity of the radiation is not Planckian but is reduced by some factor W, the lifetime will be increased by $\sim 1/W$.

4–13. The Interpretation of Natural Line Broadening. It is evident that the process of line broadening must occur in different ways in the classical and quantum theories.

Classical broadening occurs because a bound electron can oscillate in frequencies on either side of resonance, much as a radio set will receive a signal even if it is not tuned precisely to the output frequency of the broadcasting station.

There is another way of looking at classical broadening. A radiating atom can be regarded as emitting a damped wave which gradually dies out, or a wave of fixed frequency ν for a short time interval. If either kind of wave is subjected to a Fourier analysis it is found that it is not monochromatic but consists of a narrow range of frequencies whose intensity distribution is given by an expression similar to Eq. 4–80. That is, if wave trains from a classical oscillator were analyzed by a spectrograph of infinite resolving power, each spectrum line would give a broadened profile of a shape depicted in Fig. 4–11.

Although, in the quantum picture, each emitted quantum has a precise energy E and frequency ν, the atomic energy levels themselves are actually slightly fuzzy. According to the Heisenberg uncertainty principle, if

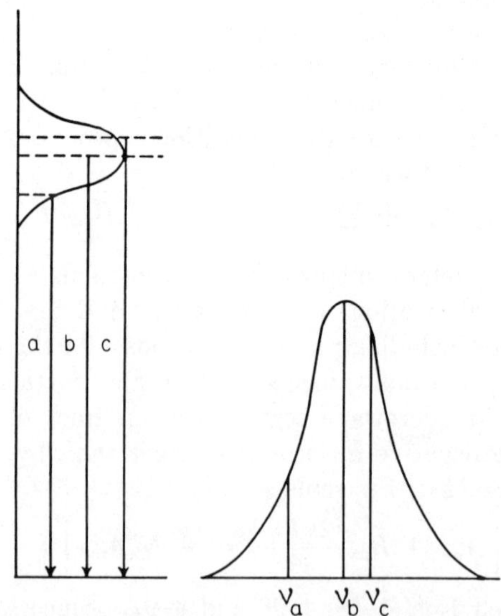

FIG. 4–13. THE RELATION BETWEEN LEVEL BREADTH AND LINE PROFILE FOR A RESONANCE LINE

The transitions from three "substates" of the broadened upper level, denoted by a, b, and c, are correlated with three frequencies ν_a, ν_b, and ν_c in the resonance line.

an atom in a certain level has a lifetime, Δt, the level necessarily will have an energy breadth ΔE, such that

$$\Delta E \, \Delta t \sim h \tag{4–109}$$

The lifetime of a ground level is very long; hence ΔE is very small and the level is quite sharp. An excited level for which $A = 10^8$ has $\Delta t \sim 1/A = 10^{-8}$ which is small. Hence ΔE will be relatively large. Since the transition may take place from any part of the broadened level, the observed spectral line will be slightly widened (see Fig. 4–13). Unlike lines emitted by classical oscillators, whose half-breadths as measured on a wavelength scale are always the same, the natural breadth of a real spectral line will depend on Γ. Lines of high transition probability will be intrinsically broader than those of low transition probability. See Fig. 4–14.

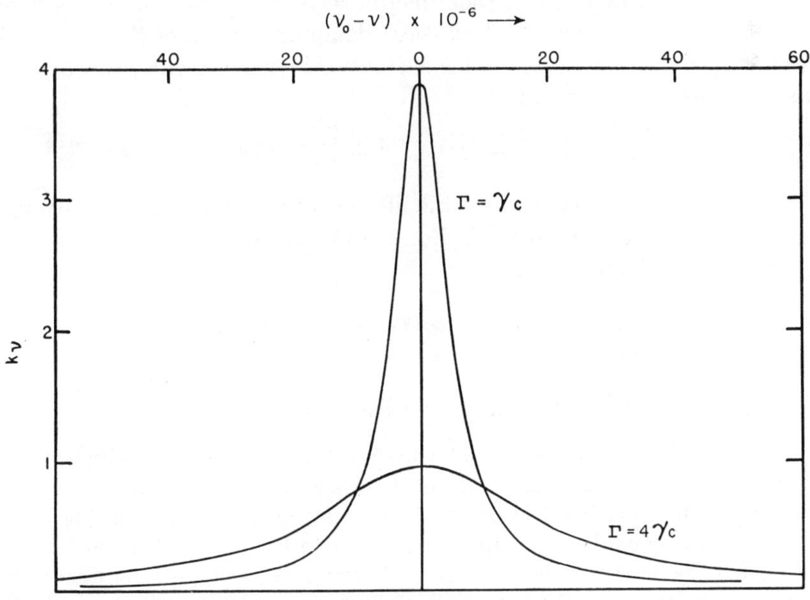

Fig. 4–14. Effect of the Damping Constant upon a Line Absorption Coefficient

4–14. The Relation Between the Einstein Coefficients and Oscillator Strengths. The relation between the f's and the Einstein A's or B's is useful for some problems. The energy absorbed per cm³ per sec by atoms in level n' is

$$N_{n'} B_{n'n} I_\nu h\nu \tag{4–110}$$

and this must equal the energy absorbed as computed in terms of the conventional absorption coefficient, viz., (see Art. 4–9)

$$N_{n'} \bar{\alpha}_\nu \, 4\pi \, I_\nu \, \overline{\Delta\nu} = 4\pi I_\nu N_{n'} \frac{\pi \varepsilon^2}{mc} f_{n'n} \tag{4–111}$$

where we have averaged the absorption coefficient α_ν over a spectral line of effective width $\overline{\Delta\nu}$. Hence

$$B_{n'n} = \frac{\pi \varepsilon^2}{mc} f_{n'n} \frac{4\pi}{h\nu} \tag{4–112}$$

and

$$A_{nn'} = \frac{g_{n'}}{g_n} \frac{8\pi^2 \varepsilon^2 \nu^2}{mc^3} f_{n'n} \tag{4–113}$$

or

$$A_{nn'} = 3 \frac{g_{n'}}{g_n} f_{n'n} \gamma_c = 3 \frac{(2J' + 1)}{(2J + 1)} f_{n'n} \gamma_c \tag{4–114}$$

$$f_{n'n} = 1.5 \times 10^{-8} \lambda^2 \frac{g_n}{g_{n'}} A_{nn'} \tag{4–115}$$

if λ is expressed in microns (one micron = 10^4A). The relation between the quantum mechanical and classical damping constants Γ_n and γ_c will be used in later discussions. For the situation in which negative absorptions can be neglected, we can write

$$\Gamma_n = \sum_{n'} A_{nn'} = 3 \sum \frac{g_{n'}}{g_n} f_{n'n} \gamma_c \tag{4–116}$$

For the more general case where the lifetime of the excited level is strongly influenced by a radiation field, Eq. 4–106 may be transformed by Eqs. 4–10, 4–96, 4–97 and 4–113 to

$$\Gamma_n = \frac{8\pi^2 \varepsilon^2}{mc} \left\{ \sum \frac{g_{n'}}{g_n} \lambda^{-2} f_{n'n} (1 - e^{-h\nu/kT})^{-1} + \sum \lambda^{-2} f_{nn''} (e^{h\nu/kT} - 1)^{-1} \right\} \tag{4–117}$$

provided photo-ionizations can be neglected.

Example: The A value for the resonance $\lambda 3933$ $4^2S_{1/2}$—$4^2P_{3/2}$ transition of Ca II is adapted from various theoretical calculations as 1.59×10^8. Hence the corresponding f is $0.738 = \frac{2}{3} \times 1.11$ by Eq. 4–115. One might suppose that the damping constant Γ would be 1.6×10^8 sec^{-1} but reference to a term diagram for Ca II shows that the $3d^2$D level lies below the 4^2P term. Transitions from this strictly metastable level to the 4^2P term give infrared lines at $\lambda 8498$, $\lambda 8542$, and $\lambda 8662$. The A value for the $3d^2D_{5/2}$ — $4p^2P_{3/2}$ $\lambda 8542$ transition is 1.2×10^7, that for the $3d^2D_{3/2}$ — $4p^2P_{3/2}$ $\lambda 8498$ transition is 1.4×10^6. Hence the damping constant for $\lambda 3933$ is

$$\Gamma = 1.59 \times 10^8 + 0.12 \times 10^8 + 0.014 \times 10^8 = 1.72 \times 10^8$$

Hence $\Gamma(\lambda 3933) = 1.19\gamma_c$ ($\lambda 3933$).

Although we shall deal almost exclusively with oscillator strengths as defined above, an emission oscillator strength is sometimes used in applications of the sum rules, thus,

$$f'_{nn'} = -\frac{g_{n'}}{g_n} f_{n'n} \tag{4–118}$$

which is given for completeness. Unless otherwise explicitly specified, whenever f values are mentioned, we refer to the ordinary (absorption) oscillator strength f.

4–15. Thomas-Kuhn Sum Rule and Theoretical f-Values. A few remarks concerning atomic f-values are pertinent at this point. A fuller discussion of this subject particularly with respect to complex atoms is found in Chapter 7. Atomic oscillator strengths obey an important relationship called the Thomas-Kuhn sum rule. If we sum f over all possible transitions between all possible configurations, we should obtain the number of electrons in the atom, viz.,

$$\sum f(nl;\ n'l') = N \tag{4–119}$$

where N is the number of electrons in the atom or ion. In iron for example, the exact sum of f-values over all x-ray and optical transitions will be 26.

If the electrons in the inner shell are so tightly bound that we need not consider interactions between them and the outer (valence) electrons, we can write for any configuration nl

$$\sum f(nl;\ n'l') = r \tag{4–120}$$

where r is the number of valence electrons and the summation is carried out over all levels to which transitions can take place. For downward transitions one most use the emission oscillator strengths defined by Eq. 4–118. If we substitute for these in terms of absorption oscillator strengths, we obtain

$$\sum_{n''} f_{nn''} - \sum \frac{g_{n'}}{g_n} f_{n'n} + \int_0^\infty f_{n\kappa}\, d\kappa = r \tag{4–121}$$

The first term written on the left represents the absorptions from the level n, and the summation is carried out over all discrete levels with $n'' > n$. The second term represents downward transitions, i.e., emissions, and the summation is carried out over all levels with $n' < n$. The third term represents bound-free absorptions in which the electron is lost from the atom. We shall explain the use of Eq. 4–121 with an application to subordinate lines of helium. Care must be exercised in the choice of the value of r, e.g., for helium r is 2 for the principal series, since there are two $1s$ electrons in the ground level and either of these may jump during a transition to a higher level. Since only one of these electrons will exist in a higher level at any one time, the f-values of subordinate lines should sum to unity as for hydrogen.

If the f-values for single lines or multiplets are given, and it is required to find the appropriate f for the whole set of transitions between the two configurations involved, we do not simply form the sum of the f's. Rather, we calculate the sum of the products of the weights of each term and the

f-value of each multiplet arising from it and then divide by the weight of the whole configuration. Thus

$$f(n'l', nl) = \frac{1}{\Sigma g_i} \Sigma g_i f(\alpha' J'; \alpha J) \tag{4–122}$$

where a particular line is represented by $\alpha' J' \rightarrow \alpha J$ and the summation is carried out over all lines. Consider a line of the Balmer series. It is composed of $2(^2S) - n(^2P)$, and $2(^2P) - n(^2S)$, $2(^2P) - n(^2D)$ doublets which coincide to form a single observable line. The weight of ^2S term is 2, that of ^2P term is 6. Hence

$$f_{2n} = \frac{1}{g_l} \sum_{l,l'} g_{2l} f(2, l; n, l') = \frac{1}{8} \left[2f(2, 0; n, 1) + 6f(2, 1; n, 0) + 6f(2, 1; n, 2) \right]$$

To illustrate some of these formulas, let us consider He I whose f values have been computed by Goldberg (1939). Reference to the He I term diagram (Fig. 2–10) will be helpful here.

The f value for the 2^3P—4^3D multiplet giving the unresolved λ4471 line is 0.129; we are to compute the A value. The statistical weights of ^3P and ^3D terms are 9 and 15, respectively. Employing Eq. 4–114 we get $A = 2.59 \times 10^7$.

The damping constant for this line is calculated as follows: Atoms may escape from $4(^3D)$ to $2(^3P)$ and $3(^3P)$ and from the lower $2(^3P)$ level to $2(^3S)$ with the emission of λ10830 (see term diagram). Calculation shows a negligible number of transitions to $3(^3P)$. Hence

$$\begin{aligned} \Gamma_{4471} &= A(4^3D - 2^3P) + A(2^3P - 2^3S) \\ &= (2.59 + 1.03) \times 10^7 = 3.62 \times 10^7 \text{ sec}^{-1} \end{aligned}$$

For the $2^3S - 2^3P$ line, $\Gamma = A(2^3S - 2^3P) = 1.03 \times 10^7$, since the lower level is metastable, i.e., it has an infinite lifetime. Similarly, $A = \Gamma$ for the resonance line $(1^1S - 2^1P)$ for which $A = 2.34 \times 10^9$. Further illustrations are worked out in Chapter 7.

The sums of the f-values provide an interesting illustration of the Thomas-Kuhn sum rule. For the $2^1S - n^1P^0$ series, Goldberg finds the sum of the f-values for discrete transitions, to be 0.666 while Huang (1948) finds the integral of the f-value over the continuum to be 0.402. Hence Σf will be 1.068 instead of 1.00. The f-values for the lines are in need of improvement.

Note that since the 2^1S level is metastable, there are no allowed downward transitions, i.e., $f' = 0$. From the 2^1P level, on the other hand, an atom may escape by going to n^1D or n^1S (with the absorption of energy) or by going to the ground ^1S level with the emission of energy. Goldberg finds

$$f(2^1P - n^1D) = 0.967 \text{ (lines)} + 0.157 \text{ (continuum)} = 1.124$$

Downward jumps to the singlet levels, 2^1S and 1^1S, must be considered. $f(2^1P - 2^1S) = 0.389$, $(2J' + 1)/(2J + 1) = 1/3$, so that the emission oscillator strength $f' = -0.389/3 = -0.130$. Similarly, for the $2^1P - 1^1S$ transition, $f' = -0.116$. Hence we have

$$0.967 + 0.157 - 0.130 - 0.116 = 0.878$$

instead of unity. The discrepancy here arises partly from the fact that Goldberg has not allowed for the 2^1P — n^1S transitions.

For atoms of simple structure it is often possible to calculate reasonably reliable f-values by quantum mechanics. Indeed, for hydrogen exact solutions can be obtained, and the principal results will be quoted here. Reliable results can be obtained for helium and certain transitions in other elements including the alkalis. For complex atoms, approximate results only can be obtained, either by theory or experiment; we defer a discussion of these problems to Chapter 7.

The f-values for hydrogen may be computed exactly by the methods of quantum mechanics. They are given by the formula

$$f_{n'n} = \frac{2^6}{3\sqrt{3}\,\pi}\,\frac{1}{g_{n'}}\,\frac{1}{\left[\dfrac{1}{n'^2} - \dfrac{1}{n^2}\right]^3}\left|\frac{1}{n^3}\,\frac{1}{n'^3}\right| g \tag{4–123}$$

Except for the factor g introduced by Gaunt, Eq. 4–123 was obtained by Kramers who used methods of the old quantum theory. This g-factor, whose precise value is tabulated by Menzel and Pekeris (1936), is not to be confused with $g_{n'} = 2n'^2$, the statistical weight of the lower level. The individual $f(n'l';\, nl)$ oscillator strengths are tabulated for the most important transitions in hydrogen by L. Green *et al.* (1957).

The Einstein A value for a transition between a level characterized by quantum numbers $nlLSJ$ (abbreviated simply as αJ) and one with quantum numbers $n'l'L'J'J'$ (denoted as $\alpha' J'$) may be expressed in terms of a parameter called the strength S_1 of the line:

$$A(\alpha J;\, \alpha' J') = \frac{1}{2J+1}\,\frac{64\pi^4\nu^3}{3hc^3}\,S_1(\alpha J;\, \alpha' J') \tag{4–124}$$

where ν is the frequency of the line, and $g = 2J + 1$ is the statistical weight of the upper level. The absolute strength S_1 depends on both the radial and angular charge distribution within the atom, and consists of two factors, viz.:

$$S_1 = S\sigma^2 \tag{4–125}$$

The A values are small in the radio-frequency region because the ν^3 factor is small, even though the S_1 factors are comparable. Here σ^2 depends on the radial charge distribution in the upper and lower levels. To the first approximation, at least, it will be the same for all lines and multiplets connecting a given pair of configurations n_1l_1 and n_2l_2.

The factor S which does not involve the frequency can be calculated for atoms in pure LS or pure jj coupling. Hence it is possible to obtain the relative strengths of all lines and multiplets belonging to a given pair of configurations $(n_1l_1;\, n_2l_2)$. In Chapter 7 the factor S is found for LS coupling. If, in addition, the σ^2 factor can be calculated or estimated by quantum mechanics, we may obtain the A or f-values for the lines in question. Further discussions of atomic line transition probabilities are given in Chapter 7. The remainder of this chapter shall be devoted to phenomena involving continuous spectra.

4–16. Continuous Absorption Coefficient of Atomic Hydrogen. For a discussion of continuous spectra of the sun and stars, of the gaseous nebulae, and of the physical state of atoms and molecules in interstellar space, it is necessary to consider continuous absorption by atoms and ions.

We have already mentioned Rayleigh and Thomson scattering as examples of processes wherein radiation is simply scattered by electrons. Transitions from one energy state of an atom to another are not involved. In the present context we are interested in processes that may be described by the general term of absorption, such as the photo-ionization of atoms from various discrete levels or the photodissociation of negative ions and molecules. The inverse process is recombination with the release of the excess energy as radiation.

The first important astrophysical example we shall discuss in some detail is continuous absorption and emission of energy by atomic hydrogen. We recall from Chapter 2 that an atom may be photo-ionized while occupying any one of its permitted levels. In a discrete transition, an atom may absorb only a particular frequency ν corresponding to the energy difference between the two levels. In photo-ionization processes, any amount of energy greater than that necessary to detach the electron may be absorbed.

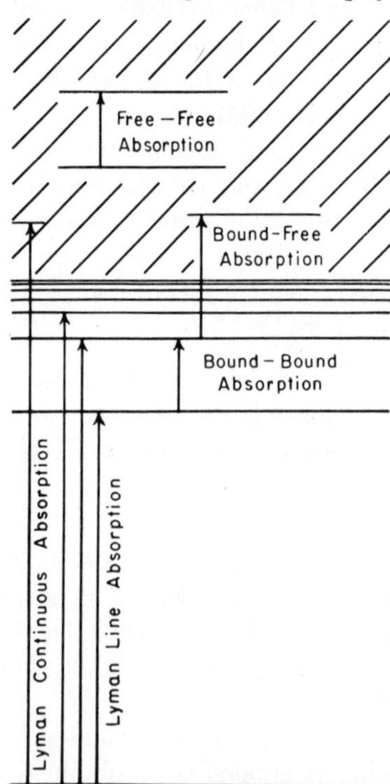

FIG. 4–15. BOUND-BOUND, BOUND-FREE, AND FREE-FREE TRANSITIONS IN HYDROGEN

For example, we have seen that the Balmer lines represent transitions from the second to higher levels. Toward the limit of this series they crowd closer and closer together until they coalesce into a continuum near λ3650. This continuous absorption beyond the Balmer limit corresponds to the photoelectric detachment of electrons from the second level. Similarly, photo-ionizations from the third level produce a continuous absorption starting at the limit of the Paschen series in the infrared and extending through the visual region of the spectrum. In the visual region of the spectra of A- and B-

type stars, the principal contribution to the continuous absorption is made by photo-ionizations from third and higher levels.

In the far ultraviolet there occurs the continuous absorption at the limit of the Lyman series which sets in at a frequency corresponding to 13.60 ev energy, i.e., $\lambda = 912$A, or $\tilde{\nu} = 109{,}679$ cm^{-1}.

Thus there is a continuous absorption associated with each of the hydrogen series, Lyman, Balmer, Paschen, Brackett, etc. In each of these continua, the absorption is strongest at the series limit and falls off gradually to shorter wavelengths. We refer to these photo-ionizations as *bound-free transitions*. See Fig. 4–15. We shall now show how the absorption coefficient per atom in an excited level n' may be computed with the aid of Eq. 4–123 which gives explicitly the f-value for a bound-bound transition. If a hydrogen atom in an excited level n absorbs a quantum of energy $h\nu$ that is greater than the energy necessary to detach the electron from the atom, the liberated electron will fly away with a velocity v given by

$$\frac{1}{2}mv^2 + \frac{hR}{n^2} = h\nu \tag{4–126}$$

since the energy of any level n is $-hR/n^2$ (referred to the ionized atom as zero). Here hR is 13.60 ev, and R is the Rydberg constant in frequency units,[5] $2\pi^2\varepsilon^4 m/h^3$, in the Balmer formula (see Art. 2–3). Then Eq. 2–12 becomes:

$$\nu = RZ^2\left(\frac{1}{n^2} - \frac{1}{n''^2}\right) \quad (n'' > n) \tag{4–127}$$

where Z is the atomic number. For a given series, n is held fixed. Menzel and Pekeris suggested that this formula be applied to the transitions involving the continuum by means of the substitution,

$$n'' \to i\kappa \tag{4–128}$$

where κ is a real but not necessarily integral quantum number. We take the real part of the resultant expression. Thus

$$\nu = RZ^2\left(\frac{1}{n^2} + \frac{1}{\kappa^2}\right) \tag{4–129}$$

is to be compared with ν as given in Eq. 4–126 and κ is defined by means of the relations

$$\frac{hRZ^2}{\kappa^2} = \frac{1}{2}mv^2, \quad \frac{-2hRZ^2}{\kappa^3}\,d\kappa = mv\,dv = h\,d\nu \tag{4–130}$$

We may calculate the absorption coefficient for the continuum on the basis of the following considerations. Since the relation between the

[5] In this and subsequent articles, we follow Menzel's notation and use R expressed in frequency units rather than in wave-number units as in Chapter 2.

absorption coefficient, α_ν, and oscillator strength f is given by Eq. 4–82 we may express α_ν as $(\pi\varepsilon^2/mc)df/d\nu$. The absorption coefficient will be continuous over a series limit. Hence, we must consider the f-value as defined for unit frequency interval. Just to the low-frequency side of a series limit, there will be Δn lines of mean oscillator strength f per unit frequency interval, whereas just on the violet side the f-value per unit frequency interval will be $f\ \Delta\kappa$. Thus the f-value corresponding to the frequency interval $d\nu$ will be

$$df = f\ d\kappa \tag{4–131}$$

Hence

$$\alpha_\nu = \frac{\pi\varepsilon^2}{mc}\frac{df}{d\nu} = \frac{\pi\varepsilon^2}{mc}\frac{df}{d\kappa}\frac{d\kappa}{d\nu} = \frac{\pi\varepsilon^2}{mc}f\frac{d\kappa}{d\nu} \tag{4–132}$$

We use $d\kappa/d\nu$ from Eq. 4–130 to obtain

$$\alpha_\nu = \frac{\pi\varepsilon^2}{mc}f\frac{\kappa^3}{2RZ^2} \tag{4–133}$$

From Eq. 4–123

$$f_{n\kappa} = \frac{2^6}{3\sqrt{3}\pi}\frac{1}{2n^2}\frac{1}{\left(\dfrac{1}{n^2}+\dfrac{1}{\kappa^2}\right)^3}\left|\frac{1}{n^3}\frac{1}{\kappa^3}\right|g \tag{4–134}$$

With the aid of Eqs. 4–129, 4–133 and the definition of R we find from Eq. 4–134:

$$\alpha_n(\nu) = \frac{32}{3\sqrt{3}}\frac{\pi^2\varepsilon^6}{ch^3}\frac{RZ^4}{n^5\nu^3}g \tag{4–135}$$

the absorption coefficient per atom in the nth level. Fortunately, the Gaunt correction factor g is always near unity. Its exact value has been tabulated by a number of workers, e.g., Karzas and Latter (1960). For the Balmer continuum, the absorption coefficient at the series limit amounts to about 1.38×10^{-17} cgs units per atom in the second level. It gradually falls to about half this value at the limit of transmission of the earth's atmosphere, λ2900. Numerically

$$\log \alpha_n(\nu) = 29.449 - 3\log\nu + \log g - 5\log n \tag{4–136}$$

We shall now show in detail how one may calculate the absorption coefficient for atomic hydrogen. We first consider the photo-ionizations from discrete levels, which produce the bulk of the absorption in spectral regions normally accessible to observation. Then we have to add the free-free contributions. Under conditions of thermal equilibrium the number of atoms in level n will be given by Boltzmann's formula (Chapter 3) and the contribution of atoms in level n to the absorption coefficient will be

$$N_n\alpha_n(\nu) = n^2N_1\,e^{-(X_1-X_n)}\,\alpha_n(\nu) \tag{4–137}$$

where N_1 is the number of atoms in level 1, $2n^2$ is the statistical weight of level n

$$X_n = \frac{h\nu_n}{kT_\varepsilon} = \frac{hRZ^2}{n^2kT_\varepsilon} = \frac{158{,}000}{n^2T_\varepsilon} \quad \text{and} \quad \nu_n = \frac{RZ^2}{n^2} \tag{4–138}$$

To compute the total absorption coefficient for a gram of neutral hydrogen at a particular temperature it is necessary to sum over the contributions of all levels that can produce the absorption at the wavelength of interest. At λ4000, for example, transitions from the third and higher levels participate in the continuous absorption. Photo-ionizations from the second level involve only quanta of wavelengths shorter than λ3650 and therefore cannot play any role at λ4000. At λ3600, contributions from $n = 2$ would have to be taken into account, whereas ionizations from the ground level, $n = 1$, need be considered only at wavelengths shorter than λ912. Our procedure will be to calculate the absorption coefficient for each level n by Eq. 4–135 and Eq. 4–136 and then to sum over all levels that can contribute to the absorption at the wavelength in question.

The number of hydrogen atoms per gram will be $1/M$. Here M denotes the mass of one hydrogen atom. The absorption coefficient per gram of neutral hydrogen in the ground level for bound-free transitions is

$$k_1(\nu) = \frac{32}{3\sqrt{3}} \frac{\pi^2\varepsilon^6RZ^4}{M\,ch^3} \frac{e^{-X_1}}{\nu^3} \sum \frac{1}{n^3} e^{X_n} g \tag{4–139}$$

where the summation is taken only over those levels for which ν exceeds ν_n (see Eq. 4–138). For the interval λ8200 – λ3650, for example, we sum over all terms, $n \geq 3$; from λ3650 – λ3000 we sum over all terms, $n \geq 2$, and so on. Unsöld (1955), Pannekoek (1936), Saito, Ueno, and Jugaku (1954), and others, have carried out detailed calculations of $k_1(\nu)$. Unsöld set $g = 1$ throughout and performed the indicated summations for levels 1 to 4, while for the higher levels he replaced the summation by an integral, more generally

$$\sum \frac{1}{n^3} e^{X_n} \rightarrow -\frac{1}{2} \int e^{X_n}\, d\left(\frac{1}{n^2}\right) = -\frac{1}{2X_1} \int e^{X_n} dX_n = \frac{(e^{X_a} - 1)}{2X_1} \tag{4–140}$$

where the summation is extended from X_a to $X_n = 0$.

In addition to the bound-free processes, free-free transitions are important at the higher temperatures. These latter transitions correspond to jumps from one unquantized level to another, and may be visualized in the Bohr picture as follows. A free electron approaching a positive ion along a hyperbolic orbit may emit a quantum of energy and fly away in a second hyperbolic orbit of less energy. Conversely, it may absorb radiant energy in the neighborhood of the ion and move on with increased speed. Since the energy changes are unquantized, the absorbed and emitted frequencies constitute a continuous spectrum. Small energy changes occur more frequently than large ones; hence free-free absorptions become particularly important in the infrared.

From the standpoint of classical theory, the ion and the electron approximate a dipole and may radiate energy in accordance with the usual electromagnetic equations. The encounter between two electrons does not constitute a dipole and neither classical theory nor quantum mechanics permits radiation to be emitted or absorbed.

In the hottest stars, free-free absorptions become important in ordinary spectral regions. Hence we must add to $k_1(\nu)$ the contribution from the free-free absorption. To calculate the absorption coefficient per proton and free electron with velocity between v and $v + dv$ we first use Eq. 4–133 for α_ν with f obtained from Eq. 4–123 wherein we make the replacements $n' \rightarrow i\kappa'$ and $n \rightarrow i\kappa$

$$f_{\kappa\kappa'} = \frac{2^6}{3\sqrt{3}\,\pi}\frac{1}{g_{\kappa'}}\frac{1}{\left[\frac{1}{\kappa^2} - \frac{1}{\kappa'^2}\right]^3}\left|\frac{1}{\kappa'^3}\frac{1}{\kappa^3}\right| g_{ff}$$
$$= \frac{2^6}{3\sqrt{3}\,\pi}\frac{R^3Z^6}{\nu^3}\frac{1}{g_{\kappa'}}\left|\frac{1}{\kappa^3}\frac{1}{\kappa'^3}\right| g_{ff} \qquad (4\text{–}141)$$

since Eq. 4–127 now becomes

$$\nu_{ff} = RZ^2\left(\frac{1}{\kappa^2} - \frac{1}{\kappa'^2}\right) \qquad (4\text{–}142)$$

The statistical weight $g_{\kappa'}$ of the free electron is calculated per unit volume and $\Delta\kappa' = 1$, viz:

$$g_{\kappa'} = g_\varepsilon \int \frac{dp_1\, dp_2\, dp_3\, dq_1\, dq_2\, dq_3}{h^3} = \frac{2}{h^3}\int dp_1\, dp_2\, dp_3 \qquad (4\text{–}143)$$

where dp's are elements of momentum and dq's are elements of coordinates, or since

$$dp_1\, dp_2\, dp_3 = p^2 \sin\theta\; dp\; d\theta\; d\phi \qquad (4\text{–}144)$$

$$g_{\kappa'} = \frac{2}{h^3}\, p^2\, dp \int_0^\pi \sin\theta\; d\theta \int_0^{2\pi} d\phi = \frac{8\pi}{h^3}\, p^2\, dp \qquad (4\text{–}145)$$

Here $g_\varepsilon = 2$ corresponding to two directions of electron spin. Hence

$$g_{\kappa'} = \frac{8\pi m^3}{h^3}\, v'^2\, dv' = \frac{16\;\pi m^2 R Z^2 v'}{h^2\, \kappa'^3}\, d\kappa' \qquad (4\text{–}146)$$

where we have used Eq. 4–130. Now set $d\kappa'$ equal to 1, substitute $g_{\kappa'}$ in Eq. 4–141, then use Eq. 4–133 to obtain the absorption coefficient per proton and free electron with velocity between v and $v + dv$

$$\alpha_\kappa\,(\nu)dv = \frac{2}{3\sqrt{3}}\,\frac{RZ^2h^2\varepsilon^2}{m^3\pi c v}\,\frac{g_{\text{III}}}{\nu^3}\, dv \qquad (4\text{–}147)$$

where g_{III} is the Gaunt correction factor for free-free transitions. If there are N_i protons cm^{-3} and $N_\varepsilon f(v)dv$ electrons cm^{-3} with a velocity v to $v + dv$, the free-free absorption coefficient per cm^3 will be

$$N_i N_\varepsilon f(v)\alpha_\kappa(\nu)\, dv = \rho\, dk_\nu^{\text{ff}} \tag{4–148}$$

where the velocity distribution of the electrons is given by the Maxwellian distribution law for speeds, viz.,

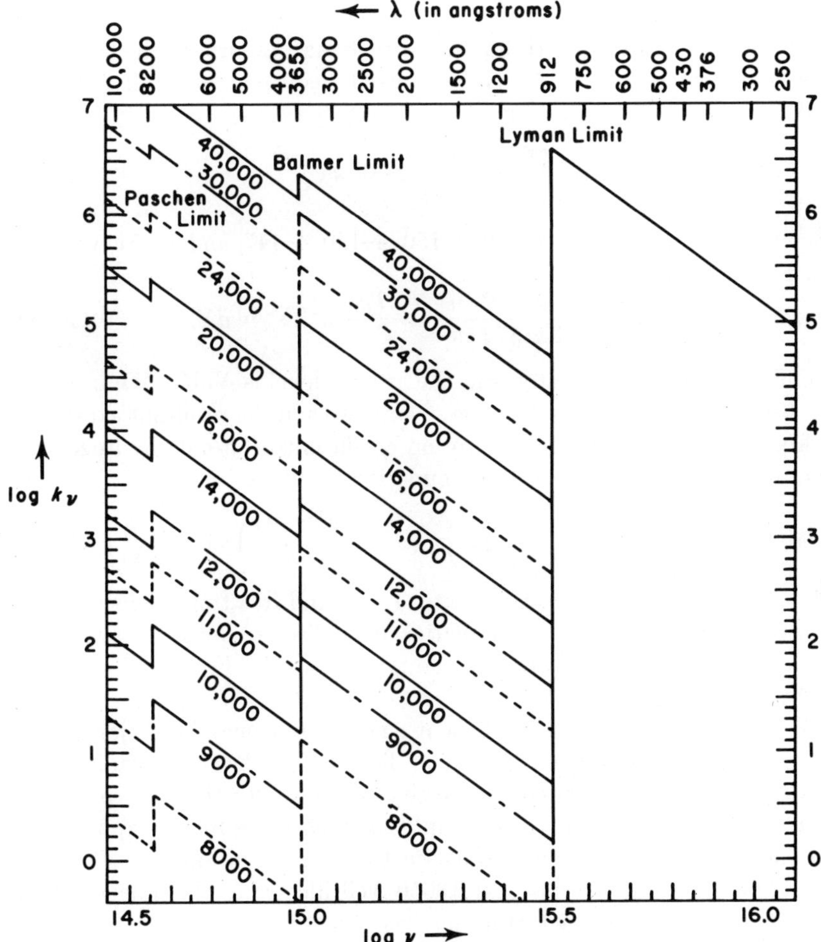

Fig. 4–16. The Absorption Coefficient of Atomic Hydrogen

The logarithm of the absorption coefficient of atomic hydrogen (not corrected for negative absorptions), calculated per gram of H atoms in the ground level, is plotted against log ν for various temperatures. The wavelength scale is indicated at the top. Notice in particular how the size of the Balmer discontinuity increases monotonically with decreasing temperature. Were atomic hydrogen the sole source of opacity in stellar atmospheres, the jump in the energy distribution at the limit of the Balmer series could become very large.

$$f(v)\,dv = 4\pi\left(\frac{m}{2\pi kT}\right)^{3/2} v^2 \exp\left(-\frac{mv^2}{2kT}\right) dv \tag{4–149}$$

Furthermore, N_iN_ε may be expressed in terms of N_1 by means of the combined Boltzmann and Saha equations (see Chapter 4). Thus, using Eq. 138, for any level n

$$N_n = \frac{N_iN_\varepsilon}{T_\varepsilon^{3/2}}\left(\frac{h^2}{2\pi mk}\right)^{3/2} n^2 e^{X_n} \tag{4–150}$$

Since electrons moving with any velocity v may absorb at the frequency ν, we must integrate Eq. 4–147 for a fixed frequency, over all velocities. Making use of Eq. 4–130 we get

$$ve^{-\frac{mv^2}{2kT}}\,dv = \frac{kT}{m}e^{-X_\kappa}\,dX_\kappa \tag{4–151}$$

Then with the aid of Eqs. 4–149, 4–150, 4–130, 4–147, and 4–151, we write Eq. 4–148 in the form

$$\rho\,dk_\nu^{\text{ff}} = \frac{8}{3\sqrt{3}}\frac{RZ^2\varepsilon^2}{hmc}\frac{N_1kTg_{\text{III}}}{\nu^3}e^{-(X_1+X_\kappa)}\,dX_\kappa \tag{4–152}$$

The density of hydrogen atoms in the ground level is N_1M. We integrate Eq. 4–152 from $X_\kappa = 0$ to $X_\kappa = \infty$. Then we add the resultant expression to Eq. 4–139 to obtain the absorption coefficient of atomic hydrogen per gram of neutral hydrogen in the ground level. Thus

$$k_1(\nu) = \frac{32}{3\sqrt{3}}\frac{\pi^2\varepsilon^6RZ^4}{ch^3M}\frac{e^{-X_1}}{\nu^3}\left\{\sum_{\nu_n<\nu}^{X_r}\frac{e^{X_n}}{n^3}g_{\text{II}} + \left(\frac{e^{X_{r+1}}-1}{2X_1}\right)\bar{g}_{\text{II}} + \frac{g_{\text{III}}}{2X_1}\right\} \tag{4–153}$$

Here g_{II} is the Gaunt correction factor for bound-free transitions (see e.g., Karzas and Latter, 1960). Summation over discrete levels is carried out to $X_r = 7$, while an integration of the type shown by Eq. 4–140 is carried out over higher levels.

A plot of the absorption coefficient against frequency has a jagged appearance (Fig. 4–16). The influence of discrete lines has not been included in this calculation. Between successive series limits the absorption coefficient falls off as ν^{-3} but rises abruptly as each series limit is approached and successive hydrogen lines coalesce to form a continuum. The confluence of the Balmer series near λ3650 well illustrates this phenomenon. Then it falls off again as ν further increases.

The continuous absorption coefficient of atomic hydrogen varies steeply with the temperature. At low temperatures nearly all hydrogen atoms are in the ground state, and practically all absorption takes place from this lowest level, i.e., in wavelengths shorter than λ912. Since under these conditions little radiation falls in this spectral region, the contribution of atomic hydrogen to the opacity, i.e., to the blocking of the outgoing radiation, is small.

At higher temperatures the upper levels become appreciably populated. Frequent photo-ionizations occur from the second and higher levels, with the consequence that considerable continuous absorption occurs in ordinary optical wavelengths.

Simultaneously, the energy distribution in the spectrum of the star is such that the fraction of radiation in the ultraviolet steadily increases. Photo-ionizations from the ground level become increasingly frequent. Atomic hydrogen then determines the opacity until it becomes practically all ionized. Only for stars somewhat hotter than the sun do the bound-free absorptions become important, and only in the very hottest stars do free-free absorptions make any appreciable contributions.

4–17. Continuous Emission of Atomic Hydrogen. Continuous emission beyond the limit of the Balmer series is observed in the solar chromosphere and in gaseous nebulae; it results from the recapture of electrons by ions. See Fig. 2–3.

The inverse of photo-ionization is the recapture of an electron by an ion. The capture coefficient, $\sigma_{\kappa n}$, for an electron of velocity v, is related to the absorption coefficient α_ν, ν and v being connected by Eq. 4–126. That is, α_ν and $\sigma_{\kappa n}$ are atomic parameters; the relation between them may be evaluated with aid of the principle of detailed balancing in thermodynamic equilibrium. Then the number of ionizations in the frequency interval $d\nu$ from a level n must equal the number of recombinations to this same level from the corresponding interval in velocity dv (see Eq. 4–130). We assume that the photo-ionization from the level n leaves the ion in its ground level.

The number of electron recaptures per cm³ from the velocity range v to $v + dv$ depends on the density of ground level ions $N_{1,i}$ the number of electrons in the relevant velocity range, $N_\varepsilon f(v)\, dv$, and the cross-section for recapture $\sigma_{\kappa n}$, all multiplied by the velocity v. The number of photo-ionizations equals the total energy absorbed per cm³ per sec, in the range $d\nu$ divided by $h\nu$. Thus

$$\frac{4\pi N_n \alpha_n(\nu) I_\nu (1 - e^{-h\nu/kT})}{h\nu}\, d\nu = N_{1,i} N_\varepsilon \sigma_{\kappa n} f(v) v\, dv \tag{4–154}$$

where the term on the left of the equal sign takes into account the negative absorptions. Here $f(v)$ is given by Eq. 4–149 and I_ν is the Planckian function, Eq. 4–10. With the aid of these relations, as well as Eqs. 4–126 and 4–130, we may solve for $N_{1,i}N_\varepsilon/N_n$ and compare the resultant expression with the combined Boltzmann and Saha formulas (Chapter 3).

In order for Eq. 4–154 to hold, $\alpha_n(\nu)$ and $\sigma_{\kappa n}$ must be connected by

$$\frac{\alpha_n(\nu)}{\sigma_{\kappa n}(v)} = \frac{m^2 c^2 v^2}{\nu^2 h^2} \frac{g_\varepsilon g_{1,i}}{2 g_n} \tag{4–155}$$

an expression first derived by Milne.

Since Eq. 4–155 is a relation between atomic constants, it depends only on the atom and level involved, the electron energy, and not at all on the existence or absence of thermal equilibrium. In the practical calculation of recombination rates, we shall assume that the velocity distribution is Maxwellian corresponding to some temperature T_ε. Hence the recombination rate will be

$$F_{\kappa n}d\nu = N_i N_\varepsilon f(v, T_\varepsilon)\, v\sigma_{\kappa n} d\nu \tag{4–156}$$

We now employ Eqs. 4–135, 4–130, 4–155, and 4–156 together with the definitions indicated by Eq. 4–138 and the statistical weights $g_{1,i} = 1$ and $g_n = 2n^2$ for hydrogen to obtain

$$F_{\kappa n}\, d\nu = N_i N_\varepsilon \frac{KZ^4}{T_\varepsilon^{3/2}} \frac{g_{\text{II}}}{n^3} e^{-h(\nu-\nu_n)/kT_\varepsilon} \frac{d\nu}{\nu} \tag{4–157}$$

where g_{II} is the Gaunt factor and the constant K is given by

$$K = \left(\frac{h^2}{2\pi mk}\right)^{3/2} \frac{8\pi^2\varepsilon^2 R^2}{mc^3} \frac{2^4}{3\sqrt{3}\pi} = 3.260 \times 10^{-6} \tag{4–158}$$

The corresponding emission per unit volume and time is

$$E_{\kappa n}d\nu \equiv F_{\kappa n}h\nu_{\kappa n}d\nu = N_i N_\varepsilon \frac{KZ^4}{T_\varepsilon^{3/2}} \frac{g_{\text{II}}}{n^3} e^{-h(\nu-\nu_n)/kT_\varepsilon}\, h\, d\nu \tag{4–159}$$

Two astrophysical applications of this formula may be mentioned. If we can measure the intensity of the continuous Balmer emission in a gaseous nebula or in a solar prominence in absolute units and make some estimate of the radiating volume, we can determine the product of the ion and electron densities. Furthermore, measures of the energy distribution with frequency ought to give an estimate of the electron temperature, i.e., the temperature appropriate to the electronic velocity distribution. Putting in numerical values for $n = 2$, we get

$$E_{\kappa 2}\, d\nu = 2.70 \times 10^{-33}\, N_i N_\varepsilon T_\varepsilon^{-3/2}\, g e^{-h(\nu-\nu_2)/kT_\varepsilon}\, d\nu \tag{4–160}$$

That is, in a continuous spectrum which arises entirely from recombination and in which self-reversal can be neglected, the energy distribution with frequency will be

$$E_\nu\, d\nu = \text{const}\, e^{-h\nu/kT_\varepsilon}\, d\nu \tag{4–161}$$

Applications to the continuous spectra of the gaseous nebulae have been made by T. L. Page and others.

Free-free transitions are the source of thermal emission in gaseous nebulae and the solar corona in the radio-frequency region. The free-free absorption coefficient per unit volume computed for electrons moving with velocities between v and $v + dv$ and corrected for the effects of negative absorption is[6]

$$N_iN_\varepsilon f(v)\,\alpha_\kappa\,(1 - e^{-h\nu/kT})\,d\nu = N_iN_\varepsilon f(v)\,\frac{h\nu}{kT}\,\alpha_\kappa\,d\nu \qquad (4\text{–}162)$$

We now proceed as in the derivation of Eq. 4–152 except that N_iN_ε is retained and Eq. 4–150 is not used. Thus, making use of Eqs. 4–147 and 4–149,

$$\rho k_\nu'^{\text{ff}} = N_iN_\varepsilon\,\frac{2}{3\sqrt{3}}\,\frac{RZ^2h^2\varepsilon^2}{m^3\pi c}\,\frac{g_{\text{III}}}{\nu^3}\,\frac{h\nu}{kT}\,4\pi\left(\frac{m}{2\pi kT}\right)^{3/2}\int ve^{-mv^2/2kT}\,dv \qquad (4\text{–}163)$$

and making use of Eq. 4–151, there results

$$\rho k_\nu'^{\text{ff}} = N_iN_\varepsilon\,\frac{8}{3\sqrt{3}}\,\frac{RZ^2h^3\varepsilon^2}{m(2\pi mkT)^{3/2}c}\,\frac{g_{\text{III}}}{\nu^2}\int e^{-X_\kappa}\,dX_\kappa = \frac{8\pi\varepsilon^6Z^2}{3\sqrt{6\pi}c(mk)^{3/2}}\,\frac{N_iN_\varepsilon}{T_\varepsilon^{3/2}}\,\frac{g_{\text{III}}}{\nu^2} \qquad (4\text{–}164)$$

as the absorption coefficient per unit volume. Putting in numerical constants

$$\rho k_\nu'^{\text{ff}} = 0.01771\,\frac{N_iN_\varepsilon}{T_\varepsilon^{3/2}}\,\frac{g_{\text{III}}}{\nu^2} \qquad (4\text{–}165)$$

The constant g_{III} may assume values considerably different from unity in the r-f region. It has been discussed by Elwert (1948) and others and has the value

$$g_{\text{III}} = \frac{\sqrt{3}}{\pi}\ln\frac{420\,T_\varepsilon}{ZN_i^{1/3}} \qquad (4\text{–}166)$$

if the "discriminant"

$$d = \frac{8.1 \times 10^{-6}\nu}{T_\varepsilon^{1/2}N_\varepsilon} \qquad (4\text{–}167)$$

is considerable less than unity. If $d > 1$

$$g_{\text{III}} = \frac{\sqrt{3}}{\pi}\ln\left[\frac{(2k)^{3/2}}{4.22\,\pi m^{1/2}\varepsilon^2}\right]\frac{T_\varepsilon^{3/2}}{Z\nu} \qquad (4\text{–}168)$$

Putting in numerical values

$$g_{\text{III}} = 0.5513\ln\left[4.97 \times 10^7\right]\frac{T_\varepsilon^{3/2}}{Z\nu} \qquad (4\text{–}169)$$

The emission per unit volume of a radiating gas will be

$$E_\nu = \rho j_\nu = 4\pi\rho k_\nu'^{\text{ff}}\,B_\nu(T) = 0.01771\,\frac{N_iN_\varepsilon}{T_\varepsilon^{3/2}}\,\frac{8\pi kT}{c^2}\,g_{\text{III}} \qquad (4\text{–}170)$$

where we substitute the Rayleigh-Jeans law for the Planck formula in this region of the spectrum. Notice that the emission per unit volume is independent of ν except for the weak dependence on ν through g_{III}.

[6] The notation k_ν' is used to indicate that we have included the negative absorptions.

We normally use Eq. 4–166 for r-f radiation from the solar corona, while for the interstellar medium Eq. 4–169 would be more appropriate. Thus

$$\rho k_\nu'^{\text{ff}} = \frac{8\,\varepsilon^6 Z^2}{3\sqrt{2\pi}c(mk)^{3/2}} \frac{N_i N_\varepsilon}{T_\varepsilon^{3/2}} \frac{1}{\nu^2} \ln\left(\frac{417\,T_\varepsilon}{Z\,N_i^{3/2}}\right) \tag{4–171}$$

is appropriate for the solar corona.

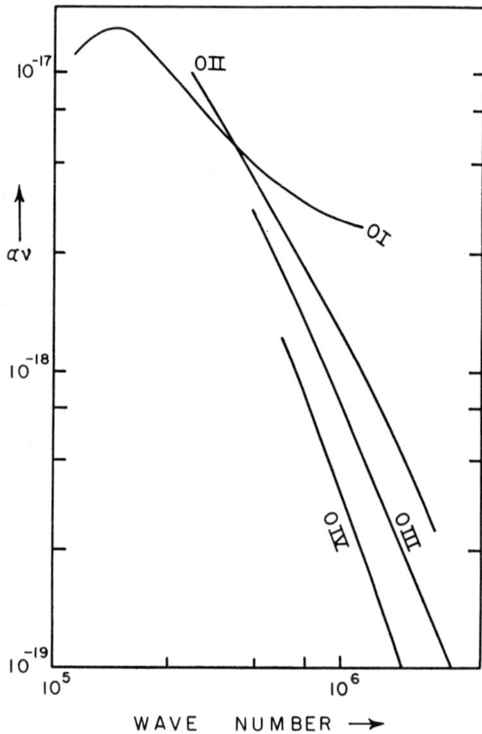

FIG. 4–17. THE CONTINUOUS ABSORPTION COEFFICIENT OF OXYGEN FOR DIFFERENT STAGES OF IONIZATION

The absorption coefficient α_ν is computed from Hartree wave functions for p^n configurations and hydrogenic d-wave functions in the continuum. The contributions of the $p^n - p^{n-1}s$ transitions to the absorption coefficient have been neglected. A logarithmic scale is employed in both coordinates. (Leo Goldberg and L. H. Aller, 1941.)

4–18. Continuous Absorption by Other Elements. Continuous absorption coefficients for bound-free transitions have been computed or measured experimentally for a number of atoms besides hydrogen (see Fig. 4–17). For the elements in the first two rows of the periodic table, we may refer especially to the recent calculation by Seaton (1958). Calculations for various elements have also been made by Green *et al.* (1949, 1950, 1951) and Biermann *et al.* (1947, 1949). The formulas are less simple than for hydrogen because of the greater complexities of the ions. Although continuous absorptions from the ground level of carbon may be important in the satellite-rocket ultraviolet spectral region of the sun, the ground level absorptions of most other abundant elements fall in the region beyond $\lambda 912$ where the hydrogen continuous absorption dominates.

In hot stars the continuous absorption of helium can become important, at least in hydrogen-depleted stars. In normal stars it is never comparable with that of hydrogen, partly because of its lower abundance and partly because of the high excitation potential of the levels that contribute to absorption in observable regions. The continuous absorption coefficients for helium have been calculated by Vinti, by Goldberg (1939), and by S. S. Huang (1948).

4–19. Continuous Absorption by the Negative Hydrogen Ion. Although absorption by neutral hydrogen explains the continuous absorption in hotter stars, it fails by many orders of magnitude for cool ones such as the sun. Also, if hydrogen is assumed to be the sole source of absorption, the break in the continuous spectrum at the head of the Balmer series will become very large as the temperature decreases, since the ratio of the absorption coefficient on the shortward side to that on the longward side increases without bound as T decreases. Actually this Balmer "jump" increases from $B0$ to a maximum at $A0$ and then decreases at still lower temperatures. Evidently some other source of continuous absorption must replace atomic hydrogen at lower temperatures.

Initially, it was assumed that metals were responsible for this absorption. If, following Pannekoek (1931), we take a hydrogen-to-metal ratio of a thousand to one, the atmospheres of cooler stars will be relatively transparent, and the amount of material above the photosphere will be measured in hundreds of grams. On the other hand, if we choose a hydrogen-to-metal ratio to fit the discontinuity in the continuous spectrum at the end of the Balmer series, we find a ratio of about 15. In either event, the metallic line intensities in cooler stars should be much greater than those observed unless some other agent is responsible for the continuous absorption. Furthermore, if metallic absorption is responsible we should observe the absorption edges of at least some metallic atom or ion.

The way out of this difficulty was found by Wildt (1938) who suggested that negative ions of hydrogen produce most of the continuous absorption in late-type stars. One interesting aspect of his proposal was that the negative hydrogen ion then had not yet been isolated in the laboratory, but had been predicted on the basis of quantum-mechanical calculations, independently, by H. Bethe and E. A. Hylleraas in 1930. R. Fuchs (1951) and Branscomb and Smith (1955) have observed the negative hydrogen ion continuum in the laboratory.

The single electron of a hydrogen atom does not completely screen the charge on the nucleus in its immediate neighborhood. An electron that passed sufficiently close to the atom would find itself in an attractive field. It would have a definite chance of becoming attached to the atom to form a negative ion, H^-. An exact calculation of the photo detachment energy shows it to be 0.754 ev and indicates that there exists only one stable energy state.

The astrophysical importance of the H^- ion lies in the fact that quanta of energy in the ordinary and near infrared spectral regions may photo-dissociate it into a neutral hydrogen atom and a free electron. The absorption spectrum of H^- in its ground state is a continuum whose edge falls at the wavelength corresponding to 0.754 ev (i.e., λ16,450), rises steadily to a maximum near λ8500, and decreases to the shorter wavelengths. In addition to the "bound-free" absorption by the H^- ion, there are also free-free transitions, corresponding to the interactions of neutral hydrogen atoms with free electrons.

Calculation of the absorption coefficient of H^- is extremely difficult, particularly for the free-free transitions. Accurate calculations have been made by Chandrasekhar and his associates (1946, 1958), by Geltman (1956), and by T. and H. Ohmura (1960). Fig. 4–18 shows the results obtained. Notice that the more recent work for the free-free absorptions indicates a smaller contribution in the infrared than had been previously obtained.

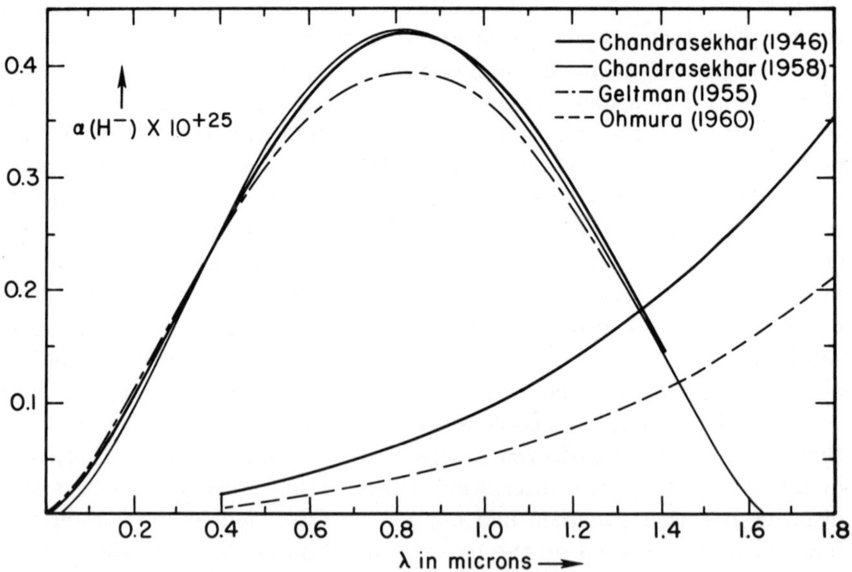

Fig. 4–18. The Absorption Coefficient of the Negative Hydrogen Ion

Chandrasekhar's (1946, 1958) data are compared with the results by Geltman (1955) and by the Ohmuras (1960). The total absorption coefficient is obtained by adding the free-free and bound-free contributions. (T = 6300°K.)

The continuous absorption curves for H^- illustrate the character of the photodetachment spectra of atomic and molecular ions. They differ from atomic photo-ionization continua in a number of significant ways. There is no accompanying line spectrum such as adjoins the Balmer limit, for example. The shape of the H^- continuous absorption coefficient is different in that it rises gradually from the threshold and reaches a maximum at a frequency about twice that of the photodetachment limit. Finally, since the photodetachment energy is only the order of a volt or even less, the photodetachment continua fall in the visual or near infrared region of the spectrum.

It is of interest to compute the fraction of hydrogen atoms that go to form H^- ions at any temperature and pressure. The H^- ion concentration obeys an "ionization" equation of the form

$$\frac{N(\mathrm{H}^-)}{N(\mathrm{H})} = \phi(T)\, P_\varepsilon \tag{4–172}$$

where

$$\log \phi(T) = -0.12 + 0.754 - 2.5 \log T \tag{4–173}$$

P_ε is the electron pressure in dynes cm^{-2}. For example, at $T = 5600°$K, $P_\varepsilon = 10$ dynes cm^{-2}, the concentration is antilog $(-7.82) = 1.51 \times 10^{-8}$ ions cm^{-3}. If α_I denotes the absorption coefficient per H^- ion, the absorption coefficient tabulated by Chandrasekhar (1946) is

$$\alpha_\lambda = \alpha_\mathrm{I}\, (\mathrm{H}^-)\phi(T) \tag{4–174}$$

and the absorption coefficient per gram of neutral hydrogen is

$$k_\lambda\, (\mathrm{H}^-) = \alpha_\lambda\, (\mathrm{H}^-)\, \frac{P_\varepsilon}{M_\mathrm{H}} \tag{4–175}$$

since M_H is the mass of the hydrogen atom. In Chandrasekhar's table $\alpha_\lambda(\mathrm{H}^-)$ and therefore $k_\lambda(\mathrm{H}^-)$ include the effect of the negative absorptions. Although the absorption coefficients calculated by Chandrasekhar (1946) are sufficiently accurate for the ordinary visual region of the spectrum λ4000–λ7000, the values in the infrared may be in need of revision in view of the corrections apparently required for the free-free absorption coefficient. It is hoped that accurate new absorption coefficients will be available shortly. To get the total absorption coefficient it is necessary to add to Eq. 4–175 the contribution of atomic hydrogen. Thus

$$k_\lambda'(\mathrm{H}) = \frac{1 - x}{M_\mathrm{H}} [\alpha_\lambda{}^\mathrm{I}\, (\mathrm{H}^-)\, P_\varepsilon + \alpha_\lambda'(\mathrm{H})] \tag{4–176}$$

is the absorption coefficient per gram of H including the effects of stimulated emissions or negative absorptions. Here x is the fraction of H that is ionized, $\alpha_\lambda'(\mathrm{H})$ is the absorption coefficient of atomic hydrogen corrected for the negative absorptions, and M_H is the mass of the hydrogen atom.

Although the main features of continuous stellar spectra are explained by continuous absorption by hydrogen, additional sources of opacity appear to occur in the violet and near ultraviolet of the sun, and in late-type stars shortward of $\lambda 4500$. Although the bulk of this absorption appears to be produced by the overlapping wings of strong absorption lines (see Liller and Lewis, 1951), it is of interest to see to what extent atomic and molecular negative ions may make significant contributions. Branscomb and Pagel (1958) have examined this question. CN^- and C_2^- have detachment energies of about 3.1 ev, OH^- has a detachment energy of 1.78 ev, while those of C^- and O^- are 1.12 and 1.465 ev, respectively. They find that O^- and O_2^- are unlikely to be important in any stars, while under special circumstances OH^- may be important in M and S stars. In hydrogen-deficient carbon stars, C^- may be important, while CN^- may be a substantial contributor to violet-region absorption ($\lambda < 4300$) in cool carbon stars. Perhaps it may be more important than C_3 in this region.

Molecular continuous absorption is important in the "rocket" ultraviolet region of the solar spectrum and is probably significant in other stars.

Recent rocket observations of stellar energy distributions indicate that below $\lambda 2000$, as yet unidentified sources of absorption cut down the emergent stellar fluxes by considerable amounts (Milligan and Stecher, 1962).

4–20. Continuous Spectra Produced by Non-thermal Processes. In addition to thermal absorption and emission processes, there occurs in certain objects, such as the Crab Nebula and perhaps also in certain solar r-f phenomena, non-thermal emission which is produced by the directed motion of charged particles — in this instance synchrotron radiation.

Consider an electron moving with a very high speed in a circular orbit under the influence of a magnetic field. It is constantly being accelerated, and therefore, in accordance with the requirements of classical electromagnetic theory, it is continuously emitting radiation. As the electron radiates it is running right behind the radiation, and the emitted wave is deformed by the Doppler effect. The radiation is polarized and is confined to a small cone in the plane of the orbit in the forward direction (see Fig. 4–19a). As seen by a fixed observer, a revolving electron will emit a series of discrete pulses separated in time by the period of the electron (see Fig. 4–19b). The resulting spectrum will be a continuous one whose intensity distribution will depend on the energy of the emitting particle (see Fig. 4–20).

The frequency distribution of the emitted energy is determined by

$$P(\nu) = \frac{\sqrt{3}\,\varepsilon^3}{mc^2}\,\mathrm{H}\tilde{\alpha}\int_{\alpha}^{\infty} K_{5/3}(\eta)\,d\eta = 2.343 \times 10^{-22}\,\mathrm{H}\,F(\tilde{\alpha})$$

where ε, m, and c have their usual meaning, H is the field strength perpendicular to the orbit plane, $\tilde{\alpha} = \nu/\nu_c$, and the function $F(\tilde{\alpha})$ has been tabulated by Oort

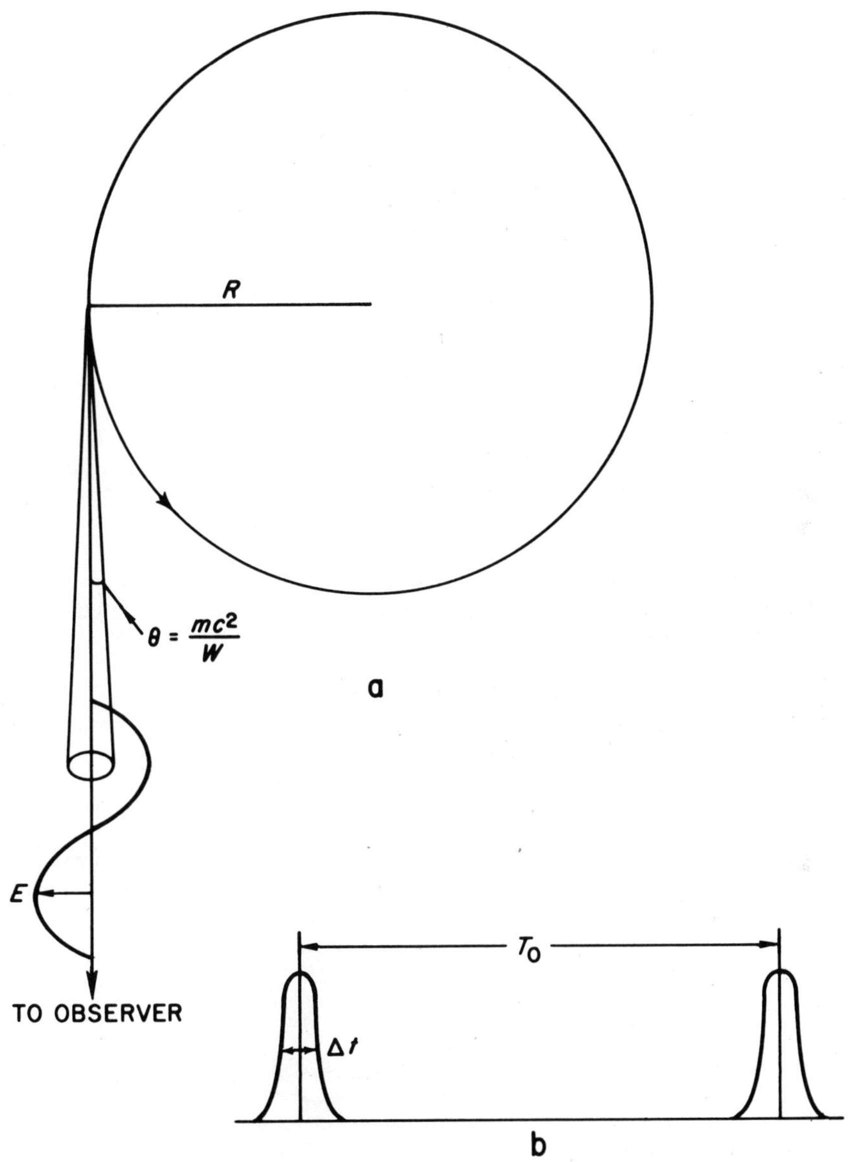

FIG. 4–19. SCHEMATIC DIAGRAM OF SYNCHROTRON RADIATION EMITTED BY AN ELECTRON MOVING WITH NEARLY THE SPEED OF LIGHT

(a) The electron moves in the plane circular orbit of radius **R** with the Larmor frequency $ceH/2\pi W$. The magnetic field **H** is perpendicular to the plane of the paper. The emitted radiation is polarized in the orbit plane and confined to the small cone $\theta = mc^2/W$, where W is the energy of the particle including its rest energy.

(b) If the intensity of the radiation is plotted against time, the radiation appears to come in discrete pulses whose width Δt is determined by the time required for the cone of width θ to sweep across the observer. The interval $T_0 = 1/\nu_0$ is simply the period of the particle in its orbit.

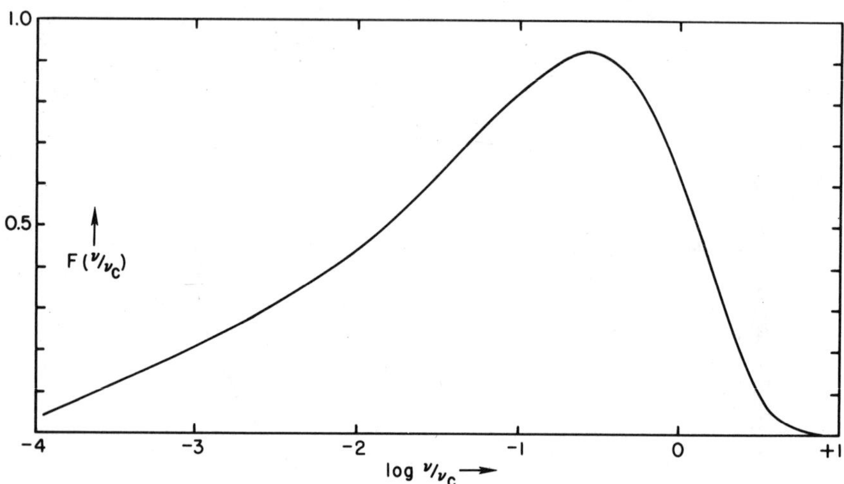

FIG. 4–20. ENERGY DISTRIBUTION IN SYNCHROTRON RADIATION

Ordinate is $F(\tilde{\alpha})$, abscissa is $\tilde{\alpha} = \nu/\nu_c$ where ν_c is the critical frequency. Notice that $F(\alpha)$ has a maximum at 0.3 and represents a broad distribution.

and Walraven. (See Fig. 4–20.) Here the critical frequency ν_c which is related to $1/\Delta t$ is given by

$$\nu_c = \frac{3c}{4\pi R}\left(\frac{W}{mc^2}\right)^3 = \frac{3\varepsilon}{4\pi mc} H \left(\frac{W}{mc^2}\right)^2$$
$$= 1.608 \times 10^{13}\, HW^2{}_{\text{Bev}}$$

where the energy W_{Bev} is expressed in billions of electron volts. Notice, however, that there is a considerable high energy extent to the spectrum.

The rate of dissipation of energy by a particle goes as W^4/R^2, or

$$\left(\frac{dW}{dt}\right)_{\text{Bev}} = -3.79 \times 10^{-6}\, H^2\, W^2{}_{\text{Bev}}$$

Hence the energy of an electron will fall to half its original value in a time

$$T = \frac{0.00835}{H^2 W_{\text{Bev}}} \text{ years}$$

Synchrotron radiation appears to be the only explanation of the strong emission from such objects as the Crab Nebula and other ex-supernovae. It seems to be the mechanism for the emission from the non-thermal radio sources. It appears to play a role in certain types of r-f radiation from the sun, although the power supply for such mechanisms remains a mystery.

PROBLEMS

4–1. Show that the Rayleigh-Jeans formula in wavelength units is

$$B_\lambda(T)d\lambda = 2kcT\lambda^{-4}d\lambda$$

4–2. If the intensity distribution of radiation emergent from the sun is given by

$$I(\theta) = \sum_{n=0}^{m} a_n \cos n\theta$$

derive an expression for the flux.

4–3. Derive the relation between the Einstein coefficients A and B defined for energy density instead of intensity.

4–4. Verify Milne's relation, Eq. 4–155.

4–5. Show that no combination of lenses or mirrors can increase the specific intensity of the radiation from an extended surface, although the flux passing through a unit area can so be increased. A lens of 20 cm diameter and 100 cm focal length is used to form an image of the sun. The angular diameter of the sun is 32′. How much brighter will the image be than a directly illuminated surface?

4–6. In the absence of an external field and with a small damping force, the equation of motion of a classical oscillator is nearly

$$m\,\frac{d^2z}{dt^2} = -m\,\omega_0{}^2 z$$

With the aid of Eq. 4–83, calculate the instantaneous rate of energy radiation, average it over a cycle, thus verifying Eq. 4–84, and derive the equation for γ (Eq. 4–85).

REFERENCES

Arts. 4–1, 4–2, 4–3, 4–4, and 4–6

Radiation laws:

Milne, E. A. 1930. Thermodynamics of the Stars, *Handbuch der Astrophysik.* Springer-Verlag, Berlin, vol. III, part 1, p. 65.

Chandrasekhar, S. 1940. *Stellar Structure.* University of Chicago Press, Chicago, chap. 5. Reprinted by Dover Publications, Inc., New York, 1957.

Derivation of Planck's law and other radiation relationships:

Richtmyer, F. K., Kennard, E. H., and Lauritsen, T. 1955. *Introduction to Modern Physics.* McGraw-Hill Book Co., Inc., New York, chap. 4.

Art. 4–5

For terms and formulas concerning antennas see:

Pawsey, J. L., and Bracewell, R. N. 1955. *Radio Astronomy.* Oxford University Press, Fair Lawn, N. J., chap. 2.

For pencil-beam antennas see:

Seeger, C. L., Westerhout, G., and van de Hulst, H. C. 1956. *Bull. Astr. Inst. Netherlands,* **13,** 89, No. 472.

Arts. 4–7, 4–8, 4–9, and 4–10

The classical formulas of dispersion and absorption are developed in:

Slater, J. C., and Frank, N. H. 1947. *Electromagnetism.* McGraw-Hill Book Co., Inc., New York.

Menzel, D. H. 1953. *Mathematical Physics.* Prentice-Hall, Inc., Englewood Cliffs, N. J.

Born, M. 1959. *Principles of Optics.* Pergamon Press, Ltd., Oxford, England.

Arts. 4–11, 4–12, 4–13, 4–14, and 4–15

For applications of quantum theory see:

HEITLER, W. 1954. *Quantum Theory of Radiation* (3d ed.). Oxford University Press, Fair Lawn, N. J.

CONDON, E. U., and SHORTLEY, G. 1935. *Theory of Atomic Spectra.* Cambridge University Press, London.

Arts. 4–16 and 4–17

For discussions of the absorption coefficient of hydrogen, see Menzel and Pekeris (1936) and Unsöld (1955).

The continuous emission of atomic hydrogen is discussed in:

MENZEL, D. H. (ed.). 1960. *Physical Processes in Ionized Plasmas.* Dover Publications, Inc., New York.

ALLER, L. H. 1956. *Gaseous Nebulae.* John Wiley & Sons, Inc., New York.

Art. 4–19

See especially Wildt (1939), Chandrasekhar (1946, 1958), Chandrasekhar and Munch (1946), and Branscomb and Pagel (1958).

Art. 4–20

The basic paper on synchrotron radiation is:

SCHWINGER, J. 1949. *Phys. Rev.* **75,** 1912.

A more elementary treatment of the basic ideas is given in:

SCHOTT, G. A. 1912. *Electromagnetic Radiation.* Cambridge University Press, London, p. 109, *et seq.*

PECKER, J. C., and SCHATZMAN, E. 1959. *Astrophysique Générale.* Masson and Co., Paris.

For applications to the Crab Nebula see:

OORT, J., and WALRAVEN, T. 1956. *Bull. Astr. Inst. Netherlands* **12,** 295.

CHAPTER 5

THE CONTINUOUS SPECTRA OF THE SUN AND STARS

5–1. The Problem of the Continuous Spectrum. Although many stars radiate roughly like black bodies, important exceptions exist. The continuous spectra of cool stars are blighted by numerous absorption lines. In hotter stars such as those of spectral class *A*, the strong wavelength dependence of the continuous hydrogen absorption plays havoc with the black-body approximation.

Thus, two effects — discrete absorptions by spectral lines and continuous absorption by atomic hydrogen — distort the continuous energy curves of many stars.

Yet, it is worthwhile to pose this question: if the continuous stellar absorption coefficient were independent of wavelength (as is actually true for electron scattering), and if each volume element of the atmosphere radiated as a "grey" body, would the stellar energy curve be Planckian? The observed energy distribution is the sum of contributions from all layers from which radiation may escape to the surface. The uppermost strata are coolest and the underlying layers become increasingly warmer. If each stratum radiates according to Kirchhoff's law, then the energy distribution contributed by each zone of constant temperature will be Planckian in character, and the resultant energy output of the star will be the sum of many black-body curves, each weighted according to the amount of radiation from that particular stratum that reaches the surface. Almost all the radiation from the upper, cooler layers reaches the surface, while the light from deeper layers is dimmed by absorption in intervening strata. From great depths little of the radiation is able to reach the surface.

In most stars the absorption coefficient is not independent of wavelength, with the consequence that in spectral regions where the absorption coefficient is high, the observed radiation emerges from the uppermost layers, whereas in other spectral regions much of the observed radiation comes from relatively hot, deep strata. The continuous absorption coefficient in an *A* star provides an illustration. To the longward side of the Balmer limit at λ3650 the continuous absorption coefficient is less than it is on the shortward side of the Balmer limit. Hence to the red of λ3650 and between the Balmer lines, the observed radiation comes from much greater depths than does radiation on the high-frequency side of the Balmer limit. At λ3600, for example, radiation reaches us only from the coolest, outermost levels of the atmosphere.

For most stars we observe only a quantity proportional to the energy distribution in the integrated light, viz., $F_\lambda(R^2/d^2)$ where R is the radius of the star and d is its distance. For the sun we can observe both the intensity distribution at the center of the disk, $I_\lambda(0)$, and the variation of the intensity distribution across the disk, $I_\lambda(\theta)/I_\lambda(0)$, the so-called darkening to the limb. This latter effect provides the most elementary argument for the existence of a temperature gradient in the sun. At the center of the disk we see to much deeper and hotter layers than we do near the limb where our line of sight enters the solar atmosphere at a glancing angle. From limb-darkening data and $I_\lambda(0, 0)$ it is possible to find empirically both the variation of the absorption coefficient with wavelength and the temperature gradient in the solar atmosphere.

In order to interpret the observations we shall derive an expression for $I_\lambda(0, \theta)$, the intensity of radiation emitted by a star in any wavelength λ, and at any angle θ with respect to a normal drawn to the surface. We may express our results in terms of either wavelength λ or frequency ν.

In Fig. 5–1, a ray reaching the observer from a point on the surface of the sun emerges at an angle θ with respect to the outward normal. This ray comprises contributions from heated strata at many different depths.

Consider a layer at a geometrical depth x and at a temperature T, and let us fix our attention upon a small cylinder of unit cross-section area and thickness

$$ds = \sec\theta\, dx \tag{5–1}$$

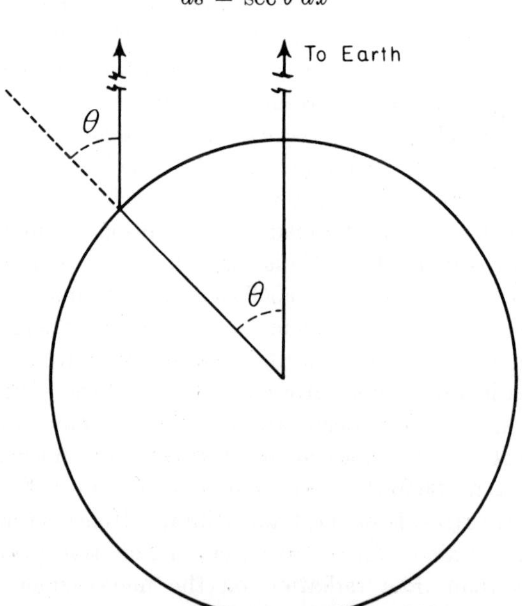

FIG. 5–1. EMERGENT RAY FROM THE SUN

(See Fig. 5–2.) The total emissivity per gram of material is j_ν and since a volume ds contains a mass $\rho\ ds$, the specific intensity of the radiation emitted by the box will be

$$dE_\nu = \frac{1}{4\pi} j_\nu\, \rho\, ds \tag{5–2}$$

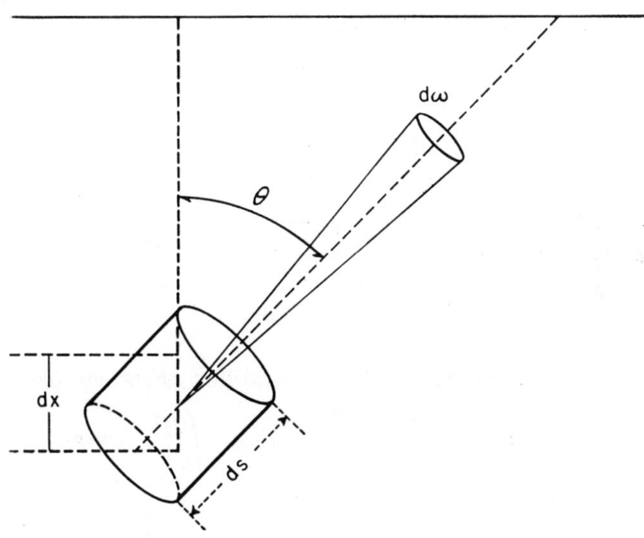

FIG. 5–2. ELEMENTARY EMITTING VOLUME

The emissivity depends on the physical conditions existing in the cylinder. If the material is in thermal equilibrium with its surroundings, (i.e., if there is no scattering) and if the emission is completely determined by the temperature, then Kirchhoff's law (Eq. 4–37) applies. Thus

$$dE_\nu = \kappa_\nu B_\nu(T)\, \rho\, dx \sec\theta \tag{5–3}$$

where κ_ν is the absorption coefficient per gram of material *including the negative absorptions.*[1] B_ν is the Planck function defined in Eq. 4–10.

The radiation emitted by the elementary cylinder in the direction of the observer is weakened by absorption in the overlying strata, such that the contribution dI_ν which actually reaches the surface is

$$dI_\nu = e^{-\tau_\nu \sec\theta}\, dE_\nu \tag{5–4}$$

where the optical depth τ_ν is defined by Eq. 4–35.

Therefore, the intensity of the emergent beam is the integral taken over all the elementary cylinders in the line of sight:

$$I_\nu(0, \theta) = \int_0^\infty B_\nu(T)\, e^{-\tau_\nu \sec\theta} \sec\theta\, d\tau_\nu \tag{5–5}$$

[1] In Chapter 4, for reasons noted in the text, k_ν was used to denote this quantity.

This is the fundamental relation expressing the connection between the emissivity of the radiating layers, the optical depth, and the observed intensity of the emergent beam. Since the absorption coefficient depends on wavelength, it is clear that the temperature will be a different function of the optical depth τ_ν for one frequency, say ν_1, than for another frequency, say ν_2. Hence, it is often desirable to define an optical depth in terms of some kind of average coefficient of absorption $\bar{\kappa}$. Thus

$$d\tau = \bar{\kappa}\rho\, dx \tag{5-6}$$

or

$$d\tau_\nu = \frac{\kappa_\nu}{\bar{\kappa}}\, d\tau \tag{5-7}$$

Frequently, however, one uses the optical depth τ_ν, at some fixed frequency, as the standard of reference. Then

$$d\tau_\nu = \frac{\kappa_\nu}{\bar{\kappa}_{\nu_0}}\, d\tau_0 \tag{5-8}$$

In most stars we can observe only the total flux πF_ν where (see Chap. 4)

$$F_\nu = 2\int_0^{\pi/2} I_\nu \cos\theta \sin\theta\, d\theta = 2\int_0^\infty B_\nu(T)\, d\tau_\nu \int_0^{\pi/2} e^{-\tau_\nu \sec\theta} \sin\theta\, d\theta \tag{5-9}$$

Now let

$$\sec\theta = y, \qquad \sin\theta\, d\theta = \frac{dy}{y^2} \tag{5-10}$$

and introduce the exponential integral function

$$E_n(x) = \int_{y=1}^\infty \frac{e^{-xy}}{y^n}\, dy \tag{5-11}$$

where E_1 is called simply E, and $E_0 = e^{-x}/x$. Then Eq. 5-8 becomes

$$F_\nu = 2\int_0^\infty B_\nu(\tau_\nu)\, E_2(\tau_\nu)\, d\tau_\nu \tag{5-12}$$

The flux (or intensity) of the emergent radiation depends on the rate at which the temperature increases with optical depth, i.e., on the temperature gradient, and on the absorptivity of the stellar material. The formula says nothing about the way in which the temperature gradient is maintained. The particular value the gradient will assume at any point will depend on the mode of energy transport, whether it is by molecular conduction, convection currents, or radiation.

Since the thermal conductivity of gas is low, e.g., that of helium is 0.344×10^{-3} cal cm^{-2} sec^{-1} deg^{-1} at N.T.P., we can easily show that thermal conduction can play no important role. In order for this process to supply the energy output of the sun, the required temperature gradient would be so high that actually the lower layers would lose more energy by thermal emission than by conduction, thus contradicting our hypothesis.

Convection currents require for their continued existence a different temperature gradient from that needed for radiative transport of energy. Although photospheric *granules* suggest that convection currents play some role in outer layers of the sun, the temperature gradient demanded for the uppermost strata by the limb-darkening observations is more nearly in accord with that appropriate to transport of energy by radiation than with that valid for a state of *convective equilibrium.* Below the radiative zone there is a zone in convective equilibrium. (See Chapter 9.) A comparison of the observed energy distributions with those predicted by theory indicate that along the main sequence at least, the atmospheric layers responsible for the production of the line and continuous spectra are in radiative equilibrium, even though the lower layers may be in convective equilibrium.

Our principal tasks in this chapter will be not only to obtain solutions of Eqs. 5–5 and 5–12, but also to show how these expressions may be employed in predicting observable features of stellar spectra.

For example, the determination of the *color temperature* of a star amounts to fitting a Planckian curve to F_ν, over a definite wavelength interval, e.g., $\lambda 4000 - \lambda 6500$. Thus, the so-defined temperature is only a parameter inserted in Planck's formula when the latter is used as a convenient interpolation device. Its relation to the physically significant *effective temperature* will depend on the temperature gradient and the absorption coefficient and is often complicated.

Our plan of attack will be as follows. First, we shall show how the temperature distribution and absorption coefficient in the solar atmosphere may be found empirically from the law of limb darkening and the energy distribution $I_\lambda(0, 0)$. Comparison between theory and observation confirms our belief that the negative hydrogen ion is the principal source of opacity in the sun. We next examine the general properties of stellar atmospheres by the crude approximation of what is called *grobanalyse* or rough analysis based on a simplified model in which the variation of absorption coefficient, ionization, etc., with depth is neglected.

A really adequate theory cannot be developed until we know how the temperature varies with depth in the star's atmosphere. Hence, the theory of the flow of radiation through a stellar atmosphere is a topic of utmost importance. We shall derive the equation of transfer and show how the temperature may be expressed in terms of the optical depth for a grey atmosphere. This is a problem of pure mathematics which has been treated by methods similar to those employed for the study of neutron diffusion in a radioactive pile, as well as by an elegant procedure suggested by Chandrasekhar. Unfortunately, the material of which stellar atmospheres are composed is not grey. Modifications are therefore required in the theory. The problem of radiative transfer in monochromatic light cannot be reduced to that of a grey body. We shall illustrate some of the procedures that have been suggested for solving the problem of the non-grey atmosphere in radiative equilibrium.

With the temperature gradient established we can then derive the variation of gas pressure, electron pressure, and continuous absorption coefficient with depth in a stellar atmosphere, i.e., the *model* of the atmosphere. Such model atmospheres will assume considerable importance in the interpretation of the dark-line spectra of the sun and stars. We shall give as an example the calculation of a model atmosphere for a B-type star and point out some of the inadequacies in these theories. In particular, the influence of strong absorption lines on the temperature distribution is not yet adequately handled. Finally, in Chapter 6 we return to the question of effective temperatures and show how measurements of energy distributions and Balmer discontinuities together may determine both the effective temperature and the surface gravity in the atmosphere of a star.

5–2. The Stellar Absorption Coefficient. Because of its overwhelming predominance, hydrogen is the leading source of opacity in stellar atmospheres. In Chapter 4 we enumerated some processes responsible for continuous absorption or extinction in stellar atmospheres. They were:

1. Photo-ionization from discrete levels of hydrogen to the continuum (bound-free transitions).
2. Free-free transitions wherein an electron traveling in one hyperbolic orbit near an atom jumps to another hyperbolic orbit with absorption of energy.
3. Electron scattering.
4. Photodissociation of the negative hydrogen ion with consequent absorption of energy.
5. Continuous absorption by molecules resulting in their dissociation.

In stars, such as the sun, the *negative hydrogen ion* primarily determines the absorption coefficient. In the hotter A and B stars, photo-ionization of atomic hydrogen and free-free transitions play the major roles, whereas in the O stars, electron scattering and even the photo-ionization of helium may make important contributions to the opacity. Tables of the continuous absorption coefficients of hydrogen and helium have been given by Saito, Ueno, and Jugaku (1954). In the coolest stars, continuous molecular absorption (corresponding to their photodissociation) and particularly the blocking of outgoing radiation by strong absorption lines becomes important. For most of the spectral sequence, from $B0$ to the late K's, absorption processes involving hydrogen primarily determine the continuous absorption.

To these sources of continuous absorption we may add:

6. Photoelectric extinction by metals (or by permanent gases other than hydrogen and helium).
7. Bound-free and free-free transitions involving H_2^+.
8. Rayleigh scattering.

Compared with processes involving hydrogen, the metallic absorption is small and can be neglected except in cool stars that are hydrogen deficient. On the other hand, the photodissociation of the H^+_2 ion and free-free absorption involving H atoms and H^+ ions can become important in A-type atmospheres (Osawa, 1956). These processes were suggested by Wildt (1947) and precise calculations of the absorption coefficient were made by Bates (1951, 1952). Rayleigh scattering is negligible in most stars, but may become significant in extended envelopes of supergiants of intermediate temperature.

There is increasing evidence that in addition to the above mentioned sources of continuous absorption in stellar atmospheres, additional agents must operate in the ultraviolet. Quasi molecules of H_2 (Stecher, 1962) in the sun and H He^+ (Stecher and Milligan, 1962) in hotter stars may exist and produce considerable ultraviolet absorption.

5-3. Empirical Determination of Temperature Distribution in the Sun. We now return to the problem of evaluating the temperature distribution in the sun from the basic relation, Eq. 5-5. In practice, we combine observations of the energy distribution at the center of the solar disk, $I_\lambda\,(0, 0)$, with those of limb darkening

$$\phi_\lambda\,(\theta) = \frac{I_\lambda\,(0, \theta)}{I_\lambda\,(0, 0)} \tag{5-13}$$

to secure not only the temperature distribution $T(\tau_\lambda)$, but also some information on the wavelength variation of the solar continuous absorption coefficient.

Let us define

$$b_\lambda\,(\tau_\lambda) \equiv \frac{B\,[T(\tau_\lambda)]}{I_\lambda\,(0, 0)} \tag{5-14}$$

Then Eq. 5-13 becomes

$$\phi_\lambda\,(\theta) = \int_0^\infty b_\lambda\,(\tau_\lambda)\,e^{-\tau_\lambda \sec\theta} \sec\theta\,d\tau_\lambda \tag{5-15}$$

Now Eq. 5-15 may be solved in a variety of ways. One may imagine the atmosphere stratified in a series of layers in each of which b_λ is constant (see, e.g., Plaskett, 1936, 1941; also Sykes, 1953). Another procedure (Przybylski, 1957) is to select some model atmosphere, calculate $I_\lambda(0, \theta)$

theoretically, and then use the discrepancies between theory and observation to correct the assumed $T(\tau_\lambda)$ and absorption coefficient κ_λ.

The essential difficulty is that one must determine a function that appears under an integral sign by inverting the integral. Whether this operation is done analytically or numerically depends on one's preferences. There are definite limitations in the determination of $T(\tau_\lambda)$ in this way. At a point on the solar disk $\mu = \cos\theta$, the radiation emerges on the average from a depth $\tau_\lambda = \mu$ (see Eq. 5–189). Hence we cannot secure any reliable data on radiation from the deep layers; e.g., only 1 per cent of the radiation comes from layers deeper than $\tau = 4.6$ and it probably suffices to assume b_λ varies linearly with τ_λ thereafter. In the shallower layers $\tau_\lambda < 0.1$, one must depend on data secured very near the limb or on profiles and central intensities of spectral lines. Analytical procedures are more convenient to apply than are strictly numerical ones. Furthermore, they fit the observational data to within their errors (see, e.g., Chalonge and Kourganoff, 1946). It is in fact impossible to determine more than three parameters in the $S_\nu - \tau_\nu$ relation, e.g., the value of the source function $S_\nu(\tau_\nu)$ at three distinct values of τ_ν, (Bohm, 1961).

Following Kourganoff, let us assume that $b_\lambda\,(\tau_\lambda)$ can be developed in terms of exponential integral functions,[2] viz.

$$b_\lambda\,(\tau_\lambda) = A_\lambda + B_\lambda \tau_\lambda + C_\lambda E_2(\tau_\lambda) \tag{5–16}$$

If Eq. 5–16 is substituted into Eq. 5–15, and properties of the exponential integrals employed (see Eqs. 5–197 to 5–204), we find

$$\phi_\lambda(\mu) = A_\lambda + B_\lambda\,\mu + C_\lambda\,[1 - \mu\, ln\,(1 + \mu^{-1})] \tag{5–17}$$

where

$$\mu = \cos\theta \tag{5–18}$$

The coefficients A_λ, B_λ, and C_λ are determined empirically by fitting $\phi_\lambda\,(\theta)$ to Eq. 5–17 by least squares. (See Pierce and Aller, 1951; Pierce and Waddell, 1961.) Consider a particular wavelength λ_0. With A_λ, B_λ, and C_λ known, one may compute $b(\tau_{\lambda 0})$ from Eq. 5–16 and $B(\tau_{\lambda 0})$ from Eq. 5–14 since $I_\lambda\,(0, 0)$ is known from measurements by Abbot, Minnaert, Peyteraux, Pierce, and others (see Chapter 6). Since $B_\lambda(\tau)$ is Planck's function of T, one finds for each value of $\tau_{\lambda 0}$ the corresponding value of T_0. If the same procedure is carried out for some other wavelength λ, the same stratum of radiating matter will be found to lie at some other value of τ_λ, for each value of T chosen. The variations of $\tau_\lambda(T)$ with λ is found to be pronounced, hence κ_λ must depend on λ and T; its variation may be established empirically as follows: From the definition of τ_λ, we have that

[2] This expression, although useful as an interpolation device, should not be construed to mean that generally we can assume b_λ is a linear function of τ_λ. Theoretical models predict otherwise.

$$\frac{d\tau_\lambda}{dT} = \rho\kappa_\lambda \frac{dx}{dT} = \rho\varepsilon_H \frac{1 - x_H}{m_H}\left[\alpha_\lambda{}^0(H^-)\, P_\varepsilon + \alpha_\lambda\,(H)\right]\frac{dx}{dT} \qquad (5\text{–}19)$$

if only negative hydrogen ions and hydrogen atoms contribute to the continuous absorption. Here $\alpha_\lambda{}^0(H^-)$ is the absorption coefficient per H-atom at unit electron pressure (corrected for negative absorptions) and $\alpha_\lambda(H)$ is the corresponding coefficient for atomic hydrogen. x_H denotes the degree of ionization of hydrogen, ε_H is the mass of hydrogen per gram of stellar material, and m_H is the mass of the hydrogen atom.

Now $d\tau_\lambda/dT = [d\tau_\lambda/db_\lambda(\tau)]\ [db_\lambda(\tau)/dT]$ may be calculated directly with the aid of Eq. 5–16 and the Planck function. For a chosen depth in the sun, ρ, dx/dT, $(1 - x_H)/m_H$, and P_ε are all fixed and the bracket in Eq. 5–19 can be computed from tables of the absorption coefficients. Although comparison of the results clearly demonstrates the negative hydrogen ion to be the principal source of continuous absorption in the solar atmosphere, there remain a number of discrepancies between the empirical κ_λ determined from $d\tau_\lambda/dT$ and the theoretical absorption coefficient. Some of these discordances are closely connected with errors in $T(\tau_\lambda)$ and may be summarized as follows:

1. Errors exist in the theoretical continuous absorption coefficient, particularly in the infrared, and in the ultraviolet where substantial revisions are indicated.

The data of Pierce and Waddell clearly demonstrate that atomic hydrogen and the negative hydrogen ion do not give a large enough absorption in spectral regions shortward of about λ4500. Some absorption must be contributed by close-packed metallic lines; some may be due to other causes such as transitions between two unstable states of a quasi hydrogen molecule (Stecher, 1962). The problem needs careful examination.

2. Errors in the observed value I_λ (0,0) probably constitute the greatest source of uncertainty (Pierce and Waddell, 1960). A 3 per cent error in I_λ can produce a 3–30 per cent error in the gradient $d\tau_\lambda/dT$.
3. Errors in the limb darkening $\phi_\lambda(\theta)$ are much less serious, except perhaps near the solar limb. Plaskett (1955) estimates that, with good data, the percentage error of the limb darkening is probably less than 2 per cent.
4. Failure of the plane-parallel approximation to the atmospheric structure is caused by temperature fluctuations and "roughness" of the surface associated with granulation and produced by turbulence.
5. Deviations from local thermodynamic equilibrium (LTE), i.e., failure of Kirchhoff's law, are probably unimportant in the continuum (see Art. 8–17; also Pagel, 1959), but can become important for spectral lines.
6. At the extreme limb, the curvature of the layers must be considered (e.g., Pagel, 1956; Proisy, 1959).

Notice that insofar as the continuum observations are concerned, the temperature distribution of the solar atmosphere is defined only for optical depths greater than 0.1. One may extrapolate $T(\tau_\lambda)$ to shallower depths with the aid of Eq. 5–16, but we must emphasize that the boundary temperature obtained in this way is not necessarily reliable. Fig. 5–3 shows the dependence of $\theta = 5040/T$ on the common logarithm of the optical depth at λ5000, as determined by several investigators. Notice that whereas all results are in good accord between $\tau \sim 0.2$ and $\tau \sim 3$, there are marked discordances in shallower layers. The Pierce-Waddell (1960) results closely corroborate an earlier Michigan solar model (Pierce-Aller, 1952). Plaskett finds an apparently too high boundary temperature; that of Przybylski seems more reliable. Pagel (1956) used continuum observations obtained by the High Altitude Observatory at the 1952 eclipse to get the solar model near the limb and in the low chromosphere, for $\tau_0 > 10^{-5}$. He calculated the darkening at the extreme limb, taking into account the curvature of the layers. As Neckel and Pagel used line data, their results may be affected by deviations from LTE.

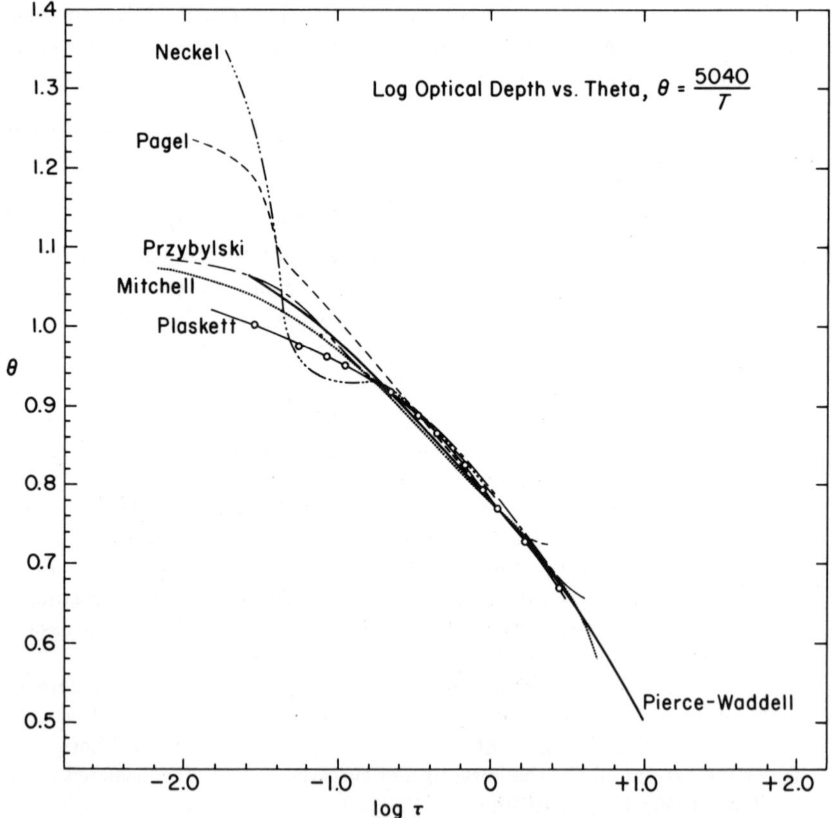

Fig. 5–3. Relation Between Optical Depth and Temperature for the Solar Atmosphere

These solar temperature distributions are strictly empirical, i.e., they involve no assumption concerning the nature of the flow of energy through the superficial layers. In Art. 5–9 the empirical $T(\tau_\lambda)$ is compared with theoretical predictions based on the flow of energy by radiation.

5–4. A Model Solar Atmosphere. With a knowledge of the coefficient of continuous absorption and of the temperature distribution with optical depth, we may now investigate the structure of the solar atmosphere. One reason for calculating the temperature and density distribution is to obtain basic data needed for precise theoretical studies of absorption line profiles. If variations of gas pressure, electron pressure, density, temperature, and absorption coefficient are known as a function of depth, the change of ionization of the element of interest and the line absorption coefficient may be calculated, provided the necessary atomic parameters for emission and absorption of energy are known. From the $T(\tau_{\lambda 0})$ relationships given in Fig. 5–3 we adopt an appropriate temperature distribution with optical depth at $\lambda_0 = 5000$A.

To make headway with this problem we suppose the atmosphere to be in hydrostatic equilibrium, i.e., at each point the gas pressure supports the weight of the overlying layers. A slab of density ρ of unit cross-sectional area, and of thickness dx, subject to an acceleration of gravity g will weigh $g\rho\,dx$; hence the increase in pressure with an increment dx in depth must be

$$dP = g\rho\,dx \tag{5–20}$$

where x is measured downward into the star. Now introduce the optical depth τ defined in terms of the absorption coefficient calculated for a wavelength λ_0, viz.

$$d\tau = \kappa\rho\,dx \tag{5–21}$$

and there results

$$\frac{dP}{d\tau} = \frac{g}{\kappa} \tag{5–22}$$

as the equation that must be satisfied to obtain the variation of gas pressure with τ.[3] For the sun, g may be taken as the gravitational acceleration; we shall temporarily assume that the levitational effects of large-scale mass motions, i.e., turbulence, can be neglected. The gas pressure, P_g, is associated with a kinetic temperature T which is customarily identified with the local temperature given by the empirical $T(\tau)$ relation described in Art. 5–3.

To determine the gas pressure, P_g, as a function of τ, it is necessary to determine κ as a function of P_g and $\theta = 5040/T$ and to solve Eq. 5–22 by numerical integration. Since κ depends on P_ε and T (see Chapter 4) we

[3] In this article we use κ and τ to denote the absorption coefficient and optical depth at the "standard" wavelength λ_0.

can find $\kappa(P_g, T)$ from $\kappa(P_\varepsilon, T)$ and $P_g(P_\varepsilon, T)$ (see Chapter 3) both of which depend in a known fashion on the assumed composition. Therefore, at the outset we must postulate the chemical composition of the sun, or at least A, the ratio of hydrogen to the metals by numbers of atoms. Also, it is necessary to specify B the He/H ratio by numbers of atoms. Helium does not contribute to the opacity but adds to the total weight of the material in the layers. The calculations in Chapter 3 were based on $\log A = 4.133$, $B = 1/7$.

One may integrate Eq. 5–22 in several ways; in place of τ we may use θ as the variable of integration. Multiply Eq. 5–22 by $2P_g\,d\tau$ and integrate to obtain

$$P_g^2 = 2g\int\frac{P_g}{\kappa(P_g, \theta)}\,d\tau = 2g\int\frac{P_g}{\kappa(P_g, \theta)}\frac{d\tau}{d\theta}\,d\theta \tag{5–23}$$

where $\theta = 5040/T$. The quantity $d\tau/d\theta$ may be computed once and for all and the equation solved quickly by iteration.

To start the integrations we suppose that all metals are singly ionized and that they alone contribute electrons. Then

$$P_g \sim P_\varepsilon A(1 + B) \tag{5–24}$$

In spectral regions near λ5000 we have that

$$\kappa_\lambda = \frac{\varepsilon_{\text{H}}}{M_{\text{H}}}(1 - x_{\text{H}})\,[P_\varepsilon\alpha_\lambda{}^0(\text{H}^-) + \alpha_\lambda(\text{H})] \tag{5–25}$$

If the negative hydrogen ion is the source of continuous absorption, we may write

$$\kappa = \frac{\varepsilon_{\text{H}}}{M_{\text{H}}}P_\varepsilon\,\alpha_\lambda{}^0\,(\text{H}^-) \tag{5–26}$$

where ε_{H} is the mass of hydrogen per gram of stellar material, M_{H} is the mass of a hydrogen atom, $\alpha_\lambda{}^0(\text{H}^-)$ is the absorption coefficient (including negative absorptions) calculated per neutral hydrogen atom. The degree of ionization of hydrogen x_{H} is small. Hence

$$P_g^2(\tau) = \frac{2g\,M_{\text{H}}\,A(1 + B)}{\varepsilon_{\text{H}}}\int_0^\tau\frac{dt}{\alpha_\lambda{}^0(\text{H}^-)} \tag{5–27}$$

yields an initial estimate for P_g. To perform a second iteration, one may find P_ε for each P_g from the $P_g\,(P_\varepsilon, \theta)$ relation, calculate κ_λ from Eq. 5–25 and integrate Eq. 5–23 directly.

An alternate procedure is to integrate in terms of a variable

$$y = \log_{10}\tau \tag{5–28}$$

Then, Eq. 5–22 may be transformed to

$$\frac{d\log P_g}{d\log\tau} = \frac{g\,\tau}{\kappa\,P_g} \tag{5–29}$$

so that with $p = \log P_g$,

$$\frac{dp}{dy} = \frac{g\,\tau\, M_H}{P_g \alpha^0(H^-) P_\varepsilon \varepsilon_H} \tag{5–30}$$

since κ is given by Eq. 5–26 throughout the relevant layers of the solar atmosphere. The integration is started by using the asymptotic form, Eq. 5–27, noting that for $y < -2$, $\alpha_\lambda{}^0(H^-)$ may be taken as constant. Once the initial values are found, Eq. 5–30 may be integrated step by step.

One reason for using y rather than τ as a variable of integration is that y is more nearly proportional to linear depth in the atmosphere than is τ. The relation between τ and linear depth x may be established as follows: From the perfect gas law

$$P = \frac{\mathcal{R}\rho T}{\mu} \tag{3–4}$$

TABLE 5–1

Model of the Solar Atmosphere

Log τ_0	θ	Log P_g	Log P_ε	x (km)
−2.0	1.073	4.290	−0.030	−115
−1.9	1.069	4.330	+0.005	−105
−1.8	1.065	4.372	+0.042	− 95
−1.7	1.059	4.417	+0.087	− 85
−1.6	1.052	4.464	+0.136	− 74
−1.5	1.044	4.512	+0.190	− 62
−1.4	1.033	4.562	+0.250	− 50
−1.3	1.021	4.612	+0.311	− 38
−1.2	1.007	4.663	+0.375	− 26
−1.1	0.991	4.715	+0.460	− 13
−1.0	0.977	4.766	+0.550	0
−0.9	0.957	4.815	+0.658	+ 12
−0.8	0.940	4.861	+0.763	+ 24
−0.7	0.922	4.906	+0.865	+ 36
−0.6	0.904	4.949	+0.971	+ 48
−0.5	0.885	4.990	+1.094	+ 59
−0.4	0.865	5.029	+1.213	+ 70
−0.3	0.845	5.066	+1.338	+ 81
−0.2	0.825	5.101	+1.478	+ 91
−0.1	0.804	5.133	+1.619	+101
0.0	0.782	5.162	+1.793	+110
+0.1	0.759	5.187	+1.983	+117
+0.2	0.736	5.209	+2.147	+125
+0.3	0.706	5.229	+2.328	+132
+0.4	0.687	5.248	+2.485	+139
+0.5	0.660	5.264	+2.718	+144
+0.6	0.631	5.277	+3.020	+149

Hydrogen/metal ratio = 7.37×10^{-5} or $\log A = 4.133$.
Fraction by weight of $H = 0.613$. The assumed abundances are given in Art. 3–10.
$\tau_0 = \tau(\lambda_0)$. $\lambda_0 = 5000$A. Pressures are expressed in units of dynes cm^{-2}.

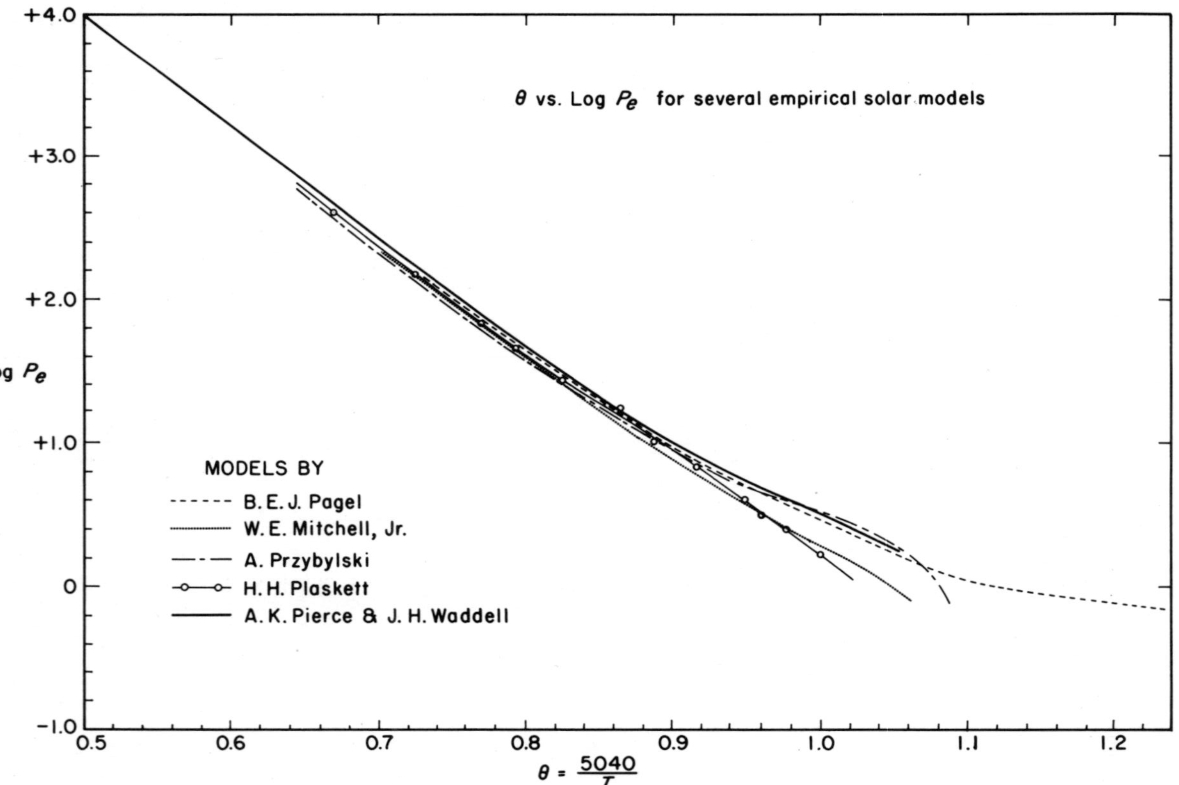

FIG. 5–4. RELATION BETWEEN $\theta = 5040/T$ AND GAS PRESSURE FOR SEVERAL EMPIRICAL SOLAR MODEL ATMOSPHERES

we note that in the solar atmosphere μ is determined by hydrogen and helium and is therefore essentially constant. Eliminate the density from Eq. 5–20 by means of Eq. 3–4 and we get

$$dx = \frac{\mathcal{R}T}{\mu g}\, d \ln P_g \tag{5-31}$$

which must be solved numerically to relate P_g with x and ultimately τ with x (last column of Table 5–1). The zero point of x is, of course, arbitrary. Table 5–1 gives the results of a numerical integration of the solar atmosphere for abundances as given in Art. 3–10. Successive columns give $\log \tau_0$, θ, $\log P_g$, $\log P_\varepsilon$, and x.

Fig. 5–4 compares the relation between θ and $\log P_\varepsilon$ for solar model atmospheres based on empirical temperature distributions as established by various authors. Notice that although divergences exist in the superficial layers, fairly good agreement is obtained in the deeper strata. The $P_g(\theta)$ or $P_g(\tau)$ relationships will differ depending on the assumed values of the He/H and metal/hydrogen ratios.

5–5: Flow of Radiation Through a Stellar Atmosphere. In calculations of a model solar atmosphere we were able to employ empirical determinations of the temperature gradient. For other stars, sufficiently accurate limb-darkening measurements cannot be obtained to permit such a study and one must rely on theoretical considerations to establish $T(\tau_\lambda)$ and ultimately the model atmosphere. As we noted in Art. 5–1, the temperature gradient depends on the mode of energy transport in the upper layers of a stellar atmosphere. Observations show that radiative transfer rather than convection plays the dominating role, at least for the sun.

We may easily see qualitatively how the intensity and directional distribution of the radiation vary as a function of depth in a star. At the surface there will be no backward flow of radiation and $I(\theta)$ will be zero for $\theta > \pi/2$. Below the surface $I(\theta)$ will be positive for all θ, i.e., there will be backward as well as forward flowing radiation, but the outward directed radiation will still predominate. See Fig. 5–5. At great depths below the surface, $I(\theta)$ will be only slightly larger for $\theta < \pi/2$ than for $\theta > \pi/2$. Almost as large a fraction of the radiation will be flowing backwards as forwards, but the same general property of the radiative field is preserved in that there is a net outward flow of radiation. Insofar as the integrated radiation is concerned, the net flux must remain constant, unless convection currents act to carry the energy.

At a point x below the surface of the star, consider a pencil of radiation of any frequency ν and of intensity $I_\nu(\theta)$ defined by a cone of solid angle $d\omega$ which makes an angle θ with respect to the outward normal (see Fig. 5–2). Let this radiation fall upon an elementary cylinder of unit cross-section and height ds, so placed that the axis of the radiation pencil is

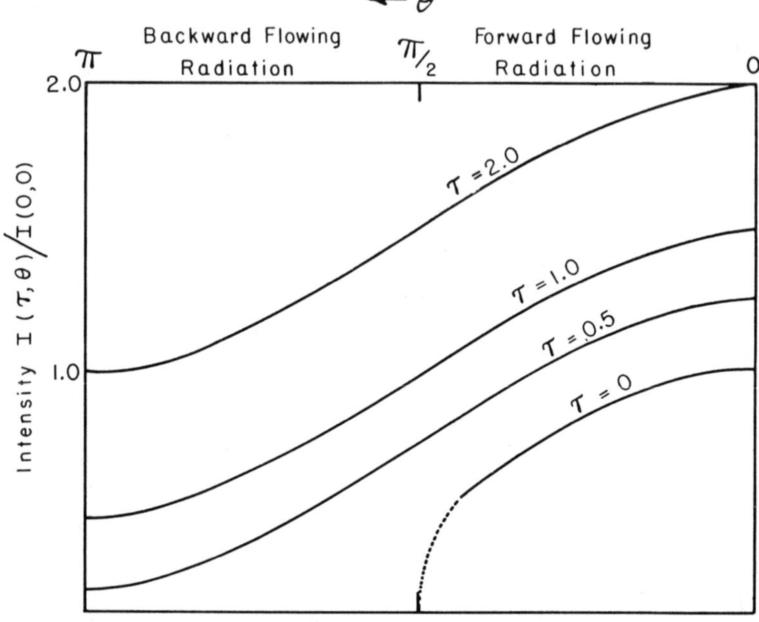

FIG. 5–5. ANGULAR DISTRIBUTION OF INTENSITY $I(\theta)$ AS A FUNCTION OF OPTICAL DEPTH

These curves are intended to show the qualitative behavior of the intensity $I(\tau, \theta)$ as a function of the angle θ at different optical depths τ.

concentric with the axis of the cylinder. In passing through the cylinder the intensity will be cut down by an amount

$$dI_\nu(\theta, x)\, d\omega = \kappa_\nu \rho I_\nu(\theta, x)\, ds\, d\omega \tag{5–32}$$

where κ_ν is the absorption coefficient including the negative absorptions and ρ is the density. Our elementary cylinder also receives radiation from all sides. Since a steady state exists, it must either scatter or diffuse this radiation, or absorb and re-emit it as thermal radiation. Both processes occur simultaneously. Let the mass emissivity of the material be denoted by j_ν. In the most general problems of energy transfer, j_ν will depend on θ as well as on x. The amount of energy emitted per second in solid angle $d\omega$ will be $(d\omega/4\pi)\, \rho\, j_\nu\, ds$. Hence, in passing through the elementary cylinder the beam suffers a net change in intensity

$$\begin{aligned} dI_\nu(\theta, x)\, d\omega &= -\kappa_\nu \rho I_\nu(\theta, x)\, ds\, d\omega + \frac{j_\nu(x, \theta)}{4\pi} \rho\, ds\, d\omega \qquad (5\text{–}33) \\ \text{(net change)} &= -\text{(energy absorbed)} + \text{(energy re-emitted)} \end{aligned}$$

We make use of Eq. 5–1, except that we now regard x as being measured positive upward while the optical depth τ_ν is measured positive downward into the star. Thus

$$d\tau_\nu = -\kappa_\nu \rho \, dx \tag{5-34}$$

and the transfer equation becomes

$$\cos \theta \frac{dI_\nu(\theta, \tau_\nu)}{d\tau_\nu} = I_\nu(\tau, \theta) - S_\nu(\theta, \tau_\nu) \tag{5-35}$$

where

$$\frac{j_\nu(\theta, \tau_\nu)}{4\pi \kappa_\nu} = S_\nu(\theta, \tau_\nu) \tag{5-36}$$

is called the source function. Before we can solve Eq. 5-35 we must know the functional form of $S_\nu(\theta, \tau_\nu)$. In an atmosphere that scatters light, extinction occurs when a pencil of radiation impinges upon a volume element and some of it is thrown out of the beam, i.e., diffused in other directions. By the same token, the radiant energy in any specified direction is augmented by the scattering of light from other directions into the particular direction under consideration. The angular distribution of the radiation scattered when a pencil of light falls upon a volume element of the medium is given by the phase function $p[\cos \Theta]$, so defined that $p[\cos \Theta]\, d\omega/4\pi$ specifies the probability of radiation being scattered in a direction Θ with respect to the direction of the incident beam. If scattering occurs according to Rayleigh's law, as with gas molecules,

$$p[\cos \Theta] = \tfrac{3}{4}[1 + \cos^2 \Theta] \tag{5-37}$$

while for isotropic scattering,

$$p[\cos \Theta] = 1 \tag{5-38}$$

In a purely scattering atmosphere with no energy losses by heating, etc.

$$\int p[\cos \Theta] \frac{d\omega}{4\pi} = 1$$

In the context of stellar continuous spectra, the situation of greatest interest is that of thermal radiation. The volume element is heated by radiation received from its incandescent surroundings, and because it is hot it radiates energy. If the total emission is of strictly thermal origin, all impinging radiation that is absorbed will be re-emitted as purely thermal energy uniformly over all directions. The emission at each frequency then will be given by Kirchhoff's law (see Eq. 4-37), and the source function will be simply

$$S_\nu(\tau_\nu) = B_\nu(\tau_\nu) = B_\nu[T(\tau_\nu)] \tag{5-39}$$

The local temperature, T, is governed by the condition that the total amount of energy absorbed over all frequencies must equal that emitted, viz.,

$$\rho ds \int_0^\infty j_\nu \, d\nu = \rho \, ds \iint_0^\infty \kappa_\nu I_\nu(\theta, x) \, d\omega \, d\nu \tag{5-40}$$

This is the equation of radiative equilibrium, which holds for the radiation integrated over all frequencies. In general, it is not valid for the monochromatic intensity, $I_\lambda(\theta)$, i.e., j_ν/κ_ν does not equal $\int I_\nu(\theta, x)\, d\omega$. The constancy of flux condition is equivalent to radiative equilibrium. We easily show this for thermal equilibrium when Kirchhoff's law holds. Multiply Eq. 5–35 by $\rho\kappa_\nu\, d\omega$ and integrate over all angles. Then integrate over all frequencies and use the definition of F_ν from Eq. 4–4:

$$\frac{d}{dx}\iint F_\nu\, d\nu\, d\omega = \rho\int \kappa_\nu \int I_\nu(\theta, x)\, d\omega\, d\nu - \rho\iint \kappa_\nu B_\nu\, d\omega\, d\nu = 0 \tag{5–41}$$

Noting that, since $j_\nu = 4\pi\,\kappa_\nu B_\nu$, we recover Eq. 5–40,

$$\iint \kappa_\nu B_\nu\, d\omega\, d\nu = 4\pi\int \kappa_\nu B_\nu\, d\nu = \int j_\nu\, d\nu = \iint \kappa_\nu I_\nu(\theta, x)\, d\omega\, d\nu$$

The local temperature T is defined in terms of the energy density at that point (see Eqs. 4–8 and 4–12),

$$u = aT^4 = \frac{1}{c}\int I(\theta, x)\, d\omega \tag{5–42}$$

Normally this temperature is identical with the gas kinetic temperature at the point in question. Under conditions of LTE, the same value of T enters in both Boltzmann's and Saha's equations.

In some stellar atmospheres, both continuous absorption and scattering may occur. Let κ_ν denote the coefficient of pure thermal absorption corrected for negative absorptions and σ_ν the coefficient of pure scattering. The loss of beam intensity in passing through a cylinder of length ds will be

$$-(\kappa_\nu + \sigma_\nu)\,\rho\, ds\, I_\nu(\theta, x)\, d\omega$$

The energy returned to the beam will consist of a contribution from thermal emission,

$$j_\nu\,\rho\, ds\,\frac{d\omega}{4\pi} = \kappa_\nu\,\rho\, ds\, B_\nu(T)\, d\omega$$

plus radiation that is simply scattered from other directions

$$ds\,\frac{d\omega}{4\pi}\int \sigma\, p(\cos\Theta)\, I_\nu\, d\omega$$

where the integration is carried out over all angles. Here Θ is the angle between the beam direction and the direction from which the radiation is scattered. Adding up the gains and losses we find for the equation of transfer,

$$\cos\theta\,\frac{dI_\nu}{\rho dx} = -(\kappa_\nu + \sigma_\nu)\, I_\nu + \sigma_\nu\int I_\nu\, p(\cos\Theta)\,\frac{d\omega}{4\pi} + \kappa_\nu B_\nu(T) \tag{5–43}$$

where $dx = ds\cos\theta$. Recall that s is measured positive towards the surface of the star.

The nature of this equation shows that absorption and scattering cannot be regarded as equivalent processes, even when scattering is isotropic, unless both κ_ν and σ_ν are independent of wavelength. Insofar as stellar continuous spectra are concerned, we are mostly involved with thermal absorption and emission according to Kirchhoff's law. Only in the hottest stars and in certain supergiants does electron scattering become important. In a rigorous treatment one must employ Eq. 5–43 and take into account the anisotropy of the radiation. Chandrasekhar has solved this difficult problem for a grey atmosphere. For many problems (particularly for non-grey atmospheres), we may take electron scattering as isotropic in order to derive the general properties of the radiation field. Let us define

$$d\tau \equiv -(\kappa_\nu + \sigma)\,\rho\,dx \tag{5–44}$$

employ Eq. 5–38 in Eq. 5–43, and recall the definition of the mean intensity J (see Eq. 4–2). If we write

$$S_\nu(\tau_\nu) = \frac{\sigma}{\kappa_\nu + \sigma} J_\nu + \frac{\kappa_\nu}{\kappa_\nu + \sigma} B_\nu = (1 - L_\nu)\,J_\nu + L_\nu B_\nu \tag{5–45}$$

where

$$L_\nu = \frac{\kappa_\nu}{\kappa_\nu + \sigma} \tag{5–46}$$

we recover Eq. 5–35, except that $d\tau_\nu$ is now defined by Eq. 5–44 rather than by Eq. 5–34. We can grasp the physical significance of the source function $S_\nu(\tau_\nu)$ perhaps a little more clearly with the aid of the following considerations. Consider our elementary radiating volume at an optical depth τ. The thermal emissivity in the continuum per unit solid angle is $(1/4\pi)\,j\rho\,ds = \kappa_\nu\,B_\nu \sec\theta\,\rho\,dx$. The light scattered by this volume element is

$$\sigma\,J\,\rho\,ds = \sigma\,J\,\rho \sec\theta\,dx$$

and the total emission from the volume element will be

$$[(1 - L_\nu)J_\nu + L_\nu B_\nu]\,d\tau_\nu \sec\theta = S_\nu(\tau_\nu) \sec\theta\,d\tau_\nu$$

The contribution from this volume element is dimmed by absorption in the upper strata, the extinction factor being $\exp(-\tau_\nu \sec\theta)$. The intensity of the emergent ray therefore is the integral over all radiating elements, viz.,

$$I_\nu(0, \theta) = \int_0^\infty S_\nu(\tau)\,e^{-\tau_\nu \sec\theta} \sec\theta\,d\tau_\nu \tag{5–47}$$

which is a generalized form of Eq. 5–5.

For many problems we need to know not only the emergent intensity but also the mean intensity J_ν (see Eq. 4–2) at a given point τ. It will be the sum of contributions from radiating elements on all sides. See Fig. 5–6. The contribution to J_ν from an elementary volume at an optical depth t_ν, in a direction θ, as seen from the point τ_ν, will be

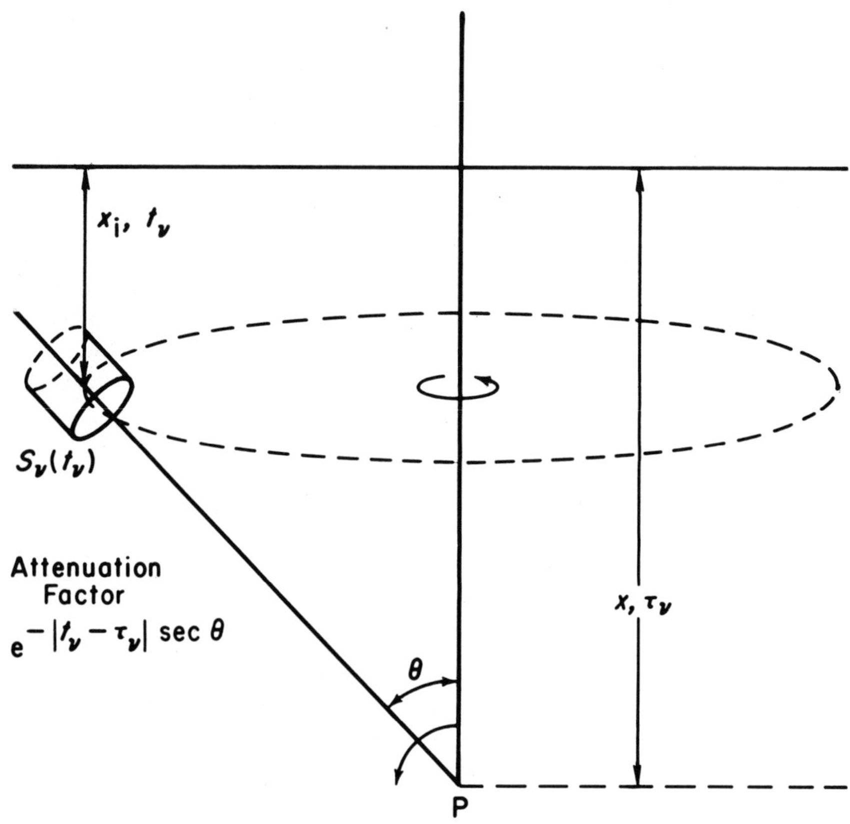

FIG. 5–6. GEOMETRICAL SITUATION FOR CALCULATION OF MEAN INTENSITY J_ν AT A DEPTH τ_ν

$$dI_\nu(\theta, \tau, t) = S_\nu(t_\nu) \exp\left[- \mid (t_\nu - \tau_\nu) \sec\theta \mid \right] \sec\theta \, dt_\nu \qquad (5\text{–}48)$$

and the contribution from all such elements integrated over all angles and to all depths will be

$$4\pi J_\nu(\tau_\nu) = 2\pi \int_{\tau_\nu}^{\infty} S_\nu(t_\nu) \int_0^{\pi/2} \exp\left[- (t_\nu - \tau_\nu) \sec\theta\right] \sec\theta \sin\theta \, d\theta \, dt_\nu$$
$$- 2\pi \int_0^{\tau_\nu} S_\nu(t_\nu) \int_{+\pi/2}^{\pi} \exp\left[-(t_\nu - \tau_\nu) \sec\theta\right] \sec\theta \sin\theta \, d\theta \, dt_\nu \qquad (5\text{–}49)$$

Now put $\sec\theta = y$ in the first integral and $\sec\theta = -y$ in the second. Notice that in the first integral $(t_\nu - \tau_\nu)$ is positive, and in the second it is negative. Making use of Eq. 5–11 we find that Eq. 5–49 may be written in the form

$$J_\nu(\tau_\nu) = \tfrac{1}{2} \int_0^{\infty} S_\nu(t_\nu) \, E_1 \left(\mid t_\nu - \tau_\nu \mid \right) dt_\nu \qquad (5\text{–}50)$$

Similarly, the flux is given by

$$F_\nu(\tau_\nu) = 2\int_{\tau_\nu}^{\infty} S_\nu(t_\nu)\, E_2(t_\nu - \tau_\nu)\, dt_\nu - 2\int_0^{\tau_\nu} S_\nu(t_\nu)\, E_2(\tau_\nu - t_\nu)\, dt_\nu \quad (5\text{–}51)$$

Now substitute Eq. 5–50 into Eq. 5–45 to obtain

$$S_\nu(\tau_\nu) = L_\nu B_\nu + \tfrac{1}{2}(1 - L_\nu)\int_0^{\infty} E_1(|t_\nu - \tau_\nu|)\, S(t_\nu)\, dt_\nu \quad (5\text{–}52)$$

Notice that Eq. 5–52 is an integral equation of the form

$$f(x) = g(x) + \int h(x - y)\, f(y)\, dy$$

where g and h are known functions, and f is to be determined. Often g and h will be given in tabular form and the equation must be solved by an iteration method. Suppose that $T(\tau_\nu)$ is known (or assumed) and we wish to calculate the corresponding radiation field, i.e., J_ν and F_ν. We must determine $S_\nu(\tau_\nu)$ by solving the integral Eq. 5–52. We follow a procedure suggested by Strömgren. Let

$$Y_\nu = J_\nu - B_\nu \quad (5\text{–}53)$$

Then from Eqs. 5–45 and 5–50 we obtain

$$Y_\nu = -B_\nu + \bar{B}_\nu + \tfrac{1}{2}\int_0^{\infty} E_1(|t_\nu - \tau_\nu|)\,(1 - L_\nu) Y_\nu dt_\nu \quad (5\text{–}54)$$

where

$$\bar{B}_\nu = \tfrac{1}{2}\int_0^{\infty} B_\nu E_1(|t_\nu - \tau_\nu|)\, dt_\nu \quad (5\text{–}55)$$

Suppose that $Y_\nu(\tau)$ can be expressed as a series of terms of the form

$$Y_\nu(\tau) = Y_\nu^0 + \Delta^1 Y_\nu + \Delta^2 Y_\nu + \cdots + \Delta^n Y_\nu \quad (5\text{–}56)$$

where

$$Y_\nu^0 = -B_\nu + \bar{B}_\nu$$

$$\Delta^n Y_\nu = \tfrac{1}{2}\int_0^{\infty} E_1(|t_\nu - \tau_\nu|)\,(1 - L_\nu)\,\Delta^{n-1} Y_\nu dt = \overline{(1 - L_\nu)\,\Delta^{n-1} Y_\nu} \quad (5\text{–}57)$$

The solution may be obtained as follows: First prepare a table of $B(\tau)$ and compute Y_ν^0. Substitute Y_ν^0 in the second integral of Eq. 5–54 to obtain $\Delta^1 Y_\nu$. Next calculate $(1 - L_\nu)\,\Delta^1 Y_\nu$, evaluate $\Delta^2 Y_\nu$, and continue the process until successive terms become negligible. Once Y_ν is found, the source function may be calculated with the aid of Eq. 5–45 and J_ν and F_ν evaluated.

Thus we need a rapid means for evaluating integrals of the form

$$\int f(t) E_1(|t' - t|)\, dt'$$

The approximation formula,

$$\frac{1}{2}\int_0^\tau f(t)E_1(\tau - t)\,dt = a_1 f(t_1) + a_2 f(t_2) \tag{5–58}$$

suggested by A. Reiz facilitates the calculation of integrals of the type of Eqs. 5–55 and 5–57. Table 5–2 lists the weights a_1 and a_2 and the points t_1 and t_2 at which $f(t)$ [which is to be identified with $B(t)$ in Eq. 5–55] is to be evaluated. Similarly,

$$\frac{1}{2}\int_0^\infty f(y)E_1(y)\,dy = b_1 f(y_1) + b_2 f(y_2)$$

TABLE 5–2

DATA FOR EVALUATION OF INTENSITY INTEGRALS

$$J(\tau) = \frac{1}{2}\int_0^\infty B(t)E_1(|\tau - t|)\,dt = a_1 B(t_1) + a_2 B(t_2) + 0.4532B(\tau + 0.292) + 0.0468B(\tau + 2.507)$$

Weighting coefficients a_1, a_2 and points t_1, t_2 at which function $B(t)$ is to be evaluated

τ	t_1	a_1	t_2	a_2
0.20	0.051	0.0851	0.169	0.1278
0.40	0.106	0.1094	0.342	0.1959
0.60	0.167	0.1198	0.520	0.2421
0.80	0.232	0.1224	0.699	0.2772
1.00	0.303	0.1214	0.881	0.3044
1.20	0.378	0.1182	1.064	0.3262
1.40	0.459	0.1138	1.249	0.3442
1.60	0.544	0.1090	1.436	0.3591
1.80	0.635	0.1040	1.624	0.3716
2.0	0.731	0.0991	1.812	0.3821
2.2	0.832	0.0944	2.002	0.3912
2.4	0.939	0.0899	2.193	0.3989
2.6	1.051	0.0857	2.385	0.4056
2.8	1.168	0.0819	2.577	0.4113
3.0	1.290	0.0783	2.770	0.4164
3.2	1.417	0.0751	2.964	0.4208
3.4	1.549	0.0721	3.158	0.4246
3.6	1.686	0.0694	3.353	0.4280
3.8	1.828	0.0670	3.548	0.4310
4.0	1.974	0.0648	3.744	0.4336
4.2	2.125	0.0628	3.940	0.4359
4.4	2.280	0.0610	4.137	0.4380
4.6	2.438	0.0594	4.334	0.4398
4.8	2.601	0.0580	4.531	0.4414
5.0	2.766	0.0567	4.728	0.4428
6.0	3.640	0.0521	5.719	0.4477
8.0	5.530	0.0482	7.710	0.4518
10.0	7.500	0.0472	9.708	0.4528

Courtesy, S. Chandrasekhar, *Radiative Transfer* (Fair Lawn, N. J.: Oxford University Press, 1950), p. 67 (reprinted by Dover Publications) for entries 0.20 to 4.8. The values 5.0 to 10.0 were kindly supplied by Jean K. McDonald in advance of publication.

where

$$y_1 = 0.292 \qquad b_1 = 0.4532$$
$$y_2 = 2.507 \qquad b_2 = 0.0468$$

TABLE 5–3

DATA FOR THE EVALUATION OF THE FLUX INTEGRALS

$$F(\tau) = 2\int_\tau^\infty B(t)E_2(t-\tau)\,dt - 2\int_0^\tau B(t)E_2(\tau - t)\,dt$$
$$= 0.8839B(\tau + 0.397) + 0.1161B(\tau + 2.723) - d_1B(t_1) - d_2B(t_2)$$

Weighting coefficients d_1, d_2 and points t_1, t_2 at which function $B(t)$ is to be evaluated

τ	t_1	d_1	t_2	d_2
0.10	0.022	0.0786	0.080	0.0889
0.20	0.046	0.1346	0.162	0.1615
0.30	0.071	0.1753	0.244	0.2247
0.40	0.096	0.2055	0.328	0.2800
0.50	0.124	0.2279	0.412	0.3289
0.60	0.152	0.2443	0.498	0.3726
0.70	0.181	0.2561	0.584	0.4118
0.80	0.212	0.2642	0.671	0.4472
0.90	0.244	0.2694	0.758	0.4792
1.00	0.276	0.2723	0.846	0.5083
1.10	0.311	0.2734	0.935	0.5349
1.20	0.346	0.2730	1.024	0.5591
1.30	0.382	0.2715	1.114	0.5814
1.40	0.420	0.2690	1.204	0.6018
1.50	0.459	0.2659	1.294	0.6206
1.60	0.499	0.2622	1.386	0.6380
1.70	0.540	0.2581	1.477	0.6540
1.80	0.583	0.2537	1.569	0.6688
1.90	0.627	0.2492	1.661	0.6826
2.00	0.672	0.2444	1.753	0.6953
2.2	0.766	0.2348	1.940	0.7182
2.4	0.865	0.2252	2.127	0.7380
2.6	0.970	0.2158	2.315	0.7553
2.8	1.079	0.2070	2.504	0.7704
3.0	1.194	0.1986	2.695	0.7836
3.2	1.314	0.1907	2.886	0.7952
3.4	1.439	0.1834	3.078	0.8054
3.6	1.569	0.1767	3.270	0.8145
3.8	1.704	0.1706	3.464	0.8225
4.0	1.844	0.1650	3.658	0.8295
6.0	3.457	0.1311	5.621	0.8684
8.0	5.324	0.1201	7.608	0.8798
10.0	7.287	0.1170	9.604	0.8830

As τ becomes larger, d_1 approaches 0.1161 and d_2 approaches 0.8839 while $(\tau - t_1)$ approaches 2.723 and $(\tau - t_2)$ approaches 0.397.

Courtesy, S. Chandrasekhar, *Radiative Transfer*, p. 68 for entries $\tau = 0.10$ to $\tau = 2.0$. The values for $\tau = 2.2$ to 3.8 are taken from a table by B. Strömgren kindly supplied in advance of publication. The values for $\tau = 4.0$ to 10.0 were calculated by Jean K. McDonald.

Thus

$$\frac{1}{2}\int_{\tau}^{\infty} f(t)E_1(t-\tau)\,dt = 0.4532f(\tau+0.292) + 0.0468f(\tau+2.507) \tag{5–59}$$

and integrals like Eq. 5–55 are the sums of Eqs. 5–58 and 5–59. For the evaluation of the flux integral Eq. 5–51, one may use an expression similar to Eq. 5–58 and calculate appropriate coefficients (see Table 5–3). The emergent flux at the surface of the star is given by

$$F_\nu(\tau=0) = c_1B(t_1) + c_2B(t_2) \tag{5–60}$$

where

$$\begin{aligned} t_1 &= 0.397 & c_1 &= 0.8839 \\ t_2 &= 2.723 & c_2 &= 0.1161 \end{aligned} \tag{5–61}$$

The selection of points t_1, t_2, t_3, etc., and coefficients a_1, a_2, b_1, b_2, etc., depends on the type of integration formula employed. Those employed here are based on the gaussian method of integration (see Chandrasekhar, 1950, for details). Other choices are possible (Kourganoff and Pecker, 1950; Reiz, 1954) and may possess certain advantages.

Cayrel (1960) suggested that equations of the type Eqs. 5–50 and 5–51 be evaluated by approximation formulas of the types

$$J(\tau_i) = \sum_{k=1}^{10} a_{ik}\,S(\tau_k), \qquad F(\tau_i) = \sum_{k=1}^{10} b_{ik}\,S(\tau_k) \tag{5–62}$$

where $S(\tau)$ and $F(\tau)$ were fixed by their values at ten points, $\tau_k = 0.01 \times 2^k (k = 1, \ldots, 10)$. Cayrel tabulates the matrices (a_{ik}) and (b_{ik}). His method is particularly well suited to computing machines. For the emergent flux he proposes the six-point formula

$$\begin{aligned} F = {} & 0.1615\,S(.038) + 0.1346\,S(0.154) + 0.2973\,S(0.335) \\ & + 0.1872\,S(0.793) + 0.1906\,S(1.467) + 0.0288\,S(3.89) \end{aligned} \tag{5–63}$$

Three-term approximation formulas often suffice for many purposes, e.g.

$$\int_0^{\infty} e^{-x} f(x)\,dx = q_1f(x_1) + q_2f(x_2) + q_3f(x_3) \tag{5–64}$$

where

$$x_1 = 0.41577,\ x_2 = 2.2943,\ x_3 = 6.290;$$
$$q_1 = 0.71109,\ q_2 = 0.27852,\ q_3 = 0.01039$$

Integrals of the type indicated by Eqs. 5–50 and 5–51 occur so frequently in astrophysical literature that we define the operators:

$$\Lambda[f(x)] = \frac{1}{2}\int_0^{\infty} f(t)\,E_1\,(|t-x|)\,dt \tag{5–65}$$

and

$$\Phi\,[f(x)] = 2\int_x^{\infty} f(t)E_2\,(t-x)\,dt - 2\int_0^{x} f(t)E_2\,(x-t)\,dt \tag{5–66}$$

5–6. Eddington's Approximation. Various approximate solutions of the transfer Eq. 5–35 illustrate the character of radiation flow. The earliest (Schuster-Schwarzschild) approximation simply divided the radiation field into an incoming and an outgoing beam. In its original form this method fails to give the correct flux.

Eddington's approximation has often been used in astrophysical transfer problems, although it gives a poor solution near the surface. He defines three quantities

$$J = \frac{1}{4\pi}\int I(\theta)\, d\omega \quad H = \frac{1}{4\pi}\int I(\theta)\cos\theta\, d\omega \quad K = \frac{1}{4\pi}\int I\cos^2\theta\, d\omega \qquad (5\text{–}67)$$

where the integrations are extended over all solid angles. J is the mean intensity of the radiation, H is the net outward flow of radiant energy, and K is the radiation pressure at each point multiplied by the velocity of light. See Eqs. 4–2, 4–4, and 4–52.

Although J, H, and K are defined here for radiation integrated over all frequencies, they may also be specified for monochromatic radiation, viz., J_ν, H_ν, and K_ν. We consider here grey material and radiation integrated over all frequencies. Making use of the fact that $S = B$ (see Eq. 5–36) and that the condition of radiative equilibrium (see Eqs. 5–39 and 5–40) is $J = B$ for thermal radiation, we first multiply Eq. 5–35 by $d\omega/4\pi$ and integrate over all angles to obtain

$$\frac{dH}{d\tau} = J - B = 0 \text{ (i.e., } H = \text{const.)} \qquad (5\text{–}68)$$

Next, multiply Eq. 5–35 by $\cos\theta$ and integrate over all angles. Since J is independent of θ in accordance with our assumption concerning thermal emission

$$\frac{1}{4\pi}\frac{d}{d\tau}\int I(\theta)\cos^2\theta\, d\omega = \frac{dK}{d\tau} = H \qquad (5\text{–}69)$$

and

$$K = H\tau + \text{const} \qquad (5\text{–}70)$$

Thus the radiation pressure varies linearly with the optical depth with H appearing as the constant of proportionality.

To this point the development has been rigorous. To make further progress, Eddington reasoned as follows: In the deep interior of the star, I must be nearly independent of θ, in which event one may take it outside the integral sign, and write:

$$K \sim \frac{I}{4\pi}\int \cos^2\theta\, d\omega \text{ or } K \sim \frac{J}{3} \qquad (5\text{–}71)$$

Eddington's approximation consists in supposing that even in superficial layers we may take $K = J/3$ so that

$$\frac{dJ}{d\tau} = 3H \tag{5–72}$$

and

$$J = 3H\tau + \text{const} \tag{5–73}$$

In the first approximation the variation of I with θ is neglected, except for a general distinction between inward and outward flow. That is

$$I(\theta) = I_1,\ 0 < \theta < \frac{\pi}{2};\quad I(\theta) = I_2,\ \frac{\pi}{2} < \theta < \pi \tag{5–74}$$

At the surface of the star, I is assumed uniform over the entire hemisphere, and $H_0 = J_0/2$ since the mean value of the cosine averaged over a hemisphere is $\frac{1}{2}$. Thus we find:

$$J(\tau) = 2H(1 + \tfrac{3}{2}\,\tau) = \tfrac{1}{2}\,F(1 + \tfrac{3}{2}\,\tau) \tag{5–75}$$

In any application of Eq. 5–5 or 5–12 to predict a theoretical energy curve for a star, a fundamental datum is the variation of temperature with optical depth. Different approximate solutions of the transfer equation differ from one another in that they predict different temperature gradients. From Eqs. 5–75 and 5–42 we see that J is related to the local temperature T and optical depth τ by

$$\frac{4\pi J}{c} = aT^4 = \frac{2\pi F}{c}\left(1 + \frac{3}{2}\,\tau\right) \tag{5–76}$$

The effective temperature T_e of a star is defined in terms of flux by

$$\pi F = \sigma T_e^{\,4} \tag{5–77}$$

From Eq. 5–76, the definition of the radiation constant a (see Chapter 4), and Eq. 5–77 it follows that

$$T_e = \sqrt[4]{2}\,T_0 = 1.19\ T_0 \tag{5–78}$$

If the effective temperature of the sun is taken as 5780°K, the boundary temperature is 4860°K. Since the source function S is here identical with J and B, we substitute Eq. 5–75 into Eq. 5–5 to obtain

$$I(0, \theta) = \int Je^{-\tau \sec\theta} \sec\theta\, d\tau = \frac{F}{2}(1 + \frac{3}{2}\cos\theta) \tag{5–79}$$

whence the limb darkening in integrated radiation for a grey body is

$$\frac{I(0, \theta)}{1(0,0)} = \frac{2}{5} + \frac{3}{5}\cos\theta, \tag{5–80}$$

in the Eddington approximation.

It is of interest to compare this predicted limb-darkening law with the observed values for the sun. Two points must be noted; first the limb darkening in integrated light must be deduced from the monochromatic values $I_\nu(0, \theta)/I_\nu(0, 0)$ and the observed energy distribution. Second, the solar material is not grey so that a grey-body temperature distribution would not necessarily be appropriate for the sun, even though an appropriate mean optical depth is chosen (Art. 5–8). Furthermore, outward radiation flow is impeded not only by continuous absorption but also by discrete line absorption in the Fraunhofer lines. The effect of line absorption on the solar temperature distribution is called the *blanketing effect* (see Art. 5–11).

Table 5–4 compares various theoretical predictions with observed values of the limb darkening in integrated light as deduced from measurements by Abbot, Aldrich, and Fowle, and by Moll, Burger, and van der Bilt which are sufficiently accurate for our present purposes. Successive columns in Table 5–4 give the values predicted by Eddington's formula, Eq. 5–80; those from Chandrasekhar's exact solution, discussed below, Eq. 5–126; Münch's results, 1945, 1946, taking the blanketing effect into account; and, finally, the observed data. Notice that the limb darkening predicted by the simple theory is too large, whereas Münch's predictions agree with the observations until $\cos\theta = 0.30$ is reached, beyond which point both theory and observation encounter difficulties.

The observed results are in harmony with the suggestion that the temperature gradient of the upper atmosphere is determined by radiative equilibrium. An atmosphere in complete convective equilibrium would be totally darkened to the limb as K. Schwarzschild showed many years ago (see Problem 5–2). At some depth below the surface, however, convective equilibrium must set in (see Chapters 9 and 10). At the present time the theory of this convection zone has not been worked out completely.

Further complications are introduced if the solar gases are not stratified in plane parallel layers, because the solar surface may have an undulatory character associated with the granules (see Chapter 9).

TABLE 5–4

Limb Darkening in the Sun

Cos θ	Eddington Approximation	Exact Solution (Chandrasekhar)	Münch (Blanketing Effect)	Observed Intensity
1.00	1.000	1.000	1.000	1.000
0.90	0.940	0.939	0.946	0.944
0.80	0.880	0.878	0.892	0.898
0.70	0.820	0.816	0.838	0.842
0.60	0.760	0.755	0.781	0.788
0.50	0.700	0.692	0.725	0.730
0.40	0.640	0.629	0.666	0.670
0.30	0.580	0.565	0.615	0.602
0.20	0.520	0.499	0.541	0.522
0.10	0.460	0.429	0.467	0.450
0.00	0.400	0.344	0.363	

5–7. The Chandrasekhar Method for the Solution of the Transfer Equation. We now give a brief sketch of the most powerful method that has been devised for the treatment of the transfer equation. For the time being we continue to confine our attention to grey material (for which the absorption coefficient is independent of wavelength). With $\mu = \cos\theta$, the equation of transfer Eq. 5–50 becomes

$$\mu \frac{dI}{d\tau} = I - \frac{1}{2}\int_{-1}^{+1} I \, d\mu \tag{5–81}$$

Our problem is to find how $I(\tau, \mu)$ depends on μ and τ, the optical depth in the star, given the appropriate boundary conditions. The appearance of the μ factor on the left-hand side complicates the solution of the equation in the neighborhood of the boundary. The equation of transfer for a grey body is formally equivalent to the problem of the diffusion of neutrons in a pile. Among the solutions of this problem was that proposed by Wick, which was amplified and extended by Chandrasekhar. In a manner analogous to that employed in the kinetic theory of gases where it is customary to resolve the motions of the molecules in a box into three equivalent streams parallel to the walls, one may divide the radiation I into several elementary pencils, each of which corresponds to a different μ. That is, we replace $I(\tau, \mu)$ which is a function of two variables by a set of $2n$ quantities, $I(\tau, \mu_i)$, $i = \pm 1, 2, \ldots, n$, each of which is a function of the one variable τ since μ is held fixed. This is an example of the gaussian method of integration, wherein the choice of the μ_j's is made so that as closely as possible

$$\int_{-1}^{+1} I(\tau, \mu)\, d\mu = \sum_{j=-n}^{+n} a_j I(\tau, \mu_j) \tag{5–82}$$

where the a_j's are the appropriate weight factors. Thus to calculate the integral we evaluate the integrand at certain specified points, multiply by pre-assigned weight factors, and add the products. Since I is symmetrical with respect to μ, it follows that

$$a_j = a_{-j}; \quad \mu_j = -\mu_{-j} \tag{5–83}$$

In order to achieve the highest accuracy with a limited number of points (at least for a function which may be well represented by a polynomial), Gauss showed that the proper procedure was to choose the μ_j's as zeros of the Legendre polynomial $P(\mu)$. When $n = 1$, $\mu_1 = -\mu_{-1} = 3^{-1/2}$. In the second approximation, the μ's are the zeros of $P_4(\mu)$; in the third the zeros of $P_6(\mu)$; in the fourth the zeros of $P_8(\mu)$, etc.[4]

Chandrasekhar replaces Eq. 5–81 by a system of $2n$ ordinary linear differential equations,

[4] For the evaluation of a, see Chandrasekhar's *Radiative Transfer* (New York: Dover Publications, 1960), p. 61. The μ_j's and a_j's are tabulated on p. 62.

$$\mu_i \frac{dI_i}{d\tau} = I_i - \frac{1}{2} \sum a_j I_j, \quad i = \pm 1, \pm 2, \ldots, \pm n \tag{5-84}$$

To solve this equation Chandrasekhar tries an expression of the form,

$$I_i = g_i e^{-k\tau} \quad (i = \pm 1, \ldots, \pm n) \tag{5-85}$$

where the g's and k's are constants to be determined. If Eq. 5–85 is put in Eq. 5–84 there results,

$$g_i(1 + \mu_i k) = \tfrac{1}{2}\Sigma a_j g_j \tag{5-86}$$

Hence,

$$g_i = \frac{\text{const}}{1 + \mu_i k} \quad (i = \pm 1, \ldots, \pm n) \tag{5-87}$$

The "constant" is independent of i. If we substitute Eq. 5–87 under the summation sign in Eq. 5–86 and employ Eq. 5–83, we obtain the equation for k,

$$1 = \frac{1}{2} \sum \frac{a_j}{1 + \mu_j k}$$

or

$$1 = \sum_{j=1}^{n} \frac{a_j}{1 - \mu_j^2 k^2} \tag{5-88}$$

Thus the k's must satisfy an algebraic equation of degree $2n$. Now

$$\sum_{j=1}^{n} a_j \mu_j^m = \int_0^1 \mu^m \, d\mu = \frac{1}{m+1} \tag{5-89}$$

and in particular,

$$\sum_{j=1}^{n} a_j = 1 \tag{5-90}$$

which means that $k^2 = 0$ is a root of Eq. 5–88. Hence Eq. 5–88 can have only $2n - 2$ distinct roots, which will appear in pairs as

$$\pm k_\alpha \quad (\alpha = 1, \ldots, n - 1) \tag{5-91}$$

That is, Eq. 5–84 will have $2n - 2$ linearly independent solutions corresponding to the $2n - 2$ distinct roots.

In addition to the solution Eq. 5–85 we notice that an expression of the form,

$$I_i = b(\tau + Q + \mu_i) \tag{5-92}$$

will also satisfy Eq. 5–84. Q and b are to be found from the boundary conditions and the total flux of radiation. The sum of Eqs. 5–85 and 5–92 — the general solution — then has the form

$$I_i = b \left\{ \sum_{\alpha=1}^{n-1} \frac{L_\alpha e^{-k_\alpha \tau}}{1 + \mu_i k_\alpha} + \sum_{\alpha=1}^{n-1} \frac{L_{-\alpha} e^{+k_\alpha \tau}}{1 - \mu_i k_\alpha} + \tau + \mu_i + Q \right\} \tag{5-93}$$

where b, Q, $L_{\pm\alpha}$ are the $2n$ constants of integration. The first term is reminiscent of the transient term in an electrical circuit. It will be important only at a small optical depth. The last three terms correspond to the steady-state solution, valid at large τ.

We must now impose the boundary conditions of the problem. Since the positive exponential would give an intensity increasing exponentially with depth in contradiction to the requirements of astrophysical theory, we must set $L_{-\alpha} = 0$. Furthermore, at the surface of the star, there is no backward flowing radiation which means

$$I_{-i} = 0 \quad \text{at} \quad \tau = 0, \quad \text{for} \quad i = 1, \ldots, n$$

With the aid of Eqs. 5–83 and 5–93, we find

$$\sum_{\alpha=1}^{n-1} \frac{L_\alpha}{1 - \mu_i k_\alpha} + Q = \mu_i \; (i = 1, \ldots, n) \tag{5–94}$$

Thus there are n equations to determine the $n - 1$ values of L and Q. This process does not fix the constant b, which we shall show to be related to the constant net flux of radiation in the atmosphere, πF. In our present notation.

$$F = 2 \int_{-1}^{+1} I\mu \, d\mu \tag{5–95}$$

We replace the integral by the sum over the $I_i\mu_i$'s and use Eq. 5–93 to obtain

$$F = 2b \left\{ \sum_{\alpha=1}^{n-1} L_\alpha e^{-k_\alpha \tau} \sum_i \frac{a_i \mu_i}{1 + \mu_i k_\alpha} + \sum_i a_i \mu_i^2 + (Q + \tau) \sum_i a_i \mu_i \right\} \tag{5–96}$$

Making use of Eqs. 5–83 and 5–89 we have

$$\Sigma a_i \mu_i^2 = \tfrac{2}{3}, \quad \Sigma a_i \mu_i = 0 \tag{5–97}$$

From Eq. 5–88 we have the identity,

$$\Sigma \frac{a_i \mu_i}{1 + \mu_i k_\alpha} = \frac{1}{k_\alpha} \Sigma a_i \left(1 - \frac{1}{1 + \mu_i k_\alpha}\right) = \frac{1}{k_\alpha} \left(2 - \Sigma \frac{a_i}{1 + \mu_i k_\alpha}\right) \tag{5–98}$$

which is equal to zero. Then

$$F = \tfrac{4}{3} b = \text{const} \tag{5–99}$$

Furthermore,

$$J = \tfrac{1}{2} \int_{-1}^{+1} I \, d\mu = \tfrac{1}{2} \Sigma a_i I_i \tag{5–100}$$

Making use of Eqs. 5–88, 5–89, 5–93, 5–97 and 5–98 we get

$$J = \tfrac{3}{4} F \left(\tau + Q + \sum_{\alpha=1}^{n-1} L_\alpha e^{-k_\alpha \tau} \right) \tag{5–101}$$

If we define

$$q(\tau) = Q + \sum_{\alpha=1}^{n-1} L_\alpha e^{-k_\alpha \tau} \tag{5–102}$$

we may write

$$J = \tfrac{3}{4}F[\tau + q(\tau)] \qquad (5\text{–}103)$$

and the temperature distribution with depth is given by

$$T^4(\tau) = \sqrt{3}T_0^4[\tau + q(\tau)] \qquad (5\text{–}104)$$

Notice that $q(\tau)$ corresponds to the constant, 2/3, in the Eddington approximation [see Eq. 5–76]. Finally, the law of limb darkening follows from a substitution of Eq. 5–101 into Eq. 5–5 to yield

$$I(0, \mu) = \frac{3F}{4}\left(\mu + Q + \sum_{\alpha=1}^{n-1} \frac{L_\alpha}{1 + k_\alpha \mu}\right) \qquad (5\text{–}105)$$

We now illustrate the calculation of J and $I(0, \mu)$ in the first two approximations.

First Approximation. Here

$$a_1 = a_{-1} = 1, \text{ and } \mu_1 = -\mu_{-1} = 3^{-1/2} \qquad (5\text{–}106)$$

Eq. 5–88 has only the one root $k = 0$ and from Eq. 5–94, $Q = \mu_1 = 3^{-1/2}$, while from Eq. 5–102, $q(\tau) = 3^{-1/2}$ and J may be found from Eq. 5–103. The law of limb darkening, Eq. 5–105, is

$$I(0, \mu) = \frac{3F}{4}\left(\mu + \frac{1}{\sqrt{3}}\right) \qquad (5\text{–}107)$$

The inward and outward beams obey the equations

$$\frac{1}{\sqrt{3}}\frac{dI_1}{d\tau} = I_1 - \frac{1}{2}(I_1 + I_{-1}), \quad -\frac{1}{\sqrt{3}}\frac{dI_{-1}}{d\tau} = I_{-1} - \frac{1}{2}(I_1 + I_{-1}) \qquad (5\text{–}108)$$

These equations are analogous to those derived by Schuster and Schwarzschild many years ago except for a factor $\frac{1}{2}$ instead of $3^{-1/2}$.

Second Approximation. Here

$$a_1 = 0.65214 \quad \mu_1 = 0.33998$$
$$a_2 = 0.34785 \quad \mu_2 = 0.86114 \qquad (5\text{–}109)$$

The summation Eq. 5–88 consists of two terms from which we may derive

$$\mu_1^2\mu_2^2k^2 = a_1\mu_1^2 + a_2\mu_2^2 = \tfrac{1}{3} \qquad (5\text{–}110)$$

Then

$$k = \frac{1}{\sqrt{3}\mu_1\mu_2} = 1.97203 \qquad (5\text{–}111)$$

and from Eq. 5–94

$$\frac{L}{1 - \mu_1 k_1} + Q = \mu_1 = 0.33998 \qquad (5\text{–}112)$$

$$\frac{L}{1 - \mu_2 k_1} + Q = \mu_2 = 0.86114 \qquad (5\text{–}113)$$

Thus

$$Q = 0.69402 \text{ and } L_1 = -0.11668 \tag{5-114}$$

Then we may calculate J from Eq. 5–101 and $I(0, \mu)$ from Eq. 5–105. The summation consists of one term.

Chandrasekhar also carried out solutions for the third and fourth approximations. The procedure should be clear from the foregoing discussion. He was also able to derive a rigorous solution of the transfer equation in a closed form. If we compare Eq. 5–105 which gives $I(0, \mu)$, the angular distribution of the radiation emergent from the surface of the star, with Eq. 5–94 which determines the constants, L_α and Q, we note that $I(0, \mu)$ which exists for $1 > \mu > 0$, is determined in terms of a function that has zeros in the complementary interval, $-1 < \mu < 0$. Let

$$S(\mu) = \sum_{\alpha=1}^{n-1} \frac{L_\alpha}{1 - \mu k_\alpha} - \mu + Q \tag{5-115}$$

The boundary condition at the surface asserts that

$$S(\mu_i) = 0 \qquad (i = 1, 2, \ldots, n) \tag{5-116}$$

while at the same time, the angular distribution of the emergent radiation, Eq. 5–105, may be written as

$$I(0, \mu) = \tfrac{3}{4}FS(-\mu) \tag{5-117}$$

Chandrasekhar shows how we may find an explicit formula for $S(\mu)$ without actually solving for the constants L_α and Q. To find such a solution, consider the product function

$$\prod_{\alpha=1}^{n-1} (1-\mu k_\alpha)S(\mu) = (1 - \mu k_1)(1 - \mu k_2) \cdots (1 - \mu k_{n-1})S(\mu) \tag{5-118}$$

which must be a polynomial of degree n, with roots, $\mu = \mu_1, \ldots, \mu_i, \ldots, \mu_n$. Thus there must be a proportionality of the form

$$\prod_{\alpha=1}^{n-1} (1 - k_\alpha \mu)S(\mu) \sim \prod_{i=1}^{n} (\mu - \mu_i) \tag{5-119}$$

We find the constant of proportionality by considering the coefficient of μ^n. For example, if $n = 3$,

$$(1 - \mu k_1)(1 - \mu k_2)\left[\frac{L_1}{1 - \mu k_1} + \frac{L_2}{1 - \mu k_2} - \mu + Q\right] = \tag{5-120}$$

$$-k_1 k_2 \mu^3 + \cdots = \text{const } (\mu - \mu_1)(\mu - \mu_2)(\mu - \mu_3)$$

from which the constant is $(-1)^3 k_1 k_2$. In general,

$$S(\mu) = (-1)^n k_1 \cdots k_{n-1} \frac{\prod_{i=1}^{n} (\mu - \mu_i)}{\prod_{\alpha=1}^{n-1} (1-k_\alpha \mu)} \tag{5-121}$$

which is the required expression. Similarly

$$S(-\mu) = k_1 \cdots k_{n-1} \frac{\prod_{i=1}^{n} (\mu + \mu_i)}{\prod_{\alpha=1}^{n-1} (1 + k_\alpha \mu)} \tag{5-122}$$

It can be shown that

$$k_1 \cdots k_{n-1} \mu_1 \cdots \mu_n = \frac{1}{\sqrt{3}} \tag{5-123}$$

whence

$$S(-\mu) = \frac{1}{\sqrt{3}} H(\mu) \tag{5-124}$$

where $H(\mu)$ is the function defined by

$$H(\mu) = \frac{1}{\mu_1 \cdots \mu_n} \frac{\prod_{i=1}^{n} (\mu + \mu_i)}{\prod_{\alpha=1}^{n-1} (1 + k_\alpha \mu)} \tag{5-125}$$

Then the angular distribution of the emergent radiation $I(0, \mu)$, Eq. 5–105, can be written as

$$I(0, \mu) = \frac{\sqrt{3}}{4} F H(\mu) \tag{5-126}$$

Chandrasekhar has considered a large number of transfer problems, for example, the scattering of light with or without absorption, and with different phase function, diffuse reflection in an extended atmosphere, scattering where polarization must be taken into account (as in the atmospheres of hot stars where electron scattering is primarily responsible for the opacity), and atmospheres of finite optical thickness, such as those of planets.

In all of these problems, the observed quantity is the emergent or reflected radiation defined for μ's in the interval $0 < \mu < 1$. The boundary conditions, however, fix the zeros of the same analytic function in the interval $0 > \mu > -1$. The result is that in any approximation, the Q and L_α constants may be eliminated and the solutions may be reduced to a closed form in the nth approximation. Except for certain constants, these equations involve $H(\mu)$ functions of the form of Eq. 5–125 where the k_α's are the positive or zero roots of a characteristic equation analogous to Eq. 5–88, viz.,

$$1 = 2 \sum_{j=1}^{n} \frac{a_j \Psi(\mu_j)}{1 - k^2 \mu_j^2} \tag{5-127}$$

As usual, the μ_i's are zeros of the Legendre polynomial, $P_{2n}(\mu)$. Here $\Psi(\mu)$ is an even polynomial in μ which must satisfy the condition

$$\int_0^1 \Psi(\mu)\, d\mu \leq \tfrac{1}{2} \tag{5-128}$$

In the transfer problem we have taken $\Psi(\mu) = \frac{1}{2}$. Other problems lead to different $\Psi(\mu)$ functions and hence to $H(\mu)$ functions that differ from one another only in the way the roots k_α are defined.

Then the $H(\mu)$ function satisfies identically the equation

$$H(\mu) = 1 + \mu H(\mu) \sum_{j=1}^{n} \frac{a_j H(\mu_j)\Psi(\mu_j)}{\mu + \mu_j} \tag{5–129′}$$

To prove this, consider first the case when the inequality holds in Eq. 5–128. Then Eq. 5–127 has n distinct positive roots. Consider the function

$$S_0(\mu) = \sum_{\alpha=1}^{n} \frac{L_\alpha}{1 - k_\alpha \mu} + 1$$

(See Eq. 5–115.) Here the constants L_α $(\alpha = 1, \ldots, n)$ are to be determined from the set of n equations

$$\sum_{\alpha=1}^{n} \frac{L_\alpha}{1 - k_\alpha \mu_i} + 1 = S_0(\mu_i) = 0 \quad (i = 1, \ldots, n)$$

Then we may write

$$\begin{aligned} S_0(\mu) &= k_1 \cdots k_n \mu_1 \cdots \mu_n \frac{(-1)^n}{\mu_1\mu_2 \cdots \mu_n} \frac{\prod_i (\mu - \mu_i)}{\prod_\alpha (1 - k_\alpha \mu)} \\ &= k_1 \cdots k_n \mu_1 \cdots \mu_n H(-\mu) = S_0(0) H(-\mu) \end{aligned}$$

since $H(0) = 1$. Thus

$$S_0(0) = k_1 \cdots k_n \mu_1 \cdots \mu_n = \sum_{\alpha=1}^{n} L_\alpha + 1$$

Since $\Psi(\mu)$ is an even function in μ, the characteristic root will satisfy an equation which can be written either as

$$1 = \sum_j \frac{a_j \Psi(\mu_j)}{1 + k\mu_j} \quad \text{or as} \quad 1 = \sum_j \frac{a_j \Psi(\mu_j)}{1 - k\mu_j}$$

Denoting a particular characteristic root as k_α, and making use of the above expressions which are satisfied by any of these characteristic roots, we obtain the identity

$$\begin{aligned} S_0(0) &= \sum_{\beta=1}^{n} L_\beta + 1 = \sum_{\beta=1}^{n} \frac{L_\beta}{k_\alpha + k_\beta} \left\{ \sum_j a_j \Psi(\mu_j) \left[\frac{k_\alpha}{1 + k_\alpha \mu_j} + \frac{k_\beta}{1 - k_\beta \mu_j} \right] \right\} + \sum_j \frac{a_j \Psi(\mu_j)}{1 + k_\alpha \mu_j} \\ &= \sum_{\beta=1}^{n} L_\beta \left[\sum_j \frac{a_j \Psi(\mu_j)}{(1 + k_\alpha \mu_j)(1 - k_\beta \mu_j)} \right] + \sum_j \frac{a_j \Psi(\mu_j)}{1 + k_\alpha \mu_j} \\ &= \sum \frac{a_j \Psi(\mu_j)}{1 + k_\alpha \mu_j} \left[\sum_{\beta=1}^{n} \frac{L_\beta}{1 - k_\beta \mu_j} + 1 \right] \end{aligned}$$

In the last expression the order of summation has been inverted and the term in brackets is $S_0(\mu)$, so that

$$S_0(\mu) = \sum_j \frac{a_j S_0(\mu_j)\Psi(\mu_j)}{1 + k_\alpha \mu_j}$$

In this expression the summation is taken over both positive and negative values of j. We have seen, however, that $S_0(+\mu_j) = 0$, so that only terms with negative j make a non-zero contribution to the above summation. Thus

$$S_0(0) = \sum_{j=1}^{n} \frac{a_j S_0(-\mu_j)\Psi(\mu_j)}{1 - k_\alpha \mu_j} \qquad (\alpha = 1, 2, \ldots, n)$$

Since $S_0(-\mu) = S_0(0)H(\mu)$, we have

$$1 = \sum_{j=1}^{n} \frac{a_j H(\mu_j)\Psi(\mu_j)}{1 - k_\alpha \mu_j} \qquad (\alpha = 1, 2, \ldots, n)$$

In accordance with this relation, the function

$$1 - \mu \sum_{j=1}^{n} \frac{a_j H(\mu_j)\Psi(\mu_j)}{\mu + \mu_j}$$

will vanish if $\mu = -1/k_\alpha$ $(\alpha = 1, 2, \ldots, n)$, since

$$1 + \frac{1}{k_\alpha} \sum_{j=1}^{n} \frac{a_j H(\mu_j)\Psi(\mu_j)}{(-1/k_\alpha) + \mu_j} = 1 - \sum_{j=1}^{n} \frac{a_j H(\mu_j)\Psi(\mu_j)}{1 - k_\alpha \mu_j} = 0$$

Therefore, if we set $\mu = -1/k_\alpha$ $(\alpha = 1, 2, \ldots, n)$, the expression

$$\prod_{j=1}^{n} (\mu + \mu_j) - \mu \sum_{j=1}^{n} a_j H(\mu_j)\Psi(\mu_j) \prod_{i \neq j} (\mu + \mu_i) = \mu_1 \mu_2 \cdots \mu_n + \text{etc.}$$

likewise vanishes. This product function is an nth degree polynomial which differs from

$$\prod_{\alpha=1}^{n} (1 + k_\alpha \mu) = 1 + \mu \sum k_\alpha + \cdots$$

only by the constant of proportionality, $\mu_1\mu_2 \cdots \mu_n$, which is established by comparing the two functions with μ set equal to zero. Hence

$$1 - \mu \sum_{j=1}^{n} \frac{a_j H(\mu_j)\Psi(\mu_j)}{\mu + \mu_j} = \mu_1 \cdots \mu_n \frac{\prod_{\alpha=1}^{n} (1 + k_\alpha \mu)}{\prod_{j=1}^{n} (\mu + \mu_i)} = \frac{1}{H(\mu)}$$

and Eq. 5–129′ follows at once.

When the integral indicated by Eq. 5–128 equals $\frac{1}{2}$, there are $(n - 1)$ positive roots and $k = 0$ is a root of the characteristic equation. The proof follows in a manner similar to the one we sketched. Eq. 5–129′ holds for any value of n, however large.

When n increases without bound, Chandrasekhar shows that $H(\mu)$ satisfies the equation

$$H(\mu) = 1 + \mu H(\mu) \int_0^1 \frac{H(\mu')\Psi(\mu')}{\mu + \mu'}\, d\mu' \qquad (5\text{–}129)$$

first found by Ambarzumian by a different line of reasoning.

Eq. 5–129 may be solved by iteration. We obtain an initial $H(\mu)$ by the approximate methods previously described, substitute it as $H(\mu')$ under the integral sign and compute a new $H(\mu)$. We continue until the $H(\mu)$ of the $(j + 1)$st approximation does not differ from that of the jth approximation.

For example, to solve the limb-darkening problem, we could start with Eq. 5–105 with the constant L and Q from Eq. 5–114, solve for $H(\mu)$ from Eq. 5–126, and substitute in Eq. 5–129. The final $H(\mu)$ obtained by iteration from Eq. 5–129 is then replaced in Eq. 5–126 to get the limb darkening.

In all approximations, the boundary temperature of the star is related to the effective temperature by

$$T_e^4 = \frac{4}{\sqrt{3}} T_0^4 \tag{5–130}$$

since $q(0) = 3^{-\frac{1}{2}}$. This is the exact solution obtained by Hopf and Bronstein some years ago. Thus

$$T_e = 1.233 T_0 \tag{5–131}$$

If the effective temperature of the sun is 5780°K, the boundary temperature is 4670°K. Further, in the expression for the source function Eq. 5–103, $q(\infty) = 0.71045$.

We have confined our attention to one simple, practical application of Chandrasekhar's method and have indicated only in sketchiest terms how his procedure may be generalized to obtain an exact solution. For details the reader should consult Chandrasekhar's *Radiative Transfer*, where a full account is given of these powerful methods and their applications to astrophysical problems, such as the scattering of polarized light in an atmosphere of a star or planet.

5–8. Other Solutions of the Transfer Equation for the Grey-Body Problem. The transfer problem for a grey body is formally equivalent to that for the diffusion of neutrons in a pile. Various physicists have published solutions of this problem. In this connection we mention the work of R. E. Marshak, G. Placzek, W. Seidel, C. Mark, J. LeCaine, and B. Davison and the earlier work of N. Wiener and E. Hopf. More recently Menzel and Sen (1949, 1951) solved the transfer equation by operational methods. For a summarizing account of the grey-body transfer problem, see Kourganoff (1952).

Some of the methods of greatest practical astrophysical interest entail the choice of an initial approximation for the source function with a subsequent improvement by an iteration or other procedure.

For example, for a grey atmosphere

$$B = J = S \tag{5–132}$$

and by an integration of Eq. 5–50 over all frequencies,

$$J = \Lambda\,(B) \tag{5–133}$$

Unsöld, who generalized a discussion by Hopf, has shown that if we start with an initial solution $B_0 = J_0$ and apply the Λ operator, $J_1 = \Lambda\,(J_0)$ will lie closer to the true solution than did J_0. We then calculate a new approximation $J_2 = \Lambda\,(J_1)$ and continue until $J_n = J_{n-1}$. In practice this method converges most rapidly in small optical depths. For larger optical depths other approximation procedures, e.g., Kourganoff's variation method, are more convenient. For example, one may assume J is given by Eq. 5–103 and expand the function in brackets in the form

$$\tau + q(\tau) = A_0 + A_1\tau + A_2E_2(\tau) + \cdots \tag{5–134}$$

where $A_1 = 1$ and E_n is given by Eq. 5–11. Kourganoff shows how the condition of constancy of flux with optical depth may be used to determine the A_n coefficients. By this method we may determine $J(\tau)$ to any desired degree of accuracy.

We shall not describe these procedures here, since the problem of temperature distribution in non-grey atmospheres is of more practical interest. Some of the methods applicable to the grey-body problem, particularly certain iteration procedures are useful for problems posed by actual stellar atmospheres.

D. Labs has suggested the representation of the function $B(\tau)$ for a grey body by a formula of the form

$$B(\tau) = \tfrac{3}{4}\,F(a + \tau - Ae^{-\alpha\tau}) \tag{5–135}$$

for which he finds $a = 0.7104$, $A = 0.1331$, and $\alpha = 3.4488$.

Turning to problems involving scattering (particularly non-isotropic scattering) or both scattering and absorption, methods such as those given by Chandrasekhar are of great value. Of particular importance is Ambarzumian's "method of invariance" (see particularly Chandrasekhar, 1950).

The properties of Chandrasekhar's $S(\mu)$ and $H(\mu)$ functions embodied in the relationships Eqs. 5–115, 5–117, or 5–124, 5–126, and 5–129 are not fortuitous. They are a consequence of the principle of invariance, one form of which may be stated in the following way: consider the radiation emergent in a direction $\mu = \cos\theta$ from a semi-infinite atmosphere stratified in plane-parallel layers. This radiation is invariant to the addition (or removal) of a layer of arbitrary thickness to (or from) the atmosphere. Similarly, in the problem of diffuse reflection as in reflection effects in close eclipsing binaries, the law of diffuse reflection by a semi-infinite plane-parallel atmosphere is invariant to the addition (or subtraction) of strata of arbitrary optical thickness to or from the atmosphere.

Instructive applications of Ambarzumian's principle have been made not only by Chandrasekhar (1950) but also, for example, by Horak (1952,

1954) and by Horak and Lundquist (1954, 1955). The value of the method is that it enables one to obtain certain essential properties of solutions of many radiation problems in a neat, concise fashion.

In connection with scattering problems, i.e., where a given quantum is passed from one atom to another without a change in frequency, one may mention Sobolev's (1951, 1954, 1956) "probabilistic method." This procedure treats radiative transfer and neutron diffusion together in a manner similar to that used for Brownian motion in a liquid or cosmic ray showers. That is, it may be described as *a stochastic theory of radiative transfer* and utilizes certain properties of the theory of the random walk. Sobolev introduces a probability $\Phi(t, \mu)$ that a photon absorbed at a point t will be re-emitted in a direction $\mu = \cos\theta$ in emergent radiation from the stellar surface. He and also S. Ueno (1957, 1958) have used this method to handle a number of classical transfer problems particularly in the theory of absorption line formation (see Chapter 8).

5–9. Definition of Mean Absorption Coefficients. A precise knowledge of the theoretical grey-body temperature distribution is of limited usefulness since most stellar atmospheres are composed of non-grey material. A strongly wavelength-dependent absorption coefficient may produce a pronounced effect on the outward flow of radiation.

We pose, first of all, this question. Is it possible to define a mean absorption coefficient $\bar{\kappa}$ in such a way that $T(\bar{\tau})$, where $\bar{\tau} = \int \bar{\kappa}\,\rho\,dx$, will be identical with $T(\tau)$, for a grey atmosphere? One may demonstrate (see Kourganoff, 1952; Krook, 1959) that such a definition is not really possible. Let us integrate the monochromatic equation of transfer (see Eqs. 5–33, 5–36, 5–39)

$$\mu\,\frac{dI_\nu(x, \mu)}{\rho\,dx} = -\kappa_\nu(x)I_\nu + \kappa_\nu B_\nu(T, x) \qquad (5\text{–}136)$$

over all frequencies. We would like to write the resultant expression in the form

$$\mu\,\frac{dI(x, \mu)}{\bar{\kappa}\,\rho\,dx} = -I(x, \mu) + B\,\{\tau(x)\} \qquad (5\text{–}137)$$

which would imply the definition

$$\frac{1}{\bar{\kappa}\,(x, \mu)}\,\frac{dI(x, \mu)}{dx} = \int_0^\infty \frac{1}{\kappa_\nu(x)}\,\frac{dI(x, \mu)}{dx}\,d\nu \qquad (5\text{–}138)$$

Notice that the mean absorption coefficient $\bar{\kappa}$ depends on μ even though $\kappa_\nu(x)$ does not, since the right-hand side of the equation involves $dI(x,\mu)/dx$. Therefore the best one could do would be to define $\bar{\kappa}(x, \mu_i)$, $i = 1, \ldots, n$, in Chandrasekhar's scheme.

The condition of radiative equilibrium Eq. 5–40 $\int \kappa_\nu J_\nu d\nu = \int \kappa_\nu B_\nu d\nu$ would suggest the definition of

1. an "intensity mean"

$$\bar{\kappa}_J = \frac{1}{J}\int_0^\infty \kappa_\nu J_\nu \, d\nu \quad \text{or} \tag{5–139}$$

2. the Planck mean

$$\bar{\kappa}_B = \frac{1}{B}\int_0^\infty \kappa_\nu B_\nu \, d\nu \tag{5–140}$$

Since J_ν depends on an evaluation of the intensity field in the stellar atmosphere, Eq. 5–139 cannot be calculated until the atmospheric structure is known. Hence, the Planck mean is more useful for many problems. It has the disadvantage that regions where κ_ν is extremely large (e.g., beyond the Lyman limit) make a huge contribution even though B_ν may be small (see, e.g., Przybylski, 1960).

A third physically significant type of mean absorption coefficient, i.e., the flux mean, may be introduced most naturally by means of the following argument: Radiation in an interval ν to $\nu + d\nu$ falling upon a slab of density ρ and thickness dx exerts a mechanical force per unit area (see Eq. 4–55).

$$dP_{r,\nu}\, d\nu = \rho\, dx \frac{\kappa_\nu \pi F_\nu \, d\nu}{c}$$

Integrating over all frequencies we obtain

$$dP_r = \rho\, dx \frac{\pi}{c} \int \kappa_\nu F_\nu \, d\nu \tag{5–141}$$

More generally, we replace κ_ν by the *total extinction coefficient*, $k_\nu = \kappa_\nu + \sigma$.

For an atmosphere in which radiation pressure is important, the equation of hydrostatic equilibrium will be:

$$dP = dP_r + dP_g = -g\rho\, dx \tag{5–142}$$

If we let

$$d\tau = -(\bar{\kappa} + \sigma)\, \rho\, dx \tag{5–143}$$

this equation becomes:

$$\frac{dP_g}{d\tau} = \frac{g}{\bar{\kappa} + \sigma} - \frac{\pi}{c}\frac{1}{\bar{\kappa} + \sigma}\int (\kappa_\nu' + \sigma)\, F_\nu \, d\nu \tag{5–144}$$

If we now define a flux mean (see Eq. 4–56)

$$\bar{\kappa} = \frac{1}{F}\int \kappa_\nu' F_\nu \, d\nu \tag{5–145}$$

the equation of hydrostatic equilibrium reduces to

$$\frac{dP_g}{d\tau} = \frac{g}{\bar{\kappa} + \sigma} - \frac{\sigma_R}{c} T_e^4 \tag{5–146}$$

where σ_R is the Stefan-Boltzmann constant. The flux mean requires a knowledge of the radiation field. In spectral regions where k_ν is large, F_ν is small (see Art. 5–11).

If $\kappa_\nu/\bar{\kappa}$ is constant with optical depth and κ_ν does not vary much with frequency (e.g., H^- ion), Chandrasekhar suggested that one may replace F_ν by the net monochromatic flux of radiation of frequency ν in a grey atmosphere. Then the Chandrasekhar mean is defined by

$$\bar{\kappa}_{\mathrm{Ch}} = \int \kappa_\nu' \left(\frac{F_\nu}{F}\right)_{\mathrm{grey}} d\nu \tag{5–147}$$

where $(F_\nu/F)_{\mathrm{grey}}$ has been tabulated by Chandrasekhar (1950). Again difficulties are encountered if κ_ν depends strongly on wavelength.

A fifth type of mean, the Rosseland mean absorption coefficient is obtained as follows.

Let us return to the equations of transfer in the Eddington notation (see Eqs. 5–69 and 5–68). We have:

Monochromatic Radiation		Integrated Radiation	
$\dfrac{dH_\nu}{k_\nu \rho dx} = J_\nu - S_\nu$		$\dfrac{dH}{\bar{k}\rho dx} = 0$	(5–148)
$\dfrac{dK_\nu}{k_\nu \rho dx} = H_\nu = \frac{1}{4}F_\nu$	(5–149)	$\dfrac{dK}{\bar{k}\rho dx} = H = \frac{1}{4}F$	(5–150)

Eq. 5–148 (our old Eq. 5–68) expresses the constancy of the total radiation flux:

$$\int H_\nu \, d\nu = H \tag{5–151}$$

also

$$\int K_\nu \, d\nu = K \tag{5–152}$$

At a given point in the atmosphere, we have from Eqs. 5–149, 5–150, and 5–151

$$\int \frac{1}{k_\nu} \frac{dK_\nu}{\rho \, dx} \, d\nu = \frac{1}{\bar{k}} \int \frac{dK_\nu}{\rho \, dx} \, d\nu \tag{5–153}$$

which defines the mean absorption coefficient or opacity. We suppose that the extinction coefficients k_ν and $\bar{k}$ contain the effects of negative absorptions and the electron scattering contribution.

Now K_ν depends on the radiation field in accordance with Eq. 5–67 and hence on the stellar atmosphere. To obtain Rosseland's mean absorption coefficient, K_ν is replaced by $\frac{1}{3} J_\nu$ (Eddington approximation) and this in turn is replaced in Eq. 5–153 by B_ν. Thus, noting

$$\frac{dB_\nu}{dx} = \frac{dB_\nu}{dT}\frac{dT}{dx}$$

we get

$$\frac{1}{\bar{k}}\int \frac{dB_\nu}{dT}\,d\nu = \int \frac{1}{k_\nu}\frac{dB_\nu}{dT}\,d\nu \qquad (5\text{–}154)$$

Now $J_\nu = B_\nu$ implies monochromatic radiative equilibrium, whereas the essence of local thermodynamic equilibrium is that there is a shift in the frequency distribution of energy with depth. Even in a strictly grey atmosphere, J_ν could not be replaced rigorously by B_ν. At large optical depths, however, replacement of K_ν by $\frac{1}{3}B_\nu$ will make little difference and the Rosseland mean coefficient will be valid. Hence one always makes use of it in calculations involving stellar interiors. Note that Rosseland's mean is a type of flux mean. Furthermore, it is a transparency mean, i.e., one calculates a weighted mean of the reciprocal of the absorption coefficient, rather than of the absorption coefficient itself. Hence as is possible for other types of means, one cannot add opacities for each mixture, i.e., each different chemical composition, but must compute the absorption coefficient *ab initio*. Nevertheless, Rosseland's mean absorption coefficient is often preferred by writers on stellar atmospheres, even though it has no obvious physical significance throughout the upper visible layers of a stellar atmosphere.

5–10. Methods for Non-Grey Atmospheres. In earlier calculations of model atmospheres, a standard procedure was to calculate Rosseland's mean absorption coefficient $\bar{\kappa}$ for the assumed composition, express a mean optical depth $\bar{\tau}$ in terms of $\bar{\kappa}$ and assume $T(\bar{\tau})$ to be given by the grey-body solution $T(\tau)$, Eq. 5–104. If an atmosphere is assumed to be in radiative equilibrium, the flux πF must be constant with optical depth. It is found, however, that this condition is not actually fulfilled, particularly for early-type stars; $F(\tau)$ may deviate sharply from the assumed initial F adopted as a parameter of the stellar atmosphere.

A number of methods have been described for handling non-grey atmospheres (see Münch, 1960). The first attempts to evaluate the influence of a strongly wavelength-dependent absorption coefficient upon the temperature distribution were made by Mustel (see Ambarzumian, 1958). Another method which is useful for getting an initial approximation was proposed by Hunger and Traving (1956). A somewhat analogous procedure was employed by Cook (1952).

Krook has developed a number of methods for deriving temperature distributions in non-grey atmospheres (1959). In one method he uses an optical depth defined in terms of the Planck mean. A systematic procedure is then developed for improving $J(\bar{\tau})$ and $T(\bar{\tau})$.

Krook (1959) subsequently discussed a class of methods based on "double moments," i.e., moments taken with respect to both angular and frequency distribution of specific intensity. Stone and Gaustad (1961) have shown that Krook's method of moments is useful for a wide range of non-grey models. Krook has also devised a perturbation procedure based on the Poincaré-Lighthill method in which the perturbations are applied not only to the dependent but also to independent variables. He concludes that a transparency coefficient has more direct physical significance for transfer problems than has the absorption coefficient. Also, he finds that it it more convenient to use the pressure as the independent variable in the model atmosphere rather than the optical depth.

A method for construction of non-grey model atmospheres without using an initial grey model has been given by Unno and Yamashita (1960). They use mean absorption coefficients defined by Eq. 5–139 and by Eq. 5–145 except that in the latter κ_ν is replaced by $k_\nu = \kappa_\nu + \sigma$.

Let us suppose that an initial temperature distribution has been obtained, either on the basis of an outright guess or by one of the more systematic procedures just mentioned. With this temperature distribution one calculates a model atmosphere and the corresponding radiation field, which yields the monochromatic source function $S_{0\nu}(\tau_\nu)$, (see Eq. 5–45) where τ_ν is the optical depth at the frequency ν. Then $\mathrm{F}_{0\nu}(\tau_\nu)$ is found from Eq. 5–51, with the definition indicated by Eq. 5–66,

$$F_{0\nu}(\tau_\nu) = \Phi[S_{0\nu}(\tau_\nu)] \tag{5–155}$$

For a given geometrical depth which corresponds to an optical depth $\bar{\tau}$ in the integrated radiation, one finds the total flux:

$$\pi F_0(\bar{\tau}) = \int_0^\infty \pi F_{0\nu}\,[\tau_\nu(\bar{\tau})]\,d\nu \tag{5-156}$$

which departs from the assumed constant flux, πF, by

$$\pi\Delta(\bar{\tau}) = \pi F_0(\bar{\tau}) - \pi F \tag{5–157}$$

Next, one must compute a correction ΔS to the source function S so that $\Delta F(\bar{\tau})$ can be made to vanish. Using the Eddington approximation and the equations of transfer, one may show that for the integrated radiation

$$-\Delta S = \frac{1}{2}\Delta F(0) + \frac{3}{4}\int_0^{\bar{\tau}} \Delta F(\bar{\tau})\,d\bar{\tau} - \frac{1}{4}\frac{d\Delta F(\bar{\tau})}{d\bar{\tau}} \tag{5–158}$$

This procedure, due to Unsöld, is particularly useful for the improvement of an initially assumed model atmosphere. It is more accurate in the deeper layers. In higher approximations and in the shallower layers it should be supplemented by a procedure that avoids the Eddington approximation as, for example, one suggested by K. H. Böhm (1954):

To the equation of transfer Eq. 5–35 apply operation $\int (\ldots) d\omega/4\pi$, that is, multiply by $d\omega/4\pi$ and integrate over all angles. Suppose further that $S_{0\nu} = B_\nu$. Noting that, by Eq. 5–65, $J_\nu = \Lambda\{B_\nu\}$, we have

$$\frac{dH_\nu}{d\tau} = \frac{1}{4}\frac{dF_\nu}{d\tau} = J_\nu - B_\nu = \Lambda\{B_\nu\} - B_\nu \tag{5–159}$$

whence for integrated radiation

$$B = \Lambda\{B\} - \frac{1}{4}\frac{dF}{d\tau} \tag{5–160}$$

Let $F = F_0 + \Delta F$, $B = B_0 + \Delta B$, where F_0 and B_0 are the actual solutions. Since

$$\Lambda(B_0 + \Delta B) = \Lambda(B_0) + \Lambda(\Delta B) \tag{5–161}$$

we get:

$$\Delta B = \Lambda(\Delta B) - \frac{1}{4}\frac{d(\Delta F)}{d\tau} \tag{5–162}$$

which enables one to correct either the monochromatic or the total radiation field. Since ΔF is known exactly for a given assumed initial atmosphere, ΔB may be found. Since the Λ operation is applied to ΔB, rather than to B itself, the procedure converges rather rapidly.

In Przybylski's variational method (1955), one assumes that the correct source function is

$$S_\nu(\tau_\nu) = S_\nu{}^0(\tau_\nu) + \Delta S_\nu(\tau_\nu) \tag{5–163}$$

where $S_\nu{}^0(\tau_\nu)$ is obtained from the initial approximation. Then

$$F^{(2)}(\tau_m) = \int_0^\infty \Phi_{\tau_{\nu,m}}\{S_\nu{}^0(\tau_\nu)\}\, d\nu + \int_0^\infty \Phi_{\tau_{\nu,m}}\{\Delta S\}\, d\nu \tag{5–164}$$

would be the corrected flux. Here $\tau_{v,m}$ is the monochromatic optical depth corresponding to the optical depth τ_m measured at the standard frequency (or the mean optical depth). We already know the first integral on the right-hand side of Eq. 5–164. For an illustration, let us consider the case where $S_\nu = B_\nu$. Przybylski simplifies the calculation by replacing the general problem where κ_ν depends on the frequency by the grey case. The second integral is the correction $\Delta F^{(1)}(\tau_m)$ which we calculate from

$$\Delta F^{(1)}(\tau_m) = \Phi[\Delta B(\tau_m)] \tag{5–165}$$

where ΔB is the correction to be applied to the total emission at optical depth τ_m. Let us write

$$\Delta B(\tau) = a_1 f_1(\tau) + a_2 f_2(\tau) + \cdots + a_n f_n(\tau) \tag{5–166}$$

where the f's are convenient functions τ^n, $e^{n-\tau}$, $E_n(\tau)$, etc., whose transforms are known. The corresponding $\Delta F^{(1)}(\tau)$ will be

$$\Delta F^{(1)}(\tau_m) = a_1 \Phi\{f_1(\tau)\} + \cdots \tag{5–167}$$

Since $\Delta F^{(1)}(\tau_m)$ is known for several values of τ_m, we solve for the coefficients a_1, a_2, etc., by least squares and obtain an improved temperature T from

$$\frac{\sigma_R}{\pi}\{T^{(2)}(\tau)\}^4 = B(\tau) + \Delta B(\tau) \tag{5–168}$$

$$= \frac{\sigma_R}{\pi} T^I(\tau)^4 + a_1 f_1(\tau) + a_2 f_2(\tau) + \cdots$$

The method is very similar to that proposed by Strömgren and successfully applied to F and G stars by Swihart (1956), to A stars by Osawa (1956), and to B stars by Kumar.

Suppose that an atmosphere is in radiative equilibrium and that the temperature distribution is expressed in the form

$$T^4 = \tfrac{3}{4} T_e^4 [a + b\tau + cE_2(\tau) + dE_3(\tau)] \tag{5–169}$$

The integrated flux at any depth is then given by

$$F_g(\tau) = \Phi_\tau [B(t)] = \frac{3\sigma}{4\pi} T_e^4 [a\phi_0 + b\phi_1 + c\phi_2 + d\phi_3] \tag{5–170}$$

where

$$\phi_0 = \Phi_\tau \{1\}, \quad \phi_1 = \Phi_\tau(t), \quad \phi_2 = \Phi_\tau \{E_2(t)\}, \quad \phi_3 = \Phi_\tau \{E_3(t)\} \tag{5–171}$$

For a grey atmosphere, $\pi F = \sigma T_e^4$ and consequently,

$$a\phi_0 + b\phi_1 + c\phi_2 + d\phi_3 = 4/3 \tag{5–172}$$

must be independent of τ. Hence, one may choose four values of τ and obtain the coefficients at once. For a non-grey atmosphere, this condition will not be fulfilled for an initially chosen temperature distribution; i.e., at each value of τ, the flux will deviate from its constant value by some amount $\delta F_g(\tau)$, where

$$\delta F_g(\tau) = \frac{3\sigma}{4\pi} T_e^4 [\phi_0 \Delta a + \phi_1 \Delta b + \phi_2 \Delta c + \phi_3 \Delta d] \tag{5–173}$$

From the value of $\delta F_g(\tau)$ at each of four different values of τ, we find the values Δa, Δb, Δc, and Δd required to restore the constancy of flux.

As an example, we consider an attempt to improve a model atmosphere composed of pure helium. In the first approximation, an initial temperature distribution was assumed, a hydrostatic model was derived, and the radiation field including the flux was computed. With $F_\nu(\tau_0)$ known, the flux mean absorption coefficient could be found and $\bar{\tau}$ evaluated from Eq. 5–145. When $F(\bar{\tau})$ was computed, it was found that the flux was not constant with depth. In the following table we give representative values of $\bar{\tau}$, $T^I(\bar{\tau})$ (first approximation), $\Delta F \times 10^{-12}$, $4\pi\Delta F/3\sigma_R T_e^4$ (see Eq. 5–170) and ϕ_0, ϕ_1, ϕ_2, and ϕ_3, which have been tabulated (McDonald, 1956). The corresponding values of the original constants a, b, c, and d are found by representing $T(\bar{\tau})$ by Eq. 5–169 by least squares. Then Δa, Δb, Δc, and Δd are calculated with the aid of Eq. 5–173 by least squares.

DATA FOR CALCULATION OF CORRECTIONS Δa, Δb, Δc, AND Δd TO THE COEFFICIENTS a, b, c, AND d FOR TEMPERATURE DISTRIBUTION IN STRÖMGREN-SWIHART METHOD

$\bar{\tau}$	$T(\bar{\tau})$	$\Delta F \times 10^{-12}$	$\frac{4\pi}{3\sigma_R}\frac{\Delta F}{T_e^4}$	ϕ_0	ϕ_1	ϕ_2	ϕ_3
0	18,530	−2.30	0.5469	1.0000	0.6667	0.4091	0.2500
. . .	. . .	. . .	. . .	. . .	. . .	. . .	. . .
0.86	26,350	−2.35	0.5588	0.2656	1.1273	−0.1561	−0.0924
1.33	28,650	−1.84	0.4376	0.1414	1.2179	−0.1320	−0.0854
. . .	. . .	. . .	. . .	. . .	. . .	. . .	. . .
3.64	34,300	−0.47	0.1118	0.0084	1.3259	−0.0172	−0.0129

Next we compare the corrections to $T(\bar{\tau})$ as obtained by the four different methods, which give results in general agreement except in the shallowest and deepest layers.

CORRECTION TO $T(\bar{\tau})$ IN A NON-GREY ATMOSPHERE

$\bar{\tau}$ =	0.00	0.32	0.86	1.33	2.41	3.64	5.33
T (°K) =	18,530	22,000	26,350	28,650	31,700	34,300	36,800
$\Delta F \times 10^{-12}$ =	−2.30	−2.32	−2.35	−1.84	−0.94	−0.47	−0.26
Correction Method	Correction ΔT(°K)						
(1) Przybylski	−1570	−2912	−2624	−2546	−2551	−1991	−1604
(2) Strömgren-Swihart	−530	−2560	−2680	−2630	−2590	−1970	−1650
(3) Unsöld	−3130	−2910	−2360	−2470	−2470	−1940	−2010
(4) Böhm	−2370	−2700	−2370	−2510	−2410	−1940	−1630

Although these methods give promising results when applied to hydrogen atmospheres at lower effective temperatures, it turns out that, in this instance, all the methods overcorrected $T(\bar{\tau})$. One reason for the difficulty is possibly that in each instance one corrects not $T(\tau_\nu)$ which was postulated at the outset, but $T(\bar{\tau})$ where the linkage between τ_ν and $\bar{\tau}$ involves the flux mean absorption coefficient. Also effects of electron scattering were not taken into account in the methods used.

The non-grey atmosphere techniques herein described have been applied mostly to situations in which departures from greyness arise from the continuous absorption coefficient. An important class of problems involves the blocking effect of absorption lines on the outgoing radiation (blanketing effect). These effects are particularly important for the Lyman lines of hydrogen in hot stars and numerous metallic lines in the sun (see Arts. 5–11 and 5–12). The blanketing effect seriously modifies the temperature distribution in the outermost atmospheric layers and at the base of the chromosphere, i.e., precisely in the regions of the atmosphere where the strong lines are formed.

5–11. Calculation of Model Atmospheres. One must recognize at the outset that all interpretations of stellar spectra are necessarily based on theoretical models. Early models were necessarily simple ones — strata at one temperature and pressure radiating as grey bodies. In later work, deviations from greyness were taken into account. Simple models of this type underlie analyses by curve of growth methods.

More sophisticated approaches entail construction of what are customarily known as model atmospheres, calculations of distributions of pressure, temperature, etc., that satisfy certain boundary conditions. These models enable one to predict observable characteristics of the outgoing radiation, e.g., energy distribution in the emergent flux $F_\nu(0)$, Balmer discontinuities (i.e., the jump in intensity at the Balmer limit), and profiles and total intensities of stellar absorption lines (Chapter 8). A comparison of predictions and observations enables one to decide whether or not the model is satisfactory. More particularly, if it is not satisfactory, one must decide if he can improve the agreement by retaining the same basic theory and simply changing the input parameters, e.g., effective temperature, surface gravity, and chemical composition, or whether he must modify the basic theoretical assumptions. For example, it may be necessary to introduce effects of stellar rotation, large scale convection currents or deviations from local thermodynamic equilibrium.

Calculation of model atmospheres may be carried out by a process of successive approximations. First, we must specify the starting parameters, effective temperature, surface gravity, and chemical composition. These quantities determine what the principal sources of opacity will be and the relation between pressure and temperature. For many purposes it is convenient to calculate a family of model atmospheres (see, e.g., Underhill, 1957, 1962), since one does not know *a priori* the effective temperature and surface gravity of a particular star in which he may be interested.

At the outset of the calculation, we must assume an initial temperature distribution $T(\tau)$. Different authors use different methods for estimating $T(\tau)$. Underhill employed a $T(\bar{\tau})$ distribution resembling that of a grey body using a $\bar{\tau}$ defined in terms of a Planck mean. Other writers, e.g., Saito (1959), argue in favor of using the Rosseland mean instead, claiming that $\bar{\kappa}_{\mathrm{F}}/\bar{\kappa}_{\mathrm{R}}$ is closer to unity than in other definitions. Here, $\bar{\kappa}_{\mathrm{R}}$ is the Rosseland mean and $\bar{\kappa}_{\mathrm{F}}$ is the flux mean. Even if an exact flux mean is employed, however, $T(\bar{\tau})$ is not that corresponding to a grey body! (See Art. 5–9.) Michard (1949) concluded that to obtain a non-grey atmosphere that most closely matches a grey atmosphere of a given effective temperature, one should use the Rosseland mean, while if we require only that the grey and non-grey atmospheres have the same temperature law and boundary temperature, one should use the Planck mean.

The boundary condition that must be satisfied is that the integrated flux is constant with optical depth. One can start with any monotonically increasing $T(\tau)$ he pleases, the "goodness" of the choice is determined by how rapidly the successive approximations approach the final constant flux. The Hunger-Traving method (1956) provides a good starting approximation for hot stars.

Second, we assume a pressure-depth law. In practice the equation of hydrostatic equilibrium is assumed unless we seek a model for a pulsating star such as a Cepheid variable. Effects of stellar rotation can also be handled in this calculation.

Third, one must postulate the nature of the interaction between radiation and matter. Processes involving thermal absorption and reemission invoke the local temperature through Kirchhoff's law, whereas those involving scattering do not. The optical depth is expressed in terms of the extinction coefficient, Eq. 5–44, specified for some particular frequency ν_0. Sometimes a mean optical depth is used, defined in terms of the Planck mean (Eq. 5–140), a Chandrasekhar mean (Eq. 5–147), a flux mean (Eq. 5–145), or the Rosseland mean defined by Eq. 5–154.

To illustrate these procedures, let us consider first the calculation of a theoretical model for an early-type star and then one for the sun.

The equation of hydrostatic equilibrium appropriate for an early-type star ($T_{\text{eff}} \sim 25{,}000$) is Eq. 5–144. At the outset of the calculation, F_ν is not known, nor is $T(\bar{\tau})$ or $T(\tau_0)$. If the temperature of the star is not too high in the first approximation, we neglect the radiation term and write

$$\frac{dP}{d\tau_0} \sim \frac{g}{\kappa_0 + \sigma} \tag{5–174}$$

where $\kappa_0 = \kappa(\lambda_0, T, P_\varepsilon)$. If the atmosphere is composed mostly of hydrogen, $P_g \sim 2P_\varepsilon$. If g is large, as in a main sequence star, σ is small compared with κ and $\kappa_0 \sim \kappa^0(\lambda_0, T)P_\varepsilon$ (see Chapter 4). Then

$$2P_\varepsilon \frac{dP_\varepsilon}{d\tau_0} \sim \frac{g}{\kappa_0\,[\lambda_0, T(\tau_0)]} \tag{5–175}$$

and

$$P_\varepsilon^2 \sim \int_0^{\tau_0} \frac{g}{\kappa_0(\lambda_0, T)}\, dt \tag{5–176}$$

In the next approximation σ is included and the iteration continued until a precise solution of Eq. 5–174 is obtained. With $P_\varepsilon(\tau_0)$ and $T(\tau_0)$ known, it is possible to calculate $\kappa(\lambda, T)$ (see Chapter 4) for each wavelength of interest, and to obtain

$$\tau_\lambda = \int \frac{\kappa_\lambda + \sigma}{\kappa_0 + \sigma}\, d\tau_0 \tag{5–177}$$

Hence $T(\tau_\lambda)$ and hence $B_\nu(\tau_\lambda)$ may be calculated for each assigned wavelength.

The boundary condition that must be imposed for each optical depth for an atmosphere in radiative equilibrium is that

$$\int_0^\infty F_\nu(\tau_0)\, d\nu = \text{const} \tag{5-178}$$

Hence we must calculate the flux $F_\nu(\tau_\lambda)$ for each selected wavelength and then, from the $\tau_0 - \tau_\lambda$ relationship, interpolate F_ν for each τ_0 of interest.

The flux $F_\nu(\tau_\lambda)$ is calculated from the source function $S_\nu(t_\nu)$ by means of Eq. 5–51, but the source function, which is defined by Eq. 5–45 is itself unknown at the start of the calculation. Hence, with the Planckian function $B_\nu(\tau_\lambda)$, known from $B_\nu(\tau_0)$ and the $\tau_0 - \tau_\lambda$ relation, we must solve Eq. 5–52 to obtain $S_\nu(\tau_\lambda)$. Calculation of $J_\nu - B_\nu$ may be carried out most conveniently by machine techniques or, if these are not available, by the Strömgren's iteration procedure. See Underhill (1960) for a comparison of accuracy of the two methods. Since $\kappa_\nu/(\kappa_\nu + \sigma)$ is known from the calculation of the absorption coefficient $\kappa_\nu(P_\varepsilon,T)$ and $\sigma = 0.666 \times 10^{-24} n_\varepsilon$ where n_ε is the number of electrons per gram, one can then obtain the source function.

With the source function known, one selects a series of values of τ_0, e.g., 0.1, 0.2, 0.4, 1.0, 1.5, 2.0, etc., and for each of these finds the corresponding value of τ_λ. Then F_ν (τ_λ corresponding to τ_0) is calculated by Eq. 5–51 and the integral Eq. 5–178 evaluated.

In practice, $F(\tau_0)$ will not be constant with τ, and the next step is to calculate an improved temperature distribution. One may now calculate a flux mean absorption coefficient from Eq. 5–145 and then define a $\bar{\tau}$. Since $T(\bar{\tau}_0)$ is known for the initial approximation, one may find $T_{\text{initial}}\,(\bar{\tau})$ and improve it by one or another of the methods described in Art. 5–10, e.g., the Unsöld procedure (see Eq. 5–158), Böhm's procedure (see Eq. 5–162), or by Strömgren's or Przybylski's method.

For the second approximation one has an estimate of the flux, so that Eq. 5–144 or 5–146 may be employed and radiation pressure taken into account. One may employ either τ_0 or $\bar{\tau}$ as independent variable. When the model atmosphere is calculated, one then proceeds as before to get the source function and the flux.

We illustrate the calculation of a model atmosphere for a B-type star, $\log g = 4.0$ and $T_e \sim 18{,}000°\text{K}$. The initially assumed model I yields the relation between $\theta = 5040/T$ and $\log P_\varepsilon$ given in Table 5–5, derived from Eq. 5–174 with an initially guessed $T(\tau_0)$ dependence. The temperature is so low that radiation pressure may be neglected. The flux corresponding to this model is given in Table 5–6, third column, while the optical depth defined in terms of the flux mean absorption coefficient is given in the second column. Note that the flux decreases with depth by about 9 per cent from the surface to $\tau_0 = 4.0$.

To obtain a second approximation the temperature distribution $T(\bar{\tau})$ is corrected by Unsöld's procedure Eq. 5–158, the source function here being taken as Planckian. The third and fourth columns of Table 5–5 give the model for this new $T(\bar{\tau})$ distribution. Notice that the boundary temperature is lowered considerably. When

TABLE 5-5

MODELS FOR A B STAR ATMOSPHERE

$\log g = 4.00 \; T_e \sim 18{,}000°K$

θ	Model I	Model II	
	Log P_ε	Log P_ε	τ_0
0.410	—	$-\infty$	0.000
0.400	—	1.698	0.017
0.390	—	1.966	0.036
0.380	$-\infty$	2.120	0.063
0.370	1.839	2.240	0.095
0.360	2.112	2.344	0.138
0.350	2.275	2.438	0.193
0.340	2.405	2.524	0.260
0.330	2.521	2.605	0.344
0.320	2.631	2.683	0.451
0.310	2.738	2.763	0.593
0.300	2.839	2.847	0.787
0.290	2.933	2.933	1.056
0.280	3.021	3.021	1.417
0.270	3.109	3.109	1.907
0.260	3.197	3.200	2.585
0.250	3.288	3.293	3.533
0.240	3.380	3.388	4.835
0.230	3.472	3.482	6.600
0.220	3.566	3.578	9.040
0.210	3.665	3.677	12.530
0.200	3.768	3.781	17.650

TABLE 5-6

INTEGRATED FLUXES $\int F_\nu(\tau_0)\, d\nu$ FOR B STAR MODELS

(Multiply F values by 10^{12})

τ_0	MODEL I		MODEL II	
	$\bar{\tau}$	F	$\bar{\tau}$	F
0.00	0	—	—	1.985
0.1	0.096	2.10	0.094	1.981
0.2	0.170	2.06	0.180	1.976
0.4	0.28	2.02	0.330	1.967
0.6	0.39	1.98	0.44	1.954
1.0	0.57	1.96	0.62	1.924
1.5	0.77	1.95	0.83	1.915
2.0	0.94	1.95	1.03	1.910
3.0	1.24	1.92	1.35	1.902
4.0	1.52	1.92	1.67	1.918

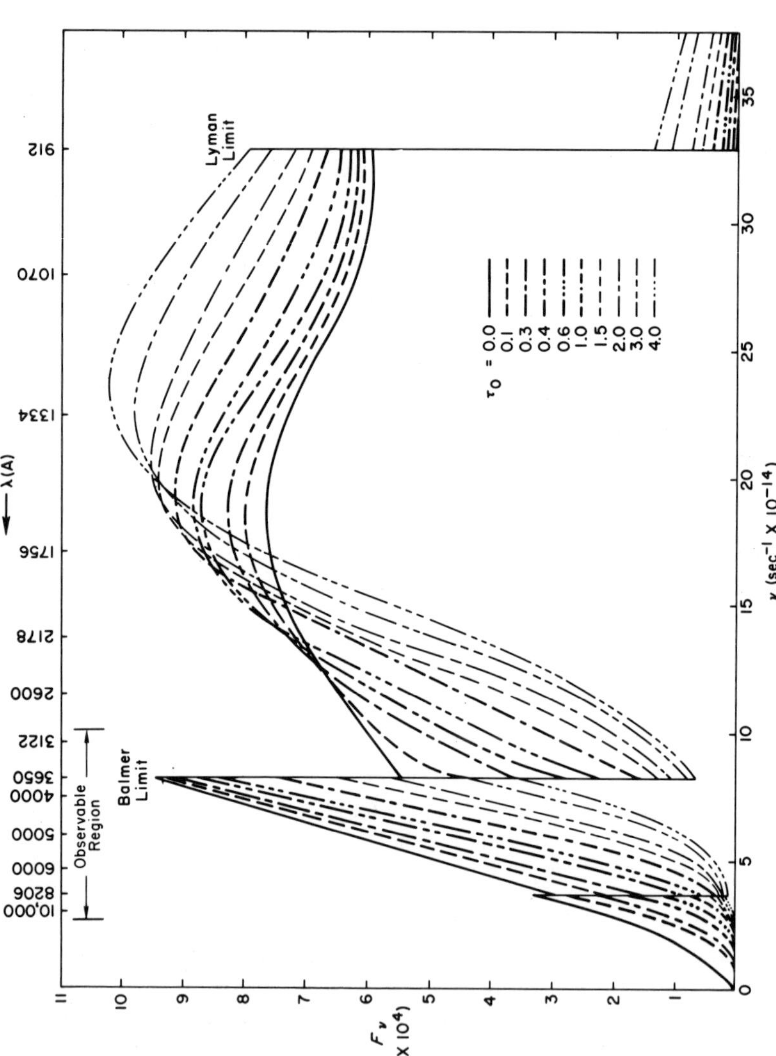

FIG. 5-7. THEORETICAL FLUX IN A *B* STAR

We plot $F_\nu \times 10^4$ as a function of ν and τ_0 (optical depth at λ5000) with a few typical wavelengths indicated. The observable region of the spectrum occupies a small domain on the left. Notice the abrupt changes in the emergent flux at the Lyman, Balmer, and Paschen discontinuities. The Brackett and higher series jumps are smoothed over as they have little influence on the emergent flux.

TABLE 5-7

MONOCHROMATIC B_ν AND J_ν FOR SELECTED τ_0

(Multiply B_ν and J_ν values by 10^{-3}

	$\theta = 0.3690$ $\tau = 0.1$		$\theta = 0.3490$ $\tau = 0.2$		$\theta = 0.3250$ $\tau = 0.4$		$\theta = 0.2920$ $\tau = 1.0$		$\theta = 0.2780$ $\tau = 1.5$	
λ	B_ν	J_ν	B_ν	J_ν	B_ν	J_ν	B_ν	J_ν	B_ν	J_ν
$14{,}588_V$	0.121	0.106	0.130	0.124	0.144	0.142	0.166	0.166	0.177	0.177
10,000	0.214	0.173	0.233	0.208	0.258	0.244	0.305	0.301	0.329	0.329
8,206	0.275	0.339	0.305	0.358	0.346	0.381	0.408	0.419	0.441	0.445
6,500	0.357	0.283	0.401	0.338	0.457	0.415	0.558	0.545	0.606	0.605
5,000	0.442	0.363	0.502	0.419	0.593	0.511	0.741	0.693	0.819	0.787
4,000	0.476	0.447	0.560	0.499	0.674	0.587	0.883	0.793	0.990	0.814
$3{,}647_R$	0.482	0.494	0.572	0.534	0.699	0.613	0.925	0.815	1.048	0.949
$3{,}647_V$	0.482	0.425	0.572	0.528	0.699	0.672	0.925	0.917	1.048	1.043
3,122	0.488	0.436	0.595	0.546	0.746	0.704	1.013	0.992	1.156	1.147
2,600	0.400	0.395	0.501	0.480	0.655	0.620	0.956	0.918	1.122	1.083
2,178	0.308	0.369	0.401	0.432	0.554	0.504	0.852	0.922	1.012	0.975
1,756	0.185	0.361	0.254	0.396	0.378	0.467	0.641	0.656	0.308	0.798
1,334	0.0611	0.317	0.0977	0.344	0.165	0.379	0.327	0.479	0.437	0.552
1,071	—	—	0.0298	0.258	0.0612	0.276	0.139	0.339	0.198	0.383
912_R	0.00504	0.243	0.00948	0.251	0.0200	0.263	0.0565	0.298	0.0871	0.322
912_V	0.00504	0.00504	0.00948	0.00948	0.0200	0.0200	0.0565	0.0565	0.0871	0.0871
833	0.00221	0.00221	0.00438	0.00438	0.00997	0.00997	0.0309	0.0309	0.0499	0.0499
708	0.000316	0.000316	0.000710	0.000710	0.000978	0.000978	0.00452	0.00452	0.0129	0.0129

the iteration is repeated (last column of Table 5–6), the flux constancy condition is more nearly fulfilled, but it might have been wiser to have used Eq. 5–162. An actual application of this equation leads to a still greater lowering of the temperature in the immediate neighborhood of the surface. Figure 5–7 shows the variation of the monochromatic flux with τ. Notice the steady shift of the maximum to higher frequencies as one goes to deeper layers. Notice also the sharp cut-off at the Lyman limit in this hydrogen atmosphere (somewhat exaggerated because the Lyman line absorption is not included). A comparison of the predicted emergent energy distribution with that of a black body that emits the same amount of energy illustrates the copious flow of energy through the "window" on the redward side of the Lyman limit.

Table 5–7 gives monochromatic values of B_ν and J_ν for selected values of τ_0. At wavelengths shorter than the Lyman limit $B_\nu = J_\nu$, but in the region just longward of this Lyman limit where much of the flow of energy occurs, J_ν does not approach B_ν except in deeper layers. These differences are to be assessed in terms of the assumptions underlying the Rosseland mean absorption coefficient, i.e., that J_ν can be replaced everywhere by B_ν (see also Underhill, 1950).

The temperature distribution $T(\bar{\tau})$ for an early-type star expressed in terms of any reasonably defined mean optical depth differs considerably from that of a grey body. The extreme non-grey character of the continuous absorption coefficient in early-type stars introduces enormous complications. Reference to the theory of the absorption coefficient, Chapter 4, shows that κ_ν varies in a jagged fashion. It is small just longward of the Lyman limit and then becomes so very large that the outgoing radiation is effectively blocked. Qualitatively the strong Lyman absorption has a blocking effect similar to that produced by the Fraunhofer lines in the solar spectrum. Many years ago, Chandrasekhar and Hopf showed that if the spectrum is crossed by a "picket-fence" of evenly spaced absorption lines, the temperature at the boundary will be depressed below that appropriate to a grey body, will then rise steeply, overshoot the grey body curve, and finally approach the latter asymptotically with increasing τ. If a_1 is the fraction of outgoing radiation blocked by the absorption lines, $x_1 = \kappa_L/\bar{\kappa}$, and $x_2 = \kappa_c/\bar{\kappa}$, where κ_L and κ_c are the absorption coefficients in the lines and between the lines, respectively, E. Hopf showed that in Eq. 5–102, $q(\tau = 0)$ becomes

$$q(0) = \frac{1}{\sqrt{3(a_1 x_1 + a_2 x_2)}} \tag{5–179}$$

Chandrasekhar (1936) derived an expression for $q(\tau)$ that enables one to correct an initially assumed temperature distribution for line absorption effects, but the strong absorption beyond the Lyman limit may be handled more conveniently by the Traving-Hunger or Krook procedures.

Model atmospheres for stars of spectral classes O and B have been calculated by many investigators, Rudkjöbing (1946), J. C. Pecker (1950),

Underhill (1950, 1951, 1957), McDonald (1953), Traving (1955, 1957), Saito and Uesugi (1959), Milligan, and others. See Münch (1960). *A*-type star models have been published by Hunger (1955), Saito (1955), Martini and Masani (1957), and particularly by Osawa (1956). In the *A* stars, we might expect (see Chapter 9) that convection would modify the temperature distribution in the atmosphere but theoretical models based on the hypothesis of radiative equilibrium are in better accord with the observations. R. Canavaggia and J. C. Pecker (1952, 1953) have published models for yellow giant stars, while an extensive calculation of model atmospheres for many spectral types has been undertaken by Neven and de Jager.

In *A*, *F*, and *G* stars the effect of the Lyman absorption is negligible since most of the radiation falls in regions of longer wavelength. Hence the atmospheric temperature distributions more closely resemble those of grey bodies. Blanketing effects modify temperature distributions in the outermost layers.

Theoretical model solar atmospheres have been calculated by numerous workers (see, e.g., Minnaert, 1953, for a summary of the earlier works). (See also Swihart, 1956.)

The most elaborate theoretical model of the solar atmosphere is Böhm's non-grey radiative equilibrium model (1954). Starting with an initial model based on earlier work, he first calculated $\tau_\nu(\tau_0)$ for each of twenty intervals into which he divided the spectrum. He then assumed that the Fraunhofer lines obeyed Kirchhoff's law and computed κ_ν (line)/κ continuum in each interval. Then he obtained $F_\nu\Delta\nu$ for each interval from Eq. 5-51 with $S_\nu = B_\nu$ and computed $\int(F_\nu\, d\nu)/F(\text{total})$ as a function of the optical depth $\bar{\tau}^*$ defined in terms of the Rosseland mean. He obtained an improvement in $B_\nu(\tau^*)$ by applying first Unsöld's Eq. 5-158 and then Eq. 5-160. The boundary temperature of 3400–3750°K corresponded to optical depths of $10^{-5} \rightarrow 10^{-4}$, i.e., to layers in the chromosphere. Discordances between predictions based on the theory of radiative equilibrium and observation occur in layers above $\tau_0 \sim 0.05$, probably because of deviations from local thermodynamic equilibrium.

Few theoretical model atmospheres have been published for main sequence stars fainter than the sun, although some calculations have been attempted by Unno and his associates. Part of the difficulty accrues from the fact that much of the atmospheric absorption comes from agencies that either have not been identified or do not permit an accurate evaluation of their absorbing properties.

Giant and supergiant stars present even more difficult problems since they often show

$$g_{\text{eff}} \sim 10^{-2} \frac{G\mathfrak{M}}{R^2} \tag{5-180}$$

Such an effect is demonstrated not only by line profiles whose shapes depend on surface gravity but also by density gradients in the atmospheres of eclipsing systems such as ζ Aurigae or 31 Cygni.

Whatever the source of levitation of atmospheres of giant and supergiant stars, calculation of models for them on the basis of hydrostatic equilibrium seems inappropriate. Radiation pressure is probably not responsible for the distention of these atmospheres. More likely there exists some mechanical cause such as the ejection of large jets of gas from deeper layers. That is, it may be necessary to employ a dynamical rather than a static model. Or, shock-wave phenomena may be important as Schwarzschild and Schatzman have suggested. In late M supergiants, actual loss of material occurs, as Deutsch noted for α Herculis.

Among the tests to which a given model atmosphere may be subjected is a comparison with observation of its prediction of the Balmer discontinuity $D = \log[F(\lambda 3650+)/F(\lambda 3650-)]$ where $F(\lambda 3650+)$ refers to the flux on the redward, i.e., long-wavelength side of the Balmer limit, $F(\lambda 3650-)$, the corresponding value for the shortward side of the Balmer limit. This parameter has the advantage that it is not affected by space absorption as are the energy distributions.

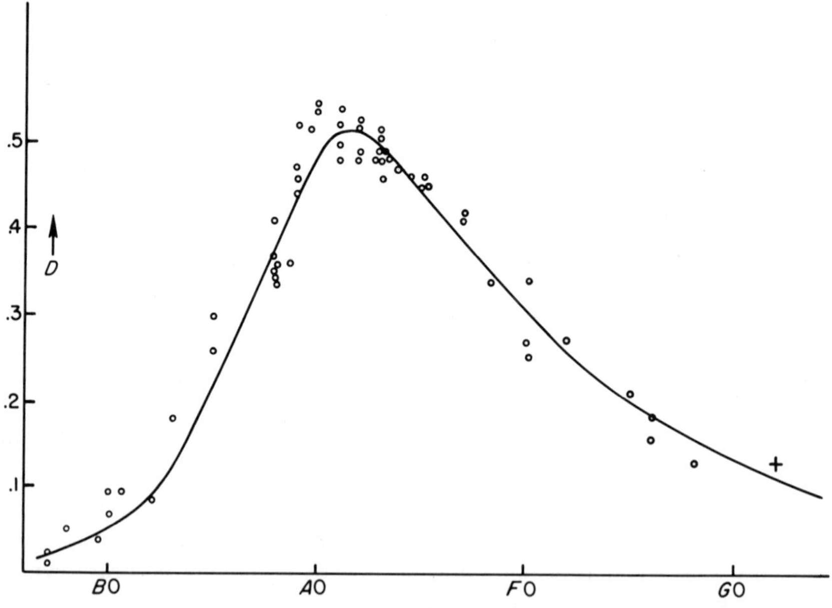

Fig. 5–8. Relation Between Balmer Discontinuity and Spectral Class for Main Sequence Stars

Ordinate is $D = \log F(3650+)/F(3650-)$ and abscissa is spectral class. Observed points are taken from Chalonge and Divan, Table 4, *Ann. d'Ap.* **15,** 1952. The cross (+) denotes the sun according to measurements by Labs and by Chalonge. Theoretical curve is based on data by Underhill, McDonald, Saito, Martini-Masani, Osawa, and unpublished University of Michigan data.

Fig. 5–8 compares theoretical and observed results for main sequence stars. The agreement is satisfactory, both qualitatively and quantitatively, thus not only providing a check on the model atmospheres but also demonstrating nicely the role of the negative hydrogen ion. If the H^- ion did not exist, the value of D would increase with decreasing temperature until the metals provided the opacity, but when that happened, metallic lines would attain an enormous strength.

Comparison of predicted energy distributions with theoretical models is more difficult. Consider first stars of spectral class *O* and early *B*. Nearly all of these stars are reddened by the interstellar medium so their energy distributions cannot be compared easily with theoretical predictions. Where comparisons can be made, the theoretical energy curves differ by an amount which is unsatisfactory.

A reference to the predicted energy distributions for early *B* stars shows that the vast bulk of the energy should flow through a "window" just longward of the Lyman limit. Hence *B* stars not affected by space reddening, e.g., γ Pegasi or α Pavonis, should be very bright in the "satellite ultraviolet" $3000 > \lambda > 912$A. Fragmentary data obtained so far from rockets fired above the earth's atmosphere (Stecher and Milligan, 1962) indicate that these stars are much fainter than expected. See Fig. 6–7. The discrepancy between theory and observation appears to be very small near spectral class *A*5 or *F*0, but increases markedly towards the earlier spectral classes. It cannot be explained by interstellar absorption, and may differ from star to star of the same spectral class. Several alternatives suggest themselves:

1. The models are seriously in error. Recent models calculated by Anne Underhill (1963) tend to remove part of the discrepancy but not all of it. The effects of spectral lines were not included, however.
2. Considerable selective absorption occurs in circumstellar envelopes. The difficulty with this explanation lies in the identification of the agent responsible.
3. Line absorption in the ultraviolet seriously modifies the emergent energy curve. Notice that the predicted energy curves neglect the influence of the Lyman lines. We see (Chapter 7) that these lines have extensive density-broadened wings that overlap as the series limit is approached, but this effect alone does not appear sufficient to modify the satellite ultraviolet flux by the required amount. In addition to the Lyman lines, there occur absorptions by resonance lines of the permanent gases and ionized metals, e.g., Ti II, Fe II, etc. (which can become very important in spectral classes *B*5 and later). Not only do these lines block outgoing radiation in the hypothetical window; they also modify the temperature distribution in the outer layers.
4. Heretofore unidentified sources of opacity in the ultraviolet may play an important role. Stecher and Milligan (1962) have suggested that transitions

between unstable states of quasi molecules of HeH^+ may produce the required continuous absorption. The quantum mechanical treatment of such transitions is difficult and no adequate theory exists at present.

These problems point out the need for accurate observations of stellar energy distributions in the ultraviolet region $\lambda3000 - \lambda912$, which can be obtained only from satellites.

For stars of spectral classes A, F, and G, good agreement between observed and predicted energy distributions has been obtained. Thus Bless finds excellent agreement for A stars with Osawa's theoretical models, while Melbourne (1960) finds that simple theoretical models reproduce observed energy distributions for A, F, and early G main sequence stars. On the one hand, one must allow for effects of lines on the observed energy distributions, and on the other for the fact that the background continuum flux corresponds to a model with a higher effective temperature T_e' (or total flux) than the true stellar effective temperature T_e because energy absorbed in the lines is put back into the stellar continuum, viz.:

$$\sigma T_e^4 = \sigma T_e'^4 - \Delta F(0) \tag{5–181}$$

where $\Delta F(0)$ is the flux absorbed in the lines. In later-type stars, models become more difficult to calculate, while effects of line absorption seriously modify the observed energy distributions, particularly in the violet and ultraviolet. For example, beyond the H and K lines in the sun we probably do not see the continuous spectrum; it is smothered by a host of strong overlapping lines, which become stronger and more pronounced towards the "rocket" ultraviolet, $\lambda2900$–$\lambda2000$, so that ultimately the "continuous" spectrum is depressed much below that of a black body at 5700°K. Liller and Lewis showed that in the neighborhood of the K line in the $K0$ dwarf ε Eridani the continuum is depressed about 35 per cent because of overlapping wings of strong lines. Calculations of model atmospheres for such objects becomes difficult because of blocking of radiation by strong lines.

5–12. Mean Atmospheric Parameters. In a preliminary reconnaissance, called a grobanalyse by Unsöld, one often assumes that certain characteristics of a stellar spectrum (most often the profiles or total intensities of absorption lines) can be calculated as though they were produced at a unique point in the stellar atmosphere characterized by a specific P_g, P_ε, T, and k_λ. This point of view underlies analyses of stellar spectra by means of the curve of growth (Chapter 8).

Consider the equation of hydrostatic equilibrium (Eq. 5–22).

$$\frac{dP}{d\tau} = \frac{g}{k} \tag{5–22}$$

Let us suppose that k is constant throughout the strata responsible for the dark lines and further that the mean temperature and level of ionization

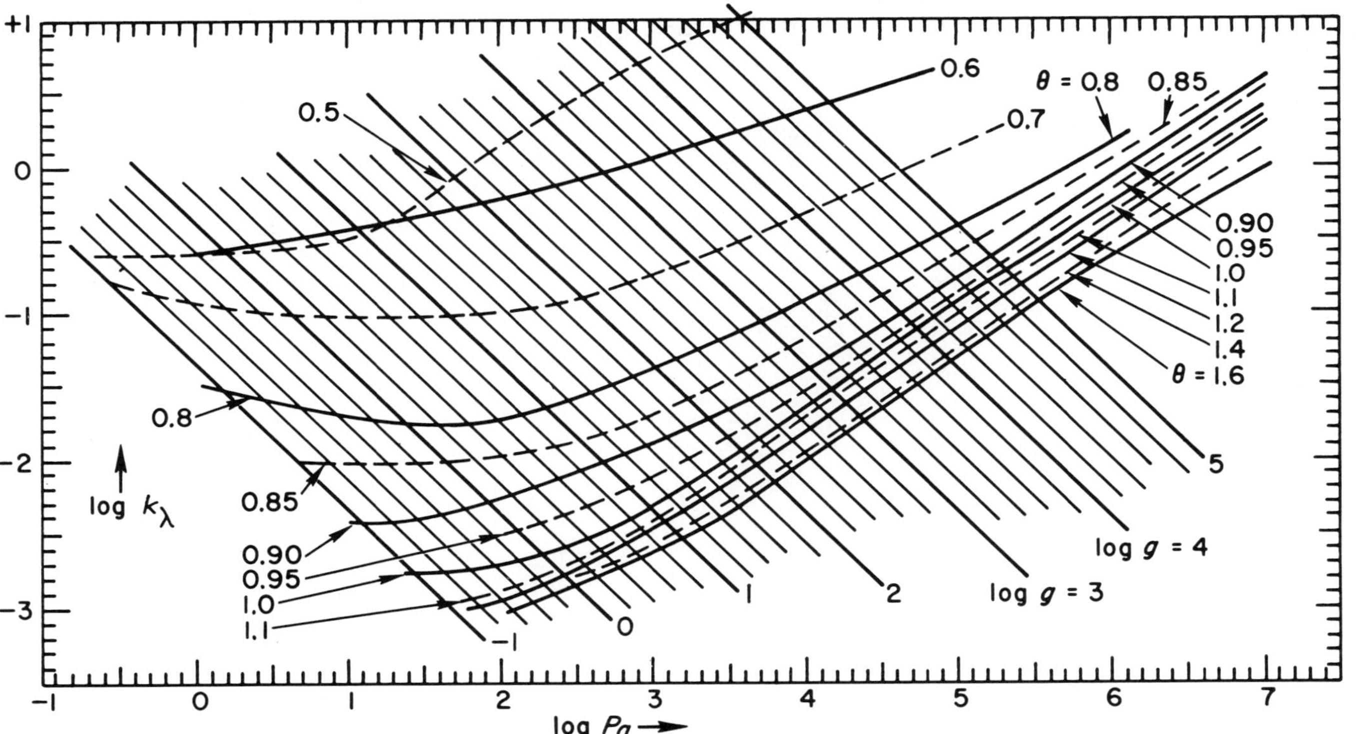

FIG. 5–9. RELATION BETWEEN GAS PRESSURE AND CONTINUOUS ABSORPTION COEFFICIENTS (λ4235) FOR DIFFERENT TEMPERATURES

We plot log k_λ against log P_g as abscissa. The straight lines with 45° slope are plots of Eq. 5–182 for different values of the surface gravity. In Figs. 5–9, 5–10, 5–11, and 5–12, we adopt a solar composition which is presumed to be appropriate for stellar population type I.

is that appropriate to an optical depth τ_0, which we tentatively adopt as 0.40.[5] In this illustrative example we adopt λ4325 as appropriate for observations in the photographic region of the spectrum; $k_\lambda(P_\varepsilon, \theta)$ is taken from calculations by Strömgren, and the composition chosen is that adopted in Chapter 4. Then

$$P_g = \frac{g}{k_0}\tau_0 \qquad \text{or} \qquad \log P_g = \log g - \log k - 0.40 \tag{5–182}$$

Now k_λ is normally calculated as a function of T (or θ) and P_ε, but with the aid of the $P_g(\theta, P_\varepsilon)$ relationship (see Art. 3–10, Fig. 3–9) the $k_\lambda(P_\varepsilon, \theta)$ relationship can be transferred to a $k_\lambda(P_g, \theta)$ relation (see Fig. 5–9). For each value of log g, Eq. 5–182 gives a line of 45° slope in the P_g-k–plane. Its intersection with a particular $k_\lambda(P_g, T)$ curve gives k_λ and log P_g for a star of temperature T and surface gravity g. Then from Fig. 3–9, we read off log P_ε for a given T.

Example: Let $\log g = 4.0$, $\theta = 0.50$. The line,

$$\log P_g + \log k = 4.00 - 0.40$$

intersects the $k(P_g, \theta = 0.5)$ curve at $\log P_g = 2.90$ and $\log k = 0.68$. From the (P_g, P_ε) relationship we find $\log P_\varepsilon = 2.47$ for $\theta = 0.5$. The amount of material "above the photosphere," i.e., above optical depth $\tau = 0.40$ is $\int \rho dx = \bar{\rho}h = \tau_0/k = 0.40/4.8 = 0.083$ gm cm^{-2}. Log $m_p = -1.08$. Next consider a giant for which $\log g = 3.0$, $\theta = 0.9$. A similar calculation shows $\log P_g = 3.99$, $\log k = -1.39$, $\log P_\varepsilon = 0.39$.

Figs. 5–10, 5–11, and 5–12, computed on the basis of the above considerations, show how the electron pressure, gas pressure, and mass above the photosphere vary with temperature for different values of the surface gravity g. Curves are drawn to indicate the values appropriate to main sequence stars, giants, and supergiants based on the (g, θ) relationships from Chapter 1. The surface gravities in giants and supergiants show a considerable range for a specified surface temperature. For a given surface gravity, notice how the electron pressure increases, while gas pressure and mass above the photosphere decrease as the temperature rises. Hydrogen is the most abundant constituent of stellar atmospheres. At lower temperatures, it is mainly neutral, and free electrons are contributed only by metals. Hence the electron pressure is quite a small fraction of the gas pressure. As the temperature rises, hydrogen becomes ionized, and both electron pressure and absorption coefficient increase. Hence the atmosphere becomes more opaque and one sees to shallower and shallower layers. The mass above the photosphere and gas pressure therefore decrease. Atomic hydrogen becomes a more effective absorbing agent than the negative hydrogen ion.

[5] For atoms that become rapidly ionized with depth, e.g., sodium in the sun, we should choose a much smaller τ_0, e.g., 0.20 or even 0.10.

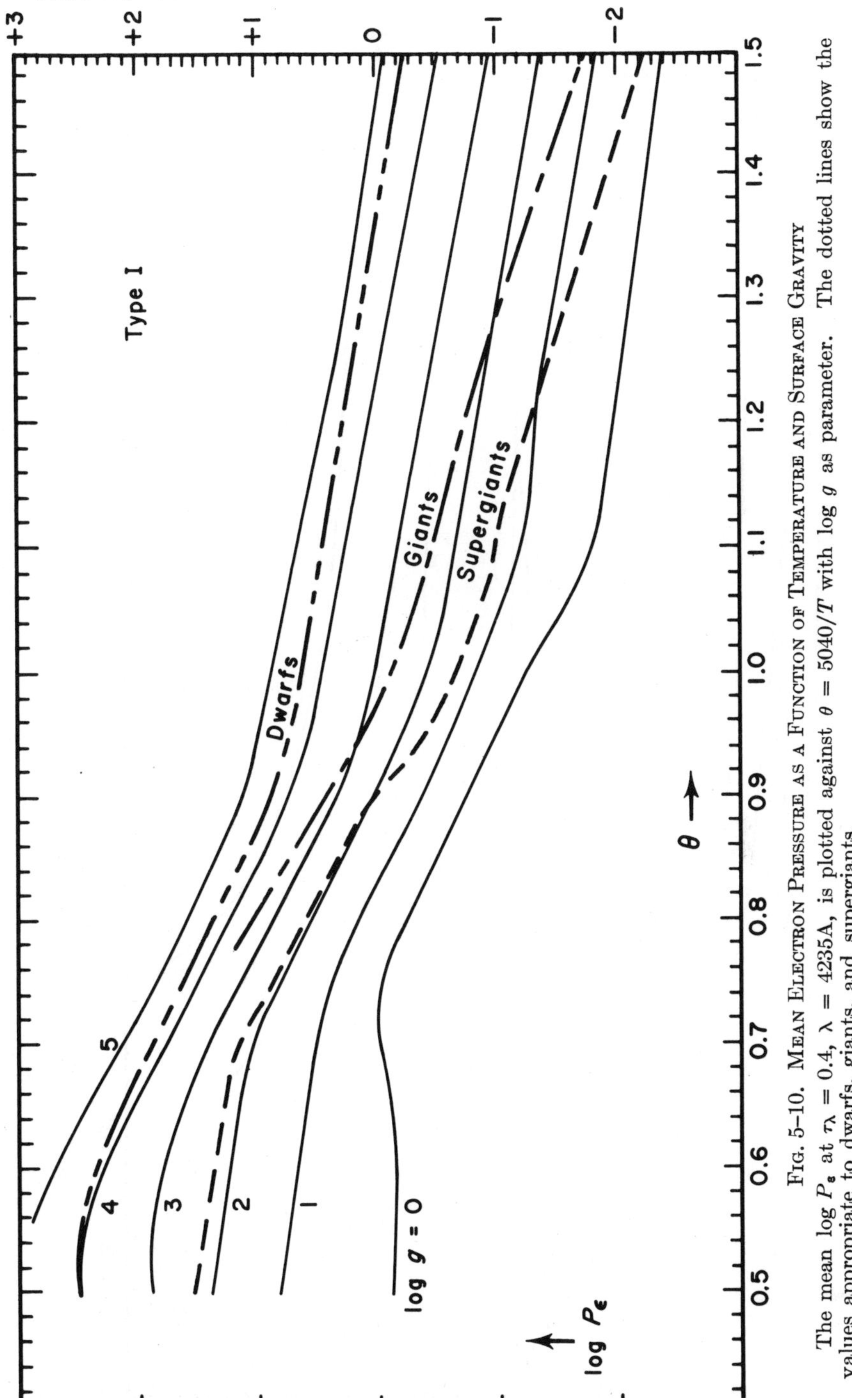

Fig. 5–10. Mean Electron Pressure as a Function of Temperature and Surface Gravity

The mean log P_ϵ at $\tau_\lambda = 0.4$, $\lambda = 4235$A, is plotted against $\theta = 5040/T$ with log g as parameter. The dotted lines show the values appropriate to dwarfs, giants, and supergiants.

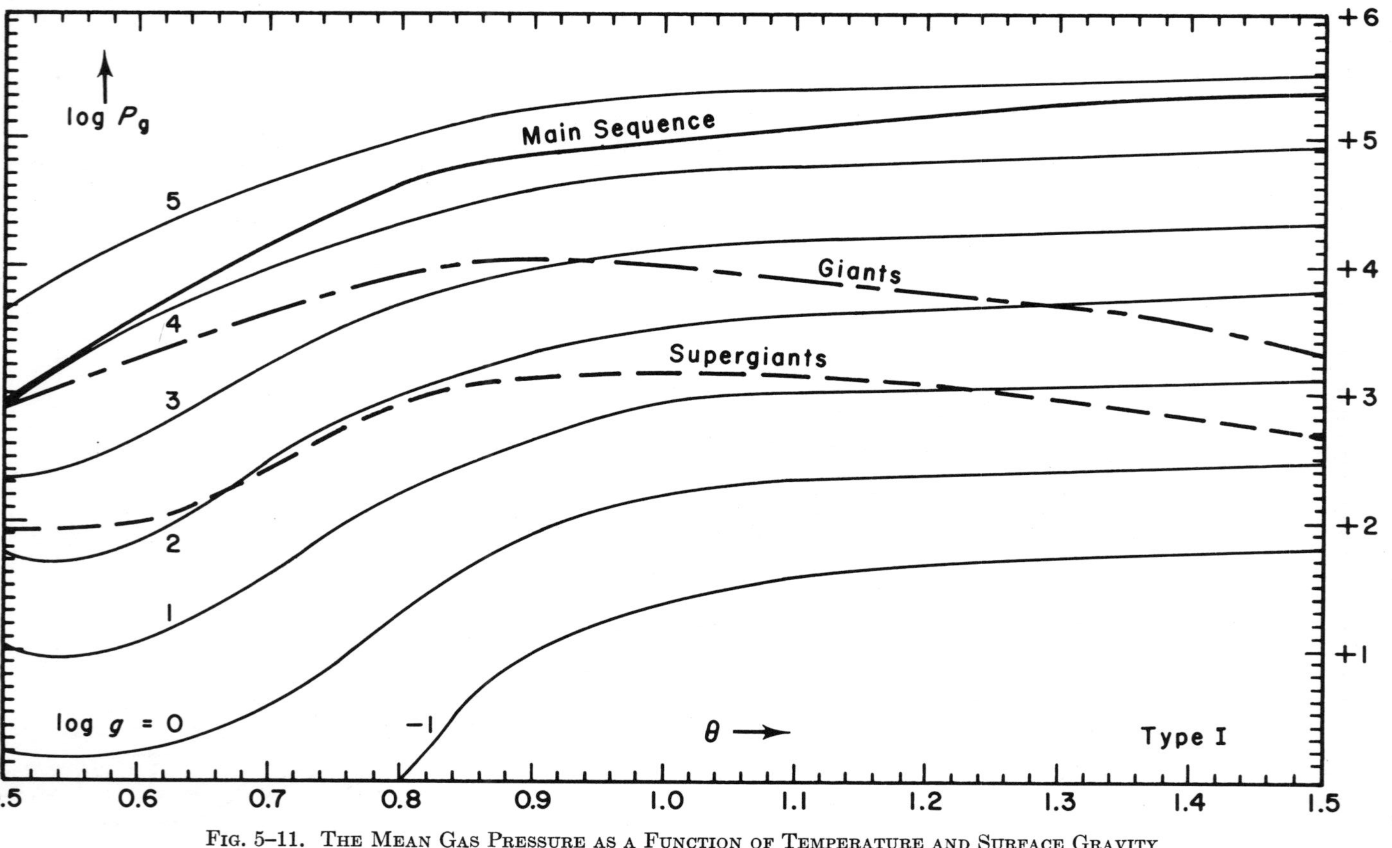

FIG. 5–11. THE MEAN GAS PRESSURE AS A FUNCTION OF TEMPERATURE AND SURFACE GRAVITY

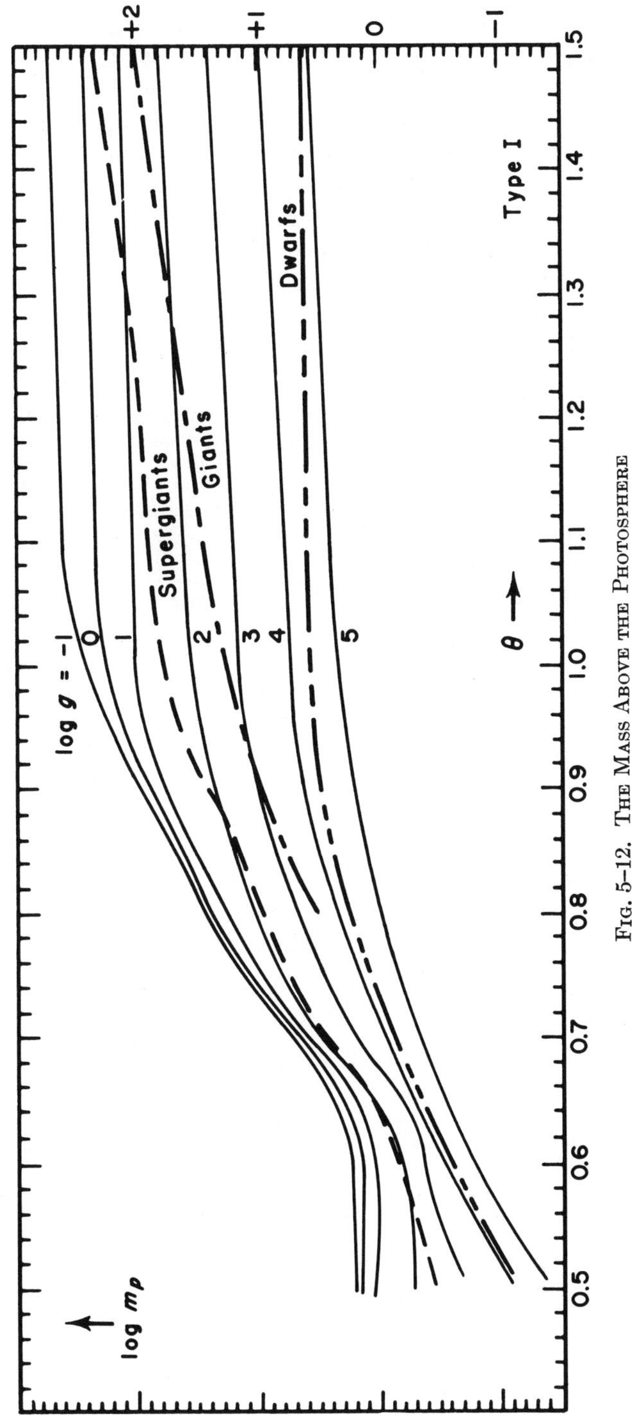

Fig. 5–12. The Mass Above the Photosphere

For example, compare Sirius ($\theta = 0.54$, $\log P_\varepsilon = 2.60$, $\log g = 4.31$) with the sun ($\theta = 0.88$, $\log P_\varepsilon = 1.00$, $\log g = 4.44$). About twenty times as much material is observed above the solar atmosphere as above that of Sirius, a result in harmony with the predictions of theory. The foregoing calculations apply to stars of population type I. Similar computations carried out for population type II stars in which the metal/hydrogen ratio is taken to be thirty times smaller than in the "standard" composition adopted in Art. 3–10 show that (1) hydrogen continues to supply most of the electrons to much lower temperatures and (2) for dwarfs cooler than the sun the gas pressure and mass above the photosphere are greater and the electron pressure is smaller in the population type II dwarfs. At low temperatures, the number of metal atoms above the photosphere may be nearly comparable in the two groups, but at higher temperatures where hydrogen supplies the electrons, the number of metal atoms above the photosphere in population type II stars is reduced more nearly in proportion to the magnitude of the metal deficiency. See Fig. 5–13.

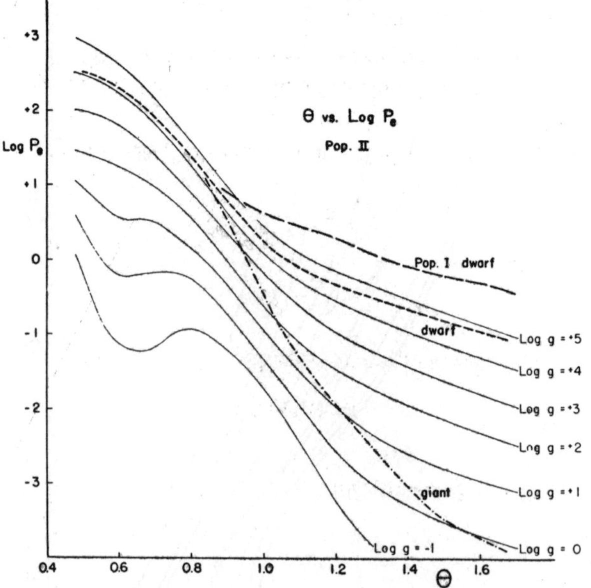

FIG. 5–13. MEAN ELECTRON PRESSURE AS A FUNCTION OF TEMPERATURE AND SURFACE GRAVITY FOR POPULATION TYPE II STARS

Compare with Fig. 5–10. The dotted lines show the values appropriate to dwarf and giant sequences in globular clusters. The assumed metal/H ratio is thirty times smaller than in population type I stars. The relation for population type I dwarf stars is shown in heavy dashed lines. (Courtesy *Astronomical Journal*, **65**, 399, Fig. I, 1960.)

PROBLEMS

5–1. Show that the temperature gradient in an atmosphere in adiabatic equilibrium

$$P = K\rho^{\gamma} \tag{5–183}$$

is constant and equal to

$$\frac{dT}{dx} = \frac{\gamma - 1}{\gamma}\frac{\mu}{\mathcal{R}}g \tag{5–184}$$

where $\mathcal{R}$ is the gas constant, γ is the ratio of specific heats, and g is the acceleration of gravity.

5–2. Consider an atmosphere in adiabatic equilibrium, and in which the absorption coefficient is proportional to gas pressure. Show that the limb-darkening law is

$$\frac{I(0, \theta)}{I(0, 0)} = (\cos\theta)^{\frac{2(\gamma-1)}{\gamma}} \tag{5–185}$$

5–3. Assume that the temperature distribution with optical depth is given by the Eddington approximation Eq. 5–76. Show that if the Planck function is expanded as $B = a + b\tau$, then

$$\frac{b}{a} = \frac{3}{8}u(1 - e^{-u})^{-1} \tag{5–186}$$

where

$$u = \frac{h\nu}{kT_0} \tag{5–187}$$

and T_0 is the boundary temperature and a, b, and B all depend on wavelength.

5–4. Prove the Eddington-Barbier relations: If the source function depends on optical depth according to the simple linear expression $B = a + b\tau$, show that

$$I(0, \theta) = B(\tau_\theta) \tag{5–188}$$

where

$$\tau_\theta = \cos\theta \tag{5–189}$$

and

$$F(0) = B(\tfrac{2}{3}) \tag{5–190}$$

What modifications are needed if

$$B = a + b\tau + ce^{-d\tau} \tag{5–191}$$

5–5. Consider an atmosphere in which both absorption and isotropic scattering occur. Show that if k and σ are both independent of wavelength, the equation reduces to the one already treated. When k and σ depend on frequency, let us define:

$$d\bar{\tau} = (\overline{k + \sigma})\rho\, dx, \quad d\tau = k\rho\, dx, \quad \text{and} \quad dt = (k + \sigma)\rho\, dx \tag{5–192}$$

Then

$$B_\lambda = a_\lambda\left(1 + \frac{b_\lambda}{a_\lambda}\frac{\overline{k + \sigma}}{k}\tau\right) \tag{5–193}$$

where the ratio b/a is given by Eq. 5–186. Solve the appropriate Eq. 5–56, with Eddington's approximation under the assumption that σ/k and $k/(\overline{k+\sigma})$ are independent of the optical depth. Show that

$$J_\lambda = B_\lambda + A_\lambda e^{-\beta t} \tag{5–194}$$

where

$$\beta = \left(\frac{3k}{k+\sigma}\right)^{1/2} \tag{5–195}$$

and the boundary conditions require that

$$A_\lambda = a_\lambda \frac{\frac{2}{3}(b_\lambda/a_\lambda)[(\overline{k+\sigma})/(k+\sigma)] - 1}{1+\frac{2}{3}\beta} \tag{5–196}$$

5–6. With the aid of the results from Problem 5–5, show that the emergent flux πF_λ is given by

$$F_\lambda = 2\int_0^\infty B(\bar{\tau})E_2(t)\,dt + 2A_\lambda \frac{\sigma}{k+\sigma}\int_0^\infty E_2(t)e^{-\beta t}dt \tag{5–197}$$

With $E_n(x)$ defined by Eq. 5–11, and

$$E_0 = \frac{e^{-x}}{x}, \quad E_1 = E \tag{5–198}$$

prove the following relations:

$$(n-1)E_n(x) = e^{-x} - xE_{n-1}(x), \quad n \geq 2 \tag{5–199}$$

$$E_n(0) = \frac{1}{n-1}, \quad n \geq 2 \tag{5–200}$$

$$\int_x^\infty E_n(x)\,dx = E_{n+1} \tag{5–201}$$

$$\frac{d}{dx}E_n(x) = -E_{n-1}(x), \quad (n \geq 1) \tag{5–202}$$

$$\int_x^\infty xE_n(x)\,dx = xE_{n+1}(x) + E_{n+2}(x) \tag{5–203}$$

$$A(\alpha, x) = \int_0^x e^{\alpha x}E(x)\,dx = -\frac{1}{\alpha}[\ln(|1-\alpha|) + E[(1-\alpha)x] - e^{\alpha x}E(x)] \tag{5–204}$$

5–7. Consider an atmosphere in which the radiative flux is no longer constant with optical depth (as in an atmosphere where convection occurs). Show that in the Eddington approximation:

$$B = \frac{F_0}{2} + \frac{3}{4}\int F\,d\tau - \frac{1}{4}\frac{dF}{d\tau} \tag{5–205}$$

REFERENCES

General Methods for Radiative Transfer

An account of the powerful new methods for the treatment of the flow of radiation through a stellar atmosphere is given in:

CHANDRASEKHAR, S. 1950. *Radiative Transfer.* Oxford University Press, Fair Lawn, N. J. Reprinted by Dover Publications, New York, 1960. See especially Chaps. 1–5, 11, and 12. An extensive bibliography is given. The discussion in our Art. 5–7 is taken largely from *Ap. J.* **100**, 76, 1944.

A general discussion of transfer problems with applications to neutron diffusion is given in:

KOURGANOFF, V. (with collaboration of Ida W. Busbridge). 1952. *Basic Methods in Transfer Problems.* Oxford University Press, Fair Lawn, N. J.

Discussion of transfer problems of both grey and non-grey atmospheres are given in:

UNSÖLD, A. 1955. *Physik der Sternatmosphären.* Springer-Verlag, Berlin.

Among the best older references are:

EDDINGTON, A. S. 1926. *Internal Constitution of the Stars.* Cambridge University Press, London.

MILNE, E. A. 1930. *Handbuch der Astrophysik.* Springer-Verlag, Berlin, vol. 3, part 1.

Applications of Ambarzumian's principle of invariance are given for example by:

HORAK, H. G. 1952. *Ap. J.* **116,** 477.

———. 1954. **119,** 640.

HORAK H. G., and LUNDQUIST, C. A. 1954. *Ap. J.*, **119,** 42.

———. 1955. **121,** 175.

BUSBRIDGE, I. W. 1953. *M. N.* **113,** 52.

———. 1955. **115,** 661.

BUSBRIDGE, I. W., and STIBBS, D. W. N. 1954. *M. N.* **114,** 2.

STIBBS, D. W. N. 1953. *M. N.* **113,** 493.

The probabilistic method is discussed by:

SOBOLEV, V. V. 1951. *A. J. USSR* **28,** 355.

———. 1954. **31,** 231.

Transfer of Radiative Energy in Atmospheres of Stars. Moscow: 1956.

UENO, S. 1957–58. *Kyoto Inst. Astrophysics Contr.* **65, 66, 67.**

See also *Ann. d'Ap.* **22,** 468, 1959.

Continuous Stellar Absorption Coefficient

In addition to extensive (unfortunately as yet unpublished) tables by Strömgren, there are the calculations by:

VITENSE, E. 1951. *Zeits. f. Ap.* **28,** 51.

SAITO, S., UENO, S., and JUGAKU, J. 1954. *Kyoto Contr. No. 43.*

Non-Grey Atmospheres

The methods we have mentioned are discussed in detail in:

AMBARZUMIAN, V. A. (ed.) 1958. *Theoretical Astrophysics* (translated by J. B. Sykes). Pergamon Press, London.

BÖHM, K. H. 1954. *Zeits f. Ap.* **34,** 182.

CHANDRASEKHAR, S. 1936. *M. N.* **96,** 21.

COOK, A. 1954. *Ap. J.* **120,** 578.

HUNGER, K., and TRAVING, G. 1956. *Zeits f. Ap.* **39,** 248.

KROOK, M. 1959. *Ap. J.* **129,** 724. **130,** 286.

PRZYBYLSKI, A. 1955. *M. N.* **115,** 650.

SWIHART, T. 1956. *Ap. J.* **123,** 139.

UNNO, W., and YAMASHITA, Y. 1960. *Publ. Astr. Soc. Japan* **12,** 157.
UNSÖLD, A. 1955. *Op. cit.*

Arguments pertaining to a choice of mean absorption coefficient are given by:

MICHARD, R. 1949. *Ann d'Ap.* **12,** 291.
SAITO, S. 1959. *Publ. Ast. Soc. Japan* **11,** 98.
PRZYBYLSKI, A. 1960. *M. N.* **120,** 3.

Approximation Formulas for Computing Flux and Intensity Integrals

KOURGANOFF, V., and PECKER, C. 1950. *Ann d'Ap.* **12,** 247.
REIZ, A. 1954. *Lund Meddelande* 184.
CAYREL, R. 1960. *Ann d'Ap.* **23,** 245.
CHANDRASEKHAR, S. 1950. *Radiative Transfer.* Oxford University Press, Fair Lawn, N. J., chap. 2.

Model Atmospheres

For an excellent general account see:

MÜNCH, G. 1960. *Stellar Atmospheres,* vol. 6 of *Stars and Stellar Systems.* J. L. Greenstein (ed.). University of Chicago Press, Chicago, chap. 1.

A useful compilation of properties of various theoretical models is given by:

MARTINI amd MASANI. 1957. *Contr. Milano-Merate* No. 119.

For models of yellow giant stars see:

CANAVAGGIA, R., and PECKER, J. C. 1952. *Ann d'Ap.* **15,** 260.
———. 1953. **16,** 47.

For *O* and *B* stars, see calculations by:

McDONALD, J. K. 1953. *Publ. Dom. Ap. Obsy. Victoria* **9,** 269.
PECKER, J. C. 1950. *Ann d'Ap.* **13,** 294, 319, 433.
SAITO, S. 1955. *Contr. Inst. Astr. Kyoto* No. 48.
SAITO, S., and UESUGI, A. 1959. *Publ. Ast. Soc. Japan* **11,** 90.
TRAVING, G. 1955. *Zeits. f. Ap.* **36,** 1.
———. 1957. *Zeits. f. Ap.* **41,** 215.
UNDERHILL, A. 1950. *Publ. Copenhagen Obsy.* No. 151.
———. 1950. *Publ. Dom. Ap. Obsy. Victoria* **8,** 357; **10,** 357. (1960) **11,** 363. (1963) *Publ. Dom. Ap. Obsy.*

For *A* stars, detailed models have been given by:

HUNGER, K. 1955. *Zeits. f. Ap.* **36,** 42.
OSAWA, K. 1956. *Ap. J.* **124,** 513.

For *A*, *F*, and *G* stars, models have been proposed by various workers, e.g.,

MELBOURNE, W. G. 1960. *Ap. J.* **132,** 101.

The older work on solar model atmospheres is summarized by:

MINNAERT, M. 1953. *The Sun.* G. P. Kuiper (ed.). University of Chicago Press, Chicago.

Among more recent models we mention that proposed by:

BÖHM-VITENSE, E. 1954. *Zeits. f. Ap.* **34,** 209.

which forms the basis for

BÖHM, K. H. 1954. Theoretical model, *op. cit.,* sec. 3.

and the theoretical model by

SWIHART, T. 1956. *Ap. J.* **123,** 143.

The empirical temperature distributions that form the basis for the model given in Table 5–1 are taken from:

PIERCE, A. K., and ALLER, L. H. 1952. *Ap. J.* **116,** 175.

PIERCE, A. K., and WADDELL, J. 1961. *Mem. Royal Astron. Soc.* **58**, 89.
PLASKETT, H. H. 1955. *Vistas in Astronomy.* A. Beer (ed.). Pergamon Press, London.
PRZYBYLSKI, A. 1957. *M. N.* **117**, 483, 600.
MITCHELL, W. 1959. *Ap. J.* **129**, 369.
PAGEL, B. 1956. *M. N.* **115**, 493.
NECKEL, H. 1958. *Zeits f. Ap.* **44**, 153, 160.

Extensive calculation of model atmospheres for all spectral classes and appropriate surface gravities have been made by Neven and de Jager (see Utrecht Observatory publications).

Attention may also be called to a recent translation from the Russian astrophysical literature:

SOBOLEV, Y. V. 1963. *Treatise on Radiative Transfer.*
Translated by S. Gaposchkin. Van Nostrand, Princeton, N. J.

CHAPTER 6

Radiation of the Stars

6–1. Significance of a Stellar Temperature. Stellar masses, radii, and luminosities are fundamental quantities. A star's spectrum, for example, is intimately related to its surface temperature and gravity. The mass and radius fix the surface gravity, whereas the temperature depends essentially upon the rate of energy generation and upon the radius. A theoretical treatment of stellar atmospheres presupposes that we know the star's boundary temperature, its surface gravity, and chemical composition. The theory of stellar structure requires the luminosity, mass, and radius of a star and some information on its chemical composition.

In Chapter 1 we mentioned how stellar masses are found from components of visual and eclipsing binaries, and radii from interferometer measures of stars of known parallax and from eclipsing binaries. In this chapter we shall concern ourselves with the determination of stellar luminosities, energy distributions, and temperatures.

The problem of stellar energy distribution is related to two broad questions: (a) stellar temperatures and (b) magnitudes and colors. A complete description of the continuous spectrum of a star would entail the measurement of the ratio $F(\lambda)/F(\lambda_0)$, i.e., the flux at all wavelengths λ referred to its value at λ_0 and then the measurement of the monochromatic flux $F(\lambda_0)$ in ergs cm^{-2} sec^{-1} per unit solid angle at the top of the earth's atmosphere. Actually such a rigorous procedure is not possible even for the sun. Our studies are hampered by limitations imposed by the small region of the spectrum we can observe through the earth's atmosphere and by the wavelength sensitivity of our detecting apparatus. In many instances we must content ourselves with measurements made in accessible spectral regions and then extrapolate to inaccessible spectral regions with the aid of model atmospheres and predicted energy distributions. Furthermore, the spectral resolution or "purity" is limited because of the faintness of most stars. Spectral energy distributions may be measured only for relatively bright stars. One must observe faint stars with broad band-pass filters to obtain stellar colors and magnitudes. Spectral lines are smoothed and information is unavoidably lost.

The temperature of a star must be defined from quantitative measurements of its line and continuous spectrum. As we have discussed in Chapter 5, there is a monotonic increase of temperature with depth in the star; for this reason alone it is impossible to assign a single number that will fully describe the spectrum. Deviations from thermodynamic equilibrium (LTE) further complicate the problem for it then becomes no longer possible even to assign a unique temperature at a given point. Astrophysicists tend to define different kinds of stellar temperatures depending on the type of observational data employed.

1. *Ionization temperatures* can be determined from the level of ionization in the atmosphere if the electron pressure is known. Saha, Milne, and Fowler, and others used this method in the early twenties to establish a stellar temperature scale.

2. *Excitation temperatures* are found from a comparison of numbers of atoms in different energy levels by Boltzmann's formula. From the equivalent width of a spectral line (see Chapter 8) it is possible to determine the numbers of atoms capable of absorbing it per gram of stellar material. Consider two levels r and r' differing in excitation energy by $\chi_{rr'}$. The relative numbers of atoms in the two levels will be given by

$$\frac{N_r}{N_{r'}} = \frac{b_r\, g_r}{b_{r'} g_{r'}}\, e^{-\chi_{rr'}/kT} \tag{6–1}$$

The factors b_r and $b_{r'}$ are corrections required for Boltzmann's formula to allow for possible deviations from local thermodynamic equilibrium LTE. If we arbitrarily set $b_r = b_{r'} = 1$ and determine T from the ratio $N_r/N_{r'}$, it will not necessarily agree with the ionization temperature or with an average temperature defined in some other way. Furthermore, different elements or even different lines of the same element will give different excitation temperatures. Determinations of excitation temperatures from equivalent widths have led often to contradictory results, partly because of poor line transition probabilities (Chapter 7), partly because different ines are formed at different depths in the stellar atmosphere, and partly because of deviations from LTE. Molecules are formed in the uppermost strata of stellar atmospheres and give excitation temperatures lower generally than those found by other means.

3. *Color temperatures.* Suppose we compare the energy distribution of a star in the wavelength region $\lambda_1 - \lambda_2$ with that of a black-body source. We may find that in this interval the distribution of energy, not its absolute value, can be represented by a Planckian function for some temperature T_c. That is

$$\log \frac{F(\lambda_1)}{F(\lambda_2)} = \left(\frac{\lambda_2}{\lambda_1}\right)^5 \frac{e^{hc/\lambda_2 k T_c} - 1}{e^{hc/\lambda_1 k T_c} - 1} \tag{6–2}$$

where $F(\lambda_1)/F(\lambda_2)$ is the ratio of the measured fluxes at λ_1 and λ_2. Over another spectral interval, $\lambda_3 - \lambda_4$, a different color temperature T_c' would be found, and so on.

Thus, a stellar color temperature is a parameter employed in Planck's function which is used as an interpolation formula to represent the slope of the energy distribution. It does not necessarily represent any physical temperature in the photospheric layers. For example, a normal $A0$ star has a photospheric temperature near 10,000°K, but the slope of its energy distribution in the region $6000 > \lambda > 4000$ corresponds to that of a Planckian curve with $T \sim 16{,}000$°K. For wavelengths shortward of $\lambda 4000$, the color temperature is near 10,000°K.

In place of color temperatures, we often use spectrophotometric gradients (Art. 6–3). The color of a star as measured with broad band-pass filters, e.g., the UBV system yields a crude color temperature (Art. 6–6).

4. *Brightness temperature.* A quantity that can be measured for a number of stars is the radiation cm^{-2} sec^{-1}A^{-1}, πF_λ, for selected points in the continuous spectrum. We can define a brightness temperature T_b, as that of the black body that would give the same energy output per angstrom as the star at wavelength λ. Brightness temperatures are measurable only for stars of known angular diameter such as eclipsing binaries of known parallax, or stars such as Sirius whose size has been measured with the photon correlation interferometer (Brown and Twiss, 1954). At $\lambda 5300$, Greaves estimates T_b's of 28,000°K for spectral type $O8$, 18,000°K for $B0$, 11,000°K for $A0$, and about 6100°K for $G2$ (the sun). The brightness temperature is really only a parameter that expresses the rate of energy radiation at certain wavelengths. It cannot be converted readily into the physical temperature of the emitting stellar surface. For example, from E. Pettit's measurements of the radiation from the whole disk of the sun, the brightness temperature of that body is found to be 6200°K at $\lambda 4500$ and 6000°K at $\lambda 6500$. At $\lambda 5263$ Labs (1957) finds $T_b = 6470$°K. High temperatures are also found in the infrared, e.g., by E. M. Lewis, but in the "rocket" ultraviolet, T_b falls to values near 4500°K.

5. *Effective temperature.* Let us suppose that a measurement of the total energy output (luminosity) L of a star of known radius R could be obtained. We define the effective temperature T_e of a star as that which a black sphere of the same radius must possess in order that its total energy output equal that of the star. That is, T_e is defined by

$$L = 4\pi R^2 \sigma_R T_e^4 \tag{6–3}$$

since $4\pi R^2$ is the surface area and $\sigma_R T_e^4$ is the black-body energy output per cm^2 (see Chapter 4). The effective temperature is a datum of great theoretical interest, but we must emphasize that it is not directly observable except perhaps for the sun whose energy can be measured from rockets and satellites above the earth's atmosphere.

To obtain the effective temperature of a star that can be observed over a limited wavelength range only, we may use model atmosphere procedures. From a model corresponding to a given surface gravity g, effective temperature T_{eff}, and chemical composition, we compute the emergent flux and compare it with observed energy distribution. In practice one uses a network of models with different surface gravities, temperatures, etc. Code and his associates have obtained effective temperatures in this way (see Chapter 5). Of course, effective temperatures obtained by this indirect method are only as reliable as the predicted emergent fluxes.

6–2. The Energy Output of the Sun. The most extended measures of stellar radiation are those that have been carried out for the sun. As indicated by the discussion in Chapter 5 the quantities wanted are $I_\lambda(0, 0)$, the specific intensity at the center of the solar disk, and $I_\lambda(0, \theta)/I_\lambda(0, 0)$, the limb darkening.

The energy distribution in the solar spectrum may be determined, in principle, in either of two ways, both of which have been used extensively. In one method, one compares the surface brightness at the center of the solar disk directly with that of a black body operating at a known temperature. This method, which seems simple in practice, is actually difficult in application. One must allow for extinction in the earth's atmosphere. Furthermore, since the black-body source is much cooler than the sun, its energy distribution will differ considerably, particularly in the visible and ultraviolet regions of the spectrum. A small error in the temperature of the black body leads to a huge error in its predicted ultraviolet intensity, I_λ.

In a second method, developed by Langley and Abbot, one measures the relative energy distribution in the spectrum with a detector whose sensitivity must be determined at each wavelength. The relative energy distribution as observed at the earth's surface is thus found, but the ordinates of this curve have to be determined from a separate observation. If the flux of radiation from the entire disk of the sun is observed, the area under the curve, $\int f_\lambda d\lambda$, must correspond to the actual energy received at the earth's surface. This energy may be measured by the heating effect it produces, with the aid of an instrument called a pyrheliometer. Several types have been devised, the type most often used employs the heating effect produced on a blackened silver disk. This device is calibrated with the aid of a fundamental instrument called a water-flow pyrheliometer in which sunlight enters a hollow, blackened chamber and heats, by a measurable amount, a stream of water which flows through a spiral tube at a known speed. The instrument is then pointed away from the sun, a resistance coil is placed in the same tube, and the amount of electric current necessary to raise the temperature of the stream of water by the same amount is measured. The heat developed by the coil ($0.24\ I^2R$ cal sec^{-1}, where I is the current in amperes and R is the resistance in ohms) must exactly equal the heat received from the sun.

To measure the solar energy curve, Abbot employed a spectrobolometer. This instrument involves the principle that when a metal wire is heated, its resistance is changed. A solar spectrum may be formed, for example, by a spectrometer with all mirror optics and a grating to avoid losses by absorption. By rotating the grating the spectrum is moved across a blackened wire. As regions of fluctuating intensity cross the wire, its temperature and therefore its resistance change, and a varying galvanometer deflection is obtained. Strong Fraunhofer lines give large depressions in the curve. When corrections are made for transparency of the spectrograph, for reflection losses in the optical system, etc., the area under the reduced bolograph curve will be proportional to the total amount of energy received provided that the bolograph curve corresponds to the total flux received, not the specific intensity at the center of the sun's disk.[1] Since the actual amount of energy is measured by the calibrated pyrheliometer, the area of the bolograph flux record is known in energy units, i.e., in calories. Hence the ordinates of the solar flux distribution are known in ergs cm^{-2} $sec^{-1}A^{-1}$ at the earth's surface for a certain zenith distance of the sun.

Because measurements are made within the earth's atmosphere, the effects of atmospheric extinction must be determined. A determination of this correction would be simple, were it not for the fact that the absorption depends strongly on wavelength. Hence both pyrheliometer and spectrobolometric measurements have to be made at different zenith distances of the sun.

If $\mathfrak{F}_{0\lambda}$ is the flux of the radiation falling on the top of the earth's atmosphere, z is the zenith distance of the sun at the instant of observation, and m is the total mass of air above the station, then the flux $\mathfrak{F}_{z\lambda}$ of the radiation reaching the observer will be

$$\mathfrak{F}_{z\lambda} = \mathfrak{F}_{0\lambda}\, e^{-k_\lambda m \sec z} \tag{6–4}$$

A similar law holds, of course, for specific intensities. At each zenith distance, z, for which $\mathfrak{F}_{z\lambda}$ has been measured, one calculates log $\mathfrak{F}_{z\lambda}$ and plots it against m sec z. If a straight line is drawn through the points and extrapolated to zero air mass ($m \sec z = 0$), we obtain log $\mathfrak{F}_{0\lambda}$, the flux the radiation of wavelength λ would have outside the earth's atmosphere. The same procedure is followed for many other wavelengths, and these extrapolated $\mathfrak{F}_{0\lambda}$ values define the solar energy curve outside the earth's atmosphere. Since pyrheliometer measurements gave the units of the $\mathfrak{F}_{z\lambda}$ curves, the area under the $\mathfrak{F}_{0z}$ curve gives the solar constant, the amount of energy received from the sun per cm^2 per min just outside the earth's atmosphere at the earth's mean distance from the sun. Of course a small correction is necessary for the radiation in the region λ2900 which can be observed only from rockets and artificial satellites.

[1] If the spectrobolometric measurements actually refer to the center of the disk, one can convert them to total flux with the aid of measurements of limb darkening.

Measurements of the solar energy distribution $I_\lambda(0, 0)$ have been obtained by numerous observers (see Minnaert, 1953; Goldberg and Pierce, 1959). Different observers have not only used different observational techniques but have measured different quantities. Some measured integrated flux πF_λ from the entire disk. $I_\lambda(0, 0)$ can be calculated from $F_\lambda(0)$ if the limb darkening is known and vice versa. Some have measured I_λ with low resolution equipment, e.g., Abbot; others have looked for "windows" between absorption lines and have published hypothetical continua based on measurements made at these points, e.g., Chalonge *et al.*, 1946, 1950; Labs, 1957.

Absorption lines are distributed unevenly throughout the solar spectrum. In the region longward of λ5200 the lines are often spaced sufficiently far apart to permit a good estimate of the background continuum. Shortward of λ5200, lines become stronger and more numerous, while below 4000 they overlap so frequently that the position of the undisturbed continuum is rarely if ever seen. Urgently needed are accurate measurements of $I_\lambda(0, 0)$ with a high spectral purity. Fig. 6–1 compares energy curves given by Pettit, 1932, 1940; Stair, 1951, 1952, 1954; and Dunkelman and Scolnik, 1952. Notice the effects of line absorption and varying spectral resolution of the equipment. Solar energy distribution compilations have been published by Mulders (1935) and by Johnson (1954), who incorporated ultraviolet results obtained by Tousey and his associates from rocket flights. Absolute intensities in the spectral range λ4010–6569 for the center of the solar disk have been published by Labs and Neckel (1962). They compared the sun with a standardized lamp with a 20A band pass. The discordances between results obtained by different observers are distressing and point up the necessity of accurate measurements of the solar energy distribution.

Precise measurements by Chalonge and his co-workers (1946–50) and by Labs (1957) have emphasized determination of solar color temperatures. Thus Labs measured $I_\lambda(0, 0)$ in various so-called "windows" between λ3300 and λ7000, comparing the sun with the crater of a carbon arc. He found $T_c = 6226°K$ $(3300 < \lambda < 3700)$, $T_c = 7540°K$ $(4000 < \lambda < 4950)$, and a Balmer discontinuity $D = 0.13$, values slightly higher than those by Chalonge and Canavaggia. Measurements in the far infrared are seriously hampered by water vapor absorption in the earth's atmosphere, but the difference in temperature between the sun and arc are not so serious. Ozone introduces severe difficulties in the photometry in the ultraviolet.

As we saw in Chapter 5, measurements of $I_\lambda(0, 0)$ combined with those of limb darkening $I_\lambda(0, \theta)/I_\lambda(0, 0)$ enable one to construct a model of the solar atmosphere. Of particular interest is the darkening at the extreme edge of the sun, i.e., for $\mu = \cos\theta \leq 0.1$. Since the pioneering work of Abbot, Fowle, and Aldrich (1912, 1913, 1914), studies have been

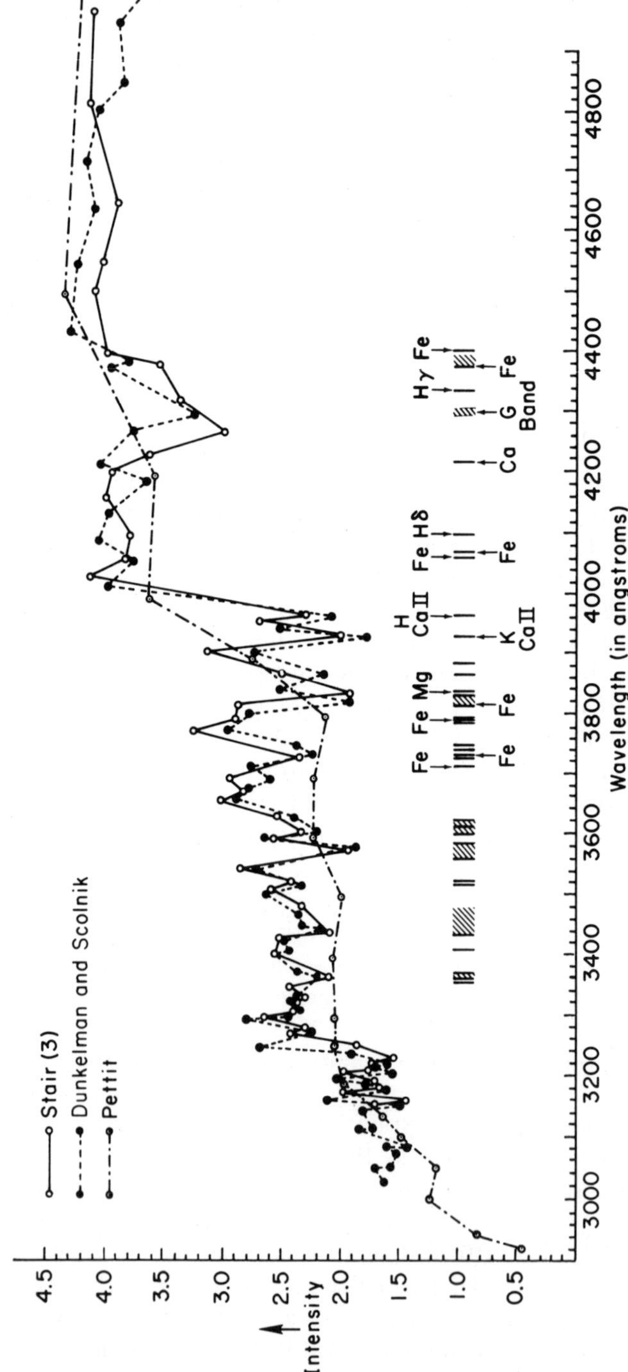

FIG. 6-1. ENERGY DISTRIBUTION IN THE SOLAR SPECTRUM λ2900–λ4900

Results obtained by Pettit, Stair, and Scolnik and Dunkelman are compared. Positions of strong Fraunhofer lines are indicated.

made by Raudenbusch (1938), by Moll, Burger, and van der Bilt (1925), by Cannavaggia and D. Chalonge (1946), by Minnaert (1949), by Peyturaux (1955), and by Pierce (1954), using observations outside of eclipse, and by Linblad and Kristenson (1951, 1955) and by Ten Bruggencate, Gollnow, and Jager (1950) using eclipse data. Goldberg and Pierce (1959) have discussed the observational difficulties inherent in finding the true limb darkening near the edge of the sun. Bad seeing, scattered light,

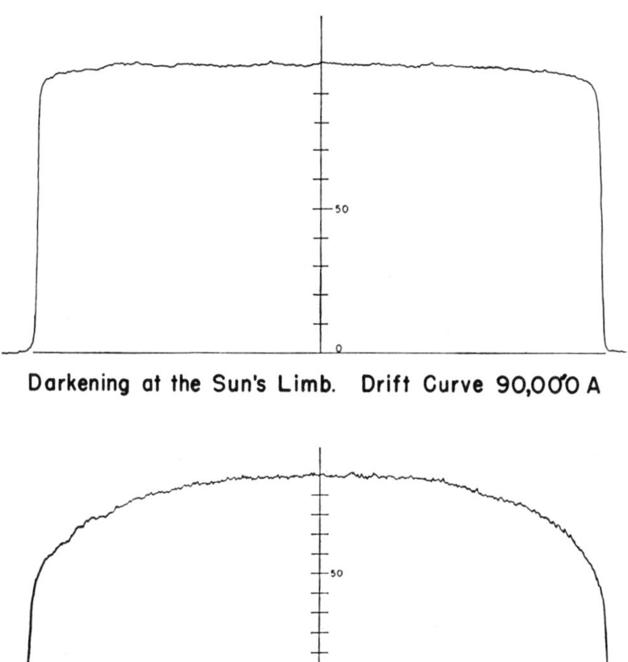

FIG. 6–2. SOLAR LIMB DARKENING

These curves show the distribution of intensity across the disk of the sun at λ90,000 and λ5970.5. In the far infrared the opacity of the solar material is so high that we receive radiation from the outermost layers both at the limb and the center of the disk; hence the curve is almost flat. At λ5970.5 the limb darkening is pronounced. (McMath-Hulbert Observatory, University of Michigan.)

diffraction, scratches on mirror surfaces, and other optical errors, as well as the finite size of scanner slot used, all complicate the problem. The measurements of Peyturaux and of Pierce seem to be the most reliable. Limb darkening depends strongly on λ, being much more pronounced in violet than in red spectral regions. See Figs. 6–2 and 6–3.

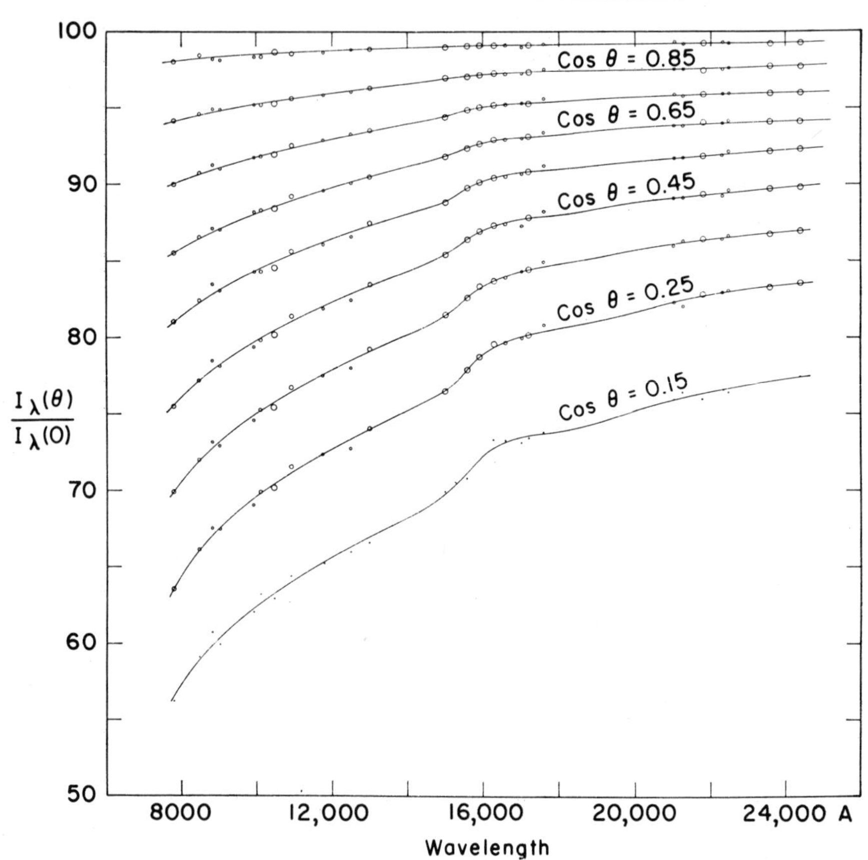

FIG. 6–3. SOLAR LIMB DARKENING AS A FUNCTION OF WAVELENGTH

Observed values of $I_\lambda(\theta)/I_\lambda(0)$ plotted as a function of wavelength. The largest circles have the greatest weight. The curves correspond to $\cos\theta = 0.15, 0.25, 0.35, 0.45, 0.55, 0.65, 0.75, 0.85$, and 0.95. (Courtesy, A. K. Pierce, *Ap. J.* **120,** 221, 1954, University of Chicago Press.)

The effective temperature of the sun may be determined from the solar constant for which Abbot found the value 1.938 cal cm^{-2} min^{-1} which amounts to 1.36×10^6 ergs cm^{-2} sec^{-1}, 0.136 w cm^{-2}, or 1.81 hp m^{-2} sec^{-1}. The total amount of energy passing through a sphere or the radius of the earth's orbit, 1.495×10^{13} cm, must equal the total energy output of the sun, 3.79×10^{33} erg sec^{-1}, or 5.08×10^{23} hp. The energy output per cm^2 per sec is 6.25×10^{10} ergs cm^{-2} sec^{-1}. The corresponding effective temperature of the sun is obtained from Stefan's law (Chapter 4). Putting in numerical values, we find

$$6.25 \times 10^{10} = 5.672 \times 10^{-5}\, T^4$$

whence $T = 5760°\text{K}$. More recently, the Smithsonian observers Aldrich and Hoover (1954) have suggested that the solar constant value is 1.940 cal cm^{-2} min^{-1}. C. W. Allen, making use of the Naval Research Laboratory rocket data, concluded that allowance for the radiation totally cut out by the earth's atmosphere would increase the solar constant to 1.970 cal cm^{-2} min^{-1}, but a more detailed discussion by F. S. Johnson (1954) leads to a final value of 2.00 ± 0.04 cal cm^{-2} min^{-1}. Further work using balloon, rocket, and satellite data is clearly indicated.

A solar effective temperature of about 5800°K is indicated, a value lower than most of the color or brightness temperatures previously cited. Whereas T_{eff} corresponds to the integral of the flux with respect to wavelength, $\int F_\lambda d\lambda$, carried out over the actual energy distribution, the brightness temperature corresponds to selected high points in the continuum between spectral lines, whereas T_c is found by fitting Planckian curves to such "windows" between lines.

No convincing evidence has been found for variations in the solar constant. If the solar energy output varied, the planets should show brightness variations when measured with respect to standard stars, but no such fluctuations have been established and one must conclude that the integrated light of the sun is very nearly constant.

6–3. Relative Energy Measurements and Spectrophotometric Gradients. Stellar spectrophotometry can be carried out either on a relative or on an absolute basis. One type involves a comparison of stellar energy distributions with one another; the other involves comparison of a stellar energy distribution with that of a laboratory source. Relative spectrophotometry achieves a much higher accuracy than does absolute spectrophotometry. Therefore, a major effort is devoted to establishing accurate absolute energy distributions for a few stars; relative energy distributions may be obtained for many.

In descriptions of stellar energy distributions in spectra it is useful to employ a quantity called the *gradient*. Let us write Planck's law in the form

$$I(\lambda, T) = \frac{c_1}{\lambda^5} \frac{1}{e^{c_2/\lambda T} - 1} \tag{6–5}$$

where $c_2 = 1.438 \times 10^8$ if λ is measured in angstroms. Suppose that for an interval λ_1–λ_2 we wish to compare a star whose energy distribution can be characterized by a color temperature T_A with a source of known temperature T_B. Measurements usually now by photoelectric photometry, give the ratio, $\log I_A(\lambda)/I_B(\lambda)$, which is a linear function of $1/\lambda$ even for large spectral intervals. We define the relative gradient:

$$G_{AB} = -\frac{d\,[\ln I_A(\lambda)/I_B(\lambda)]}{d(1/\lambda)} = -2.30\frac{d \log I_A(\lambda)/I_B(\lambda)}{d(1/\lambda)} \tag{6–6}$$

for the wavelength range involved. By hypothesis, we suppose that the energy distributions of A and of B resemble those of black bodies at T_A and T_B, respectively, in the interval λ_1–λ_2. Hence, from Eq. 6–5

$$G_{AB} = \frac{c_2}{T_A}\left(1 - e^{-c_2/\lambda T_A}\right)^{-1} - \frac{c_2}{T_B}\left(1 - e^{-c_2/\lambda T_B}\right)^{-1} \tag{6–7}$$

If we define

$$\phi = \frac{c_2}{T}\left(1 - e^{-c_2/\lambda T}\right)^{-1} = 5\bar{\lambda} - \ln 10\,\frac{d \log I_\lambda}{d\left(\frac{1}{\lambda}\right)} \tag{6–8}$$

then

$$G_{AB} = \phi\,(T_A) - \phi\,(T_B) \tag{6–9}$$

If T is not large, λT will be small, and G_{AB} will approach

$$G_{AB} = c_2\left(\frac{1}{T_A} - \frac{1}{T_B}\right) \tag{6–10}$$

The observational quantity G_{AB} gives the relative energy distribution in the star and source over a selected wavelength range. It is a more fundamental and useful quantity than the color temperature. Suppose we compare a certain number of stars with the same primary standard (star or laboratory source), for which a temperature T_B is assumed. If an error δT_B is subsequently found in T_B, the ϕ_A's will all be affected by the same amount $\delta\phi_B$, but the changes in color temperature will be much greater for hot stars than for cooler ones. The transformation from gradient to color is most easily done with Kienle's (1941) monograms.

Relative spectrophotometry is most accurately done with the aid of a photoelectric spectrum scanner of the type employed by Whitford and Code, Liller, Oke, and others. Such instruments employ all-mirror optics with a grating as a dispersing element, and are usually equipped with antimony-caesium or multi-alkali photomultiplier tubes for the region $\lambda < 6000$, and caesium oxide on silver photomultiplier tubes for 6000–11,000A. As the grating rotates, the spectrum moves across the photocell. The purity of the spectrum can be controlled by appropriate slots.

The observer chooses a small group of stars, preferably of early type, well distributed over the sky as primary standards and compares the program stars with these objects. It is necessary to observe at least one star over a sufficient range of zenith distance to accurately evaluate atmospheric extinction. Whitford and Code scanned spectra of sixty-seven stars with resolutions of 10A in blue and 20A in infrared and published relative fluxes at sixteen wavelengths from 3400A to 10,000A (see Code, 1960). Oke (1960) published accurate fluxes for six stars at twenty-four wavelengths in the continuum between λ3390 and λ6000. Among the older relative spectrophotometric studies we mention particularly the Greenwich work (Greaves *et al.*, 1940) which were extended to the southern skies by Gascoigne (1950) and the work of John Hall (1941). The Greenwich observers measured the relative gradients (referred to the mean gradient for an *A*0 star) for 250 stars from *Oe*5 to *G*0 using a baseline from λ4300 to λ6300. With the aid of gratings, Hall measured relative gradients of sixty-seven bright stars (λ4560 – λ10,300) with a caesium-oxide-on-silver cell.

A number of teams have measured energy distributions in a few stars absolutely and then have compared numerous stars with these fundamental standards, e.g., Kienle and associates (1938, 1940), Barbier and Chalonge (1940).

6-4. Absolute Spectrophotometry. The comparison of a laboratory source with a star entails severe difficulties, particularly if a long wavelength range is involved: (1) the laboratory source must be constant with time, must be reproducible, and both its brightness temperature and energy distribution must be known; (2) the laboratory source and the stars must be observed over the same optical paths; and (3) we must observe the stellar spectrum at several altitudes, i.e., through different air masses in order to estimate the reddening effect of the earth's atmosphere upon the observed energy distribution. Photocells are commonly used as detectors, but lead sulfide cells may be used in the infrared. Thermocouples and bolometers are so insensitive that they can be used only on the brightest stars with the largest telescopes.

The temperature of the primary source must be accurately established. Below the solidification point of Pd at 1552°C, a gas thermometer may be used; above the gold solidification point at 1063°C one may use the radiation laws. The black-body radiation from a source of molten platinum at 1769°C (which is just at the point of solidification) is often taken as the primary standard although some workers use the gold point.[2] The melting points of Ir, Pd, and Ni provide other fixed points. Secondary standards include the tungsten ribbon filament lamp (operated near 2800°K), and the xenon high pressure lamp which has a color temperature near 5600°K and is very stable. The positive crater of a pure graphite electrode of a burning carbon arc can be used as a standard emitter whose temperature is near 3996°K ± 15°.

Special care is necessary to insure that the light path from the source

[2] In fact, Stebbins and Kron suggest that the largest errors come not from comparing stars with cool terrestrial sources, but from uncertainities in the gold point and the constant c_2 in the Planck equation.

and the star are exactly the same, and that the source is properly calibrated *in situ*. Atmospheric extinction has to be determined with great care.

Among the older fundamental investigations carried out by photographic photometry were those of Robley Williams (1938, 1939), who measured the energy distribution in Vega; of Kienle, and associates; and of Barbier and Chalonge and their co-workers (1940). Kienle *et al.* measured fundamental energy distributions in the range λ4000–λ6500 with great care for a set of early-type stars and then measured relative energy distributions for thirty-nine stars from B to M. Barbier and Chalonge covered the range λ3150 to λ4600. In their early work, these observers compared stellar spectra photographed with an objective prism with various lamps and with a hydrogen discharge tube. One trouble was the large number of steps necessary to compare stars with an ultimate standard lamp. A further difficulty is that the hydrogen source was bluer than the stars themselves. Later, Chalonge and Divan (1952) used fluorescent powders excited by a mercury lamp as secondary calibrations with an improved spectrograph.

Chalonge and associates tabulate the gradients ϕ_b (defined for λ4600–λ4000) and ϕ_{UV} (defined for the region beyond the Balmer limit λ3700–λ3150). They also give the discontinuity D at the Balmer limit, defined by

$$D = \log \frac{I(> \lambda 3650)}{I(< \lambda 3650)} \qquad (6\text{–}11)$$

Here $I_>$ refers to the intensity computed from a Planckian curve fitted to the longward side of the Balmer discontinuity, whereas $I_<$ refers to the intensity derived from a curve drawn through the energy distribution on the shortward side. For an $A0$ star such as Vega, Chalonge, and Divan get $\phi_b = 0.96$ ($T_c \sim 17{,}000°$K) and $\phi_{UV} = 1.43$ ($T_{UV} \sim 10{,}000°$K). They also give for each star a parameter λ_1, a wavelength that characterizes the position of the Balmer discontinuity (see Fig. 6–4). This parameter is useful in distinguishing between dwarfs and supergiants among early-type stars. Thus, in the $A0$ dwarf, Vega, the Balmer lines merge at a longer wavelength ($\lambda_1 = 3770$) than in the $A2$ supergiant, α Cygni ($\lambda_1 = 3703$).

Chalonge and his co-workers have shown how the three quantities λ_1, D, and ϕ_b can be closely correlated with the Morgan-Keenan spectral and luminosity classification. The absolute gradient ϕ_b is affected by space absorption, but λ_1 and D are not. One may use ϕ_b to separate the objects into dichotomies earlier than $A2$ or later than $A2$, and then λ_1 and D to fix spectral class and luminosities wth a precision perhaps greater than the MK classes themselves.

Spectral scans can also be calibrated to give luminosity classes. Strömgren (1958) showed how one can measure parameters that depend essen-

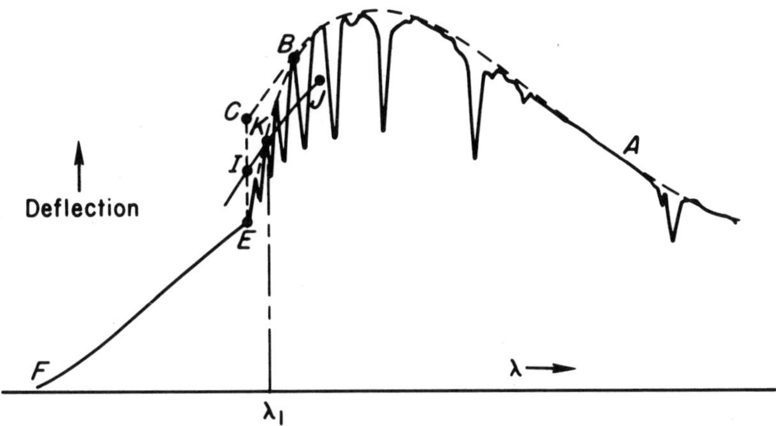

FIG. 6–4. DEFINITION OF CHALONGE'S PARAMETER λ_1

In this scanner tracing of the spectrum of α Gruis $B2$V, the dotted curve ABC denotes the "blue" continuum, the curve EF the "ultraviolet" continuum. The jump CE corresponds to the Balmer discontinuity $D = \log I(C)/I(E)$. BKE is drawn as the envelope of the observed continuum from B to E. Its position is determined partly by the true overlap of lines, partly by lack of resolution. The curve IK passes through the midpoint of the Balmer jump D and intersects the envelope BE at the "half-intensity" point K which is at a distance $D/2$ below ABC. Abscissas are proportional to λ. Since dispersion is linear, the deflection is $I_\lambda S_\lambda \exp(-\kappa_\lambda{}^A \sec z)$, where I_λ is the true intensity, S_λ is the sensitivity of the cell multiplied by the transmissivity of the optics, $\kappa_\lambda{}^A$ is the atmospheric extinction coefficient, and z is the zenith distance.

tially on the Balmer discontinuity and the total intensity of $H\beta$ with narrow band pass filters and a photoelectric cell and use these data to determine accurate spectrum and luminosity classes. See Chapter 8.

Greaves (1956) reports the zero point of the Greenwich relative gradients as $\phi_0 = 1.10 \pm 0.05$, $\lambda = 5000$A, $A0$V, while Cayrel de Strobel (1957) concluded that the Chalonge-Divan (1952) results should be decreased by 0.10. A comparison of Kienle's and Chalonge's gradients suggests a large correction from B to $A5$, but a small one beyond $A7$. Table 6–1 gives ϕ_b ϕ_{UV} and D as adopted from a comparison of data of Kienle, Greaves, and Chalonge. See also Code (1960).

TABLE 6–1

SPECTROPHOTOMETRIC GRADIENTS AND BALMER DISCONTINUITIES OF MAIN SEQUENCE STARS

Spectral Class	ϕ_b	ϕ_{UV}	D	Spectral Class	ϕ_b	ϕ_{UV}	D	Spectral Class	ϕ_b	ϕ_{UV}	D
$O7$	0.65	0.60	0.04	$B5$	0.76	1.01	0.27	$A5$	1.60	1.70	0.46
$O9$	0.65	0.65	0.06	$B8$	0.85	1.20	0.38	$A7$	1.45	1.77	0.41
$B0$	0.66	0.70	0.08	$B9$	0.90	1.33	0.47	$F0$	1.65	1.88	0.33
$B1$	0.65	0.75	0.10	$A0$	0.95	1.40	0.51	$F2$	1.28	1.93	0.25
$B2$	0.64	0.80	0.14	$A2$	1.06	1.55	0.52	$F5$	2.04	1.98	0.18
$B3$	0.63	0.84	0.21	$A3$	1.15	1.60	0.49	$F8$	2.30	2.03	0.13

Sizable discordances still exist between work of various observers. Further fundamental work must be undertaken to provide reliable basic data. Accurate recent work indicates that the energy distribution λ3800 – λ5000 cannot be represented precisely by a black-body curve. Recent photoelectric spectrophotometry by Code and his associates (1960) and by Oke (1960) should provide us with definitive standards.

6–5. Stellar Magnitudes and Colors. Spectral scans can be obtained only for relatively bright stars. See Figs. 6–5A, B. As long as we can obtain actual energy distributions, we may compare stars by means of their monochromatic fluxes at one point in the continuum. Woolley and Gascoigne (1948) compared the sun and Sirius at four different wavelengths.

In dealing with faint stars such as appear in globular clusters and nearby

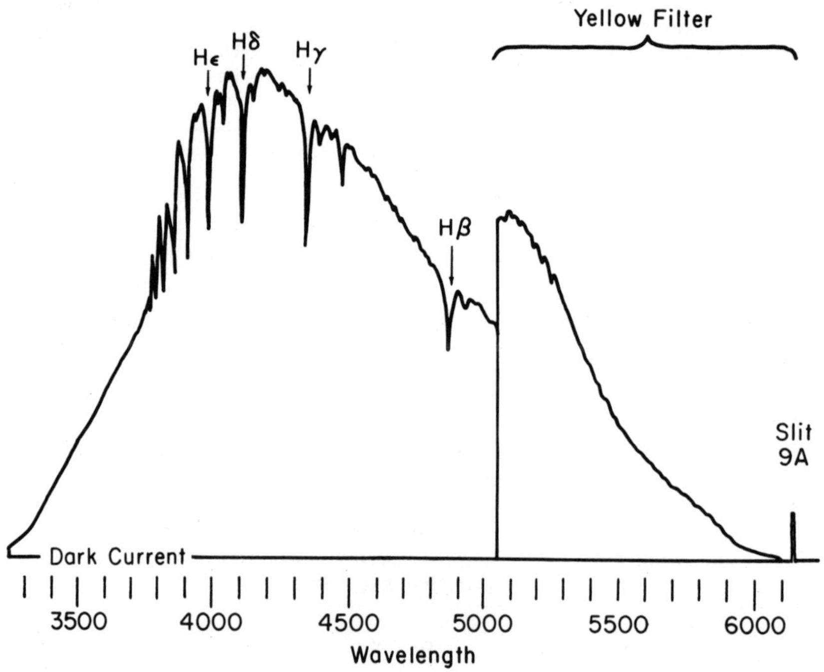

FIG. 6–5A. TYPICAL SPECTRAL SCANS: PHOTOELECTRIC SPECTRUM TRACING OF ALPHA PAVONIS

Deflection is proportional to intensity multiplied by the transmissivity of the atmosphere, optics, and sensitivity, of the cell. The yellow filter is introduced to eliminate the effects of overlapping orders. (Mount Stromlo Observatory, Australian National University.)

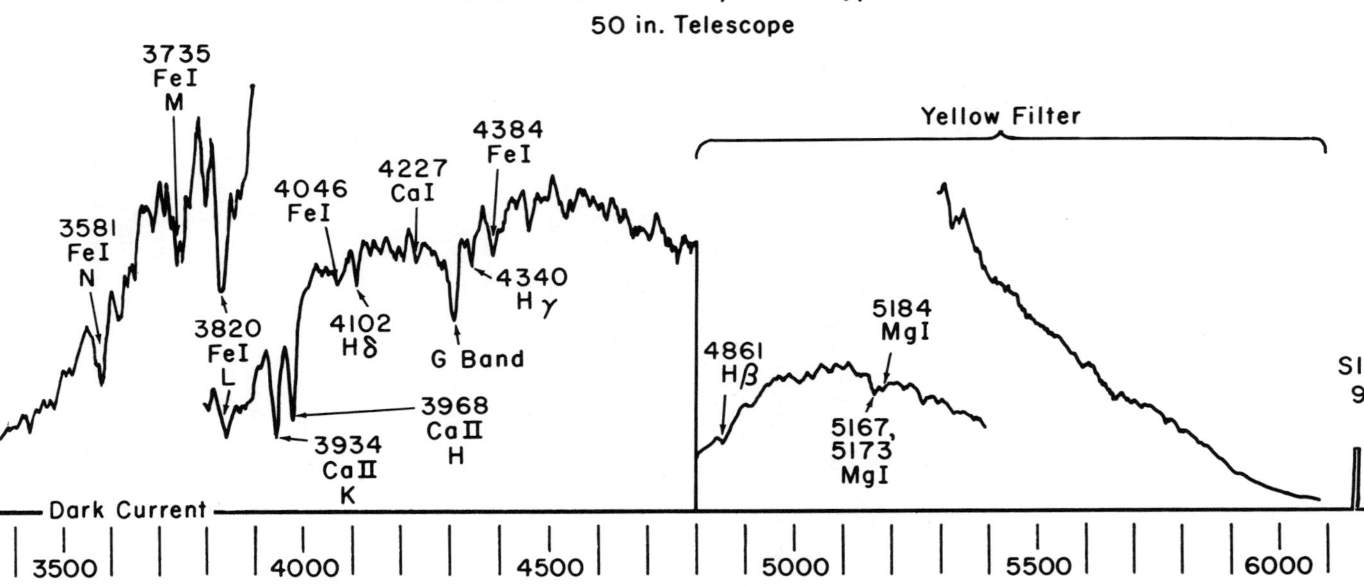

FIG. 6–5B. TYPICAL SPECTRAL SCANS: PHOTOELECTRIC SPECTRUM TRACINGS OF THE BRIGHTER COMPONENT OF ALPHA CENTAURI

Because of its complexity, this spectrum was scanned at a slower speed. The breaks are produced by changes in adopted sensitivity. With a resolution of 9A, all but the strongest lines are smoothed over. (Mount Stromlo Observatory, Australian National University.)

external galaxies, spectral scans are out of the question and one is restricted to measurement of magnitudes and colors that involve energies integrated over hundreds of angstroms.

The "brightness" of a star, defined as total luminous flux effective on the receiver and expressed in magnitudes, is obtained in a heterochromatic photometric system. This measured brightness involves the sensitivity function of the detector (e.g., photographic plate or photoelectric cell and filters), $R(\lambda)$, the transmission of the optics $T(\lambda)$, the energy distribution of the star $F(\lambda)$, its diameter and distance, and the effects of space absorption, A_λ. Finally, since the star is observed through the earth's atmosphere with a telescope of aperture a_0, atmospheric extinction and light collecting power are involved. Hence the measured brightness $b(z, a_0)$ will be

$$b(z, a_0) = \int b(\lambda) A_\lambda\, T(\lambda)\, R(\lambda)\, A(\lambda, z)\, \pi a_0^2\, d\lambda \tag{6–12}$$

where

$$b(\lambda)\, d\lambda = \pi \left(\frac{R}{r}\right)^2 F_\lambda d\lambda \tag{6–13}$$

is the flux of energy in the interval λ to $\lambda + d\lambda$ impinging upon each unit area outside the earth's atmosphere. Here R is the radius of a star of distance r, and $\pi F_\lambda\, d\lambda$ is the flux radiated between λ and $\lambda + d\lambda$.

Now $A(\lambda, z)$ is the fraction of energy transmitted by the earth's atmosphere at wavelength λ and zenith distance z, viz.:

$$A(\lambda, z) \sim e^{-(k_0 + k_1\lambda^{-4}) \sec z} \tag{6–14}$$

where k_0 is independent of λ and may vary from site to site or even from night to night. The coefficient of Rayleigh scattering, k_1, is proportional to barometric pressure. The observer's first task is to eliminate the effects of atmospheric extinction. With monochromatic magnitudes this task is simple, but with broad band pass filters it is necessary to observe in two or more colors and determine atmospheric extinctions in both magnitudes and colors. The thus corrected brightness b is related to the magnitude m by

$$m = -2.5 \log b + c_0 \tag{6–15}$$

where we postulate a Pogson scale (i.e., each magnitude step corresponds to a 2.5-fold ratio in brightness). The constant c_0 fixes the zero point of the system.

Standard magnitude and color systems correspond to certain functional forms of T and R. Now T and R vary from one instrument photocell filter combination to another; hence each observer obtains his own private color and magnitudes and the problem becomes one of reducing these individualistic magnitudes to a well-defined standard system. Such reductions are practical only if certain basic conditions are fulfilled. Over the λ-range covered by the cell-filter combination the energy distribution must be represented by a gradient corresponding to a unique temperature T_c. In early-type stars one must use cell-filter combinations that do not straddle the Balmer limit. See Table 6–2.

TABLE 6–2

Summary of Principal Features of Various Magnitude Systems

System	Detector	Zero Point	Approximate Wavelength Range (in angstroms)	Remarks
Visual	Eye	Essentially set in classical antiquity	4500–7500	Now replaced by V magnitudes of U, B, V, system in accurate work.
Photographic	Blue-sensitive plates	$m_{pg} = m_V$ for $A0$ stars	3700–5000	Essentially defined by north polar sequence.
U,B,V	Photocell + filters	$U = B = V$ for $A0V$ stars	See Fig. 1–4	H. L. Johnson and W. W. Morgan *Ap. J.* **117,** 313, 1953.
Six-color	Photocell + narrow band-pass filters	See reference	3000–12,000	J. Stebbins and A. E. Whitford, *Ap. J.* **102,** 273, 1945.
Radiometric	Thermocouple	$m_r = m_V$ for $A0$ stars	All radiation that reaches detector	E. Pettit and S. B. Nicholson *Ap. J.* **68,** 279, 1926.
Bolometric	—	$\Delta m_{\text{bol}} = -0.11$ at $G2$V	$0 \rightarrow \infty$	A derived quantity, not directly observable.

If these conditions are fulfilled it is often possible to reduce the colors and magnitudes on the observer's systems to colors and magnitudes on a standard (i.e., UBV) system by linear equations.

$$c_s = a + b\,c_0 \qquad m_s = m_0 + gc_s + h \tag{6–16}$$

where m_0 and c_0 and m_s and c_s are magnitudes and colors on the observer's and the standard system, respectively. One measures standard stars over a large range in color and magnitude with his equipment and solves for the empirical constants a, b, g, and h by least squares.

In general, precise transformations between different color and magnitude systems will be non-linear. If radiation below λ3800 is involved in either system, the transformation may be both non-linear and multivalued. If only radiation redward of λ3800 is involved, the transformation will be reasonably simple. Whenever possible, an observer should determine magnitudes of standard stars in three colors in widely separated wavelength regions. He should select objects with a large range in color in order to find the constants necessary to reduce his results to the standard systems. When a star is photographed or observed photoelectrically, the recorded light will cover a considerable range of wavelength. The interval embraced will depend on the spectral sensitivity of the receiver and the filters employed. Unless the star shows abrupt variations of energy distribution in its spectrum, caused, for example, by intense emission lines or strong molecular bands, the magnitude of the star will be nearly the same as it would be if all the recorded light were lumped together at one wavelength. Therefore, it would be helpful if we could determine a wavelength λ_0 such that the *heterochromatic magnitude* of a star would be equivalent to its *monochromatic magnitude* m_{λ_0} defined for some λ that would depend only on the telescope-photocell-filter combination and not on the stellar energy distribution. Such a definition is strictly possible only for narrow band passes. We might, however, define a *constant energy wavelength* (Strömgren, 1937)

$$\lambda_0 = \frac{\int \lambda ATR\, d\lambda}{\int ATR\, d\lambda} \tag{6–17}$$

Wesselink showed that if $d^2b/d\lambda^2$ is sufficiently small over the band pass to be neglected, we may write

$$b = \int_0^\infty b(\lambda)ATR\, d\lambda = Db(\lambda_0) \quad \text{where } D = \int_0^\infty ATR\, d\lambda \tag{6–18}$$

Thus the heterochromatic brightness b differs only by a constant factor from $b(\lambda_0)$. Hence the heterochromatic magnitude is identical with the monochromatic magnitude at λ_0 except for the zero-point, provided narrow band pass filters are used.

The neglect of $d^2b/d\lambda^2$ amounts to ignoring spectral lines; hence the approximation is good for regions with relatively few lines and relatively narrow band pass filters. Detailed analysis (Code, 1960) shows that it is best to define λ_0 by

$$\frac{1}{\lambda_0} = \frac{\int_0^\infty \frac{1}{\lambda} T(\lambda)R(\lambda)\, d\lambda}{\int_0^\infty T(\lambda)R(\lambda)\, d\lambda} \tag{6–19}$$

In older work frequent reference was made to a quantity called the *effective wavelength* which differs from the constant energy wavelength in that it includes the stellar energy distribution, viz.:

$$\lambda_{\text{eff}} = \frac{\int \lambda b_\lambda ATR \, d\lambda}{\int b_\lambda ATR \, d\lambda} \tag{6-20}$$

Hence, the effective wavelength changes in going from hot to cool stars in the sense that λ_{eff} becomes greater for the cooler stars. The effect will be more serious the greater the wavelength range covered in a given filter and plate combination. For a small range of wavelength and of spectral type, λ_{eff} usually may be regarded as constant. For any combination of telescope, plate, and filter, the effective wavelength can be determined with the aid of photographs secured with a wire grating placed over the objective. The effective wavelength for conventional photographic magnitudes is about λ4250. In photoelectric work, λ_{eff} is found from the filter transmission and cell sensitivity.

The classical photographic-photovisual magnitude system was set up by photographic photometry. It suffered from a number of important defects: (a) the zero points were set by the condition that the photographic magnitudes of white (*A*0) stars between 5.5 and 6.5 equaled the Harvard visual magnitudes of these stars. Spectral class *A*0 includes stars of different luminosity classes and colors and shows a wide spread in intrinsic color. (b) The standard sequence was set up at the North Pole, a region affected by space absorption and containing no *O* and *B* stars. Furthermore, the North Pole is inaccessible or at least inconvenient to many observers, so secondary sequences had to be set up. (c) The wavelength ranges embodied in the photovisual and particularly in the photographic magnitudes are too broad. The photographic band pass straddles the Balmer limit; hence great complications are encountered.

The difficulties encountered in the older work have been overcome by recourse to photoelectric photometry and the use of narrower band pass filters. Stebbins and Whitford (1945) employed a photocell and filter combination that gives effective wavelengths at λ3350, λ4220, λ4880, λ5700, λ7190, and λ10,300 in their six-color photometry of various stars. The zero point must be established from other data, such as relative energy distributions.

From plots of 4220A – 10,000A colors against spectral class, Stebbins and Whitford showed that main sequence, giant, and supergiant stars defined separate curves. Furthermore, the λ3530 – λ4220 or λ3530 – λ4880 base lines (color indices) include effects of the Balmer discontinuity, whereas the λ4880 – λ10,000 index does not. Hence a plot of these two sets of colors against one another shows strikingly the effects of space reddening and Balmer discontinuities. This six-color system has been applied successfully to a number of problems by Stebbins and Kron. Although the base line is long, which is advantageous for many problems, the band passes are narrow and the measurements are restricted to relatively bright stars.

The requirements of a good magnitude system are (Johnson and Morgan, 1953): (a) The set of stars selected to define the standard system should include examples of most spectrum and luminosity classes. Furthermore, it should include stars of different chemical composition (e.g., subdwarfs, white dwarfs, high velocity stars) and both reddened and unreddened stars. (b) Observations should be made in three or more widely spaced colors so observers can easily reduce their results to the standard system. (c) Colors and magnitudes must be measured accurately. (d) Standards should be conveniently located over the sky. (e) The sensitivity functions, etc., should be known so that from fundamental energy distribution measurements made upon the standard stars one can reproduce the observed colors and magnitudes. In particular, the zero point of color indices should be given in terms of a star whose spectral energy distribution can be predicted accurately from its spectrum and luminosity class, e.g., *A*0V.

H. L. Johnson and W. W. Morgan devised a magnitude system which attempts to meet these requirements. Their *V* magnitudes can be regarded as equivalent to visual magnitudes, but their *B* magnitudes differ from the international photographic magnitudes because they have tried to avoid the Balmer limit. The zero point of their system of visual magnitudes is based on the revised values of Stebbins, Whitford and Johnson. Then the zero points of the *B* and *U* magnitudes are fixed by the requirement the U-B and B-V color indices are both equal to zero for the mean of six bright stars of spectral class *A*0V. The relation between *V* magnitudes and colors has already been shown in Chapter 1. Fig. 6–6 shows the relation between U-B and B-V colors for main sequence stars and supergiants. Notice that the color indices of a star depend not only on its spectral class but also on its luminosity class so one can sometimes use colors to separate stars by luminosity. Further, as had been previously demonstrated by W. Becker (1938), the use of three colors instead of two enables one to assess effects of space reddening in star clusters.

Finally, it must be emphasized that the color of a star is affected by the presence of lines in the spectrum. As long as one compares stars of roughly the same hydrogen/metal ratios, color-magnitude arrays can be matched against one another to evaluate distance moduli, space reddening, and so on. If this H/metal ratio differs substantially, there will be pronounced effects on the color indices. That is, two stars of different H/metal ratio but the same effective temperature may differ markedly in color. Subdwarfs (hydrogen deficient stars) of the same effective temperature as the sun are brighter in the ultraviolet than stars of normal (i.e., solar) composition. Hence their U-B colors will differ. Melbourne (1960) and also the Burbidges, Sandage, and Wildey (1960) showed that if one eliminated

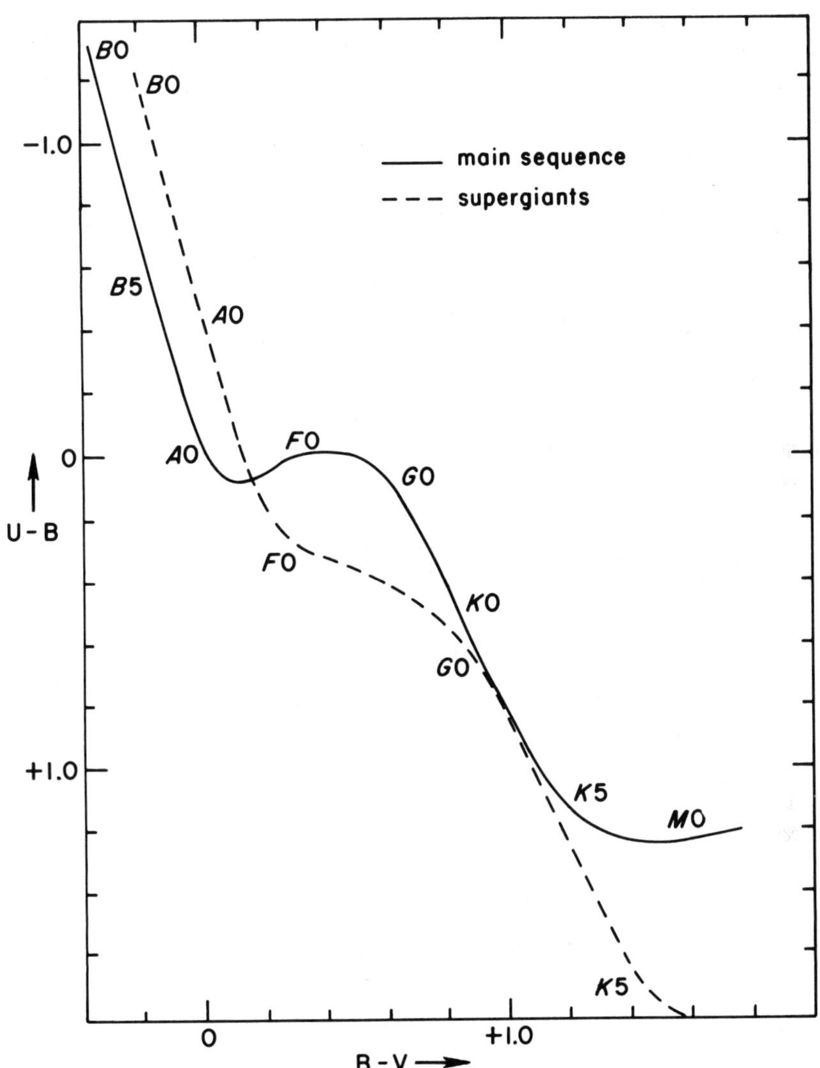

Fig. 6–6. Relation Between U-B and B-V Colors
(After Johnson and Morgan. Adapted from diagrams in *Ap. J.* **117,** 313, 1953; courtesy, University of Chicago Press.)

the effects of spectral lines, these stars would have the same colors as similarly corrected normal stars of the same effective temperature.

Many workers have considered the influence of lines on stellar colors and spectral scans. One may start with scans of the energy distribution in the spectrum, measure the total absorption of lines observed with high resolution spectra and obtain the position of the background continuum much as Chalonge and Labs attempted to do for the sun. Such a procedure was carried out by Melbourne (1960), who compared theoretical and observed stellar energy distributions for both normal stars and subdwarfs. Some years ago Milford (1950) considered effects of lines on measured stellar magnitudes and derived corrections to get the photovisual magnitude of the continuum. He also derived absolute fluxes of stars from $B3$ to $K5$ for the continuum at $\lambda 5450$. If appropriate sensitivity and optical transmissivity curves for the UBV magnitude systems were known it would be possible to correctly predict stellar colors from energy scans. So far such a procedure has not proven possible and the UBV system must be regarded as empirically defined in terms of measurements made on certain standard stars. Part of the fault may lie in the energy distributions, part may lie in uncertainties in the basic response curves of the photoelectric cell plus filter plus telescope.

A fact important to the practicing spectroscopist is that stellar colors are often available when energy distributions are not. Therefore, we must use colors to obtain checks on stellar temperatures. Since color measurements embrace broad band passes it is necessary to use such data carefully, particularly when dealing with stars of abnormal H/metal ratios.

6–6. Relation Between Stellar Radii and Temperature. The emission per cm^2 per angstrom by a stellar surface will depend on its brightness temperature at that wavelength. Hence the monochromatic absolute magnitude, M_λ, stellar radius, R, and brightness temperature must all be related. For the moment, let us suppose we can regard the heterochromatic magnitude (e.g., V magnitude) as equivalent to a monochromatic magnitude measured at some effective wavelength λ_e.

The impression upon the photographic plate or response of the photocell will be proportional to $r^{-2}\pi R^2 E(\lambda_e, T) Q \Delta\lambda$, where R is the radius of the star whose distance is r and E is the emission at the effective wavelength λ_e over a wavelength interval $\Delta\lambda$. Q is a factor which depends on the sensitivity of the apparatus and the transparency of the optical system. We assume that atmospheric extinction has been determined and taken into account. Further, let us suppose that the stellar emission may be represented by Planck's function for some temperature T. That is

$$E(\lambda_e, T) = \pi B(\lambda_e, T) \tag{6–21}$$

Ideally, T is the brightness temperature at λ_e, but within the framework of our present approximation we regard it as the stellar "temperature." If we express the measured brightness in magnitudes, we obtain

$$\begin{aligned} m_\lambda = &-2.5 \log \pi^2 Q - 2.5 \log c_1 \lambda_e^{-5} \Delta\lambda - 5 \log R + 5 \log r \\ &+ \frac{2.5\, c_2}{\lambda_e T} \log e + 2.5 \log (1 - e^{-c_2/\lambda T}) \end{aligned} \tag{6–22}$$

which can be written in the form

$$m_\lambda = a - 5 \log R + \frac{1.561}{\lambda_e T} + x + 5 \log r \qquad (6\text{–}23)$$

Here a represents the first two terms and x is the last term on the right of Eq. 6–22. Since distance, apparent magnitude, and absolute magnitude are related by

$$m_\lambda = M_\lambda - 5 + 5 \log r \qquad (6\text{–}24)$$

we can write

$$M = C_\lambda - 5 \log R + \frac{1.561}{\lambda_e T} + x \qquad (6\text{–}25)$$

The correction factor x, has been tabulated by Russell-Dugan-Stewart[3] as follows:

$\frac{1.561}{\lambda_e T}$	1.0	2.0	3.0	4.0	5.0
x	−0.55	−0.19	−0.07	−0.03	−0.01

For visual magnitudes, $\lambda_e \sim 5480$. We may evaluate the constant $C_\lambda = 5 + a$, with the aid of the known absolute visual magnitude of the sun, + 4.84 (Kron and Stebbins, 1957) and solar temperature (5800°K). We express radii in terms of the sun's radius. Then $C_\lambda = -0.08$ and we have

$$M_V = -0.08 - 5 \log R + \frac{28{,}500}{T} \qquad (6\text{–}26)$$

as the relation between absolute visual magnitude, radius, and temperature for the stars like the sun. Thus, given the gradient and the absolute visual magnitude, we may estimate the stellar diameter.[4] It is evident that the procedure is rough. For B magnitudes $\lambda_e = 4360$, hence

$$M_B = C_B - 5 \log R + \frac{35{,}800}{T} \qquad (6\text{–}27)$$

Let us assume that the same value of the brightness temperature holds at λ4360 and λ5480 (perhaps a rather bold assumption for many stars). Then we can correlate the color index of a star undimmed by space absorption, (B-V) = ($M_B - M_V$), with temperature. For an $A0$ star, $m_V = m_B$ and $M_B = M_V$ by definition. If T is taken as 10,000°K, C_p will be −0.81. Hence color index (B-V) is related to temperature T by

$$T = \frac{7300}{(\text{B-V}) + 0.73} \qquad (6\text{–}28)$$

[3] *Astronomy* (Boston: Ginn & Co., 1927), II, p. 732.

[4] Michelson's stellar interferometer has been employed to measure angular diameters of red giant and supergiant stars while "photon correlation" interferometry may be applied to bright white stars such as Sirius (Brown and Twiss, 1954). Thus, if the angular diameter of a star is known and its distance can be measured, R and M_{vis} are determined. From Eq. 6–26 one can then find the temperature.

In view of the crudeness of the assumption that stars radiate like black bodies, we have neglected the small correction factor x in Eqs. 6–25, 6–27, and 6–26. Furthermore, the temperature determined from Eq. 6–28 is rather loosely defined, correlated with, but not identical with, either the effective or color temperature.

6–7. Bolometric Magnitudes. All magnitude systems in common use entail the measurement of stellar radiation only over a portion of a star's spectrum. Sensitive detectors such as photocells respond only to a limited range in wavelength. A thermocouple responds to all radiation that reaches it but lacks sensitivity. The minimum detectable signal under normal operating conditions is about 10^{-3} erg sec^{-1} so it can be used only on bright stars with large telescopes. Nevertheless, Pettit and Nicholson (1928, 1933) used this instrument successfully to measure radiometric magnitudes for bright red stars, most of whose energy falls in the far infrared where thermocouples respond and other detectors do not. Corrections for absorption by water vapor and other molecular constituents of the earth's atmosphere are troublesome. In order to make them it is usually necessary to assume a form for the infrared energy distribution. Much more work can be done in the near infrared with photocells to λ12,000 and with photoconductive cells for longer wavelengths.

Magnitudes measured with a thermocouple are called radiometric magnitudes. The difference between the visual and radiometric magnitude of a star,

$$m_{\text{visual}} - m_{\text{radiometric}}$$

is called its *heat index*. The zero point of the radiometric magnitude scale is so chosen that the heat index is zero at class $A0$. Heat indices are positive and large for cool, red stars. Since hot stars emit most of their energy in the ultraviolet beyond the atmospheric cut-off at λ2900 it will be necessary to observe them from instruments flown in artificial satellites.

In problems of stellar structure and evolution one is concerned with the total power output of a star, in kilowatts or in units of the power output of the sun (3.9×10^{23} kw). We express a star's total luminosity in terms of its *bolometric magnitude. Each magnitude step corresponds then to 4 db in power output.* Bolometric magnitudes are not observed quantities but must be derived from a combination of observation and theory. The *bolometric correction* is defined as the difference between the bolometric and visual (V) magnitude.

$$\text{BC} = m_{\text{bol}} - m_{pv} \tag{6–29}$$

The zero point of the bolometric scale is so adjusted that bolometric corrections are small for stars like the sun (i.e., -0.11 at $G2$V). They become numerically large for hot stars which radiate most in the far ultraviolet and for cool ones whose principal energy output is in the infrared. Now BC is calculated by the expression

$$BC = 2.5 \log \frac{\int p_\lambda F_\lambda \, d\lambda}{\int F_\lambda \, d\lambda} + \text{const} \tag{6–30}$$

where p_λ is the sensitivity function of the eye, πF_λ is the flux and the constant is determined by the condition that for the sun, BC = −0.11. In modern work one replaces p_λ by the sensitivity function R_λ for the V-magnitude system.

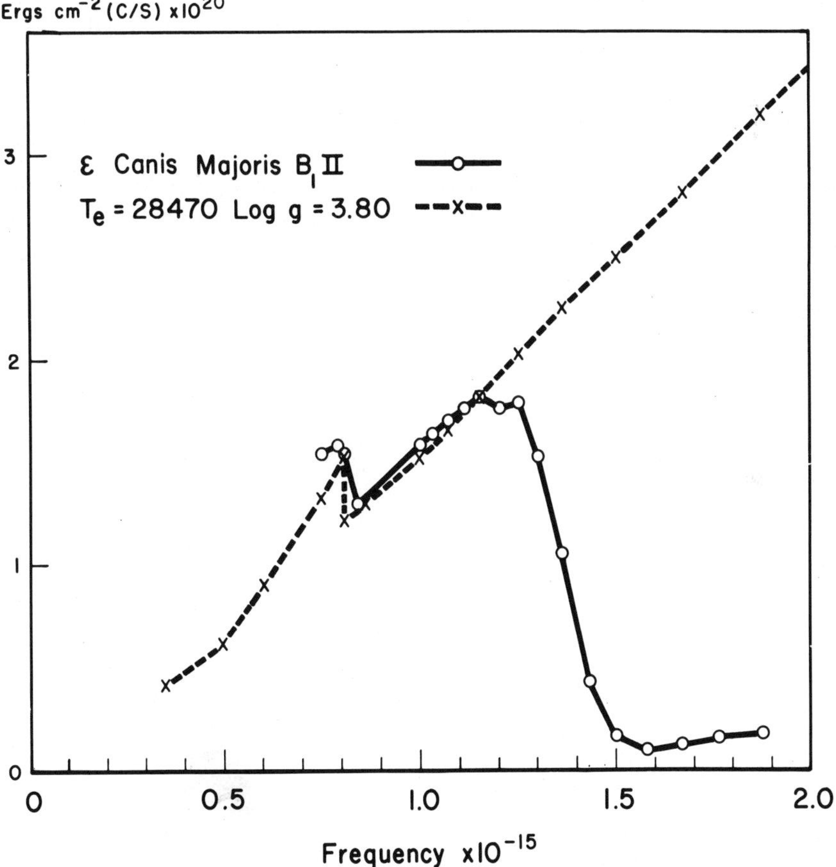

FIG. 6–7. THE OBSERVED ENERGY DISTRIBUTION IN ε CANIS MAJORIS, B1II

The solid curve gives the energy distribution observed with detectors flown in rockets above the earth's atmosphere. The dotted curve is that predicted (Underhill, 1957) for an atmospheric model with T_{eff} = 28470°K and log g = 3.80, and fitted to the observed curve at λ2600. The discordance is reduced in Miss Underhill's more recent calculations, which predict an effective temperature nearer 26,000°K for this spectral class. (Courtesy T. P. Stecher and J. E. Milligan, N.A.S.A.)

Bolometric corrections are obtained in the following way. For the region from λ3200 to about λ12,000 one may rely on spectral scans to obtain F_λ. Spectral regions outside this range are important both for very cool and for very hot stars. For cool stars one may use thermocouple results (Pettit and Nicholson, 1928; Emberson, 1941) to estimate infrared energy. Usually one assumes that these stars radiate like black bodies in the infrared. Admittedly this approximation is unsatisfactory, but we have no suitable cool star atmospheric models, nor are the blanketing effects by vibration, rotation bands, etc., known. The spectral regions of cool stars normally observed are distorted by strong line and band absorption, but the bulk of radiation passes outward in infrared spectral regions where blocking by line and band absorption is less marked. Consequently, the atmospheric structure would appear to be determined mostly by what happens in regions we do not observe.

The difficulty for hot stars may be appreciated best by reference to Fig. 6–7, where we compare the energy distribution in ε Canis Majoris observed from rockets fired above the earth's atmosphere (Stecher and Milligan, 1962) with that predicted by a model atmosphere for the same spectral class (Underhill, 1957). Although more recent calculations (Underhill, 1963) reduce the discordance, they do not remove it (see Chapter 5), and the theory must be regarded as unsatisfactory.

TABLE 6–3

TENTATIVE BOLOMETRIC CORRECTIONS FOR HIGH TEMPERATURES

$T_e(°K)$	ΔM_{bol}	$T_e(°K)$	ΔM_{bol}	$T_e(°K)$	ΔM_{bol}
12,000	−0.55	20,000	−1.6	30,000	−2.8
14,000	−0.75	22,000	−1.9	35,000	−3.1
16,000	−1.03	24,000	−2.2	40,000	−3.3
18,000	−1.35	26,000	−2.4	45,000	−3.5

Accordingly, in Table 6–3 we are guided by the rocket observations in our estimates of bolometric corrections, but we must regard these figures as tentative. Table 6–4 gives bolometric corrections for luminosity classes I, III, and V and spectral classes $A0$ or later. These were obtained from spectral scans, calibrated by Code's monochromatic magnitudes and supplemented by model atmosphere predictions where necessary. Bolometric corrections for M dwarfs are due to Limber (1958).

If one compares Table 6–4 (which is based mostly on spectral scans interpreted by Eq. 6–30) with Popper's (1959) careful analysis based on radiometric and visual magnitudes, he will note some systematic differences. The two methods for deriving BC's should agree, particularly for spectral classes where the bulk of the energy output can be measured by both thermocouples and photoelectric cells. The discordances may partly

TABLE 6-4
SEMIEMPIRICAL BOLOMETRIC CORRECTIONS

Spectral Class	Luminosity Class			Spectral Class	Luminosity Class		
	V	III	I		V	III	I
A0	−0.40			K0	−0.29	−0.43	−0.75
A2	−0.34			K2	−0.37	−0.62	−0.93
A5	−0.27			K3	−0.44	−0.75	−1.00
A7	−0.22			K4	−0.56	−0.90	−1.02
F0	−0.16			K5	−0.68	−1.05	−1.10
F2	−0.13		−0.13	K6	−0.80		−1.17
F5	−0.10		−0.08	M0	−1.26		−1.50
F8	−0.08		−0.13	M1	−1.40		−1.60
G0	−0.09	−0.12	−0.23	M2	−1.62		−1.80
G2	−0.11	−0.15	−0.31	M4	−2.24		
G5	−0.16	−0.23	−0.42	M6	−2.81		
G8	−0.21	−0.37	−0.58	M8	−3.80		

arise from difficulties in calibrating magnitudes in terms of spectral energy distributions.

The giant star Aldebaran, $K5$III, has an apparent magnitude of 1.06. The bolometric correction is −1.05 and, hence, the star has a bolometric magnitude of $1.06 - 1.05 = +0.01$. Bolometrically, the irregular variable, α Orionis, $M21$, magnitude $0.1 - 1.2$, becomes comparable with Sirius ($m = -1.58$, $T = 10{,}000$°K), since the bolometric correction of α Orionis is −1.80.

6-8. Scale of Effective Temperatures. Closely related to the problem of bolometric magnitudes is that of effective temperatures (see Eq. 6-3). Greatly improved accuracy in measurements of stellar energy distributions and in their interpretations, theoretical treatments, and data from rockets have made possible considerable progress since Kuiper's (1938) discussion of stellar effective temperature scales. Important contributions have been made by Limber (1958), Popper (1959) and Code and his associates. See Table 6-5.

For red dwarfs probably Limber's (1958) scale is the most reliable. Luminous cool stars have been measured with thermocouples, but temperature scales are difficult to determine for the same reasons that bolometric corrections are hard to find. The effective temperature of the sun is known well enough so that temperature scales proposed by various authors in the range $F5$–$K5$ are in reasonably good accord. From $F5$ to $A0$ different workers have obtained temperature scales showing a considerable spread. We rely here on the work of Popper (1959) who used (B-V) colors and adopted as fixed points Sirius, whose temperature was obtained from interferometer measurements by Brown and Twiss (1954), and σ Bootis, whose temperature was obtained by Code from a comparison of observed and predicted energy distributions. Bless obtained somewhat lower tem-

TABLE 6–5

SCALE OF EFFECTIVE TEMPERATURES FOR MAIN SEQUENCE STARS
(Luminosity Class V)

Spectral Class	T_e (°K)	Spectral Class	T_e (°K)	Spectral Class	T_e (°K)
*O*5	37,500	*A*0	9450	*K*0	5170
*O*7	34,000	*A*2	8900	*K*1	5000
*O*9	31,300	*A*5	8220	*K*3	4660
*B*0	28,000	*A*7	7800	*K*5	4350
*B*1	22,500	*F*0	7200	*K*7	4000
*B*2	20,300	*F*2	6850	*M*0	3520
*B*3	18,800	*F*5	6500	*M*1	3400
*B*4	17,500	*F*6	6400	*M*4	3150
*B*5	16,500	*F*8	6200	*M*5	3050
*B*6	15,300	*G*0	5970	*M*6	2950
*B*7	14,300	*G*2	5780	*M*8	2700
*B*8	13,000	*G*5	5570		
*B*9	11,500	*G*8	5350		

peratures from a comparison of photoelectric tracings with fluxes predicted by Osawa's model atmospheres. The adopted temperatures are based on Bless's and Popper's results.

For stars of extremely high temperature one must rely on a judicious combination of ionization temperatures, theoretical models, and fragmentary data secured from rockets. The rocket data would suggest that ionization temperatures such as those derived by Petrie (1950) are closer to the effective temperatures than are those obtained from many models so far computed.

REFERENCES

Art. 6–2

MINNAERT, M. 1953. *The Sun*, G. P. Kuiper (ed.). University of Chicago Press, Chicago, chap. 3.

GOLDBERG, L., and PIERCE, A. K. 1959. *Handbuch der Physik*. S. Flügge (ed.). Springer-Verlag, Heidelberg, vol. 51, p. 11.

Arts. 6–3 and 6–4

See especially the review article by:

CODE, A. 1960. *Stellar Atmospheres*, vol. 6 of *Stars and Stellar Systems*. J. L. Greenstein (ed.). University of Chicago Press, Chicago, chap. 2.

BEER, A. (ed.). 1960. *Vistas in Astronomy*. Pergamon Press, London. Articles by D. Chalonge, p. 1328; H. Kienle, p. 1321; W. M. H. Greaves, p. 1309; B. Strömgren, p. 1336.

Art. 6–5

An historical survey is given by:

WEAVER, H. F. 1946. *Pop. Astr.* **54,** 211, 287, 339, 451, 504.

See also:

PECKER, J. C., and SCHATZMAN, E. 1959. *Astrophysique Générale*, Masson & Cie., Paris.

SEARES, F. H. 1938. *Publ. Astron. Soc. Pac.*, **50,** 5.
———. 1943. *Ap. J.*, **98,** 302.
BEER, A. (ed.). 1956. *Op. Cit.*
WOOLLEY, R. v. D. R., p. 1095, and STOY, R. H., p. 1099.

Photoelectric magnitude scales have been established by:

JOHNSON, H. L., and MORGAN, W. W. 1953. *Ap. J.* **117,** 313.
STEBBINS, J., and WHITFORD, A. E. 1945. *Ap. J.* **102,** 273.

The influence of line absorption and relation between scanner traces and UBV magnitudes is given by:

MELBOURNE, W. 1960. *Ap. J.* **132,** 101.

See also:

WILDEY, R. L., SANDAGE, A. R., BURBIDGE, E. M., BURBIDGE, G. R. 1962. *Ap. J.* **135,** 94.

Arts. 6–7 and 6–8

Thermocouple observations and radiometric magnitudes are discussed by:

PETTIT, E., and NICHOLSON, S. B. 1928. *Ap. J.* **68,** 279.
———. 1933. **78,** 320.
EMBERSON, R. M. 1941. *Ap. J.* **94,** 427.

Effective temperatures and bolometric corrections are discussed by:

KUIPER, G. P. 1938. *Ap. J.* **88,** 429.
LIMBER, N. 1958. *Ap. J.* **127,** 363. For *M* stars.
POPPER, D. M. 1959. *Ap. J.* **129,** 659. For stars of intermediate temperatures.
STECHER, T., and MILLIGAN, J. 1962. *Ap. J.* **136,** 1.

CHAPTER 7

STRENGTHS AND BREADTHS OF SPECTRAL LINES

7–1. Importance of Transition Probabilities and Line-Broadening Mechanisms. The essential data obtainable for an interpretation of a stellar atmosphere are the energy distribution in its continuous spectrum and the equivalent widths or profiles, and displacements of its spectral lines. Observation and interpretation of stellar continua have been discussed in Chapters 5 and 6. In order to interpret line profiles and intensities, we must know (in addition to some model of a stellar atmosphere) (a) the transition probability of a line (f- or A-value) and (b) quantitative nature of line broadening.

An interpretation of intensities and profiles of emission lines in spectra emitted by hot, ionized gases (plasmas) in laboratory sources likewise depends on these same parameters. Hence they are of interest whenever one wishes to deduce the physical state of a radiating gas from its spectrum.

The importance of the f-value is obvious. A sodium atom in its ground state has a 70-times greater chance of absorbing a quantum of one of the D-lines (for this $3^2S—3^2P$ doublet $f = 0.98$) than it has of absorbing a quantum of the next doublet of the principal series for which $f = 0.014$. Line broadening is important because if the absorptivity of a given transition is spread over a great range in wavelength, a large number of absorbing atoms can remove much more energy from an emerging beam than is possible if the line is narrow.

The discussion in Chapter 4 gave the following relation between the Einstein coefficient $A_{nn'}$ and the Ladenburg $f_{n'n}$:

$$A_{nn'} = \frac{g_{n'}}{g_n} \frac{8\pi^2\varepsilon^2\nu^2}{mc^3} f_{n'n} \tag{7–1}$$

where n' denotes the lower, n the upper level, the g's are the corresponding statistical weights. Numerically (see Eq. 4–115),

$$f_{n'n} = 1.5 \times 10^{-8} \lambda_\mu{}^2 \frac{g_n}{g_{n'}} A_{nn'} \tag{7–2}$$

where λ is given in microns. The Einstein A-value is also related to another quantity called the "strength" of the line, S_1, (see Eq. 4–124).

$$A(\alpha J; \alpha' J') = \frac{1}{2J+1} \frac{64\,\pi^4\nu^3}{3\,hc^3} S_1(\alpha J; \alpha' J') \tag{7–3}$$

where α and α' denote the term designations for the upper and lower levels of the transition. The absolute strength is related to the relative strength, S, by

$$S_1 = S\,\sigma^2(nl;\, n'l') \tag{7–4}$$

where σ depends on the radial charge distribution in each energy level. It can be calculated for some atoms by quantum mechanics.

7–2. Relative Line Strengths in *LS* Coupling. If an atom is in good LS coupling, relative line strengths and relative f-values can be obtained for all lines in a whole transition array. In a calibration of absorption line intensities in terms of numbers of atoms, it is useful to have a large number of lines whose relative f-values are known, even if their absolute f-values cannot be obtained.

We calculate S in two steps. First we find the strength of the line in question referred to that of the whole multiplet, $s/\Sigma s$, and then we compute the strength of the entire multiplet itself.

Even before the advent of quantum mechanics, Sommerfeld, Honl, Russell, and Kronig derived formulae for relative strengths within an LS multiplet, tabulated relative strengths for different values of L with J as argument, and then grouped the individual tables according to different values of spin. Table A–1, based on their work, gives log $\Sigma s/s$, where $s/\Sigma s$ is the strength of each line in terms of that of the whole multiplet.[1]

As an example, let us compute relative strengths for a ^{4}P—^{4}D multiplet. The terms are quartets, $S = \frac{3}{2}$, hence we use the table with spin $= \frac{3}{2}$ and choose the box with $L_1 = 1$, $L_2 = 2$. The values of log $\Sigma s/s$ from Table A–1 are:

		^{4}D			
		$\frac{1}{2}$	$\frac{3}{2}$	$\frac{5}{2}$	$\frac{7}{2}$
^{4}P	$\frac{1}{2}$	1.08	1.08		
	$\frac{3}{2}$	1.77	0.97	0.68	
	$\frac{5}{2}$		2.00	1.05	0.40

e.g., for $^4P_{3/2}$—$^4D_{5/2}$, log $s/\Sigma s = -0.68 = \bar{9}.32$ or $s/\Sigma s = 0.21$. That is, the $\frac{3}{2}-\frac{5}{2}$ line contributes 0.21 of the total strength of this multiplet.

For checks on strength calculations, *sum rules* are of great importance. For a given multiplet these are:

1. The sum of the strengths of all lines of a multiplet that end on a common final level is proportional to $2J + 1$ for this level.
2. The sum of strengths of all lines of a multiplet that start from a common initial level is proportional to the weight, $2J + 1$, of the initial level.

As an example, consider a ^{4}P—^{4}D multiplet. In terms of $s/\Sigma s$, the box now looks like the following:

[1] These tables may also be used to compute the strengths of lines in jj coupling since the Sommerfeld, Russell, etc., formulas remain valid if we replace L by j_2 and S by j_1, and take j_1 as the quantum number that does not change during a transition.

		^{4}D					
		$\frac{1}{2}$	$\frac{3}{2}$	$\frac{5}{2}$	$\frac{7}{2}$	Sum	Weight $(2J+1)$
^{4}P	$\frac{1}{2}$	0.083	0.083			0.166	2
	$\frac{3}{2}$	0.017	0.107	0.21		0.334	4
	$\frac{5}{2}$		0.010	0.09	0.40	0.500	6
Sum of s-values		0.10	0.20	0.30	0.40		
Weight $(2J+1)$		2	4	6	8		

Notice that the sums of the strengths taken along a row or column are proportional to $2J + 1$ in accordance with the sum rules. When the multiplet is in good LS coupling the relative strengths, s, computed by theory are in good agreement with experiment. Relative strengths of multiplets in a transition array may be found from Goldberg's (1935, 1936) tables. Table A–2 is adapted from extensive calculations he has published. As an example of application of his tables, we shall consider multiplets of the transition array, $2s^22p^23s$—$2s^22p^23p$ in O II. The s^2 electrons play no role and from Table A–2 we read relative multiplet strengths of the p^2s—p^2p transition as follows:

$$p^2s(^3\mathrm{P})\,^4\mathrm{P}—p^2s(^3\mathrm{P})\,^4\mathrm{D} = 20$$
$$^4\mathrm{P}— \qquad ^4\mathrm{P} = 12, \text{ etc.}$$

To obtain the relative strengths s of any particular line we first obtain $s/\Sigma s$ from Table A–1, and then the strength of the multiplet from Table A–2.

As an example, let us compute the strength of λ4317.16 of O II, $2p^23s(^3\mathrm{P})\,^4\mathrm{P}_{1/2}$—$2^2p3p(^3\mathrm{P})\,^4\mathrm{P}_{3/2}$. For a quartet, $S = \frac{3}{2}$. With $L_1 = 1$, $L_2 = 1$, $J_1 = \frac{1}{2}$, $J_2 = \frac{3}{2}$, Table A–1 gives $s/\Sigma s = 0.14$. From Table A–2, the relative strength of the multiplet is 12. Hence the strength of the line is $0.14 \times 12 = 1.68$.

To obtain the actual f-value of the line we now employ Eqs. 7–1, 7–3, and 7–4. Let the total strength S_1 be measured in atomic units, $a_0\varepsilon^2$, where a_0 is the radius of the first Bohr orbit and ε is the electronic charge. Let λ be measured in angstrom units (10^{-8} cm). Then

$$f = \frac{304}{g_1\lambda} S_1 \tag{7–5}$$

where

$$S_1 = S_0 \frac{s}{\sum s} \sigma^2 \tag{7–6}$$

Here S_0 is the multiplet strength given in Table A–2, $s/\Sigma s$ is obtained from Table A–1, and σ^2 must be obtained by quantum mechanical arguments or estimates based on them.

Approximate values of σ^2 can be calculated for certain transitions between high levels in light atoms. Bates and Damgaard (1949) have given tables from which σ may be computed from the effective quantum numbers of the upper and lower levels of a transition. For the 4317.16 O II line, $g_1 = 2$, and $\sigma^2 = 5.5$. Since $S = 1.68$, we find $f = 0.325$.

Goldberg's tables cover most of the multiplets of interest. Other examples can be handled with the aid of certain sum rules. All the lines of a transition array that originate in (or terminate on) a given level constitute the *J-file* of that level. The sum of the strengths of these lines is the strength of the file. If the jumping electron is not equivalent to any other electron in either the initial or the final configuration, the sum of the strengths of all the multiplets originating from, or ending in, a term $nl^k n'l'$ of the transition array, $nl^k n'l' - nl^k n''l''$ is

$$(2S + 1)(2L + 1)(l' + 1)(2l' + 3)\sigma^2 \quad \text{if } l' = l'' - 1$$

or

$$(2S + 1)(2L + 1)(l')(2l' - 1)\sigma^2 \quad \text{if } l' = l'' + 1$$

When the jumping electron is equivalent to others in either the initial or the final configuration, we may employ other sum rules such as those given by Menzel (1947) or Rohrlich (1959b).

As an example of the J-file sum rule consider the $2s2p3s'^4\text{P} - 2s2p3p'^4\text{P}$, multiplet of N III at λ3360. The only term in the $2s2p3s$ configuration with which the $3p^4\text{P}$ term can combine is the ^4P term, as transitions to ^2P terms are excluded in good LS coupling. Hence the sum of the strengths of all transitions of the $2p3s - 2p3p$ array originating in the $3p^4\text{P}$ term is simply the strength of the λ3360 multiplet. For the upper level $S = \frac{3}{2}$, $L = 1$, and $l = 1$; hence $S = 12$. Goldberg's tables are incomplete in that they omit transition arrays such as $s^2p - sp^2$ or $s^2p^3 - sp^4$. Also, if two similar terms originate from a configuration (e.g., a d^3 configuration gives two ^2D terms), these tables give only the sum of the strengths. These difficulties are overcome with the aid of general formulas for multiplet strengths involving fractional parentage and certain parameters known as *Racah coefficients* and given recently by Rohrlich (1959a). See also Kelly and Armstrong (1959).

Theoretical calculations of f-values are resolved into two separate problems: (1) deviations from LS coupling and (2) evaluation of the σ^2 factor.

The validity of the relative f-values derived by foregoing methods depends on how well the atom in question follows LS coupling. For certain transitions in light atoms such as carbon or oxygen, the results are probably good first approximations although, e.g., the $3p - 3d$ array in O II appears to show substantial departures from LS coupling. For other atoms such as neon, this is no longer true. Heavier elements usually show marked deviations from LS coupling and often display strong lines connecting terms of different multiplicity (intercombination lines).

In good LS coupling, spin-orbit interaction is not important, i.e., the splitting of a term of a configuration is small compared with the separation of the terms (see Art. 2–7). Also, the configuration interaction is negligible. When spin orbit interaction becomes significant, deviations from LS coupling occur and we have intermediate coupling. Relative multiplet strengths in a configuration array show effects of deviations from LS coupling first. Relative line strengths within a multiplet, however, become sensibly distorted only when the atomic levels show appreciable deviations from LS coupling.

The theory of line strengths in intermediate coupling is well known, but each array for each atom must be considered individually with parameters established empirically from observed positions of energy levels, etc. Gottschalk (1948) obtained intermediate coupling (IC) line strengths for the $3d^7 4s - 3d^7 4p$ transition array in Fe I that were a marked improvement over the LS coupling theory. Garstang (1954) has carried out IC calculations for ions of O, Ne, S, and A of astrophysical interest, but much work remains to be done.

Evaluations of the σ^2 factor present a much more formidable problem.[2] Different methods of calculations often lead to substantially different results even for relatively simple one-electron systems such as Na I or Ca II, particularly for transitions involving higher levels. For methods of calculation involved, see Bethe and Salpeter (1959) and for a brief summary of the status of the problem see Garstang (1955). Extensive calculations of astrophysical interest have been undertaken by Czyzak, by Green, by Seaton, by Biermann and their associates.

The mean f-value, $\bar{f}$, for an entire transition array is computed by the rule

$$\bar{f} = \frac{1}{\Sigma g_i} \Sigma g_i f(\alpha J; \alpha' J') = \frac{8}{3} \frac{\pi^2 mc}{h\varepsilon^2} \frac{1}{\Sigma g_i} \Sigma \frac{S_1}{\lambda} = \frac{304}{\Sigma g_i} \Sigma \frac{S_1}{\lambda} \tag{7–7}$$

where we have used Eq. 7–5. Here Σg_i is the sum of the weights of all the levels in the lower configuration. Some authors (e.g., Unsöld) employ a special type of f-value, **f**, defined by

$$\mathbf{f}(\alpha J; \alpha' J') = \bar{f} \frac{S}{\Sigma S} \frac{s}{\Sigma s} = \frac{(2J' + 1)}{\Sigma g_i} f(\alpha J; \alpha' J') \tag{7–8}$$

Here $S/\Sigma S$ is the strength of the multiplet in terms of that of the whole configuration array. Note that

$$N_{J'} f(\alpha J; \alpha' J') = N_{\alpha'} \mathbf{f}\,(\alpha J; \alpha' J') \tag{7–9}$$

where $N_{J'}$ is the number of atoms in level J', $f(\alpha J; \alpha' J')$ is the *true* f-value for the transition, and $N_{\alpha'}$ is the number of atoms in the whole lower configuration.

7–3. Experimental Determinations of f-Values. In complex atoms, such as iron, configuration interaction occurs and deviations from LS coupling become very large. Accurate f-values cannot be calculated by theoretical means, and recourse must be had to empirical determinations of line strengths and f-values. Many methods, involving various techniques and goals, have been developed. The type of method employed depends on whether one is determining absolute or relative f-values,

[2] The required solutions of the appropriate radial wave equation all entail extensive, difficult calculations. Effects of electron exchange, polarization, and configuration interaction all complicate matters. A further difficulty is caused by the manner in which the wave functions have to be computed. Wave functions that adequately represent the energy levels may fail utterly for transition probabilities. Recent work by Layzer and by Varsavsky point out some need for improvements in the basic theory.

whether one is studying resonance transitions or subordinate transitions, or absorption or emission spectra.

Some astrophysical problems, e.g., a determination of the excitation temperature of the sun, require only *relative* f-values. Determinations of relative f-values require a knowledge of the temperature of the emitting (or absorbing) source of line spectrum, but the total number of participating atoms in the line of sight need not be known. Absolute f-value determinations, however, usually require not only a knowledge of the excitation temperature but of that of the number of atoms in the line of sight as well.

Methods Involving Absorption Lines

In a typical experiment, light from a source producing a continuous spectrum, e.g., a tungsten filament lamp, passes through a tube containing the vapor of the element to be examined. Ideally, the temperature of the vapor should be constant in the tube, but actually the gas is cooler at the ends near the windows and due allowance must be made for this effect.

The influence of the gas upon the beam passing through it may be studied in a number of ways. Magneto-rotation effects (rotation of the plane of polarization in a medium placed in a magnetic field) and measurements of the index of refraction of a gas in the neighborhood of an absorption line (anomalous dispersion) yield values of hNf where Nh is the number of atoms in the vapor column capable of absorbing the line. If Nh is known, f may be determined.

R. B. King and A. S. King and their co-workers measured relative f-values from absorptions produced in thin layers of an absorbing gas. Then the amount of energy subtracted from the beam by the gas is very nearly proportional to the f-value.

Consider a weak absorption line upon a continuous spectrum. We express its intensity as an equivalent width W, which is the width of a perfectly black line of rectangular profile that would remove exactly the same amount of energy from the spectrum. We measure I_ν/I_0 at each point in the line profile and define

$$W_\nu = \int \frac{I_0 - I_\nu}{I_0}\, d\nu \tag{7–10}$$

the *equivalent width in frequency units.* A beam of light passing through an absorbing medium of length h and density ρ with an absorption coefficient κ will be extinguished according to the law

$$I = I_0\, e^{-h\kappa\rho} = I_0\, e^{-hN\alpha} \tag{7–11}$$

where N is the number of atoms cm^{-3} in the tube. Both κ and α depend on the frequency. If the optical thickness of the layer is small,

$$I = I_0\,(1 - hN\alpha + \text{small terms}) \tag{7–12}$$

then since (Eq. 4–82)

$$\int \alpha_\nu \, d\nu \doteq \frac{\pi \varepsilon^2}{mc} f$$

$$\rho h \int \kappa d\nu = hN \int \alpha_\nu d\nu = hN \frac{\pi \varepsilon^2}{mc} f = \int \frac{I_0 - I_\nu}{I_0} d\nu = W_\nu \quad (7\text{–}13)$$

In wavelength units,

$$W_\lambda = \frac{\lambda_0^2}{c} W_\nu \quad (7\text{–}14)$$

and

$$W_\lambda = hN\lambda_0^2 \frac{\pi \varepsilon^2}{mc^2} f \quad (7\text{–}15)$$

Hence if N can be found from the vapor pressure in the tube and W_λ can be measured, we may obtain the absolute value of f. If N cannot be determined, only relative f-values may be found. We emphasize that weak lines must be used.

In practice, the ideal condition of a layer of absorbing vapor whose optical depth at all wavelengths is much less than 1 is difficult to fulfill, and allowance must be made for deviations from linearity in the W–Nf relationship (*curve of growth*), i.e., for the higher order terms in Eq. 7–12. Fortunately this relationship is well known (see Chapter 8). It may be shown that very strong lines yield an estimate of $Nf\Gamma$; hence the damping constant Γ must be estimated as well as N, or if N and f are known from other data we can evaluate Γ. The fact that Γ depends on the density complicates the problem.

R. B. King determined absolute f-values for iron lines from specially purified samples of this metal heated in a quartz tube in an electric furnace at an accurately known temperature. The vapor pressure of iron is known as a function of temperature; hence the number of gaseous iron atoms in the quartz tube may be found.

A better procedure is the *atomic beam method*, employed by Kopfermann and Wessel (1951) and refined by R. B. King and his associates whose experimental set up is illustrated in Fig. 7–1. The metal to be studied is placed in a small crucible (1), about 7/8 in. long, which has a small aperture at the top through which the vapor escapes into a chamber (3), evacuated to a pressure of about 5×10^{-5} mm of Hg. The crucible which must be made of some substance that will stand temperatures up to 2200°C and not react chemically with the metal being investigated is placed in an electric furnace (2). The angular spread of the emergent beam is restricted by the rectangular opening in the chamber (3). The beam then passes to the plate (4) that contains a number of holes, through one of which (5) the temperature of the crucible orifice is monitored by an optical pyrometer.

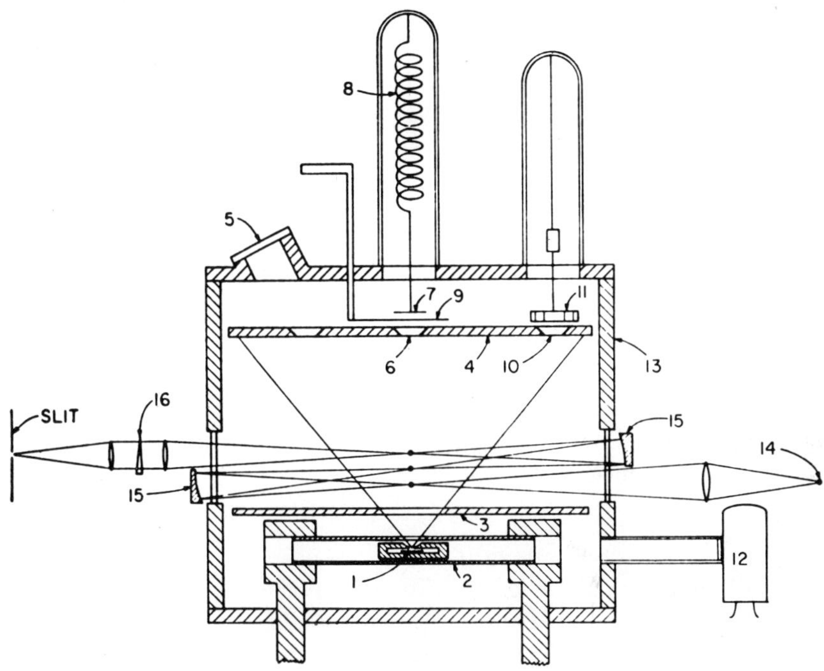

FIG. 7–1. SCHEMATIC DIAGRAM OF ATOMIC BEAM APPARATUS
(Courtesy, Bell, Davis, King, and Routly, *Ap. J.* (University of Chicago Press) **127,** 782, 1958.)

The method of operation is the following: A continuous spectrum light source (14) is focussed in the chamber and refocussed on the slit of the spectrograph. Mirrors (15) are employed to lengthen the light path to increase the effective number of absorbing atoms. The prism (16) serves to isolate the spectral region and minimize scattered light in the grating. When the crucible is heated, the escaping vapor atoms produce weak resonance absorption lines on the continuous spectrum as photographed in the spectrograph. Some escaping vapor atoms pass through the aperture (6) and collect on the thin aluminum plate (7) where the deposit is weighed by the thin quartz spring balance (8). The shutter (9) prevents atoms from reaching the disk until measurements are commenced. A quartz torsion balance (11) detects any change in the flux or velocity distribution of atoms passing through the aperture (10) during the course of the experiment.

Thus the equivalent width of a resonance line is measured directly and the number of atoms involved can be computed from gas-kinetic theory and the known temperature of the crucible, the rate of deposition of material on the pan suspended from the spring balance, and the geometry. The experiment, although simple in principle, is difficult to perform with the required accuracy. Since the density in the beam must be so low that multiple collisions are negligible, only strong resonance lines can be observed. Hence, although one may measure absolute f-values for resonance lines in atomic beam experiments, the f-values of subordinate lines must be determined from other experiments, in which only relative f-values are found.

In electric furnace experiments when the temperature can be controlled accurately, relative f-values can be measured for resonance lines and lines arising from low levels. Transitions between high energy levels cannot be measured in absorption experiments and recourse must be had to arcs and other sources of emission line spectra.

Methods Involving Emission Lines

1. *Lifetime measurements.* Since the lifetime of a level is related to the Einstein A coefficients by

$$T = \frac{1}{\sum_{n'} A_{nn'}}$$

where the summation is carried out over all lower levels, a measurement of T gives $\Sigma A_{nn'}$. If the relative transition probabilities are known or if the state decays only by one particular transition, the f-values can be found.

The great advantage of lifetime measurements is that one need not know the temperature or density of the gas (i.e., parameters necessary to compute the population of the upper level). Several methods are available. In one type of experiment atoms are excited to an upper level by collisions or by illumination by resonance radiation. The source of excitation is cut off suddenly, and the number of photons emitted by spontaneous de-excitation at a time, Δt, after the excitation is measured (see e.g., Heron *et al.*, 1954, 1956; Ziock, 1956). One may also excite selected upper levels by a modulated beam and observe the time lag between the modulation and the radiation. By using polarized light, and magnetic resonance, it is possible to excite particular Zeeman states if desired. In yet another method, a steady source of light is employed to excite an upper level which decays in a cascade transition emitting two frequencies ν_1 and ν_2 which are isolated with interference filters and observed with two photomultipliers. Each photon ν_1 and ν_2 produce a pulse; those of frequency ν_1 are delayed by an amount Δt and coincidences with pulses ν_2 are observed (see, e.g., Brannen *et al.*, 1955).

Decay times can be measured to a high accuracy by coincidence counter techniques, and the like, developed for nuclear physics. On the other hand, the decay constants are not necessarily lifetimes. Resonance quanta, in particular, may be scattered from one atom to another so the radiation may die away much more slowly. The measured decay constant then depends on gas density, the geometry of the experiment, etc.; consequently one must run the experiment for different densities and possibly geometries. Cascade phenomena produce further complications, particularly with subordinate lines. Eventually, lifetime measurements may provide the most accurate technique for getting transition probabilities of high level lines.

2. *Ordinary arcs.* Ordinary arcs have been used as sources of emission spectra in which quantitative measurements have been made. For example, Allen and Asaad, and also Corliss at the National Bureau of Standards, used an arc struck between two copper electrodes with which were alloyed small, known amounts of other metals. Lines for which absolute f-values were known from other experiments were used to fix electron densities and ionization temperatures. For example, Corliss found an arc temperature of $5100 \pm 110°K$ and an electron density of 2.4×10^{14} cm^{-3}. Then, from measured intensities of lines of program metals, one could determine their f-values on the assumption that the mixture in the vapor was the same as in the arc and that a unique excitation temperature could be assigned to the emitting volume of the arc. Extensive tables for lines of 70 elements have been published by the Bureau of Standards; the f-values are probably good only to within a factor of 2, but they form a homogeneous set of material.

3. *Stabilized arcs.* Since ordinary arcs have a tendency (as usually operated) towards temperature and density instability, most f-value studies have been carried out with stabilized arcs operating under carefully controlled conditions.

In the water-stabilized arc developed by Lochte-Holtgreven and his associates at Kiel University, the current flows down the axis of a rapidly rotating tube driven by a water turbine. Water flows down the wall of the tube, the vapor becomes dissociated, ionized, and raised to a high temperature near the tube axis. If the emitting column is observed end-on through a hole in the electrode, one can measure intensity variations in the lines of hydrogen and of oxygen in various stages of ionization and of hydrogen across the tube. With the aid of the Saha and Boltzmann equations one can then assign to each distance from the tube axis a unique temperature and density. The broadening of the hydrogen lines may be used to fix the electron density. By mixing small quantities of other elements in the form of soluble salts or liquids or by blowing in gaseous mixtures whose proportions are known, one can obtain f-values for other elements. Unfortunately, whirling-water arcs tend to be unstable and erratic. To overcome these difficulties the wall-stabilized arc has been developed. Current flows through an aperture on the axis of a stack of copper plates that are insulated from one another by spacers that permit gas to flow in freely from the sides. Mastrup and Wiese (1958) used this device to measure f-values for nitrogen.

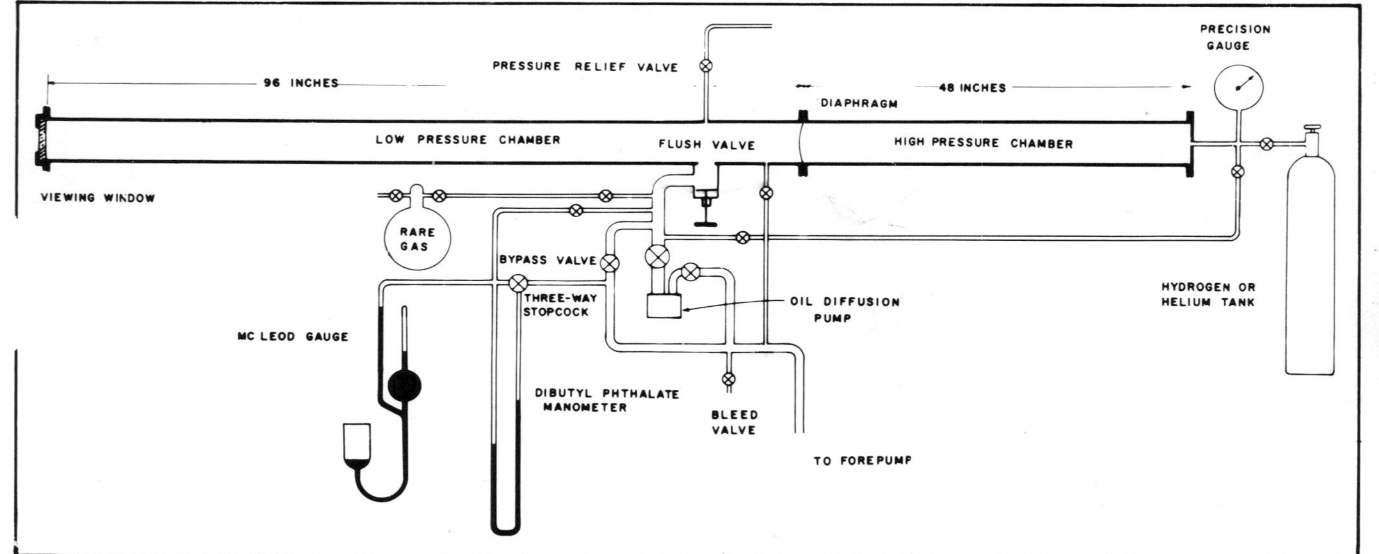

FIG. 7–2. SCHEMATIC DIAGRAM OF LUMINOUS SHOCK TUBE

Driver gas from the hydrogen or helium tank is introduced into the high-pressure chamber on the right. The low-pressure chamber is evacuated by a fore pump and oil-diffusion pump and the rare gas whose spectrum is to be studied is introduced in a known amount into the low-pressure chamber. Its pressure is measured with the McLeod gauge. When the diaphragm is pierced, the gas in the high-pressure chamber rushes to the left, producing a shock wave which is reflected from the left end wall of the tube. (Courtesy E. B. Turner. University of Michigan Physics Laboratory.)

Noble gases present a relatively easy problem since they can be run in pure form in an arc at a known pressure and easily determined temperature. H. N. Olsen measured argon f-values in an arc at atmospheric pressure.

4. *Flames.* Flames have been used, for example, by Huldt and Lagerquist (1952) who measured the absolute emission intensity of the unresolved resonance triplet of Mn excited in a flame. They obtained the concentration of atoms from the flame temperature, the dissociation energy, and thermochemical arguments.

5. *Shock tubes.* The luminous shock tube, described by Kantrowitz and Laporte, has been developed as a tool for the study of high temperature gases. The shock tube employed at Laporte's laboratory at the University of Michigan (Fig. 7–2) consists of a long tube, one segment of which is heavily reinforced to stand internal pressures of the order of a hundred atmospheres. This high pressure chamber is separated from the low pressure chamber (that contains a gas at a few millimeters pressure) by a diaphragm that may be punctured at will. The end of the low pressure chamber contains a strongly reinforced viewing box that permits the experimenter to observe the reflection of the shock wave from the end of the tube. The phenomena are observed with cameras and spectrographs equipped with rapidly rotating film holders.

The gas whose f-values are to be studied is mixed with a noble gas and introduced in the low pressure chamber at a few millimeters pressure. If one wishes to study a metal, for example, a volatile compound is prepared and mixed with the noble gas. Hydrogen is normally employed as the driver gas at a pressure exceeding 10^8 dynes cm^{-2}.

As soon as the diaphragm is pierced, the rapidly expanding gas from the high pressure chamber produces a strong shock wave that pushes ahead of it the gas originally contained in the low pressure chamber. When the shock wave hits the end of the tube it is reflected. Then a conversion of the ordered motion of the advancing gas into random thermal motion produces a sudden rise in temperature of the noble gas and its impurity.

Radiation effects are extremely short-lived but excitation lasts long enough for thermal equilibrium to be established in the radiating volume, so that one may employ the Saha and Boltzmann equations. The great advantage of the shock tube technique lies in the fact that temperatures and pressures can be calculated from the known composition and initial pressures of the gas with the aid of measures of the shock speed and well-known hydrodynamical relations. Spectroscopic measurements are not involved at all except as a check.

Transition probabilities for neon and argon have been measured by L. Doherty, while Wilkerson has measured f-values for neutral and ionized chromium.

Special forms of shock tubes are required for high temperatures, from

20,000° to 50,000°K. With a magnetically driven shock, Kolb and his associates have measured helium lines at temperatures up to 40,000°K. McLean *et al.*, (1960) have demonstrated that equilibrium is attained very quickly in a shock plasma; McLean has checked the method by measuring helium *f*-values. The ionic concentration in the plasma may be found to within ±10% from measurements of the continuum, the emission line profiles, and the index of refraction. The temperature can be found to within ±2 per cent from the ionization equilibrium and the continuum intensity. To obtain transition probabilities for lines of other highly ionized atoms, one may use the ionized helium gas (helium plasma) as a thermometer. The gas to be studied is mixed in a known concentration with the helium and the line intensities are measured photoelectrically.

7-4. Comparisons of Different Methods for *f*-Value Determination. Although wavelengths of spectral lines may be measured to an accuracy of one part in a million, transition probabilities cannot be found with an accuracy much better than ten or fifteen per cent except under exceptional circumstances. In many instances the errors are much larger. Even relative *f*-value measurements by different workers are often sadly discordant, probably because of errors in the excitation temperature. Examples are compiled for instance by Goldberg, Müller, and Aller (1960) and by C. W. Allen (1960).

As an example of *f*-value determinations for a resonance line, let us consider first Mn I λ4031, which has been measured by a number of experimenters as follows:

0.062 ± 50%	Huldt and Lagerquist (1951) flame
0.045 ± 60%	Allen and Asaad (1957) copper arc
0.056 ± 2%	Ostrovsky and Penkin (1957) anomalous dispersion
0.061 ± 0.005	Bell, Davis, King, and Routly (1959) atomic beam

For iron λ3720 satisfactory results are also obtained, viz.:

0.030	R. B. King (1938) absorption tube, corrected vapor data
0.043 ± 20%	Kopfermann and Wessel (1951) atomic beam
0.046 ± 30%	Ziock (1957) lifetime of upper levels
0.036 ± 60%	Allen and Asaad (1957) copper arc
0.032 ± 12%	Bell, Davis, King, and Routly (1958) atomic beam

Unfortunately, the situation is less happy for other metals, e.g., titanium. Furthermore, resonance lines of abundant elements are usually of less use for abundance studies than are subordinate lines.

In particular, investigations of deviations from thermodynamic equilibrium require a knowledge of accurate *f*-values for lines arising from levels embracing a considerable range in excitation potential. For the measurement of *f*-values of subordinate lines, and lines of ions, the atomic beam method cannot be used and one must resort to lifetime measurements, arc, or shock-tube studies. Lifetime measurements must be carried out on an individual basis for selected transitions; they are not at present amenable to production of quantities of data as are shock tube or arc studies.

The big advantage of the arc is that it can be sustained for a long period of time permitting the temperature and electron pressure to be measured. On the other hand, the long lifetime of the discharge is itself a cause of trouble in that thermal diffusion, for example, may produce ionic concentration gradients and composition differences. Even if a homogeneous gas is introduced, it will not so remain. Suppose one uses a water stabilized arc, takes the known f-values for hydrogen, and uses them to get P_ε and T from the hydrogen line intensities with the aid of the Boltzmann and Saha equations. Then from the measured intensities of the oxygen lines he obtains the f-values for O. Then the arc is run in air to get the f-values for nitrogen. With the f-values for oxygen and nitrogen known, one tries to get those for carbon by running CN and CO in the arc. The results will not check because in the whirling-water arc the chemical composition depends on temperature and cannot be assumed uniform. Furthermore, the liquid arc does not even have the advantage of stability, a very serious fault, since one must assume T and P are independent of time.

Hence, determinations of f-values from arcs requires that not only previously described cross-checks be applied but that also the same f-values must be obtained when different densities and temperatures are used in the operation of the arc. Insofar as relative f-values are concerned, the arc offers a number of advantages. Since it remains steady with time, longer exposures may be secured with a spectrograph and weaker lines observed. One can produce a steady state in an arc even if thermal equilibrium is not attainable. On the other hand, faint lines cannot be observed with a shock tube. One of the disadvantages of the arc is that there is a change of ionization and excitation with distance from the axis so that it is necessary to construct a model in order to interpret the experimental results. Furthermore, most arcs are temperature limited, and it is usually not possible to get above 25,000°K. Although it would be possible to go to higher temperatures by confining the arc with the aid of a magnetic field, it is not possible to attain temperatures as high as those that can be produced in a shock tube.

The shock tube has two considerable advantages as a device for producing ionized gases under controlled or known conditions: (1) In shock tubes such as the one employed at Michigan, the geometry is simple (depending on only one dimension). Hence it is not necessary to correct for effects of cylindrical asymmetry as in an arc. (2) The chemical composition of the heated gases can be accurately fixed, since there is no time for concentration gradients to be set up. Pressures and temperatures can be predicted from shock tube hydrodynamics and checked to see if the assumptions underlying the spectroscopy are fulfilled.

At high temperatures, complications are sometimes observed. For example, Kolb and Griem found that helium gas in front of the shock wave is ionized and excited by UV radiation from the discharge driving the shock wave, but the problem of such precursor radiation can be handled. McLean measured relative *f*-values in He I to within about 3 per cent. At lower temperatures such troubles do not exist, e.g., *f*-values for O_2 measured at Cornell and at Avco Laboratory agreed to about 10 per cent (the numbers are probably good to within 20 per cent when allowance is made for the basic errors in the experiment).

One disadvantage of the shock tube is that faint lines cannot be measured, although good *f*-values can be obtained. Time resolution is often technically difficult. On the other hand, transition probabilities for lines of highly excited ions can be measured only in shock tubes and hot plasma.

7–5. Collisional Broadening of Spectral Lines. The intensity of a stellar absorption line clearly must be related to the number of atoms acting to produce it; the stronger the line the greater the number of participating atoms. At first, one might suppose the equivalent width, W, of a line to be simply proportional to the effective number of atoms, Nf, and this is indeed true for very weak lines. Lines of intermediate and large equivalent width are not related to numbers of effective atoms in such a simple way. The character of the line broadening plays the decisive role, such that the exact relationship will depend on the importance of the Doppler effect, radiation damping, broadening caused by charged ions, etc. Consider what happens as an increasing number of atoms is added to an atmosphere or optical path. (We can get the same effect by considering lines of the same atom but of different oscillator strength f, since Nf, not just N, is the relevant factor.) When the number of atoms is small, the amount of energy subtracted from the continuous spectrum is small and is proportional to the number of atoms, i.e., W varies as Nf. As Nf increases, the energy subtracted in the center of the line is depleted and further absorption has to take place in the wings. Thus the total amount of energy subtracted from the outgoing radiation as Nf increases will depend upon how much energy the atoms can absorb at great distances from the line center, i.e., on how the absorption coefficient depends on the distance $\lambda - \lambda_0$ from the line center.[3] Hence we must first consider sources of line broadening and their influence on the absorption coefficient.

Absorption lines in stellar spectra are intrinsically broadened by the following causes:

[3] Analogous considerations hold for emission lines. The central intensity of a line builds up to a value given by Planck's formula, and the line increases in total intensity by becoming broader.

1. Doppler effect arising from random kinetic motions of atoms. To this may be added a possible "turbulence" broadening caused by large scale motions of huge masses of gas, especially in atmospheres of giant and supergiant stars.
2. Radiation damping, which is a consequence of the finite lifetimes of excited levels. Classically, it corresponds to the fact that the finite wave train emitted by a radiating atom is non-monochromatic (see Chapter 4).
3. Collisional broadening. A radiating atom may be perturbed by its neighbors and emit a broadened spectral line. The perturbing particles may be neutral atoms, ions, or electrons.
4. Hyperfine structure is responsible for the broadening of certain lines in the solar spectrum (e.g., lines of Mn).
5. Zeeman effect. Lines produced in sunspots or magnetic stars are broadened or split by the magnetic field.

Finally, there are *extrinsic* causes of broad lines in stellar spectra. An absorption line profile in a rapidly spinning star is a composite of contributions from different parts of the disk, some of which are approaching and some of which are receding. The net result will be a broad, dish-shaped line. Expanding shells in novae, peculiar stars, or Cepheids also produce broadened lines. If large-scale convection currents occur, widened lines may also result.

The absorption coefficient of a line broadened by thermal motions alone (see Eq. 3–90 has a bell shape, falling off as $\exp[-\text{const}\ (\delta\lambda)^2]$ in the wings. Radiation damping gives a narrow absorption coefficient with broad wings in accordance with Eq. 4–80. In resonance lines the quantum mechanical damping constant Γ may be of the order of the classical damping constant γ. Observations show, however, that damping constants of five, ten, or even fifty times the classical value are required to explain profiles of certain strong Fraunhofer lines. Collisional broadening appears to be responsible for these large damping constants.

There are several distinct collisional broadening situations that arise in problems of astrophysical interest:

1. Collisions of the radiating atom with neutral particles of another element, e.g., broadening of metallic lines by collisions with neutral hydrogen atoms.
2. Collisions of a radiating atom with a neutral atom of the same kind, i.e., self-broadening. In astrophysical applications this type of broadening is important only for Hα (see Traving and Cayrel, 1960) — a line which is also broadened by Doppler and Stark effects.
3. Perturbations by static ion fields. (a) First order (linear) Stark effect exists if the levels are degenerate (e.g., hydrogen-ionized helium). (b) Second order (quadratic) Stark effect exists if the levels are not degenerate.
4. Perturbations by rapidly moving electrons, which can usually be described in the impact approximation. Two physical situations occur: (a) The electron collisions may cause phase shifts. (This is called the adiabatic

impact theory in the older work.) (b) The electron collisions may cause transitions between close-lying levels; these inelastic collisions correspond to non-adiabatic processes.

Detailed studies (e.g., Griem *et al.*, 1962) show that if electrons induce transitions between levels of different J, these non-adiabatic effects are as large or larger than pure phase-shift effects.

Two or more sources of collisional line broadening may operate simultaneously, e.g., hydrogen lines are always broadened by both ions and electrons while metallic lines may be broadened by collisions with both neutral and ionized particles and electrons.

We may examine the collisional broadening of spectral lines from two limiting points of view, corresponding to the "impact limit" and the "static limit."

In the impact theory we regard the emitting and absorbing atoms as being disturbed by separate, i.e., discrete encounters with other particles. In 1906 Lorentz gave a phenomenological theory: a wave emitted by a radiating atom is interrupted by collisions with perturbing particles, and the intensity distribution in the emitted line is obtained by a Fourier analysis as a function of temperature and density. The impact theory was developed notably by Lenz, Weisskopf, Anderson, Lindholm, and Foley. A general impact theory, which contains the results of Anderson, Lindholm, and Foley as a special case and is in harmony with the quantum mechanical theory of Kivel, Bloom, and Margenau, has now been developed (Baranger, 1958; Kolb and Griem, 1959). In the discrete encounter or impact approximation[4] the time dependence of the perturbation is the strategic factor.

In the "static" theory, following Stark, Debye, Holtsmark, or Margenau, we can ask what will be the instantaneous value, at a given atom, of the electric field, F, produced by surrounding charged particles. The direction and magnitude of F will fluctuate as the configuration of the particles change, but we can compute the probability, $W(F/F_0)$, for a field of strength F at a given instant. F_0 is the time average of the field. Since to each F there corresponds a shift, $\delta\nu$, in the frequency of the emitted radiation, we can compute for what fraction of time the spectral line is shifted by amounts, $\delta\nu_1$, $\delta\nu_2$, ... Then, by summing over all shifts according to their statistical probabilities, we get the shape of the broadened emission line and hence the wavelength dependence of the absorption coefficient, since both must have the same form (see Chapter 4). In the center of the line, one must consider corrections to the Holtsmark theory

[4] Most line broadening theories (which apply to densities of astrophysical interest) use the classical path approximation (CPA) in which a perturbing atom or ion is regarded as moving in a path like a classical particle, i.e., the wave nature of the particle is ignored.

due to ion-ion correlations, but such effects are not important at astrophysical densities (Mozer and Baranger, 1960).

In the limiting extreme of no relative motion between radiating atom and perturber, Holtsmark's theory would be valid. When there is relative motion, we shall see that the central part of the line is governed by the discrete encounter theory and the statistical theory applies in the wings. Under conditions prevailing in stellar atmospheres, usually one or the other of these two extremes applies for a given line, e.g., for the broadening of Ca II λ3933 or Ca I λ4227 the discrete encounter theory is appropriate, whereas for the ionic contribution to the broadening of hydrogen lines we apply Holtsmark's theory.

To decide whether the discrete encounter (impact) or statistical approximation applies in a given instance we proceed as follows: Consider a perturber moving in a straight line with the velocity v such that the closest distance to a radiating atom is ρ (Fig. 7-3). Suppose that the perturbation, $P(t)$, obeys the law

$$P(t) = D_n\, r^{-n}(t) = D_n\, [v^2t^2 + \rho^2]^{-n/2} \tag{7-16}$$

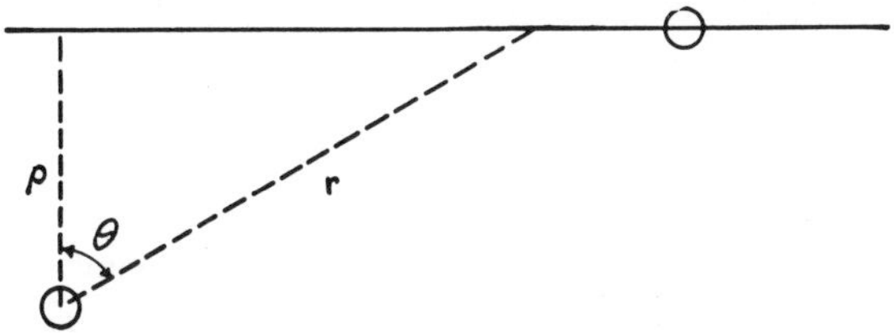

FIG. 7-3. GEOMETRY OF AN ENCOUNTER

The perturbing particle whose velocity is v passes the radiating atom at a minimum distance ρ.

We now define the "characteristic time" t_D for the encounter as that interval during which $P(t)$ exceeds

$$\tfrac{1}{2}\, P_{\max} = \tfrac{1}{2}\, D_n\, \rho^{-n} \tag{7-17}$$

That is,

$$t_D = \frac{2\rho}{v} \tag{7-18}$$

Now consider a circular frequency shift $\Delta\omega = 2\pi\Delta\nu$, and a time interval, Δt, connected by the condition

$$\Delta\omega\, \Delta t \sim 1 \tag{7-19}$$

If $\Delta\omega$ is sufficiently small that

$$\Delta t \gg \frac{t_D}{2\pi} \tag{7-20}$$

the discrete encounter approximation is valid. Since Δt must appreciably exceed the duration of the collision, this approximation is valid near the line center (small $\Delta\omega$). The range in $\Delta\omega$ increases with the speed of the perturber.

Spitzer (1939a) showed that for general perturbations of the type described by Eq. 7–16, the discrete encounter approximation holds if

$$(\Delta\omega) \ll \left[\frac{v^n}{D_n}\right]^{1/(n-1)} \tag{7–21}$$

It may be shown (Holstein, 1950) that the statistical approximation is valid if

$$\Delta\omega \gg \left(\frac{v^n}{D_n}\right)^{1/(n-1)} \left(\frac{n}{6}\right)^{n/(n-1)} \tag{7–22}$$

Stating the situation in qualitative terms: A very small time interval Δt is connected with a very large frequency interval $\Delta\omega$, i.e., the distant parts of a line correspond to very short time differences during which the spatial configurations of the disturbing particles change very little. Therefore the far wings of a line are broadened in accordance with the statistical theory.

A numerical application to $H\gamma$ shows that insofar as the ions are concerned, the statistical theory is valid for values of $\Delta\lambda$ greater than 0.5*A* at T = 20,000°K, but that electrons broaden according to the impact (discrete encounter) theory.

Before we derive an expression for the shape of a broadened emission line (which gives the corresponding absorption coefficient), let us consider the line broadening by static ion fields from the point of view of the quantum theory. An undisturbed atom will radiate a frequency, ν_0, given by $h\nu_0 = W_2^0 - W_1^0$. If this atom is perturbed by an intruder, its energy levels will be shifted. A plot of W_1 and W_2 against r, the separation between radiating atom and perturbing particle, yields curves resembling the po-

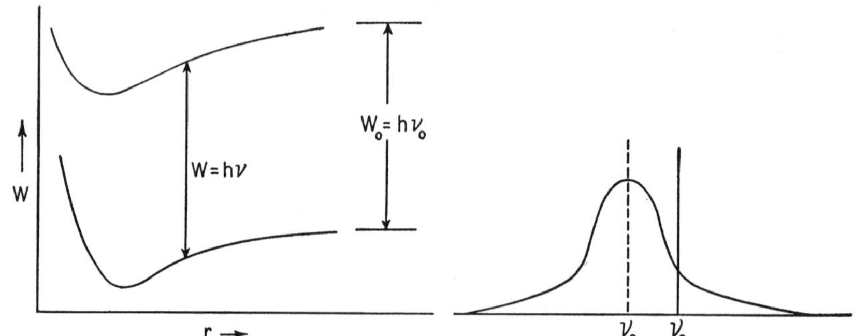

Fig. 7–4. Collisional Displacement of a Spectral Line

The left part of the figure depicts the distortion of the energy levels as a function of the separation r between the atom and perturber. The undisturbed frequency is ν_0. The resultant spectral line (right) not only is broadened because encounters take place at different r values but is shifted as well.

tential energy curves of a diatomic molecule. If the atom radiates when the separation is r, then $h\nu = W_2(r) - W_1(r)$. This energy will differ from that ordinarily radiated since $W_2 - W_1$ changes with r and the position of the spectral line will be shifted. Throughout a radiating volume, emissions will occur at differing r values and the resultant spectral line will not only be shifted but broadened as well. See Fig. 7–4.

We shall now sketch the elementary theory of impact broadening, which has underlain interpretations of much astrophysical data in the past. Quantitatively, the theory is inadequate in that although it gives the correct order of magnitude for the line widths, the predicted shifts are wrong. Nevertheless, it forms a background for a more complete theory which includes effects of collision-induced transitions which are very important for line broadening.

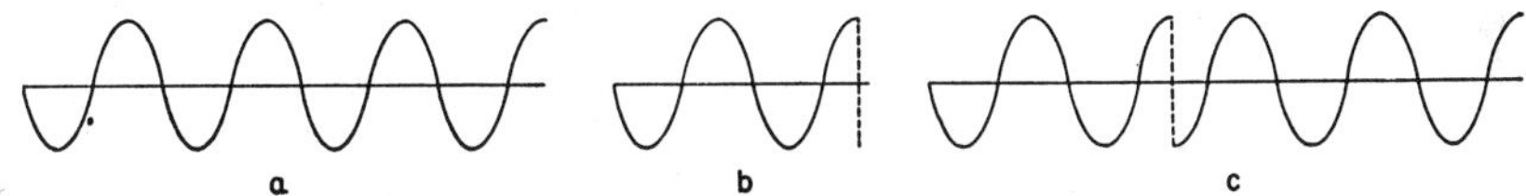

FIG. 7–5. QUENCHING OF A WAVE TRAIN IN AN ENCOUNTER

(a) Normal classical wave of finite duration, (b) quenched wave, (c) collision with a phase shift of 180°.

In the classical discrete-encounter or "impact" picture, an atom radiating a wave train suffers an encounter with another particle which quenches the wave or changes its phase so appreciably that the net effect is as though two distinct wave tracings were emitted. In all encounters the phase of the wave will be advanced or retarded; sometimes the phase shift is large (see Fig. 7–5); at other times it is so small that the net effect is negligible. For this reason discrete encounter theories may be called "phase shift" theories if collision induced transitions are negligible.

Let us suppose, following Weisskopf, that the phase shift depends on an inverse power of the separation, r^{-n}, viz.,

$$\delta\omega = \frac{2\pi C}{r^n} \tag{7–23}$$

where ω is measured in circular frequency units and C is a constant. The total phase shift is found by integrating $\delta\omega$ over the duration of the encounter. That is,

$$\eta = \int_{-\infty}^{+\infty} \delta\omega\, dt = 2\pi C \int_{-\infty}^{+\infty} \frac{dt}{(\rho^2 + v^2t^2)^{n/2}} = \frac{2\pi C}{v\rho^{n-1}}\, a_n \tag{7–24}$$

since $r = \rho \sec\theta$ (see Fig. 7–3); whence

$$a_n = \int_{-\pi/2}^{+\pi/2} \cos^{n-2}\theta\, d\theta = \frac{\sqrt{\pi}\,\Gamma\left(\dfrac{n-1}{2}\right)}{\Gamma\left(\dfrac{n}{2}\right)} \tag{7–25}$$

where Γ here denotes the gamma function. Here a_n is related to n by:

$$\begin{array}{llllll} a_n = & \pi & 2 & \frac{\pi}{2} & \frac{4}{3} & \frac{3\pi}{8} \\ n = & 2 & 3 & 4 & 5 & 6 \end{array} \tag{7-26}$$

Weisskopf supposed, arbitrarily, that if a phase shift was greater than a critical value, η_0, which he set equal to 1, an atom would be so disturbed by the collision that the emissions before and after the event should be regarded as independent processes, insofar as their influence on the shape of the spectral line is concerned.

Suppose that an atom radiates with an undisturbed frequency, $\nu_0 = \omega_0/2\pi$, for an interval $-T/2$ to $+T/2$ between collisions. We may analyze this finite wave train, $e^{i\omega_0 t}$, of time duration T by means of the Fourier integral theorem. We write

$$F(t) = \int_{-\infty}^{+\infty} e^{i\omega t}\, d\omega \int_{-a}^{+a} F(u)\, e^{-i\omega u}\, du \tag{7-27}$$

where $F(t) = 0$ outside the interval $-a$ to $+a$. With $a = T/2$, and $F(t) = e^{i\omega_0 t}$, we find

$$e^{i\omega_0 t} = \int_{-\infty}^{+\infty} e^{i\omega t}\, d\omega \int_{-T/2}^{+T/2} e^{-i(\omega-\omega_0)y}\, dy \tag{7-28}$$

which means that the finite wave train $-T/2$ to $+T/2$ is equivalent to the superposition of infinite wave trains, $e^{i\omega t}$, having an amplitude

$$\int_{-T/2}^{+T/2} e^{-i(\omega-\omega_0)y}\, dy$$

and therefore an intensity dependence on frequency given by

$$I(\nu) \propto \left[\int_{-T/2}^{+T/2} e^{-i(\omega-\omega_0)t}\, dt\right]^2 = \left[\frac{\sin\left(\dfrac{\omega-\omega_0}{2}\right)T}{\dfrac{\omega-\omega_0}{2}}\right]^2 \tag{7-29}$$

The time between encounters T, or flight time, depends on the mean free path of the particle. It is the velocity divided by the free path. If T_0 is the mean flight time (sometimes called mean free time) interval between collisions, the probability $F(T)$ of a flight time T is $(1/T_0)\, e^{-T/T_0}$. To obtain the intensity distribution in a spectral line we must sum over all flight times T. Then

$$I(\nu) = A \int_0^\infty \frac{1}{T_0}\left[\frac{\sin \pi(\nu-\nu_0)T}{\pi(\nu-\nu_0)}\right]^2 e^{-T/T_0}\, dT = \frac{A}{2\pi^2}\, \frac{1}{(\nu-\nu_0)^2 + \left(\dfrac{1}{2\pi T_0}\right)^2} \tag{7-30}$$

where A is determined by the condition that

$$\int I(\nu)d\nu = I_{\text{Total}} \tag{7-31}$$

Since the shape of the absorption coefficient is the same as that of the corresponding emission line, the absorption coefficient per atom is

$$\alpha_\nu = \frac{\pi\varepsilon^2}{mc} f \frac{1}{2\pi^2 T_0} \frac{1}{(\nu - \nu_0)^2 + \left(\dfrac{1}{2\pi T_0}\right)^2} \tag{7-32}$$

This expression shows the same frequency dependence as the radiation damping formula if the constant γ be replaced by $2/T_0$.

Weisskopf supposed that only those collisions that change the phase, η, by an amount greater than η_0 contributed to line broadening. Let us suppose that a distance of closest encounter, ρ_0, corresponds to a phase shift, η_0. Then the collision cross-section is $\pi\rho_0^2$ for line-broadening encounters. Let N_b denote the number of particles cm^{-3} responsible for the broadening. For problems such as the widening of metallic lines in the solar spectrum by the action of neutral atoms, N_b will be very nearly the number of hydrogen atoms cm^{-3}. That is, the number of perturbers will vastly exceed the number of emitting particles. The number of line-broadening collisions sec^{-1} for each radiating particle will be

$$S = \frac{1}{T_0} = \pi\rho_0^2 N_b V \tag{7-33}$$

where V, the mean relative speed, is given by kinetic theory as

$$V = \sqrt{\frac{8kT}{\pi M_0}\left(\frac{1}{A_1} + \frac{1}{A_2}\right)} \tag{7-34}$$

M_0 is the mass of a particle of unit atomic weight while A_1 and A_2 are the atomic weights of the radiating and perturbing atoms. Since the density of stellar atmospheres is always low, the mean distance between particles $r_0 = \sqrt[3]{3/4\pi N} >> \rho_0$, and we may neglect multiple collisions.

With $\eta = \eta_0$, ρ_0 follows from Eq. 7–24, viz.,

$$\rho_0 = \left[\frac{2\pi C a_n}{V\eta_0}\right]^{1/(n-1)} \tag{7-35}$$

The collisional damping constant will be

$$\Gamma_{\text{coll}} = \frac{2}{T_0} = 2\pi V N_b \rho_0^2 = 2\pi N_b \left[\frac{2\pi C a_n}{\eta_0}\right]^{\frac{2}{n-1}} V^{1-\frac{2}{n-1}} \tag{7-36}$$

or

$$\Gamma_{\text{coll}} = 2^{\left(\frac{5n-7}{2n-2}\right)} C^{\left(\frac{2}{n-1}\right)} \pi^{\frac{n+5}{2(n-1)}} a_n^{\frac{2}{n-1}} \left[\frac{kT}{M_0}\left(\frac{1}{A_1} + \frac{1}{A_2}\right)\right]^{\frac{n-3}{2n-2}} \eta_0^{-\frac{2}{n-1}} N_b \tag{7-37}$$

if the force law is of the form r^{-n} and if inelastic collisions are negligible.

Atoms other than hydrogen and ionized helium, perturbed by ions and electrons would have a quadratic Stark effect, whilst the collisional broadening that arises when other absorbing atoms suffer encounters with neutral hydrogen atoms would involve a van der Waals type of force for which $n = 6$. We must emphasize that Eq. 7–37 can give only the order of magnitude of the damping constant since it does not take into account the inelastic collisions.

Another difficulty besides that with inelastic collisions lies in the choice of η_0. Weisskopf and Lorentz made the simple assumption that no relation existed between the phases and after collision and that $\eta_0 = 1$. Actually, in some collisions that influence line broadening, there does exist a relation between initial and final phases. Foley (1946) and Lindholm (1946) gave a more detailed discussion of this problem wherein they took into account the cumulative effect of numerous encounters in which the phase is changed by varying amounts. See Fig. 7–6. There are few collisions in which η is large but many in which η is small. In agreement with experiment and in contrast with Eq. 7–32, they found a frequency shift as well as a line broadening so that the intensity distribution is given by

$$I(\nu) = I_0\left(\frac{\gamma}{4\pi^2}\right)\frac{1}{(\nu - \nu_0 + \beta')^2 + (\gamma/4\pi)^2} \tag{7–38}$$

in agreement with predictions of quantum theory. Encounters for which $\eta \geq 1$ are most important for line broadening, whereas encounters for which $\eta < 1$ play the more important role in line shifts. Then the Lindholm-Foley theory gives:

$$\Gamma_{el} = 38.8\, C_{el}^{2/3}\, V^{1/3}\, N_e \tag{7–39}$$

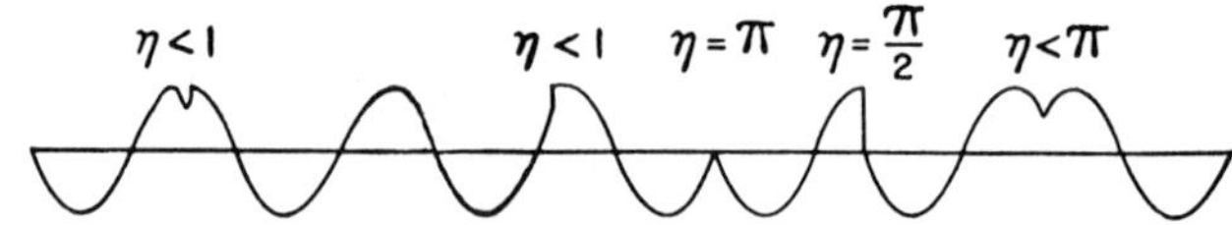

Fig. 7–6. Phase Shifts in Encounters

for the electronic contribution to the quadratic Stark effect. More recent calculations (Griem *et al.*, 1962) show that because of the inelastic collisions the so-calculated Γ_{el} may be two or three times too small even if C_{el} is exact.

The corresponding expression for interactions of the van der Waals type is

$$\Gamma_W = 17.0\, C^{2/5}\, V^{3/5}\, N \tag{7–40}$$

The predicted ratios of damping constants to line shifts are

$$\begin{aligned} \gamma/\beta &= 1.16 \quad (n = 4) \\ &= 2.80 \quad (n = 6) \end{aligned} \tag{7–41}$$

i.e., they are constant; but the more exact theory shows that if inelastic collisions are included the shifts can vary over a considerable range but are usually smaller than predicted by the Lindholm-Foley theory.

The values of η_0 to be employed in Eq. 7–35 to get ρ_0 are

$$\begin{aligned} \eta_0 &= 0.64 \quad (n = 4) \\ &= 0.61 \quad (n = 6) \end{aligned} \tag{7–42}$$

The electron-impact broadening of spectral lines (which are already subject to the quadratic Stark effect of ions in static fields) is especially important in an ionized gas and generally dominates the broadening due to static ion fields. The above discussion, based on the elementary Lindholm-Foley theory), where we considered only the phase shifts and neglected the collision-induced transitions, corresponds to interactions with $n = 4$. Note that for hydrogen and ionized helium, both the ionic and electronic contributions are important; these lines are subject to the first-order Stark effect.

In a preliminary reconnaissance, where we use the Lindholm-Foley theory, it is necessary to evaluate the constant C_{el}, which can be found from measured displacements and splittings of the spectral lines in an electric field; for an example, see Allen (1940).

When experimental data are not available, the necessary interaction constants can be computed sometimes by theory of the second-order Stark effect. If an atom is placed in a homogeneous electric field, E, an energy level W_i will be displaced by an amount

$$\Delta W_i = -E^2 \sum_j \frac{|(i|P_z|j)|^2}{W_i - W_j} \tag{7–43}$$

where P_z is the electric dipole moment and the summation extends over all states j which will combine with i under the selection rules. The summation must be taken out over all Zeeman states and appropriate averages taken. It turns out (see, e.g., Hunger, 1960) that if E is expressed in kv cm^{-1} and λ in microns, the shift of the level in wave number units will be

$$\Delta\tilde{\nu}(\text{cm}^{-1}) = -2.0 \times 10^{-6} E^2_{\text{kv}} \Sigma \pm f \lambda_\mu^2 \tag{7–44}$$

The plus sign and conventional f's are used for higher perturbing terms, the minus sign and emission f's for the lower terms. With $\Delta\tilde{\nu}$ for a given value of E we may compute C from

$$C\,\Delta\tilde{\nu} = \frac{C}{\varepsilon^2}\left(\frac{E_{\text{kv}}}{0.3}\right)^2 \tag{7–45}$$

or, numerically,

$$C = 6.21 \times 10^{-10} \frac{\Delta\tilde{\nu}}{E^2_{\text{kv}}} \tag{7–46}$$

For a transition $(nl) \rightarrow (n'l')$,

$$\Delta\tilde{\nu}_{\text{net}} = \Delta\tilde{\nu}(n, l) - \Delta\tilde{\nu}(n', l') \tag{7–47}$$

$$C = C(n, l) - C(n'\, l') \tag{7–48}$$

Hence C may be calculated for each transition of interest.

Example: Calculate $\Delta\nu$ for the $3s^4P$ term 83330 cm^{-1} in N I for a field of 100 kv cm^{-1}. We consider all perturbing terms of importance, i.e., terms to which transitions can take place neglecting the higher terms.

Perturbing Terms	χ (cm^{-1})	$f\lambda_u^2$
$3p^3\ {}^4S^0$	0	$-\ 0.447 \times 10^{-2}$
$3p\ {}^4D^0$	94830	$+39.35$
$3p\ {}^4S^0$	96752	$+\ 6.50$
$4p\ {}^4D^0$	106820	$+\ 0.27$
		$\Sigma(\pm f\lambda^2) = 45.67 \times 10^{-2}$

$$\Delta\tilde{\nu}\ (\text{cm}^{-1}) = -2.0 \times 10^{-14}\,(100)^2 \times 0.457 = 9.5 \times 10^{-3} \qquad (7\text{–}49)$$

If we use Eqs. 7–39 and 7–34, note that $1/A_2 = 1836$, and express N_ε in terms of P_ε and T we get

$$\log \Gamma_e = 19.38 + \tfrac{2}{3}\log C_{el} + \log P_\varepsilon - \tfrac{5}{6}\log T \qquad (7\text{–}50)$$

If contributions of both ions and electrons are to be included, the constant term depends on the mass of the radiating atom and whether the ions are protons or heavier particles. Note that, as remarked above, Eq. 7–50 gives a *lower limit* to the electronic damping constant, since the non-adiabatic effects are not taken into account.

Collisions between radiating atoms or ions and neutral hydrogen atoms are characterized by the van der Waals force law ($n = 6$). Then $\eta_0 = 0.61$, and by Eq. 7–24

$$\rho_0 = \left(\frac{3\pi^2}{4V}\frac{C}{\eta_0}\right)^{1/5} = \left(12.5\,\frac{C}{V}\right)^{1/5} \qquad (7\text{–}51)$$

where C, the *van der Waals constant of interaction*, can be found by experiment in some instances or calculated by quantum mechanics in others. With hydrogen as the principal perturbing atom, we may write C as

$$C = \frac{\varepsilon^2}{h}\,\alpha R_k^2 \qquad (7\text{–}52)$$

where $\alpha = 6.63 \times 10^{-25}$ cm^3 is the *polarizability of the hydrogen atom* (computed from quantum mechanics), and R_k^2 is the mean square radius appropriate to the upper level of the broadened absorption line. If R_k is measured in units of a_0, the radius of the first Bohr orbit,

$$C = 6.46 \times 10^{-34}\,\frac{R^2}{a_0^2} \qquad (7\text{–}53)$$

If quantum mechanical calculations of R_k^2 are not available, we may often use the approximate formula,

$$\frac{R_k^2}{a_0^2} = \frac{n^{*2}}{2Z^2}\,\{5n^{*2} + 1 - 3l(l+1)\} \qquad (7\text{–}54)$$

valid for the hydrogen-like levels of many light atoms. Here n^* is the effective quantum number, l is the azimuthal quantum number, $Z = 1$ for neutral atoms, 2 for singly ionized atoms, etc. In the approximation employed by Unsöld, one replaces the brackets by $5n^{*2}$ where $n^{*2} = 13.5Z^2(\chi_r - \chi_{r,s})^{-1}$ and obtains

$$C_s = 1.61 \times 10^{-33} \left(\frac{13.5\, Z_e}{\chi_r - \chi_{r,s}}\right)^2 \tag{7-55}$$

where Z is the effective nuclear charge, χ_r is the ionization potential of the atom, and $\chi_{r,s}$ is the excitation potential of the level s. The interaction constant for a transition $s \rightarrow s'$ is simply $C_s - C_{s'}$.

Interaction constants differ from one gas to another, for helium the coefficient in Eq. 7–55 is replaced by 0.50×10^{-33}. Weidemann and Unsöld proposed getting interaction constants from experiments involving noble gases; then using the above theory to estimate what the constants would have been for hydrogen.

The derivation of Eq. 7–52 requires that the energy level scheme of the perturbing atom has a big gap between the ground level and excited level near the ionization limit, and further that the energy difference between the level of the radiating atom and nearer combining levels are small compared with ionization energy of the perturbing atom. These conditions are fulfilled by hydrogen. We find from Eq. 7–40 that $\gamma_{\text{H}}/\gamma_{\text{He}} \sim 1.7$ and if we assume that $N(\text{H})/N(\text{He}) \sim 7$, $\overline{\gamma_{\text{H}} + \gamma_{\text{He}}} = 1.06\, \gamma_{\text{H}}$, and $\gamma_{(\text{H+He})} = 18.0\, C^{2/5}\, v_{\text{H}}{}^{3/5}\, N_{\text{H}}$, where v_{H} is the velocity with respect to H. Then

$$\log \gamma_6 = 19.61 + \tfrac{2}{5} \log C_{\text{H}} + \log P_g - \tfrac{7}{10} \log T \tag{7-56}$$

The validity of the Lindholm theory for van der Waals' interactions has been considered by a number of workers. Kusch (1958) measured the pressure broadening of iron lines by atomic and molecular hydrogen. Hindmarsh (1959) studied the broadening of λ4227 (calcium). He found the experimental value of $2\gamma/\beta$ to differ considerably from the theoretical value of 2.76 and concluded that Lindholm's theory was tolerable for line broadening but poor for line shifts. This discrepancy is to be expected because of the neglect of the inelastic collisions in the Lindholm-Foley type of theory, see Griem *et al.* (1962).

As an example, consider the broadening of the H and K lines of Ca II, for which we take $R_k{}^2 = 23a_0{}^2$, whence $C = 1.49 \times 10^{-32}$. From Eq. 7–34 we obtain $V = 1.11 \times 10^6$ cm sec^{-1} since $A_1 = 40$, $A_2 = 1$, and $T = 5700$°K. The target radius for collisional broadening in the Lorentz-Weisskopf sense is then, from Eq. 7–51, $\rho_0 = 4.40 \times 10^{-8}$ cm $= 4.40A$. At a typical point in the line-forming region in a G-dwarf star like the sun we may take $\log P_g = 4.86$, and $T = 5700$°K so that

$$S = \pi V \frac{P_g}{kT} \rho_0{}^2 = 6.2 \times 10^8$$

is the number of broadening collisions per second.

Another type of broadening which can become important in the laboratory when lines are broadened because of collisions of atoms with others of their own kind is self-broadening. The damping constant is given by (see Cayrel and Traving, 1960; Breene, 1957):

$$\gamma_n = \frac{4}{3}\frac{\varepsilon^2}{m\nu_0} f_{1n} N_0 \tag{7–57}$$

where f_{1n} is the f-value for the transition from level 1 to level n and N_0 is the number of atoms in the ground level.

Cayrel and Traving (1960) find that self-broadening is important for Hα in the solar spectrum and for singlet lines of helium in a white dwarf star that displays a pure He I spectrum.

In sunspots and certain peculiar A stars, macroscopic magnetic fields produce a Zeeman splitting of lines. A single line becomes replaced by a number of components whose separation depends on the field strength. Each component is broadened by pressure and Doppler effects and the individual patterns overlap. Since different groups of components are differently polarized, the broadening of a line depends not only on the magnitude of the field but on its direction as well.

7–6. Line Absorption Coefficient for Combined Doppler, Natural, and Collisional Broadening. Except for those of hydrogen and helium, spectral lines are widened by Doppler effects, natural broadening, and collisional broadening in accordance with the impact theory. Collisional broadening by neutral atoms and radiation damping can be handled by a suitable definition of the damping constant. Broadening by collisions with electrons usually requires a more sophisticated theory (Griem *et al.*, 1962).

An atom moving with velocity v toward the observer will emit a line centered on a frequency ν' given by

$$\nu' = \nu_0 + \frac{v}{c}\nu_0 \tag{7–58}$$

and the absorption coefficient for this atom will be

$$\alpha_\nu = \frac{\pi\varepsilon^2}{mc} f \frac{\Gamma}{4\pi^2} \frac{1}{\left(\nu_0 + \frac{v}{c}\nu_0 - \nu\right)^2 + \left(\frac{\Gamma}{4\pi}\right)^2} \tag{7–59}$$

To obtain the total absorption coefficient per atom in unit frequency interval at ν, we must multiply Eq. 7–59 by the fraction of atoms moving with a velocity v to $v + dv$ and then integrate over all velocities. The number moving in a velocity range v to $v + dv$ is given by Maxwell's equation for velocities (see Eq. 3–80):

$$dN = N\sqrt{\frac{M}{2\pi kT}}\, e^{-\frac{Mv^2}{2kT}}\, dv \tag{7–60}$$

where $M = AM_0$ is the mass of the atom in question. A is the atomic weight and M_0 the mass of an atom of unit atomic weight. To obtain the absorption coefficient α_ν at a frequency ν we now integrate over all velocities that may contribute to the absorption coefficient at this frequency. Thus from Eqs. 7–59 and 7–60

$$\alpha_\nu = \frac{\pi\varepsilon^2}{mc} f \frac{\Gamma}{4\pi^2} \int_{-\infty}^{+\infty} \frac{\left(\frac{M}{2\pi kT}\right)^{1/2} e^{-\frac{Mv^2}{2kT}}}{\left(\nu_0 + \frac{v}{c}\nu_0 - \nu\right)^2 + \left(\frac{\Gamma}{4\pi}\right)^2} dv \tag{7–61}$$

We transform this integral with the aid of the following substitutions. First let

$$\Delta\nu_0 = \frac{\nu_0}{c}\sqrt{\frac{2kT}{M}} \tag{7–62}$$

$$\Delta\nu = \frac{\nu_0}{c} v \tag{7–63}$$

so that

$$dv = \frac{c}{\nu_0} d(\Delta\nu) \tag{7–64}$$

Then let us define

$$y = \frac{\Delta\nu}{\Delta\nu_0} \tag{7–65}$$

$$u = \frac{\nu - \nu_0}{\Delta\nu_0} \tag{7–66}$$

$$\delta' = \frac{\Gamma}{4\pi} \tag{7–67}$$

$$a = \frac{\delta'}{\Delta\nu_0} \tag{7–68}$$

where Γ is called the effective damping constant. The absorption coefficient at the line center for zero damping (sometimes called the fictitious absorption coefficient) is:

$$\alpha_0 = \frac{\pi\varepsilon^2}{mc} f \frac{1}{\Delta\nu_0} \frac{1}{\sqrt{\pi}} \tag{7–69}$$

or

$$\alpha_0 = \frac{g_n}{g_{n'}} A_{nn'} \frac{\lambda^2}{8\pi^{3/2}} \frac{1}{\Delta\nu_0} \tag{7–70}$$

where $A_{nn'}$ is the probability Einstein coefficient for the transition, and g_n and $g_{n'}$ are the weights of the upper and lower levels, respectively.

$$\alpha_\nu = \alpha_0 \frac{a}{\pi} \int_{-\infty}^{+\infty} \frac{e^{-y^2}}{a^2 + (u - y)^2} dy \tag{7–71}$$

When both radiation damping and collisional broadening occur, we can write

$$\delta' = \delta_1 + \delta_2 = \frac{\Gamma_{\text{rad}}}{4\pi} + \frac{S}{2\pi} \tag{7–72}$$

where Γ_{rad} is the sum of the reciprocal lifetimes of levels n and n' for radiative processes, and S is the number of damping collisions $\sec^{-1}$. Then

$$\delta' = \frac{1}{4\pi}\,\Sigma\, A_{nn'} + \frac{1}{2}\,N_b\,\rho^2 V \tag{7–73}$$

To obtain the damping constant of any line, we must know the number of broadening particles cm^{-3}, N_b, and the damping collisional cross-section for the two levels of the transition in question.

At a fixed temperature and pressure, a is constant and the integration may be carried out over y. Helpful tables for the calculation of α_ν/α_0 as a function of v and a have been given by Hjerting, by Mitchell and Zemansky, and by D. Harris (1948) who writes α_ν/α_0 as

$$\alpha_\nu = \alpha_0\, H(a, u) \tag{7–74}$$

which may be expanded in the form

$$\frac{\alpha_\nu}{\alpha_0} = H_0(u) + aH_1(u) + a^2H_2(u) + a^3H_3(u) + \cdots \tag{7–75}$$

He tabulates H_0, H_1, etc. (see Table 7–1).

Example: Calculate the line absorption coefficient for λ3933 of Ca II, $4^2S_{1/2}$ — $4^2P_{3/2}$ at $T = 5700$°K. The A-value for this transition is taken as 1.59×10^8 sec $^{-1}$. Since transitions may occur from the $^2P_{3/2}$ level not only to $^2S_{1/2}$ but also to $3d^2D_{5/2}$ and to $3d^2D_{3/2}$,

$$\begin{aligned}\Gamma_{\text{rad}} &= A(4p^2P_{3/2} - 4s^2S_{1/2}) + A(4p^2P_{3/2} - 3d^2D_{5/2,\,3/2})\\ &= 1.59 \times 10^8 + 0.12 \times 10^8 + 0.01 \times 10^8 = 1.72 \times 10^8\end{aligned}$$

From Eq. 7–62 we calculate $\Delta\nu_0 = 0.392 \times 10^{10}$ $\sec^{-1}$ and since

$$\Delta\lambda_0 = \frac{c}{\nu_0^2}\,\Delta\nu_0 = \frac{\lambda}{c}\sqrt{\frac{2kT}{M}} \tag{7–76}$$

we get

$$\Delta\lambda_0 = \frac{3933}{3 \times 10^{10}}\sqrt{\frac{2 \times 1.38 \times 10^{-16} \times 5700}{40.07 \times 1.67 \times 10^{-24}}} = 0.0202A,$$

or since

$$\begin{aligned}u &= (\lambda - \lambda_0)/\Delta\lambda_0 \text{ (dimensionless)}\\ (\lambda - \lambda_0) &= 0.020\, u \text{ (angstroms)}\end{aligned} \tag{7–77}$$

Likewise

$$a = \frac{\Gamma}{4\pi\Delta\nu_0} = \frac{1.72 \times 10^8}{4\pi \times 0.392 \times 10^{10}} = 0.0035$$

TABLE 7–1*

THE FUNCTIONS H_0, H_1, H_2, H_3, AND H_4

u	$H_0(u)$	$H_1(u)$	$H_2(u)$	$H_3(u)$	$H_4(u)$
0.0	+1.000 000	−1.128 38	+1.000 0	−0.752	+0.50
0.1	+0.990 050	−1.105 96	+0.970 2	− .722	+ .48
0.2	+0.960 789	−1.040 48	+0.883 9	− .637	+ .40
0.3	+0.913 931	−0.937 03	+0.749 4	− .505	+ .30
0.4	+0.852 144	−0.803 46	+0.579 5	− .342	+ .17
0.5	+0.778 801	−0.649 45	+0.389 4	− .165	+ .03
0.6	+0.697 676	−0.485 52	+0.195 3	+ .007	− .09
0.7	+0.612 626	−0.321 92	+0.012 3	+ .159	− .20
0.8	+0.527 292	−0.167 72	−0.147 6	+ .280	− .27
0.9	+0.444 858	−0.030 12	−0.275 8	+ .362	− .30
1.0	+0.367 879	+0.085 94	−0.367 9	+ .405	− .31
1.1	+0.298 197	+0.177 89	−0.423 4	+ .411	− .28
1.2	+0.236 928	+0.245 37	−0.445 4	+ .386	− .24
1.3	+0.184 520	+0.289 81	−0.439 2	+ .339	− .18
1.4	+0.140 858	+0.313 94	−0.411 3	+ .280	− .12
1.5	+0.105 399	+0.321 30	−0.368 9	+ .215	− .07
1.6	+0.077 305	+0.315 73	−0.318 5	+ .153	− .02
1.7	+0.055 576	+0.300 94	−0.265 7	+ .097	+ .02
1.8	+0.039 164	+0.280 27	−0.214 6	+ .051	+ .04
1.9	+0.027 052	+0.256 48	−0.168 3	+ .015	+ .05
2.0	+0.018 3156	+0.231 726	−0.128 21	− .010 1	+ .058
2.1	+0.012 1552	+0.207 528	−0.095 05	− .026 5	+ .056
2.2	+0.007 9071	+0.184 882	−0.068 63	− .035 5	+ .051
2.3	+0.005 0418	+0.164 341	−0.048 30	− .039 1	+ .043
2.4	+0.003 1511	+0.146 128	−0.033 15	− .038 9	+ .035
2.5	+0.001 9305	+0.130 236	−0.022 20	− .036 3	+ .027
2.6	+0.001 1592	+0.116 515	−0.014 51	− .032 5	+ .020
2.7	+0.000 6823	+0.104 739	−0.009 27	− .028 2	+ .015
2.8	+0.000 3937	+0.094 653	−0.005 78	− .023 9	+ .010
2.9	+0.000 2226	+0.086 005	−0.003 52	− .020 1	+ .007
3.0	+0.000 1234	+0.078 565	−0.002 10	− .016 7	+ .005
3.1	+0.000 0671	+0.072 129	−0.001 22	− .013 8	+ .003
3.2	+0.000 0357	+0.066 526	−0.000 70	− .011 5	+ .002
3.3	+0.000 0186	+0.061 615	−0.000 39	− .009 6	+ .001
3.4	+0.000 0095	+0.057 281	−0.000 21	− .008 0	+ .001
3.5	+0.000 0048	+0.053 430	−0.000 11	− .006 8	.000
3.6	+0.000 0024	+0.049 988	−0.000 06	− .005 8	.000
3.7	+0.000 0011	+0.046 894	−0.000 03	− .005 0	.000
3.8	+0.000 0005	+0.044 098	−0.000 01	− .004 3	.000
3.9	+0.000 0002	+0.041 561	−0.000 01	− .003 7	.000
4.0	+0.000 0001	+0.039 250	0.000 00	−0.003 3	0.000

u	$H_1(u)$	$H_3(u)$	u	$H_1(u)$	$H_3(u)$
4.0	+0.039 250	−0.003 29	8.0	+0.009 0306	−0.000 15
4.2	+ .035 195	− .002 57	8.2	+ .008 5852	− .000 13
4.4	+ .031 762	− .002 05	8.4	+ .008 1722	− .000 12
4.6	+ .028 824	− .001 66	8.6	+ .007 7885	− .000 11
4.8	+ .026 288	− .001 37	8.8	+ .007 4314	− .000 10
5.0	+ .024 081	− .001 13	9.0	+ .007 0985	− .000 09
5.2	+ .022 146	− .000 95	9.2	+ .006 7875	− .000 08
5.4	+ .020 441	− .000 80	9.4	+ .006 4967	− .000 08
5.6	+ .018 929	− .000 68	9.6	+ .006 2243	− .000 07
5.8	+ .017 582	− .000 59	9.8	+ .005 9688	− .000 07
6.0	+ .016 375	− .000 51	10.0	+ .005 7287	− .000 06
6.2	+ .015 291	− .000 44	10.2	+ .005 5030	− .000 06
6.4	+ .014 312	− .000 38	10.4	+ .005 2903	− .000 05
6.6	+ .013 426	− .000 34	10.6	+ .005 0898	− .000 05
6.8	+ .012 620	− .000 30	10.8	+ .004 9006	− .000 04
7.0	+ .011 8860	− .000 26	11.0	+ .004 7217	− .000 04
7.2	+ .011 2145	− .000 23	11.2	+ .004 5526	− .000 04
7.4	+ .010 5990	− .000 21	11.4	+ .004 3924	− .000 03
7.6	+ .010 0332	− .000 19	11.6	+ .004 2405	− .000 03
7.8	+ .009 5119	− .000 17	11.8	+ .004 0964	− .000 03
8.0	+0.009 0306	−0.000 15	12.0	+0.003 9595	−0.000 03

From Eq. 7–70 with $g_n = 4$, $g_{n'} = 2$, $A_{nn'} = 1.59 \times 10^8$, we compute

$$\alpha_0 = \frac{4}{2} \times 1.59 \times 10^8 \frac{(3.933 \times 10^{-5})^2}{8 \times \pi^{3/2}} \times \frac{1}{0.392 \times 10^{10}} = 2.82 \times 10^{-12}$$

Computation of α_ν/α_0 from Table 7–1 is straightforward; e.g., for $u = 0\,4$,

$$\alpha_\nu/\alpha_0 = 0.85214 - 0.0035 \times 0.80346 + (0.0035)^2 \times 0.5795 = 0.84934$$

If both radiation and collision damping have to be taken into account, we must use Eq. 7–72 with an appropriate expression for S. For example, if $P_g = 7.2 \times 10^4$ dynes cm^{-2}, we find

$$\delta' = 0.137 \times 10^8 + 0.99 \times 10^8 = 1.127 \times 10^8$$

and

$$a = \frac{\delta'}{\Delta\nu_0} = \frac{1.127 \times 10^8}{0.392 \times 10^{10}} = 0.0287$$

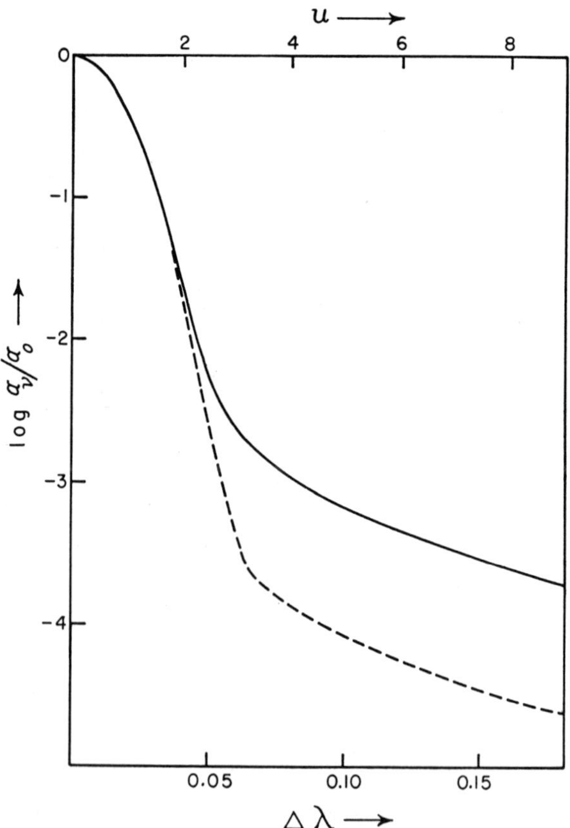

FIG. 7–7. THE LINE ABSORPTION COEFFICIENT FOR CA II λ3933

We plot log α_ν/α_0 against u and $\Delta\lambda$ as abscissa for $T = 5700°$K. The solid curve applies to combined radiation and collisional broadening at a gas pressure of 7.2×10^4 dynes; the dotted curve applies for radiation damping alone.

for the combined effects of natural damping and collisional broadening. The resultant values are given in Fig. 7–7. Notice how radiation damping plus collision damping gives a much larger absorption coefficient at great distances from the line center than does radiation damping alone. Near the line center, Doppler broadening alone determines the absorption coefficient. At distances much larger than $\Delta\lambda_0$, the Doppler contribution is small; there, collision and radiation damping fix α_ν/α_0.

To summarize: Doppler motions of the atoms determine the shape of the absorption coefficient near the line center, whereas radiation and collision damping fix the absorptivity at greater distances. In the sun and similar stars, collisions primarily determine Γ, so that an accurate knowledge of the collision damping parameter ρ and the gas pressure P_g is needed for a precise specification of α_ν throughout those layers that play a role in producing absorption lines. Profiles of weak lines in a stellar spectrum are determined entirely by absorption in the Doppler core. Absorption in line wings becomes important only for moderately strong lines so that, whereas an interpretation of weak lines requires a knowledge of their f-values, that of strong lines necessitates knowledge of radiative and collisional damping constants as well.

7–7. Broadening of Hydrogen Lines. It seems paradoxical that the simplest atom, hydrogen, should have one of the most complicated types of line broadening. The reason is that outside the narrow Doppler core the broadening arises from perturbations produced by ions which broaden according to the statistical theory and electrons which broaden according to the impact (discrete encounter) theory. Indeed, fully satisfactory theoretical treatments have been given only recently (Baranger, 1958; Kolb and Griem, 1958; Griem, Kolb, and Shen, 1959).

We may recapitulate the physical situation briefly as follows: Slowly moving ions in the neighborhood of a radiating hydrogen atom produce a net electric field E whose magnitude and direction changes slowly with time. This field resolves the degenerate level n into a number of individual states whose separation from one another depends on the field strength. Although an individual atom, in jumping from one particular state of the upper level to another particular state of the lower level, may emit a relatively sharp line, the net effect of many contributions by a radiating volume is to produce a blurred line.

In addition to the effects of slowly moving ions, one must consider the contribution to broadening by fast-moving electrons which act according to the discrete encounter theory. The effects produced by the electrons are complicated by the fact that they may cause an atom to jump from one degenerate state, β, to another, β'. Hence one must sum over the effects of all such radiationless collisions on the line broadening. Consequently the calculation of the electronic contribution to the line broadening is complicated.[5]

Let us first consider the contribution of the ions to line broadening effects: In 1913 Stark showed that when a field of the order of 10^5 v cm^{-1} is applied to incandescent hydrogen, the Balmer lines split into a number of components. Lo Surdo, soon thereafter, observed the splitting of lines emitted in the cathode dark space of the discharge tube. The change in energy of a given level is expressible by an equation of the form

$$\delta T = aE + bE^2 + cE^3 + \cdots \tag{7–78}$$

If δT is measured in cgs units and E in electrostatic units, the value of the first-order Stark coefficient will be

$$a = \frac{3h^2}{8\pi^2 m\varepsilon} n(n_2 - n_1) \tag{7–79}$$

where n is the total quantum number and n_1 and n_2 may assume values from 0 to $n - 1$. If E is measured in kv cm^{-1}, and λ in cm, the shift in wavelength for a particular Stark component will be

$$\delta\lambda = 0.0643\,\lambda^2 X_{\alpha a} E \tag{7–80}$$

where

$$X_{\alpha a} = \{n(n_2 - n_1)_\alpha - n'(n_2' - n_1')_a\} \tag{7–81}$$

where the upper state α, belonging to the quantum level n, is specified by three quantum numbers n_1 n_2, and m subject to the conditions noted above and $|m| = n - 1 - n_1 - n_2$. The lower state a belonging to quantum number n' is specified by n_1', n_2', and m'. Individual Stark components are polarized parallel to the field if $m = m'$, and perpendicular to the field if $m - m' = \pm 1$. Strengths of individual components may be obtained by quantum mechanics (see, e.g., Underhill, 1951). Thus for the kth component

$$S_k = \frac{f_k}{\Sigma f_k} \tag{7–82}$$

The number of components and absolute value of the Stark splitting increase towards higher members of the series.

These expressions were derived for atoms radiating in a uniform constant field E. Electric fields in stellar atmospheres, however, arise from rapidly moving charges and are neither uniform nor constant with time. What will interest us in a calculation of line absorption coefficients is the probability that perturbations from surrounding charges will displace a given

[5] Experimental checks obtained by Turner and his associates in Laporte's laboratory at The University of Michigan, by N. N. Sobolev at Moscow, and particularly by Wiese *et al.* at N.B.S. and Berg *et al.* at the University of Maryland (1962) showed that the Kolb, Griem, and Shen theory gave a more satisfactory representation of the measured hydrogen line profiles than did the theory based on ion broadening alone.

Stark component by an amount between $\delta\lambda$ and $\delta\lambda + d\lambda$. In hydrogen, we need to consider usually only the first order or linear Stark effect which produces a symmetrical splitting of the line. We shall be particularly interested in large shifts which determine the absorption in the wing. Within a sphere of radius R, let there be K charged particles. The probability of finding all K of these within a radius r is simply

$$\left[\frac{4\pi}{3} r^3 \Big/ \frac{4\pi}{3} R^3\right]^K$$

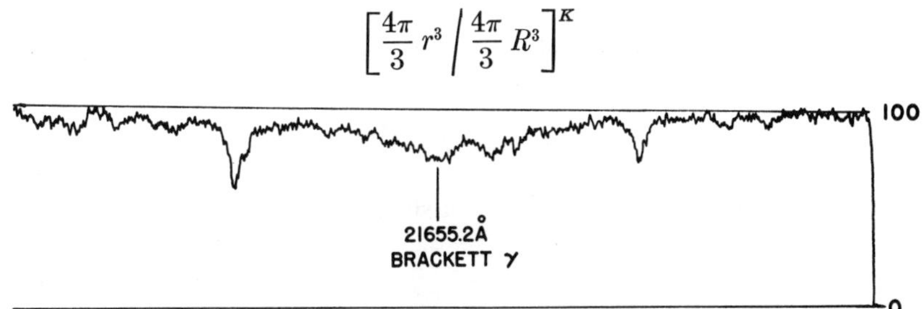

FIG. 7-8. BRACKETT γ IN THE SOLAR SPECTRUM

Notice the extremely broad wings and shallowness of the (4–7) hydrogen transition. Stark effect is pronounced and since the line is formed close to the mechanism of local thermodynamic equilibrium, it has a high central intensity. (McMath-Hulbert Observatory, University of Michigan.)

and the probability that no one of the K-charged particles lies within the sphere of radius r is $[1 - (r/R)^3]^K$. Then the probability that at least one of the K particles falls within the sphere of radius r is

$$p = 1 - \left[1 - \left(\frac{r}{R}\right)^3\right]^K \sim 1 - \exp\left(-\frac{4\pi}{3} Nr^3\right)$$

where we have compared the expansions $(1 - x)^n$ and e^{-xn} and have noticed that $\frac{4}{3}\pi R^3 N = K$, where N is the number of perturbing charges cm^{-3}. Hence the probability that at least one particle falls in the shell r to $r + dr$ is found by differentiation:

$$dp = e^{-y}\, dy \tag{7-83}$$

where

$$y = \left(\frac{r}{r_0}\right)^3 \text{ and } \frac{4\pi}{3} r_0{}^3 N = 1 \tag{7-84}$$

determine the mean separation r_0 of the two perturbing particles. The "average" field corresponding to the distance r_0 will be

$$E_0 = \frac{\varepsilon}{r_0{}^2} = 2.60\, \varepsilon N^{2/3} \tag{7-85}$$

The above treatment (due to Russell and Stewart) has been amplified by a more detailed discussion by Holtsmark which takes into account the simultaneous action of several particles, and shows that the factor 2.60

should be replaced by 2.61. Here N refers to the number of heavy charges (ions). Hence

$$E_0 = 46.8 \left(\frac{P_i}{T}\right)^{2/3} \tag{7-86}$$

where the ionic pressure P_i is expressed in dynes cm^{-2} and E is given in esu. Under most circumstances the ionic pressure P_i will equal the electron pressure P_ε.

A charged particle approaching a distance r from the atoms produces a displacement of the kth Stark component,

$$\Delta\nu = a_k E = \frac{C_k}{r^2} \tag{7-87}$$

In particular, a field E_0 will produce a shift

$$\Delta\nu_0 = \frac{C_k}{r_0^2} \tag{7-88}$$

where

$$\frac{\Delta\nu}{\Delta\nu_0} = \frac{\Delta\lambda}{\Delta\lambda_0} = \frac{r_0^2}{r^2} = \frac{E}{E_0} = \beta \tag{7-89}$$

Hence the statistical probability of a displacement $\Delta\lambda$ of a certain component is found from Eqs. 7–83 and 7–89 to be

$$W(\beta) = \tfrac{3}{2}\beta^{-5/2} \exp(-\beta^{-3/2}) \tag{7-90}$$

or, more precisely, in accordance with the exact Holtsmark theory,

$$W(\beta) = 1.496\, \beta^{-5/2}\, (1 + 5.106\, \beta^{-3/2} + 14.43\, \beta^{-3} + \cdots) \tag{7-91}$$

for large β. For other values of β we may use the exact theory (see Table 7–2 and Traving, 1960).

TABLE 7–2

THE HOLTSMARK FUNCTION $W(\beta)$*

β	$W(\beta)$	β	$W(\beta)$	β	$W(\beta)$
0	0.00	2.0	0.339	6.0	0.024
0.5	0.095	2.5	0.257	8.0	0.0105
1.0	0.271	3.0	0.166	10.0	0.0055
1.5	0.364	4.0	0.080	15.0	0.0019

* From Verweij, *Pub. Astron. Inst. Amsterdam* no. 5.

In a uniform field, E_0, every hydrogen line will be resolved into a number of symmetrically spaced components, each of which has a unique displacement $\Delta\lambda_0(n_1n_2n'_1n'_2)$. In a stellar atmosphere, each of these components will be spread out into a broadened distribution about the static field position, $\lambda_0 + \Delta\lambda_0(n_1n_2n'_1n'_2)$. Insofar as ionic effects are concerned, the final absorption coefficient will be proportional to the properly weighted sum of the individual curves. The necessary calculations for Balmer lines up to H XVIII have been carried out by Underhill and Waddell (1959). Our chief interest is in the line wings, which at least for the Balmer series, are produced by momentary fields corresponding to β values much greater than 1. When we sum over the components of a Balmer line, we may write the asymptotic form of the ionic contribution to the absorption coefficient per atom as (see Underhill, 1951; Unsöld, 1955):

$$\alpha_i = 3.54 \times 10^{-15}\, \lambda^5 f_D\, E_0^{3/2} \sum \frac{S_k\, X_k^{3/2}}{\Delta\lambda^{5/2}} \tag{7–92}$$

where f_D is the total strength of all *displaced* components; it is identical with the f-value of the line for Hβ, Hδ, . . . , lines which have no central Stark component. For lines such as Hα or Hγ one must add to $\alpha(\Delta\lambda)$ the contribution of the thermally broadened central component, viz.,

$$\alpha_0 = \frac{\pi\varepsilon^2}{mc} f_0 \frac{H(a, u)}{\sqrt{\pi}\Delta\nu_0} \tag{7–93}$$

where $H(a, u)$ is given by Eq. 7–74 and f_0 is the f-value of the undisplaced component.

One may readily calculate $\Sigma S_k X_k^2$ from the known strengths and displacements of the Stark components. We may write:

$$\alpha(\Delta\lambda) = c_n E_0^{3/2}(\Delta\lambda)^{-5/2} = 321\; c_n \frac{P_\varepsilon}{T} (\Delta\lambda)^{-5/2} \tag{7–94}$$

where c_n has been tabulated by various workers. We give here a compilation due to Traving.

At the very high temperatures of the O stars, Underhill finds that the shape of Hγ is determined largely by thermal Doppler effect, by radiation and collision damping of the dispersion type, and by turbulence if any exists. Stark effect is dominant, however, for main sequence stars cooler than 25,000°K.

We must now add the contribution of the electrons to the absorption coefficient. The final expression for the wings may be written in the form

$$\alpha(\Delta\lambda) = \frac{c_n\, E_0^{3/2}}{\Delta\lambda^{5/2}} [1 + R(N, T)\sqrt{\Delta\lambda}] \tag{7–95}$$

due to Griem, Kolb, and Shen (1959) who have tabulated $R(N, T)$ for Hα, Hβ, Hγ, and Hδ for ranges of temperature and density of laboratory and astrophysical interest. This correction factor $R(N, T)$ varies slowly with temperature and density. In Eqs. 7–94 and 7–95 and Tables 7–3 and 7–4, c_n and $R(N, T)$ are both tabulated for $\Delta\lambda$ in angstrom units. In the line wings, the electronic and ionic contributions to the absorption coefficient

TABLE 7–3

PARAMETERS FOR CALCULATION OF STARK EFFECT
(After G. Traving)

(E_0 is given in cgs units and $\Delta\lambda$ in angstroms)

n (upper level)	$\log c_n + 17$	n (upper level)	$\log c_n + 17$	n (upper level)	$\log c_n + 17$
3	1.495	9	0.327	15	0.239
4	0.944	10	0.301	16	0.234
5	0.646	11	0.282	17	0.230
6	0.504	12	0.267	18	0.225
7	0.417	13	0.256		
8	0.364	14	0.248		

TABLE 7–4

STARK BROADENING OF *H* LINES IN PLASMA

Ratio $R(N, T)$ ($A^{-\frac{1}{2}}$) of electron to ion contribution to the absorption coefficient for the wings of the Balmer lines*

$N(\text{cm}^{-3})\backslash T(°\text{K})$	$\frac{1}{2} \times 10^4$	10^4	2×10^4	4×10^4	$\frac{1}{2} \times 10^4$	10^4	2×10^4	4×10^4
		Hα				Hβ		
10^{10}	1.50	1.05	0.79	0.60	1.39	1.05	0.80	0.60
10^{11}	1.34	0.93	0.71	0.54	1.21	0.93	0.71	0.54
10^{12}	1.17	0.82	0.63	0.48	1.04	0.81	0.62	0.48
10^{13}	1.01	0.70	0.54	0.42	0.86	0.68	0.54	0.42
10^{14}	0.85	0.59	0.46	0.36	0.69	0.56	0.45	0.35
10^{15}	0.68	0.47	0.38	0.30	0.51	0.44	0.36	0.29
10^{16}	0.52	0.35	0.30	0.25	0.34	0.31	0.27	0.23
10^{17}	0.35	0.24	0.22	0.19	0.17	0.19	0.19	0.17
10^{18}	—	0.12	0.14	0.13	—	0.07	0.10	0.11
		Hγ				Hδ		
10^{10}	1.79	1.37	1.04	0.79	2.17	1.66	1.27	0.96
10^{11}	1.56	1.20	0.92	0.70	1.87	1.45	1.12	0.85
10^{12}	1.32	1.03	0.80	0.62	1.57	1.24	0.97	0.75
10^{13}	1.08	0.87	0.68	0.53	1.27	1.03	0.82	0.64
10^{14}	0.84	0.70	0.57	0.45	0.97	0.81	0.67	0.54
10^{15}	0.61	0.53	0.45	0.37	0.67	0.60	0.52	0.43
10^{16}	0.38	0.36	0.33	0.28	0.37	0.39	0.37	0.32
10^{17}	—	0.20	0.21	0.20	—	—	—	—
10^{18}	—	—	—	—	—	—	—	—

* Courtesy H. R. Griem, A. C. Kolb, and K. Y. Shen, *Physical Review* **116,** 4, 1959.

are comparable and both must be taken into account.[6] In the far wings, Eq. 7–95 must sometimes be modified where the usual impact approximation fails for electrons (Griem, 1962b). For most stars this region is of no importance.

[6] Exact calculations for the profile of Hβ have been made by Griem, Kolb and Shen (1962). This treatment takes into account the modifications of the Holtsmark distribution noted by Mozer and Baranger (1960). For astrophysical applications, Eq. 7–95 should suffice since we are nearly always concerned only with the wings.

The Kolb-Griem theory has been applied to the calculation of stellar line profiles by Elste, Jugaku, and Aller (1956), and subsequently by other workers (Osawa, 1956; van Regemorter, 1958; Cayrel and Traving 1960), all of whom have found that the newer theory gives hydrogen line profiles in better accord with observations.

In the sun, and particularly in subdwarfs where $P_g/P_\varepsilon > 10^4$ the influence of self-broadening must be considered in addition to effects produced by ions and electrons. Cayrel and Traving (1960) find the effect to be important in the sun for Hα but not for the higher Balmer lines, and write for the line mass absorption coefficient

$$l(\Delta\lambda) = n_{0,2}P_\varepsilon T^{-1}\zeta_1 \left\{1 + \left[R(N_\varepsilon T) + \frac{n_0}{n_\varepsilon}\zeta_2\right]\Delta\lambda^{1/2}\right\}\Delta\lambda^{-5/2} \qquad (7\text{–}96)$$

	Hα	Hβ	Hγ	Hδ
$\zeta_1 \times 10^{13}$	1.01	0.284	0.142	0.099
$\zeta_2 \times 10^4$	4.269	0.7716	0.3542	0.1917

Here $n_{0,2}$ denotes the number of second-level hydrogen atoms per gram of material, n_0 is the number of H atoms per gram, n_ε the number of electrons per gram, and the other symbols have their usual meaning.

For calculations of profiles of higher members of the Balmer series, one may use approximate formulae proposed by Griem (1960).

The confluence of the lines near the limit of the Balmer series shows the influence of the Stark effect. The hydrogen lines in a stellar spectrum gradually coalesce toward the higher series numbers, and for some maximum value of n, they will cease to be distinguishable from one another. The interionic Stark effect and electron broadening widen the lines until they overlap and cannot be resolved with any spectrograph. Between the last resolvable line, whose quantum number is n_m, and the series limit the spectrum looks continuous. If both ionic and electronic contributions are taken into account, the electron density N is related to the quantum number of the last resolvable line by the Inglis-Teller (1939) relation,

$$\log 2N = 23.26 - 7.5 \log n_m \qquad (7\text{–}97)$$

as modified (by a factor 2) by Griem (1962).

Example: In the spectrum of the supergiant χ Orionis, $n_m \simeq 23$. Hence $\log N_\varepsilon = 12.76$. If $T = 20{,}000°$K, $P_\varepsilon = 16$ dynes cm^{-2}, much smaller than for a main-sequence star of the same spectral class.

The broadening of ionized helium lines follows a theory essentially similar to that of hydrogen (Griem and Shen, 1961). Stark patterns are more complicated because the transitions observed correspond to Paschen and Brackett lines. See Fig. 7–8 for Brackett γ.

7–8. Broadening of Helium Lines. Whereas the hydrogen lines in stellar spectra show symmetrical profiles that can be explained in a straight-

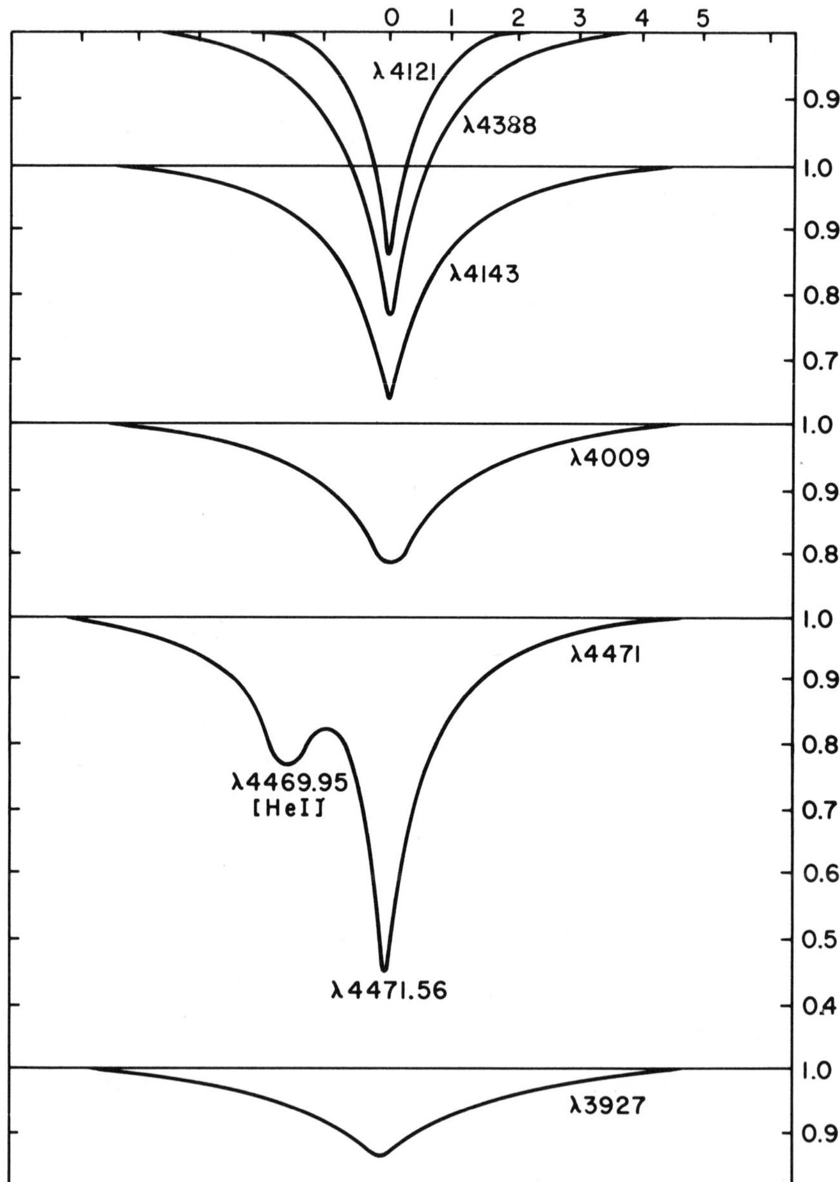

FIG. 7–9. HELIUM LINE PROFILES IN THE *B* STAR *HD* 36959

Notice the asymmetry of λ4471 and the appearance of the forbidden component λ4470.

forward way, those of helium exhibit a variety of shapes. Fig. 7–9 reproduces profiles measured in the spectrum of HD 36,959 (*B*2V). Notice that whereas lines such as λ3820, λ4121, λ4143, and λ4388 show relatively

TABLE 7–5

CALCULATED LINE BROADENING PARAMETERS FOR SOME HELIUM LINES OF ASTROPHYSICAL INTEREST*

		10,000°K	20,000°K	40,000°K
$2^1S - 4^1P$	w	102	92.5	80.9
3965A	d/w	−0.54	−0.45	−0.37
$C = +39.4$	α	0.89	0.96	1.06
$C' = +37.3$	$\log \sigma$	2.05	1.86	1.65
$2^3P - 5^3S$	w	98.4	108	111
4121A	d/w	+1.14	+0.91	+0.73
$C = -15.7$	α	0.49	0.45	0.45
	$\log \sigma$	2.00	1.89	1.75
$2^1P - 5^1S$	w	168	176	172
4438A	d/w	+0.99	+0.79	+0.64
$C = -28.1$	α	0.56	0.54	0.55
	$\log \sigma$	2.17	2.04	1.88
$2^3P - 4^3S$	w	43.0	49.0	51.9
4713A	d/w	+1.25	+1.02	+0.82
$C = -3.14$	α	0.33	0.30	0.28
$C' = -2.9$	$\log \sigma$	1.52	1.43	1.30
$2^1S - 3^1P$	w	37.9	35.2	31.9
5016A	d/w	−0.58	−0.49	−0.40
$C = +4.14$	α	0.49	0.52	0.56
$C' = +4.3$	$\log \sigma$	1.41	1.23	1.04
$2^1P - 4^1S$	w	75.5	81.7	82.3
5048A	d/w	+1.09	+0.88	+0.71
$C = -5.79$	α	0.38	0.36	0.36
$C' = -5.2$	$\log \sigma$	1.71	1.59	1.44
$2^3P - 3^3D$	w	17.6	18.1	18.0
5876A	d/w	−0.33	−0.13	−0.02
$C = +0.39$	α	0.19	0.18	0.18
$C' = +0.67$	$\log \sigma$	0.94	0.81	0.65
$2^1P - 3^1D$	w	40.5	36.6	32.9
6678A	d/w	+0.60	+0.56	+0.51
$C = -2.43$	α	0.48	0.52	0.56
$C' = -2.6$	$\log \sigma$	1.19	1.00	0.80

*All quantities are calculated for $N_\varepsilon = 10^{18}\text{cm}^{-3}$. For $N < 10^{18}\text{cm}^{-3}$, widths of lines without forbidden components are found by multiplying tabulated values by $N/10^{18}$, the asymmetry parameter α is found by multiplying by $\sqrt[4]{N/10^{18}}$, and the parameter σ by multiplying by $(N/10^{18})^{2/3}$. If hydrogen is the broadening agent, σ must be multiplied by $\sqrt{M(\text{H})/M(\text{He})} = \frac{1}{2}$.

The electron impact half-width w is given in angstroms; the static Stark coefficients C (computed) and C' (measured) are in units of cm^{-1} per $(100\ \text{kv/cm})^2$. The ratio of shift to half-width is d/w. The absorption coefficient profiles are evaluated with the aid of Table 7–6.

Courtesy Griem, Baranger, Kolb, and Oertel, *Phys. Rev.*, **125**, 177, 1962.

deep symmetrical profiles, those of λ4009 and λ3927 are broad and shallow, while λ4026 and especially λ4471 are asymmetrical. Helium exhibits both linear and quadratic Stark effects and even displays lines that are normally forbidden.

The near coincidence of many energy levels in helium makes the theory extremely complicated. If an electric field is applied to an atom whose energy levels are well separated from one another, its spectral lines will be broadened by the quadratic Stark effect. Hydrogen (where levels of the same quantum number n are all coincident) shows a linear Stark effect. Consider now two levels which are separated by an energy difference ΔE. If the Stark splitting is less than ΔE, lines connecting this level with others of a similar nature tend to follow the quadratic pattern, whereas if Stark splitting greatly exceeds ΔE, line broadening will follow the linear effect.

With fields of the order of 15–25 kv cm^{-1}, certain lines such as λ4713.3, λ4437.5, and λ3613.6 show the quadratic Stark effect, but the problem is complicated by the inelastic collision effects. Collisional broadening by electrons produces a symmetrical profile, whereas broadening by encounters with ions is responsible for asymmetry.

Fortunately, a detailed, accurate theory of helium line broadening is now available (Griem *et al.*, 1962) and calculations have been carried out for several lines of astrophysical interest. The behavior of strong lines in the diffuse subordinate series, e.g., 2^3P − 3^3D λ5876 is more complicated than that of λ4437 or λ4713. The generalized impact theory for isolated lines also leads to a dispersion line shape but the damping constant is larger than that given by the Foley-Lindholm theory. There is a pronounced asymmetry due to ionic effects.

Table 7–5 gives theoretical line broadening parameters for a number of frequently observed helium lines. Let w denote the electron impact half-width; v the relative velocity of the radiating and perturbing particles; and ρ_m denote the "mean distance" between electrons given by

$$\frac{4\pi}{3}\,\rho_m{}^3\,N = 1 \tag{7–98}$$

Griem *et al.* define the parameter σ by

$$\sigma = w\,\rho_m/v \tag{7–99}$$

and the asymmetry parameter by

$$\alpha = (2\pi\,C_4/u\,\rho_m{}^4)^{3/4} \tag{7–100}$$

Where C_4 is the quadratic Stark coefficient, the ratio of shift to impact parameter is d/w.

Table 7–6 gives the profile shape $j(x,\alpha,\sigma)$ in reduced units. For any given value of N and T we can calculate w, α, and σ (see footnote of Table 7–5) and convert from x to A. Hence we can derive, for any helium line for which the necessary parameters are tabulated, the line absorption coefficient $\alpha(\lambda-\lambda_0)$.

TABLE 7-6

REDUCED LINE PROFILES $j(x, \alpha, \sigma)$ AND $j_H(x, \alpha)$ FOR VARIOUS VALUES OF THE ION BROADENING PARAMETERS α AND σ*

α	σ \ x	−2.0	−1.5	−1.0	−0.7	−0.5	−0.3	−0.2	−0.1	0.0	0.1	0.2	0.3	0.5	0.7	1.0	1.5	2.0	2.5	3.0	4.0	5.0
0.1	1.0	0.052	0.073	0.111	0.148	0.173	0.200	0.214	0.226	0.239	0.251	0.259	0.265	0.261	0.242	0.204	0.142	0.096				
	∞	0.049	0.072	0.114	0.153	0.186	0.221	0.238	0.254	0.267	0.277	0.282	0.283	0.270	0.244	0.199	0.132	0.089				
	j_H	0.050	0.073	0.115	0.154	0.186	0.218	0.235	0.247	0.260	0.270	0.274	0.274	0.263	0.240	0.198	0.134	0.092				
0.2	0.6	0.044	0.058	0.083	0.103	0.121	0.141	0.150	0.161	0.172	0.182	0.194	0.204	0.219	0.223	0.211	0.171	0.127				
	1.0	0.039	0.056	0.084	0.110	0.130	0.153	0.166	0.180	0.193	0.205	0.217	0.227	0.238	0.231	0.215	0.167	0.121				
	∞	0.038	0.054	0.083	0.110	0.134	0.162	0.177	0.195	0.207	0.222	0.234	0.243	0.253	0.248	0.221	0.163	0.115				
	j_H	0.039	0.056	0.088	0.116	0.143	0.169	0.183	0.196	0.208	0.220	0.228	0.234	0.238	0.228	0.206	0.156	0.114				
0.3	0.3	0.041	0.055	0.075	0.091	0.105	0.122	0.130	0.139	0.147	0.155	0.162	0.171	0.186	0.199	0.212	0.206	0.166				
	0.6	0.036	0.049	0.071	0.089	0.104	0.121	0.130	0.139	0.149	0.158	0.167	0.175	0.192	0.207	0.219	0.198	0.158				
	1.0	0.032	0.045	0.067	0.087	0.103	0.123	0.133	0.144	0.155	0.163	0.172	0.183	0.201	0.214	0.218	0.190	0.145				
	∞	0.030	0.042	0.062	0.082	0.098	0.118	0.130	0.142	0.154	0.170	0.183	0.195	0.214	0.225	0.222	0.184	0.139				
	j_H	0.032	0.045	0.069	0.091	0.109	0.131	0.143	0.155	0.166	0.178	0.188	0.195	0.204	0.206	0.197	0.164	0.128	0.100	0.078	0.047	0.030
0.4	j_H	0.026	0.038	0.056	0.073	0.086	0.104	0.113	0.123	0.131	0.141	0.150	0.157	0.170	0.175	0.176	0.160	0.134	0.110	0.090	0.058	0.036
0.5	j_H	0.022	0.032	0.047	0.058	0.072	0.090	0.094	0.102	0.110	0.117	0.127	0.133	0.144	0.154	0.156	0.148	0.130	0.113	0.096	0.065	0.040

* To obtain line profiles on a wavelength scale, multiply each x-value by the electron impact width (taken from Table 7–5 and reduced to density considered). To retain the normalization the "intensity" (given in the body of the table) is multiplied by $1/w$. The zero point of the wavelength scale is then shifted by the electron-impact shift d from the "zero" position of the line.

If $\alpha > 0.3$ it suffices to use $j_H(x, \alpha)$. For smaller values of α, interpolate x for the proper values of α and σ from the table.

Courtesy Griem, Baranger, Kolb, and Oertel, *Phys. Rev.*, **125,** 189, 1962.

Neutral helium lines are observed with great strength in *B* stars. Now that a reliable quantitative theory of line broadening is available (Griem *et al.*, 1962), we can use the helium line profiles as an independent check on the structure of the atmosphere and to obtain the H/He ratio (see Jugaku, 1958).

The higher members of the diffuse subordinate series $2^3P - 4^3D$, $2^3P - 5^3D$, etc., show pronounced broadenings and asymmetries. Fluctuating interatomic electric fields 10^3–10^4 volts cm^{-1} not only produce a displacement of the line to the red; they also result in the appearance of lines forbidden by ordinary selection rules, e.g., $2p - 4f$, $2p - 5f$, etc., which fall to the violet of the normal diffuse series lines.

Many years ago Struve noticed that λ4471 ($2^3P - 4^3D$) was flanked by the forbidden line, λ4469.9 ($2^3P - 4^3F$), in the spectra of certain *B* stars. This line appears in laboratory discharge tubes when the electron density is appreciable. Simultaneously, the $p - d$ lines are broadened while others are unaffected. Since stellar helium lines showed the same behavior, Struve concluded there was good evidence for an interatomic Stark effect arising from charged particles. Subsequent work by Goldberg, Foster, and Douglas; Unsöld, M. K. Krogdahl, and Kolb; and Griem has confirmed this idea. (See Traving, 1960).

Laboratory experiments by J. S. Foster showed that with increasing electric field strength, the permitted λ4471 ($2^3P - 4^3D$) line shifts to the red, while λ4470 is displaced to the violet. The intensity of the latter grows with increasing field strength at the expense of λ4471, until at fields of 100 kev cm^{-1} it approaches equality with λ4471. For fields less than 15 kev cm^{-1}, the Stark effect is quadratic, whereas for fields greater than 15 kev cm^{-1} the effect is linear. Thus, λ4471 displays "discrete" collisional broadening (quadratic Stark effect) plus Doppler broadening near the line core, and statistical Holtsmark broadening in the wings where the linear Stark effect prevails. The wavelength of the stellar forbidden helium line λ4469.92 corresponds to a vanishingly small field strength when the line is produced in a static laboratory field. To match the observed stellar intensity, an appreciable static field would be required and this field would shift the line appreciably in position. In other words, the stellar line appears at the zero static-field position but with an intensity that corresponds to an appreciable static-field strength. The answer seems to be that in an ionized gas where the time-dependent electrostatic field comes from the ions and free electrons, its effect on the energy levels differs from that of a static laboratory field. In contradiction to what is observed in the laboratory, the space between λ4470 and λ4471 tends to be filled because of the broadening effects of collisions, as Struve pointed out in 1938.

The line broadening theories of Kolb, Griem, Baranger, and their associates may be applied to other atoms as well as helium; it would be of great astrophysical interest to study silicon in various stages of ionization, and perhaps also carbon, nitrogen, and oxygen.

In summary, intensive laboratory studies and theoretical work are needed to supply the basic parameters, f-values, and collision broadening parameters for transitions of astrophysical interest. Until this information is obtained, vast quantities of spectroscopic data cannot be interpreted.

PROBLEMS

7–1. Use the J-file sum rule to verify the strengths of the following transitions:

$$S(s - p) = 6 \quad S(p - d) = 60 \quad S(d - f) = 210$$
$$S(f - g) = 504 \quad S(g - h) = 990$$

7–2. Verify Eq. 7–92.

7–3. Show that in the far wings of a line, $u \geq 8$,

$$\alpha_\nu = \frac{\pi\varepsilon^2}{mc} f_{n'n} \frac{1}{4\pi^2} \frac{\Gamma}{(\nu - \nu_0)^2} = \frac{2\pi\varepsilon^4}{3m^2c^4} \frac{\nu_0^2}{(\nu - \nu_0)^2} f_{n'n} \frac{\Gamma}{\gamma_c} \tag{7–101}$$

and

$$\alpha_\lambda = 16.5 \times 10^{-26} f_{n'n} \frac{\Gamma}{\gamma_c} \frac{\lambda_0^2}{(\lambda - \lambda_0)^2}$$

7–4. For the sodium "D" lines, $\lambda5889$ and $\lambda5896$, $\rho = 4.6A$, ($T = 5700$°K). With $A = 6.8 \times 10^7$, calculate α_0, $\Delta\lambda_0$, δ', and a for $\log P_g = 5.00$. Then compute the absorption coefficient for values of u from 0 to 9 to obtain a curve similar to that illustrated in Fig. 7–7.

7–5. Show that the absorption coefficient in the wing of a Balmer line in a high temperature star is:

$$\begin{aligned}\log l_\lambda = \log n_{0,2}\,\alpha_\lambda &= 3.59 + \log c_n + 3.38\,\theta - 3.5 \log T + 2 \log P_\varepsilon \\ &\quad -2.5 \log(\Delta\lambda) + \log[1 + R\sqrt{\Delta\lambda}] + \log n_1\end{aligned} \tag{7–102}$$

where $\Delta\lambda$ is given in angstroms and n_1 is the number of ionized hydrogen atoms per gram of stellar material.

REFERENCES

General References

UNSÖLD, A. 1955. *Physik der Sternatmosphären.* Springer-Verlag, Berlin, chaps. 12, 13, 14.

Transition Probabilities (Theory)

An extensive quantum mechanical treatment of radiation formulas, line strengths, and related topics will be found in:

BETHE, H., and SALPETER, E. 1959. *Handbuch der Physik,* S. Flügge (ed.). Springer-Verlag, Berlin, vol. 35.

CONDON, E. U., and SHORTLEY, G. 1935. *Theory of Atomic Spectra.* Cambridge University Press, London.

RACAH, G. 1942. *Phys. Rev.* **61,** 186; **62,** 438.

———. 1943. **63,** 367.

———. 1949. **76,** 1352.

Tables and formulas for actual calculation of line strengths are given by:

Goldberg, L. 1935. *Ap. J.* **82,** 1.
———. 1936. **84,** 12.
Rohrlich, F. 1959. *Ap. J.* **129,** 441, 449.
Bates, D. R., and Damgaard, A. 1949. *Phil. Trans. Roy. Soc. A,* **242,** 101.

Transition Probabilities (Experiment)

For descriptions of various techniques see, e.g.:

Arcs

Allen, C. W., and Asaad, A. S. 1957. *Mon. Not. R. A. S.* **117,** 36.
Corliss, C. H., and Bozman, W. R. 1962. "Experimental Transition Probabilities for Spectral Lines of 70 Elements," National Bureau of Standards Monograph.

Water-Stabilized Arc

Lochte-Holtgreven, W. 1957. *Colloque de Spectroscope d'Amsterdam, 1956.* Pergamon Press, Ltd., Oxford, England, p. 111.

Electric Furnace

King, R. B., and Stockbarger, D. 1940. *Ap. J.* **91,** 488.
King, R. B., and King, A. S. 1938. *Ap. J.* **87,** 24.

Lifetime Measurements

Mitchell, A., and Zemansky, M. 1934. *Resonance Radiation and Excited Atoms.* Cambridge University Press, London, pp. 145–47.
Heron, S., McWhirter, R. W. P., and Roderick, R. 1956. *Proc. Roy. Soc. A,* **234,** 565.
Brannen, E., Hunt, F. R., Adlington, R. H., and Nicholls, R. W. 1955. *Nature* **175,** 810.
Ziok, K. 1957. *Zeit f. Physik* **147,** 99.

Flames

Huldt, L., and Lagerquist, A. 1952. *Ark. Phys.* **5,** 1, 2.

Shock Tube

Aller, L. H. 1954. *Sky and Telescope* **14,** 59.
Kolb, A. C., and Griem, H. 1962. In *Atomic and Molecular Processes,* D. R. Bates (ed.). Academic Press, New York.

Atomic Beam

Bell, G. D., Davis, M. H., King, R. B., and Routly, P. M. 1955. *Ap. J.* **127,** 775.

Bibliographies of f-Values

A most useful summary of the data available to 1962 is given by:

Corliss, C. National Bureau of Standard Monograph. In press.

See also compilations and bibliographies by:

Minnaert, M. 1960. *Trans. Int. Astr. Union,* **10,** 220.
———. 1962. **11A,** 120.
Goldberg, L., Müller, E., and Aller, L. H. 1960. *Ap. J. Suppl. No. 45.*
Allen, C. W. 1960. *M. N.* **121,** 299.
van Regemorter, H. 1959. *Journ. Phys. Radium* **20,** 907.

Line Broadening (Review Articles)

Breene, R. G. 1957. *Rev. Mod. Phys.* **29,** 94.
Chen, S., and Takeo, M. 1957. *Rev. Mod. Phys.* **29,** 20.
Margenau, H. 1959. *Rev. Mod. Phys.* **31,** 569.
Traving, G. 1960. *Mitt. Ast. Geselt. Sonderheft No. 1.* G. Braun, Karlsruhe, Germany.

For the general impact theory see:

Baranger, M. 1962. In *Atomic and Molecular Processes,* D. R. Bates (ed.). Academic Press, New York.

The theory of broadening of helium lines is given by:

Griem, H. R., Baranger, M., Kolb, A. C., and Oertel, G. 1962. *Phys. Rev.* **125,** 177.

CHAPTER 8

The Fraunhofer Spectrum

8–1. Introduction. The fundamental problem of stellar atmospheres is the following: given the intensity distribution in its continuous spectrum and the displacements (Doppler shifts) and profiles or total intensities of its absorption lines, to determine the chemical composition, dynamical and physical state of the atmosphere and the surface gravity of the star. In Chapter 5 we discussed the interpretation of stellar continuous spectra. Line displacements due to large scale-motions in an atmosphere can be observed in detail only in the sun; for other stars it is possible usually to get only the integrated effects over an entire stellar disk.

In this chapter we shall be concerned with the interpretation of spectral line intensities. First of all we need to define what we mean by the "intensity" of a line. Rowland made eye estimates on an arbitrary scale of what he called the intensities of the dark lines in the solar spectrum and these estimates, when properly calibrated, are useful for some problems. Unfortunately, no single parameter can completely describe an absorption line. A Fraunhofer line may be represented by an intensity curve, with a certain depth, width, and shape. Since every one of these features arises from a definite physical cause, an adequate theory should be able to account for the complete intensity distribution or *profile* of the line. It is true that spectral lines have certain common characteristics as regards shape. Usually they are symmetrical (notable exceptions are certain helium lines in hot dwarf stars) and possess a single minimum. But in other features they may differ markedly. The strong lines show broad "wings" — extensions on either side of the minimum. Weak lines do not have such wings, and if they are photographed with an instrument of sufficient resolving power, they show a bell-shaped profile. Also some lines, e.g., $\lambda 4227$ of Ca I, are nearly black at the center; others have appreciable central intensities. Thus it is not possible to write an empirical equation containing a single adjustable parameter capable of representing all lines in the solar spectrum, for example. These differences in shape arise from differences in numbers of atoms acting to produce the lines, from differences in temperature, density and large scale mass motions in those strata where lines are formed, and from the fact that processes important for strong lines, e.g., collisional broadening are not significant for weaker ones.

Observational determinations of line profiles are difficult for stars because the resolving power of the spectrograph is often comparable with the intrinsic line widths. Hence it is usually possible to observe profiles of only strong stellar lines. With modern instrumentation such as the vacuum solar spectrograph at the McMath-Hulbert Observatory of the University of Michigan it is practical to secure observations with a resolving power of about 700,000. The instrumental blurring of about 0.01A is substantially smaller than the Doppler width of the line in the solar atmosphere, so that shapes of even very weak lines can be observed across the disk of the sun. Scattered light amounts to less than 2 per cent in this instrument. Other powerful solar spectrographs are under construction or in operation at Oxford, in France, at the Crimean Astrophysical Observatory, and at the American National Optical Observatory.

Future solar spectroscopic studies are likely to emphasize line profiles whereas stellar spectroscopic studies will continue (largely) to be restricted to "equivalent widths." The equivalent width of a line is defined as the total amount of energy subtracted from the background continuum expressed in angstroms of this continuum. By the "intensity" of an absorption line we shall mean, unless otherwise indicated, its equivalent width.

Line profiles and equivalent widths may be measured both photographically and photoelectrically. Consider solar observations first. With a photographic plate one can observe simultaneously the intensity and displacement of a given spectrum line over a distance of several tens of thousands of kilometers on the sun's disk. A photoelectric scan of the spectrum yields a greater accuracy but one integrates along a strip of the sun several thousand kilometers long.

Furthermore, each point in the scan corresponds to a different time. Hence, if one is interested in behaviors of spectral lines averaged over time and distances large compared with small scale solar phenomena (such as granules, spicules, etc.) it is advantageous to scan the spectrum with a photoelectric detector. If one wishes to study time and space variations of solar lines produced by mass motions of material (see Figs. 8–1, 8–2), as far as image tremors due to "seeing" will permit one should use photographic recording. In stellar spectroscopy, the source is nearly always faint and one wishes to record large ranges of the spectrum simultaneously. Hence one uses usually a coudé grating spectrograph. A photoelectric spectrum scanner can observe only one line at a time but it yields results of much higher accuracy than does a photographic plate. Hence measurements made with it can be used as a check on the photographic photometry. For studies of limited wavelength regions a Fabry-Perot etalon is probably superior to a grating. Lack of space precludes lengthier discussions of these experimental and observational techniques.

Our problem will be to describe what can be learned from equivalent widths and profiles of absorption lines. We present first of all the "classi-

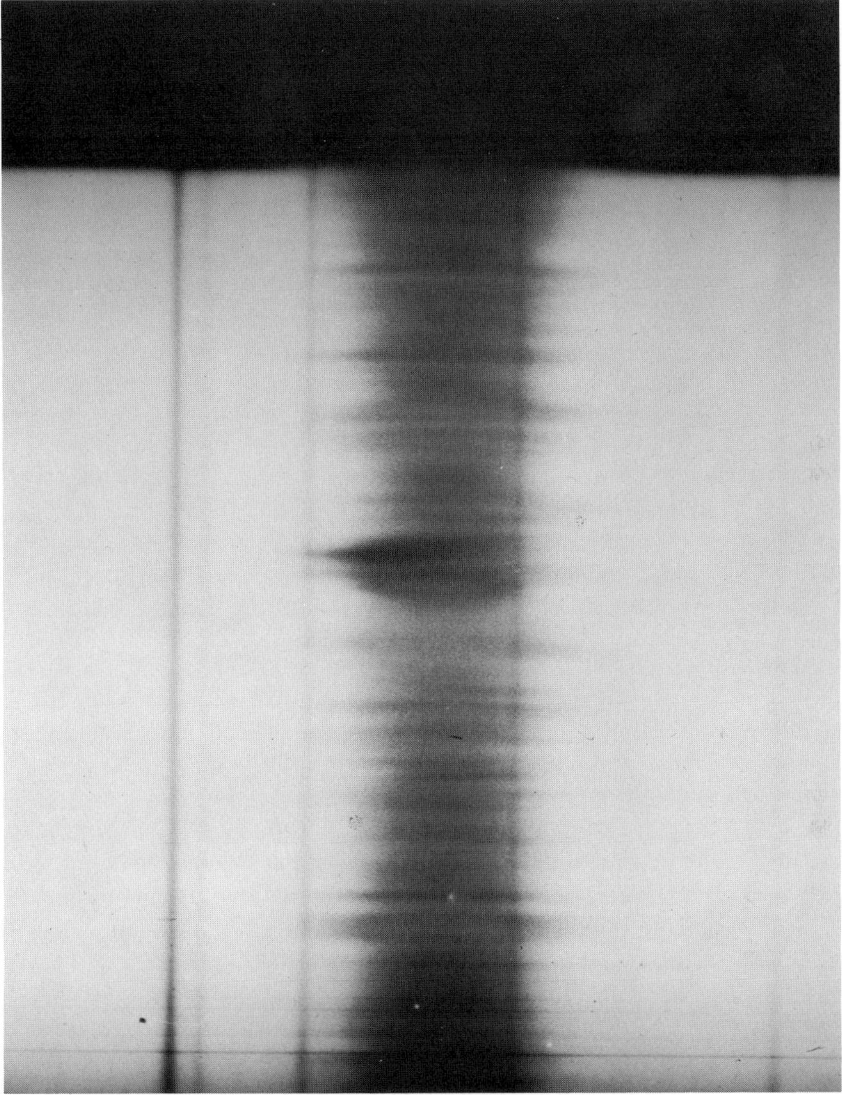

FIG. 8–1. CORE OF Hα IN THE SOLAR SPECTRUM

The jagged appearance of the line arises from differential vertical motions of the radiating gases. Photographed with the vacuum spectrograph of the McMath-Hulbert Observatory, August 3, 1955. (Courtesy, The McMath-Hulbert Observatory of the University of Michigan.)

cal" theory of spectral line formation as presented by K. Schwarzschild, Eddington, and Milne, then certain of the refinements introduced by Chandrasekhar and by Strömgren. Next we describe how spectral line shapes and equivalent widths may be calculated for the sun and stars,

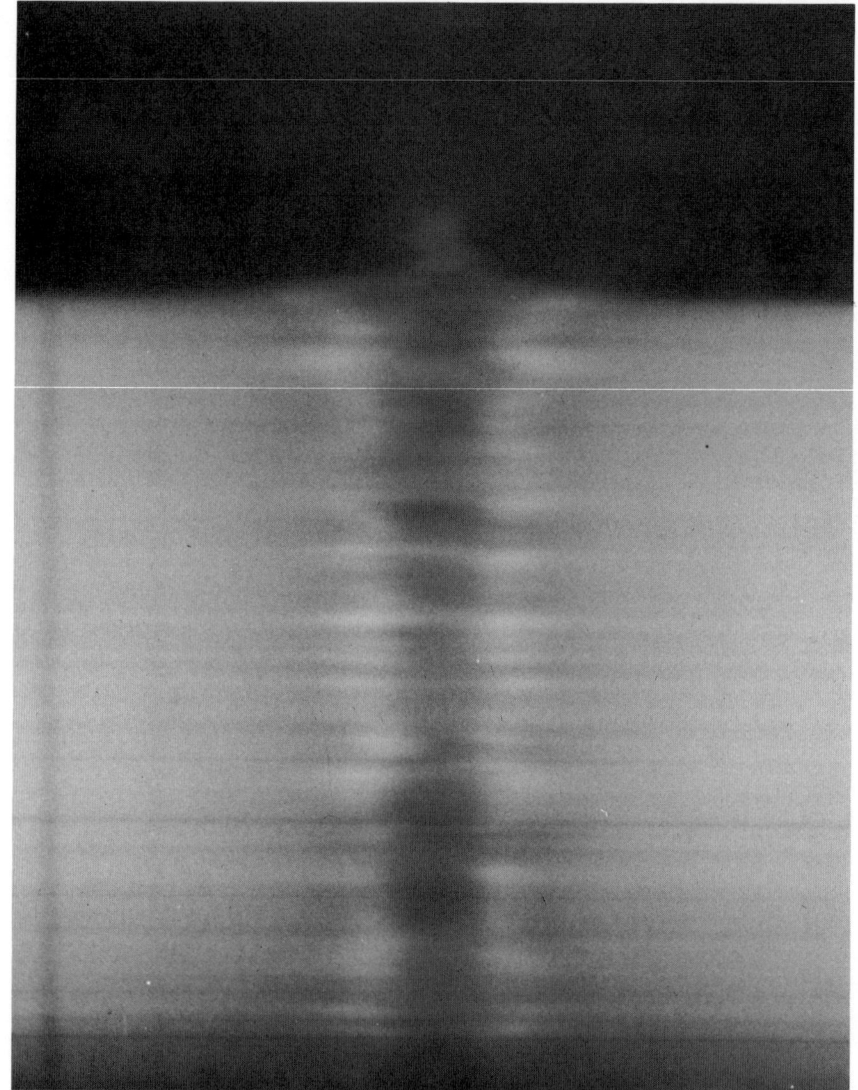

Fig. 8–2. Core of the K Line in the Solar Spectrum

Notice the bright K_2 emission and the darker central K_3 core. Compare Fig. 9–14. A prominence is seen at the edge of the disk. Photographed with the vacuum spectrograph at The McMath-Hulbert Observatory, August 2, 1955. (Courtesy, The McMath-Hulbert Observatory of the University of Michigan.)

and give, in some detail, a simple but useful technique for analyzing stellar spectra — the curve of growth. Other applications to stellar astronomy (luminosity effects, stellar rotations, magnetic fields and compositions) are briefly described. Finally, we re-examine the basic assumptions underlying the physical theory of absorption line formation. We defer a discussion of the fine structure (Figs. 8–1, 8–2), produced by vertical motions, to Chapter 9.

We must first of all discuss two basic questions: in what layers are absorption lines formed and what are the physical mechanisms involved? The same strata that produce continuous absorption also contribute to Fraunhofer line formation. The ratio of line absorption coefficient, l_ν, to continuous absorption coefficient, κ_ν, varies with depth in the atmosphere in a different way for different elements, for different levels of the same element, and for different parts of a given line.

For neutral metals in the sun, as Ca I and Na I, this ratio will diminish with optical depth because of increased ionization with increased temperature. The approximation is sometimes made of regarding such lines as formed in a "reversing layer" which overlies a photosphere that produces a pure continuous spectrum. This schematic model in which the upper layer produces line absorption or line scattering only, whereas all continuous radiation comes from the photosphere below, is called the *Schuster-Schwarzschild*, SS, model. It is the one we have in mind when we speak of the "number of atoms above the photosphere," i.e., the number of atoms required in an hypothetical reversing layer to produce lines of the same intensities as those observed in real stars. In the *Milne-Eddington* (ME) atmospheric model, on the other hand, l_ν/κ_ν is constant with optical depth, i.e., all strata are equally effective in producing line and continuous absorption. This approximation is reasonably good for ionized metals in the sun Ca II, Fe II, etc., for which second ionization is not important in the relevant layers. The true situation for any line must lie between the two extremes. The "reversing layer" and photosphere merge gradually into one another, and the factor that distinguishes the latter is a gradual increase in the continuous absorption opacity. In calculation based on model atmosphere methods one takes into account the actual stratification of the absorbing atoms but in preliminary analyses (as by a curve of growth treatment) one must use either the ME or SS model. The choice will depend on the atom or molecule one investigates. We may regard the physical processes of line formation from two extreme points of view. One might, for example, suppose that a unique temperature completely determines all emission and absorption processes in a given volume element, i.e., Kirchhoff's law holds. This condition is called *local thermodynamic equilibrium* (LTE) and is sometimes referred to as "absorption." From this point of view, the radiation from the center of a strong line, therefore, will correspond to the temperature of the uppermost stratum since l_ν at this wavelength is large and radiation reaches us only from the surface. In the nearby continuum, the bulk of the radiation comes from hotter, deeper layers. Toward the limb of the sun, the emergent radiation arises only from the uppermost layers both in the continuum and in the lines and the latter should disappear at least in principle. In the other extreme, atoms are not in temperature equilibrium with the radiation field at all, but simply scatter quanta reaching them from (mostly) greater depths. Thus a particular light quantum may be absorbed and re-emitted many times on its way through the atmosphere, and since it may be thrown either forward, sideways, or backward, its chance of reaching the surface is small. A line formed according to this mechanism of *scattering*[1] will have a black center unless it is weak. Most Fraunhofer lines are neither black at the center nor invisible at the limb. Some lines, e.g., the resonance line λ4227 of Ca I, have low central intensities, about 3 to 5 per cent, whereas the infrared subordinate lines of O I weaken as the limb is approached. There seems to be a tendency for resonance lines to favor the scattering mechanism and high level subordinate lines to lean towards the LTE mechanism. Some writers have favored the LTE mechanism even for strong lines, however. It is supposed that the sun has an extremely steep temperature gradient at optical depths $\tau < 0.05$. The lines will still appear at the observable edge of the sun since the effects of the lowered temperature would be visible only within less than a second of arc from the geometrical limb. For example, Neckel deduced the temperature distribution in the solar photosphere from the central intensities of Fraunhofer lines of known f-value and found a temperature decline in the outer layers out to log $\tau = -5$ where a boundary temperature of about 3600°K is indicated. This result appeared to be substantiated by an analysis of limb darkening at the extreme edge of the sun. Investigators find that different lines give different boundary temperature. Most likely no unique temperature can be invoked to describe the degree of ionization, population of levels, etc. in this region where severe departures from thermodynamic equilibrium must exist.

Scattering requires that the absorption of a quantum be followed by the re-emission of the same quantum. The absorption of a strong resonance line, e.g., λ5896 Na I, or λ4227 Ca I, is likely to be followed by the re-emission of the same quantum. Sometimes, however, an atom may be immediately ionized from the higher level and a resonance quantum disappears. Conversely, an ion may recapture an electron in the upper level and cascade downward with the rebirth of a resonance quantum. In denser strata, collisions may populate and depopulate levels. These always tend to restore the condition of local thermodynamic equilibrium at the gas kinetic temperature of the colliding particles. The emission and absorption of subordinate quanta will not follow the scattering mechanism since upper levels are populated to a large extent by transitions from

[1] Milne labeled this process *monochromatic radiative equilibrium* (MRE). We prefer to call it simply *scattering*.

the ground level, by cascade from higher levels, by recombination and in deeper layers by collisions. For example, the $(n = 2) \rightarrow (n = 6)$ absorption transition in hydrogen Hδ is more likely to be followed by a Paschen or a Lyman transition than by a Balmer transition. Whenever the re-emission of the particular quantum absorbed is improbable, scattering is unimportant and the LTE mechanism prevails.

In the hydrogen spectrum, for example, Lyman α tends to follow the scattering mechanism, whereas the Balmer lines, and more particularly the Paschen and Brackett lines, will follow the LTE scheme. If the upper levels of an atom are populated largely by radiative processes that follow the mechanism of scattering, we could expect that the numbers of atoms in these levels would be controlled by a non-grey radiation field and that deviations from local thermodynamic equilibrium would exist, at least in the higher atmospheric strata. Hence the population of a level, s, would be

$$N_s = b_s \frac{g_s}{u^*(T)} e^{-\chi_s/kT_\varepsilon} \quad u^*(T) = \sum_s b_s g_s e^{-\chi_s/kT_\varepsilon} \tag{8-1}$$

where b_s is a measure of the deviation of the level from LTE, and T_ε is the local temperature as measured by the gas kinetic velocity of the electrons. Deviations from LTE are treated in Art. 8-17.

8-2. The Equation of Transfer for Line Radiation. The theoretical study of absorption line formation was initiated by Schuster and Schwarzschild many years ago and has been developed in more recent times by Milne, Eddington, Unsöld, Pannekoek, Minnaert, Strömgren, Chandrasekhar, and others. A simple model atmosphere, stratified in plane parallel layers, and subject to no large scale or turbulent motions is postulated. Usually the formation of a given line is treated as though one can ignore all other lines in the spectrum.

The attack on the problem consists of two phases: (1) a determination of the appropriate absorption and emission coefficients for the line and continuum from data of atomic physics, and (2) a study of the outward flow of radiation through the atmosphere of a star of assumed chemical composition, surface gravity, and effective temperature.

Line intensity calculations can attain any level of complexity depending on the assumed initial conditions. First we need a standard solution for line formation under simplified self-consistent conditions. This standard solution can serve then as a basis for more elaborate treatments, wherein deviations from our idealized model are taken into account.

Let us first follow the classical approach suggested by Eddington. His derivation of the transfer equation is idealized in that he does not introduce specific cross-sections for collisional processes, etc., but treats radiation and collision processes from an essentially macroscopic point of view.

Later (see Art. 8–17) we shall look at the same problem from the microscopic point of view — treating detailed mechanisms.

We suppose that the continuous spectrum is formed purely by absorption processes — electron scattering is neglected. Let κ denote the coefficient of continuous absorption in the neighborhood of a line and l the coefficient of line absorption. Both κ and l depend on the frequency or wavelength; l varies sharply across a line; κ varies so slowly that it may be taken as constant in the neighborhood of the line.

To derive the equation of transfer, we again fix our attention on events occurring in an elemental cylinder of unit area and length ds placed at a depth x below the surface (see Fig. 7–2). In the distance ds, $I(\theta)$ suffers a loss from both continuous and line absorption, viz.,

$$dI\ (\theta) = -\kappa I(\theta)\ \rho ds - lI(\theta)\ \rho ds \tag{8–2}$$

If the line radiation is simply re-emitted in the same frequency, simple scattering obtains. Then the problem is simply one of diffusion of a light quantum through the atmosphere. Often, however, while an atom lingers in the excited level, it may collide with a passing electron which carries away the excess energy in a super-elastic collision (collision of the second kind). The atom may become ionized from the upper state, or further excited by radiation or inelastic collisions, or it may cascade to a lower level other than the initial one.

We shall assume that of all quanta initially absorbed, the portion $(1 - \varepsilon)$ is re-emitted as scattered radiation. The total number of ν-quanta absorbed in the cylinder is

$$l\rho ds \int I(\theta)\ d\omega$$

of which the fraction scattered in $d\omega$ will be

$$(1 - \varepsilon)l\,\frac{d\omega}{4\pi}\,\rho\ ds \int I(\theta)d\omega = (1 - \varepsilon)lJ\rho\ ds\ d\omega$$

where J is the mean intensity averaged over all angles.

We must take into account the fact that additional line emission is provided by inelastic collisions, by electron recaptures on higher levels, and by cascading. In practice, the rate of these processes will depend on the local temperature T. The amount of energy emitted is assumed by Eddington to be $\varepsilon lB(T)\rho\ ds$, where $B(T)$ is Planck's function. If we set $dx = -\sec\theta\ ds$, and add up the gains and losses in our elemental cylinder, we obtain *Eddington's transfer equation:*

$$\cos\theta\,\frac{dI}{\rho\,dx} = (\kappa + l)I - \kappa\,B(T) - (1 - \varepsilon)lJ - \varepsilon lB(T) \tag{8–3}$$

An alternative form of this equation is often useful. We think of absorption occurring in the line as being compounded of two processes: a straight scattering of radiation, and a thermal absorption at the local temperature $T(x)$, followed by emission at this same temperature. We define a scattering coefficient

$$s = (1 - \varepsilon)l \tag{8-4}$$

and a line absorption coefficient, $a = \varepsilon l$, which refers to these thermal processes. Thus $l = a + s$, and

$$\cos\theta \frac{dI}{\rho\, dx} = (\kappa + s + a)I - sJ - (\kappa + a)B \tag{8-5}$$

Let us define the optical depth in the line as

$$dt = (\kappa + l)\,\rho\, dx \tag{8-6}$$

or

$$dt = (1 + \eta)\,\rho\, dx \tag{8-7}$$

where $\eta = l/\kappa$ is the ratio of line to continuous absorption. If we introduce [2]

$$L = \frac{1 + \varepsilon\eta}{1 + \eta} \tag{8-8}$$

Eq. 8-51 becomes

$$\cos\theta \frac{dI}{dt} = I - LB(T) - (1 - L)J \tag{8-9}$$

The *source function* (see Eq. 5-45) will be

$$S = LB(T) + (1 - L)J \tag{8-10}$$

The transfer equation then reduces to the conventional form

$$\cos\theta \frac{dI}{dt} = I - S \tag{8-11}$$

8-3. Solution of the Transfer Equation for a Constant Ratio of Line to Continuous Absorption. Although numerical methods must be employed to solve Eq. 8-9 if $L(t)$ is arbitrary, an exact solution may be found if L is constant with optical depth, i.e., Milne-Eddington model. For some purposes, it suffices to divide the radiation field into two streams — an outgoing and an ingoing beam. This is the Schuster-Schwarzschild approximation, which we use in the modified form proposed by Chandrasekhar.

Analogous to Eq. 5-108 we now have:

$$\frac{1}{\sqrt{3}} \frac{dI}{\rho\, dx} = (\kappa + l)I - (\kappa + \varepsilon l)B - (1 - \varepsilon)l\, \tfrac{1}{2}(I + I') \tag{8-12}$$

$$-\frac{1}{\sqrt{3}} \frac{dI'}{\rho\, dx} = (\kappa + l)I' - (\kappa + \varepsilon l)B - (1 - \varepsilon)l\, \tfrac{1}{2}(I + I') \tag{8-13}$$

[2] Notice that t, L, and J all depend sharply on the distance from the line center $(\nu - \nu_0)$. The subscript ν is here omitted to simplify the notation.

Here I, I', B, κ and l all depend on the wavelength. The flux πF and mean intensity J are defined with the aid of

$$F = 2\,\Sigma\, I\mu_j\, a_j = \frac{2}{\sqrt{3}}\,(I - I') \tag{8–14}$$

and

$$J = \tfrac{1}{2}(I + I') \tag{8–15}$$

respectively, since $n = 1$ (see Eq. 5–106).

These equations must be solved subject to the boundary condition that J does not increase exponentially with t, and that

$$I' = 0 \text{ when } t = 0 \tag{8–16}$$

since there is no inward flux on the outer boundary. Also we shall suppose that to an adequate approximation we can write

$$B = B_0 + B_1\tau \tag{8–17}$$

where τ is the optical depth in the continuum in the neighborhood of the line and B_0 and B_1 are constants.

Furthermore, if $\eta = l/\kappa$ is constant with optical depth

$$B = B_0 + pt = B_0 + \frac{B_1}{1 + \eta}\,t \tag{8–18}$$

Subtract Eq. 8–13 from Eq. 8–12 to obtain

$$\frac{1}{\sqrt{3}}\,\frac{d(I + I')}{dt} = (I - I') \tag{8–19}$$

using Eq. 8–6. Now add Eqs. 8–12 and 8–13 and employ Eqs. 8–6, 8–8, 8–19, and 8–15. We get

$$\frac{d^2J}{dt^2} = 3L(J - B) \tag{8–20}$$

Now employ Eqs. 8–16, 8–14, and 8–15 to obtain

$$F_0 = \frac{4}{\sqrt{3}}\,J_0 \tag{8–21}$$

for $t = 0$. Similarly from Eqs. 8–19 and 8–15 we find at $t = 0$

$$\frac{1}{\sqrt{3}}\left(\frac{dJ}{dt}\right)_0 = J_0 \tag{8–22}$$

Now let

$$y = J - B \tag{8–23}$$

Then from Eq. 8–20 and the condition imposed on B by Eq. 8–18, we find

$$\frac{d^2y}{dt^2} = 3\,Ly \tag{8–24}$$

or

$$y = y_0\, e^{-\sqrt{3L}\,t} \tag{8–25}$$

From Eqs. 8–18, 8–22, 8–23, and 8–25 there results

$$\frac{1}{\sqrt{3}}\left(\frac{dy}{dt}\right)_0 = -\sqrt{L}y_0 = B_0 - \frac{1}{\sqrt{3}}p + y_0 \tag{8–26}$$

Solving for y_0 and employing Eqs. 8–23 and 8–18 we get

$$J = B_0 + pt - \frac{(B_0 - \frac{1}{\sqrt{3}}p)}{1+\sqrt{L}} e^{-\sqrt{3L}\,t} \tag{8–27}$$

Using Eq. 8–21 the emergent flux is computed from

$$F = \frac{\frac{4}{3}B_0\sqrt{3L} + \frac{4}{3}p}{1+\sqrt{L}} \tag{8–28}$$

In the continuum $\eta = 0$, $L = 1$, and $F = F_0$.

Let us write

$$M = \frac{1}{1+\eta} \tag{8–29}$$

Hence the residual intensity in the line is

$$r = \frac{F}{F_0} = \frac{2}{\sqrt{3} + \frac{B_1}{B_0}} \frac{\sqrt{3L} + M\frac{B_1}{B_0}}{1+\sqrt{L}} \tag{8–30}$$

which is the solution of the problem. For an adopted mean temperature and electron pressure we calculate first κ for the wavelength in question. Then we find l, η, and L as functions of $(\lambda - \lambda_0)$ if ε is known. Finally with the aid of Eq. 8–30 it is possible to compute r for selected values of $(\lambda - \lambda_0)$ and draw the profile.

Chandrasekhar has given an exact solution of the transfer equation for L = constant, in terms of his H functions and their first and second moments.[3] In place of Eq. 8–30, the expression for the residual intensity for pure scattering becomes:

$$r(\lambda) = \frac{L^{3/2}}{\frac{1}{3} + \frac{1}{2}\frac{B_0}{B_1}}\left[\alpha_2 + \frac{B_0}{B_1}\frac{\alpha_1}{L} + \frac{1-L}{2\sqrt{L}}\alpha_1^2\right] \tag{8–31}$$

where α_1 and α_2 are respectively the first and second moments of the H functions (see Table 8–1). For the center-limb variations, Chandrasekhar gives the following expression for the residual intensity:

$$r_\mu(\lambda) = \frac{L^{3/2}}{\mu + B_0/B_1} H(\mu)\left[\mu + \frac{B_0}{LB_1} + \frac{1-L}{2\sqrt{L}}\alpha_1\right] \tag{8–32}$$

where $\mu = \cos\theta$.

[3] See his *Radiative Transfer*, p. 321 *et seq.*, especially his Eqs. (71) and (70). $H(\mu)$ is tabulated on p. 125.

TABLE 8–1

THE FIRST AND SECOND MOMENTS OF CHANDRASEKHAR'S $H(\mu)$ FUNCTION*

L	α_1	α_2
1.0	0.500000	0.333333
0.9	0.515609	0.344357
0.8	0.533154	0.356787
0.7	0.553123	0.370985
0.6	0.576210	0.387466
0.5	0.603495	0.407030
0.4	0.636636	0.430922
0.3	0.678674	0.461423
0.2	0.735808	0.503218
0.1	0.825318	0.569449
0.075	0.858734	0.594404
0.050	0.901864	0.626785
0.025	0.964471	0.674134
0.00	1.154701	0.820352

*Courtesy, S. Chandrasekhar, *Ap. J.* (University of Chicago Press) **106**, 151, 1947, Table 1.

As we shall see, Eqs. 8–31 and 8–32 lie at the basis of Wrubel's calculation of the theoretical curve of growth for the Milne-Eddington model. One difficulty frequently encountered in applications of Eqs. 8–30, 8–31, and 8–32 is the assumption that B varies linearly with optical depth τ. One may estimate B_0/B_1 from the observed or computed law of limb darkening for the sun or star or one can determine it from the variation of the Planck function with optical depth using either the first derivative computed for some representative optical depth or an empirical value from a plot of B against τ; in actual practice one often finds that B varies steeply with τ in shallower layers and then more slowly in deeper layers so that a proper choice of B_0/B_1 is often difficult.

8–4. Solution of the Transfer Equation for η Variable with Optical Depth.

a. *Method based on choice of a mean value of* $\overline{L}$ *and* $\sqrt{\overline{\overline{L}}}$

We might expect the assumption η = const to hold for such a line as λ3933 of Ca II where both l and κ increase with increasing optical depth. Actual calculation shows that even for this line, η varies with τ. If the variations in η are not too large, Strömgren showed that one can employ Eq. 8–30 with a $\sqrt{\overline{\overline{L}}}$ and $\overline{M}$ defined by

$$\sqrt{\overline{\overline{L}}} = \int_0^\infty \sqrt{L}\, e^{-2z}\, 2dz \qquad (8\text{–}33)$$

and (see Eq. 8–29)

$$\overline{M} = \int_0^\infty M\, e^{-z}\, dz \qquad \overline{L} = \int_0^\infty \overline{L}\, e^{-z}\, dz \tag{8–34}$$

where

$$z = \left(\frac{3}{L_0}\right)^{1/2} \tau \tag{8–35}$$

The best procedure is to choose $\sqrt{\overline{\overline{L}}} = \sqrt{L_0}$ by a process of iteration.[4] We may employ Eqs. 8–31 and 8–32 with the following substitutions:

$$L^{3/2} \to \overline{L}\sqrt{\overline{\overline{L}}};\ (1 - L) \to \left[1 - \left(\sqrt{\overline{\overline{L}}}\right)^2\right];\ \frac{B_0}{LB_1} \to \frac{B_0}{\overline{L}B_1} \tag{8–36}$$

Given $\eta(t)$, one computes $\overline{L}$ or $\sqrt{\overline{L}}$ for several points in the line profile by Eqs. 8–33 and 8–34, and then obtains $r(\lambda)$ from Eq. 8–31 with the replacements as noted in Eq. 8–36. Center-limb variations follow from Eqs. 8–36 and 8–32. In calculations of center-limb variations we note that $\overline{L}$ and $\sqrt{\overline{L}}$ will vary with μ. Toward the limb, radiation is emitted from successively higher layers. Hence in Eq. 8–35, τ must be replaced by τ/μ. The integrals Eqs. 8–33 and 8–34 may be evaluated conveniently by use of approximation formulas, e.g., Eq. 5–64. Profiles computed by this method are sensitive to the choice of L_0. Also variations of η with optical depth must be small. The method fails, for example, for the hydrogen lines in the sun although it succeeds reasonably well for lines of ionized metal atoms. The usual difficulties with the choice of the constants B_0 and B_1 also conspire to limit the usefulness of the method.

b. *Method of direct numerical integration*

A number of years ago Pannekoek suggested a method of direct integration of the transfer equation based on the Eddington approximation. We divide Eq. 8–5 by κ, substitute $d\tau = +\,\kappa\rho\, dx$, multiply by $d\omega/4\pi$, integrate over all solid angles, and use the Eddington J, H, and K with the understanding that these quantities here depend on the wavelength (Eq. 5–67). We obtain:

$$\frac{dH}{d\tau} = \left(1 + \frac{a}{\kappa}\right)(J - B) \tag{8–37}$$

If we multiply Eq. 8–5 by $\cos\theta\, d\omega/4\pi$ and integrate we find

$$\frac{dK}{d\tau} = \left(1 + \frac{a + s}{\kappa}\right) H \tag{8–38}$$

In the Eddington approximation we assume

$$K = \tfrac{1}{3} J \tag{8–39}$$

[4] One selects η for $t = 0.25$, calculates the corresponding value of L which is taken as the first approximation to L_0 for computation of z by Eq. 8–35. Then $\sqrt{\overline{L}}$ is evaluated by Eq. 8–33 and the process repeated until $\sqrt{\overline{L}} = \sqrt{L_0}$.

Hence,

$$\frac{dJ}{d\tau} = 3\left[1 + \frac{a + s}{\kappa}\right] H \tag{8–40}$$

Here κ denotes the coefficient of continuous absorption in the neighborhood of the line. Also,

$$a = \varepsilon l \tag{8–41}$$

is the coefficient of line absorption as distinct from the coefficient of line scattering

$$s = (1 - \varepsilon) l \tag{8–42}$$

Let us express J, H, and B in units of B_0 (the Planck function evaluated for the boundary temperature). Denote the corresponding quantities by primes. Then Eqs. 8–37 and 8–40 are replaced by the expressions:

$$\Delta J' = 3(1 + \eta) H' \Delta\tau \tag{8–43}$$

$$\Delta H' = (J' - B') \Delta\tau \tag{8–44}$$

for lines formed according to the mechanism of pure scattering. To start the integrations we adopt an arbitrary J_0' at the surface of the star. The boundary condition

$$H_0' = \frac{1}{2} J_0' \tag{8–45}$$

or

$$H_0' = \frac{1}{\sqrt{3}} J_0' \tag{8–46}$$

(depending on whether we use the Eddington or exact condition) will fix H_0'. We then compute the increments $\Delta J'$ and $\Delta H'$ from Eqs. 8–43 and 8–44 for a small step $\Delta\tau$, find new values of H' and J', and continue the process. If J_0' had been chosen correctly, J' would approach B' deeper in the atmosphere. Usually J will deviate sharply from B and new J_0' values will have to be tried until a solution is found. The corresponding H_0' will be the correct flux in the line. Finally,

$$r(\lambda) = \frac{H_0'}{H_{0c}'} \tag{8–47}$$

where H_{0c}' the flux in the continuum, is obtained from the solution of Eqs. 8–43 and 8–44 with $\eta = 0$. The method has the advantage that variations of B, ε, and η with τ may be taken into account exactly and is readily adapted to modern electronic computational techniques. One objection which may be raised against this method is the use of Eddington's approximation which is inaccurate near the surface. A superior procedure would be to adapt Chandrasekhar's method to the problem. Even the first approximation would give a better result than the Eddington approximation.

c. *Solution of the radiative integral equation by iteration*

Instead of integrating the differential equation of radiative transfer directly, one may follow a procedure suggested by Strömgren and attack the corresponding integral equation (see Art. 5–5). This method, although advantageous in treatments of continuous spectra where scattering and absorption are of comparable magnitude converges very slowly if the scattering coefficient is large compared with the absorption coefficient.

d. *Line formation with the mechanism of pure absorption*

If a spectral line is formed according to the strict mechanism of pure absorption, its profile may be calculated directly without any reference to method a, b, or c. The absorption coefficient in the line is $(\kappa_\nu + l_\nu)$ and the corresponding element of optical depth is

$$dt_\nu = (\kappa_\nu + l_\nu)\,\rho\,dx \tag{8–48}$$

Assuming that the continuum is formed according to the mechanism of pure absorption, i.e., electron scattering is unimportant

$$S_\nu(t_\nu) = B_\nu(t_\nu) \tag{8–49}$$

The residual intensity in the line is the ratio of the specific intensity in the line to that in the continuum. Using Eq. 5–5 we obtain:

$$r_\nu\,(\theta) = \frac{I_\nu\,(0, \theta)}{I_\nu^c\,(0, \theta)} = \frac{\int_0^\infty B_\nu\,(t_\nu)\,e^{-t_\nu \sec\theta} \sec\theta\,dt_\nu}{\int_0^\infty B_\nu\,(\tau_\nu)\,e^{-\tau_\nu \sec\theta} \sec\theta\,d\tau_\nu} \tag{8–50}$$

Similarly, for the integrated flux from the entire star we get

$$r_\nu = \frac{\int_0^\infty B_\nu(t_\nu)\,E_2(t_\nu)\,dt_\nu}{\int_0^\infty B_\nu(\tau_\nu)\,E_2(\tau_\nu)\,d\tau_\nu} \tag{8–51}$$

where $E_2\,(\cdots)$ is the exponential integral function. When $B_\nu\,(\tau_0)$ is given we calculate $B_\nu\,(\tau_\nu)$ from the structure of the model atmosphere and evaluate t_ν in terms of τ_ν.

Fig. 8–3 shows the results of calculations carried out for the Ca II λ3933 "*K*" line. We assume $A = 1.59 \times 10^9$ sec^{-1}, $\Gamma_{\text{rad}} = 1.72 \times 10^8$ sec^{-1}, and calculate Γ_{coll} by methods of Art. 7–5. Calculations are carried out for an average point in the atmosphere, $\log \tau_0 = -0.5$ (see Eq. 8–32) by the method of Art. 8–4a (see Eqs. 8–32 to 8–36) and by Eq. 8–50. We adopt the solar model atmosphere described in Chapter 5. Notice that the three assumptions give similar profiles for curves A1, A2, and A3, but with small, detectable differences.

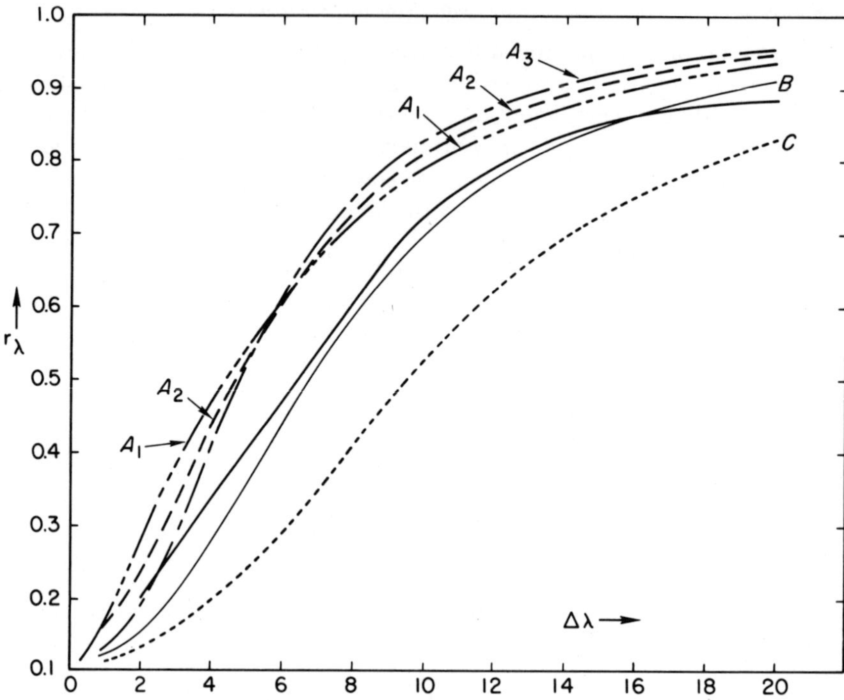

FIG. 8–3. THEORETICAL INTENSITY PROFILE OF THE K LINE IN THE SOLAR SPECTRUM

The assumed values of log $[N(\mathrm{Ca})/N(\mathrm{H})]f$ are -5.983 (A), -5.682 (B), and -5.380 (C). Curves A3, B, and C are calculated for pure absorption with our adopted solar model (Chapter 5), including the influence of the upper layers (see Eq. 8–50). Curve A1 is computed for a homogeneous atmosphere (Eq. 8–32) with the following atmospheric parameters $\log \tau_0 = -0.5$, $\theta = 0.885$, $\log P_g = 4.99$, and $\log P_\varepsilon = 1.094$. Curve A2 is calculated by Eq. 8–32 with mean values calculated by Eqs. 8–35 and 8–36 and our adopted model atmosphere.

The observed line profile, as adapted from the work of Houtgast, is indicated by the solid curve.

8–5. Precise Calculation of a Line Profile. The precise calculation of the theoretical profile of a line involves the following steps:

1. Calculate a model atmosphere for an assumed effective temperature, surface gravity, and chemical composition. The dependence of temperature on optical depth τ_0 at some wavelength λ_0 is usually fixed by the condition of radiative equilibrium. For the sun $T(\tau_0)$ is given from limb darkening measurements. This model atmosphere gives the gas pressure, electron pressure, and absorption coefficient as a function of τ_0.

2. As the theory of the continuous absorption coefficient gives $\kappa_\lambda(P_\varepsilon,T)$, one may calculate the relation between the optical depth τ_λ at any wavelength λ and the optical depth τ_0 at λ_0.

3. Since P_ε and T are known as functions of τ_λ, one may calculate the

fraction of all atoms of a given element capable of absorbing the line in question, viz., N_r/N_{total} with the aid of the ionization and Boltzmann equations. If one has a theory for deviations from local thermodynamic equilibrium he may introduce the b_n correction factor at this point. See Eq. 8–1.

4. Before we can calculate the line absorption coefficient l_λ and the ratio $\eta = l_\lambda/\kappa_\lambda$ we must assume an abundance for the element in question. In practice one usually assumes several values of the abundance (or the product Nf) and calculates the corresponding profiles.

5. We must now adopt the form of the line absorption coefficient. For hydrogen lines one employs the Kolb-Griem theory, for helium the Griem-Baranger-Kolb-Oertel theory (Art. 7–8), while for lines of other elements one may adopt the combined Doppler plus dispersion-type broadening theory (Art. 7–6). We calculate the fictitious absorption coefficient at the center of the line, α_0, by Eqs. 7–69 and 7–70. Then the damping constant for radiation is computed from the Einstein probability coefficients for spontaneous emission. The collisional damping constant depends on the gas and electron pressure. See Eqs. 7–56 and 7–50.

6. Next calculate the parameters a and $u = (\lambda - \lambda_0)/\Delta\lambda_0$ (defined by Eqs. 7–68 and 7–66, respectively) and find α_ν/α_0 from Eqs. 7–74 and 7–75 as a function of optical depth.

7. With α_λ/α_0 and $N_r\alpha_0$ both known as a fraction of optical depth, one may calculate the line absorption coefficient $l_\lambda = N_r\, \alpha_0(\alpha_\lambda/\alpha_0)$.

Since $\kappa_\lambda(\tau)$ is already known from the model atmosphere, one may calculate $\eta_\lambda = l_\lambda/\kappa_\lambda$ and hence $dt_\lambda = (1 + \eta_\lambda)\, d\tau_\lambda$.

8. The method to be employed for calculating a line profile then depends on the assumed mechanism. If the line is formed by pure absorption we may proceed to a direct numerical calculation of Eq. 8–50 or 8–51. See Fig. 8–3 for an illustrative example. If the line is formed according to the mechanism of scattering the calculations become more involved. When the variation of η with optical depth is small, one may employ method (a); whereas if η varies over a large range in the relevant atmospheric layers (e.g., the sodium D lines in the sun) (a) cannot be used, and we may employ (b) or (c), or the method of weighting functions [see (6)]. Perhaps some modification of method (b) is best suited for calculation with an electronic computer, although Strömgren's iteration method (which is superior, in that the conditions are more rigorously handled) might be adapted to machines.

When a theoretical profile (or equivalent width) computed on the basis of a specific model atmosphere is compared with observations for a particular star, discrepancies are usually found indicating that the model does not correspond to the star. Similar calculations may be carried out for a range of models with different effective temperatures, surface gravities, and even chemical compositions (particularly different hydrogen/metal ratios). Then, by a process of interpolation one may find a model that represents not only the continuous spectrum of the star, but its line spectrum as well.

We may illustrate the procedure by two examples, the sun and a B-type star. The temperature distribution in the solar atmosphere is known empirically at least in the upper layers. Furthermore, spectral lines may be observed, not only at the center of the disk but also elsewhere so that center-limb variations may be investigated. Consider the infra-red transition of nitrogen and oxygen studied by Faulkner and Mugglestone (1962). These subordinate lines arise from levels 9–10 v above the ground level. Hence they are formed predominantly in deep layers by the mechanism of absorption and a study of their behavior across the disk should yield valuable clues to the temperature distribution in the deeper layers of the solar atmosphere where limb darkening measurements give little help. Furthermore, deviations from LTE should be unimportant.

A machine program for the direct calculation of equivalent widths of solar absorption lines has been developed by J. P. Mutschlecner. Fig. 8–4 illustrates the schematic flow diagram which shows the plan of the calculation. The left-hand column gives basic input data. The first box gives the model atmosphere data, i.e. the variation of $\theta = 5040/T$, the gas pressure, electron pressure and turbulent velocity, ξ, with the logarithm of optical depth at the standard wavelength, $\lambda 5000$. That is

$$y = \log \tau(\lambda_0) \tag{8–52}$$

Provision is made for the situation where deviations from LTE occur, i.e., the excitation temperature differs from the local gas kinetic temperature.

The second box includes selected values of parameters for the calculation — i.e., distances $\Delta\lambda_i = (\lambda - \lambda_0)_i$ from the line center at which the depth in the line is to be computed, the points on the solar disk where the profile is to be evaluated $\mu_i = (\cos\theta)_i$, and choices of the product of abundance and f-value, $(N_{el}f_{r,s}/N_{\rm H})_i$, the *van der Waals interaction constant* $C_{{\rm H}i}$ and damping constant, $\Gamma_{\rm rad}$. The third box gives numerical constants pertinent to the element involved, i.e., the ionization potentials for the neutral and ionized atom, the mass (atomic weight units) and the partition functions, u_0, u_1, and u_2.

The fourth box gives the atomic constants appropriate to the line in question, i.e., wavelength, excitation potential, statistical weight $g_{r,s}$, and parameters related to the continuous spectrum in the neighborhood of the line.

The fifth box gives physical and numerical constants needed in the calculation h, e, π, etc. while the sixth box gives tables for the calculation of the continuous absorption coefficient and the line broadening function $H(a, u)$ defined by Eq. 7–74

The machine operations are summarized in the center of the figure. In the lower left-hand box the collisional damping constant Γ_6 (see Eq. 7–56) is calculated, given the van der Waals interaction constant. Then, given the radiation damping constant we calculate the effective damping constant which is the sum of radiative and collisional contributions. The next step is to compute the parameter $a(y)$, in Eq. 7–68, where $\Delta\lambda_0(y)$ is now defined by

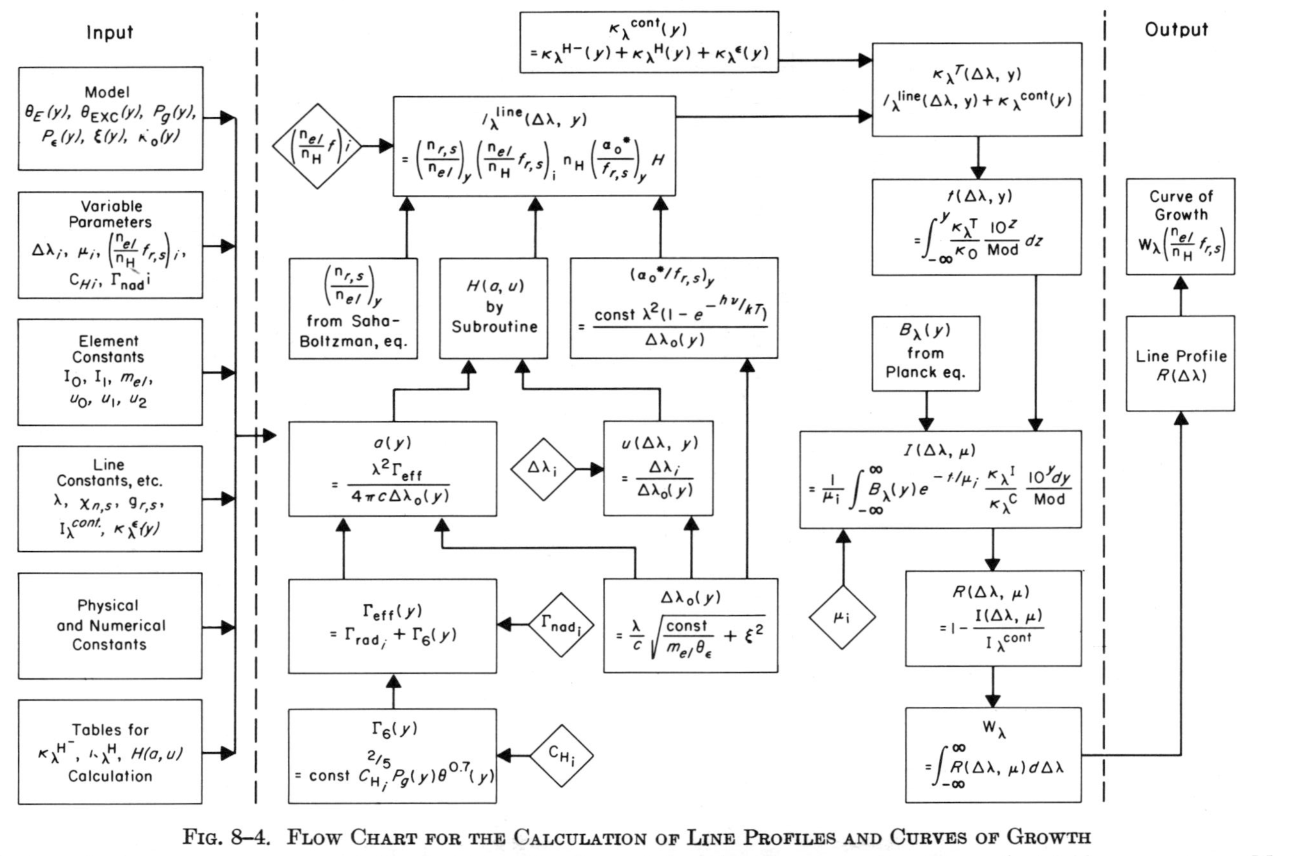

Fig. 8-4. Flow Chart for the Calculation of Line Profiles and Curves of Growth

Rectangular boxes represent computational steps while diamond-shaped boxes represent settings of variable parameters. Most symbols have their usual meanings within the line-formation theory (see text). C_H is the van der Waals broadening constant, ξ the turbulent velocity, κ_λ^ϵ an empirical absorption coefficient necessary in addition to those of atomic hydrogen and the negative hydrogen ion in certain spectral regions. (Courtesy, Paul Mutschlecner.)

$$\Delta\lambda_0 (y) = \frac{\lambda}{c}\sqrt{\frac{2kT}{M} + \xi^2} \tag{8–53}$$

instead of Eq. 7–76, since we must include effects of large scale mass motions in the atmosphere or "turbulence" (see Art. 8–10). There ξ is the rms velocity of the turbulent elements.

For selected values of $\Delta\lambda_i$, Eq. 7–66 is used to calculate $u(\Delta\lambda, y)$ from $\Delta\lambda_i$ and $\Delta\lambda_0$. Then the "fictitious" absorption coefficient at the center of the line divided by the f-value is calculated by Eq. 7–69. The line absorption coefficient is then given in the form

$$l(\Delta\lambda, y) = \left[\frac{N_{r,s}}{N_{el}}\right]\left[\frac{N_{el}}{N_{\rm H}} f_{r,s}\right]_i N_{\rm H}\left(\frac{\alpha_0{}^*}{f_{r,s}}\right) H(a, u) \tag{8–54}$$

where the first factor which is the ratio of the number of atoms capable of absorbing the line in question to the total number of atoms of that element, is calculated from the Saha and Boltzmann equations. The second factor embodies the assumed abundance of the element relative to hydrogen. $H(a, u)$ is calculated by appropriate methods from the integral involved in Eq. 7–71.

The continuous absorption coefficient may be evaluated from tables of $\kappa_\lambda(\mathrm{H}^-)$ and κ_λ(H-atomic) for the sun. In addition to these well-recognized hydrogenic sources of continuous absorption, there also appears (particularly in the ultraviolet solar spectrum) a background absorption produced by overlapping wings of metallic lines. This additional contribution may be assessed in the following way: we suppose that

$$\kappa_\lambda\,(\mathrm{emp}) = A_\lambda\,\mathfrak{D}\,(\tau_0)$$

where $\mathfrak{D}(\tau_0)$ describes the variation of the contributors to absorption with depth. One calculates the intensity at the center of the disk $I(0, 0)$ and the limb darkening for different values of A and assumptions concerning the depth dependence of $\mathfrak{D}(\tau_0)$ until the observed data are represented in a satisfactory way. Detailed calculations show that $\mathfrak{D}(\tau_0)$ is proportional to the gas pressure, suggesting that overlapping line wings produce the absorption. Mutschlecner has developed a program for the calculation of κ_λ (emp).

The second column from the right-hand side shows the relevant steps in the calculation. First one evaluates the sum of the line and continuous absorption coefficients, $l_\lambda(\Delta\lambda, y) + \kappa_\lambda(y)$. Then we compute the optical depth at a point $\Delta\lambda$ in the line profile (see Eq. 8–48)

$$t(\Delta\lambda, y) = \int_0^\tau [1 + \eta]\frac{\kappa_\lambda}{\kappa_0}\,d\tau_0 = \int \frac{l_\lambda + \kappa_\lambda}{\kappa_\lambda}\frac{10^y}{\mathrm{mod}}\,dy \tag{8–55}$$

where y is defined by Eq. 8–52 and mod means the modulus of natural logarithms. The intensity at a distance $\Delta\lambda$ from the line center and at a point μ on the disk becomes

$$I(\Delta\lambda, \mu) = \frac{1}{\mu}\int_{-\infty}^{+\infty} B_\lambda(y)\, e^{-t/\mu}\left(\frac{l_\lambda + \kappa_\lambda}{\kappa_\lambda}\right)\frac{10^y}{\mathrm{mod}}\,dy \tag{8–56}$$

where B_λ is the Planck function. The corresponding intensity in the continuum, $I_\lambda{}^c(\mu)$ (see Eq. 5–5) is known from previous calculations. Eq. 8–56 is evaluated for as many points on the solar disk as are of interest. Next, one calculates the line depth

$$R(\Delta\lambda, \mu) = 1 - \frac{I(\Delta\lambda, \mu)}{I_\lambda{}^c} \tag{8–57}$$

and evaluates the equivalent width of the line

$$W_\lambda = \int_{-\infty}^{+\infty} R(\Delta\lambda, \mu)\, d\,\Delta\,\lambda \tag{8–58}$$

In practice $W_\lambda(\mu)$ is calculated for different values of μ and of the product $(N_{el}f_{r,s}/N_{\rm H})$. A plot of W_λ/λ against $N_{el}f_{r,s}$ or some precisely defined function thereof is called a *curve of growth,* although the term is more often used (as we shall illustrate in Art. 8–7) for relations between W_λ and Nf for special atmospheric models, e.g., the Schuster-Schwarzschild and Milne-Eddington models.

To calculate theoretical equivalent widths for stellar spectra, we want results referring to the flux from the entire stellar disk. Such fluxes may be computed from:

$$\frac{W_\lambda}{\lambda} = \int_0^1 \frac{W_\lambda(\mu)}{\lambda}\,\frac{I_\lambda{}^c(\mu)}{F_\lambda{}^c}\, 2\mu\, d\mu \tag{8–59}$$

where $\pi F_\lambda{}^c$ is the flux in the continuum.

Methods of direct calculation of line intensities on the electronic computer have been developed by a number of investigators. Important contributions have been made by Heiser (1957), Cayrel (1956), Neven and de Jager, and others. The elaborate character of the computations required means that even with a versatile, fast machine such as the IBM 704 several minutes are required for the calculation of each value of W_λ.

Hence methods employing various approximations and assumptions have been employed. The most successful of these methods and the one involving the fewest objectionable assumptions is the method of weighting functions (see Art. 8–6). It is practical only when data justify it and enough is known concerning the star to permit one to calculate a meaningful stellar model atmosphere. The simple curve of growth method is certainly the one to apply in initial stages of an analysis.

8–6. The Method of Weighting Functions. Many years ago Milne showed that in a stellar atmosphere where P_g, P_ε, and T varied with optical depth one could assign to each elementary stratum a weight $G(\tau)$ which determined how much that layer contributes to the emergent flux, whether in the line or in the continuum. This idea has been subsequently developed by Unsöld and by Minnaert for the interpretation of weak lines or the far wings of strong ones, and by Pecker and others for lines of moderate strength.

From the definitions $a = \varepsilon l$ and $s = (1 - \varepsilon)l$ (see Eq. 8–4), Eqs. 8–10, 8–14, 5–50, and 5–52 we can write the integral equation for the source function in the line as

$$(\kappa + s + a)\, S(t) = (\kappa + a)B(t) + \frac{s}{2}\int_0^\infty S(t')\, E_1\,(t - t')\, dt' \tag{8–60}$$

Consider first a weak line or the far wings of a strong line where

$$(a + s)/\kappa \ll 1$$

For the zeroth approximation we neglect the contribution of scattering and set $S(t) = B(t)$ under the integral sign on the right-hand side of Eq. 8–60. Then

$$S_1(t) = \frac{\kappa + a}{\kappa + a + s} B(t) + \frac{s}{2(\kappa + a + s)} \int_0^\infty B(t')\, E(|t - t'|)\, dt' \qquad (8\text{–}61)$$

Using Eqs. 5–51 and 5–55 we compute the emergent flux from

$$\begin{aligned} \tfrac{1}{2} F(0) &= \int_0^\infty S_1\,(t)\, E_2\,(t)\, dt \\ &= \int_0^\infty \frac{\kappa + a}{\kappa + a + s} B\,(t)\, E_2\,(t)\, dt + \int \frac{s}{\kappa + a + s} \overline{B}\,(t)\, E_2\,(t)\, dt \end{aligned} \qquad (8\text{–}62)$$

Now

$$dt = (\kappa + a + s)\, \rho\, dx = d\tau + d\tau_a + d\tau_s \qquad (8\text{–}63)$$

Expand in a Taylor series to obtain:

$$\begin{aligned} E_2(t) &= E_2\,(\tau) + (\tau_a + \tau_s)\, dE_2/d\tau + \cdots \\ &= E_2\,(\tau) - (\tau_a + \tau_s)\, E(\tau) \end{aligned} \qquad (8\text{–}64)$$

where we have noted that $(a + s)/\kappa \ll 1$ and have used (see Problem 5–6) the property of the $E(\tau)$ functions

$$\frac{d}{d\tau} E_n\,(\tau) = -E_{n-1}\,(\tau) \qquad (n \geq 1) \qquad (8\text{–}65)$$

Then Eq. 8–62 reduces to

$$\begin{aligned} F(0) &= 2 \int_0^\infty B(\tau) E_2(\tau) d\tau + 2 \int_0^\infty B(\tau) E_2(\tau) d\tau_a \\ &\quad - 2 \int_0^\infty (\tau_a + \tau_s) B(\tau) E(\tau) d\tau + \int_0^\infty E_2(\tau) d\tau_s \int_0^\infty B(\tau) E(|\tau - \tau'|) d\tau' \end{aligned} \qquad (8\text{–}66)$$

The first term on the right represents $F_c(0)$, i.e., the flux in the continuum divided by π. Define

$$f(x) \equiv \int_x^\infty B(x) E(x)\, dx \qquad (8\text{–}67)$$

and notice that by integration by parts

$$\int_0^\infty \tau_a B(\tau) E(\tau)\, d\tau = \tau_a f(\tau) \Big|_0^\infty - \int_0^\infty f(\tau)\, d\tau_a = - \int_0^\infty f(\tau)\, d\tau_a, \text{ etc.} \qquad (8\text{–}68)$$

We obtain for the depth of the line (in the flux):

$$R = 1 - r = \frac{F_c(0) - F(0)}{F_c(0)} = \int_0^\infty G_1(\tau)\, d\tau_a + \int_0^\infty G_2(\tau)\, d\tau_s \qquad (8\text{–}69)$$

$$= \int_0^\infty \frac{a}{\kappa} G_1(\tau)\, d\tau + \int_0^\infty \frac{s}{\kappa} G_2(\tau)\, d\tau$$

where $G_1(\tau)$ and $G_2(\tau)$ represent the weighting functions for the flux for absorption and scattering processes respectively. Thus:

$$G_1(\tau) = \frac{2}{F_c(0)} [f(\tau) - B(\tau)\, E_2(\tau)] \tag{8–70}$$

and

$$G_2(\tau) = \frac{1}{F_c(0)} \left[2f(\tau) - E_2(\tau) \int_0^\infty B(\tau')E(|\tau - \tau'|)\, d\tau' \right] \tag{8–71}$$

The important property of these weighting functions is that they may be calculated once and for all as a function of wavelength for a given model atmosphere. They do not depend on the atom or ion involved. The line absorption or scattering coefficients (which do depend on the depth in the atmosphere) enter through the factors a/κ and s/κ. Notice that one can include both scattering and absorption contributions in the mechanism of line formation.

In the following discussion we shall emphasize the absorption weighting function; the expressions can be easily modified for scattering processes when it is required. Note that Eq. 8–70 can also be written in the form

$$G_\lambda{}^0(\tau_\lambda) = \frac{2}{F_c(0)} \int_{\tau_\lambda}^\infty \frac{dS_\lambda(\zeta_\lambda)}{d\zeta_\lambda} E_2(\zeta)\, d\zeta_\lambda \tag{8–72}$$

which is evaluated by integrating from deeper layers outwards. Here ζ_λ denotes the optical depth appearing as a variable of integration.

Weighting functions may be defined in terms of intensity at a point on the disk, μ, rather than in terms of the integrated flux, viz.

$$g_\lambda{}^\mu(\tau_\lambda) = \frac{\displaystyle\int_{\tau_\lambda}^\infty S_\lambda(\zeta) \exp\left(-\frac{\zeta_\lambda}{\mu}\right) \frac{d\zeta_\lambda}{\mu} - S_\lambda(\tau_\lambda) \exp\left(-\frac{\tau_\lambda}{\mu}\right)}{I_\lambda{}^c(\mu)} \tag{8–73}$$

or alternately

$$g_\lambda{}^\mu(\tau_\lambda) = \frac{1}{I_\lambda{}^c(\mu)} \int_{\tau_\lambda}^\infty \frac{dS_\lambda(\zeta)}{d\zeta_\lambda} \exp\left(-\frac{\zeta_\lambda}{\mu}\right) d\zeta_\lambda \tag{8–74}$$

The methods just described permit us to handle only weak lines. Fortunately, Pecker has developed a theory that permits one to calculate equivalent widths of lines of moderate strength.

We shall not attempt to reproduce the rather lengthy derivations here but shall merely state the results, explaining the physical significance of each factor involved. Consider first the equivalent width of a line at a point, μ, on a stellar disk. It is given by (see Elste, Aller, and Jugaku, 1957):

$$\frac{W_\lambda{}^\mu}{\lambda} = \int_{-\infty}^{+\infty} \frac{M}{\mu} Z(y) \Psi\left(\frac{Y}{\mu}, a\right) g_\lambda{}^{*\mu}(y)\, dy \tag{8–75}$$

Here $y = \log \tau_0$ (Eq. 8–52) is the independent variable. The expression under the integral sign is called the *contribution function* and indicates the extent to which each layer contributes to the formation of a particular line. It is therefore of interest in assessing effects of pressure and temperature stratification upon numerical values of equivalent widths.

Here $g_\lambda{}^*(\mu)$ denotes the weighting function (of Eq. 8–73) multiplied by the factor $[1 - \exp(-h\nu/kT)]$ to allow for the diminution of the absorption coefficient by stimulated emissions. It depends only on μ, λ, and y for a given model atmosphere.

Of the remaining factors, M is a constant for each line and assumed abundance, $Z(y)$, expresses the variation of the population of a given level with depth in the atmosphere, Ψ is called the *saturation function* which expresses the physical result that strong lines tend to be formed in the upper atmospheric layers.

Let us examine these quantities in turn. In practical applications of Pecker's method it is convenient to introduce sort of an "excitation" temperature for the atmosphere, denoted by T_a. Let

$$\theta_a = 5040/T_a \tag{8–76}$$

Then (following Elste) one may write

$$\log M = \log \frac{N_E}{N_{\mathrm{H}}} + \log g_{r,s} f_{J',J} \lambda_A k_a + \theta_a \Delta\chi \tag{8–77}$$

Here N_E, N_{H} denote the assumed number of atoms of element E and of hydrogen respectively, per gram of stellar material; $g_{r,s}$ denotes the statistical weight of the sth level in the rth stage of ionization; $f_{J',J}$ is the f-value for the observed transition $(\alpha'L'S'J' \rightarrow \alpha LSJ)$ from the sth to some higher level; and λ_A denotes the wavelength expressed in angstroms. We have defined θ_a by Eq. 8–76 while $\Delta\chi$ depends on the excitation potential of the (r, s) level. Consider an element that is represented in a given atmosphere in three stages of ionization. A good example is provided by silicon in early B-type stars, where lines of Si II, Si III and Si IV may be observed simultaneously in a given spectrum. Then

$$\begin{aligned} \Delta\chi &= (\chi_{r-1} - \chi_{r-1,s}) && \text{for level } (r-1, s) && \text{e.g., Si II} \\ \Delta\chi &= -\chi_{r,s} && \text{for level } (r, s) && \text{e.g., Si III} \\ \Delta\chi &= -(\chi_{r+1,s} + \chi_r) && \text{for level } (r+1, s) && \text{e.g., Si IV} \end{aligned} \tag{8–78}$$

The constant k_a depends only on the molecular weight of the un-ionized gas, μ^0 and on numerical and physical constants, viz.,

$$k_a = \frac{\pi\varepsilon^2}{mc^2} \frac{10^{-8}}{\text{mod}} \frac{1}{M_{\mathrm{H}}\mu^0} \tag{8–79}$$

where mod is the modulus of natural logarithms.

For a "normal" stellar composition, $\log k_a \sim 3.84$.

The variation of the population of a given level (r, s) with depth in the stellar atmosphere is described by the function

$$Z(y) = \frac{1}{\kappa_\lambda{}^{\text{eff}}} \frac{N_{r,s}}{N_E} \frac{10^{y-\theta_a\Delta\chi}}{g_{r,s}} \tag{8–80}$$

where $\kappa_\lambda^{\text{eff}}$ is the continuous absorption coefficient (including effects of electron scattering if necessary). The ratio of the population of the (r, s) level to the total for the element is obtained from the Saha and Boltzmann equations.

The saturation function (in the normalized form suggested by Elste) is

$$\Psi\left(\frac{Y}{\mu}, a\right) = \frac{2}{\sqrt{\pi}} \int_0^\infty H(a, u) \exp\left[-H(a, u)\frac{Y}{\mu}\right] du \tag{8–81}$$

where u is given by Eq. 7–77, $H(a, u)$ is defined by Eqs. 7–74 and 7–71 and

$$\frac{Y}{\mu} = \frac{M}{\mu}\frac{c}{\sqrt{\pi}}\frac{1}{V}\int_{-\infty}^{y} Z(w)\,(1 - e^{-h\nu/kT})\,dw \tag{8–82}$$

where $Z(w)$ is obtained from Eq. 8–80. Since the abundance enters directly into Eq. 8–82 through the factor M, we have to evaluate the factor Y for several choices of M and selected values of y in the atmosphere. Then the integrand of Eq. 8–75 is computed for each of these choices.

For small values of $N_{r,s}f$, i.e., small Y, $\Psi(Y, a) \rightarrow 1$ and we have in fact the weak line theory already described. For large values of Y/μ, Ψ approaches zero. Since Y increases with y, this means that the effect of Ψ is to cause the upper layers to make the largest contribution to the integral, Eq. 8–75.

The expression for W/λ for flux rather than intensity is more complicated than Eq. 8–75. It may be written in the form:

$$\frac{W_\lambda}{\lambda} = M\mathfrak{L}_\lambda^* = M\int_{-\infty}^{+\infty} Z(y)\Psi(Y, a)\left[\frac{G_\lambda^m(y)}{G_\lambda^0(y)}\right] G_\lambda^0(y)\,(1 - e^{-h\nu/kT})\,dy \tag{8–83}$$

where $G_\lambda^m(y)$ is defined by an equation analogous to Eq. 8–72 with $E_2(y)$ replaced by $E_{m+2}(y)$ where m depends in a known way on the intensity of the line (Elste, Aller, and Jugaku, 1957). For the limiting situation of weak lines, we may write

$$\left(\frac{W_\lambda}{\lambda}\right)_0 = ML_\lambda^* = \int_{-\infty}^{+\infty} M\,Z(y)\,G_\lambda^0(y)\,(1 - e^{-h\nu/kT})\,dy \tag{8–84}$$

or defining

$$\log C = \log gf\lambda k_a + \Delta\chi\theta_a + \log L_\lambda^* \tag{8–85}$$

we may write:

$$\log\left(\frac{W}{\lambda}\right)_0 = \log C + \log\left(\frac{N_E}{N_{\text{H}}}\right) \tag{8–86}$$

For an atom in a given stage of ionization, one proceeds to calculate the theoretical equivalent widths as follows: first, one selects a model atmosphere and calculates the weighting functions for selected wavelengths in the spectral range of interest. Then for each configuration of energy levels, $Z(y)$ may be calculated, and for each line $\log gf\lambda k_a + \Delta\chi\theta_a$ is computed. The integral L^* defined by Eq. 8–84 is evaluated for each atomic configuration and wavelength of interest. Then the quantity $\log C$ can be computed for each line. Calculation of the integrand in Eq. 8–83 is more complicated because of the Ψ factor and the correction to the weight function. For each selected value of M, we obtain a value of $(W/\lambda)_0$ from Eq. 8–84 and W/λ from Eq. 8–83. A plot of $\log W/\lambda$ against $\log C$ constitutes a *curve of growth* for the particular lines involved — a different curve is obtained for each group of energy levels for each atom in each stage of ionization.

To summarize, the Pecker theory may be applied to the calculation of theoretical line intensities where one wishes to take into account the effects of the detailed structure of the atmosphere on the formation of the lines. The procedure has definite advantages over the direct integration method described in Art. 8–5, because certain quantities, such as the saturation functions, weighting functions, etc., may be calculated once and stored for future reference. Thus the problem is broken down into an evaluation of a contribution function that is made up of several factors; $G_\lambda{}^0(y)$ or $G_\lambda{}^m(y)$ that depend only on y; a numerical constant M; a function $Z(y)$ that depends on the energy level involved; and a function Y (plus a function Ψ derived from it) that depends on the product MZ.

In practice, one is usually confronted with the problem of interpreting the spectrum of a particular star. As a specific illustration we shall take a B-star. The observational data consist of (1) measurements of the energy distribution in the continuous spectrum, perhaps expressed in terms of

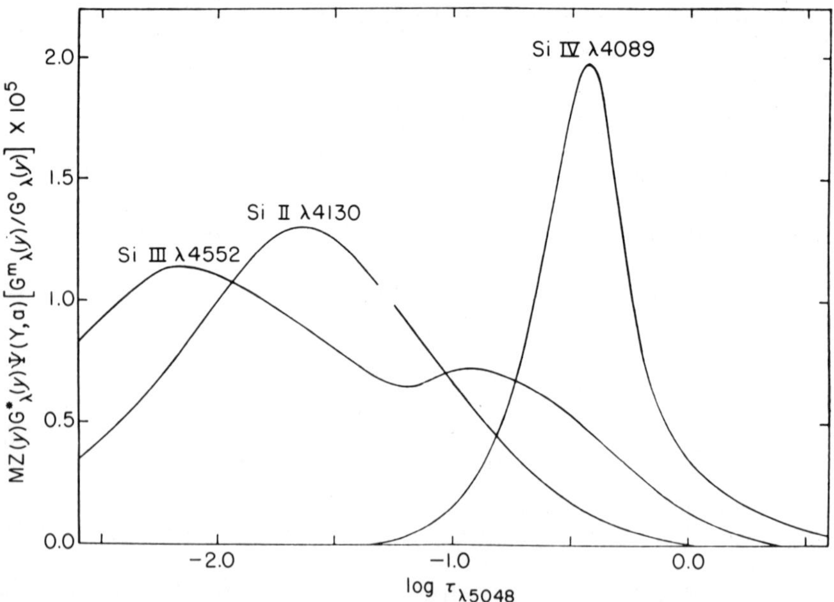

Fig. 8–5. Contribution Function for Lines of Silicon in an Early Type Stellar Model

Here the contribution functions (integrand of Eq. 8–83), evaluated for a silicon/hydrogen ratio of 5×10^{-5} and plotted against $\log \tau_{5048}$ are displayed for Si II λ4130, Si III λ4552, and Si IV λ4089 for Underhill's model II corresponding to spectral class $B2.5$V, $\log g = 4.20$. (From *Ap. J. Suppl.* no. 38, **4**, 109, 1959, Fig. 1; courtesy, University of Chicago Press.)

color temperatures above and below the Balmer limit — and the Balmer discontinuity, (2) equivalent widths of absorption lines, and (3) profiles of hydrogen and helium lines.

Theoretical model atmosphere calculations enable us to compute the flux $\pi F_\lambda(0)$ in the emergent continuous spectrum and by the Pecker method (or some other procedure) we can calculate theoretical line intensities. For the hydrogen line profiles one may employ Eq. 8–5 with the line absorption coefficient in Eq. 8–48 obtained from the Kolb-Griem theory of Chapter 7. In practice one often finds that none of the models at hand will represent the spectral data for the star in question. Then one may interpolate between the models. It is often found that one cannot simultaneously represent all the observed data. In γ Pegasi (for example, Underhill, 1957; or Aller and Jugaku, 1959), one may interpolate a model that represents the intensities of the Si II and Si IV lines, but the Si III lines will then be too strong. The continuous energy distribution often is difficult to interpret because of effects of space reddening. On the other hand, in spectral classes *B* and *A*, hydrogen line profiles are much more sensitive to surface gravity than to temperature. Hence one may use the Si II, Si III and Si IV line intensities to obtain the effective temperature of the model atmospheres that most closely represents the observed line spectral characteristics, whereas the hydrogen line profiles enable us to fix the surface gravity.

Calculation of theoretical intensities of silicon lines in a particular stellar model may be discussed briefly. Fig. 8–5 shows contribution functions for three well-known silicon lines in a model atmosphere calculated by Underhill (1957). The temperature of the atmosphere which corresponds to that of a *B*2.5V star is such that higher energy levels of ions of Si^{++} are favored over those of Si^{+++} and relevant levels of Si^{+}. The Si IV lines are formed in deeper layers than those of Si II, whereas the Si III lines are formed over a huge range of optical depth. Notice that the Si III curve has two maxima. The deeper one corresponds to the maximum of the product $ZG_\lambda{}^0$ whereas the higher one displays the influence of the saturation factor which strongly favors the higher layers. A substantial contribution probably is made by the chromospheric layers. With a lower Si/H ratio the ordinates of the curves would be decreased and their maxima would be shifted to deeper layers.

Theoretical curves of growth for silicon lines in Underhill's model II are shown in Fig. 8–6. Notice that the different ions give different theoretical curves; that for Si IV lies considerably below the curves for Si III and Si II because the Si IV lines are formed much deeper in the atmosphere. Stating the result in another way — a good share of the contribution to the continuous spectrum arises in layers above these primarily responsible for the Si IV lines.

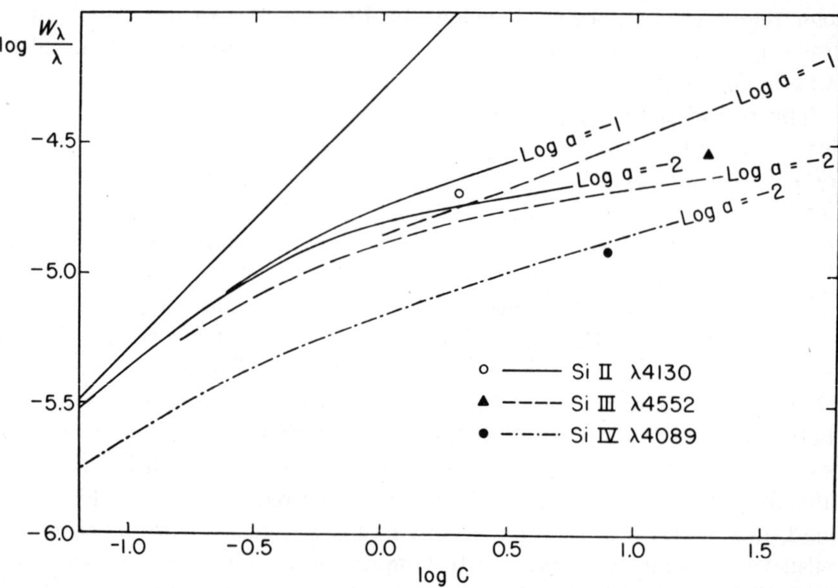

FIG. 8–6. THEORETICAL CURVES OF GROWTH FOR SILICON LINES IN UNDERHILL'S MODEL II

Ordinates are log W/λ, abscissas are log C (defined by Eq. 8–85). Theoretical curves for different damping constants are compared with intensities obtained by Underhill (1957) from direct calculations of line profiles. (From *Ap. J. Suppl.* no. 38, **4,** 109, 1959, Fig. 2; courtesy, University of Chicago Press.)

Underhill (1957), Cayrel (1956), and others have calculated line intensities in early-type stars by methods of direct integration rather than by the Pecker theory. Comparison with the latter generally shows fair agreement (Fig. 8–6); the discrepancies between the theoretical calculations generally are less than those between either theory and observation. The existence of these discrepancies show that faults still lie in our model atmospheres.

A number of possible causes for the discordances may be cited: (1) errors in adopted f-values, (2) errors in the damping constants, (3) errors in observed line intensities, (4) the models are wrong, and (5) deviations from thermodynamic equilibrium.

Difficulties (1) and (2) will be removed as reliable experimental data and theoretical procedures become available. Errors in the measured line intensities cannot account for the discrepancies, although improvements in the measured profiles and equivalent widths are urgently needed.

The models are inadequate in that they do not allow for the "blanketing" effect produced by lines upon the temperature distribution in the uppermost layers, nor upon the influence of such lines upon the flux. The rocket data (Fig. 6–7) indicate that the theoretical fluxes are grievously in error in the ultraviolet. Further, the effects of stellar rotation should be considered. Huang and Struve (1956) note that rotational effects on stellar spectra can be important! Deviations from thermodynamic equilibrium are particularly nasty to handle (see Art. 8–17). Theoretical calculations are reliable only when the entire radiative and hydrodynamical structure of the atmosphere is known, although empirical methods have been employed for the sun by Pecker and his associates. He shows that if the excitation temperature T_{ex} does not agree with the local temperature T defined by the gas kinetic motions, one must replace Eq. 8–73 by

$$g_{\lambda,i}{}^{\mu}(\tau_\lambda) = \frac{\displaystyle\int_{\tau_\lambda}^{\infty} B_\lambda(T)\, e^{-\zeta_i/\mu}\, \frac{d\zeta_i}{\mu} - B_\lambda(T_{ex})\, e^{-\tau_\lambda/\mu}}{I_\lambda{}^c(\mu)} \tag{8–87}$$

where $I_\lambda{}^c(\mu)$ is calculated as before from Eq. 5–5 in terms of T, but T_{ex} may vary from one level to another in an atom or ion. Here one notes that $B_\lambda(T)$ is the source function for the continuum, and $B_\lambda(T_{ex})$ that for the line.

Model atmospheres appropriate for theoretical continuum and line spectrum calculations have been published by a number of investigators (see also Chapter 7). We may mention especially the models calculated by Cayrel, Saito (1954–56), McDonald (1953), Pecker (1951), and Underhill (1957) (B stars); those of Hunger (1955), Osawa (1956), Saito (1956), Martini and Masani (1957), Przybylski (1953) and Hack (1956) (A–F stars); and the extensive series for many spectral types and surface gravities by de Jager and Neven. Pecker has attempted to calculate models for giant stars. With increasing use of high speed electronic computers and improvements in basic astrophysical theory we may expect that good theoretical models will become available for the entire main sequence from early O to K and eventually M stars. Models for giant atmosphere will be more difficult to construct. Concurrent with these developments we may expect extensive calculations of theoretical intensities of many lines of interest for determining pressures (or surface gravities) and temperatures in stellar atmospheres. One must keep in mind that such calculations can never be any better than the validity of the basic underlying theory, including the atmospheric models.

8–7. The Curve of Growth. Methods that involve the use of model atmospheres and detailed theories of line formation are called *fine analysis* methods. The elaborate, lengthy calculations required are justified only when accurate observational data are available and the temperature, surface gravity and chemical composition of the atmosphere are known to a good approximation.

For the first reconnaissance of a stellar spectrum it behooves the astronomer to apply a curve of growth analysis of a conventional type in which one has a precise relationship between W/λ and the number of atoms acting to produce the line in question. Various representations of the model atmosphere (usually Milne-Eddington or Schuster-Schwarzschild) may be employed and mechanisms of line formation (usually pure scattering or pure absorption) postulated. In addition, it is assumed that there exists a specific value of the electron pressure, of the gas pressure and of the temperature that will describe adequately the state of ionization of each particular element and the coefficient of continuous absorption (ME model) or permit one to compute the numbers of atoms above the photosphere (SS model). It is sometimes supposed (particularly in work on solar type stars) that there exist excitation temperatures T_{ex} [that describe the population of levels by Boltzmann's formula] that differ from the ionization temperature. This non-equality of T_{ex} and T_{ion} does not imply necessarily deviations from LTE; it may merely be a consequence of the temperature stratification in the atmosphere. Furthermore, it is assumed that all lines of the same equivalent width have exactly the same shape and usually each line follows the same curve of growth as every other line. Provision is made, however, for the possibility that different lines may have different damping constants.

First of all, let us look at the problem of the curve of growth from the simplest point of view: How will the equivalent width (intensity) of a spectral line change as the number of atoms acting to produce it increases? The significant parameter is not simply N, the number of atoms, but Nf, the number of atoms times the f value. When Nf is small, we observe only a small depression in the continuous spectrum. As it increases, the line center deepens and widens until, when Nf becomes very large, prominent "wings" begin to appear. We depict the process in Fig. 8–7a, which shows the growth of a line formed by atoms in a thin layer (Schuster-Schwarzschild model) according to the mechanism of pure scattering. The number, N_0, is so chosen that for this number of atoms above the photosphere, the optical depth at the line center, which we call X_0, will be 1. When N is less than $100N_0$, the Doppler effect is dominant in fixing the line shape, and the line profile is bell-shaped. For a time, W grows very slowly with N; then damping wings begin to appear. At $1000N_0$ the line shows a combination of a bell-shaped Doppler profile and incipient "wings." Beyond $N = 10{,}000N_0$, these wings are marked, and finally they alone fix the value of W. In the calculation of the profiles depicted in Fig. 8–7a we have assumed radiation damping only. If collision damping occurs, wings become developed for smaller numbers of atoms.

We could construct a relation between W and numbers of atoms, i.e., the curve of growth, from a diagram such as Fig. 8–7a by measuring the areas subtracted from the continuum by the profiles and plotting the resultant log W against log Nf. For example, we could calculate the profile of λ3933 by the method of Art. 8–5 for different values of log N, measure the area of the profile W, and plot it against the number of atoms. The

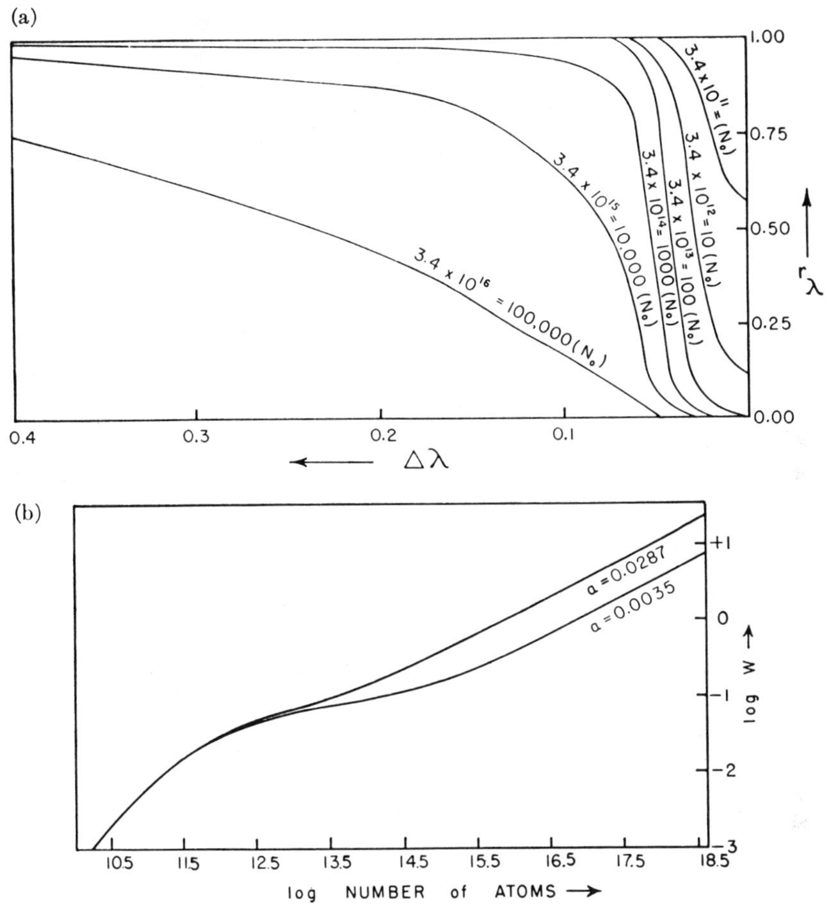

Fig. 8–7. The Curve of Growth for λ3933

(a) Theoretical profiles calculated for the Schuster-Schwarzschild model and pure radiation damping show how the shape of the line changes as the number of absorbing atoms increases. The number $N_0 = 3.4 \times 10^{11}$ is so chosen that the optical depth at the center of the line for N_0 atoms, X_0, will be 1. N denotes the number of atoms above the photosphere.

(b) From the integration of the profiles of Fig. 8–7a we obtain log W which we plot against log N, the number of atoms above the photosphere. Curves are given for $a = 0.0035$ and 0.0287 (see Eqs. 7–68 and 7–72).

results of such a calculation are shown in Fig. 8–7b where we have carried out computations for: (1) pure radiation camping with $a = 0.0035$ and (2) collision plus radiation damping with $a = 0.0287$. Notice that for small numbers of atoms, W is proportional to N; later the curve flattens out and finally W varies as $\sqrt{N}$ or more precisely as $\sqrt{Nf\Gamma}$.

We emphasize that the curve of growth is a plot of W against Nf, not simply N. In practice, we measure the equivalent widths of lines whose relative f values are known, and plot them against f. This operation defines an empirical curve whose comparison with the theoretical curve of growth may serve to fix N, the number of atoms acting to produce the line. In practice it is necessary to know also the effective damping constant. Fortunately, to a sufficient degree of accuracy, we can calculate the curve of growth once and for all analytically for a given set of assumptions and adopted mechanism of line formation. The curve is relatively insensitive to the atmospheric model and to the details of the line shape. The equivalent width of a line is given by

$$W = \int (1 - r)\, d\lambda = \Delta\lambda_0 \int (1 - r)\, du \tag{8-88}$$

where r is the residual intensity, $\Delta\lambda_0$ is defined by Eq. 7–76 and u is defined by Eq. 7–77, i.e.,

$$\Delta\lambda_0 = \frac{\lambda}{c}\sqrt{\frac{2kT}{M}} \quad \text{or} \quad \Delta\lambda_0 = \frac{\lambda}{c}\sqrt{\frac{2kT}{M} + \xi^2} \tag{8-89}$$

if turbulence is present (see Art. 8–10).

$$u = (\lambda - \lambda_0)/\Delta\lambda_0 \tag{8-90}$$

Various workers have calculated curves of growth with the aid of different assumptions about r. Unsöld and his associates have used Minnaert's empirical formula,

$$\frac{1}{R} = \frac{1}{R_c} + \frac{1}{N_{r,\,s}\,\alpha_\nu}, \; R = 1 - r \tag{8-91}$$

where R_c is the central depth of the line and $N_{r,s}$ is the number of atoms above the photosphere, although it can also be redefined to apply to model atmosphere calculations. Menzel (1936) employed the Schuster-Schwarzschild model and took

$$r_\nu = [1 + N_{r,s}\,\alpha_\nu]^{-1} \tag{8-92}$$

although Barbara Bell wrote

$$(1 - r) = \frac{R_c N \alpha_\nu (\kappa_0/\kappa_\nu)}{1 + N\alpha_\nu(\kappa_0/\kappa_\nu)} \tag{8-93}$$

The factor R_c allows for the fact that spectral lines have a finite central intensity $r_c = 1 - R_c$ and the factor κ_0/κ_λ allows for the variation of the depth of the photosphere with wavelength.

Although Huang (1952) concluded that Unsöld and Menzel's curves of growth were satisfactory for practical use, a number of objections can be raised against Eqs. 8–91, 8–92, or 8–93. For example, Münch (1958) concludes that in hot stars where R_c is of the order of 0.5 or 0.6 instead of 1, Eq. 8–91 predicts line intensities that are systematically in error. Pecker and van Regemorter (1955) find from a more precise treatment of the

Schuster-Schwarzschild curve of growth that although small departures from the shape of Menzel's curve are indicated, the principal discordance comes from the fact that Menzel's curve corresponds to taking the weighting function $G(0)$ equal to 1 at the surface, whereas actually it is less than one. Corrections are also indicated for Unsöld's curve of growth. See also Barbier (1956) and van Regemorter (1959) who discuss curve of growth theory from points of view of weighting functions. Hunger (1956) has discussed analyses of curves of growth based on Minnaert's formula, Eq. 8–91 (1949, 1950). The most elaborate calculations are those of Wrubel, which are based on the mechanism of scattering, the Milne-Eddington model and Chandrasekhar's exact solution of the transfer equation (cf. Eqs. 8–31 and 8–32) for the integrated flux and for different parts of the solar disk. More recently Wrubel (1954) has calculated exact curves of growth for scattering in the SS model. He has also considered the curve of growth for pure absorption in the ME model (1960).

Here we shall discuss Wrubel's curves of growth for the total flux based on the mechanism of scattering and the ME model. He supposes that the Planckian function B depends linearly upon the optical depth at the wavelength in question in accordance with Eq. 8–17. The coefficients B_0 and B_1 may be estimated in several ways as noted in Art. 8–3. If one supposes a grey-body distribution of temperature, we can show (Problem 5–5) that

$$\frac{B_0}{B_1} = \frac{8}{3}\frac{\kappa_\lambda}{\bar{\kappa}}\frac{[1 - \exp(-h\nu/kT_0)]}{h\nu/kT_0} \tag{8–94}$$

where T_0 is the boundary temperature of the star, $\bar{\kappa}$ is a mean absorption coefficient, and κ_λ is the absorption coefficient in the neighborhood of the line.

Wrubel calculated curves of growth for ratios $B_0/B_1 = 1/3$, $2/3$, $4/3$, and $10/3$ and for values of $\log a = -1.0, -1.4, -1.8, -2.2, -2.6$, and -3.0. Then for each value of B_0/B_1 he tabulated the curve of growth for different values of the damping constant. Interpolation of curves of growth for values of B_0/B_1 and a different from those tabulated is straightforward.

As ordinate one employs

$$\log\frac{W}{\lambda}\frac{c}{V} = \log\frac{W}{\Delta\lambda_0}$$

where V is the most probable velocity of the atoms. It may contain contributions from both thermal and large-scale (turbulent) motion (see Art. 8–10 and Eq. 8–89). See Fig. 8–8.

As abscissa one uses

$$\log\eta_0 = \log N_{r,s} + \log\alpha_0 - \log\kappa_\lambda \tag{8–95}$$

where $N_{r,s}$ is the number of atoms per gram of stellar material capable of absorbing the line in question, α_0 is the fictitious absorption coefficient at the center of the line (Eqs. 7–69 and 7–70).

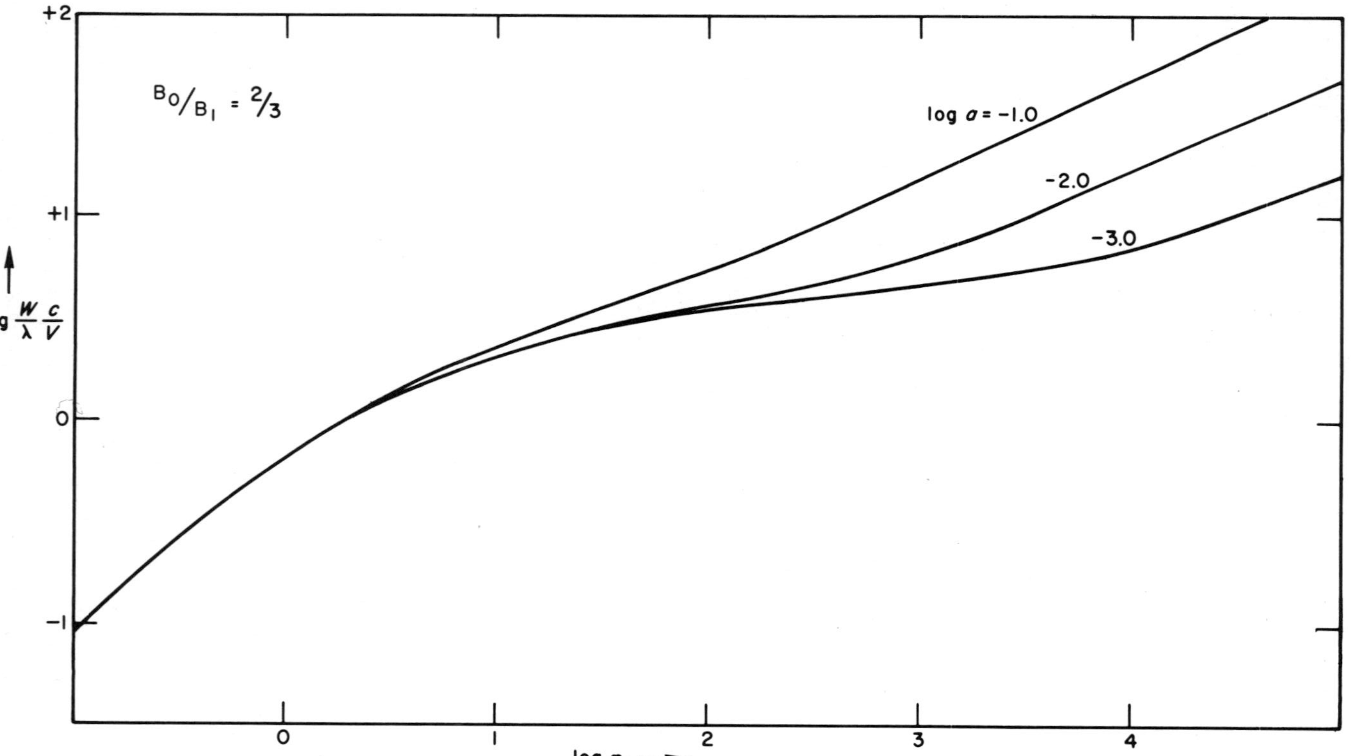

Fig. 8–8. Wrubel's Curve of Growth for Pure Scattering in the Milne-Eddington Model with $B_0/B_1 = 2/3$

To construct a curve of growth we usually employ lines covering a considerable wavelength range. Hence a single value of B_0/B_1 will not be valid (see Eq. 8–94). In practice we choose a particular set of curves, e.g., the family for $B_0/B_1 = 2/3$ and apply for each line a correction $\Delta \log \eta_0$ to η_0 depending on the value of B_0/B_1 at the wavelength in question.

How do curves of growth based on different theoretical hypotheses compare with one another? Superficially, most such curves look alike but closer examination shows a number of important differences. First, curves based on the hypothesis of pure absorption (ME model) differ considerably from those based on the scattering hypothesis since a line formed by the latter mechanism is always black at the center whereas a line formed by absorption has a central intensity corresponding to $B_\nu(T_0)$ where T_0 is the boundary temperature. Furthermore the approximation (Eq. 8–17) gives a poor representation for stars such as the sun. Differences also exist between the SS and ME models; for example, limb darkening has a greater

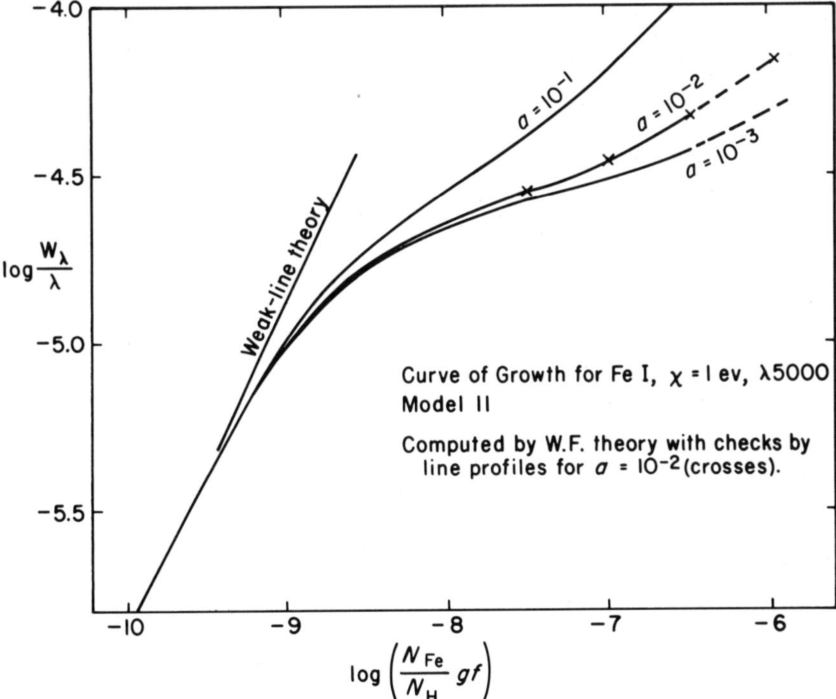

Fig. 8–9. A Theoretical Curve of Growth for Iron Lines in the Solar Spectrum

Ordinates are log W_λ/λ, abscissae are log $\{[N(\text{Fe})/N(\text{H})]\,gf\}$. P. Mutschlecner calculated these curves for a model solar atmosphere by the weighting function method for an iron line of lower excitation potential 1 v, falling at λ5000. The values for $a = 0.01$ were checked by direct calculation by line profiles. The mechanism of pure absorption was assumed.

effect in the latter and the shape of the curves for log $W_\lambda(\theta)c/\lambda V$ change more markedly across the disk (Wrubel, 1954).

Accordingly, results may depend on the kind of curve of growth used. For example, in an analysis of early-type stars, different abundances values may be obtained from pure absorption curves of growth than from pure scattering ones in the ME model.[5] A theoretical curve of growth suitable for solar type stars is not necessarily best for high temperature stars.

Fig. 8–9 shows a theoretical curve of growth for the sun, calculated for the center of the disk. It gives a fairly good representation of the observational data for V, Ti and Fe. One procedure, which is gaining favor, is to calculate a theoretical curve for a neutral metal, e.g., Fe, and another for an ionized metal (e.g., Fe^+), making use of a model atmosphere for the star and some such procedure as is described in Art. 8–5. Then, in *a differential comparison* (see Art. 8–9) *of stars of nearly the same spectral and luminosity class* one may use the "neutral" curve of growth for all neutral lines and the "ionized" curve for all ionic lines.[6] Further refinements can be introduced, but one must decide whether the basic data justify the extra effort involved. To utilize a theoretical curve of growth we measure the equivalent widths of a group of lines whose f-values are known or can be estimated. From the spectral type and UBV, six-color photometry, or spectral energy distribution we can usually estimate the temperature and electron pressure sufficiently well to obtain a good working value of $V_{\text{thermal}} = \sqrt{2kT/M}$ and κ (see Chapter 5). Then for each line we may calculate $\log \eta_c/N$ and $\log Wc/\lambda v$ and fit a plot of the points on our chosen theoretical curve by a horizontal shift.[7] The amount of the horizontal shift then determines $N_{r,s}$ the number of atoms capable of absorbing the line. Now

[5] See, e.g., *Handbuch der Physik* **51**, 324, 1958.

[6] The curves shown in Fig. 8–9 may be used for lines of neutral metals in solar type stars. They were calculated by the weighting function method and checked by actual calculation of line profiles. Care must be taken to include properly contributions from the uppermost layers which are important for strong lines.

$\log \frac{N_{\text{Fe}}}{N_H} gf$	$\log W/\lambda$		
	$\log a = -3$	-2	-1
-10.00	-5.87	-5.87	-5.87
-9.50	-5.41	-5.41	-5.40
-9.00	-5.03	-5.03	-5.01
-8.50	-4.78	-4.78	-4.73
-8.00	-4.65	-4.64	-4.55
-7.50	-4.57	-4.55	-4.39
-7.00	-4.52	-4.47	-4.20
-6.50	-4.43	-4.33	-3.96

[7] A vertical shift is sometimes required. This means that to the gas thermal motion one must add the contribution of turbulence (see Eq. 8–89) and Art. 8–10.

$N_{r,s}$ will depend on the ionization equilibrium, the temperature, and the abundance of the element in question.

8-8. An Illustration of the Curve of Growth Method for the Sun. To explain one practical use of the curve of growth, let us interpret the equivalent widths of solar lines of aluminum with the aid of Wrubel's curve of growth for scattering in the ME approximation.

At the outset one must choose certain mean values of the temperature, electron, and gas pressures, viz., $\overline{T}$, $\overline{P}_\varepsilon$, $\overline{P}_g$, from which we can obtain parameters needed for the curve of growth. For a typical star we might use the graphs discussed in Chapter 5, but for the sun we may be guided by results of model atmosphere calculations. Thus we choose

$$\bar{\tau} = 0.3 \qquad \theta = \frac{5040}{T} = 0.9 \qquad P_\varepsilon = 10 \text{ dynes cm}^{-2}$$

The abscissa of the curve of growth is given by Eq. 8-95. If $N_{r,s}$ can be expressed in terms of Boltzmann's equation:

$$\log N_{r,s} = \log N_r + \log g_{r,s} - \log u(T) - \theta\chi_{r,s} \tag{8-96}$$

where $u(T)$ denotes the partition function (see Chapter 3). Numerically,

$$\alpha_0 = \frac{\sqrt{\pi}\varepsilon^2}{mc} f \frac{1}{\Delta\nu_0} = 0.0150\, f\lambda/V, \quad V^2 = \left(\frac{2kT}{M}\right) + \xi^2 \tag{8-97}$$

Then, from Eqs. 8-95, 8-96, and 8-97

$$\log \eta_0 = -1.824 + \log N_r + \log gf\lambda - \theta\chi_{r,s} - \log Vu(T) - \log \kappa_\lambda \tag{8-98}$$

is the expression for the abscissa for the curve of growth.

To calculate V we adopt a turbulence velocity $\xi = 1.3$ km sec^{-1}. The thermal velocity for aluminum atoms, $\sqrt{2kT/M}$ is 1.9 km sec^{-1}, so $V = 2.3$ km sec^{-1} (see Eq. 8-97). Hence $\log c/V = 5.115$. At solar temperatures $u(T) = 6$ for Al I. At λ5000 the continuous absorption is assumed to be due to the negative hydrogen ion only. If hydrogen contributes 0.613 by weight, and if the absorption coefficient per hydrogen atom at $\theta = 0.9$, $P_\varepsilon = 1$ dyne cm^{-2} is 0.515×10^{-25} (Chandrasekhar), then

$$\kappa(\lambda_0) = 0.515 \times 10^{-25} \times 10 \times 6.025 \times 10^{23} \times 0.613 = 0.19$$

per gram of solar material. Substituting $\kappa_\lambda = (\kappa_\lambda/\kappa_{\lambda 0})\,\kappa_{\lambda 0}$ in Eq. 8-98 along with appropriate numerical values we obtain

$$\log \eta_c + \theta\chi - \log N_r = \left[-12.24 + \Delta \log \eta - \log \frac{\kappa_\lambda}{\kappa_{\lambda 0}}\right] + \log gf \frac{\lambda(A)}{1000} \tag{8-99}$$

where η_c denotes the value of η_0 corrected by $\Delta \log \eta$ so that all points may be fitted to the curve of growth for $B_0/B_1 = \frac{2}{3}$. Table 8-2 gives the data relevant for the problem. The first two columns give the wavelengths of the lines, the transitions and J-values involved. The third column gives the excitation potential

TABLE 8–2

CURVE OF GROWTH DATA

λ	ΔJ	χ	$\log gf\dfrac{\lambda}{1000}$	$\log\dfrac{W}{\lambda}\dfrac{c}{V}$	$\dfrac{B_0}{B_1}$	$\log\dfrac{\kappa_\lambda}{\kappa_{\lambda_0}}$	$\Delta\log\eta_0$	$\log\eta+\theta\chi$ $-\log N_r$
^2P–^2S								
3961.54	3/2–1/2	0.00	+0.33	+1.48	0.28	−0.04	+0.13	−11.74
3944.02	1/2–1/2	0.00	+0.03	+1.34				−12.04
^2S–^2P								
6696.03	1/2–3/2	3.13	−0.52	−0.19	0.65	+ .10	−0.00	−12.86
6698.67	1/2–1/2	3.13	−0.82	−0.18				−13.16
^2S–^2P								
13123.32	1/2–3/2	3.13	+1.35	0.53	0.67	−0.15	−0.00	−10.73
13151.02	1/2–1/2	3.13	+1.06	0.41				−11.03
^2S–6^2P								
5557.07	5/2–3/2	3.13	−1.35	−0.93	0.49	+0.05	+0.04	−13.60
^2D–^2F								
8773.91	5/2–5/2, 7/2	4.00	+0.95	+0.17	0.88	+0.12	−0.06	−11.47
8772.77	3/2–3/2	4.00	+0.77	+0.08				−11.65
^2D–^2F								
7836.13	5/2–5/2, 7/2	4.00	+0.56	+0.03	0.80	+0.12	−0.04	−11.84
7835.32	3/2–5/2	4.00	+0.38	−0.15				−12.02
^2D–^2F								
7362.29/	5/2–5/2, 7/2	4.00	+0.39	−0.11	0.75	+0.12	−0.03	−12.00
7361.55/	3/2–5/2	4.00	+0.21	−0.26				−12.18
^2P–^2D								
16719.12	1/2–3/2	4.08	+1.53	+0.44	0.50	−0.30	+0.06	−10.36
16750.65	3/2–5/2	4.05	+1.79	+0.56				−10.09
16763.48	3/2–5/2	4.05	+0.83	+0.09				−11.06

of the lower levels. (See Charlotte Moore, *Revised Multiplet Table*.) Column 4 gives $\log gf[\lambda(A)/1000]$. The f-values were taken from Biermann and Lübeck (1948) or computed from Bates-Damgaard tables (Chapter 4). Column 5 gives $\log (W\,c/\lambda\,V)$ as obtained from the observed equivalent widths of the solar aluminum lines ($V = 2.3$ km sec^{-1}).

The calculation of B_0/B_1 for column 6 is carried out with the aid of Eq. 8–94 with $T_0 = 4600°$K and with $\bar{\kappa} = 1.10\,\kappa$ ($\lambda 5000$). Then $\log \kappa_\lambda/\bar{\kappa}_{\lambda 0}$ is calculated from tables of the absorption coefficient of the negative hydrogen ions, and $\Delta\log\eta_0$ is taken from Table 8–3. Finally, the right-hand side of Eq. 8–99 is tabulated in the last column.

TABLE 8–3

VALUES OF Δ LOG η TO BE ADDED TO LOG η TO REDUCE TO THE CURVE OF GROWTH FOR $B_0/B_1 = 2/3$

(After M. Wrubel)

B_0/B_1	$\log\dfrac{Wc}{\lambda V} = -1.0$	−0.5	0	+0.5	+1.0
1/3	+0.08	+0.09	+0.11	+0.15	+0.11
4/3	−0.10	−0.11	−0.13	−0.15	−0.12
10/3	−0.24	−0.25	−0.29	−0.31	−0.26

We now separate the lines into three sets according to their χ values, 0, 3.13, and 4 v. For each of these sets we now plot log $(W\,c/\lambda\,V)$ against $\log \eta_c/N_r + \theta\chi$, i.e., the entries in column 5 against the corresponding numbers in column 9. In order to fit these empirical plots for the λ3944 and λ3961 lines to the theoretical curve of growth for $B_0/B_1 = \frac{2}{3}$ we must obtain the damping constant for these lines.

From our solar model atmosphere we adopt $P_g = 8.9 \times 10^4$ dynes cm^{-2} (we could have used the $\bar{P}_g\bar{P}_\varepsilon$ curves from Chapters 3 and 5). Using Eq. 7–56, we get $\Gamma_{\text{coll}} = 1.12 \times 10^9$, while $\Gamma_{\text{rad}} = A_{nn'} = 1.1 \times 10^8$ (Allen). Hence $\Gamma_{\text{total}} = 1.2 \times 10^9$ and $\log a = -1.8$.

Fig. 8–10*a*, *b*, *c* exhibit the manner of fitting empirical points to the theoretical curve. The amount of shift required depends on χ. Thus we get:

Set	χ	$\log N_r - \theta\chi$
(a)	0.00	16.10
(b)	3.13	12.80
(c)	4.0	12.13

We plot $(\log N_r - \theta\chi)$ against χ and pass a straight line through the three points. The slope of the curve gives θ and the intercept with the $\chi = 0$ axis the value of N_r. In this way we find $\theta = 1.0$ or an excitation temperature of 5000°K, somewhat lower than T_{ion}. The corresponding $\log N_r$ is 16.10. This result is sensitive to the choice of damping constant for λ3944 and λ3961. If we assume that $\theta = 0.9$ and omit these two strong lines, we find $\log N_r = 15.60$. Actually, one should use only lines of comparable intensity for determinations of excitation temperatures. Excluding the resonance lines, we find $\log N_r = 15.60$ for neutral aluminum. If $T = 5600$°K, $P_\varepsilon = 10$ dynes cm^{-2}, the ionization equation yields $\log N_1/N_0 = 2.06$ whence $\log N(\text{Al}) = 17.66$. With $\log N(\text{H}) = 23.57$ we get $\log N(\text{Al})/N(\text{H}) = -5.91$.[8]

Since the continuous absorption depends on the concentration of negative hydrogen ions, while line absorption depends on $N_{r,s}f$, the ratio, η, depends on the abundance ratio of the element in question to hydrogen. Furthermore, the determination of the aluminum abundance is insensitive to the choice of $\bar{P}_\varepsilon$. The concentration of neutral aluminum atoms is proportional to $\bar{P}_\varepsilon$, while at the same time the negative hydrogen ion absorption $\kappa_\lambda(\text{H})$ is proportional to $\alpha_0{}^0\,(\text{H}^-)P_\varepsilon$. On the other hand the abundance of an element calculated from its enhanced lines would be sensitive to the adopted electron pressure.

In principle one may use the preceding method with lines of neutral and ionized atoms and establish the ionization equilibrium. In theory one can get both T and P_ε this way but in practice it is better to adopt the temperature from the intrinsic color or energy distribution in the star, and use ionization equilibrium of Si I and Si II, Sr I and Sr II, etc. to get P_ε. The chief difficulty here is that one often does not have the necessary f-values for the ion-lines.

[8] An analysis of the same data with the SS model and Menzel's curve of growth gives $\log N = 16.09$ for the total number of aluminum atoms above the photosphere. (See Aller, 1961.) The depth of the photosphere depends on $\kappa_\lambda\,(\text{H}^-)$ and thus also on the concentration of H and the electron pressure.

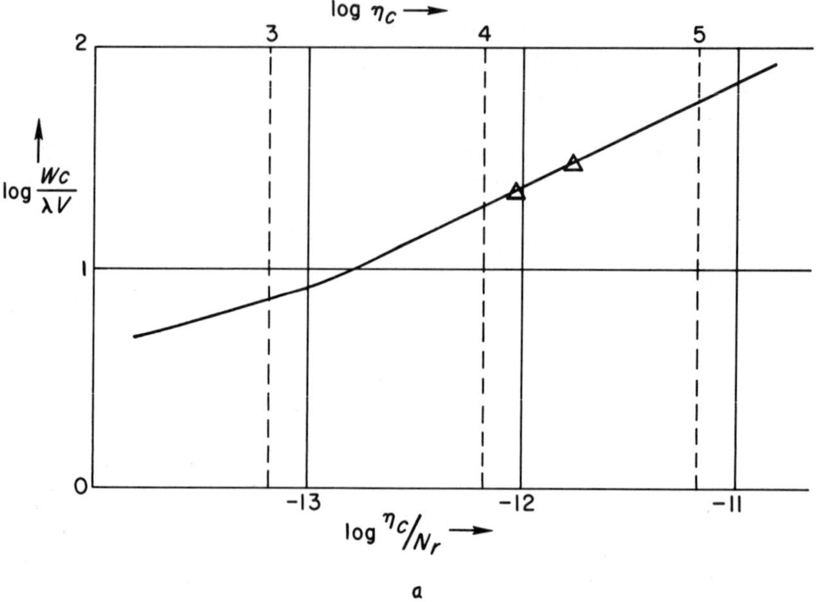

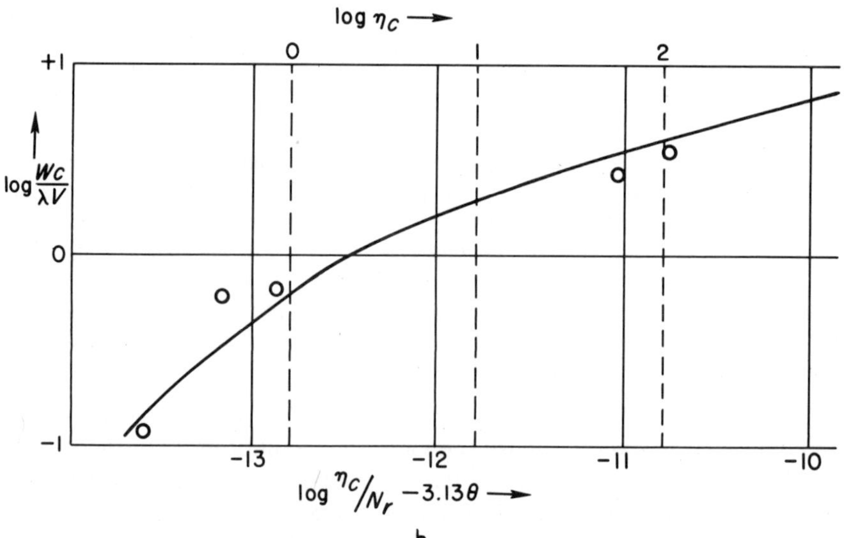

FIG. 8–10. CONSTRUCTION OF AN EMPIRICAL CURVE OF GROWTH

The theoretical curve is that due to Wrubel, $B_0/B_1 = 2/3$ and $\log a = -1.8$. (a) The empirical fit of the points for $\chi = 0$ to the theoretical curve. The upper scale gives

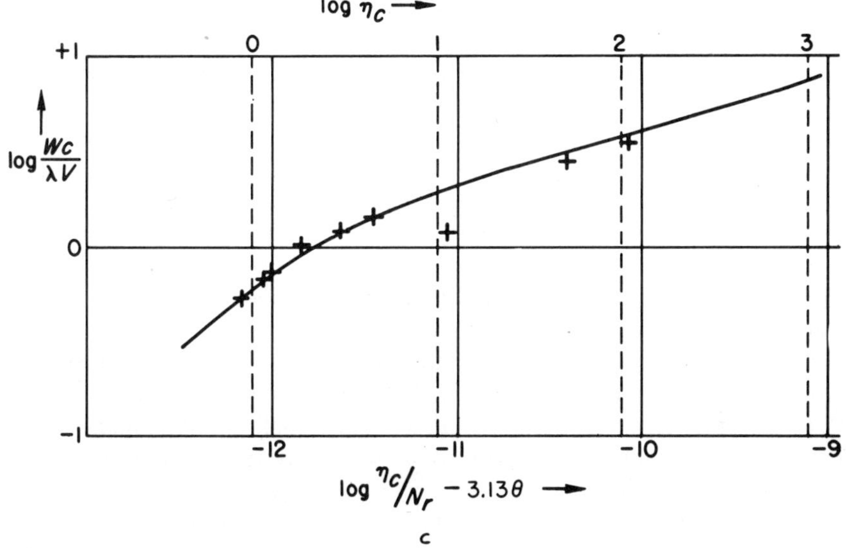

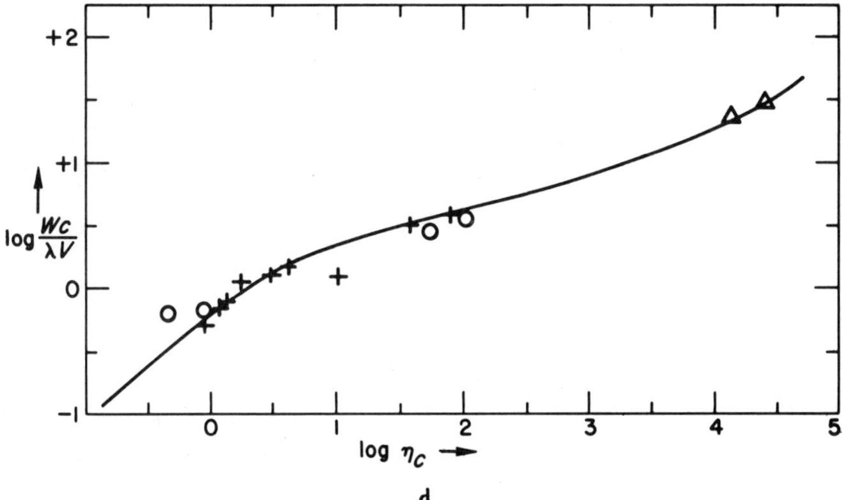

abscissas of the theoretical curve, the lower scale those of log η_c/N_r; ordinates are log $Wc/\lambda V$. (b) Data for χ = 3.13 ev. The lower scale of abscissas is log η_c/N_r −3.13θ. (c) Data for χ = 4.00 ev. Lower scale is log η_c/N_r − 4.00θ. (d) The curve of growth for aluminum.

8–9. A "Differential" Curve of Growth Method. In the method we have just described, each star is treated essentially independently of any other star. That is, one decides on a particular mechanism of line formation, atmospheric structure, set of f-values and proceeds from more or less first principles. The difficulty with this fundamental approach is that first of all one must know the f-values for all transitions used. Secondly, one must decide on the mechanism of line formation (scattering or absorption or a combination of the two), whether the SS or the ME model best represents the situation for the lines in question, and then select a theoretical curve that fits these considerations. Blends and the failure of a particular stellar atmosphere to fit the expected model will all affect the individual results.

On the other hand, in the differential method one compares a program star with a standard star for which the chemical composition, $\overline{T}$, $\overline{P}_\varepsilon$ and $\overline{P}_g$ are presumably known. The method can be applied only to stars which are sufficiently similar in temperature as to display the same spectral lines with not too dissimilar excitations.

To illustrate the procedure let us consider a star whose temperature does not differ greatly from that of the sun. Using the superscripts O and S to denote quantities for the sun and star respectively, we obtain for each spectral line from Eq. 8–98

$$\log \frac{\eta_0{}^O}{\eta_0{}^S} = \log \frac{N_r{}^O}{N_r{}^S} - \chi_{r,s}\,(\theta_O - \theta_S) - \log \frac{V^O u^O(T)}{V^S u^S(T)} - \log \frac{\kappa_\lambda{}^O(P_\varepsilon{}^O, T)}{\kappa_\lambda{}^S(P_\varepsilon{}^S, T)} \tag{8–100}$$

Notice that not only numerical constants but also the $gf\lambda$'s cancel out. Now $\eta_0{}^O$ and $\eta_0{}^S$ may be read for each line in the sun and star from the curve of growth with $\log Wc/\lambda V$ as ordinate. This curve is assumed to be the same for sun and star. To simplify the notation, let us write for any quantity, ζ

$$\log \zeta^O/\zeta^S = [\zeta] \tag{8–101}$$

Then Eq. 8–100 is written in the form

$$[\eta] = [N_r] - \chi_{r,s}(\theta^O - \theta^S) - [Vu(T)] - [\kappa_\lambda] \tag{8–102}$$

Again, for each element, we group the lines according to excitation potential and for each of these groups we determine the shift necessary to fit the empirical curve to the theoretical curve both for the sun and the star. The difference in the $\log \eta$ shifts thus provides for each of these groups the value of $[\eta]$. If T^S and T^O do not differ greatly $[\kappa_\lambda]$ will be virtually independent of λ and will depend mostly on the $P_\varepsilon{}^O/P_\varepsilon{}^S$ ratio.

By the method described in Art. 8–8 for determining the excitation temperature, we can find $(\theta_O - \theta_S)$, since the terms $[N_r]$, $[Vu(T)]$ and $[\kappa]$ are all independent of λ. Now $T_S = 5040/\theta_S$ should be close to but is not

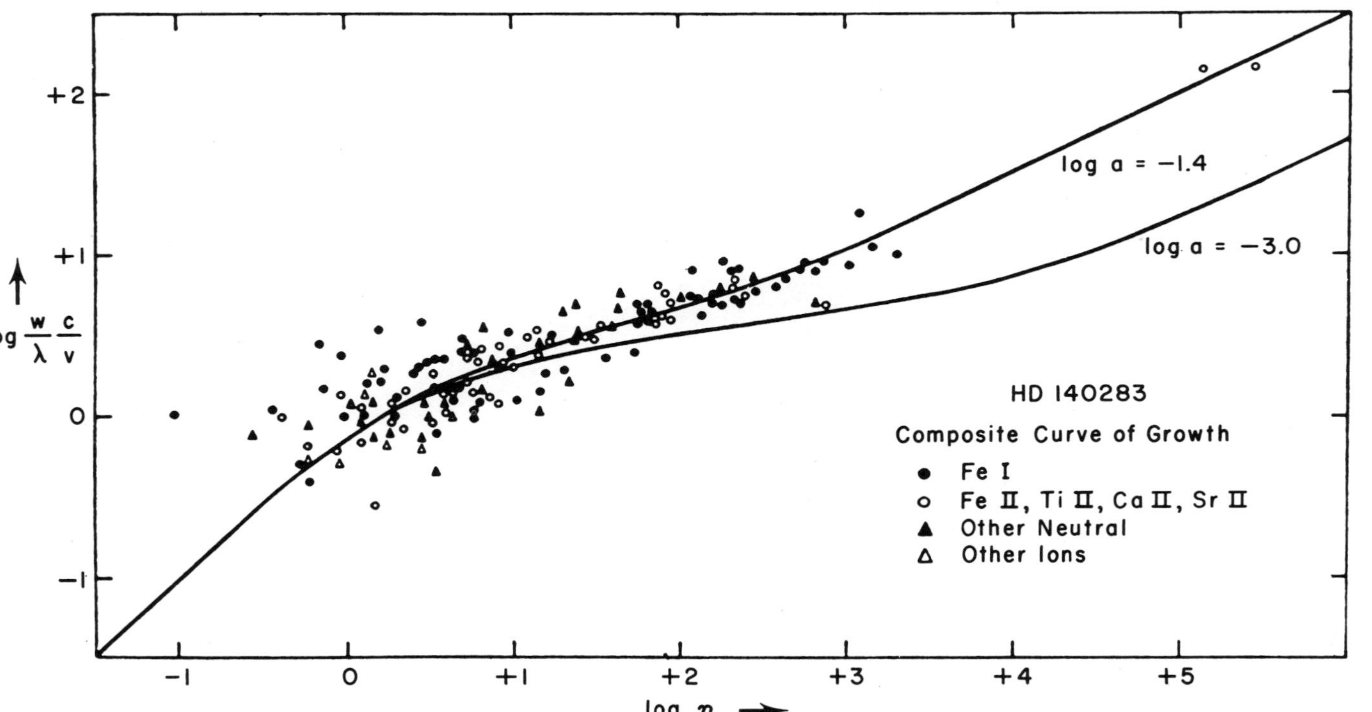

FIG. 8–11. CURVE OF GROWTH FOR THE SUBDWARF *HD* 140283
(From Aller and Greenstein, *Ap. J. Suppl. No. 46*, **5**, 139, Fig. 2, 1960; courtesy University of Chicago Press.)

necessarily identical with the mean temperature of the atmosphere. Furthermore, V may contain a turbulent contribution (see Eq. 8–89) which would require a vertical shift. Often we can estimate T from the color or energy distribution in the star and thus evaluate the term $[Vu(T)]$. On the other hand, the ratio $P_\varepsilon^O/P_\varepsilon^S$ may be determined from the ionization equilibrium. In other words, given P_ε^O, T^O, κ_λ^O, and ξ^O, we must determine P_ε^O, T^S, κ_λ^S, and ξ^S before we can interpret $[\eta]$ in terms of $[N_r]$, and $[N_r]$ then in terms of actual abundance ratios.

These procedures have been applied by various investigators, for example by K. O. Wright, by the Burbidges, and by Greenstein, Wallerstein, and their co-workers to determine composition differences between stars. As an illustration let us compare the subdwarf HD 140283 and the sun (Aller and Greenstein, 1960; see Fig. 8–11). Wrubel's curve of growth for scattering in ME model (Art. 8–8) was used to analyze the observations. First one selects a group of unblended lines, takes the solar equivalent widths (Allen, 1935; *Utrecht Atlas*), calculates $\log (W c/\lambda V)$ and reads off from the solar curve of growth the corresponding values of $\log \eta_0{}^c$. Similarly from measurements of the stellar equivalent widths one obtains $\log W/\lambda$. Preliminary trials show that turbulence is negligible in this star; so $\log (Wc/\lambda V) = \log (W/\lambda)(c/\sqrt{2kT/M})$.

At the outset of a curve of growth analysis one knows the temperature accurately enough from the star's color to compute gas thermal velocities, etc., but not accurately enough for calculating the ionization equilibrium. Thus the temperature of this star as deduced from its color and level of excitation of the spectrum is close to that of the sun. From the stellar $\log (Wc/\lambda V)$ values we read $\log \eta_0{}^S$ from the curve of growth. Thus we obtain $[\eta]$ and may attempt to determine the excitation temperature. As a consequence of the weakness of the lines in HD 140283 and other subdwarfs, determination of $(\theta_O - \theta_S)$ is very uncertain; in fact, one cannot be sure that the temperatures differ at all. Consequently, throughout the remainder of the analysis we assume $\theta_O = \theta_S$ so that also $u(T_O) = u(T_S)$ so Eq. 8–102 becomes

$$[\eta] = [N_r] - [\kappa V] \tag{8–103}$$

The velocities differ in the two stars because turbulence exists in the sun but not in the subdwarf. Further, the electron pressures and ionization temperatures differ. The adopted effective temperature $T_{\text{eff}}{}^S$ (assumed equal to $T_{\text{ion}}{}^S$) corresponds to that of a model atmosphere that reproduces the spectral energy distribution as measured by Melbourne (1960) with a photoelectric scanner, due allowance being made for distortion of the energy curve by lines. Usually such excellent supplementary data are not available and one must rely on UBV or preferably six color photometry. With the temperature chosen one must next obtain P_ε.

Suppose that the neutral and ionized lines of an element E fall in the same spectral region and are produced in the same atmospheric strata. Then for the neutral 0 and ionized 1 stages respectively we have

$$[\eta]_0{}^E = [N]_0{}^E - [\kappa V] \qquad [\eta]_1{}^E = [N]_1{}^E - [\kappa V] \tag{8–104}$$

We may then compute for each observed (neutral atom-ion) pair the quantity:

$$[\eta]_1 - [\eta]_0 = [N]_1 - [N]_0 = \log \frac{N_1{}^O}{N_0{}^O} - \log \frac{N_1{}^S}{N_0{}^S} = \Delta \qquad (8\text{–}105)$$

Now $\log N_\varepsilon{}^O/N_0{}^O$ may be calculated for each element as soon as P_ε and T are chosen for the sun. We chose $\log P_\varepsilon{}^O = 1.30$ and $\theta = 0.89$ on the basis of earlier model atmosphere results. From the computed $\log N_1{}^O/N_\varepsilon{}^O$ and Δ one finds $\log N_1{}^S/N_0{}^S$ while from the ionization equation and T we readily compute $\log N_1{}^S P_\varepsilon{}^S/N_0{}^S$ whence P_ε is found at once.

TABLE 8–4

CALCULATION OF LOG P_ε FROM CURVE OF GROWTH FOR *HD* 140283.*

Element	$\log \eta_0{}^0/\eta_0{}^s$	$\log N_1{}^0/N_0{}^0$	Δ	$\log N_1{}^s/N_0{}^s$	$\log N_1{}^s P_\varepsilon/N_0{}^s$	$\log P_\varepsilon$
Ca I	+1.66	2.62	0.61	3.23	3.56	+0.33
Ca II	+1.05					
Ti I	+1.15	2.04	0.32	2.36	2.93	+0.57
Ti II	+0.83					
Cr I	+1.44	1.72	1.24	2.96	2.61	−0.35
Cr II	+0.20					
Fe I	+1.48	1.00	0.65	1.65	1.84	+0.19
Fe II	+0.83					

*From L. H. Aller and J. L. Greenstein, *Ap. J. Suppl. No. 46*, **5**, 139, 1960. Courtesy University of Chicago Press.

Table 8–4 illustrates the details of the calculation. Successive columns give $\log \eta_0{}^O/\eta_0{}^S$, $\log N_1{}^O/N_0{}^O$, Δ, $\log N_1{}^S/N_0{}^S$, from Eq. 8–105 and the data of columns 1, 2, and 3, $\log N_1{}^S P_\varepsilon/N_0{}^S$ from the ionization equation for $\theta_{\rm eff} = 0.94$, and finally $\log P_\varepsilon{}^S$. The adopted $\overline{P}_\varepsilon{}^S = 2.2$ dynes cm^{-2}, whence the corresponding $\kappa_\lambda{}^O/\kappa_\lambda{}^S$ is 6.6 (as obtained from the tables of Allen or Strömgren). With $\overline{P}_\varepsilon{}^S$ and $\theta_{\rm eff}{}^S$ we may calculate the ratio $N_T{}^S(E)/N_r{}^S(E)$ where $N_r{}^S(E)$ is the observed number of atoms in the rth stage of ionization and $N_T{}^S(E)$ is the total number per gram of stellar material of atoms of a particular element E.

From Eq. 8–103 we have that

$$\log \frac{N^O}{N^S} = \log \frac{N^O}{N_r{}^O} - \log \frac{N^S}{N_r{}^S} + [\eta] + [V] + [\kappa] \qquad (8\text{–}106)$$

With $[V]$ and $[\kappa]$ known, $[\eta]$ obtained from the curves of growth, and the first two terms from the ionization equation with θ^O, $P_\varepsilon{}^O$, θ^S, $P_\varepsilon{}^S$ we can obtain N^O/N^S the ratio of abundances at once. In this way, the abundances given in the last column of Table 8–5 are obtained.

Notice that the abundance of metals in this subdwarf is two orders of magnitude smaller than in the sun! This subdwarf displays an extremely low metal abundance, although a number of similar stars are known in globular clusters. We have chosen an extreme example in which the temperature and level of excitation in the atmosphere is close to that obtaining in the sun, and the chief difference arises from an actual chemical composition difference. Of course differences as great as this are immediately obvious upon a visual inspection of the stellar spectrum with its prominent hydrogen lines and weak lines of *both* ionized and neutral metals. This star has also been studied by Baschek (1959) who used the actual f-values and both a curve of growth and model atmosphere method, but since he employed different observational data the results are not strictly comparable. Strictly, subdwarfs are only metal-deficient dwarfs.

TABLE 8–5

DETERMINATION OF COMPOSITION OF *HD* 140283

Atom or ion	Log η_0^0/η_0^s	Solar N^0/N_r^0	Stellar (N_r^s/N^s)	Log (v^0/v^s)	Log (N^0/N^s)
CH	+1.70				3.40
Mg I	+1.50	23.4	0.0136	0.07	1.87
Al I	+2.45	64.1	0.0040	.08	2.73
Si I	+1.74	3.8	0.103	.09	2.23
Ca I	+1.66	418	0.00059	.18	2.02
Ca II	+1.05	1.01	1.00		2.03
Sc II	+1.35	1.005	1.00	.21	2.34
Ti I	+1.15	111	0.0025	.22	1.72
Ti II	+0.83	1.01	1.00		1.85
V II	+0.73	1.015	1.00	.23	1.76
Cr I	+1.44	53.5	0.00537	.24	1.94
Cr II	+0.20	1.02	1.00		1.25
Mn I	+1.45	25	0.0122	.25	1.99
Fe I	+1.48	11.0	0.0306	.25	2.06
Fe II	+0.83	1.10	0.97		1.91
Co I	+0.88	8.4	0.039	.25	1.45
Ni I	+0.50	6.1	0.0555	.26	1.42
Sr II	+0.92	1.00	1.00	.35	2.03
Y II	+1.48	1.00	1.00	.35	2.59
Zr II	+1.14	1.00	1.00	.36	2.30
Ba II	+1.44	1.015	0.782	0.45	2.59

*From L. H. Aller and J. L. Greenstein, *Ap. J. Suppl. No. 46*, **5**, 139, 1960. Courtesy University of Chicago Press.

Caution must be exercised in employing this method to compare giants or supergiants with dwarf stars or stars of rather dissimilar spectral classes. Application of any differential method implies that curves of growth of the same shape are valid for both stars. The greatest value of the method lies in comparing stars suspected of having different chemical compositions but otherwise of comparable surface temperature and luminosity. Nevertheless, blends and other factors must be considered. For example, in a subdwarf such as HD140283, it is relatively easy to locate the continuum shortward of λ4000 where the solar spectrum is crowded with strong lines. Hence one cannot use the differential method to compare solar and subdwarf abundances of nickel whose lines fall in this region. A fundamental shortcoming of the curve of growth method is that it makes no use of line profiles, handling the data within a rigid framework of assumptions. Its great advantage is that it permits a rapid assessment of the physical state and chemical composition of a stellar atmosphere.

8–10. Turbulence and Extended Atmospheres. A curve of growth consists of three parts, the Doppler section wherein the equivalent width W is proportional to Nf, the flat portion where W increases slowly with Nf, and the damping part where W varies as $\sqrt{Nf}$. For a star like the sun we adopt a most probable kinetic velocity, v, on the basis of its effective temperature, calculate log W/λ and plot it against log η_0/N. We then try to fit the empirical curve to the theoretical curve by a horizontal shift, a procedure that is often satisfactory for main sequence stars like the sun.

Struve and his co-workers found that empirical curves of growth derived for giants would not fit the corresponding theoretical curves unless one supposed that v was much larger than the effective temperature of the star would suggest. Observational curves of growth for 17 Leporis (T_e = 10,000°K), and ε Aurigae (T_e = 6500°K) resemble theoretical curves, except that the values of v required would correspond to kinetic temperatures of 30,000,000°K and 2,000,000°K. Thus, either the kinetic temperature of the gas is unaccountably high or there is a large-scale motion of the gas; the latter effect is often called *turbulence*. The kinetic velocity v would be replaced by V defined by

$$V^2 = \left(\frac{2kT}{M}\right) + \xi^2 \tag{8–107}$$

where ξ is interpreted as the "turbulent" velocity, the most probable speed of large jets or masses of the gas. Struve found ξ to be 67 km sec^{-1} for 17 Leporis, and 20 km sec^{-1} for ε Aurigae. The line profiles lend further support to the idea that large bodies of gases in these stellar atmospheres are moving with speeds much greater than the thermal motions of the atoms themselves. The weak lines are broad and shallow. The strong lines, unlike those in dwarf stars, show no wings but are bell-shaped.

From a conventional curve of growth of the spectrum of δ Canis Majoris, Steel determined a turbulent velocity of about 5 km sec^{-1}, while Struve determined that the profiles of the strong lines require a turbulent velocity of 30 km sec^{-1}. In the spectrum of η Aquilae, M. and B. Schwarzschild and W. S. Adams found a turbulent velocity of 4 km sec^{-1} from the curve of growth, whereas the line profiles give a turbulent velocity of 12.2 km sec^{-1}, if we interpret the broadening as arising from large-scale gas motions and not from rotation. Hence turbulent velocities relevant to line profiles are not necessarily the same as those of most importance in curve of growth studies. These discrepancies are interpreted as follows:

It is usually assumed that the emitting "blobs" of gas move with a random distribution of speed, the number with velocities between U_x and $U_x + dU_x$ along the line of sight being given by

$$N(U)\,dU = \text{const}\exp\left(-U^2/\xi^2\right)dU \tag{8–108}$$

Furthermore, the sizes of the blobs (which may be regarded as turbulent eddies) are not uniform. Different eddy sizes may have different characteristic velocities, ξ. We may distinguish two limiting cases.

a. The individual moving elements of gas are small compared with the mean free path of a light quantum, i.e., small compared with the depth of the photosphere. Hence masses of gas moving with different velocities absorb at different distances from the center of the line and a broadened profile and increased equivalent width results. The flat portion of the curve of growth will lie above that predicted for lines broadened by thermal motions only.

b. The individual moving elements of gas are large compared with the mean free path of a light quantum. That is, the linear dimensions of the turbulent elements exceed the depth of the effective photospheric layers. The profile of the line is broadened, but its equivalent width is not increased. Hence there is no influence on the curve of growth.

A situation described by (a) is sometimes called *microturbulence,* that described by (b) is similarly called *macroturbulence.*

It appears that in many stellar atmospheres eddies of many different sizes occur simultaneously. The velocities characteristic of the small eddies (whose diameters are less than the thickness of the layers responsible for the absorption lines) will fix the turbulent velocity found from the curve of growth, whereas the line profile may reflect the velocities of larger masses in the atmosphere. Huang and Struve (1952, 1955) showed that by considering both the equivalent widths of lines and the Doppler half-width of their profiles, or the correlation between central intensity and equivalent width one could get the turbulent velocities characteristic of eddies of types (a) and (b). If a large range of eddy sizes exists, the problem is extremely complicated. Since different parts of a line are formed at different depths, condition (a) may be fulfilled in the wings but not near the center of the line, and very strong lines may give different results from lines of moderate intensity.

Many workers have studied "turbulence" in the sun both from curve of growth and profile studies. Center-to-limb measurements enable one to assess variations of these random turbulent velocities with depth and to search for possible anisotropies. Studies by van de Hulst, Reichel, Voigt, Suemoto, Wehlau, and Waddell have been discussed by de Jager (1959), who concluded that the mean turbulent velocity increases with height. Waddell found no depth dependence but concluded the motion was anisotropic. From an application of Goldberg's (1958) method to lines of moderate intensity, Unno (1959) finds that turbulent velocity *decreases* with height from a value of 1.5 km sec^{-1} at $\tau_0 = 0.6$ to 0.75 km sec^{-1} at $\tau_0 = 0.2$.

Generally, we expect curves of growth for giants and dwarfs to differ as shown in Fig. 8–12. In giants, turbulence often raises the flat portion, although the damping constant may remain comparable with the classical damping constant, γ, because of the low atmospheric density. In dwarfs, on the other hand, the flat portion frequently corresponds to a kinetic temperature close to the effective temperature, while the damping portion indicates a Γ much greater than γ because it is determined by collisions. Also, as expected from theory (see Chapter 7) damping constants determined from lines arising from high levels tend to be systematically larger than those found from resonance and low-level lines.

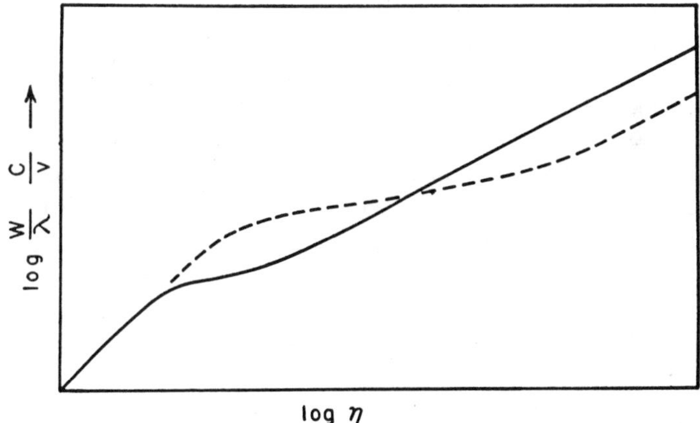

FIG. 8–12. SCHEMATIC CURVES OF GROWTH FOR GIANTS AND DWARFS

In the (dashed) curve of growth for the giant, the flat portion is raised by turbulence but the damping constant is small. In the (solid) dwarf curve, turbulence is small but damping is large.

The prevalence of turbulent motions in giant and supergiant atmospheres may be connected with their lower surface gravities and with the fact that some of the stellar envelopes appear to be at the limit of stability or even actually unstable. Stratification effects are probably more marked than in dwarf atmospheres, as one looks through a greater range in pressure and temperature as well as geometrical depths.

For example, K. O. Wright found indications that in the atmosphere of α Persei, the lines of neutral metals, which seemed to show greater damping constants and excitation temperatures than those of ionized metals, were formed in deeper layers than were the lines of the ionized metals.

Many years ago Paddock found semiregular variations in the radial velocity of the supergiant Deneb, and careful studies have been made for similar stars by Abt (1957). The supergiant Canopus (*F*0 Ib–II) also shows semiregular variations of extremely short period. These effects may be manifestations of irregular pulsations produced in the stellar interior, rather than purely atmospheric phenomena. Nevertheless these quasi-periodic variations add to difficulties in interpretations of stellar atmospheres.

Some of our best clues to structures of supergiant atmospheres will come from studies of eclipsing systems such as ζ Aurigae or 31 Cygni, where a *K* supergiant eclipses a *B*-type companion, or VV Cephei where an *M* supergiant hides a much smaller *F* star. Olin Wilson found that the density gradient in the atmosphere of ζ Aurigae is much less steep than the hypothesis of hydrostatic equilibrium would predict. In other words, the atmosphere of the giant *K* star cannot be in mechanical equilibrium at its effective temperature. From curves of growth constructed for different levels of the atmosphere, Wilson found some evidence for an increase of turbulence with height, but the increased turbulent velocities themselves do not suffice to support the distended atmosphere as McCrea had suggested long ago for the solar chromosphere. Evidently we must discard the picture of an atmosphere in static equilibrium and adopt in its stead some kind of a dynamical picture. The behavior of the chromospheric *K* line in the *K*-type component of 31 Cygni illustrates this point. As the *B* star emerges from eclipse, the wings of the *K* line produced in the chromosphere of the cool star rapidly weaken and the line shows a square profile cor-

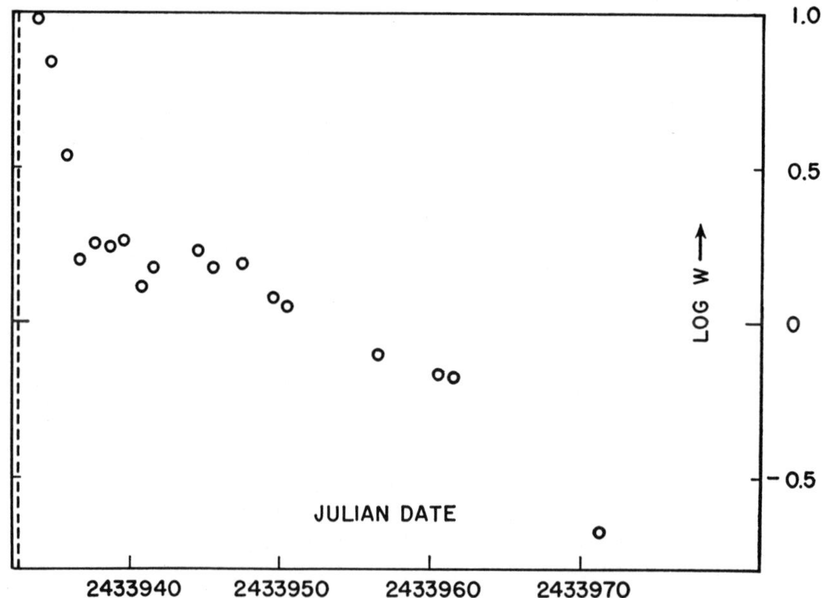

Fig. 8–13. Changes in Intensity of the Chromospheric *K* Line During the Egress of the 1951 Eclipse of 31 Cygni

The logarithm of the equivalent width of the *K* line, as measured upon plates secured at the University of Michigan Observatory, is plotted against the Julian date. Note the rapid decline after the end of the geometrical eclipse (October 14, 1951) and the subsequent fluctuations. The *K* line became very weak about November 21 (J.D. 2433971) (compare Fig. 8–14) and subsequently became stronger before disappearing. The density of the absorbing gas is estimated to be comparable with that found in the lower strata of the solar corona.

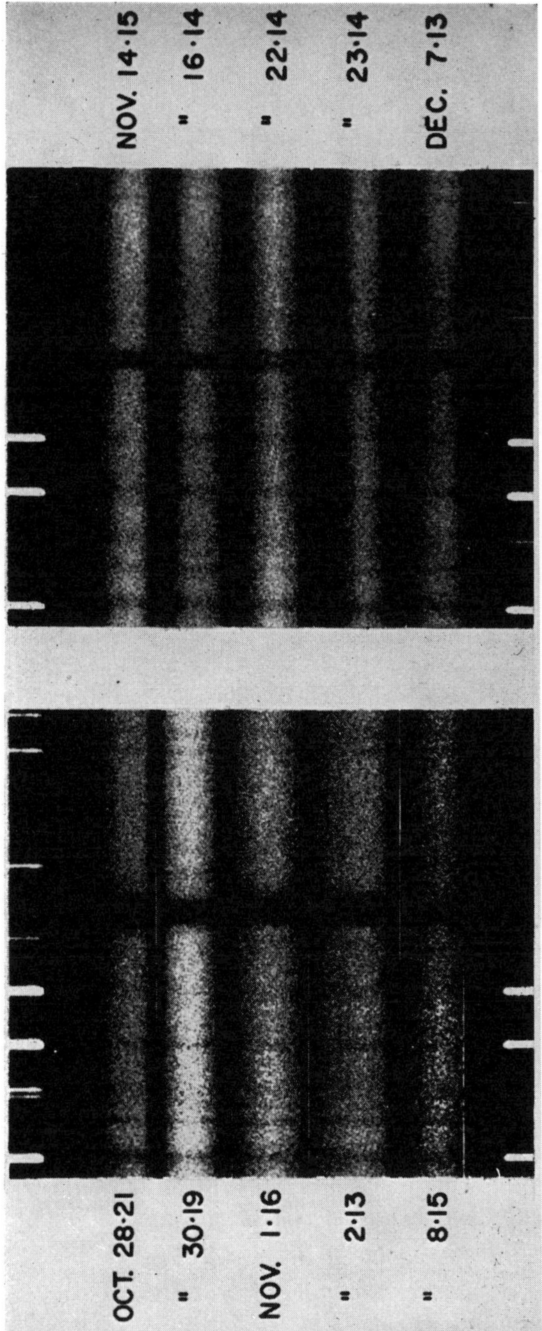

Fig. 8–14. Structure in the Chromospheric K Line During the Egress of the 1951 Eclipse of 31 Cygni

Shortly after the end of the geometrical eclipse the K line showed a broad bell-shaped profile which gradually narrowed until the line finally became resolved into two components whose total and relative intensities changed in an erratic fashion as the light from the B star passed through varying thicknesses of absorbing gases in the extended chromosphere of the K-type supergiant. (Courtesy, Andrew McKellar and Graham Odgers, Dominion Astrophysical Observatory, Victoria, B. C.)

responding to a turbulent velocity of about 20 km sec^{-1}. As the line further weakens it shows a structure consisting sometimes of a single line, at other times of a line and a companion of variable intensity and velocity displacement. The equivalent width of the main component varies irregularly. See Fig. 8–13. Probably the extended atmosphere of the giant star consists of a multitude of prominences in more or less rapid motion with respect to one another. See Fig. 8–14.

This picture finds support in Wilson's and Abt's (1954) conclusion that their observations of ζ Aurigae can be understood only in terms of an inhomogeneous atmosphere of detached masses of differing sizes and densities.

As our final example of complexities exhibited by giants and supergiants, we shall mention the circumstellar envelopes identified by Deutsch (1956).

The third-magnitude *M* supergiant, irregular variable, α Herculis, has a fifth-magnitude *G*-type companion which is a spectroscopic binary. High dispersion spectrograms show a group of stationary lines, attributable to a circumstellar envelope surrounding the *G*-star system and at some distance from it. Since other *G* stars show no such phenomena and *M* supergiants are known to have extended atmospheres, these lines must be attributed to a huge envelope involving the whole system. The absorption lines arise from the lowest energy levels and appear as stationary lines in the spectrum of the binary and as violet displaced lines (appropriate to an expanding shell) in the *M*-system spectrum. Deutsch estimated the minimum radius of the envelope as 200,000 that of the sun, its minimum mass as about three times that of the earth. It is not uniform, but consists of discrete patches or clouds filling only about a ten millionth of the volume. The envelope expands at a rate of about 10 km sec^{-1} and represents a mass loss for the *M* supergiant of about 3×10^{-8} solar masses per year.

Besides α Herculis, several other binaries with late-type giants exhibit evidence of cool envelopes many times larger than the star. Among the most interesting of these is Antares *M*1 (Ib) which has a *B*4V companion that excites a nebula produced from the layers ejected from the supergiant component. Another example is the long period variable Mira whose hot faint, variable companion may be an incipient white dwarf. Deutsch has established that virtually all *M* giants and supergiants exhibit circumstellar absorption lines at *H* and *K*. Circumstellar lines are found in some *K*-type supergiants and red variables. Complex emission features due to circumstellar envelopes are also observed.

Giant and supergiant stars represent fairly advanced stages in stellar evolution. As they evolve, their atmospheres increase in size until they escape into space and the core settles down as a white dwarf. This transitional stage appears to be represented by remarkable combination variables such as Z Andromedae.

8–11. Spectral Classification, Luminosity, and Chemical Composition Effects. The Henry Draper spectral classification is a one-dimensional system, based on the level of excitation and given in terms of criteria that are not too precisely described. Modern systems of spectral classification are at least two-dimensional — one quantity relates to the level of excitation of the spectrum or color temperature of the star; the other refers to its intrinsic luminosity or surface gravity.

Empirical studies first established the important fact that the spectra of the stars include important clues to their true luminosities. Antonia Maury noticed that stars of the same Henry Draper spectral class often showed marked differences in line sharpness. From a study of the proper motions, Hertzsprung showed that the sharp-lined stars Maury had indicated by *c* were intrinsically brighter than the broader-lined objects indicated as *b* or *a*.

Adams and Kohlschutter firmly established the spectroscopic differences between giants and dwarfs and laid the foundation for a determination of what came to be known as spectroscopic parallaxes. Subsequently, Adams and Joy and their colleagues determined spectroscopic parallaxes for many hundreds of stars. Important advances were made by Lindblad and his associates in Sweden, by Morgan and Keenan and their co-workers, by Chalonge and collaborators, by Strömgren and by others who have employed a variety of spectrographic and photometric techniques.

The observed spectrum of a normal star will depend on three parameters: (a) the surface temperature which is the main factor in controlling ionization and excitation in the atmosphere, (b) the surface gravity which controls the density in the atmosphere, and (c) the chemical composition which may influence the appearance of the spectrum both directly and indirectly. As long as we deal with stars of nearly "normal" composition like the sun and population type I stars, the influence of the composition will usually be small, but this factor must be taken into account for objects such as subdwarfs, members of globular clusters and certain late-type stars. For stars hotter than the sun, the most important factor is the metal/H ratio. In cool giant stars, the C/O ratio and the relative abundances of the metals of the first and second long periods (titanium group)/(zirconium group) are of critical importance. Spectral and luminosity classification criteria that are valid for stars of "standard" (i.e., solar composition) may be worthless for metal deficient objects.

A variety of methods for spectrum and luminosity classifications have been developed because not only the line spectrum of a star, but also its continuous spectrum, or more crudely its color, depend on T, g, and composition. These methods can be described as "subjective" (i.e., they depend on eye estimates of intensities, line ratios, etc.) or "objective" (i.e., depending on quantitatively measured features of the spectrum plus line intensities or energy distributions. They can also be considered from the point of view of the dispersion employed from $\sim$ 30,000A mm^{-1} in a technique due to Morgan, Meinel, and H. M. Johnson to about 10A mm^{-1}

in "*K*-line" method of Wilson and Bappu. Some methods, notably those developed by Strömgren and his co-workers do not use spectra at all, but employ narrow band pass filters that isolate lines or features sensitive to luminosity or composition differences.

The establishment of absolute magnitude criteria in practical application depends somewhat on the type of instrument employed. The most suitable criteria may differ from one dispersion to another. The use of low dispersion limits the number of lines that can be employed but enables the observer to reach fainter stars. W. W. Morgan has demonstrated that good absolute magnitudes can be obtained with low dispersion (110A mm^{-1}) spectra.

Let us first consider stars of the same chemical composition but of different luminosity. Their spectra will differ in four characteristics: (1) Color effects or energy distribution; giants are redder than dwarfs of the same spectral class. (2) Line sharpness, or "*c*" characteristics. (3) Intensities of lines not used in spectral class determination. (4) Intensities of bands such as CN. Interstellar lines are also sometimes helpful in early-type stars. Color alone can be employed when we can eliminate effects of space reddening. Line sharpness is a qualitative rather than a quantitative criterion. With spectra of intermediate and high dispersion, criterion (3) is most important, while (4) can sometimes be used with very low dispersion spectra.

In one method, a determination of luminosity classes or absolute magnitudes of stars of the same spectral class proceeds as follows: (1) From an intercomparison of stars of known luminosity classes or absolute magnitude, lines are chosen that appear to vary with intrinsic stellar luminosity. (2) The intensity of each luminosity-sensitive line with respect to a neighboring non-variable line is estimated. (3) From calibration curves derived from stars of known absolute magnitude, the ratio of the intensities of variable to non-variable lines are converted to absolute magnitudes. In practice one employs a procedure similar to Argelander's method for variable stars. One estimates intensity differences on a step scale and reads the corresponding absolute magnitude from a calibration curve. For example, the ratio of 4215 Sr II to 4250 Fe I is often employed as a luminosity criterion in classes *F* to *K*.

An alternative and somewhat better method is that involved in the Morgan-Keenan system of spectral classification. It has the enormous advantage that spectrum and luminosity classes are defined in terms of specific stars, which the astronomer can observe with his own particular equipment. Better results are obtained if one employs nearly the same dispersion as used in the MK Atlas. A further advantage is that the system employs spectrum and luminosity classes, with steps chosen as small as can be effectively utilized. The observer first secures spectrograms of MK standard stars and then matches his program stars against the standards. With good-quality, properly exposed spectrograms, reliable luminosity and spectral classifications can be made rapidly.

The first attempts to apply quantitative measurements to the problem of spectral classification were those by Lindblad and developed by him and his associates at Stockholm and Uppsala. The idea was to replace eye estimates of intensity by quantitative measurements of integrated intensities of lines and slopes and discontinuities in the energy curves with a view of applying it to low dispersion material. It is possible in this way to establish criteria that are excellent for obtaining absolute magnitudes even though equivalent MK classes cannot always be established unambiguously. The Swedish workers developed appropriate criteria for both cool *G*, *K*, and *M* stars and hot *B* and *A*.

Lindblad showed that with accurate spectrophotometry, the cyanogen band intensity is an excellent absolute magnitude criterion; the more luminous the star, the stronger the bands. In his earlier work he measured the intensity in terms of exposure ratios needed to give the cyanogen band and neighboring spectral regions the same blackening. Later, techniques of photometric calibration were developed and intensities were measured at several points in the spectrum to obtain a number of criteria. With the cyanogen criterion one can employ dispersions so low that the atomic lines are invisible.

Later, Keenan showed that with good standard spectra and accurate spectral classes one can attain an accuracy as great as in methods using much higher dispersion. With a dispersion as low as 425 A mm^{-1}, so that the bands looked like atomic lines, he was able to distinguish between different luminosity classes.

The CN criterion works reliably only for stars of population type I. In stars of population type II, the CN bands are often abnormally weak and one must use atomic lines to obtain a reliable luminosity classification.[9]

In *M* stars one uses the TiO bands to get the spectral class. The CaH λ6385 band is strong in red dwarfs (Ohman), but for most purposes the most important criterion is Ca I λ4227, which is the strongest line in the spectra of dwarfs. In stars of increasing luminosity it progressively weakens until in supergiants it is less intense than the absorption lines on either side.

For lines on the square-root portion of the curve of growth, i.e., lines for which the wings determine the total intensity, pressure broadening may play an important role since α_ν in the wings varies as $Nf\Gamma$. In dwarfs the damping constant, Γ, being essentially the collisional damping constant, is proportional to the density. Even if the number of atoms above the photosphere of a supergiant and a dwarf were the same, the vastly greater density of the dwarf atmosphere would so increase the damping constant in the wings of the line as to augment substantially the total intensity. In the sun the damping constant for resonance lines

[9] Suppose that in a cool star, where metals supply the electrons, all elements but H are cut down by a factor x in abundance. Since atmospheric opacity depends on P which is diminished by a factor x, the mass above the photosphere is likewise increased by a factor x. Hence metallic lines, or bands of CH or OH are little affected in intensity. The concentration of a molecule like CN will be cut down by a factor x^2, so it will suffer a net weakening of roughly a factor x.

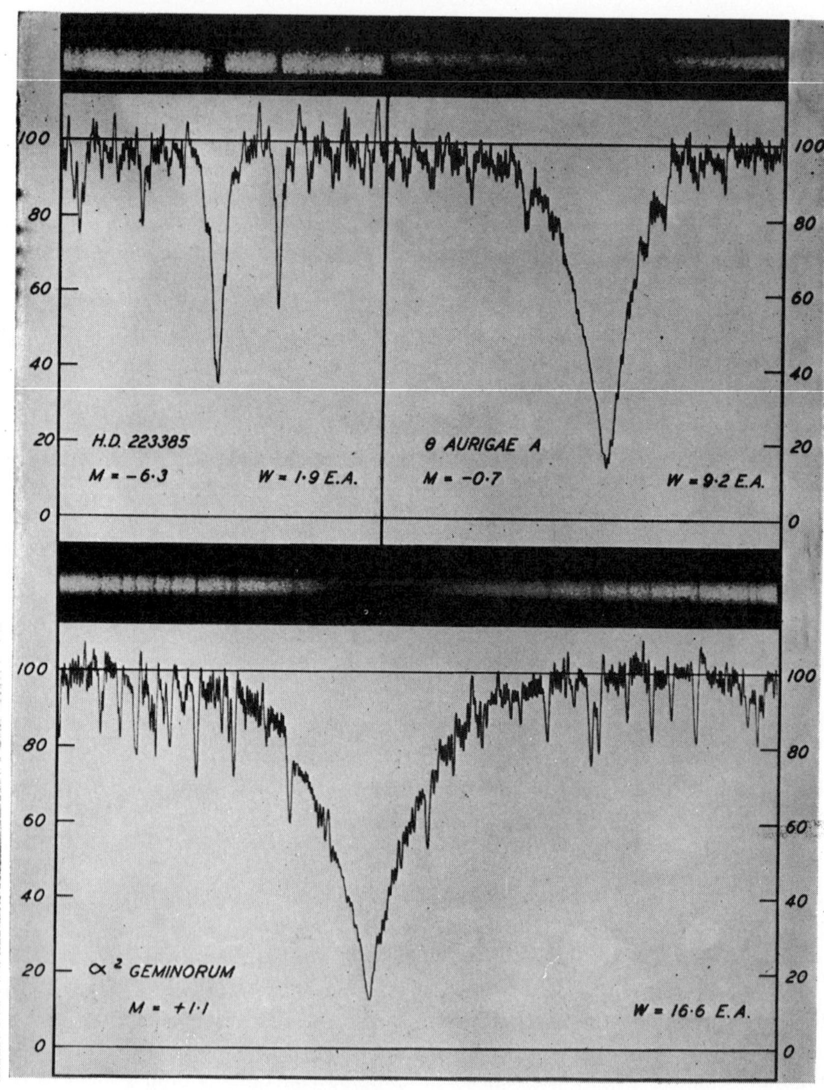

Fig. 8–15. Luminosity Effects at Hγ in *A*-Type Spectra

Notice the weakness of the Hγ line in *HD* 223385, an *A*2 supergiant similar to α Cygni and its great strength in the dwarf, α^2 Geminorum. The microphotometer tracings show Hγ to be much wider in α^2 Geminorum than in θ Aurigae *A*, although the differences are not conspicuous to the eye. (Courtesy, R. M. Petrie.)

is of the order of ten times the classical value, and in *M* dwarfs it may be much greater. Lindblad found the λ4227 line in red dwarfs to be asymmetrically broadened, probably because of the formation of quasi-molecules of Ca_2 such as are observed in calcium vapor at high pressures.

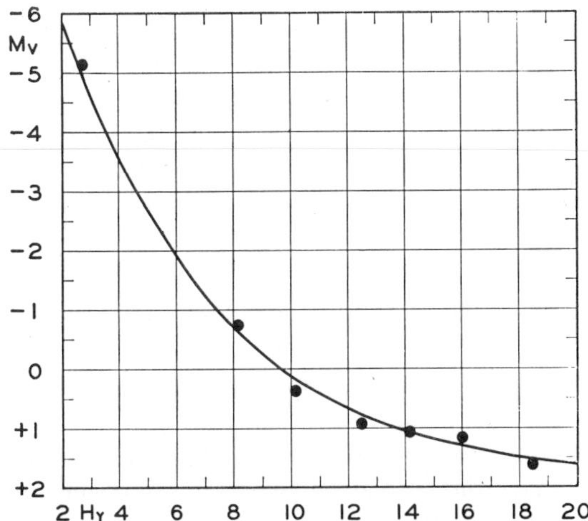

FIG. 8–16. THE MEAN RELATION BETWEEN THE EQUIVALENT WIDTH OF Hγ IN ANGSTROM UNITS AND ABSOLUTE MAGNITUDES
(Courtesy, R. M. Petrie, Dominion Astrophysical Observatory, Victoria, B. C.)

The *G*-band, a composite structure of molecular bands and atomic lines, is a useful criterion in class *M*. The break in the spectrum at λ4308 corresponding to the red end of the *G* band becomes much more pronounced as the luminosity increases (Nassau and Keenan).

Among the recent Swedish investigations of *G*, *K*, and *M* stars, we may mention particularly those of Elvius (1955–1956) and of Westerlund (1951–1953) who used the *G* band, the CN band and the intensity of λ4227. For example, Westerlund found the intensity of the CN band (λ3835) to be particularly useful for dwarfs. He also used the intensity ratio of hydrogen/*G* band, I (λ4227), and discontinuities at the *G* band and at Ca + *K* line.

An important advantage of microphotometric techniques is that, as Ramberg showed, one can use unwidened spectra and thus reach a fainter limiting magnitude with a given telescope. Yoss (1961) found that with objective prism spectra one could reach considerably fainter stars than was possible by conventional techniques. A considerable increase in labor is involved.

For stars of spectral class *G* and *K*, Hossack (1953) and Heard (1958) employed an oscilloscopic microphotometer in which one displays traces of two spectrograms in a screen one above the other. The amplitude of either pattern can be varied and line ratios measured directly. The luminosity can be measured with a precision of about a seventh of a luminosity class.

The value of the total intensities of the hydrogen lines as luminosity criteria for hot stars was demonstrated by Linblad and his associates, by E. T. R. Williams, Anger, S. Gunther, E. G. Williams, P. Rudnick, and particularly by Petrie and his associates. See Figs. 8–15 and 8–16.

Theory predicts and observation confirms that in stars hotter than 10,000°K, the number of atoms above the photosphere capable of absorbing the Balmer lines will be the same in supergiants and dwarfs. The appearance of the hydrogen lines will differ markedly; in the dwarfs the earlier members will be broad, whereas in the supergiants they will be relatively sharp because of the much greater broadening due to ions and electrons in the denser star (see Chapter 7). Over a fair range of spectral class, the equivalent widths of the Balmer lines longwards of $\lambda 3889$ depends strongly on absolute magnitude.

The Balmer lines provide one example in which theory gives a quantitative explanation of observed phenomena — the actual line shapes as a function of surface gravity. Even the early version of the Kolb theory, which was checked experimentally by Turner and Doherty at Michigan, was shown by Elste, Jugaku, and the author in 1955 to give an adequate interpretation of line profiles in B stars whose surface gravities were known from eclipsing binary data. Various workers have since confirmed nicely the validity of the Kolb theory (and the more accurate version given by Kolb, Griem and Shen) for the prediction of stellar H-line profiles.

Luminosity effects and the establishment of accurate absolute magnitudes for B stars are extremely important for galactic structure studies. The spiral arms of our galaxy are defined by emission nebulosities and clouds of ionized hydrogen which are excited by O and B stars. Hence a knowledge of the absolute magnitudes of these stars is urgently needed. Morgan and Keenan (1951) set up line intensity criteria for B-star luminosity classes and established absolute magnitudes by using galactic stars clusters of known distance, trigonometric and statistical parallaxes, galactic rotation, and interstellar lines. R. M. Petrie (1956) used the measured strength of the H lines and calibrated them with the aid of galactic clusters, visual binaries, and eclipsing binaries. This distance calibration agreed with that by E. G. Williams, but discordances between the Victoria and Yerkes system persist for main sequence objects. The discrepancies between these two investigations, both of which were carefully carried out, point out the need for further fundamental work on this problem. The problem is further complicated by stellar evolution effects which are most marked among the brightest stars. With spectrograms of high resolution, one may also use the number of resolvable Balmer lines to separate supergiants and dwarfs (see Chapter 7).

A relation between the Balmer discontinuity and the central intensity of Hδ has been employed by Hack (1953) to obtain a spectrum-luminosity classification for stars between $O6$ and $F8$. Heterochromatic photometry of the Balmer discontinuity was suggested by Barbier and by Westerlund about 1951–1953.

Among spectrum-luminosity methods whose criteria depend on properties of the hydrogen atom and its negative ion are those of Chalonge (1956) and of Strömgren (1958) (and variations thereof). Chalonge uses

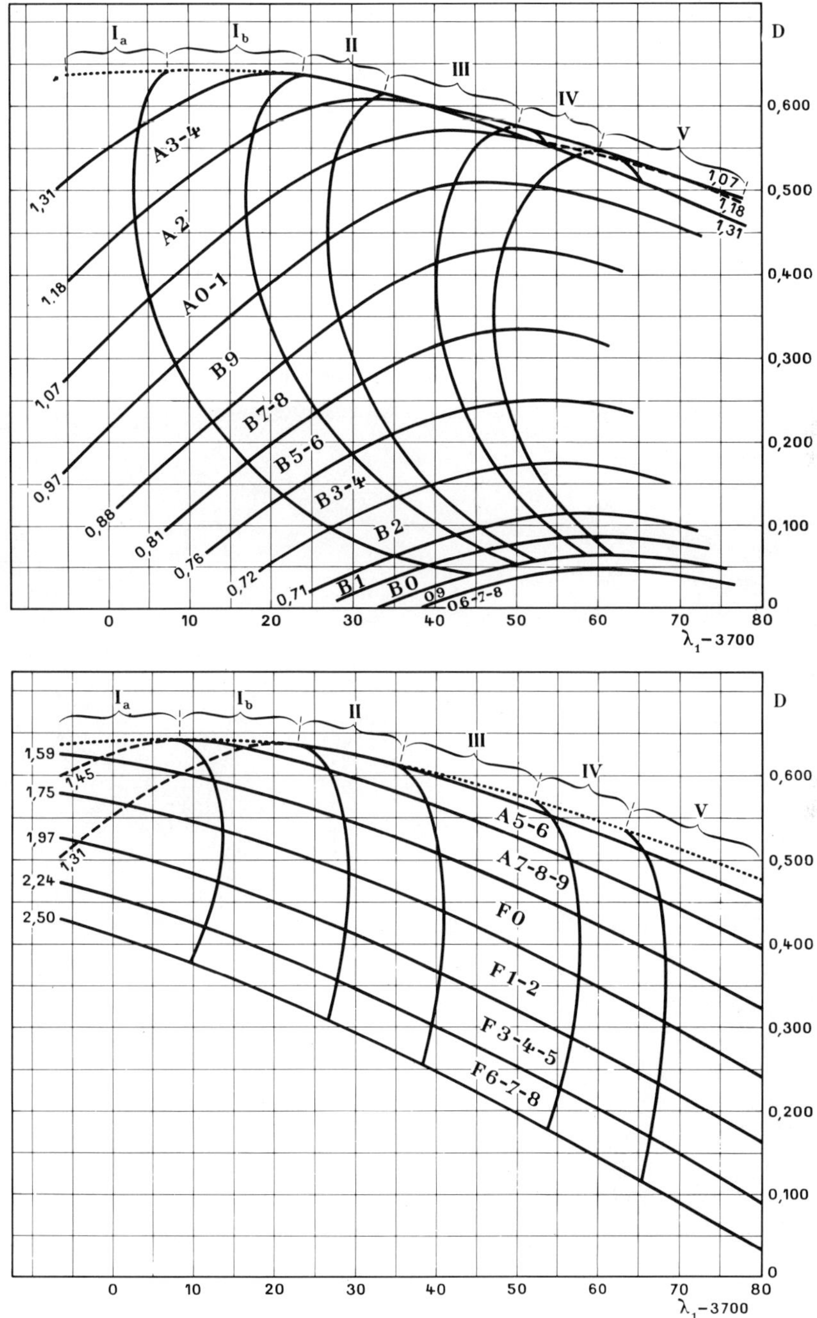

Fig. 8–17. Relation Between λ_1 and D and the MK Spectral Classes

measured properties of stellar spectra, Strömgren employs photoelectric techniques with narrow band-pass filters. Previously (see Chapter 6) we have seen how Chalonge and his co-workers measured the spectrophotometric gradients $\phi_b(3700 < \lambda < 4500)$, $\phi_{uv}(3100 < \lambda < 3700)$, the Balmer discontinuity D and the wavelength λ_1, defining the position of the discontinuity. Now ϕ_b and ϕ_{uv} vary with spectral class but are also affected by space absorption; hence they are not by themselves too reliable indicators of spectral class and luminosity (see Figs. 8–17 and 8–18). Since the relation is not single-valued (two stars may have the same value of λ_1 and D, yet one may be an F star, the other a B star), one must use the gradient ϕ_b to separate stars earlier than $A3$. Thus, the three observed quantities (λ_1, D, ϕ_b) define a three-dimensional figure, and one can correlate

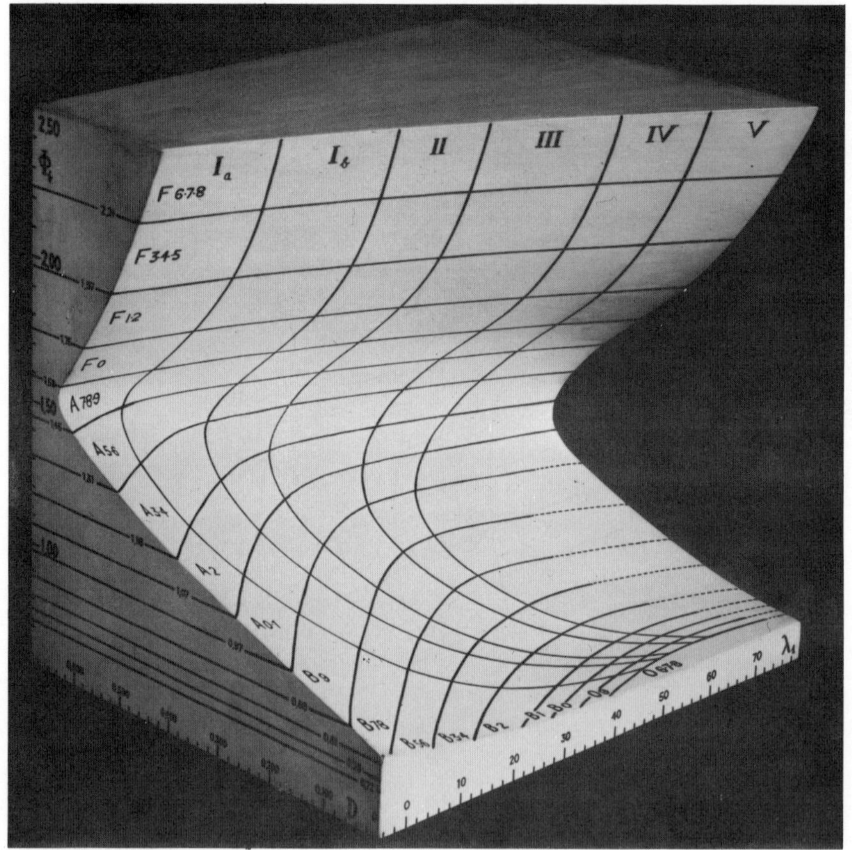

Fig. 8–18. The (λ_1, D, ϕ) Surface

If a three-dimensional plot of λ_1, D, and ϕ_b is constructed, normal stars fall on a two-dimensional surface whose shape is shown by this plaster of Paris figure. (Courtesy, D. Chalonge, *Ann. d'Ap.* **19**, 256, Fig. 3, 1956.)

the MK spectral classes with small areas on its surface. For example, a star for which $\lambda_1 = 3710$, $D = 0.45$ (earlier than $A3$) has an MK class $A2$Ib.

We notice, however, that the three parameters are not really independent and for most purposes we need employ only λ_1 and D (once we have decided that the star is earlier or later than $A3$). When we can eliminate effects of space reddening, we can use λ_1, and ϕ_b to define an HR diagram, since for normal stars ϕ_b is independent of luminosity class from O to $F8$, whereas λ_1 depends only on absolute magnitude. Although the points for normal stars fall on Chalonge's (λ_1, D, ϕ_b) surface, metallic line stars fall off the surface on one side, and subdwarfs on the other (concave) side — the more abnormal the metal/H ratio the greater the distance from the surface.

In practice, it is usually easier to decide whether a star is earlier or later than $A3$. Then one may use the (λ_1, D) parameters to construct a diagram analogous to the HR or color-luminosity diagram without any correction for distance or interstellar absorption. The technique is useful in clusters where some components are located in absorbing clouds.[10] Various reassuring checks on Chalonge's three-dimensional classification may be applied, e.g., one may use close binaries (where the difference M_v in absolute magnitudes is accurately known), (Berger, 1958). The relation between the Chalonge system and the UBV photometry has been discussed by Rozis-Saulgeot (1959, 1960). The Chalonge technique can be applied from spectral class O to $F5$, beyond which metallic lines become so strong as to subdue the hydrogen spectrum.

For early-type stars, Strömgren's photoelectric narrow band pass photometry offers many advantages for determining not only effective temperature and luminosity but also mass and age. From color magnitude arrays for galactic clusters one can construct (see Art. 1-5) the relation between luminosity and color for the "zero-age" main sequence, i.e., the main sequence defined by young stars that have just started to operate on the hydrogen-burning mechanism. The amount by which a star of a given luminosity departs from this line depends on its age. Hence accurate measurements of color and luminosity should be capable of indicating the initial luminosity L_0 (and therefore the mass) and also the age of the star (i.e., the time interval since the star started shining on the main sequence provided the evolutionary tracks are known either by theory or empirically).

In the B stars, for example, the equivalent width of Hβ indicates the luminosity, whereas the Balmer discontinuity indicates the surface temperature. To measure these quantities Strömgren employed six filters whose

[10] Also by plotting the components of multiple stars, one may use the positions of the points with respect to curves for clusters of known age to obtain the age and stage of evolution of the particular multiple stars.

wavelengths of maximum transmission and half-widths were: (a) λ5000 (190A), (b) λ4861 (35A), (c) λ4900 (100A), (d) λ4500 (80A), (e) λ4030 (90A), and (f) λ3600 (350A). He then combined the corresponding intensities $I(a)$, $I(b)$, $I(c)$, etc., to get two indices,

$$c = \text{const} + 2.5\,[2 \log I(e) - \log I(d) - \log I(f)] \qquad (8\text{–}109)$$

which measures the Balmer discontinuity and

$$l = \text{const} + 2.5\,[\tfrac{1}{2} \log I(a) + \tfrac{1}{2} \log I(c) - \log I(b)] \qquad (8\text{–}110)$$

which measures the strength of Hβ. Some photometers have been devised to measure $I(a)$, $I(b)$, $I(c)$ simultaneously with the aid of beam splitters, hence avoiding difficulties with atmospheric transparency. One can thus plot the index c against l to obtain the equivalent of a color magnitude array. Absolute magnitudes are obtained from the c-l classification with a precision of about $0^m.2$, whereas the probable error of the c-index expressed in terms of spectral class is about 0.01 of a spectral class.

Strömgren was particularly interested in using the position of a star in the c-l diagram to establish its age. One can do this by comparing main sequences of galactic clusters of different ages with the zero-age main sequence, or with theoretical calculations. A relation between c and effective temperature T_e can be obtained from model atmosphere work (see Chapter 5). The mass is known as a function of luminosity for stars on the main sequence from binary star data (Chapter 1). From L and T_e one may deduce R, and from M and R the surface gravity g. Evolutionary tracks for stars of mass M, and therefore initial luminosity L_0, have been computed by Taylor and by Kushwaha so that one can obtain $R(t)$ and $L(t)$ and from these data the values of g, T_e, whence we can derive the indices c and l from model atmospheres and the Kolb-Griem theory of hydrogen line broadening. On the theoretical c-l diagram, one can draw a grid of curves corresponding to different masses and to different ages, e.g., 25, 50, and 75 million years, and thus read off for each star its age and mass. Thus Strömgren found that it was possible to track the path of an evolving star in the c-l diagram, starting from the zero-age line and to indicate from one point to another the age of the star. See Fig. 8–19. The accuracy of the measured age is lower for fainter stars whose speed of evolutionary age is slower.

Applications of this technique to star clusters and association shows that differences in chemical composition among B stars are not pronounced. A small change in the He/H ratio could be detected since helium contributes weight but not continuous absorptivity, and hence increases pressure, and therefore the broadening of hydrogen lines for stars of a given surface gravity.

For A stars, Strömgren used the Hβ index and $(B\text{-}V)_0$, the standard color index corrected for space absorption in order to make age determi-

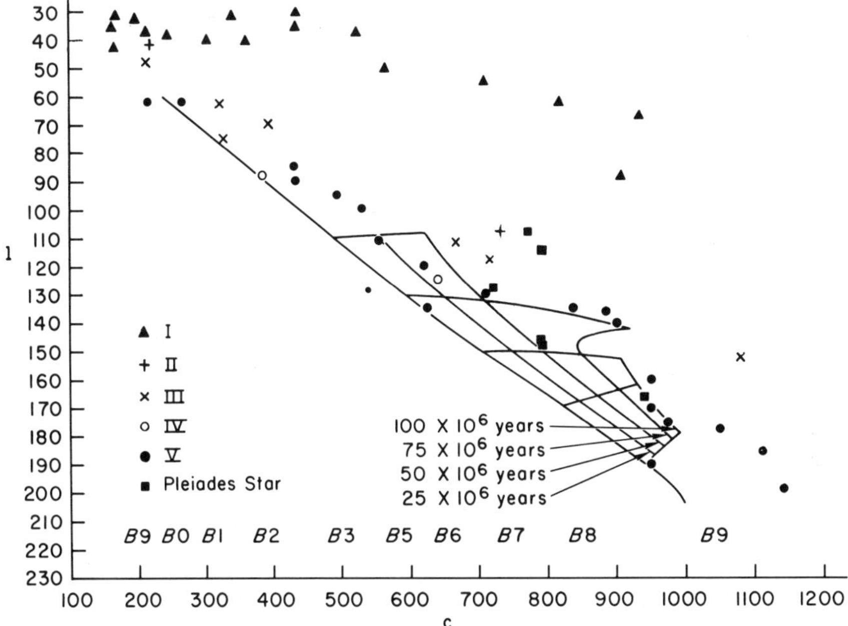

FIG. 8–19. STRÖMGREN'S c–l DIAGRAM AND THE DETERMINATION OF STELLAR AGES

The index c is obtained from narrow band-pass filters centered at $\lambda 4500$, $\lambda 4030$, and $\lambda 3600$ or from UBV photometry (see Eq. 8–109), whereas the l index is measured from Hβ photometry (see Eq. 8–110). Theoretical calculations yield a grid of curves corresponding to different masses and ages. When the observed values of c and l for a given star are plotted on the diagram, its age may be read off by interpolation between theoretical curves corresponding to different ages but the same mass.

nations. For the population type-I F stars he calibrated the c-l diagram in terms of $(B\text{-}V)_0$ and absolute normal magnitude M_v.

Effects of composition differences, or more particularly the metal/hydrogen ratio have been studied in A, F, and G stars by Johnson and Morgan; Roman, Barbier, and Chalonge; Sandage, Wildey, and the Burbidges (1960); and others. Stars with reduced metal content will have weaker lines and the effects will be particularly marked in the ultraviolet where metallic lines tend to crowd. Thus the effect is more pronounced on the $(U\text{-}B)_0$ index than on the $(B\text{-}V)_0$ index. Sandage and Eggen (1959) found that if all lines were eliminated from the solar spectrum, the changes in the B-V and U-B indices would be 0.17 and 0.32, respectively. Certain subdwarf stars have almost these line-free colors.

Hence Strömgren introduced the m-index which is sensitive to the metal to hydrogen ratio, viz.:

$$m = \text{const} - 2.5\,[\log I(a) + \log I(e) - 2 \log I(d)]$$

In F stars the metallic line absorption reduces $I(a)$ and $I(d)$ by about equal amounts, where $I(e)$, which refers to $\lambda = 4030$, is reduced more markedly. The m-index increases with increasing metal/H ratio but is not sensitive to variations in T_e and g. For extreme subdwarfs, for which the metal/H ratio is about 10^{-5}, Strömgren finds an m-index of $0.^m02$ as compared with $0.^m14$ for stars with a normal ($\sim 10^{-4}$) metal/H ratio and a value of $0.^m19$ for 20 Can Ven for which Savedoff found an abnormally high metal/H ratio.[11]

In B stars the influence of the metal/H ratio upon colors and energy distributions is not great. It is important to realize that the c-index may be replaced by U, B, V colors so that a combination of the Hβ index with conventional three-color photometry will yield critical information on hot stars and their evolutionary status.

Strömgren and Gyldenkerne (1955) also developed narrow band-pass (80–120A) criteria for G and K spectral classes, viz.:

$$k = 2.5 \log I(3920)/I(4070) \qquad \text{break at Ca}^+ \ K \text{ line}$$
$$n = 2.5 \log I(4280)/I(4210) \qquad \text{CN absorption}$$
$$g = 2.5 \log I(4370)/I(4280) \qquad \text{break at } G \text{ band}$$

They separated giants and dwarfs from CN absorption equivalent and spectral type from the break in the G band, Gyldenkerne (1958).

Spectral energy distributions secured with a photoelectric spectrum scanner may also be employed to obtain spectral classes and luminosities. Much more information is obtained in a spectral scan than in colors and magnitudes measured with broad band-pass filters, but the observations necessarily require more time and can be undertaken only for fewer stars. Nevertheless they are necessary for obtaining a more complete understanding of stellar atmospheres and luminosity effects.

As was pointed out by Johnson and Morgan, the U, B, V, colors themselves can be used to obtain absolute magnitude information among early-type stars. The U-B colors give clues to the amount of the Balmer discontinuity. If one plots the U-B color indices against the B-V indices for main sequence stars in a cluster, for example, it will be found that the points fall along a well-defined curve (see Fig. 6–6). Giants and supergiants will not fit this standard curve, nor will subdwarfs or other metal deficient stars. Space absorption will shift the main sequence with respect to the standard curve and can be evaluated by comparing the empirical with the plot for zero space absorption.

The six-color photometry of Stebbins and Kron covers a wide base line in color so it is suited to obtaining very accurate space absorption and luminosity data.

[11] Baschek has attempted to derive a quantitative relation between Strömgren's abundance index and the hydrogen/metal ratio. If A denotes the ratio of hydrogen to the metals by numbers of atoms, $m - m_0 = 0.05 \log A(\text{sun})/A(*)$.

It also enables the observer to recognize unusual stars whose UBV colors might seem relatively normal.

Mention must be made of the important five-color photometry of the Walravens (1960) who measured photoelectric colors of *O* and *B* stars in five spectral regions, three of which corresponded closely to the UBV system, one in the far ultraviolet *W* at λ3220, and another, *L*, is at λ3900 where the Balmer lines crowd.

One final, very accurate example of luminosity criteria will now be given. In 1957, Wilson and Bappu called attention to the existence of a relation between the width of the reversed *H* and *K* lines in late-type stars and their absolute magnitudes. Specifically, the logarithm of the width of the *H* and *K* reversal depends linearly on the absolute magnitude. It is independent of the intensity of the reversal and of the spectral type of the star. Remarkably, the correlation extends from stars as bright as $M_v = -6$ to faint main sequence stars, i.e., over a range of 15 magnitudes. Wilson (1960) has used measurements of the widths of bright reversals in *H* and *K* in the sun and yellow giants of the Hyades to correlate log $(\Delta\lambda)_{EM}$ and M_v. For giants and supergiants the probable error of a single measurement is ± 0.26 magnitude, whereas the probable error due to intrinsic scatter is equal to $\pm$ 0.20 magnitude; possibly the scatter is caused by stellar rotation. In spite of the fact that the method is limited to high dispersions, and therefore requires a large telescope and relatively bright stars, it is the most powerful method yet devised for studying luminosity and evolutionary characteristics of late-type stars in the neighborhood of the sun.

The task of the theoretical astrophysicist is to give a rational explanation of absolute magnitude effects in terms of surface gravity, effective temperature and composition differences if any. Although some observed effects, particularly the hydrogen line behavior, have been explained by theory, others have not been interpreted and calibrations must be carried out empirically. We do not yet know enough about stellar atmospheres to predict precisely the change in intensity of an arbitrary line in going from a dwarf to a supergiant. One of the biggest difficulties, of course, is that we do not know enough about the structures of supergiant atmospheres to permit reliable predictions of line profiles and intensities. Eventually, absolute magnitude effects may be studied theoretically by constructing synthetic spectra for various model atmospheres. This approach, suggested by van Albada and others may be aided by families of model atmospheres such as those calculated by Neven and de Jager and by modern computational techniques.

There are a number of stars that do not fit into ordinary classification schemes. These include the metallic line (ML) stars and the magnetic stars. The temperatures and B-V colors of ML stars are essentially the same as those of main sequence stars of the same hydrogen line intensity.

The K line indicates an earlier spectral class and other metallic lines a later spectral class. Hack (1959) suggests that ML stars are located on a sequence that runs parallel to the main sequence but about half a magnitude above it, and attributes the strengthening of the metallic lines to an actual increase in abundance although some agency operates to produce abnormal ionization. The Jascheks (1959, 1960) found that in most ML stars the elements behaved normally except Ca, Sc, Ni, Sr, and Y, while Bohm-Vitense (1960) suggests that the peculiar features of the spectra can be explained by an atmosphere that is distended in its outer layers, possibly by magnetic fields. She finds that in high atmospheric layers, the electron pressure is lower, whereas in deeper atmospheric layers it is about equal to the electron pressure in main sequence stars of the same temperature. Abt finds all ML stars to be spectroscopic binaries. Among late B and early A stars are also found examples of "silicon" and "manganese" stars which show lines of these elements with abnormal strength.

8–12. Molecules in Stellar Spectra. The astrophysical importance of molecules is being increasingly appreciated. They are observed in all stars later than about $F5$, and become more prominent in cooler ones. We have already mentioned how certain molecular bands, those of CN, are important as luminosity indicators, or sometimes as population type indicators, since they tend to be weakened in stars of population type II. Molecules also serve as indicators of excitation temperature, either vibrational or rotational. Abundance of certain elements in the sun and other stars can be obtained only from bands of molecules involving them. The evidences for compositional differences in cool stars depends largely on molecular data (see Chapter, 3). Furthermore, slight composition differences that would defy detection in a study of atomic lines, may sometimes appear in molecular bands. Stellar isotope ratios, particularly of carbon, can be investigated only with the aid of molecular bands. The sun offers a particularly fruitful field for studies of molecular spectra. The structure of the uppermost photosphere can be investigated from the center-limb variations of lines of molecules such as CO (Newkirk, 1958), while the excitation temperature of the lower chromosphere has been obtained from emission bands of λ3883 CN (D. V. Thomas, 1958), and of C_2 (Parker, 1955).

Molecules can also have an influence on the structure of a stellar atmosphere. The rotation-vibration bands of polyatomic molecules may be important as an opacity source in the late-type stellar atmospheres (Swings, 1958), while the dissociation of the hydrogen molecule may have a profound influence on the structure of the convection zones in late-type stars (Wildt, Vardya 1960).

The sun can be observed with higher dispersion than any star, the intensity variations of the individual molecular lines across the disk can be investigated, and finally effects of the lowered temperature in sunspots can be examined.

Identification of solar bands is complicated by their superposition on the rich atomic spectrum of the sun, by blends among themselves, and by inadequate laboratory data. Many molecules have been studied only at low temperatures where their spectra differ vastly from their appearance in the sun. Dissociation constants, f-values and other parameters are often poorly known. The band spectrum of even a single electronic transition may be bewilderingly complex. Furthermore, certain radicals (partially dissociated molecules), not yet observed in the laboratory, may be found in the sun. Unidentified molecular absorption bands in sunspots are more numerous than those which are recognized. Under the best conditions, much of the sunspot spectrum shows no true continuous background at all. Rather, it consists of many faint absorption features arising from the superposition of delicate lines too faint to be seen separately, which are probably molecular in origin. The Zeeman effect, however, is helpful in separating atomic and molecular lines.

Spectra of sunspots often differ from one another and are difficult to observe because of scattered light from the disk.

The forbidden "A" band of atmospheric oxygen illustrates the dependence of the visibility of a band head upon the abundance of the responsible compound. With an air path of 100 km when the sun is low, the head of the band is sharply defined on the violet edge, but with an air path of only three meters, the absorption spectrum shows only two doublets in the P branch of the band, 40A to the red of the head. The lines of the R branch are weaker than those of the P branch, and thus disappear first with decreasing numbers of molecules. Thus absence of a given band head doesn't always mean the molecule in question is necessarily absent. Many faint lines in the solar spectrum may be remnants of bands arising from the ground state of rare molecules or abundant molecules in excited levels.

The compounds now identified in the sun are OH, NH, O_2, CH, CN, CO, C_2, MgH, SiH, TiO, and CaH, the last four being best represented in spots. The data for BH, ScO, BO, AlO, ZrO, YO, and SiF are not conclusive. It is of interest that molecules give us our only clues to the existence of boron and the halogens in the sun. Compounds which are very abundant in the sun include CO whose pure rotation-vibration bands are observed in the infrared, and O_2 whose transitions from the vibrational levels 18–21 to 0, 1, 2 were observed by H. D. Babcock (1945). Other compounds which must be abundant in the sun but whose lines fall in the far ultraviolet are H_2, NO, SiO, and N_2. Of these compounds only NO has so far been observed in the ultraviolet solar spectrum. Babcock suggests that under favorable conditions lines arising from high vibrational levels of NO and N_2 may become visible in sunspots.

The stars can be observed only with lower resolution. On the other hand, one can choose a much greater range in temperature and density and also can select stars of different compositions. For example, in the supergiant β Pegasi, D. N. Davis (1947) found the bands of TiO and VO to be the strongest, but in addition to the compounds observed in the sun (excepting CO, and NO) she noted AlH, CrO, BO, SiF, and SiN.

Molecular bands sometimes prove useful for excitation temperature determinations. Two methods are used: (a) a comparison of different vibration bands of the same band system; (b) a comparison of individual rotational lines of a given electronic vibrational band.

With the dispersions ordinarily applied in stellar spectroscopy, the component lines of an individual band are not resolved and method (a) is the only one available. For bands of such low intensity that curve of growth effects can be neglected, we may write

$$W_\nu(v'' - v') = N_0 \frac{\pi \varepsilon^2}{mc} f(v'' - v') \frac{e^{-E''/kT}}{\sum e^{-E''/kT}} \tag{8-111}$$

where the energy of the low vibrational level E'' is given by Eq. 2–52. The transition probability $f(v'' - v')$ has been calculated by Hutchisson for symmetric molecules such as C_2, and his formulas are still quantitatively applicable to molecules such as CN which are almost symmetrical. They are not valid, however, for molecules such as CH, TiO, ZrO, etc., in which the component atoms have appreciably different masses.

Method (b) may be employed with high dispersion material, such as is obtainable for the sun. If the bands are weak, the intensities of the individual rotational components will be

$$W_\nu(J) \sim N_0 \left(\frac{\pi \varepsilon^2}{mc}\right) f(J) \frac{g(J)\, e^{-E(J)/kT}}{u(T)} \tag{8-112}$$

where $f(J)$, the f-value for a line characterized by J, is known from theory, $E(J) = BJ(J + 1)$ is the energy of level J, $g(J) = 2J + 1$ is its statistical weight, and $u(T)$ is the partition function of the molecule $\Sigma\, g(j)\, e^{-E(J)/kT}$. Since u, f, and E are all known, measures of equivalent widths as a function of J will permit a determination of the excitation temperature. Physicists and chemists have employed this method to find the temperatures of arcs, flames, and glow discharges.

In practice, both methods are often difficult to apply. Even with adequate spectral resolutions (which means we are restricted to bright stars with large telescopes) there are still difficulties in locating the background continuum. Also, curve of growth effects have to be assessed as is illustrated, for example, in the work of Climenhaga in the study of carbon isotopes in the R stars. For example, Phillips (1952) obtained a rotation temperature of about 3000°K for β Pegasi and lower values for cooler M stars, whereas for R stars he found rotational temperatures that were discordant with those derived from the vibration bands. From 5 diatomic molecules observed in the spectrum of β Pegasi, C_2, CH, AlH, MgH, and SiH, Wyller (1961) derived a rotational temperature 3362°K in fair agreement with the effective temperature obtained with a thermocouple for this star (Pettit and Nicholson, 1928).

Rotation temperatures from solar molecular bands are given in Table 8–6. These excitation temperatures are comparable with those found from atomic lines; they refer to the uppermost portions of the photosphere.

TABLE 8–6

ROTATION TEMPERATURES (°K) FROM SOLAR MOLECULAR LINES
(The values are rounded off to the nearest even hundred)

Molecule	Temperatures
C_2	4900 (h), 4600 (a) 4900 (l)
CH	4400 (h), 4500 (l)
CN	4500 (h), 4300 (rb), 4500 (lb), 4700 (l)
CO	4300 (g, m)
NH	4700 (h), 4600 (l)
OH	4800 (h), 4600 (r), 4800 (l)
MgH	4600 (l)

(a) M. G. Adams (1938) *M. N.* **98,** 544
(rb) R. T. Birge (1922) *Ap. J.* **55,** 273
(lb) L. Blitzer (1940) *Ap. J.* **91,** 421
(h) J. Hunnaerts (1947) *Ann d'Ap.* **10,** 237
(g, m) L. Goldberg, E. A. Müller (1953) *Ap. J.* **118,** 397
(r) Roach, F. E. (1939) *Ap. J.* **89,** 99
(l) Laborde, G. (1961) *Ann d'Ap.* **24,** 89

As noted briefly in Chapter 3, composition differences in cool stars produce profound effects upon their spectra and particularly upon the compounds found therein. Merrill (1956) pointed out that spectra of cool giant and supergiant stars are dependent on (1) temperature, (2) oxygen/carbon abundance ratio, and (3) Ti group/Zr group abundance ratio. In the *M* stars the oxygen abundance exceeds that of carbon, which is tied up in carbon monoxide, whereas in the carbon (*C* or *R, N*) stars the oxygen is tied up in CO. In the *S* and heavy metal stars, the elements of the second long period are abundant. Frequently *S* stars show TiO and ZrO bands, along with bands of LaO, YO, SiH. They also show strong lines of Y, Zr, Nb, and lanthanide elements, so one uses the ratios of atomic lines of Zr group to Ti group to assess the degree of "*S*-ness" of a particular star, whereas the sum of the intensities of molecular bands ZrO and TiO is used to get the spectral class, i.e., a parameter that depends on temperature (Keenan, 1948, 1954). Closely related to the *S* stars are hotter objects that show strong atomic lines of Ba, Sr, etc. Although *S* stars are rare, they are of great astrophysical interest because they display results of element building *in situ*. Merrill (1952) found lines of the unstable element technetium in their spectra with intensities comparable with those of neighbor elements. Presumably this element whose lifetime is about 200,000 years, is built by neutron capture from the neighboring element molybdenum.

The complexity introduced by composition differences on spectral classification is not better illustrated than in carbon stars. The (R, N) spectral classification was unsatisfactory in that it did not give a smooth dependence of temperature on spectral class. Keenan and Morgan (1941) noted that the way out of the difficulty was to use atomic line ratios and intensities of which the equivalent width of the sodium D lines was the most useful. Their spectral classes $C0 - C7$ run parallel with the normal sequence from $G4$ to $M4$; $C1$, $C2$, $C3$ correspond to $R0$, $R3$, and $R5$, but the N types do not fit and the R, N classification scheme should be discarded. Although Bouigue (1954) found molecular vibration temperature to decline steadily with advancing spectral class and McKellar and Buscombe found vibration temperatures for R stars in good agreement with their colors, Wyller (1959) found not only systematically lower vibrational temperature but also values that showed no correlation with spectral class.

The carbon/oxygen ratio not only determines the character of the spectrum in a general way but may also influence the TiO/ZrO ratio in S stars. The ratios, Ti group/Zr group, O/C, C/Zr group, may all vary independently. The abundance of nitrogen may also be variable. Carbon stars also show abnormal isotope ratios, e.g., the C^{12}/C^{13} ratio which is about a hundred for the earth and normal stars may be as low as five or six for certain carbon stars (Climenhaga). A few carbon stars show extreme abundances of lithium, an element particularly vulnerable to nuclear reactions and which is preferentially destroyed in thermonuclear reactions.

Because they show the influence of element-building mechanisms in stars, the red giants and supergiants deserve very careful spectroscopic study. Unfortunately, for lack of space we cannot pursue these questions here. We shall mention some of the ways in which molecular problems differ from atomic problems in stellar atmospheres.

In the sun molecular bands tend to be found in the highest photospheric layers; quantitatively they must be handled by model atmosphere methods. Although models of main sequence stars down to about $M1$ probably could be calculated with presently known opacity data and some start could be made for giant and supergiant stars, little work has been done along these lines. Stratification effects are particularly important for molecules and quantitative abundance results will require that they be taken into account.

8–13. Stellar Rotation. Spectral lines in many early-type stars are broad and diffuse. For example, among A stars the lines in α Aquilae or α Piscis Austrini are fuzzy compared with the lines in α Lyrae or α Canis Majoris. The reasons for believing these lines to be washed out by rapid axial rotation are: (1) The line widths are proportional to wavelength which suggests a Doppler origin to the broadening, i.e., the observed line shape represents the superposition of line profiles from different parts of the disk of a spinning star. At longer wavelengths, the lines become wider and shallower and lines which are normally strong may even become practically invisible. Thus Shajn found the strong visual Fe II lines to be "washed out" in the spectrum of Aquilae. (2) The relative intensities of lines in multiplets are the same as in non-rotating stars. This relation would not always hold if lines were widened by turbulence. (3) The lines in spectroscopic binaries of short period and large amplitude are always broad and diffuse. (4) Consider an eclipsing system just before primary minimum. If the orbital revolutions and axial rotations are in synchronism, as is usually true for close binaries, the uneclipsed segment of the bright star has an additional motion away from us because of its rotation. The rotational motions measured in this way (Rossiter and McLaughlin), as well as the line profile variations (discussed by Struve and Elvey for Algol), give speeds very similar to those deduced from an analysis of the broadened lines outside of eclipse.

The measured velocities usually range up to 250 or 300 km sec^{-1}, while speeds as high as 400 or 500 km sec^{-1} are found in stars such as ϕ Persei. The corresponding kinetic energy of rotation,

$$E = \tfrac{1}{2} K^2\omega^2 M \tag{8–113}$$

where K is the radius of gyration, ω is the angular velocity, and M is the mass, is of the order of 10^{47} ergs. This quantity is comparable with the total amount of radiant energy stored within the star.

To deduce the rotational velocity from the broadened profile, Struve and Elvey supposed that the intrinsic profile radiated by a small area of the surface is essentially the same as that emitted by a non-rotating star of the same spectral class and luminosity. Each element of the surface radiates a profile $r_i(\lambda - \lambda_0)$ shifted by the Doppler effect from the undisplaced position by an amount $v\lambda/c$.

$$r_\lambda = \iint r_i\left[\lambda - \lambda_0 \pm \frac{v}{c}\,\lambda\right] dx\, dy \tag{8–114}$$

is obtained from an integration over the entire disk of the star. Now v is a function of x and y, and reaches its maximum, v_r when $y = 0$ and x is the radius of the star. For a constant x, the component of the rotational velocity in the line of sight is constant. Hence it suffices to divide the apparent disk of the star into a number of vertical strips and sum over a set of displaced profiles, each multiplied by the area of the appropriate strip. For each trial rotational velocity a profile is constructed and compared with the observed profile. The excellent agreement between the observed and computed profiles strongly suggests that the diffuse lines in η Orionis, α Aquilae, α Virginis, α Piscis Austrini, and other such stars, are broad because of rapid axial rotation. Carroll has shown that both the rotational velocity, v, and the true profile can be deduced in principle from the measured profile since a rotationally distorted line still retains certain features of the undisturbed profile.

Struve points out that the centrifugal force at the surface of certain of these spinning stars approaches that of gravity, a fact which supports the idea that the shells observed around some of the fastest rotating *B* stars arise from rotational instability. In accordance with theoretical expectations such shells are unstable and soon disintegrate. For example, that of γ Cassiopeiae disappeared after a few years. The shells around other stars, apparently not too different from that of γ Cassiopeiae, however, have lasted many years, e.g., *HD* 193182. Not all rapidly rotating stars show shells or bright lines, e.g., η Ursae Majoris, α Virginis, or α Aquilae show no signs of such appendages. Main sequence stars of spectral classes *O*, *B*, *A*, and early *F* show the largest rotations.

Slettebak and Howard (1955) find that stars from *B*5V to *B*7V show the fastest rotation $>$ 200 km sec^{-1}. The late *F*'s rotate slowly and in the *G*'s rapid rotation occurs only in spectroscopic binaries.

Herbig and Spalding (1955) found no stars later than *G*5 with rotationally broadened lines except for spectroscopic binaries. Supergiants of all types and normal giants of type *F* and later, never show rapid rotations. Further, *F*0–*G*0 stars of intermediate luminosity show greater speeds of axial rotation than do main sequence stars. These observations have been interpreted by stellar evolution theory. A hot *B* star evolves into a supergiant. A fast rotating main sequence star can evolve into an *F* star of intermediate luminosity which still rotates faster than a main sequence *F* star. As a spinning star expands in diameter after leaving the main sequence, conservation of an angular momentum requires that its rotational velocity decrease. Hence rapidly rotating supergiants are not found. From a statistical study of line broadening in *O*5–*B*5 stars, Slettebak (1956) concluded that the broad lines in main sequence stars were produced by rotation, those in supergiants by turbulence. The space velocities of the stars are not correlated with their rotational speeds; the axes appear distributed at random. Abt (1957, 1958) found the measured line profiles in bright giants and Ib supergiants to be consistent with the idea that these stars evolved from *B* stars, provided that the giants and supergiants do not rotate as rigid bodies.

The observed rotation of a star is not correlated with its galactic latitude, and average rotational velocities in different physical groups may differ markedly. Among the *B* stars in the Pleiades there is a greater proportion of rapidly spinning objects than among *B* stars chosen at random. Treanor (1960) finds that the rotations of the Hyades and Praesepe stars tend to be somewhat larger than those of the general field in the neighborhood of spectral class *F*0.

The origin of stellar rotation is probably to be understood in terms of conservation of angular momentum in the clouds of dust and gas from which the proto stars were formed. A contracting, spinning dust cloud may undergo fission into two stars or even evolve into a solar system. Enough residual angular momentum may remain to permit the remaining star to rotate rapidly. Collisions of gas clouds and proto stars may contribute further angular momentum to the system in initial stages of its formation (Huang and Struve, 1954).

8–14. Stellar Magnetic Fields. Magnetic fields of thousands of gauss exist in sunspots, although large-scale magnetic fields on the solar surface of the order of 10–200 gauss are confined to plage areas. The suggestion that solar magnetic fields were in some way connected with the sun's rotation would imply that large-scale magnetic fields might appear in rapidly rotating stars. Thus Horace Babcock guessed that one might observe appreciable magnetic fields in rapidly spinning *A* and *F* stars, and accordingly selected sharp-lined *A* stars which were likely to be objects in rapid rotation with axes directed towards the observer. It is possible to observe a Zeeman splitting only if the stellar lines are widened to not more than 0.3A by Doppler motions (stellar rotation or turbulence). Further, you would observe no recognizable magnetic effect if the stellar surface were covered with large regions of alternating magnetic polarity.

Babcock secured coudé spectrograms with a suitable analyzer in front of the slit so that two parallel spectra of the star, analyzed for right- and left-handed circular polarization, respectively, are photographed simultaneously. If the field is similar to that of a uniformly magnetized sphere and the axis is pointed toward the observer, the observed spectral lines will tend to be split into two groups of components, of opposite circular polarization. The separation will depend on the magnitude of the magnetic field, the atomic transition, and the wavelength of the line. Babcock found a field of 1500 gauss in 78 Virginis while a *K*0 control star showed no such effect. Magnetic fields were subsequently found in nearly every other sharp-lined *A* star.

Most *A* stars rotate rapidly and it seems well established that most of those with sharp lines are seen pole-on. These same sharp-lined stars often show "peculiar" spectra characterized by abnormally strong lines of various metals, e.g., Mn, Eu, Cr, and Si, which sometimes vary periodically in intensity. The spectral peculiarities and strong magnetic fields characteristic of the stars are a consequence of their rapid rotation. With increasing obliquity of view, the lines become "washed out" by rotation; Zeeman effects and spectral peculiarities are no longer observable. The equivalent widths of lines and energy distribution in the continuum depend on latitude. With decreasing obliquity of view, the star would become bluer, brighter, and display a more highly enhanced spectrum. Thus the peculiar A stars tend to fall on the upper edge of the main sequence. In a rapidly spinning star probably certain spectral peculiarities are produced in the polar regions. Sharp-lined stars with no magnetic fields or spectral peculiarities, e.g., γ Geminorum, are probably slow rotating objects.

Among the A stars Babcock has distinguished three types of magnetic variables. The prototype of the α type is α^2 Canum Venaticorum. They show essentially periodic (P = 4–9 days), variations in magnetic field with large amplitude in a range in H of the order of 4000 gauss, symmetric reversals of polarity and variations in intensities of certain lines that are synchronous with the magnetic charges. The lines of Cr II in α^2 Can Ven vary in antiphase with those of the rare earth. The β type (e.g., β Coronae Borealis) usually show spectral variations, $\overline{H} \sim 2000$ gauss, and fields with irregular fluxes and reversals in polarity. The γ type (e.g., γ Equulei) have irregular magnetic variations, $\overline{H} \sim 1100$ gauss, and show the same polarity.

The strongest magnetic field known in nature was found by Babcock in HD 215,441 whose field of 34,000 gauss suffices to resolve lines into their individual π and σ groups of components with an over-all spread of 1A. The sharpness of the components show the field to be very regular; evidently it is essentially dipolar and the axes of the rotation and magnetic dipole have only a small obliquity to the line of sight. The sharpness of the lines further shows that the dominance of magnetic forces in the outer layers suppresses turbulence. The field is so strong that its influence on stellar structure cannot be neglected. Sharp H and K lines of Ca^+ arise in a circumstellar envelope, their Doppler shifts show a relative outward velocity of 5 km sec^{-1} indicating that the star is losing mass.

All sharp-lined A stars classified as Ap (peculiar) show magnetic fields. The stars of earlier classes $A0p$ and $A2p$ show the most rapidly varying strong fields and exhibit a number of abnormally strong lines of various elements. The slower magnetic variations in $A5$, $A7$, and $F0$ stars (e.g., γ Equulei whose period is 226^d) accentuate only the lines of Sr, Cr, and the rare earths. Those peculiar A stars that are spectrum variables have periodic magnetic fields. The irregular variables of Babcock's classes β and γ are not spectrum variables, but there is a small group of magnetic stars with irregular spectral charges that are not correlated with magnetic changes.

The lines of Sm II and Eu II tend to show the largest amplitude, while lines of Sr II, Fe, Mg, and Ca II tend to show little or no change. Metallic line A stars tend to have stronger metallic lines than do peculiar A stars.

Although stellar magnetic fields have been studied most intensively in A stars, they are by no means confined to objects of this type. Babcock found them in five metallic line F stars, in three M stars, two S stars, the pulsating variable RR Lyrae, a subdwarf and the variable AG Pegasi. He concludes that since stars with mixed magnetic polarity (e.g., the sun) probably far outnumber those with uniform or coherent fields, magnetic fields of significant strength may occur in nearly all stars.

The prevalence of magnetic fields in *A* stars may be due to some kind of regenerative dynamo process which depends on axial rotation in combination with an outer convective layer in which Coriolis forces played a part (Elsasser, 1956; Parker, E.N., 1955). For a simple dipole field Babcock showed that the effective field strength as measured by the Zeeman effect can be multiplied by a factor of about 3 to yield the true field intensity. In many stars, however, the magnetic field distribution on the surface is complex, variable, and subject to irregular variations. Furthermore, the sharp edges of the line profiles suggest that one is observing a region of apparently uniform intensity covering a larger area of the surface. No evidence is found for small super sunspots with fabulously strong magnetic fields. Stars near *A*0 might possess magnetized areas covering a portion of the surface; possibly they are analogous to large plage areas, except that the fields are much stronger.

The abnormal line intensities arise from actual abundance differences between magnetic and normal stars, while the spectrum changes are due to actual abundance differences over the surface. It seems most likely that those differences are produced by magnetic segregation effects rather than by nuclear reactions in the superficial layers, as suggested by the Burbidges, Fowler, and Hoyle (1957).

A strong magnetic field may produce some effects on observed equivalent widths but cannot account for the huge intensification of certain lines, particularly those of the rare earths. Attempts have been made to use observed line intensities to deduce strength of magnetic fields (Stawikowsi, 1959). A magnetic field first of all influences the atomic absorption coefficient. Consider, as an example, an iron line such as $\lambda 4325(^3F_2 - {}^3G_3^0)$. When viewed along the direction of the magnetic field, this line will be split into ten circularly polarized Zeeman components. The absorption coefficient of each one of these components will be washed out by the Doppler effect and instead of a half-width of about 0.03A, the over-all spread of the pattern will yield an effective half-width of about 0.12A. Secondly, the equations of transfer become much more complicated, see discussions by Unno (1956), Hubenet (1954), Warwick (1955), and Stepanov (1958, 1960).

Various theories have been proposed to account for magnetic fields in stars. Babcock suggested that the mechanism was related to that responsible for the solar magnetic cycle. The sun possesses a "general" field confined to high latitudes and a "toroidal" field connected with the sunspot zones; possibly both are of the same origin (see Chapter 9). The α-type variables may undergo magnetohydrodynamic oscillations similar to those proposed for the sun (see Chapter 9) but on an exaggerated scale. One alternative theory developed particularly by Deutsch (1956, 1958) attributes the principal phenomena to the rotation of an oblique rotator. In its simplest form the oblique rotator consisted of a rigid sphere whose axis of rotation and magnetic dipole field were inclined to one another. This model was inadequate and was replaced by more complex formulations. In the periodic variables, the atmosphere may be organized by the field in distinct characteristic regions whereas in the irregular variables the fields may be subject to less well-defined and perhaps erratic or random fluctuations. Since the electrical conductivity of the gas is high, any magnetic field will be dragged along by the gas in its motion.

8–15. Abundances of Elements as Derived from Astronomical Data. In preceding articles we have alluded to chemical composition differences between stars. Roughly speaking these are of two types. In one type of stars we observe composition differences presumably due to differences in the composition out of the interstellar medium from which the stars were found. In subdwarfs and certain high velocity stars we observe metal/hydrogen ratios appreciably smaller than those found in stars like the sun. There also may exist stars such as 20 Can Ven (Savedoff and Santirocco) where the metal/hydrogen ratio is actually greater than in the sun. According to modern theories of stellar evolution, as the galaxy ages, dying stars add an increasing proportion of heavy elements to the interstellar medium from which new stars are formed. Hence the younger the star the greater the metal/hydrogen ratio. There also exists the possibility that the composition of the interstellar medium may differ at any one time in different places. Further, it seems very likely that the rate of heavy element building in stars was much greater during the early life of the galaxy than at present.

A second type of composition differences is to be attributed to nuclear processes actually occurring in the star concerned. Thus the trifurcation of the spectral sequence into *M*, *R*, and *S* stars, the abnormal abundance of lithium in certain stars, and the existence of helium stars, such as *HD* 124448, all point to composition differences that have been produced after the star was formed.

We concern ourselves here with a more restricted problem, namely, the abundances in a "normal" star such as the sun whose composition is believed to reflect that of the solar system. The data in Table 8–7 are obtained from the following sources: (a) early-type stars whose spectra show lines of elements not observed or not usable in the sun, (b) an analysis of the sun by Goldberg, Müller, and Aller (1960) using the method of weighting functions, and (c) data from chondritic meteorites. Certain elements such as helium, neon, and argon are all observed in the sun only in the corona or chromosphere under conditions of physical excitation (see Chapter 9) such that they cannot be used for abundance determinations. We use data from *B* stars which show for elements that can be observed in the sun abundances that are closely comparable with solar results. The Michigan analysis of the sun gave abundances of the following elements with respect to hydrogen, viz.: Li, Be, C, N, O, Na, Mg, Al, Si, P, S, K, Ca, Sc, Ti, V, Cr, Mn, Fe, Co, Ni, Cu, Zn, Ga, Ge, Rb, Sr, Y, Zr, Nb(Cb), Mo, Ru, Rh, Pd, Ag, Cd, In, Sn, Sb, Ba, Yb, and Pb. Results for some elements are highly uncertain because

TABLE 8-7

RELATIVE ATOMIC ABUNDANCES

Log $N(0) = 10.00$

	Log N		Log N		Log N		Log N		Log N
H	13.05	K	6.03	Rb	3.40	Ba	3.13	Re	1.95
He	12.25	Ca	7.10	Sr	3.75	La	2.15	Os	2.45
Li	4.55	Sc	3.98	Y	3.50	Ce	2.34	Ir	2.25
Be	3.85	Ti	5.87	Zr	3.85	Pr	1.71	Pt	1.75
B	3.93	V	4.87	Nb	2.55	Nd	2.41	Au	1.70
C	9.65	Cr	6.40	Mo	2.93	Sm	1.95	Hg	1.80
N	9.09	Mn	6.20	Ru	2.49	Eu	1.53	Tl	1.59
O	10.00	Fe	8.00	Rh	1.85	Gd	2.11	Pb	2.73
F	7.06	Co	5.69	Pd	2.31	Tb	1.29	Bi	1.45
Ne	9.75	Ni	6.95	Ag	1.87	Dy	2.13	Th	1.03
Na	7.35	Cu	4.85	Cd	2.50	Ho	1.43	U	0.75
Mg	8.48	Zn	5.25	In	1.80	Er	1.89		
Al	7.30	Ga	3.50	Sn	1.47	Tm	1.13		
Si	8.55	Ge	4.25	Sb	2.00	Yb	1.83		
P	6.45	As	3.15	Te	3.09	Lu	1.11		
S	8.35	Se	4.38	I	2.40	Hf	1.45		
Cl	7.30	Br	3.69	Xe	3.11	Ta	1.80		
Ar	7.93	Kr	4.25	Cs	2.20	W	1.65		

of the weakness of the observed lines, inaccuracies in f-values, and inadequacies in term analysis for some elements. Lines of many elements are not observed in accessible regions of the solar spectrum, e.g., Ne, As, Hg, and Bi. Hence the solar data have to be supplemented by information from other sources. Except for a few special abundance ratios, the earth's crust is an unsatisfactory source of quantitative data. It has been subjected to extreme fractionization and chemical separation and bears no more resemblance to the composition of the earth as a whole than does the composition of the slag in a smelter crucible bear to the composition of the original ore. Meteorites which are believed to be fragments of a defunct planet probably represent the best sample we can get of the non-volatile constituents of the solar system. Of these, the type known as carbonaceous chondrites seems to give the best clues to the original material. Unfortunately they are fragile and rare since few have survived the passage through the earth's atmosphere, and analyses have been inadequate, particularly for the rarer so-called trace elements. The ordinary chondrites, the most abundant type of stony meteorite, give reliable abundances for most elements that are not themselves volatile nor form volatile compounds. Compilations of abundances based largely on meteorites were proposed by Goldschmidt, more recently by Harrison Brown and particularly by Suess and Urey who made a careful study of the entire problem of elemental abundances. More recent work has been summarized by Ehmann (1961). Where they can be compared, the chondritic meteorites yield abundances of non-volatiles that generally are in good agreement with the solar data. One outstanding discordance must be mentioned, iron for which the f-values, etc., are particularly reliable. Nevertheless the solar abundance of iron seems to be about three times smaller than the meteoritic result which appears to be substantiated by other considerations, viz.: density of planets and satellites, etc. Chondritic meteorites, however, show a considerable spread in abundances of rare elements.

The data of Table 8–7 must be considered as tentative, i.e., as a working compilation only. Further improvements will be made as accurate analyses of carbonaceous chondrites become available, as the structure of the solar atmosphere becomes better understood, and as basic physical parameters such as f-values, etc., are reliably measured. Yet the problem of the composition of the solar system will be only solved by space research, by obtaining actual samples of the moon, asteroids, comets, and particularly material from remote parts of the system, e.g., satellites of Saturn. The inspiration for abundance studies comes from our hopes to understand the origin not only of the earth and solar system, but of the elements themselves and the stages of stellar evolution therein involved.

8–16. Interlocking, Non-Coherent Scattering, and Center-to-Limb Variations of Spectral Lines. In most of the preceding discussions we have considered absorption lines in the framework of simple assumptions embodied in curves of growth or line profiles formed by the mechanism of pure absorption. In the latter instance we can ignore the presence of all other levels than αJ and $\alpha' J'$ in considering a transition $\nu(\alpha' J', \alpha J)$. If lines are formed, however, by the mechanism of scattering, three ugly questions immediately arise: (1) What happens in a level n will not only depend on temperature and density (T, ρ) but also on the population in other levels, n'', whence atoms may jump to level n. (2) Since the population of a level n no longer depends on the local T, ρ, deviations from local thermodynamics equilibrium (LTE) occur (see Art. 8–17). (3) Is the scattering coherent? That is, if a quantum is absorbed at a particular frequency in the line, will it be re-emitted at the same frequency?

Consider a subordinate line such as Hα or Mg I (λ5167) (see below). The upper level can be populated by captures from the continuum, by cascading from higher levels, or by collisional excitations. Collisions are not likely to be important in the upper atmosphere of the sun, but recaptures in higher levels and cascading can become important for subordinate lines as Strömgren and others have emphasized. The Balmer lines provide a good illustration. An atom may be raised from the ground level by the absorption of Lyman β, for example. As it cascades to the second level it emits a λ6563 quantum and if this process occurs more often than its inverse, Hα will tend to be filled in. The Hα line may be said to be interlocked with the Lyman β and other transitions occurring in the hydrogen atoms. Woolley 1934 suggested that this interlocking effect might be important in determining the observed central intensities of the Balmer line. (See Fig. 8–20.)

A number of workers have considered the problem of interlocked lines. For example, Busbridge and Stibbs (1954) have solved the transfer problem

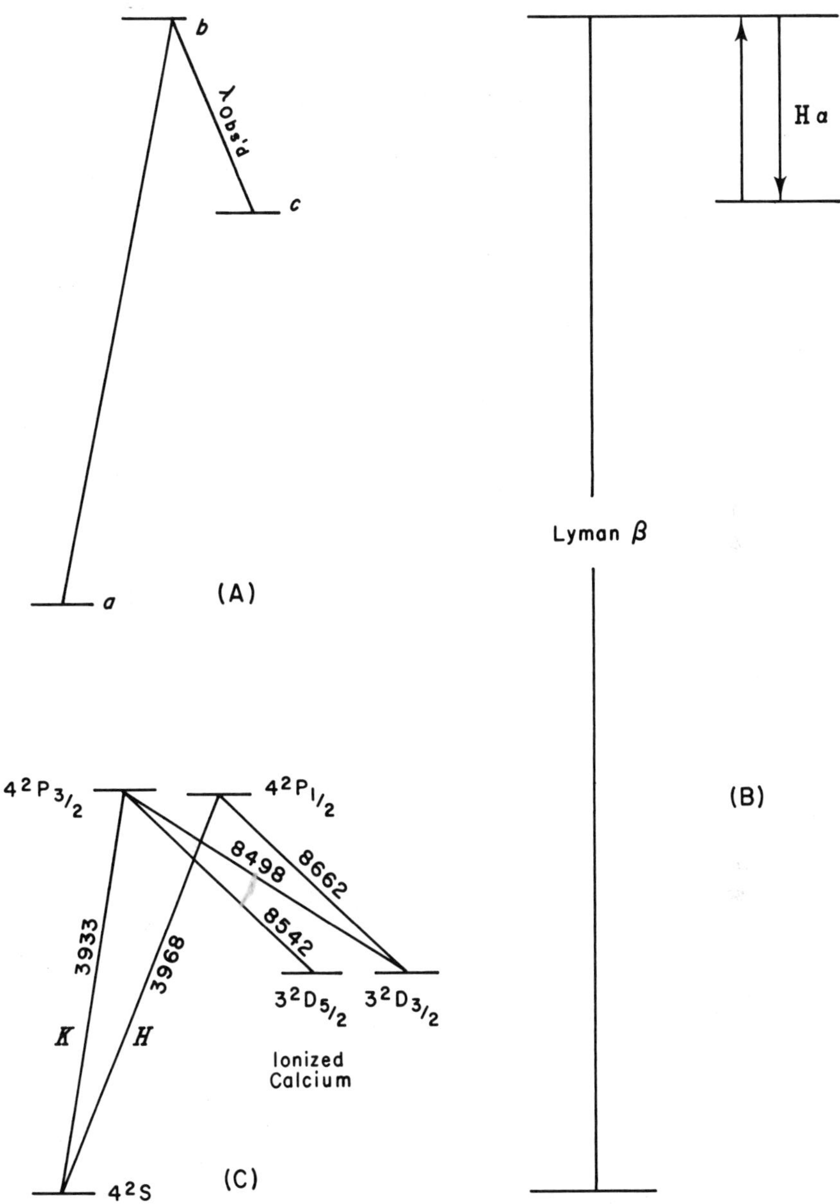

FIG. 8–20. INTERLOCKING BETWEEN SPECTRAL LINES

(A) In a typical situation, an upper level is excited by transitions from the ground level a as well as from level c. The number of transitions b-to-c does not necessarily equal the number c-to-b.

(B) The third level in hydrogen can be reached from the ground level by the absorption of Lyman β and from the second level by the absorption of Hα. Under conditions of non-LTE, the number of absorptions may not equal the number of Hα emissions.

(C) Ionized calcium provides an especially interesting situation in that the ^{2}D term is metastable. Atoms typically escape to the ground term by transitions to ^{2}P followed by cascade.

for interlocked lines with a common upper level and the ME approximation, e.g., Mg I λ5167, λ5173, and λ5184 ($3s3p^3P_{012}$—$3s3p^3S_1$).

The infrared Ca II lines λ8498, λ8662, and λ8542 ($3d^2D$—$4p^2P^0$) present a particularly interesting situation. Their upper levels are the same as the upper levels of the great *H* and *K* lines, whereas the lower levels are metastable. They have high central intensities 0.26–0.34 because of strong interlocking with the resonance lines. Although the *H* and *K* lines tend to follow the scattering mechanism, these lines more closely approach the absorption mechanism because of the metastability of the lower levels. Their centers are probably of chromospheric origin (Miyamoto, 1954). Interlocking effects are likely to be very important in the cores of strong lines formed in chromospheric regions (see, e.g., Jefferies, 1960).

We must now consider the important question of whether light absorbed in one part of a line is re-emitted at precisely the same part of the line. If this is true, as we usually assume in line profile calculations, the darkening in each frequency is independent of what happens in other frequencies. On the other hand, the re-emitted frequency may be correlated with, but not determined by, the absorbed frequency. Under this condition, which is called *non-coherent scattering*, *n. c. s.*, re-emission is not the exact inverse of absorption. When a transition occurs between two broadened levels, the atom is not likely to return to the same spot in the broadened lower level from which it started. There are varying degrees of non-coherency, and it is necessary to consider correlated, partially correlated, and uncorrelated non-coherent scattering.

Doppler broadening leads to correlated non-coherent scattering as follows: If a moving atom absorbs a quantum from one direction and emits it in another, the absorbed and emitted frequencies are not equal, even if the scattering as viewed by an observer traveling with the atom were coherent. Rather, the frequency shift is determined by the geometry of the encounter. In the far wings, for frequency shifts that correspond to the rms velocity, the scattering is often assumed two-thirds non-coherent and one-third completely coherent.

Other types of non-coherent scattering may occur with natural broadening and particularly with collisional broadening.

Following a discussion by Spitzer, let us first describe non-coherent scattering by atoms subject to natural broadening only. In the first example (Fig. 8–21A), the ground level is sharp and the atom must emit the same frequency as it absorbs. If as for subordinate lines, however, both upper and lower levels are fuzzy, as indicated in Fig. 8–21B, the atom may emit a different amount of energy from what it absorbs. Eventually, the sum of the energies of all absorbed quanta must equal the sum of the energies of all emitted quanta. If the atom, upon returning to the lower level, emits a quantum of frequency less than the frequency of the quantum

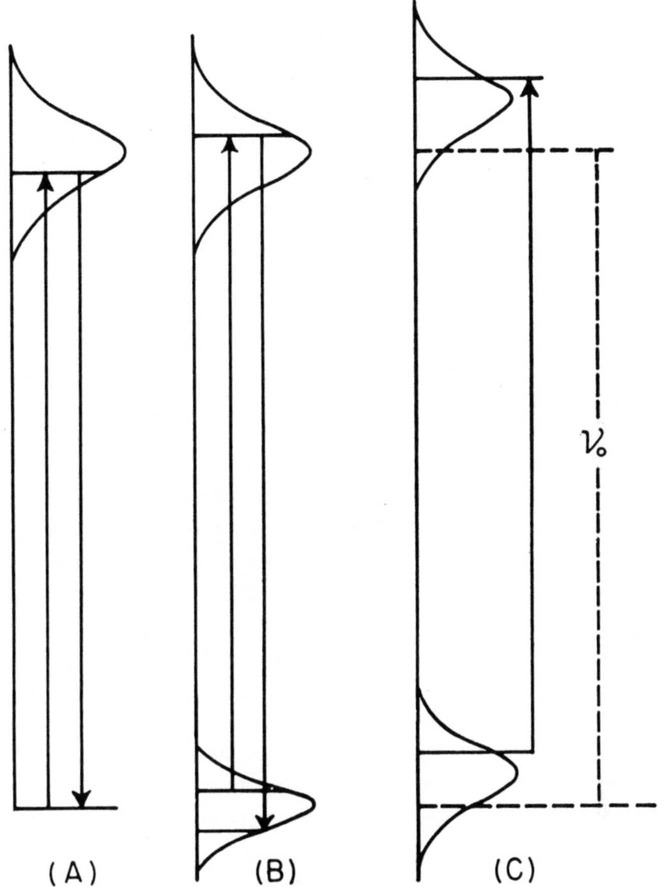

FIG. 8–21. NON-COHERENT SCATTERING

(A) A resonance line arising from a perfectly sharp lower level follows coherent scattering if the radiating atoms are at rest. (B) Non-coherent scattering occurs if both upper and lower levels are broadened by radiation damping, or by collisions as in (C) where ν_0 denotes the undisturbed frequency.

previously absorbed, then in the next capture of a quantum, the emitted frequency will tend to be greater than the absorbed frequency, i.e., the atom acts as though it had a memory. In the end, the atom tends to smooth out all energy gains and losses.

Pressure broadening is the most important source of non-coherent scattering. Suppose that the velocities of the perturbing atoms are small, so that the profile of the broadened absorption coefficient reflects the statistical distribution of the perturbers. Then, if the electron jumps to a higher level when a disturbing atom is nearby, the absorbed frequency ν will differ from that of the undisturbed atom, ν_0. When the atom returns to the lower level, the perturber may be far away and the atom will emit ν_0. The perturbing atom absorbs the excess energy to increase its own velocity.

W. Orthmann and P. Pringsheim showed that when mercury vapor is exposed to radiation of a narrow resonance line from an auxiliary mercury lamp, the profile of the re-emitted line is independent of perturbing atoms in the auxiliary lamp and is determined only by the Lorenz widening in the gas exposed to the radiation. Hence, we may conclude damping by collision produces at least some degree of non-coherent scattering.

The degree of non-coherency will depend on how much the perturbation changes between absorption and emission of a quantum. If the time interval is small enough the scattering will tend to be coherent, whereas if the time interval is long, and the perturbation changes appreciably, the scattering will be uncorrelated non-coherent scattering. In impact broadening theories the total effect of a collision cannot be expressed in terms of the time interval between absorption and re-emission, whereas in statistical broadening theories the concept of such a time interval is admissible.

Zanstra (1941, 1946) suggested that the problem could be handled by considering a classical oscillator whose scattered radiation consisted of a coherent and an uncorrelated non-coherent component in the ratios $\Gamma_{\text{rad}}/\Gamma_{\text{coll}}$, where Γ_{rad} and Γ_{coll} are the damping constants for radiation and collision, respectively. Edmonds (1955) has treated the problem from the point of view of a simplified quantum mechanical formulation of non-coherent scattering and has shown that Zanstra's ratio of coherent to non-coherent scattering is too large by an order of magnitude. Further, where statistical broadening is important we can neglect any coherent contribution to scattering to within an error of 5 per cent.

The physical theory of the redistribution of the re-emitted radiation is not too well understood. The simplest hypothesis is that the energy re-emission is proportional to the absorption coefficient and independent of the frequency distribution of the incident radiation.

If J_ν is the re-emission of radiation in the line per gram per unit solid angle, I_ν is the mean intensity of the radiation with respect to angle and l_ν is the line scattering coefficient, then

$$\int j_\nu \, d\nu = \int J_\nu l_\nu \, d\nu = \overline{J}_\nu \int l_\nu \, d\nu = n_{rs} \frac{\pi \varepsilon^2}{mc} f \overline{J} \tag{8–115}$$

where n_{rs} is the number of atoms capable of scattering the line per gram of stellar material, then

$$j_\nu = l_\nu \overline{J} \tag{8–116}$$

Thus, for a situation in which the continuum is formed by pure absorption and the line is formed by absorption and non-coherent scattering, Eq. 8–3 is replaced by

$$\cos\theta \frac{dI_\nu}{\rho dh} = (\kappa + l_\nu) I_\nu - (\kappa + \varepsilon l_\nu) B_\nu(T) - (1 - \varepsilon) l_\nu \frac{\int J_\nu l_\nu \, d\nu}{\int l_\nu \, d\nu} \tag{8–117}$$

where κ varies slowly with ν, but l varies rapidly. Non-coherent scattering grievously complicates the theoretical calculation of line profiles since it is necessary to solve a transfer equation whose last term involves an integration of the quantity sought, I_ν, over all frequencies across the line.

Before enumerating some of the mathematical techniques that have been proposed for solving Eq. 8–117, let us discuss briefly some of the qualitative effects of non-coherent scattering on line profiles. If we solve the equations of line transfer for pure scattering (see Art. 8–3, note especially Eqs. 8–17, 8–18, and 8–27), we find that the intensity (averaged over all solid angles) at the center of the line, $J_{\nu 0}(\tau)$, while small at the surface, rapidly rises to the black-body value with increasing optical depth. Away from .the line center where the absorption coefficient is smaller, J_ν approached B_ν much more slowly. Non-coherent scattering cannot directly alter the total net flux in the whole line, but it can transfer energy from one frequency to another. Let us suppose that the mean intensity of the center of the line, $J_{\nu 0}(\tau)$, computed for coherent scattering can be used as a starting approximation in the treatment of non-coherent scattering. Then, at large optical depths, radiation taken up by the line will be re-emitted in the wings at a rate proportional to $\alpha_\nu B_\nu$ for each atom, since the redistribution is proportional to α_ν and the radiation field $J_{\nu 0}$ is B_ν at the center of the line. In other words, the re-emission is determined by what is happening at the center of the line where the energy density is $B_\nu(\tau)$, rather than $J_\nu(\tau)$, the average intensity at the frequency ν in the line wing. Toward the line center J_ν approaches $J_{\nu 0}$ and the deviations from the coherent scattering picture are not great. Near the surface, the energy absorbed in the wings tends to be put back near the line center and there is extinction with no re-emission in the wings. Hence non-coherent scattering should show up as a combination of absorption and extinction in the far wings of absorption lines.

This very feature of non-coherent scattering makes it extremely difficult to decide between the mechanisms of pure absorption and of scattering on purely empirical grounds. Thus, Unsöld has pointed out that line profiles formed by pure absorption in an atmosphere with a steep temperature fall near its surface would closely resemble the observed profiles.

Let us discuss briefly the variations of profiles of strong lines across the solar disk, and the relation of these data to the problem of non-coherent scattering. At different distances from the center we observe radiation from effectively different depths of the solar atmosphere. Hence line pro-

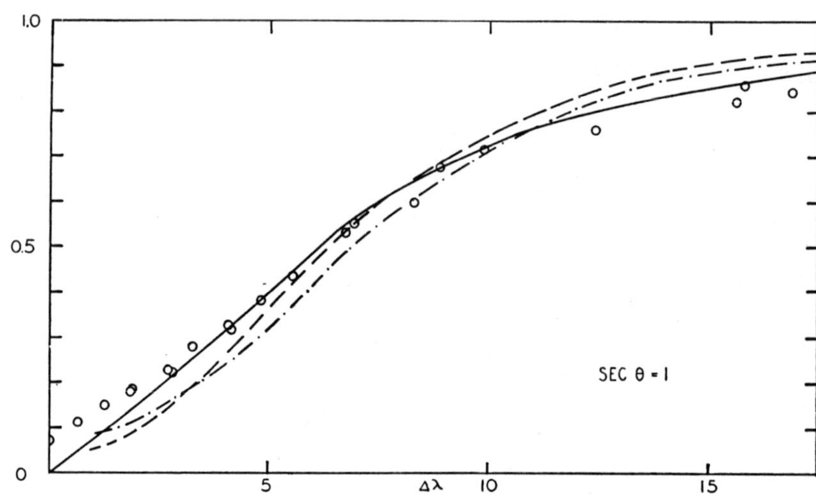

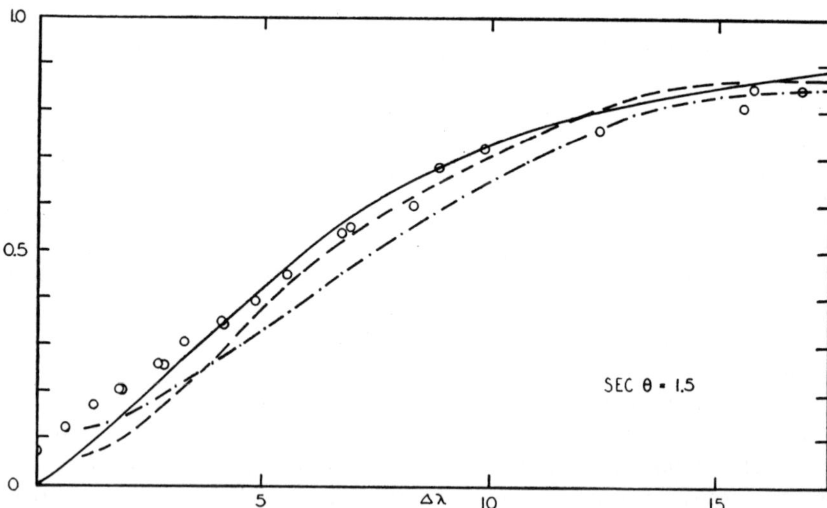

Fig. 8–22. The Center-Limb Variation in λ3933

We compare Houtgast's observations (circles) with theoretical curves. At sec θ = 1.5 and 3, the plotted points are interpolated from the observations. The solid curves give the profile computed by Eq. 8–32 with the modifications suggested by Eq. 8–36. The dashed curve is computed by the iteration method, while the dot-dash curve gives the latter corrected for non-coherent scattering. Notice that although

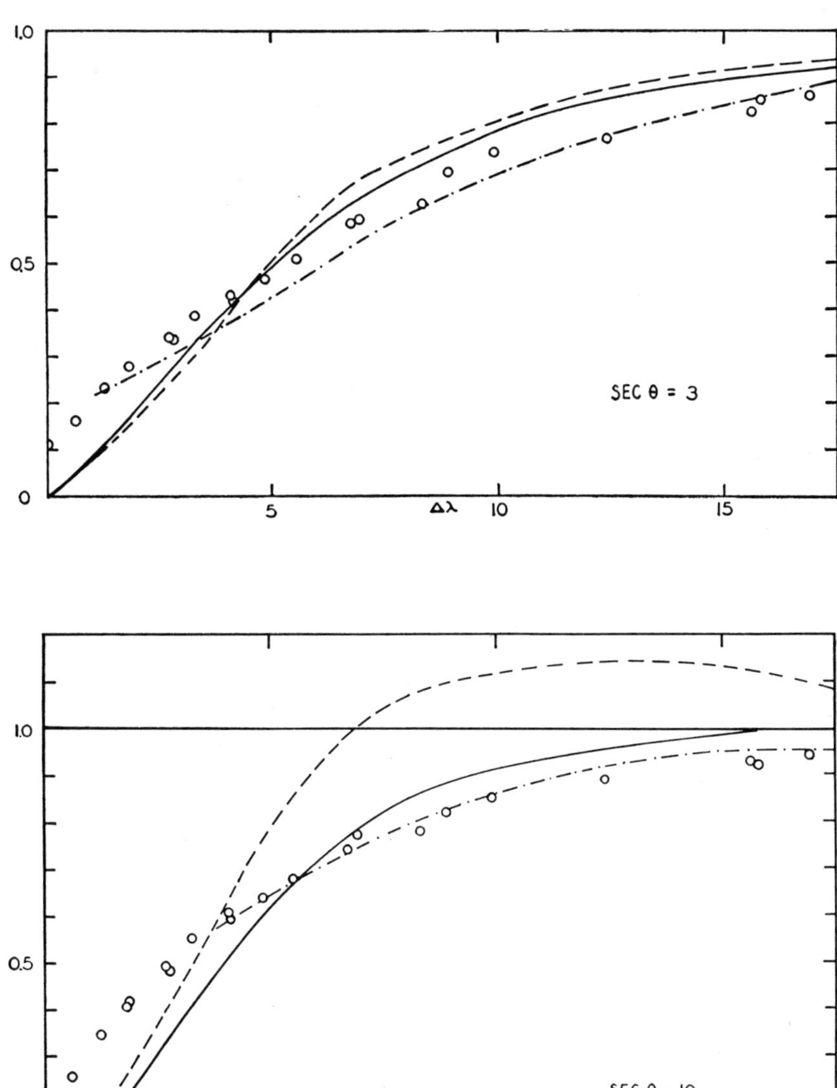

there is fair agreement between the theoretical curves near the center of the disk, the curves show significant departures from one another toward the limb. The non-coherent scattering curve seems to give a better representation of the center-limb variations than does the completely coherent theory. These calculations, carried out in 1949, are based on an earlier solar model with a slightly different temperature distribution, $T(\tau)$, from that given in Chapter 5.

file variations should reveal something of the stratification of the radiating atoms. Theory predicts that the behavior of the line wings is insensitive to whether scattering or absorption prevails but is sensitive to the variation of η with depth. At the extreme limb itself, the wings would vanish for pure absorption, whereas lines formed in strict monochromatic radiative equilibrium may actually appear in emission there. (See Fig. 8–22.) If we represent $B_\nu(\tau)$ by a linear expression (Eq. 8–17),

$$B_\nu = B_0 + B_1\tau$$

we find that to the violet of the energy maximum B_1 will be so large that B_ν will rise rapidly with τ. A volume element in local thermal equilibrium near the surface will absorb a greater amount of radiation of frequency ν than it will emit as thermal energy of the same frequency, since the emission depends on the local temperature in accordance with Kirchhoff's law. On the other hand, purely scattering atoms will emit all the radiation absorbed at ν in the same frequency ν. Thus they may emit more energy than the atoms responsible for the continuous spectrum, provided that the flux in the line is comparable with that in the continuum, a condition which is fulfilled for the far wings of the line. Hence the wings may actually appear in emission. On the other hand, to the red of the energy maximum in the spectrum, scattering may act to intensify the dark wings.

Among older center-limb variation studies of line profiles we may mention those of K. Schwarzschild, A. Unsöld, H. H. Plaskett, G. Righini, T. Royds, and A. L. Narayan, M. Minnaert, R. O. Redman, C. W. Allen, D. S. Evans, E. Cherrington, and C. D. Shane. The most extensive series was that of Houtgast who studied twenty-three representative lines of Fe I, Mg I, Ca I, Ca II, and Na I across the solar disk in an effort to assess the role of non-coherent scattering in the sun.

The central intensities of the strong lines simply mirror the center-limb variations in the continuous background. Variations in the line wings are of great interest in evaluations of the relative roles of scattering, absorption, and non-coherency. Towards the limb, scattering must predominate in the wings, since radiation is received from only the uppermost layers. It follows that since it appears necessary to introduce an absorption coefficient to explain the profiles near the limb, we are really dealing with non-coherent scattering which transfers energy from the wing towards the center of the line! Towards the line center, in accordance with our expectations, a formal representation by coherent scattering becomes possible. In order to explain the measured profiles, Houtgast found it necessary to invoke non-coherent scattering.

On the basis of rather idealized assumptions concerning η (Eq. 8–7) and the variation of B with depth, Munch (1949) and Savedoff (1952) interpreted the center-limb variations of the K lines by non-coherent scattering.

Fig. 8–22 compares Houtgast's observations with theoretical profiles of the K line calculated with the aid of a model atmosphere, both for coherent scattering (mean-value and iteration methods) and for non-coherent scattering. It would appear that the center-limb behavior of the wings can be explained by non-coherent scattering at least in a quantitative way. Unfortunately, the behavior of the computed profile near the line center is unsatisfactory. The immediate neighborhood of the line center is formed in the chromosphere where we would expect no agreement, but the deviations for $\Delta\lambda \sim 2$–3A probably arise partly from inadequacies of the 1949 solar model atmosphere. Sobolev (1954) concluded that the observational data favored non-coherent scattering, but Miyamoto (1954) found that the green magnesium "b" lines gave no clear-cut choice between absorption and non-coherent scattering. Harris (1949) found some evidence for non-coherent scattering in his studies of the D lines.

On the other hand, studies of the sodium D lines by Priester (1953) and by Hitotuyanagi and Inaba (1954) indicated that the profiles could be accounted for by the mechanism of pure absorption without invoking non-coherent scattering. Bray (1956) concluded that for the sodium "D" lines the argument for non-coherent scattering is not convincing.

The problem of non-coherent scattering effects in line profiles can be attacked from another point of view, developed by H. Zanstra. A resonance line like λ4227 of Ca I should behave like a classical oscillator with negligible true absorption. The scattered light should be polarized such that the intensity of the component along the solar radius, I_r, differs from that parallel to the solar limb, I_L. The difference, $I_L - I_r$ expressed in units of the nearby continuum depends on the position on the disk and the limb-darkening coefficient. Redman found a polarization effect in λ4227 ten times smaller than the value predicted by Zanstra for pure scattering. The conclusion is that collisional damping rather than radiation damping prevails for the line. The scattering is non-coherent; only a tenth of the radiation absorbed in the line wings is re-emitted in the same frequency; the rest is distributed over the entire line. Edmonds' re-examination of Zanstra's calculation suggests that the observations can be interpreted in terms of uncorrelated non-coherent scattering, the contribution of actual absorption is negligible.

There are good theoretical reasons for believing that strong resonance lines are formed according to the mechanism of scattering. Since the contribution of the coherent component is of negligible importance compared to the non-coherent component, theories of resonance line formation must emphasize non-coherent scattering.

Since Eddington's original suggestion (1929) and the pioneer theoretical work of Spitzer (1936), Woolley (1938), Henyey (1940), and Zanstra (1941, 1946), many writers have examined the problem from different points of view. Edmonds

(1955) has critically examined the basic physical theory. The theory of spectral line formation has occupied the attention of many writers who have used different mathematical techniques and assumptions, Suemoto (1949), Savedoff (1952), Stibbs (1953), Sobolev (1949, 1951, 1954), Busbridge (1953, 1955), Warwick (1955), Inaba (1955), Ueno (1955, 1957, 1958), Labs (1954), and Miyamoto (1955). Among methods used are principle of invariance, Laplace transformations, discrete ordinates, and probabilistic methods. Various assumptions are made, such as the ME approximation, $B(\tau)$ dependence as in Eq. 8–17, and replacement of the line absorption coefficient by a step function. In a quantitative discussion of non-coherent scattering, Labs compares pure absorption, coherent scattering, and non-coherent scattering for different values of the limb-darkening coefficient and line strength. He finds that pronounced differences occur near the limb and that true absorption and non-coherent scattering behave in different ways.

Clearly, a decision concerning the importance of non-coherent scattering rests on a better understanding of the structure of the upper solar atmosphere and observations of line behavior at the extreme limb of the sun.

The wings of strong lines are broadened by collision for which the relevant cross sections are not well enough known. Hence in recent years the emphasis has been on the center-to-limb variations of weak and moderately strong lines. Interpretations of such measurements are complicated by our ignorance of the variation of turbulent velocity with depth, by the granulation of the solar surface, or by deviations from LTE. As previously remarked, Allen (1949) and Waddell (1958) found turbulence to be asymmetrical, i.e., the vertical and horizontal velocity components differ. Furthermore, Suemoto (1957) and Unno (1959) find that turbulence decreases with height in the atmosphere while de Jager draws the opposite conclusion.

The existence of the photospheric granulation (Chapter 9) shows that the solar surface has temperature fluctuations and that the plane parallel approximation fails. Theoretically, the problem can be approximated by assuming an undulation in the photospheric surface (Redman) or by replacing a plane-parallel solar atmosphere by one consisting of alternate columns of different temperature (two or even three stream models have been proposed), e.g., Böhm (1954), Schroter (1957), Voigt (1956, 1959). Although some success has been encountered in representing some of the line intensity changes by such models, the required temperature differences $\Delta T \sim 400°C$ are inconsistent with the temperature differences $\sim 100°C$ found from granulation studies (Schwarzschild). These and other studies, however Elste (1955), Bretz (1956), Mitchell (1959), indicated that there remained discrepancies between theory and observation that suggested deviations from local thermodynamic equilibrium, a topic to which we now turn our attention.

8–17. Deviations from Local Thermodynamic Equilibrium. Throughout most of our discussion we have assumed that the state of excitation, dissociation and ionization in a gas can be described in terms of some unique temperature T, even though the radiation field J_ν does not necessarily correspond to the Planckian radiation field $B_\nu(T)$. We must now examine to what extent our use of the Boltzmann and Saha equations is justified.

The question is *not* whether deviations from thermodynamic equilibrium exist but how large are they, and what are their influences on observable parameters, e.g., intensity distribution in the continuum, profiles and central intensities of lines, and their equivalent widths.

Even for a grey atmosphere, the existence of a temperature gradient and consequent non-isotropy of the radiation field requires that the source function S_ν (τ) be non-Planckian (Wildt, 1957). Actually (Chapter 5) stellar material is colored (i.e., the absorption coefficient depends on wavelength), so $I_\nu(\theta, \tau_\nu)$ depends on both direction and frequency. Calculations presented in Chapter 5 show that the intensity averaged over all angles, J_ν, may depart substantially from the Planckian function B_ν corresponding to the energy density of radiation at the same point. In treating deviations from LTE, the non-greyness of the radiation field must be considered already in the first step. Further $(J_\nu - B_\nu)/B_\nu$ depends not only on frequency but also on optical depth, decreasing with increasing depth in the stellar atmosphere.

We might expect deviations from LTE to show up in its effects on the following:

1. Distribution law of velocities.
2. Ionization equation.
3. Boltzmann equation.

Departures from LTE may affect (a) processes producing the continuum and (b) processes producing lines — whose wings and cores are formed at different depths; hence effective deviations may differ substantially from one part of a line to another.

Let us first consider effects of deviations from LTE on the Maxwellian velocity distribution. Processes tending to destroy a Maxwellian distribution for electrons are (a) inelastic collisions with atoms, (b) photoionizations, (c) free-free emissions, and (d) recaptures; whereas processes tending to restore it are elastic collisions of electrons with ions, atoms, and particularly with one another. Even under such extreme conditions as those existing in a tenuous gaseous nebula electrons will maintain a Maxwellian distribution. The reason is that electron-ion and particularly electron-electron encounters are so enormously more frequent than inelastic encounters that deviations from a Maxwellian distribution are negligible (D. Bohm and Aller, 1947).[12]

Even though electrons and ions follow Maxwellian distributions it is

[12] If an electron field is present so that an atom may pick up a fair amount of energy between each encounter, noticeable deviations from a Maxwellian distribution may occur.

not immediately obvious that each distribution would necessarily correspond to the same gas kinetic temperature. An analysis by Bhatnagar, Krook, Menzel, and Thomas (1955) shows that T (electron) $= T$ (ion) over a wide range in temperature and density because energy exchange between electrons and heavier particles proceeds much faster than energy dissipation in inelastic collisions, free-free emissions, etc.

Departures from thermodynamic equilibrium may be studied most neatly in gaseous nebulae (e.g., Aller, 1956; Seaton, 1960). Here the dilution of the radiation is so great the transfer problem may be ignored for all except resonance transitions, the hydrogen, helium and carbon lines are excited by photo-ionization followed by recapture, and strong (forbidden) lines of permanent gases are all excited by inelastic electron collisions.

In stellar atmospheres and particularly in the chromosphere and extended envelopes the problem is much less simple and one must proceed by successive approximations.

Deviations from LTE may affect the Saha and Boltzmann equations in a complicated manner depending on the radiation field, density and chemical composition of the gas. Collisional processes always tend to restore LTE whereas radiative processes tend to produce departures since the mean intensity does not correspond to the Planckian function at the local temperature.

Consider first the ionization equation. The number of radiative ionizations per cm³ sec⁻¹ for $N_{r,s}$ atoms in each unit where in the sth level of the rth ionization state will be:

$$F_{r,s}(\text{ion}) = N_{r,s} \int_{\nu_0}^{\infty} \frac{4\pi J_\nu}{h\nu} \alpha_\nu'(s)\, d\nu \tag{8–118}$$

where $\alpha_\nu'(s)$ is the atomic continuous absorption coefficient (corrected for negative absorptions) for s-level atoms. The number of collisional ionizations will be:

$$\mathfrak{F}_{r,s}(\text{ion}) = N_{r,s}N_{\varepsilon}\mathfrak{Q}_{r,s}{}^{\text{ion}} = N_{r,s}N_{\varepsilon}\int_{v_0}^{\infty} vf(v)\sigma_{r,s}{}^{i}(v)\, dv \tag{8–119}$$

where $\sigma_{r,s}{}^i(v)$ denotes the cross-section for collisional ionization and $f(v)$ denotes the velocity distribution function. The big difficulty in the calculation is that $\sigma(v)$ is not accurately known and one must use approximations. On the basis of a classical approximation to σ, K. H. Böhm (1960) estimated

$$\frac{F_{r,s}(\text{ion})}{\mathfrak{F}_{r,s}(\text{ion})} \sim 10^{-22}\, \frac{T^{3/2}\nu_0{}^2}{N_\varepsilon} \left(\frac{h\nu_0}{kT}\right) \frac{E_1(h\nu_0/kT)}{E_2(h\nu_0/kT)} \tag{8–120}$$

and concluded that throughout relevant layers of the solar atmosphere, the radiative ionizations of a typical metal like iron are a hundred times as frequent as collisional ionizations. Hence one can neglect collisional ionizations in comparison with radiative ionizations, except for the higher atomic levels,

For all practical purposes we can write the ionization equation in the following form for LTE

$$N_{r+1}{}^{o}\, N_{\varepsilon}{}^{o} a^{o}(T) = N_{r}{}^{o} \sum \frac{N_{r,s}{}^{o}}{N_{r}{}^{o}} \int \frac{4\pi B_{\nu}}{h\nu}\, \alpha_{\nu}'(s)\, d\nu \qquad (8\text{–}121)$$

where the superscript ^{o} refers to LTE values and $a^{o}(T)$ is the recombination coefficient. In the stellar situation:

$$N_{r+1} N_{\varepsilon} a(T) = N_{r} \sum \frac{N_{r,s}}{N_{r}} \int \frac{4\pi}{h\nu}\, J_{\nu}\, \alpha_{\nu}'(s)\, d\nu \qquad (8\text{–}122)$$

If $a(T) = a^{o}(T)$, i.e., the departures from LTE are not so severe that the recombination coefficient is seriously affected, we can write

$$\frac{N_{r+1}N_{\varepsilon}}{N_{r}} \left[\frac{N_{r+1}{}^{o} N_{\varepsilon}{}^{o}}{N_{r}{}^{o}}\right]^{-1} = \sum \frac{N_{r,s}}{N_{r}} \int J_{\nu}\alpha_{\nu}(s)\, \frac{d\nu}{\nu} \left\{ \sum_{s} \frac{N_{r,s}{}^{o}}{N_{r}{}^{o}} \int B_{\nu}\alpha_{\nu}(s)\, \frac{d\nu}{\nu} \right\}^{-1} \qquad (8\text{–}123)$$

in order to assess the effects of deviations from LTE on the ionization equilibrium. Hence it is clear that the departure from LTE will involve two factors, (1) the difference between J_{ν} and B_{ν} and (2) the departure of the population of the (r, s) level from a Boltzmann distribution. One can get some idea of the departure if he assumes a Boltzmann distribution appropriate to the local temperature and uses the J_{ν} valid for the layers in question. For the ionization of iron in the sun, Böhm found a ratio of 2.88 for a continuum optical depth of about 0.01 as a consequence of a considerable departure of J_{ν} from B_{ν} in these upper layers. That is, large departures from LTE exist in optical depths shallower than 0.1 and it is precisely in these regions that the cores and even much of the wings of strong Fraunhofer lines are produced. From a study of the center-limb variations of weak to medium-strong metallic lines, Bretz (1956) found deviations from LTE in the sense of increased ionization in the upper layers. In early-type stars (see Fig. 8–23) one can anticipate that the main contribution to the integral of $\alpha_{\nu}J_{\nu}$ comes from the region where J_{ν} most closely approaches B_{ν}.

What will be the effects of deviations from LTE on the continuous spectrum? In early-type stars one might expect departures of the populations of the higher levels from a Boltzmann distribution with consequent effects on $\kappa_{\nu}(T, P_{\varepsilon})$. Hence J_{ν} would be modified and an iteration procedure would have to be employed.

In the sun the problem is much simpler since the main source of continuous absorption is the negative hydrogen ion. This ion is destroyed by photo-ionization and electron impact. It is also necessary to consider associative detachments as in the reaction $H + H^{-} \rightarrow H_2 + \varepsilon$, but this

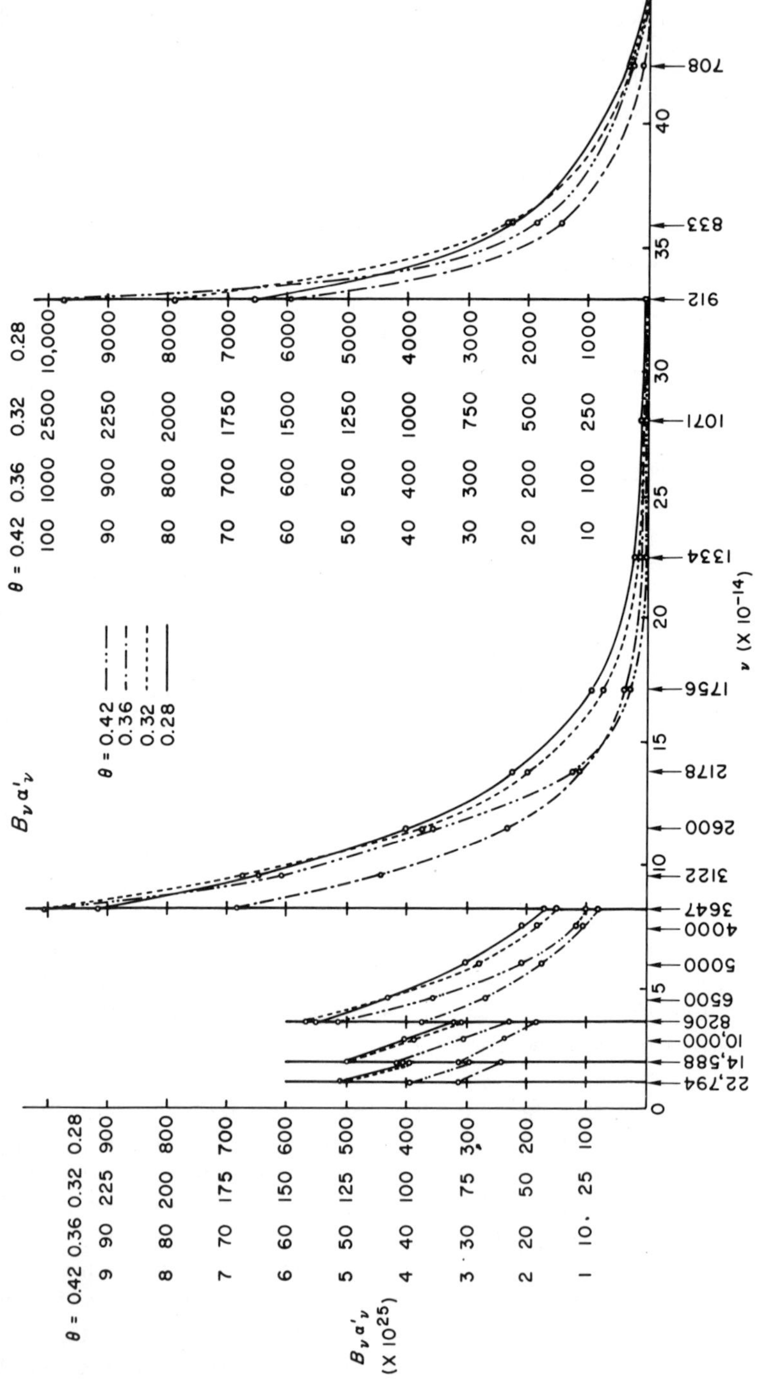
$B_\nu a'_\nu$ (X 10^{25})
θ = 0.42 0.36 0.32 0.28
$B_\nu a'_\nu$
θ = 0.42
0.36
0.32
0.28
ν (X 10^{-14})
22,794
14,588
10,000
8206
6500
5000
4000
3647
3122
2600
2178
1756
1334
1071
912
833
708

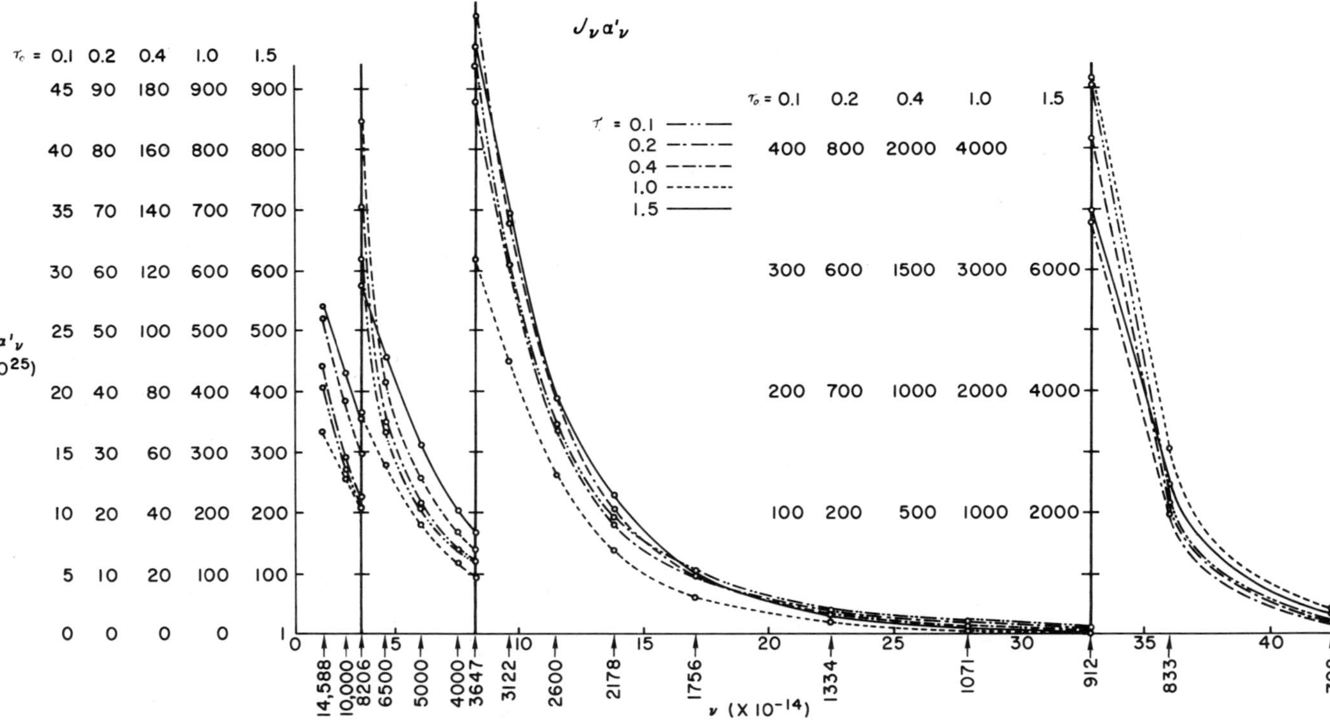

FIG. 8-23. COMPARISON OF $\alpha_\nu' B_\nu$ AND $\alpha_\nu' J_\nu$ FOR THE ATMOSPHERE OF A B STAR

(a) We plot $\alpha_\nu' R_\nu$ ($\times 10^{25}$) against ν for different values of $\theta = 5040/T$ corresponding to different optical depths in the star. The effects of lines are not included. Notice the pronounced maxima at the series limits. (b) We plot $\alpha_\nu' J_\nu$ ($\times 10^{25}$) against ν for different values of the optical depth $\tau_0 = 0.1, 0.2, 0.4, 1.0$, and 1.5. Different wavelengths are also indicated on the abscissa scale.

Note the different scales on the left-hand side applying to the different curves and the abrupt scale change at the Lyman limit. The stellar atmosphere model employed is that described in Chapter 5.

(like all collision processes) tends to maintain LTE. Pagel (1959) finds that the $N(H^-)/N(H)$ ratio is appropriate to LTE throughout the photospheric layers relevant to the formation of the continuous spectrum (see also Mädlow, 1962).

We must consider that although the usual ionization equation is apparently satisfactory in those layers that produce the continuous spectra of stars like the sun, we can expect departures in strata that produce strong or medium-strong lines, or even relatively weak lines. The physical reason for these departures is that radiative processes are much more important than collisional ones and the radiation field may depart substantially from a Planckian. For line transitions the departure of J_ν from B_ν becomes even worse.

We must now examine the question of the validity of Boltzmann's equation. Collisions (both inelastic and superelastic) tend to maintain a Boltzmann distribution, whereas radiative processes tend to destroy it. Following Woolley and Stibbs (1953) one can compare radiative and collisional processes involving the excitation of levels and verify that in the normally studied region of the solar spectrum collisional excitation is much less important than radiative excitations. As far as observed transitions in hot stars are concerned, the shuffling between levels by collisions may be more important than radiative processes, but these levels are often primarily populated by radiative transitions from the ground level. Nevertheless, infrared spectral lines in early type stars should admit of a good approximation to LTE.

Spectral line formation under conditions of non-LTE can present difficult problems. In Art. 8–2 we derived the equation of transfer from the macroscopic point of view. We now re-derive it from the microscopic point of view, but in order to simplify the discussion we shall suppose that for the frequency concerned, the continuous absorption is negligible compared with the line absorption (Thomas, 1957).

Let there be $N_{n'}$ atoms in level n' per unit volume. The amount of energy absorbed by atoms in the elementary volume of unit area and ds in the elementary cone $d\omega$ will be

$$N_{n'} I_\nu (\theta)\alpha_\nu \, ds \, d\omega$$

Here $I_\nu(\theta)$ is the intensity in the frequency range ν to $\nu + d\nu$ and α_ν is the *atomic* absorption coefficient which is related to the Einstein $B_{n'n}$ by

$$\alpha_\nu = B_{n'n} \frac{h\nu}{4\pi} \phi_\nu \tag{8–124}$$

(see Eq. 4–113). Here ϕ_ν describes the shape of the absorption coefficient. Near the center of the line it will be determined by the Doppler motions of the atoms.

The amount of energy re-emitted in $d\omega$ in transitions ν (nn') will be

$$N_n A_{nn'} \frac{h\nu}{4\pi} \psi_\nu \, d\omega + N_n B_{nn'} \frac{h\nu}{4\pi} I_\nu(\theta)\phi_\nu \, d\omega$$

The first term refers to spontaneous emission, the second to induced emissions. We suppose that the absorption coefficient shows the same frequency distribution, ϕ_ν for both normal and negative absorptions (induced emissions). The shape of the emission distribution function ψ_ν need not, a priori, be the same for spontaneous emission as for absorption, i.e., ψ_ν may differ from ϕ_ν. The equation of transfer now becomes:

$$\cos\theta \frac{dI_\nu}{dx} d\omega = -N_{n'} I_\nu(\theta) B_{n'n} \frac{h\nu}{4\pi} \phi_\nu \, d\omega + N_n \frac{h\nu}{4\pi} d\omega \left[A_{nn'}\psi_\nu + B_{nn'} I_\nu(\theta)\, \phi_\nu\right] \quad (8\text{–}125)$$

Making use of the relations between the Einstein coefficients (see Eqs. 4–96 and 4–97), we obtain

$$\cos\theta \frac{dI_\nu}{-dx} d\omega = B_{n'n}\, I_\nu(\theta) \frac{h\nu}{4\pi} \phi_\nu \left[N_{n'} - \frac{g_{n'}}{g_n} N_n\right] d\omega - \frac{g_{n'}}{g_n} \frac{2h\nu^3}{c^2} B_{n'n}\, N_n\, \psi_\nu \frac{h\nu}{4\pi} d\omega \quad (8\text{–}126)$$

Let us now define

$$d\tau_\nu = -N_{n'}\, B_{n'n}\, \phi_\nu \frac{h\nu}{4\pi} \left[1 - \frac{N_n}{N_{n'}} \frac{g_{n'}}{g_n}\right] \quad (8\text{–}127)$$

and let (see Eq. 8–1)

$$S_\nu = \frac{2h\nu^3}{c^2} \left[\frac{N_{n'}}{N_n} \frac{g_n}{g_{n'}} - 1\right]^{-1} \frac{\psi_\nu}{\phi_\nu} = \frac{2h\nu^3}{c^2} \left[\frac{b(n')}{b(n)} e^{h\nu/kT} - 1\right]^{-1} \frac{\psi_\nu}{\phi_\nu} \quad (8\text{–}128)$$

denote the source function. It follows that the equation of transfer reduces to

$$\cos\theta \frac{dI_\nu}{d\tau_\nu} = I_\nu\,(\theta) - S_\nu \quad (8\text{–}129)$$

which is the standard form. Notice that S_ν reduces to the Planckian function if and only if $\psi_\nu = \phi_\nu$ which is true for pure absorption *or* completely non-coherent scattering, and if $b(n') = b(n) = 1$ for all energy levels, i.e., Boltzmann's equation holds. For partially coherent scattering, additional terms will appear in the equation as interlocking may occur between two substates dE_n and $dE_{n'}$ in any energy level n.

If $\psi_\nu/\phi_\nu = 1$, but $b_n \neq 1$, we can still define an "excitation temperature," T_{ex}, for any pair of levels.

Thomas (1957) shows that for instances of astrophysical interest the frequency dependence of the emission coefficient closely follows that of the absorption coefficient, the extent of the difference amounting to a factor of only four over the Doppler core, while over the same frequency interval the absorption coefficient varies over a range of 10,000. In problems of line formation in the upper photosphere and chromosphere one may take advantage of the fact that the absorption coefficient follows a Doppler shape over a core about three Doppler widths in extent. The mechanism of line formation is essentially pure absorption or n.c.s.

In practice, the problem is horrid because one must also set up the equations of statistical equilibrium and solve them simultaneously with the transfer equation. That is, we must write down explicitly the condition that a steady state exists for each level. Consider, for example, a three-state atom consisting of a ground level (1), an excited level (2), and a continuum (3). The equation of statistical equilibrium (steady state) for the ground level is:

$$N_2A_{21} + N_2B_{21}I_{21} + N_2N_\varepsilon Q_{21} + N_3N_\varepsilon^2 Q_{31} + N_3N_\varepsilon \int f(v)\sigma_{31}(v)\, dv$$
$$+ N_3N_\varepsilon \int_0^\infty f(v)\, \sigma_{3,1}{}^iI_{31}\, v\, dv = N_1B_{12}I_{12} + N_1 \int B_{13}\, I_{13}\, d\kappa + N_1N_\varepsilon Q_{12} + N_1N_\varepsilon Q_{13} \tag{8–130}$$

where the first term on the left-hand side denotes spontaneous emissions from 2 to 1, the second term the induced emissions, the third term represents the superelastic de-excitations of the 2nd level, the fourth term represents the collisional recaptures, the fifth term the normal recombinations, and the sixth term the induced recombinations. On the right-hand side the terms represent successively radiative excitations to the second level, photo-ionizations, collisional excitations, and collisional ionizations. As more energy levels are considered, the problem becomes correspondingly more involved.

Plaskett (1955) suggested one could solve the problem of non-LTE by introducing several distinct temperatures, T_{rad}, T_ε, and T_{ex} that could differ from one another. The trouble is that no unique excitation temperature can describe the population distribution in even *one* atom, to say nothing of several. Pecker and his associates proposed an empirical procedure for finding the b_n's (see Eq. 8–1) from the profiles and central intensities of Fraunhofer lines.

The most elegant treatment is a perturbation-type method proposed by Henyey (1946) and applied by him and Grasberger (1955) to some situations of astrophysical interest. They consider only the first order deviations from LTE, a Maxwellian distribution of velocities, and neglect the collisional excitations and ionizations. They also suppose that the population of atoms is distributed among the various substates (ΔE) of an energy level E in the same manner as for LTE, i.e.

$$N_n\,(\Delta E)/N_n{}^0\,(\Delta E) = N_n/N_n{}^0 \tag{8–131}$$

An approximation is made that

$$N_n\,(\Delta E)/N_n{}^0\,(\Delta E) = 1 + \xi_n\,(\Delta E) \tag{8–132}$$
$$N_iN_\varepsilon/N_i^0N_\varepsilon^0 = 1 + \xi, \quad J_\nu = B_\nu\,[1 + \eta_\nu]$$

where the superscript zero denotes the equilibrium value. Grasberger considered an atmosphere of hydrogen and calculated the deviations of the central intensities $R_c(\text{H}\alpha)$, $R_c(\text{H}\beta)$, and $R_c(P_\alpha)$ from the values they would have had under conditions of LTE for atmospheres corresponding to spectral classes $A0 \rightarrow A5$. For Hα and Hβ they amount to 10–12 per cent; for P_α they amount to 15–20 per cent.

The difficulty in these calculations are that the assumed deviations are small, a linear approximation to $B(T)$ is adopted, and the deviations from the Saha equation are presumed small.

The theory of deviations from thermodynamic equilibrium for the chromosphere is extremely difficult (see Thomas and Athay, 1960). Since deviations are large, successive approximations are needed. Schematic examples are helpful in getting an idea of the situation involved, but a final solution of the problem depends on knowing the actual parameters for collision and ionization processes and taking into account the detailed transfer problem.

The source function as defined by

$$S_\nu = \frac{2h\nu^3}{c^2}\left[\frac{N_{n'}}{N_n}\frac{g_n}{g_{n'}} - 1\right]^{-1}$$

depends on what physical processes dominate in fixing the ratio $N_n/N_{n'}$ and this in turn depends on the positions of the energy levels involved, ionization potential and radiation field. Thus, Thomas (1957) found that the source function for the ionized metals in the upper photosphere and chromosphere tends to be dominated by collisions and therefore depends only on N_ε and T_ε. For neutral metals and non-metals the source function is dominated by the photoelectric ionization term, so there tends to exist a capture spectrum for at least the lower temperatures. Thomas concluded that in the general case the source function does not lie strictly between the case of pure coherent scattering, where $S_\nu = \bar{I}_\nu$ and pure absorption $S_\nu = B_\nu(T_\varepsilon)$ as is sometimes stated, but rather is intermediate between a value fixed by photoelectric transitions and a value fixed by collisional transitions. When radiation terms dominate, a solution of the radiative transfer equation is necessary.

How serious are deviations from LTE in photospheric layers? Some writers, e.g., Pecker, claim that huge deviations exist. Others assert that deviations are small except for resonance lines formed mainly in chromospheric layers. Empirical studies are difficult. Measurements of center-limb variations can give only differential effects in LTE, but if the deviations from LTE depend strongly on optical depth we would expect to detect them. Estimates of the ratio of the scattering coefficient to the absorption coefficient may be made (see, e.g., Mitchell, 1959) and will give an idea of how large the deviations might be. The lower level of the infrared oxygen lines λ7772–4–6 lie only 2.92 v below the ionization limit, so radiation shortward of λ4220 can ionize atoms from this level. Even at the limb, these lines appear to be formed deep in the atmosphere where the continuum originates.

Some of the simplest examples of deviations from LTE are provided by gaseous nebulae and the solar corona. In gaseous nebulae, hydrogen is ionized by quanta from a high temperature star whose energy output beyond the Lyman limit is given by

$$I_\nu = W_\nu \frac{2h\nu^3}{c^2}\left[e^{h\nu/kT} - 1\right]^{-1} \tag{8–133}$$

where the dilution factor W_ν is of the order of 10^{-12} to 10^{-14}. All ionizations take place from the ground level at a rate

$$\int F_{1\kappa}\, d\nu = \int_{\nu_1}^{\infty} B_{1\kappa}\, I_\nu\, N_1\, d\nu = \int_{\nu_1}^{\infty} 4\pi\, N_1\, I_\nu \frac{\alpha_\nu}{h\nu}\, d\nu \tag{8–134}$$

where N_1 is the number of atoms cm^{-3} in the ground level, $B_{1\kappa}$ is the Einstein coefficient and α_ν is the coefficient of continuous absorption. In a steady state, this rate equals the rate of recombinations cm^{-3} sec^{-1}.

$$N_1\, N_\varepsilon \sum_{n=1}^{\infty} a_n\,(\tau)$$

where $a_n(T_\varepsilon)$ is the recombination coefficient for level n. It depends only on T_ε. Hence the ionization of hydrogen depends on W_ν and T_ε in a straightforward way (see Aller, *Gaseous Nebulae* (1956), p. 138). Note that recombinations may occur on all levels even though photoionizations took place only from the ground level.

In the solar corona, atoms are ionized by collisions only so that

total number of collisional ionizations = total number of recombinations

(See, e.g., Woolley and Allen, 1948.) The number of such ionizations is

$$\begin{aligned}\mathfrak{F}_{i,\,i+1} &= N_i\, N_\varepsilon \int_{v_1}^{\infty} \sigma(v)\, vf(v)\, dv = N_i N_\varepsilon C_{i,\,i+1} \\ &= N_i N_\varepsilon\, 4\pi\left(\frac{m}{2\pi k T_\varepsilon}\right)^{3/2} \int_{v_1}^{\infty} v^3 \sigma_i(v)\, e^{-mv^2/2kT_\varepsilon}\, dv\end{aligned} \tag{8–135}$$

which must equal the number of radiative recombinations

$$F_{i+1,i} = N_{i+1}\, N_\varepsilon \sum_n \int_{v=0}^{\infty} f\,(v,T)\, v\sigma_{\kappa n}\, dv \tag{8–136}$$

where one sums over the bound levels, n. Since

$$\mathfrak{F}_{i,\,i+1} = F_{i+1,\,i} \tag{8–137}$$

the electron density N_ε cancels out.

Burgess, van Regemorter, and Seaton find for the rate of collisional ionizations

$$C_{i,i+1} = 1.15 \times 10^{-8}\left[3.1 - \frac{1.2}{Z} - \frac{0.9}{Z^2}\right]\frac{s_0 T^{1/2}}{\chi^2}\, 10^{-\theta\chi}\ \text{cm}^{-3}\ \text{sec}^{-1} \tag{8–138}$$

χ = ionization potential in volts, s_0 = number of electrons in outer shell, T is the temperature in °K, and Z is the ionic charge after ionization. The rate of recombinations (see Elwert, 1952) is

$$0.97 \times 10^{-12}\, \chi\, n_0\, g T^{-1/2}\, N_{i+1}\, N_\varepsilon\ \text{cm}^{-3}\ \text{sec}^{-1} \tag{8–139}$$

where n_0 is total quantum number of the ground level and the correction factor g is of the order of 3 (most metals) and 4 for iron. Hence the ionization ratio is

$$N_{i+1}/N_i = C_0 \times 10^4\, T\chi^{-3}\, 10^{-\theta\chi}\, sn_0^{-1} \tag{8–140}$$

where $C_0 = 0.75$ for iron, 1.0 for other metals. Illustrative calculations based on equations are given by C. W. Allen (1961). See also pp. 540, 541.

PROBLEMS

8–1. Consider a line formed according to the mechanism of pure absorption in an atmosphere whose opacity is produced both by thermal absorption and by electron scattering. Show that the source function is given by the integral equation

$$S_\nu(t_\nu) = \frac{\kappa_\nu + l_\nu}{\kappa_\nu + l_\nu + \sigma} B_\nu(t) + \frac{\sigma}{\kappa_\nu + l_\nu + \sigma} \frac{1}{2} \int S_\nu(t') \, E_1 (|t - t'|) \, dt'$$

where l_ν, κ_ν, and σ are the coefficients of line absorption, continuous absorption and electron scattering respectively while t is the optical depth in the line.

8–2. Show that for very weak lines, such that the residual intensity is

$$R = \int_0^\infty G(\tau) \, dt_\nu, \text{ then } W = G(0) \frac{\pi \varepsilon^2}{mc} \lambda^2 Nf$$

where N is now the number of atoms above τ^*, where $\tau^* \sim \frac{2}{3}$ for pure absorption. More accurately

$$\tau^* = \frac{1}{G(0)} \int_0^\infty G(\tau) \, d\tau$$

8–3. Consider a weak line which is formed in the wing of a stronger one. If η is constant with optical depth (ME model) and both lines are formed by the mechanism of pure scattering, compare the profile of the blended line with that which it would have if the strong line were absent. Suppose that the depth of the strong line at the center of the profile of the weak one is R.

8–4. Molecular bands are normally formed in the highest coolest layers of a stellar atmosphere. Examine the question of deviations from local thermodynamic equilibrium on the equations of dissociative equilibrium, noting the influence of the deviations on rotational, vibrational and radiative terms.

8–5. Consider Eq. 8–130 for the statistical equilibrium of a 3-state atom in non-LTE, assume that the cross-section for collisional ionization can be written in the form

$$\sigma_{AB} = \frac{1}{g_A} \frac{h^2}{4\pi m^2} \frac{\Omega(A, B)}{v^2} \qquad v > v_0, \ \Omega = \text{const}$$

make use of Eq. 4–155, assume $I_\nu = WB_\nu(T)$, and derive an expression for the stimulated captures. Then write out the explicit expression for the source function.

REFERENCES

General References

AMBARZUMIAN, V. A. and others. 1958. *Theoretical Astrophysics.* Pergamon Press, Ltd., Oxford, England. Translated by J. B. Sykes.

BARBIER, D. 1958. *Encyclopedia of Physics.* S. Flügge (ed.). Springer-Verlag, Berlin, vol. **50**, 0. 274.

CHANDRASEKHAR, S. 1950. *Radiative Transfer.* Oxford University Press, Fair Lawn, N. J., chap. 12.

HYNEK, J. A. and others. 1951. *Astrophysics, a Topical Symposium.* McGraw-Hill Book Co., Inc., New York.

PECKER, J. C., and SCHATZMAN, E. 1959. *Astrophysique Générale.* Masson and Co., Paris.

UNSÖLD, A. 1955. *Physik der Sternatmosphären.* Springer-Verlag, Berlin.
WOOLLEY, R. v. d. R., and STIBBS, D. W. 1953. *The Outer Layers of a Star.* Oxford University Press, Fair Lawn, N. J.
GREENSTEIN, J. L. (ed.). 1960. *Stellar Atmospheres*, vol. 6 of *Stars and Stellar Systems.* University of Chicago Press, Chicago. The best general reference for this chapter.

Art. 8–1

High dispersion solar and stellar spectrographs are discussed respectively by:

MCMATH, R. R. 1956. *Ap. J.* **123,** 1.
DUNHAM, T. 1955. *Vistas in Astronomy.* A. Beer (ed.). Pergamon Press, Ltd., Oxford, England.

Arts. 8–2, 8–3, and 8–4

See texts and review articles listed under general references. Strömgren's method is described by him in *Ap. J.* **86,** 1, 1937; see also *Zeits. f. Ap.* **10,** 237, 1935.

Art. 8–5

As example of calculation of a line profile on electronic computers see:

HEISER, A. M. 1957. *Ap. J.* **125,** 470.
Symposium on "The Profiles of Stellar Spectral Lines." 1960. C. de Jager (ed.). *Annales d'Astrophysique* **23,** No. 6.

Art. 8–6

The method of weighting functions is discussed by Pecker and Schatzman, *op. cit.*: Unsöld, *op. cit.*, and in vol. 6 of *Stars and Stellar Systems*, J. L. Greenstein (ed.). For practical applications see, for example:

ALLER, L. H., ELSTE, G., and JUGAKU, J. 1957. *Ap. J. Suppl.* No. 25.
ALLER, L. H., and JUGAKU, J. 1959. *Ap. J. Suppl.* No. 38.

Art. 8–7

The curve of growth theory is given by:

MENZEL, D. H. 1936, *Ap. J.* **84,** 462.
UNSÖLD, A. 1955. *Ap. J.* **84,** 462.
WRUBEL, M. H. 1949. *Ap. J.* **109,** 66.
———. 1950. *Ap. J.* **111,** 157.

See also other writers mentioned in the text and, for example:

GREENSTEIN, J. L. (ed.). 1960. *Stellar Atmospheres*, vol. 6 of *Stars and Stellar Systems.* University of Chicago Press, Chicago.

Arts. 8–8 and 8–9

Examples of curve of growth method are numerous in the astrophysical literature. A good illustrative example of the differential method is:

HELFER, H. L., WALLERSTEIN, G., and GREENSTEIN, J. L. 1959. *Ap. J.* **129,** 700.

An example of the use of absolute *f*-values in curve of growth construction is:

BASCHEK, B. 1959. *Zeits f. Ap.* **48,** 95.

See also:

WRIGHT, K. O. 1948. *Publ. Dom. Ap. Obs.* **8,** 1.
———. 1950. *Publ. Dom. Ap. Obs.* **8,** 281.
———. 1954. *Publ. Dom. Ap. Obs.* **10,** 1.
———. 1944. *Ap. J.* **99,** 249.

Art. 8–10

For a general discussion of turbulence in stellar atmosphere see the review article by

WRIGHT, K. O. 1955. *Transactions of International Astronomical Union*, **9,** (Dublin meeting).
UNDERHILL, A. B. 1961. *Suppl. Nuovo Cimento* **22,** 1.

HUANG, S. S., and STRUVE, O. 1960. *Stellar Atmospheres*, vol. 6 of *Stars and Stellar Systems*. J. L. Greenstein (ed.). University of Chicago Press, Chicago, Chap. 8.

The complexities of the atmosphere of the supergiant 31 Cygni are best exhibited in a series of papers by:

McKELLAR, A., WRIGHT, K. O., and associates. 1958, 1959. *Publ. Dom Ap. Obs.* **11,** 1, 35, 59, 77.

For a discussion of ζ Aurigae and related problems, see:

WILSON, O. C., and ABT, H. A. 1954. *Ap. J. Suppl.* No. 1.

WILSON, O. C. 1960. *Stellar Atmospheres*, vol. 6 of *Stars and Stellar Systems*. J. L. Greenstein (ed.). University of Chicago Press, Chicago, Chap. 11.

The loss of mass from supergiant atmospheres is discussed by

DEUTSCH, A. 1956. *Ap. J.* **123,** 210.

———. 1960. *Stellar Atmospheres*, vol. 6 of *Stars and Stellar Systems*. J. L. Greenstein (ed.). University of Chicago Press, Chicago, Chap. 15.

WILSON, O. C. 1960. *Ap. J.* **131,** 75; **132,** 136.

Art. 8–11

An enormous literature exists in spectral classification and luminosity effects. For a summarizing review with bibliography see:

FEHRENBACH, CHARLES. 1958. *Encyclopedia of Physics.* S. Flügge (ed.). Springer-Verlag, Berlin, vol 50, p. 1.

Also see:

KEENAN, P. C. 1958. *Encyclopedia of Physics.* Vol 50, p. 93.

The classical references on spectral and luminosity classification are:

CANNON, A. J. *Henry Draper Catalogue.* Annals of Harvard College Observatory **91–99,** 1918–1924 and H. D. extension in Harvard Annals, **100, 105** (1937), **112** (1949).

MORGAN, W. W., KEENAN, P. C., and KELIMAN, E. 1942. *An Atlas of Stellar Spectra.* University of Chicago Press, Chicago.

For the use of the *K* emission lines as a source of luminosity data and applications thereof see:

WILSON, O. C. 1957. *Ap. J.* **126,** 525.

———. 1961. *Ap. J.* **130,** 496, 499; **133,** 457.

WILSON, O. C., and VAINU BAPPU, M. K. 1957. *Ap. J.* **125,** 661.

The determination of stellar surface gravities from the profiles of the hydrogen lines and the Kolb-Griem theory is described in:

ALLER, L. H., and JUGAKU, J. 1958. *Ap. J.* **128,** 616.

———. 1959. *Ap. J.* **130,** 469.

Strömgren's technique is described in the Halley lecture reported in *Observatory* **78,** 1958.

Art 8–12

A summarizing account of molecules in stellar spectra, with bibilography, is given by:

SWINGS, P. 1958. *Encyclopedia of Physics.* S. Flügge (ed.). Springer-Verlag, Berlin, vol. 50, p. 109.

Symposium on "Les Molécules dans les Astres." 1957. *Mem. Roy. Soc. Liége* **18,** 215 *et seq.*

Art. 8–13

Among examples of recent work on stellar rotation we mention especially:

SLETTEBAK, A., and HOWARD, R. F. 1955. *Ap. J.* **121,** 102.

HERBIG, G. F., and SPALDING, J. F. 1955. *Ap. J.* **121,** 118.

HUANG, S. S., and STRUVE, O. 1960. *Stellar Atmospheres*, vol. 6 of *Stars and Stellar Systems*. J. L. Greenstein (ed.). University of Chicago Press, Chicago, Chap. 8.

Art. 8–14

The basic characteristics of magnetic fields in *A* stars are described in:

BABCOCK, H. W. 1958. *Ap. J.* **128,** 228.

———. 1960. *Stellar Atmospheres,* vol. 6 of *Stars and Stellar Systems.* J. L. Greenstein (ed.). University of Chicago Press, Chicago, Chap. 7.

For theoretical discussions of stellar magnetic fields see articles in:

Symposium on Cosmical Electrodynamics. 1958. B. Lehnert (ed.). Cambridge University Press. International Astronomical Union Symposium No. 6.

DEUTSCH, A. J. 1958. *Encyclopedia of Physics.* S. Flügge (ed.). Springer-Verlag, Berlin, vol. 51, p. 689.

Art. 8–15

The basic modern compilation of abundances of elements is that by:

SUESS, H., and UREY, H. C. 1956. *Rev. Mod. Phys.* **28,** 53.

See also the report of the Symposium on Cosmochemistry in *Journ. of Chemical Education* **38,** 51, 1961.

The problem is summarized from the astronomical point of view in:

ALLER, L. H. 1961. *Abundances of the Elements.* Interscience Publishers, New York.

Art. 8–16

The theory of non-coherent scattering and interlocking is summarized in Woolley and Stibbs (*op. cit.*). See also:

BUSBRIDGE, I. 1953. *M. N.* **113,** 52.

———. 1955. *M. N.* **115,** 661.

EDMONDS, F. N. 1955. *Ap. J.* **121,** 418.

HENYEY, L. G. 1940. *Proc. Natl. Acad. Sci.* **26,** 50.

SPITZER, L. 1936. *M. N.* **96,** 794.

———. 1944. *Ap. J.* **99,** 1.

STIBBS, D. 1953. *M. N.* **113,** 493.

UENO, S. 1955–58. *Kyoto Astr. Contr.* 58, 62, 63, 64, 67.

ZANSTRA, H. 1941. *M. N.* **101,** 273.

———. 1946. *M. N.* **106,** 225.

Art. 8–17

Deviations from thermodynamic equilibrium are discussed extensively in current astrophysical literature. See, for example, papers by J. C. Pecker, *Annales d'Astrophysique* **22,** 499, 1959, and especially:

THOMAS, R. N., and ATHAY, R. G. 1960. *Physics of the Solar Chromosphere.* Interscience Publishers, New York.

BÖHM, K. H. 1960. *Stellar Atmospheres,* vol. 6 of *Stars and Stellar Systems.* J. L. Greenstein (ed.). University of Chicago Press, Chicago.

UNSÖLD, A. 1963. *Zeits. f. Physik* **171,** 44.

Basic Tables and Atlases

BABCOCK, H. D., MINNAÉRT, M., and MOORE, C. E. In press. *Revised Rowland Table.* Mt. Wilson Observatory. Gives wavelengths, identifications, and equivalent widths of principal solar lines.

MINNAERT, M., MULDERS, G. F. W., and HOUTGAST, J. 1940. *Photometric Atlas of Solar Spectrum.* I. Schnabel, Amsterdam. Gives profiles of solar lines on a true intensity scale.

MOHLER, O. C., PIERCE, A. K., MCMATH, R. R., and GOLDBERG, L. 1951. *Photometric Atlas of the Infrared Solar Spectrum from* λ8465 to λ25,242. University of Michigan Observatory, Ann Arbor.

HILTNER, W. A., and WILLIAMS, R. 1946. *Photometric Atlas of Stellar Spectra.* University of Michigan Observatory, Ann Arbor.

MERRILL, P. W. 1956. *Lines of the Chemical Elements in Astronomical Spectra.* Carnegie Institution, Washington, D. C., Publ. No. 610.

CHAPTER 9

The Sun: Closeup of a Dwarf Star

9–1. Introduction. Heretofore we have been almost exclusively concerned with studies of stars by means of their spectra and energy distributions. The spectra of stars such as the sun can be understood in terms of a fairly simple model, an atmosphere stratified in plane parallel layers with a steady increase of temperature with depth.

When we examine the sun itself, however, we are confronted with phenomena of bewildering complexity, only a trace of which is hinted by the dark-lined Fraunhofer spectrum. As the variety and power of our observational techniques increase so do the number and intricacy of the problems presented.

The outer strata of the sun comprise three distinct layers that differ in structure, density and temperature. The bright disk of the sun, which is seen in white light and which produces the familiar Fraunhofer spectrum, is called the *photosphere*. It has a sharp edge (see Fig. 9–1).

At times of solar eclipse, when the bright photosphere is hidden by the moon, we observe the outer envelopes that surround the sun, the chromosphere, a fiery-looking ring above which often extend delicate filamentary protuberances called prominences, and the much fainter, pearly corona, which fades away into the zodiacal light millions of miles from the sun.

First, let us enumerate some of the ways in which the sun can be observed:

1. White light observations (secured by projection of the solar image on a screen or photographs). Direct photographs of the sun show that the solar disk is not uniformly bright but has a pronounced *limb darkening*. In Chapter 5 we explained how this darkening was a consequence of a temperature gradient in the solar atmosphere; light from the limb reaches us from higher and cooler layers than does radiation emergent from the center of the disk. Under conditions of good seeing we find that the bright surface or photosphere is resolved into small bright granules of the order of 200–1200 miles in diameter which have been likened in appearance to rice grains, and which cover 50 to 60 per cent of the surface (see Figs. 9–2 and 9–3). The most striking features of all are the sunspots, refrigerated regions that possess strong magnetic fields. Unlike the granules which occur everywhere on the solar surface they are concentrated in well-defined zones from 5° to 40° on either side of the equator. Near the limb we often

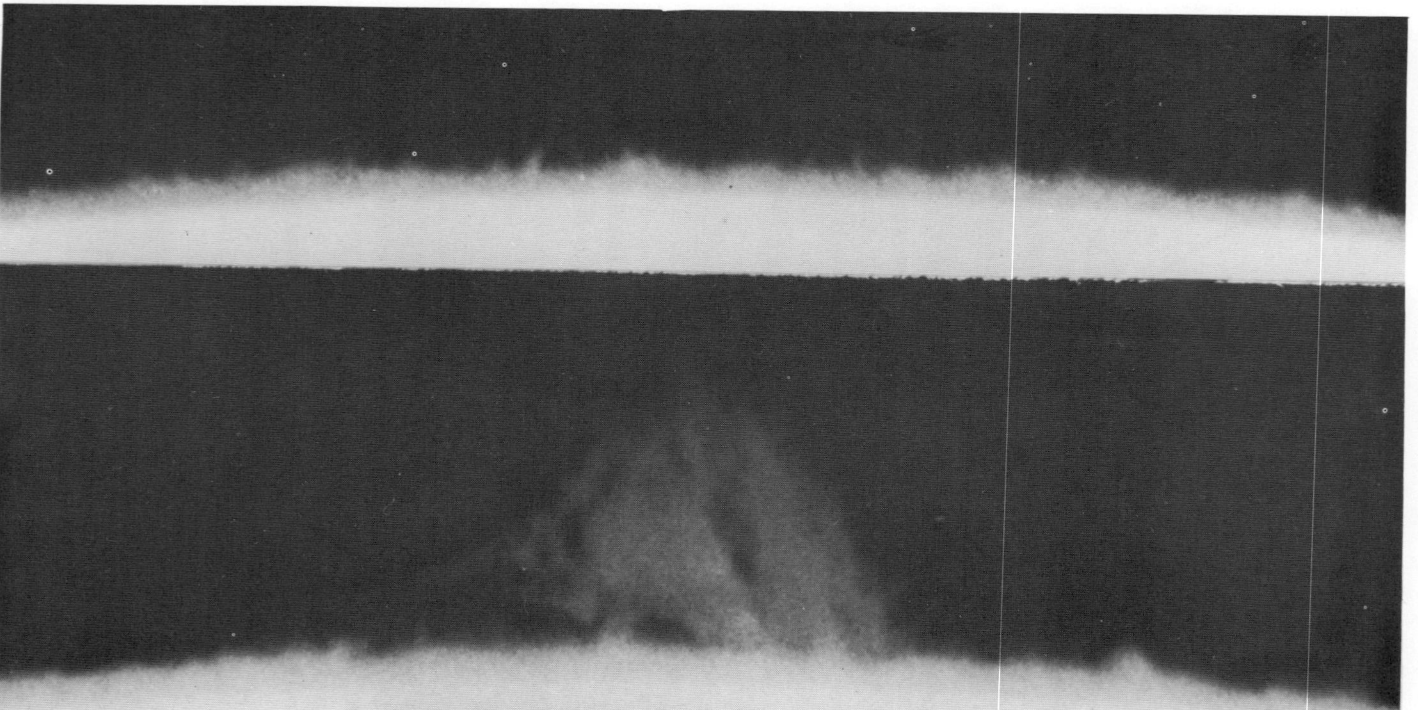

FIG. 9–1. SEPARATION OF THE CHROMOSPHERE AND THE PHOTOSPHERE

This illustration, which shows the line of demarcation between the photosphere and the chromosphere, was taken from a motion picture film obtained by Bernard Lyot at the Pic du Midi in 1942 with a 0.38 m refractor and a polarizing monochromatic filter of 1.5A band-pass centered on Hα. The upper strip shows an undisturbed portion of the chromosphere; the lower part shows a prominence rising above the chromosphere.

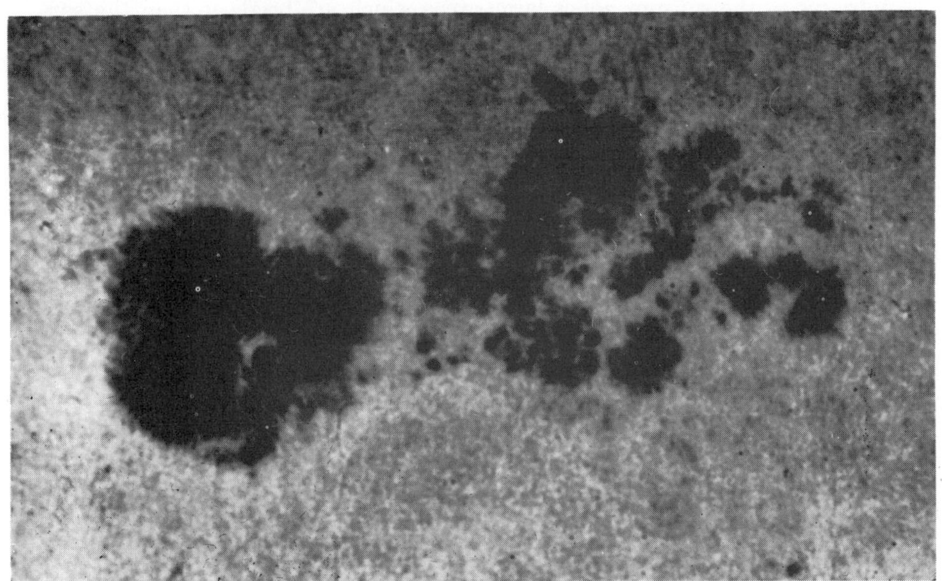

Fig. 9-2. Sunspots and Granules
Photographed by Janssen, June 22, 1885. (Courtesy, Meudon Observatory.)

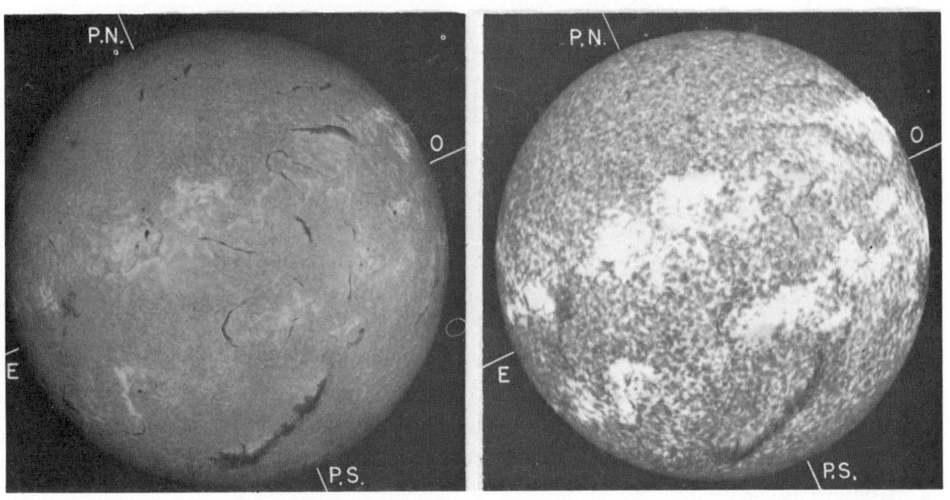

Fig. 9-3. Spectroheliograms in Hydrogen and Ionized Calcium

Left: The monochromatic picture in the light of Hα shows the plages and prominences projected upon the disk as filaments.

Right: Notice that the plages in the light of the center of the *K* line are much more brilliant than in Hα. The fine structure of the background appears coarser.

October 5, 1947, 7^h 32^m G.C.T. (Courtesy, L. d'Azambuja, Meudon Observatory.)

observe in the sunspot zones large areas which are brighter than the surrounding photosphere. They are called *faculae.*

2. Monochromatic observations can be secured with a spectroscope, the spectroheliograph (or spectrohelioscope), and narrow band pass filters such as the *polarizing monochromator.* The spectroheliograph, devised independently by Hale and Deslandres, enables one to photograph the solar disk in light emitted in one of the dark Fraunhofer lines. For example, one may use the H and K lines of ionized calcium, the Hα line, the $D3$ helium line, or any other sufficiently strong line. Spectroheliograms are usually obtained in the light of Hα or Ca "K" and differ in appearance depending on the part of the line utilized. In examining the sun with a spectroheliograph at the center of a strong line we are really looking at the chromosphere in the light of the particular element involved. As we shift away from the line center we look deeper and deeper into the chromosphere and finally into the photosphere. The monochromatic appearance of the solar disk differs strikingly from the white light appearance (Fig 1–1). Near sunspot groups are large bright areas called "plages faculaires" by Deslandres. They are sometimes called bright *flocculi* in older work. The faculae, observed in white light near the limb, coincide with the most intense portions of the plages.[1] Elsewhere on the disk are often seen large sinuous dark *filaments* (or *dark flocculi* in older literature), prominences seen in projection (see Fig. 9–3).

3. The coronagraph, invented by Lyot, permits prominences and even the brightest portion of the inner corona to be studied under specially transparent atmospheric conditions.

4. The magnetograph, invented by H. W. Babcock, and modifications of spectroheliograph techniques (Leighton) or narrow band pass filter methods (Giovanelli) permit one to observe large magnetized areas on the solar surface.

5. Radio techniques have revolutionized studies of the outer solar envelopes and their attendant transient phenomena. Although the angular resolution in radio frequency (r-f) studies is usually small, ingenious interferometric techniques (e.g., Mills, Christiansen) permit one to locate disturbed areas on the solar surface. Dynamic r-f spectra (Wild) show the frequency-time characteristics of transient (e.g., flare) phenomena.

Further, r-f observations differ from optical observations in other ways besides resolving power. A radio wave of frequency ν will travel in an ionized medium as long as

$$\nu^2 > \frac{N_e}{\pi}\frac{\varepsilon^2}{m} \tag{9–1}$$

[1] In the light of the center of the K line the plage is larger than the facula. Spectroheliograms taken a short distance away from the center of the line show "faculae" over the whole disk very similar to the white light form observed at the limb.

where N_ε is the electron density. A wave whose frequency exceeds 10^5 mc could penetrate the chromosphere, $N_\varepsilon < 10^{12}$ and its intensity should be correlated with the effective temperature of the sun, as is observed to be true. For wavelengths of the order of 5 m, the critical electron density is of the order of 10^8 cm^{-3}. Such waves must originate in the corona: the high intensity of r-f radiations from the quiet sun in the meter range gives one of the strongest arguments for the high temperature of the corona. In the neighborhood of the critical frequency, $\varepsilon \sqrt{N_\varepsilon/\pi m}$ the refractive index differs appreciably from unity and the rays suffer considerable curvature. If a magnetic field is present, there is also double refraction and the waves become circularly polarized.

6. Finally, the earth's atmosphere acts as a "counter" for certain solar events. Enhanced ultraviolet or x-ray radiation at times of solar activity manifests itself on the sunlit hemisphere through increased ionization in the ionosphere or a sudden ionospheric disturbance (SID) and induction effects in cables, which is recognized through its effects on short-wave radio transmission. Particle emission from the sun leads to much more complicated interactions because for a given angle of incidence and geomagnetic latitude the earth's magnetic field segregates charged ions according to their energies. Further, the Van Allen belt acts on an incoming stream of particles in a way that cannot yet be accurately evaluated. Hence geomagnetic storms, aurorae, and cosmic rays all serve as indices of solar activity, but detailed interpretations are difficult because of effects produced by the earth's field and the interplanetary medium in the earth's neighborhood.

One cannot escape the feeling that basic solar phenomena are controlled by (1) the existence of a convection zone in the outer layers of the sun, and (2) magnetic fields manifested in sunspots, plages, and in the weak general field. It is commonly believed that the convection coupled with solar rotation produces the observed magnetic effects (Elsasser, Parker, Babcock). Spectacular solar phenomena such as flares, prominences, etc., that are associated with cosmic rays, SID's, magnetic storms, "polar blackouts," etc., probably owe their origin to convective and magnetic phenomena not yet fully understood.

In attempting to construct models for convective zones and magnetized areas in other stars, we must be guided by phenomena observed in the sun. In this chapter we are primarily concerned with the sun and its outer envelopes from the point of view of stellar problems. In the next chapter we emphasize the sun from the point of view of solar-terrestrial relationships. Magnetic fields are fundamental to all the activity, flares, sunspots, plages, etc., associated with the active sun.

9-2. Solar Rotation. Observations of sunspots and other phenomena establish that the sun does not rotate as a rigid body. The rotation period is shortest at the equator and increases towards the higher latitudes. The daily angular motion of solar rotation as determined from the motions of faculae, plages, sunspots and filaments, and by radial velocity measures of Hα, λ4227, and the lower level lines is plotted against heliographic latitude in Fig. 9–4.

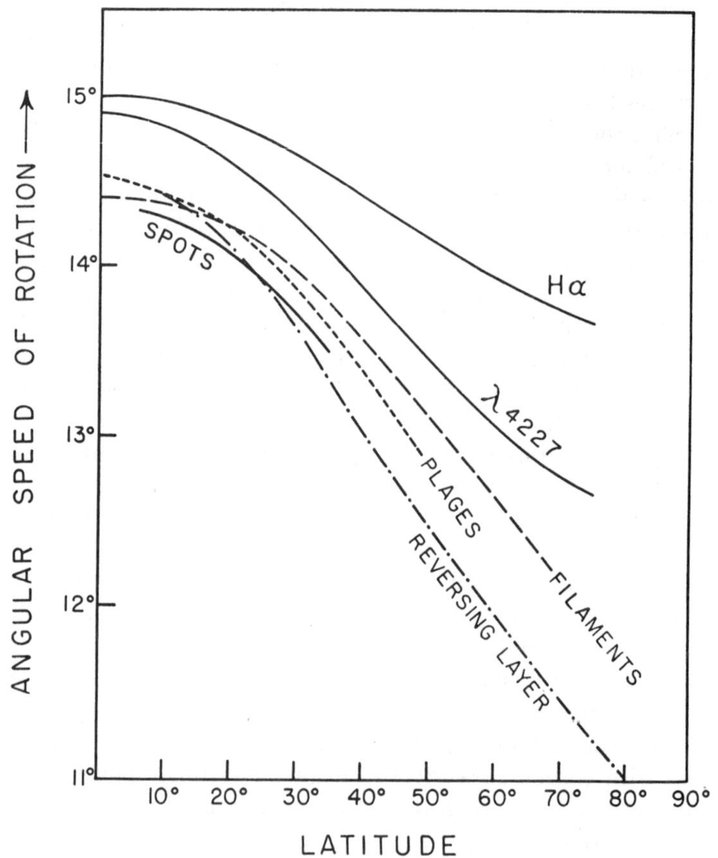

Fig. 9–4. The Rotation of the Sun

Measurements of the solar rotation from the position of spots on the disk were made many years ago by Scheiner, Carrington, Sporer, Faye, and others. Chevalier and Maunder observed the rotation of the faculae, and Fox measured that of the flocculi from the spectroheliograms obtained at Yerkes. The d'Azambujas have determined the solar rotation from the long-lived filaments. W. S. Adams, Duner, and others carried out extensive measures of the solar rotation from the Doppler shift in the Fraunhofer spectrum at the limb of the sun. If opposite limbs of the sun are observed, the lines are shifted to the violet on the approaching limb and to the red on the receding limb. Rotation measurements by the Doppler shift can be carried to latitudes as high as 75°; the solar rotation diminishes smoothly from a period of 24.6 days at the equator to 32.0 days at a latitude of +75°. The rotational speeds of the centers of the strong Hα and λ4227 (Ca I) lines (which originate in the upper layers of the solar atmosphere) appear to diminish more slowly with latitude. Lyot's measures of the wavelength of the green coronal line on the east and west edges of the sun show that the corona rotates in the same direction as the rest of the sun and with the same speed. Studies of individual corona features, centers of activity and r-f data amply confirm this result. It is of interest, however, to establish just how far out the solar corona rotates as a rigid body. The solar rotation period does not vary with time (Newton and Nunn). Such effects as have been found are probably due to large scale motions in the solar atmosphere (Hart, Plaskett).

9–3. Convection in the Solar Atmosphere. In Chapter 5 we showed how solar limb darkening indicated that the outermost portions of the photosphere were in radiative equilibrium. Precise empirical determinations of temperature distribution in the deeper layers of the solar atmosphere, (e.g., Pierce and Aller, Sykes) show that small but significant departures from the radiative temperature gradient do occur. The theoretical criterion for stability against convection is

$$\left[\frac{d\,(\ln T)}{d\,(\ln P)}\right]_{\text{photosphere}} < \left[\frac{d\,(\ln T)}{d\,(\ln P)}\right]_{\text{adiabatic}} \tag{9–2}$$

That is, if the temperature gradient required to supply the actual flux by means of radiative transfer exceeds that corresponding to adiabatic equilibrium, instability may occur.

Plaskett used Sykes' temperature distribution to construct a model atmosphere for the deeper layers. In this empirical model he found that for τ_0 greater than about 0.8, the gradient becomes unstable. This instability is not connected with that produced in the deep convection zone where hydrogen becomes ionized. Rather, it is produced here by a steep temperature gradient. Instability may occur in two ways: (a) If τ_0 is greater than about 0.8 the gradient becomes unstable because

$$[d\,(\ln T)/d\,(\ln P)]_{\text{radiative}}$$

or (dT/dh) becomes large as a consequence of a rapid increase of opacity in these layers. Here κ depends strongly on the temperature. (b) In much deeper layers, instability may set in because γ becomes small (see Eq. 9–3).

If, following Plaskett, we take $h = 0$ at $\tau_0 = 3$, instability sets in at $h - h_0 = 60$ km. Hence we may regard the photosphere as composed of two layers: an upper zone $(h - h_0) > 60$ km where the Fraunhofer lines arise and a lower zone $(h - h_0) < 60$ km, characterized by a high temperature gradient and instability.

The structure of the convection zone in the sun and similar stars is certainly quite complicated. K. Schwarzschild's criterion for stability, Eq. 9–2, is established by a straightforward argument. If a given temperature gradient exists in an atmosphere and a mass of gas density ρ and temperature T which is in equilibrium at a pressure P with its surroundings is suddenly displaced upwards to a point where the ambient temperature is T', pressure is P' and density is ρ', it will rapidly attain pressure equilibrium with its surroundings. Its internal temperature and density will be determined by the adiabatic gas law since it was displaced suddenly. If its density ρ_e exceeds ρ' it will sink back and such an atmosphere will tend to be stable against convection. If $\rho_e < \rho'$ the element will continue to rise and the corresponding atmosphere would be unstable against convection. The adiabatic temperature gradient corresponds to $\rho_e = \rho'$ and hence to a limiting equilibrium.

$$\left[\frac{d\,(\ln T)}{d\,(\ln P)}\right]_{\text{ad}} = \frac{\gamma - 1}{\gamma} \tag{9–3}$$

where γ is the ratio of specific heats. For a normal monatomic gas $\gamma = 5/3$ but deep below the photospheric surface hydrogen becomes ionized. When this happens $\gamma = C_p/C_v$ falls below 5/3 and does not rise again to this value until the hydrogen is *all* ionized.[2] In this hydrogen ionization zone, criterion (2) cannot be satisfied, energy transport is probably mainly by convection, and the temperature gradient eventually approaches that corresponding to adiabatic equilibrium, in deeper layers.

Nevertheless, we must emphasize that when convective equilibrium is established it does not necessarily set up an adiabatic temperature gradient. A rising element of heated gas must be subject to a buoyancy force sufficient to overcome viscosity, and furthermore it radiates while it ascends. The uppermost, observable, part of the solar convection zone has a temperature gradient approaching that of a radiative equilibrium more closely than that of convective equilibrium.

These hydrogen convective zones are of enormous importance for theories of stellar structure, not only for dwarf stars like the sun but more particularly so for giant and supergiant stars, whose radii depend on the structures of their convection zones.

The solar hydrogen convective zone, whose thickness may be of the order of 60,000 km, is extremely inhomogenous. Böhm (1958) finds that in the uppermost 500 km, the opacity changes rapidly, $[(dT/dh) - (dT/dh)_{\text{ad}}]$ is about a hundred times higher than in the deeper layers, and the density may be ten thousand times lower. He concluded that in the uppermost strata, you would expect circulation patterns with sizes of the order of the thickness of the most unstable layers, i.e. 500 km.

[2] The reason for this decrease in γ is that ionization acts like an additional degree of freedom; energy supplied to the gas is used up not only in increasing the mean speeds but also in ionizing the gas.

Older calculations were often based on Prandtl's mixing length theory. In unstable, turbulent convection it is supposed that a rising blob of material moves a distance comparable with its diameter, dissolves into its surroundings. New blobs are then formed and energy gradually flows upward. This mixing length theory has been widely used in calculations concerning the structure of convection zones in stars. It fails in laboratory experiments because of effects of viscosity which probably are not important in stellar atmospheres. To calculate a convection zone for a star by Rayleigh-type methods it is necessary to have an initial starting model which can be provided by the mixing length theory. Hence, although mixing length theories eventually must be replaced by more accurate formulations, they may play useful roles in the zeroth order approximation.

The most promising method for handling convection problems appear to be those based on a procedure proposed by Rayleigh (1916) to explain a convection experiment done by Benard. Suppose a thin (1 mm or less) layer of a liquid is placed on a carefully leveled plate and protected from extraneous disturbances while a uniform temperature T_1 is maintained on the lower surface and a lower temperature T_2 on the upper one. Then, Benard found that if the temperature gradient dT/dh exceeded a certain critical value, convection would be set up in a definite cellular pattern. For example, a pattern of beautifully regular hexagons could be established.

Lack of space prevents reproduction of Rayleigh's elegant discussion here. We can state only the results. One sets up the equation of state for the fluid under consideration, the equations of motion, and the equation of continuity and then examines what would happen if a disturbance were created by shifting a small volume element. Would this volume element continue to move or not? Let $A(x, y, z, t)$ denote the amplitude of a displacement. We assume a solution of the form

$$A(x, y, z, t) = A_0 e^{qt} \sin \eta_1 x \sin \eta_2 y \sin \eta_3 z \tag{9-4}$$

and evaluate q. Here η_1, η_2, and η_3 are proportional to the reciprocal wavelengths of the disturbing displacements. If $q < 0$ convection will not occur, whereas if $q > 0$ it certainly will. Furthermore, one can find a combination of η_1, η_2, and η_3 that corresponds to a minimum temperature gradient that would produce instability corresponding to convection. The condition for stability is not that of $q > 0$ for a given combination of η_1, η_2, η_3, but rather that for a given temperature gradient, viscosity, thickness of layer, etc., there exists no mode of disturbance, i.e., no combination of η_1, η_2, η_3 for which it is unstable.

Rayleigh found that convection would occur if $\beta > \beta_c$ where

$$\beta = \frac{dT}{dh}, \qquad \beta_c = \left|\frac{dT}{dh}\right|_{\text{critical}} = R\,\frac{\kappa\nu}{ga}\,d^{-4} \tag{9-5}$$

Here g is the acceleration of gravity, d is the thickness of the layer, while

a, κ, and ν are, respectively, the coefficients of volume expansion, thermal conductivity, and kinematic viscosity. R is a non-dimensional quantity called the *Rayleigh number*. Although Rayleigh considered simplified boundary conditions and physically correct ones were introduced only later by Jeffreys, Low, Pellew, and Southwell, the essential properties of the solution remain unchanged.

One considers a horizontal, two-dimensional periodic disturbance of wavelength λ and wave number $\tilde{\eta} = 2\pi/\lambda$ (i.e., $\eta_1 = \eta_2 = \tilde{\eta}$, $\eta_3 = 0$) and solves for the values of $R(\tilde{\eta})$ such that $q \geq 0$. In this way, one finds a value $\tilde{\eta}_0$ such that $R(\tilde{\eta}_0)$ is a minimum. This is the critical value R_c, for if $R < R_c$, convection cannot occur, while if $R > R_c$, there exists some mode of motion for which convection is possible. Naturally, the value of R_c depends on boundary conditions. If both surfaces are rigid, $R_c = 1708$; if the bottom is rigid and the top is free, $R_c = 1100$ and if both are free $R_c = 635$. Schmidt and Milverton verified the results for rigid boundary surfaces. See also Malkus, 1954.

If convection is only just initiated, only small velocities develop and the motion is almost entirely in discrete Benard-type cells corresponding to the most favorable wave numbers. The amount of heat transportable in this fashion is somewhat limited and this strictly orderly type of motion prevails only over a restricted range of the driving temperature gradient. If R continues to increase substantially above R_c, i.e., the temperature gradient becomes much larger than what is necessary to initiate convection, the layers become unstable over a wide range of wave numbers. That is, there occur abrupt transitions of the convective flux, corresponding to definite values of R. Finally, for rigid surfaces at R about 18,000 the flow becomes turbulent. The motion in modes corresponding to low wave numbers becomes so fast that the pattern gets disorderly as energy is fed from lower to higher wave numbers until dissipated by viscosity. Most of the energy of motion may be still associated with the lower wave numbers, but small scale turbulence may occur.

In the theoretical treatment we look for solutions with cellular type patterns such as are observed in the laboratory. In a given unstable situation, one cell tends to grow faster than others; this is the "mode of maximum instability." We could add this solution to the old steady state one (in the fashion of usual perturbation treatments) and do the problem over again. Thus by successive approximations we get a fully developed theory which predicts the convective heat transport accurately, the sudden breaks in the relation between energy flux and temperature gradient, and the experimentally observed onset of turbulence.

Chandrasekhar (1957, 1961) has emphasized that instability can set in in two ways — as convection as described above or in an oscillatory fashion for which the time dependent part is

$$a(t) = A_0\, e^{-qt} \sin pt \tag{9–6}$$

This situation is an example of Eddington's "overstability" where restoring forces overshoot equilibrium and build up steadily increasing oscillations.

Rotation can hinder the onset of convection which can set in as overstability when $\nu/\kappa \ll 1$. Chandrasekhar's predictions were beautifully verified by the experiments of Nakagawa, Frenzen, and Flutz.

Since magnetic lines of force tend to act as though glued to the material of a highly conductive medium, a magnetic field will tend to inhibit convection. Matter which is a good electrical conductor can flow freely along the lines of force but cannot easily flow perpendicular thereto.

Thus the critical Rayleigh number for the onset of stability depends on both magnetic field, H, and electrical conductivity, σ. Chandrasekhar finds that for an incompressible fluid

$$R_c \sim \frac{\pi^2\mu^2H^2\sigma^2}{\rho\nu} \tag{9–7}$$

where μ is the magnetic permeability. The corresponding critical temperature gradient (see Eq. 9–5) is

$$\beta_c \sim \pi^2 \frac{\mu^2H^2\sigma\kappa}{ga} d^{-2} \tag{9–8}$$

Notice that viscosity ν disappears because ohmic dissipation rather than viscous dissipation is the main factor stopping convection. The magnetic field gives the fluid a pseudoviscosity. Chandrasekhar's predictions received a beautiful confirmation in experiments by Nakagawa. The simultaneous action of both rotation and magnetic fields was considered.

Chandrasekhar's investigations, although confined to incompressible fluids, represent a fundamental advance towards understanding problems of astrophysical interest, even though the latter are not amenable to exact treatment.

Astrophysical problems involve the following complexities: (1) In stellar atmospheres, density changes strongly with height; (2) the unstable layers interact with overlying ambient stable layers; (3) the gases radiate energy so radiative transfer must be considered; (4) in early-type stars stellar rotation is important; and (5) magnetic fields are important in sunspots, and to a lesser degree in plage areas and also in certain early-type stars.

For a compressible fluid one must replace

$$\beta = \frac{dT}{dh} \text{ by } \beta^* = \left\{ -\left(\frac{dT}{dh}\right) + \left(\frac{dT}{dh}\right)_{\text{ad}} \right\} \tag{9–9}$$

which is the Schwarzschild condition. Accordingly, the ordinary Rayleigh number (see Eq. 9–5)

$$R = \frac{ga}{\kappa\nu} d^4 \left|\frac{dT}{dh}\right| \tag{9–10}$$

has to be replaced. Unsöld (1955) finds

$$R = \frac{gd^4}{K\nu} \frac{\gamma - 1}{T} \Delta(\nabla T) \tag{9–11}$$

where d is here taken as the "scale height" of the layer, i.e., the depth across which the density increases by a factor of about e, ν is the viscosity, K is the radiative conductivity, and $\Delta(\nabla T)$ is the excess of temperature gradient over the adiabatic gradient.

In the upper, highly unstable convective layers of the solar atmosphere, Schwarzschild adopted from the work of Böhm-Vitense (1955) and of Edmonds (1956) the following values:

$$\begin{aligned} &T = 8200°K && K = 2 \times 10^{12} \\ &\rho = 3 \times 10^{-7}\, g\ \text{cm}^{-3} && d = 300\ km \\ &\log P = 5.2 && (\gamma - 1)T^{-1}\,\Delta(\nabla T) = 2.5 \times 10^{-4}\ \text{km} \end{aligned} \tag{9–12}$$

Convection problems of astrophysical interest have been considered by a number of writers. Among more recent efforts we mention especially those of Skumanich (1955), Unno (1958), Böhm and Richter (1959, 1960), Spiegel (1960), and of Schwarzschild, Ledoux, and Spiegel (1961). Skumanich applied a Rayleigh-type method to a fluid with a marked density variation, Böhm and Richter considered the combined effects of a strong density variation and radiative exchange on convective instability, Unno considered the effect of convective motion upon an overlying stable zone, while Spiegel considered effects of radiative exchange on a gray atmosphere with no large changes of density in the relevant layers.

Ledoux, Schwarzschild, and Spiegel (1961) considered non-linear interactions in an effort to investigate size distribution of convective cells in the upper part of the solar convection zone. Consider the distribution of cell sizes from the point of view of velocities and of temperature fluctuations. Both distribution functions have maxima at the same element size which is established by the density scale height, but the distribution function for temperatures falls off much more steeply than that for velocities. The reason is, that since radiative losses tend to smooth out temperature differences, there will be no small scale temperature fluctuations even though velocity fluctuations corresponding to physically small cell sizes may yet exist.

Studies of convection in the sun, based on mixing-length theories have been made by Siedentopf (1932, 1933, 1935), Woolley (1941–1943), Vitense (1953), de Jager (1959), and others. Problems of convection zones in other stars, particularly from the point of view of stellar structure, have been considered by Schwarzschild (1958), Kippenhahn, St. Temesvary, and Biermann (1958), and others. A systematic study of the outer convection zones in stars of different effective temperatures and surface gravities from the point of view of mixing length theories and including the effects of radiative exchange has been carried out by Böhm-Vitense (1958).

9–4. Large-Scale Motion in Solar Atmosphere. The convective instability of the deeper layers of the solar atmosphere produces (a) velocity fluctuations in the photospheric layers and (b) granulation. Let us consider first the motions. Mass motions of the radiating gases affect equivalent widths and profiles of lines in the sun and stars. Under conditions of good seeing, high resolution spectra of the photosphere show "wiggly lines."

In Art. 8–10 we considered effects of large-scale motions denoted as "turbulence" on equivalent widths and line profiles. If the characteristic motions had a scale greater than the mean free path of a light quantum the profile would be affected but not the equivalent width. If the characteristic motion is much smaller than the mean free path of a light quantum equivalent widths and profiles both are affected. Since equivalent widths and profiles may be observed at sundry points on the solar disk, it should be possible not only to separate large-scale motions (macro-turbulence) from small-scale motions (microturbulence), but also to investigate the dependence of turbulent velocities on depth and any anisotropy that may be present. Unfortunately, results found by various workers are inconsistent. Although the evidence seems fairly good that the motions are anisotropic (Allen, 1949; Waddell, 1958); there is no agreement on the dependence of turbulence on optical depth.

If one can measure accurately the profiles of two lines of the same multiplet wherein relative f-values are accurately known, the wavelength dependence of the absorption coefficient, and often the depth dependence of the Doppler width $\Delta\lambda_D$ can be found by a neat method due to Goldberg (1958). By this technique Unno found the turbulent velocity to increase with depth in the solar atmosphere. This is what we would expect: convective motions in unstable layers below optical depth 0.8 would overshoot into the overlying photospheric layers and their effects would gradually die away with increasing height.

Under conditions of good seeing, spectral lines on the solar disk display a wiggly appearance. Thus Richardson and Schwarzschild at Mount Wilson from direct measurements of displacements in the lines give a mean random velocity of 0.37 km sec^{-1} for structures of the order of 2″ or 1500 km diameter. More recent observations, such as those obtained with the vacuum spectrograph at the McMath-Hulbert Observatory, show Fraunhofer lines to exhibit pronounced Doppler shifts. (See Fig. 8–1.) The fine structure of the pattern (2″) appears to be limited to atmospheric seeing; hence the random turbulent velocities obtained from the line wiggles tend to be lower limits to the true values. Note that ξ_1 derived from wiggly lines refers to large-scale or macroturbulence, whereas that obtained from line profiles displays effects of both large and small-scale motions. Not only does ξ_1 vary across the solar disk, it also tends to increase with height in the chromosphere. (See Art. 9–16.)

No solar atmosphere stratified in plane parallel layers could explain these line shifts nor could such a model explain the observed center-limb variations of many spectral lines. Hence model atmospheres with ascending and descending columns of gases at different temperatures have been suggested. See de Jager (1959), Böhm (1954), Voigt (1956), and Schröter (1957). These various models differ not only from one another in differences ΔT between hot and cold columns, but also in the assumed variation of ΔT with depth, and in deviations from LTE.

The wiggly-line structure observed in the stronger lines appears to be primarily of *chromospheric* rather than of photospheric origin. The main results obtained from high dispersion studies with the McMath-Hulbert vacuum spectrograph have been summarized by O. C. Mohler as follows: the velocities are at random and range from about 0.2 to 1 km sec^{-1}, the strongest Balmer lines and λ3933 of Ca^+ show no dependence of velocity on height, while random velocities derived from metallic lines increase with height in the chromosphere. The sizes of elements of chromospheric structure appear to be of the order of 6″–20″; they seem to be correlated with photospheric granulation in some continuous fashion. At low levels in the chromosphere, the motions tend to be arranged in rising and falling currents, while at higher levels, the motions tend to be disorderly. The chromosphere is continually in motion with respect to the underlying photosphere.

The Michigan observers found very little correlation between the brightness of areas corresponding to individual wiggles of lines and the velocities. In this connection, we mention moving pictures of high dispersion solar spectra obtained by J. W. Evans and R. Michard. They found most of the radial velocity effect at the center of the solar disk to consist of oscillatory and roughly sinusoidal motion. Their interpretation of these observations is that a bright granule "starts a wave" in the photosphere and then decays (without indication of oscillatory change in brightness) while the upper photospheric material oscillates for one or two periods (period about 260 sec). Comparison of phases of strong lines that originate high in the atmosphere with weak lines that originate low in the atmosphere indicates that the wave (which has a horizontal length of 1000–5000 km) is progressing upward. Hence the velocity is only related "historically" to the brightness of the underlying surface — a fact which explains the low correlation found in the past between line displacement and continuum brightness.

Thus the instabilities in the sub-photospheric layers seemingly cause two modes of macroscopic motion, a convective mode which produces the observed brightness fluctuations in the granulations, and a wave mode shown by the phenomena of the wiggly lines. The wave mode may have two components; a horizontal component that does not seem to depend on depth and a vertical component that has a period of the order of four minutes. Other large-scale oscillations of the chromosphere may also exist.

The existence of such progressive waves was suggested on theoretical grounds by Alfvén, by Biermann, and by Schwarzschild. As such waves pass through gases of lower and lower density in the chromosphere, the velocity steadily increases. Again, this conclusion seems to be substantiated by the work of Unno who found that the mean random velocity increased with height in the atmosphere.

A few remarks concerning large-scale motions in the solar atmosphere may be added. Hart (1954, 1956) has described long-lived local velocity fields in the photosphere, while Plaskett (1959) concluded that photospheric motions were not random. There may exist meridional currents and asymmetries in the solar rotation suggesting a photospheric circulation analogous to that shown by the earth's atmosphere. These motions may originate in the solar convection zone or they may represent large-scale, randomly excited, free oscillations of outer solar layers.

Using an adaptation of a technique developed for studying solar magnetic fields with the aid of specially processe dspectroheliograms, Leighton, Noyes, and Simon (1962) measured line of sight velocities in the photospheric layers. They found evidence for two distinct velocity fields in the photosphere — one a large-scale, long-living cellular pattern of horizontal currents. The other was a smaller-scale, quasi-oscillatory pattern of vertical (and possibly also horizontal) motion. The larger cells had diameters of about ten thousand miles, lifetimes of the order of hours, while matter flows from the center toward the periphery with a speed of about 500 m $\sec^{-1}$. These large cells are related geometrically to structures observed in the chromosphere. They suggest that the cellular patterns and chromospheric network are surface manifestations of a supersystem of convection currents originating deep in the convective layers. The smaller scale, oscillatory, velocity field appears closely related to the granulation; apparently these are the same type of oscillations as observed by Evans and Michard. Bright elements move upward at low levels and downward at higher levels; in the chromosphere the velocity depends strongly on altitude. Energy needed to maintain oscillations is supplied from the convection zone.

According to the general theory of relativity, spectral lines produced in a quiescent, homogeneous, solar atmosphere should show a red shift given by

$$\frac{\Delta\lambda}{\lambda} = \frac{G\mathfrak{M}}{Rc^2} = 2.12 \times 10^{-6} \tag{9–13}$$

i.e., about 0.3 km $\sec^{-1}$. Normally, such shifts should be difficult to detect, because of a large-scale mass motion and the necessity of measuring very small deflections freed of systematic errors. J. E. Blamont developed a magnetic scanning method which enabled him to observe a line profile with a resolution of 10^{-3}A. At the center of the solar disk, he found a red shift which corresponded to the predicted relativistic value.

9–5. Solar Granules. Perhaps nowhere in astronomy is the necessity of radically new observational techniques better illustrated than in studies of fine structure of the solar surface, particularly the granulation. The difficulty with much of the older work on granules was that even the largest ones were near the limit of resolution. Under such circumstances it may be misleading to treat measurements made on the image as though they were made on the object itself. Special attention must be paid to the way an optical system of a given aperture handles information about structure of an object; some data may be completely suppressed and others handled in a metamorphosed form (see Fellgett, 1959). Uberoi (1955) noted you could get a discrete granular structure if a "random" brightness distributed were photographed with limited resolution and high contrast and concluded that measurements made up to 1954 were consistent with a completely turbulent atmosphere. More recent observations, Thiessen (1955),

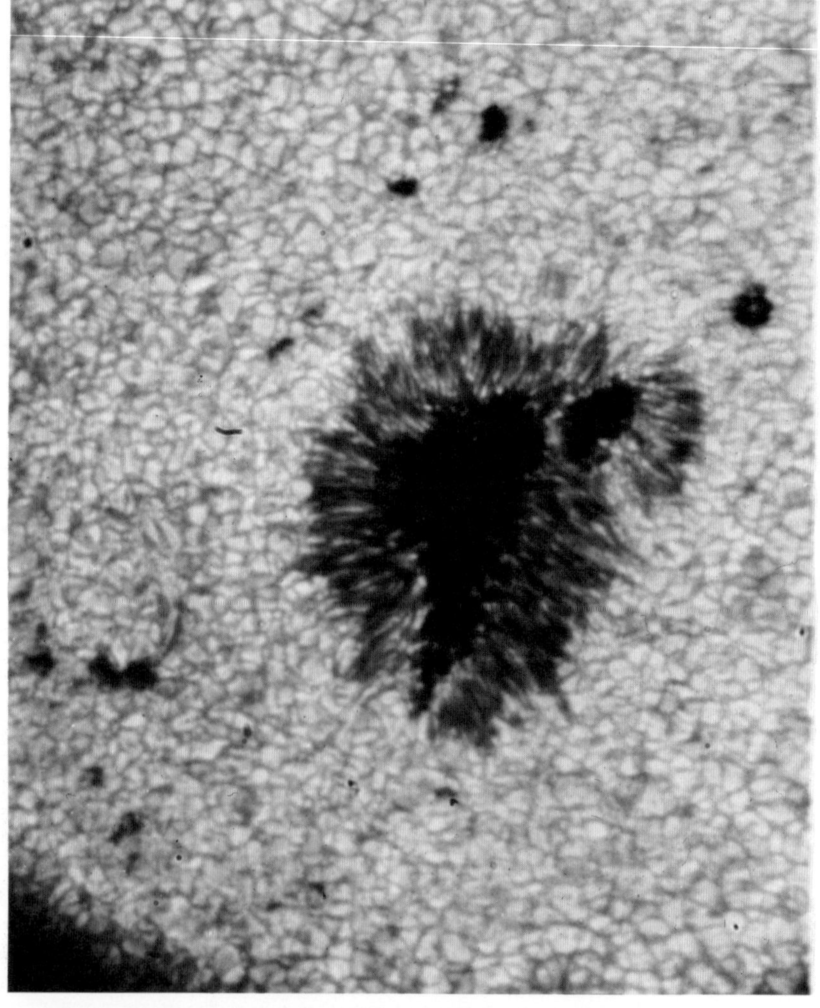

FIG. 9–5. SOLAR GRANULATION AS PHOTOGRAPHED FROM A BALLOON
(Photographed August 17, 1959, 11:53 A.M. C.D.T. Courtesy M. Schwarzschild.)

Leighton (1959), and Rösch (1959), indicate the presence of non-random features on the sun. From a balloon at 18,000 ft Blackwell, Dewhirst, and Dollfus (1959) observed granules and confirmed the complex structure of the photosphere, but A. H. Mikesell's studies of seeing during a night balloon ascension indicated it is necessary to go to much greater heights to eliminate effects of bad seeing.

Schwarzschild's observations (1959) secured from an unmanned balloon at 80,000 ft seem to give the definitive answer. See Fig. 9–5. Solar granules appear to range in size from 300 km or less to about 1800 km. The mean diameter as measured by an impersonal statistical procedure is about 700 km. Bright granules appear as simple blobs, but the dark regions between them seem to form a net of narrow strips so there exists a conspicuous asymmetry between bright and dark areas.

In shape the granules simulate irregular polygons. Their range in size, shape and short lifetimes all indicate they are not Benard cells, but on the other hand they are not entirely random phenomena either. Schwarzschild suggests they represent a state of non-stationary convection in which a fair range of cell sizes exist, the cells have irregular shapes and there is a pronounced geometrical difference between the rising columns of material and the narrow areas representing sinking material. Thus the solar granules lie between the domain of stable convection and classical turbulence.[3]

A correlation analysis of the photospheric intensity variations gives a mean fluctuation of 4.6 per cent of the mean value, which after correction for instrumental profiles indicates a rms temperature difference $\sqrt{(\Delta T)^2}$ of about 92°C between the hot and cold elements (Bahng and Schwarzschild, 1961).[4]

Granules cover the entire disk of the sun. Rösch estimates the total number on the solar disk to be about 3,500,000. The fact that they seem to appear within 10″ of the solar limb suggests that their influence extends to the uppermost photospheric layers of the sun. Macris finds that both size and number vary with the solar cycle.

Granules are even found in the central dark area or umbra of a sunspot, and granule-like structures are found in the somewhat brighter surrounding area or penumbra (see Figs. 9–5 and 9–6). The penumbra is resolved into a system of bright, narrow, predominantly radial filaments with diameters

[3] One difficulty is that if numerical values are substituted in Eq. 9–11, one finds a value of R indicative of turbulent motion. If, however, in place of the ordinary gas viscosity one supposes there exists a "turbulent" viscosity due to a few small-scale vortices in the medium, R is lowered and the observed result is understandable.

[4] Blackwell, Dewhirst, and Dollfus (1959) obtained $\Delta T = 520°K$ from their observations, but Gaustad and Schwarzschild from a discussion of the same data found a rms fluctuation of 7.3 per cent and $\overline{\Delta T} = 89°K$. Rösch finds a mean diameter $d = 2.''6$ from the mean distance of centers of neighboring granules and a temperature difference $\Delta T = 250°K$ from the intensity ratios of brightest and least bright points. Hence the d's and ΔT's given by Schwarzschild and Rösch refer to different things.

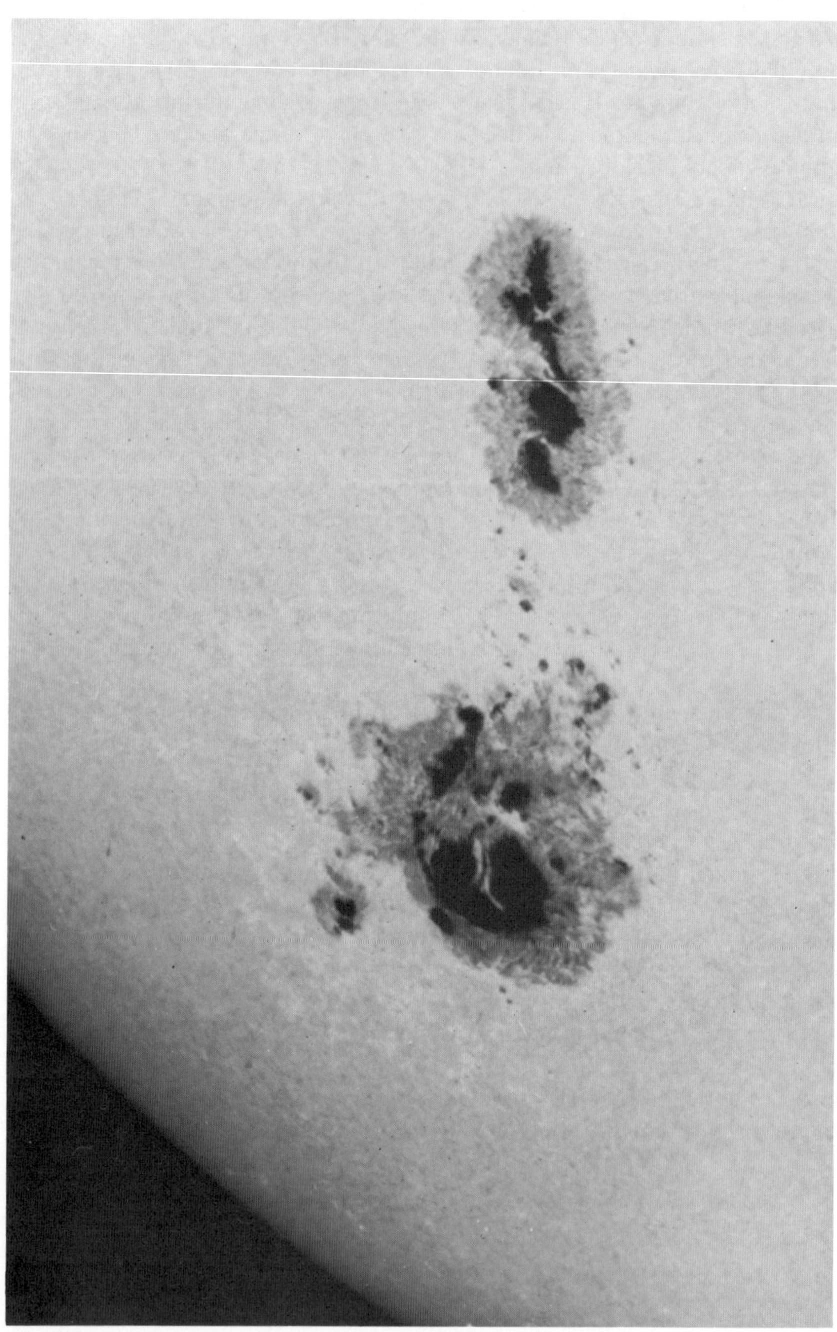

FIG. 9–6. THE GREAT SUNSPOT GROUP OBSERVED ON FEBRUARY 2, 1946
(Courtesy, Seth B. Nicholson and J. O. Hickox, Mount Wilson Observatory.)

of 300 km or less and lengths of about 5000 km (Danielson, 1961). From an application of Chandrasekhar's theory (1952) of convection in a magnetic field, Danielson finds that these filaments can be interpreted as long narrow convection cells or "rolls" provided the magnetic field is nearly horizontal.

The lifetimes of granules are summarized in Table 9-1.

TABLE 9-1

LIFETIMES OF PHOTOSPHERIC AND SPOT FINE STRUCTURE

Feature	Lifetimes	Observers
Photospheric granules	$7\text{–}8^m$	Macris (1953)
	8^m	Bahng and Schwarzschild (1960)
	10^m	Bray and Loughhead (1958)
	10^m	Rösch and Hugon
Umbral granules	$15\text{–}30^m$	Bray and Loughhead (1959, 60)
Penumbral filaments	40^m	Danielson (1961)
	2^h	Bray and Loughhead (1958)
Facular granules	2^h	Bray and Loughhead, Waldmeier, ten Bruggencate

Rösch finds that once a granule is formed, its diameter increases until it reaches about 2″, whereupon it disintegrates into several small granular fragments that disappear at the same spot as the granule originated. In view of the inhibiting effect of a magnetic field on the establishment of convection, the increased lifetimes of granules in spots is of particular interest. Loughhead and Bray found umbral granules to be closer together than photospheric granules, to last longer (10 per cent of them survive for more than 2 hours) and to show no change in brightness or shape during their lifetimes as detectable structures. The penumbral filaments have lifetimes of the order of hours.

Spectroheliograms show fine details sometimes reminiscent of granules, but these structures depend not only on the spectral line employed but also the part of its profile. Thus they depend on the depth in the chromosphere to which one is looking and the structure changes steeply with height. Chromospheric and photospheric fine structure must be related, but the exact nature of the linkage has not yet been worked out.

9-6. Measurement of Solar Magnetic Fields. Magnetic fields on the sun have been found in spots, in plages, and elsewhere on the disk of the sun. The coronal structures observed at sunspot minimum are indicative of a solar poloidal field.

Earlier we described how a spectrum line is split into a number of polarized components under the influence of a magnetic field. Each level is resolved into a number of Zeeman states whose displacement in wave number units from the field-free position is given by

$$\Delta T = \frac{\varepsilon}{4\pi mc^2} HgM \text{ cm}^{-1} \tag{9–14}$$

where M is the magnetic quantum number and g is the Lande factor, which is different for each level (see Chapter 2). A practical difficulty that arises is that for an average spot field, the Zeeman splitting is the same order of magnitude as the intrinsic (Doppler + damping) line width. Hence even with high resolution equipment the magnetic components are properly separated only near centers of large spots.

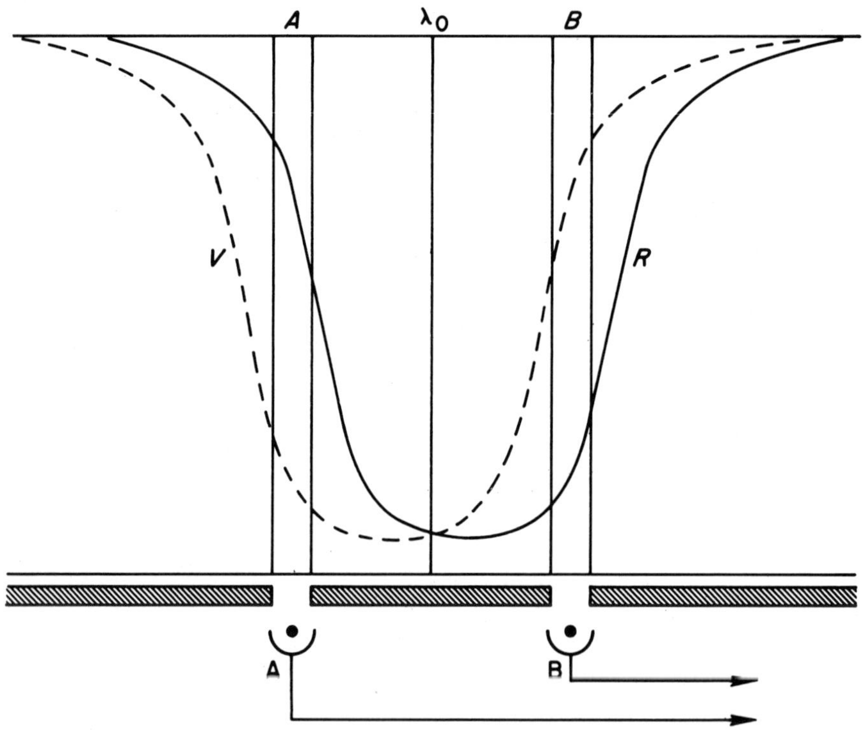

FIG. 9–7. ANALYSIS OF BROADENED PROFILES OF LINES IN A MAGNETIC FIELD (H. W. Babcock's Method)

The two circularly polarized components of a line whose undisplaced center is at λ_0 are denoted as R (solid line) and V (dotted line). In the incoming beam H. W. Babcock places a Nicol prism and a crystal of ammonium dihydrogen phosphate, which acts as a quarter-wave plate for either right- or left-handed circularly polarized light depending on the direction of the voltage. If an alternating current is applied, the profile flickers from position V to R depending on which of the two components is transmitted. The spectrum is imaged on a screen which contains two parallel slits so separated that they normally fall on the steep sides of the profile. The light from each slit is then fed into a separate multiplier tube. If no magnetic field exists the line profile will be unchanged as the analyzer transmits first the right- and then the left-hand circularly polarized beam. If a field exists the two photocells A and B alternately see an enhanced intensity as the profile shifts from R to V. The difference in the outputs of the two photocells is rectified, then amplified and its magnitude and phase as displayed on a cathode-ray tube give the magnitude and polarity of the field.

Hence it is necessary to use an analyzer of polarized light. Looking in the direction of the field the p-components disappear while the s-components are circularly polarized, one group is left-handed and the other, which is shifted in the opposite direction from the undisplaced position, is right-handed. With a Nicol prism and a quarter-wave plate one can suppress one line or the other, depending on the orientation of the Nicol. Hale used a mosaic of quarter-wave plate strips so that on the photographed spectrum, alternate components appear in successive strips. Both the polarity and magnitude of the field could be established by these measurements.

For weak fields the displacements are so small that they cannot be seen. To overcome this difficulty H. W. Babcock (1953) devised a magnetograph, which permits the measurement of fields as weak as 0.3 gauss. Consider the profile of a line formed in a weak field. The separation of the components is much smaller than the natural width of the line so the line is virtually unchanged in appearance. If we observe the line with an analyzer for circularly polarized light with two parallel slits placed at the steepest parts of the profile (see Fig. 9–7) we can detect a slight change in intensity as the analyzer is changed from a position where left-hand circularly polarized light is transmitted, to its inverse. We assume the observer is looking along the lines of force; otherwise, he measures the component of the field in the line of sight.

In Babcock's magnetograph the image of the λ5250.216 Fe I line or the 5247 Cr I line is placed on double slits in the focal plane of a spectrograph with a dispersion of 11 mm per 1A and a spectral resolving power of 600,000. The analyzer is placed before the entrance slit of the spectrograph, and as the line image flickers back and forth, photocells behind each slit detect a change in intensity whose phase and amplitude depend on the polarity and magnitude of the field. Fig. 9–8 illustrates the principle of the magnetogram and its interpretation. In earlier magnetograms, the deflections sometimes overlapped. In an improved revision (as illustrated in Fig. 9–8) tilted lines are used in place of vertical deflections. Very strong fields are indicated by horizontal lines. The direction of tilt indicates polarity, and the brightness of the trace indicates intensity. See Fig. 9–9. Usually fields from 1 to 50 gauss are measured with an angular resolution of about 23″, although for detailed studies, e.g., flares, a higher resolution 5″ or 10″ is possible.

Leighton (1959) obtained trick spectroheliograms in which areas with a field greater than about 20 gauss appear darker or lighter depending on the polarity. The method was to set the second slit of the spectroheliograph on a steep part of the line. An analyzer is then put in the system and two spectroheliograms are obtained, one picture is taken with right-hand circularly polarized light, the other with left-hand circularly polarized

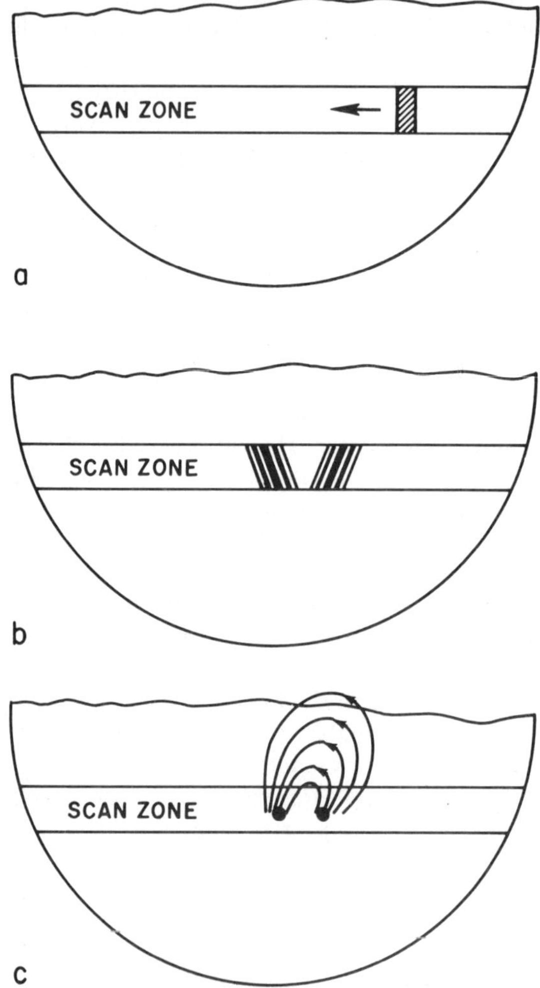

FIG. 9–8. RELATION BETWEEN THE MAGNETOGRAPH SCAN PATTERN AND THE INFERRED FLUX LOOPS

(a) The analyzing slit, usually taken as 0.04 of the solar diameter, moves slowly across the disk. (b) When the field is directed toward the observer, the trace is tilted to the right; its strength determines the brightness of the trace. (c) The flux pattern inferred from the strips in (b) indicates a bipolar magnetic region, BMR. (After H. W. Babcock.)

light. If a field is present, one spectroheliogram will correspond to radiation that is shifted closer to the line center than the other and the two photographs will differ. By making a positive of one film, a negative of the other and printing through the pair, areas with no field will be medium density, those with fields will be darker or lighter.

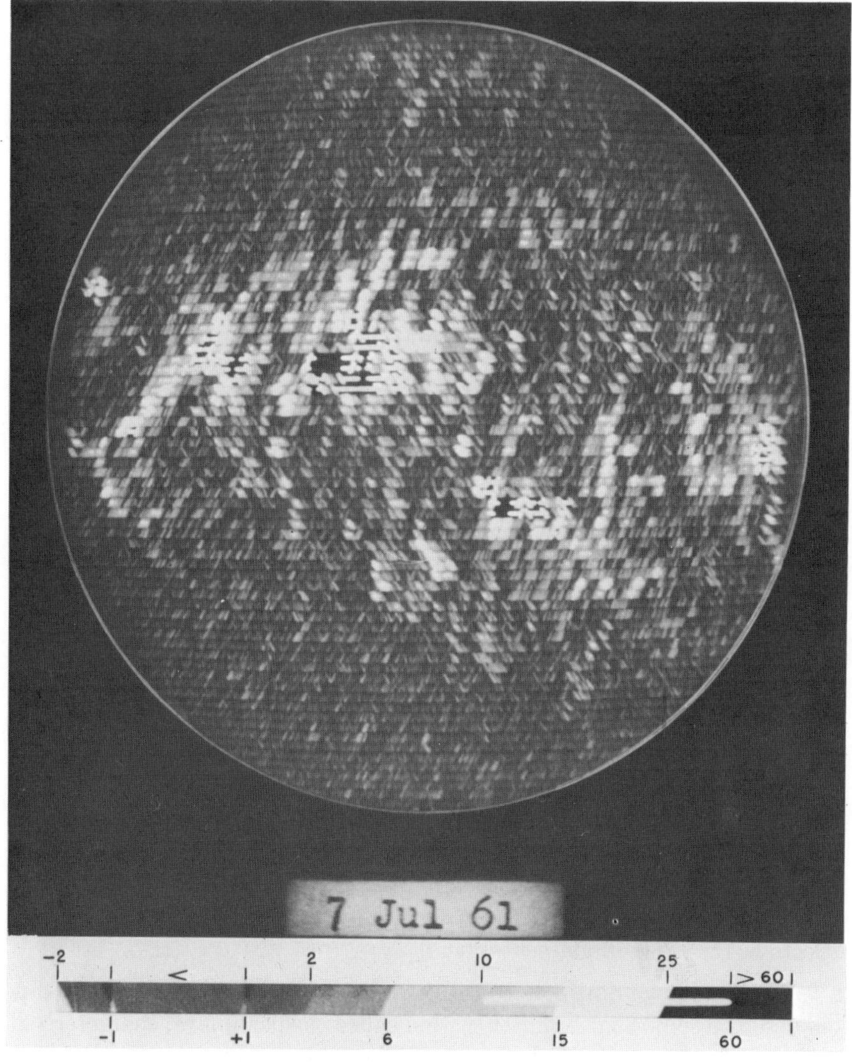

Fig. 9–9. Magnetogram of the Sun

To make this record, the magnetograph automatically scanned the disk of the sun in 80 parallel traces, with a resolution of 23 seconds of arc, taking about one hour. A high-dispersion spectrograph, with photoelectric detectors, analyzed the sensitive lines λ5247 Cr I and λ5250 Fe I for the Zeeman effect. A recording circuit brightened or modified the trace of the cathode-ray tube to indicate seven distinct levels of magnetic-field intensity (1, 2, 6, 10, 15, 25, and 60 gauss), as shown on the calibration strip. Magnetic polarity is shown by the slant of the recording trace to right (+) or left (−). North is to the top, east to the right. Note the three characteristic bipolar magnetic regions (BMR's); also the weak fields of the north and south polar caps. (Mount Wilson and Palomar Observatories.)

Giovanelli used a monochromatic filter with a very narrow band pass $\frac{1}{8}$A and photographic subtraction techniques in a method for obtaining motions as low as 50 meters/sec in photospheric layers. The technique can also be adopted for use with analyzers to measure weak magnetic fields. Extensive work on solar magnetic fields is also carried out by Severny and his associates in the Crimea.

Babcock's method detects extremely weak fields and can be calibrated precisely. It employs a low angular resolution ($\sim 20''$) unless one wants to take a long time to scan the disk. Leighton's technique gives high angular resolution but is limited to fields greater than 20 gauss, in the area over which a differential photographic technique can be employed.

For particular problems a number of specialized techniques have been devised. For example, to measure the direction of a magnetic vector in a sunspot, Treanor (1960) used a Babinet compensator to study polarization fringe systems in components of normal Zeeman triplets.

The curved arches often observed in prominences in neighborhoods of sunspots suggests the presence of magnetic fields but give no quantitative data. Likewise the polar rays of the minimum corona strongly suggested the existence of a poloidal solar field and even present rough estimates of its magnitude (van de Hulst). Finally, circularly polarized radio-frequency emission from certain disturbed areas on the sun permit one to infer the existence and magnitude of magnetic fields where none could be observed otherwise.

In summary, the solar magnetic fields may be divided into two components, both of which exhibit time dependent characteristics: (a) a toroidal field associated with sunspots and magnetic regions of the sunspot zones (e.g., plages); (b) a poloidal field associated with the weak "general" magnetic field of the sun and the coronal spikes. A recent theory due to Babcock (Art. 9–10) suggests that both of these are merely aspects of the same fundamental field.

9–7. Sunspots. Sunspots consist of a dark central area, or umbra surrounded by a somewhat brighter section called the penumbra, characterized by filaments of granules radiating from the umbra. The sizes and shapes of the spots vary greatly, but the percentage of large to small spots is nearly constant with the phase of the cycle.

Usually, spots of apparently common origin appear in a group. A typical prominent group will first appear as small spots, apparently formed from small pores between the granules, extending 3° or 4° in longitude. Two main spots develop, grow rapidly, and separate in longitude to a distance of 10° or more. The following spot attains a maximum size in three or four days, while the leader (which acquires the larger size) takes a week or nine days for its full development. Often smaller spots develop within the group. After attaining maximum size, a typical spot declines more slowly, taking weeks or even months. The following spot breaks up into several spots which shrink and disappear in a few days or weeks while the leader survives as a single spot which gradually fades away in a few weeks or months.

One of the largest spot groups ever observed on the sun (January to May, 1946) has been described by Nicholson and Hickox of Mount Wilson (see Fig. 9–6). It consisted of two large spots with complicated umbrae and a number of smaller spots. The group was so large that two days were required for solar rotation to bring it into view. Both of the large spots were roughly oval in shape; the preceding spot was 62,000 miles long and 35,000 miles wide, and the following spot was 90,000 miles long and 60,000 miles wide. They were 40,000 miles apart; hence the total length of the group was nearly 200,000 miles. The total area of the group, six billion square miles, comprised 0.0054 of the sun's visible surface. When the group returned to the east limb on the next rotation its area had diminished but had spread out to over 36° in longitude, or more than 220,000 miles. This great spot group made four transits across the disk, and was seen last on May 8, 1946 on the west limb, 99 days after it was first seen on January 29. The last remnant was a small stable spot, 10,000 miles in diameter.

Sunspot groups are usually oriented in such a way that the preceding spot lies closer to the equator than the following spot. The magnitude of this effect is greatest at 30°–40° and decreases for pairs nearer the equator (Brunner, 1930). Furthermore, sunspot activity tends to be repeated in intervals of longitudes where there has been previous activity. Likewise, regions free of activity favor certain longitude intervals. The fact that unequal numbers of sunspots are found in the eastern and western halves of the solar disk was interpreted long ago by Maunder by supposing that the axes of the spots were not perpendicular to the solar surface, but were tilted forward.

Many years ago Hale noticed vortical forms of gaseous filaments on Hα spectroheliograms, which suggested material flowing along penumbral filaments into the spots. St. John's measurements of displacements of the hydrogen lines confirmed an inward and downward motion of material, although he found no cyclonic motions. Moderately strong lines of Rowland intensity 10 to 40 showed no shifts. On the other hand, Evershed's measurements of lines of intensity -1, -2, or -3 indicated material flowing out of the spot tangent to the solar surface at rates up to 2 km sec^{-1}. The "vortices" observed in Hα spectroheliograms are structures located much further from the spots than normal Evershed motions.

At the center of a strong line, e.g., Hα, we see to small optical or geometrical depths, i.e., we observe radiation from great heights in the atmosphere. In a weak line, on the other hand, radiation reaches us from strictly photospheric levels. Centers of lines of intermediate strength represent radiation from intermediate layers. Hence high and low level atoms have different motions near sunspots. Recent work indicates that the motions are probably not "vortical" near normal spots, while in complex spots the motions are very complicated. The "vortices" of the type observed by

Hale appear to follow the sense of terrestrial cyclones in four-fifths of all examples, i.e., the direction of rotation is clockwise in the Southern Hemisphere and counterclockwise in the Northern Hemisphere. This fact suggests that the motions arise from Coriolis forces rather than from some other cause such as magnetic or electrical forces.

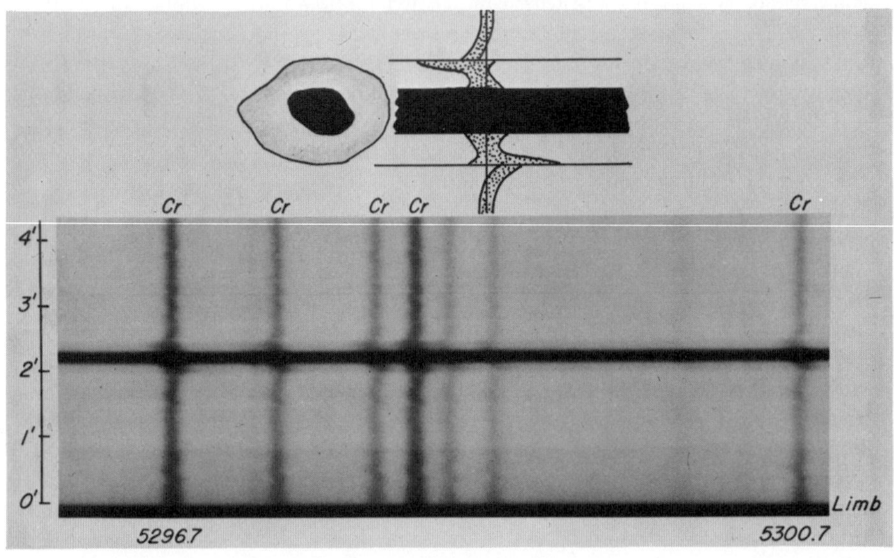

Fig. 9–10. The Evershed Effect

The sketch at the top schematically depicts the Evershed effect of the chromium aed other lines that are photographed across the spot. The thick, dark, horizontal band is the underexposed image of the spot umbra. On either side lies the greyish penumbra which is bordered by a thin, slightly brighter thread corresponding to the image of a narrow, slightly brighter ring that surrounds the penumbra. Although the velocity displacement increases to a maximum of about 4 km sec^{-1} at the edge of the penumbra, it decreases rapidly outside the spot. Photographed with the vacuum spectrograph of the McMath-Hulbert Observatory, September 2, 1955. (The McMath-Hulbert Observatory of the University of Michigan.)

Fig. 9–10 shows the Evershed effect as observed for a large spot with the vacuum spectrograph at the McMath-Hulbert Observatory. Note the distortion of the lines in the penumbral region and the rapid establishment of normal line positions outside the spot. Thus the Evershed effect clearly depends on the distance from the center of the spot. It also depends on the size of the spot (Kinman, 1952, 1953), while Michard (1951) finds that it depends on the magnetic field, the stronger the field the faster the outflow of material in the photospheric layers. Kinman gave a theoretical treatment of the energy balance and concluded that the material flowing out of the spot emitted energy at the edge of the penumbra which may thus produce bright rings sometimes seen at the spot edge.

The intensity distributions across spots (Richardson, 1933; Das and

Ramanathan, 1953) suggest steep transition regions between the umbra and the penumbra and the latter and the photosphere. Better measurements can be obtained from balloons. Are spaces between bright penumbral filaments any brighter than spot umbrae themselves?

The temperature ratio of sunspot to photosphere may be obtained from a comparison of their respective rates of emission of energy, the intensity distribution in their respective continua, and changes in line intensity from spot to photosphere.

Pettit and Nicholson compared the fluxes of spot and photosphere by means of a thermocouple. They found a flux ratio of 0.47 which would imply a spot temperature of about 4700°K. The true ratio may be actually smaller, because of scattered light from the photosphere, in which event an even lower temperature is implied.

The brightness of the umbra and penumbra varies from spot to spot. Larger spots have a tendency to be darker, but the size of a spot does not tell the whole story. Korn (1940) found the intensities of umbrae to range from 0.03 to 0.85 that of the photosphere, corresponding to a temperature range from 3920°K to 5610°K. Intensities in the penumbra range from 0.52 to 0.82 that of the photosphere, implying a temperature range from 5210°K to 5580°K. Makita and Morimoto 1960 find umbral temperatures of 4100°K and penumbral temperatures of 5500°K at $\tau = \frac{1}{2}$ from measurements made in the continuum.

The ratio of umbral to photospheric continuum intensities has been tabulated by Allen (1955), while de Jager (1959) suggests that one can derive a model for a sunspot by simply adding

$$\Delta\theta = \left(\frac{1}{T'} - \frac{1}{T}\right) \times 5040 = 0.25$$

to corresponding values for a photospheric model. Here T is the photospheric temperature, T' is that of the spot.

Finally, we may estimate excitation and ionization temperatures for a spot from a comparison of selected spectral lines in the spot and photosphere (see Fig. 9–11). The scattering of photospheric light into the spot by the earth's atmosphere, the appearance of many weak molecular lines in the spot spectrum, and the splitting and polarization of lines in the magnetic field of the spot, all serve to complicate the problem. Moore used eye estimates of intensities interpreted with the aid of the Rowland scale calibration. From a comparison of the same lines in spot and photosphere she found $\delta\theta = 0.19 \pm 0.01$. Similarly, the lines of neutral and ionized atoms, employed in connection with the Saha equation, give the electron pressure variation from spot to photosphere. Moore found the ratio of P_ε (spot) to P_ε (photosphere) to be about 0.6. Somewhat lower ratios were found by Krat, and by P. ten Bruggencate and von Kluber

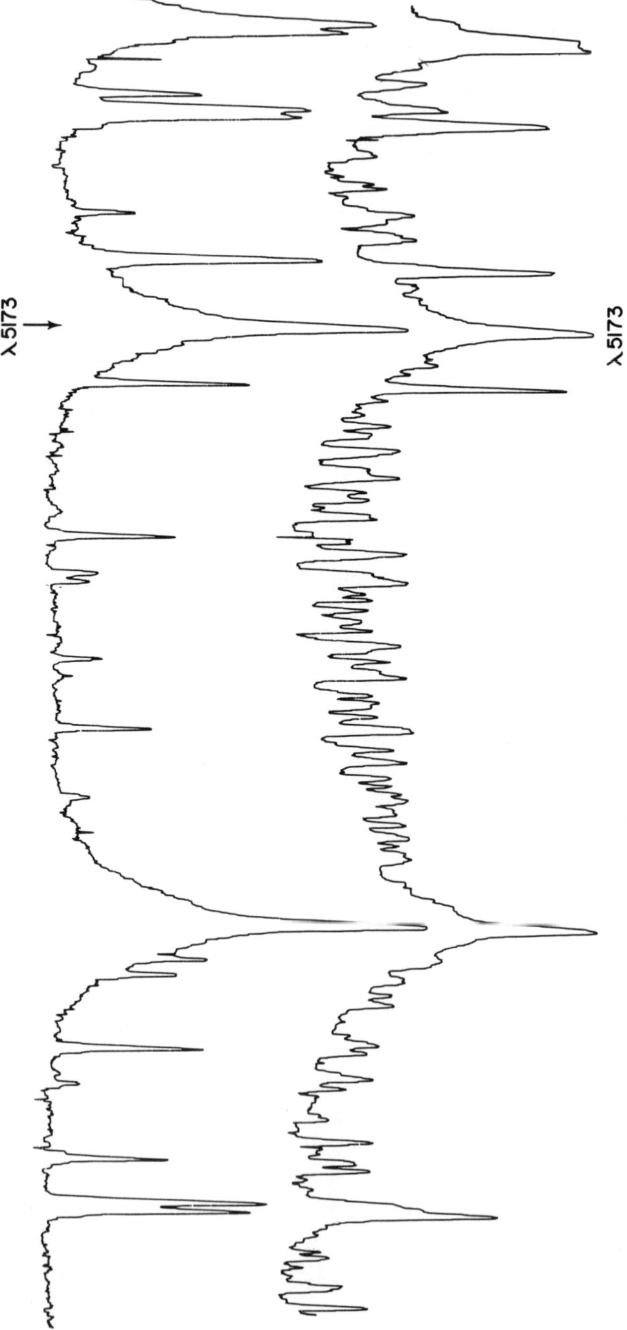

FIG. 9–11. COMPARISON OF THE PHOTOSPHERIC AND SUNSPOT SPECTRA IN THE NEIGHBORHOOD OF THE MAGNESIUM "B" LINES

Upper: Trace of photospheric spectrum. *Lower:* Trace of sunspot spectrum.

Notice the extreme complexity of the sunspot spectrum. Many of the lines are due to fragments of molecules in high-energy states of rotation and vibration not normally observed in the laboratory. (Courtesy, O. C. Mohler, McMath-Hulbert Observatory, University of Michigan.)

who used a curve of growth method. The "most likely" value appears to be about 0.2 for the line-forming layers but is rather sensitive to the assumed $\delta\theta$. Excitation temperatures in the neighborhood of 3900°K were found by these workers and by Zwaan (1960) who studied neutral lines of titanium and iron. Using the rotational structure of bands of OH, NH, CH, C_2 and Mg H, Laborde (1961) obtained spot temperatures ranging from 3600°K to 3900°K.

The derived temperatures refer to different layers, $\tau_0 \sim 0.8$ for the continuum, $\tau_0 \sim 0.2$ for line spectra and ~ 0.07 for molecular bands. Much more work remains to be done on this problem which is very difficult both observationally (because of scattered photospheric light) and theoretically.

Attempts to construct models for spots have been made by Michard (1953) and by Mattig (1958). The results are not in good accord because of different assumptions concerning the metal/hydrogen ratio, consequently there are considerable differences in the P_ε (spot)/P_ε (photosphere) ratios, and in the height scales. Mattig and de Jager conclude that spots are less transparent than the photosphere, whereas Michard, Bray, and Loughhead, and others have concluded that the opposite is true.

Difficulties attend any attempt at detailed quantitative treatment of spots. To add to the usual observational difficulties, one spot differs from another so a complete series of observations should be obtained for each spot. The theory of line formation is complicated by Zeeman splitting in a magnetic field (Hubenet, 1954; Warwick, 1951, 1955, 1957; Unno, 1956, 1957, 1959). One might expect some intensification because of the broadening of the line absorption coefficient in a magnetic field. Precise line profile and equivalent-width calculations by Pecker's method for lines formed in an inhomogeneous magnetic field indicate that with increasing field strengths the central intensities and half-width change and there is an increase in asymmetry and equivalent width (Mattig 1958). The intensification of the lines is however insufficient to account for the fact noted by ten Bruggencate and von Kluber that the flat portion of the curve of growth is raised in the spot with respect to the curve for the photosphere. Hence the difference in the curve of growth for the photosphere and the spot has been attributed to increased turbulence in the latter, in spite of the well-known fact that magnetic fields tend to suppress turbulence.

A further difficulty in interpretation of spot spectra is that the source of opacity in the ultraviolet is not well known (although Michard showed that the H^- ion accounts for the opacity for $\lambda > 4500$). Formation of the negative hydrogen ion is favored by the lower temperature but the lowered electron temperature acts in the opposite direction. Because of the great increase in the number of lines, particularly those of molecules, blending poses a serious problem. Furthermore, in constructing models for spots we cannot be sure that pressure equality exists. The refrigeration in the spots certainly occurs as a consequence of their magnetization, not vice versa. Spots may owe their low temperatures to the circumstance that they are in radiative equilibrium to very deep strata, whereas the nearby photosphere layers are in convective equilibrium for $\tau_0 > 1$, the temperature gradients being such that the surface in the spots is much cooler than in the photosphere.

The most obvious, but not the most satisfactory, definition of the solar cycle is provided by the sunspot members. Although the periodicity in spot members was discovered by Schwabe in 1843, Wolf in 1856 was the first to establish the eleven-year cycle and put the observations of sunspots on a systematic basis. The first problem is what do we count — the numbers of individual spots, groups, or total areas? As a compromise Wolf proposed the quantity $(10g + s)$ where g is the number of groups, and s the number of individual spots. The system has been retained in order to secure uniformity in statistics. In practice, each observer compares his counts with those made at Zurich and determines a correction factor K so that $K(10g + s)$ will agree with the Zurich system.

Fig. 9–12 shows the yearly variation of sunspot members. There are in addition short-period fluctuations from month to month that are often quite marked.[5] In each cycle, the mean latitude of the spots decreases according to Spörer's law (see Fig. 9–13), which is illustrated by the Maunder "butterfly" diagram. This phenomenon can now be explained.

9–8. Magnetic Fields of Sunspots. The vortical structure observed on Hα spectroheliograms suggested to Hale that it would be worthwhile to look for magnetic fields in the spots.[6] By methods previously described he was able to demonstrate that each spot possessed a magnetic field, and that these fields were large by terrestrial standards — of the order of 3000 to 4000 gauss. One may measure not only the magnitude but also the direction of the field. An outward-field direction would mean that a north-seeking or positive pole would be urged outwards. Hale denoted sunspot polarities as R or V. For R polarity a particular polarized component of a line in a spot of positive polarity is displaced towards the red. With the same apparatus the same line component in a spot of negative polarity would be shifted toward the violet. We now use N(north) and S(south) polarities instead of R and V.

Hale and Nicholson classified spots in three classes, *unipolar*, *bipolar*, and *complex*. Unipolar spots are single spots or groups of spots with the

[5] A periodicity of 80 years may be superposed on the 11-year cycle and both longer and shorter cycles have been suggested.

[6] Later work suggested these vortices to be analogous to cyclones on the earth, i.e., purely hydrodynamical phenomena, although the Crimean observers suggest that the Hα structure around spots follows the lines of force.

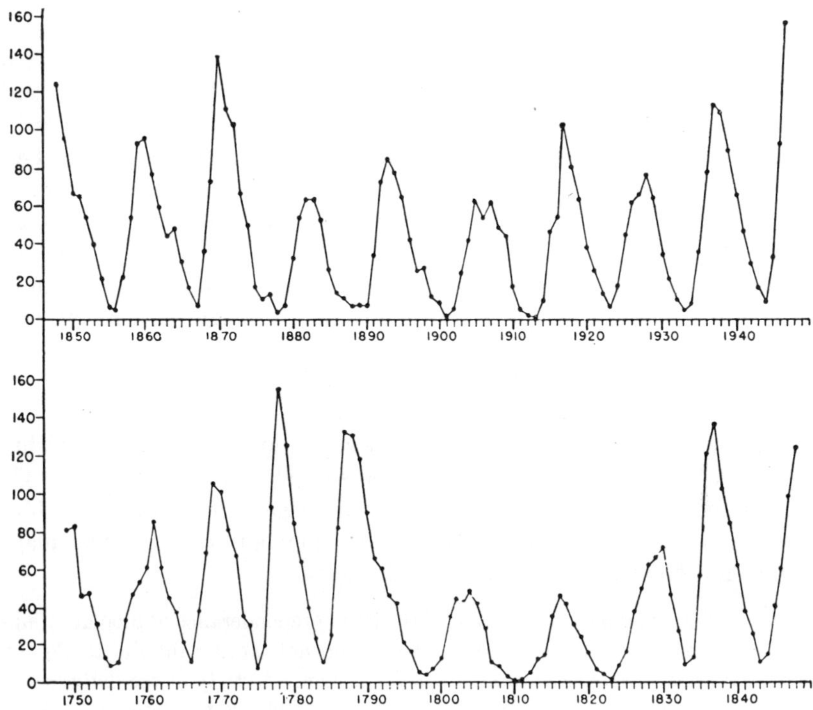

FIG. 9–12. SUNSPOT NUMBERS

The most conspicuous feature of the solar cycle is the variation of the number of sunspots with a period of 11.5 years. Note the unequal heights of the various sunspot maxima. (Courtesy, D. H. Menzel, *Our Sun*, Cambridge: Harvard University Press, 1949.)

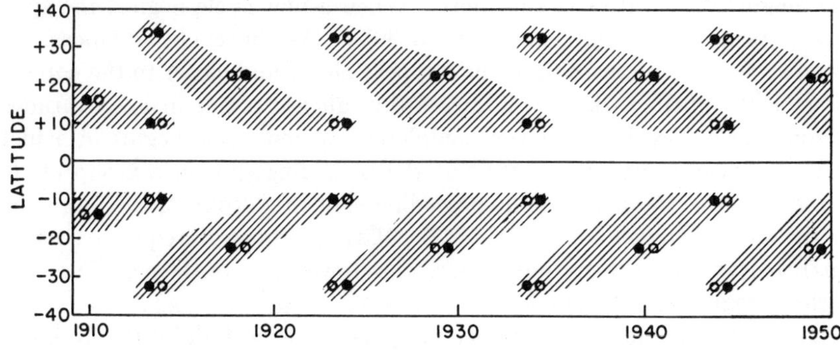

FIG. 9–13. THE LAW OF SUNSPOT POLARITY

The diagram shows the approximate variation in latitude and polarities of sunspots observed at Mount Wilson from June, 1908 to January, 1950. The preceding spot is indicated on the right. Filled circles denote spots of north-seeking polarity; open circles denote those of south-seeking polarity. The cross-hatched area indicates roughly the distribution of the spots at any one time. Notice the decrease in the average latitude of the spots as the cycle advances (Spoerer's law), and the reversal of polarity in each hemisphere at the end of each cycle.

same magnetic polarity. In their simplest form, bipolars consist of two spots of opposite polarity; the line joining them makes a small angle with the equator. The preceding spots of such pairs are of opposite polarity in the two hemispheres. For example, if the preceding spot in the northern hemisphere is positive (N or north-seeking), the preceding spot in the southern hemisphere will be negative (S or south-seeking). More often a bipolar group will consist of a stream of spots, whose preceding and following members have opposite polarity. Direct telescopic observation cannot distinguish between spots of opposite polarity, although polarization tests in their spectra can tell them apart at once. In complex spot groups, spots of opposite polarity are mixed together. The bipolar appears to be the fundamental type.

Hale and Nicholson employed both the appearance of the bright calcium plages and the magnetic polarities in their spot classification scheme. The plages play an important role. Although plages are observed without sunspots, no spot has ever been caught without an accompanying plage.

Hale and Nicholson remark:[7]

> The most striking fact brought out is the large preponderance of bipolar groups, and the tendency of unipolar groups, even when not lapsing intermittently into the bipolar state, to hover continually on its margin. A study of the daily history of the spots further emphasizes this tendency as it shows the frequent passage, back and forth, from one type to the other and the development from time to time of invisible spots in unipolar groups which never visibly attain the bipolar condition.

Of considerable importance is the relation between the solar cycle and the magnetic polarities of the spots. When a new cycle begins, the spots first appear in high latitudes (up to 60°). As the cycle advances, the number of spots increases and they appear closer and closer to the equator (Fig. 9–13) until, as the cycle dies away, all are found in low latitudes. Before the old cycle dies away completely, a new cycle begins in a high latitude, with reversed polarity. If the preceding spot in a group of the dying cycle has N polarity, the preceding spot in a group of the new cycle in the same hemisphere will have S polarity. Since it requires twenty-three years for the status quo to be restored, we think of the complete solar cycle as requiring 2×11.5 years.

Furthermore, the total number of spots tends to be alternately larger and smaller in successive eleven year periods. In higher maxima, the peak is reached earlier than in low maxima and the average latitude of the spots will tend to be greater.

The classical investigations of magnetic field directions were those of Nicholson at Mount Wilson, who found that the angle θ between a normal

[7] *Ap. J.* **62,** 203, 1925.

drawn to the surface and the lines of force emerging from a spot could be closely represented by an equation of the form

$$\theta = \frac{\pi}{2}\left(\frac{r}{b}\right) \tag{9-15}$$

where b is the radius of the spot to the outer edge of the penumbra and r is the distance from the center. The field strength of the spot can be represented fairly well by the formula

$$H = H_0\left(1 - \frac{r^2}{b^2}\right) \tag{9-16}$$

proposed by Broxon (1942), or a more elaborate expression suggested by Mattig (1953). More recent work by Treanor indicates that the field is more strongly outward directed than Eq. 9–15 would indicate. He finds that 4 per cent of the total flux of the spot is funneled into unit solid angles. Treanor finds good agreement between his observations and Eq. 9–16 if for b we substitute r_0, the outer radius of the strongly magnetic area.

Attempts to estimate the gradients of magnetic fields were made from (a) center limb variations in field strengths of spots, (b) comparisons of weak and strong lines, (c) from divergence of the lines of force, and (d) from asymmetry of lines. Houtgast and van Sluiters (1948) find 5.7 gauss km^{-1} from (a), 2.5 gauss km^{-1} from (c). Mattig (1958) concludes method (d) will work if the gradient exceeds about 1 gauss km^{-1}. Treanor concludes that the gradient must be less than 1 gauss km^{-1} while Schluter and St. Temesvary (1958) derive 0.4 gauss km^{-1} from theoretical considerations.

The area of a long-lived (60^d) spot increases rapidly at first, reaches its maximum in about 10 days, and then gradually decreases. The magnetic field, on the other hand, reaches a maximum near 3000 gauss and remains roughly constant for about 30 days. Thereafter it declines, slowly at first and then more rapidly, as though the lines of force were retracted below the surface of the photosphere. Thus the total magnetic induction varies with the area.

The magnetic field in a spot does not remain strictly constant with time. Even quiet spots show variations of the order of 20 to 30 gauss an hour and in the course of several days variations of hundreds of gauss may occur. The variations do not seem to be more frequent or pronounced in magnitude when flares are observed in the spot group.[8]

Following S. Chapman we may discuss the magnetic fluxes, dipole moments, and mechanical forces involved in spots. If we integrate the vertical components

[8] At the Crimean Observatory, however, Severny and his associates find indications that significant magnetic field changes are associated with flares.

of the magnetic field $H \cos\theta$ over the area of a circular spot (see Eq. 9–15), the emergent flux is found to be

$$F = \tfrac{1}{3}\,\pi\, b^2 H_0 \tag{9–17}$$

which is related to the pole strength m of the spot by

$$F = 2\pi m \tag{9–18}$$

For example, the spot MW 6618, September 21, 1939, had a field strength H_0 of 3600 gauss, a radius b of 28,000 km, and thus a magnetic flux of 4×10^{22} gauss cm^2.

The magnetic dipole moment M corresponding to a bar magnet of pole strength m and length d is $M = md$; e.g., the bipolar spot MW 6725 consisted of two spots each of radius $b = 20{,}000$ km, and of field strength, 3900 gauss, separated by 175,000 km. Hence the magnetic moment is

$$M = \frac{F}{2\pi} \times d = 5 \times 10^{16}\, \Gamma\ \text{km}^3 \tag{9–19}$$

or 600,000 times the dipole moment of the earth. The dipole moments of sunspots are usually nearly parallel to the equator, although the leading spot is often a little closer to the equator than the following spot.

Further, it is to be expected that two spots of opposite polarity in a bipolar group should attract one another. Following Chapman, consider two spots, each of the pole strength $(4 \times 10^{22}/2\pi)\ \Gamma$ cm^2, and separated by 125,000 km. If we treat them as point poles the mutual attraction is

$$\frac{(4 \times 10^{22})^2}{2\pi}\,\frac{1}{(1.25 \times 10^{10})^2} = 2.6 \times 10^{22} \text{ dynes (or } 260 \times 10^{12} \text{ tons weight)}$$

During the growth of a spot group, the two components tend to move apart against this strong magnetic attraction, apparently under the influence of powerful hydrodynamical forces.

The mechanical forces caused by these large magnetic fields must be considerable. The lines of force may be regarded as subject to stresses, consisting of a tension $H^2/8\pi$ along the lines of force, and equal pressure p_{mag} normal to them. Since lines of force tend to repel one another, we might expect the sunspot to expand continually in area during the course of its life.

Alfvén suggested that a spot was maintained in mechanical equilibrium by a difference in gas pressure inside and outside the spot such that

$$p_s + p_{\text{mag}} = p_s + \frac{H^2}{8\pi} = p_0 \tag{9–20}$$

where p_0 is the external gas pressure and p_s is the gas pressure inside the spot. We cannot, however, be sure that sunspots are actually in mechanical equilibrium.

The magnetic energy stored in a sunspot is

$$\frac{1}{8\pi}\int H^2\, dv \tag{9–21}$$

integrated over the entire volume of the spot. At the center, where the field is

3000 gauss, the energy density is 3.6×10^5 ergs cm^{-3} and since the energy radiated at the center of the spot is about 1.2×10^{10} ergs cm^{-2}, the radiation per second is equal to the amount of magnetic energy contained at any one time in a layer $\frac{1}{3}$km thick.

Insofar as surface phenomena give any indication, the total amount of magnetic energy stored in the sun appears negligible compared with that residing in other forms. If we suppose that magnetic fields in spots are produced by electrical currents flowing in circular paths about their axes, as in a solenoid, the total flow through the radial half-plane passing through the axis of the magnet is $\bar{H}/4\pi$ emu $= 10\bar{H}/4\pi$ amperes per centimeter length along its axis. Thus if $\bar{H}$ is 3000 gauss, the current is 2400 amp cm^{-1} depth below the surface. The direction of electron flow around the axis of a north-seeking polarity is clockwise as seen from the earth.

Some years ago Cowling showed that the magnetic field cannot grow and decay with the appearance of a spot, but must exist before the spot becomes visible and must persist after it disappears. He calculated that about 300 years would be required for the electromagnetic decay of a sunspot field. Or, conversely, this is the time required for the field, if due to electric currents, to increase to its steady state value against the inertial effects of self-induction. More recent work indicates that even longer times are required.

Before discussing sunspot theories we must examine the nature of magnetic fields in their neighborhood. We might suppose that the flux from one member of a bipolar spot simply flowed into the other spot of the group, whereas magnetic flux from a unipolar spot presumably must return to the solar surface somewhere in an unspotted region. The same phenomenon must also occur in bipolar and complex groups where the total north-seeking flux and south-seeking flux from recognized spots do not balance. The plage sometimes suggests where missing magnetic flux returns to the sun. They precede the birth of a spot group and persist after it has vanished. Thus, they usually indicate that a particular unipolar spot is a remnant of a defunct bipolar group. As a sunspot group decays, the plages fade also, and, in the end, they symmetrically surround the expiring remnant of the group. Hale and Nicholson found magnetic fields in plages around a spot, and called such regions of appreciable magnetic flux, but no darkening, "invisible spots."

The question might then be asked whether plages, in general, have magnetic fields which in many instances are too weak to be measured by classical methods. The answer is that they do and that magnetic fields appear to play an essential role in production and maintenance of plages and faculae.

9-9. Faculae, Plages, and Magnetic Regions. Sunspots are the most easily observable indication of disturbed areas on the sun, wherein originate certain types of prominences, such as surges, flares and streams of charged particles, and electromagnetic non-thermal radiation. Such areas, called centers of activity, are described further in Chapter 10. Here we describe plages, faculae, and magnetic regions.

The amount of information obtainable from direct photographs of the solar disk is limited. One recognizes that facular areas tend to be associated with sunspots and that they precede the development of spots and persist after the spots have disappeared.

Under conditions of good seeing, the faculae have been resolved into grains of about the same size as the photospheric granules. According to Waldmeier and P. ten Bruggencate and to Bray and Loughhead the facular grains have lifetimes of the order of two hours. Rogerson (1961) proposes a model for faculae in which these regions are about 900°C hotter than normal in regions above an optical depth of about 0.3. On photographs of excellent definition showing the granules well, it is hard to trace faculous regions since the contrast is only 10 per cent, although very bright faculae may be 40 or 45 per cent brighter than the background.

Faculae may be called white light plages. They fall into two distinct classes. Those near the pole are small (Saito and Tanaka, 1960). Those connected with the sunspot zones have a greater extension in latitude than do the spots but show a progression in latitude with solar cycle similar to that shown by spots. They appear before the spots sometimes by several days and they often connect several successive spots. They gradually disappear after the spots are gone but they may last for several months. One of the most outstanding single characteristics of the faculae and particularly of the calcium plages is their long life. The average lifetime of a plage is about six rotations from its first appearance to its break-up into a widely dispersed group of faint features. Faculae not associated with spots are fainter and shorter-lived, and occupy only about 10 per cent of the total plage or facular areas.

Spectroheliograms show these faculae to coincide with brighter portions of the plages and yield much more detail than direct photographs. The aspect of the spectroheliogram will depend on the line utilized and the part of the line. The strong Ca II "K" line has a central emission component called K_2, upon which appears a narrow absorption feature K_3. See Fig. 9–14.

Photographs taken in the light of the wings K_1 correspond to the upper levels of the photosphere and (see Fig. 9–15, *left*) show a mottled structure with bright plages around the sunspots. The facular regions tend to have sharp boundaries. In the light of K_3, which corresponds to a high level in the chromosphere (see Fig. 9–15, *right*) notice that some spots are hidden, that the plages are prominent and that there appear long sinuous dark filaments. The latter, however, are seen better in Fig. 9–3. In K_2 the filaments disappear but the plage areas are still larger than those observed

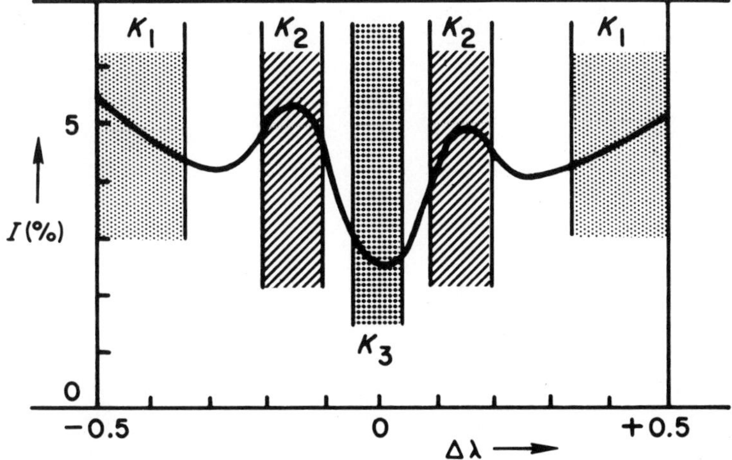

Fig. 9–14. Profile of the Center of the K Line

Intensity in percentage of the nearby continuum is plotted against wavelength over the central angstrom of the K-line. The K_1 spectroheliograms are taken in the "wings" at distances of greater than 0.4A, whereas the K_2 spectroheliograms are obtained in the emission (which is usually asymmetrical as depicted) and the K_3 spectroheliogram is taken in the central reversal. Compare Fig. 8–2.

in white light. In the outer part of the line K_1, the appearance of the spectroheliogram approaches that of a direct photograph.

Fig. 9–15 illustrates the fact that the structures observed in spectroheliograms are essentially of two fundamental types, (1) those connected closely with the photosphere, granules, etc., (2) those originating in the high altitudes above the solar surface, the dark filaments, plages and bright flocculi.

Fig. 9–3 permits a comparison of a spectroheliogram taken at the center of the Hα line with a Ca II "K_3" image. Notice that the plages are less prominent in the hydrogen photograph, whereas the filaments are more pronounced and the spots are sometimes visible. The "fine structure" of the background in the hydrogen photograph appears more delicate; the individual grains are five or six times larger than the photospheric granules and appear dark on a bright background. Hydrogen spectroheliograms also often show a vortical structure around spots, presumably due to cyclonic motions, but under conditions of exquisite seeing they also sometimes show many dark, thin "fibrils" (Lippincott, 1955; Kiepenheuer, 1957) whose orientations suggest that they are connected with magnetic fields. Michard suggests there is no difference between the Hα "vortical" structure and the structures made up by the fibrils.

Spectroheliograms taken in the wings of the Hα line more nearly resemble K_1 spectroheliograms. Fig. 9–16 compares spectroheliograms taken in successive lines of the Balmer series. Notice the weakening of the plages and the increased prominence of the spots, toward the higher members of the series, as we look more and more closely down to the photospheric layers.

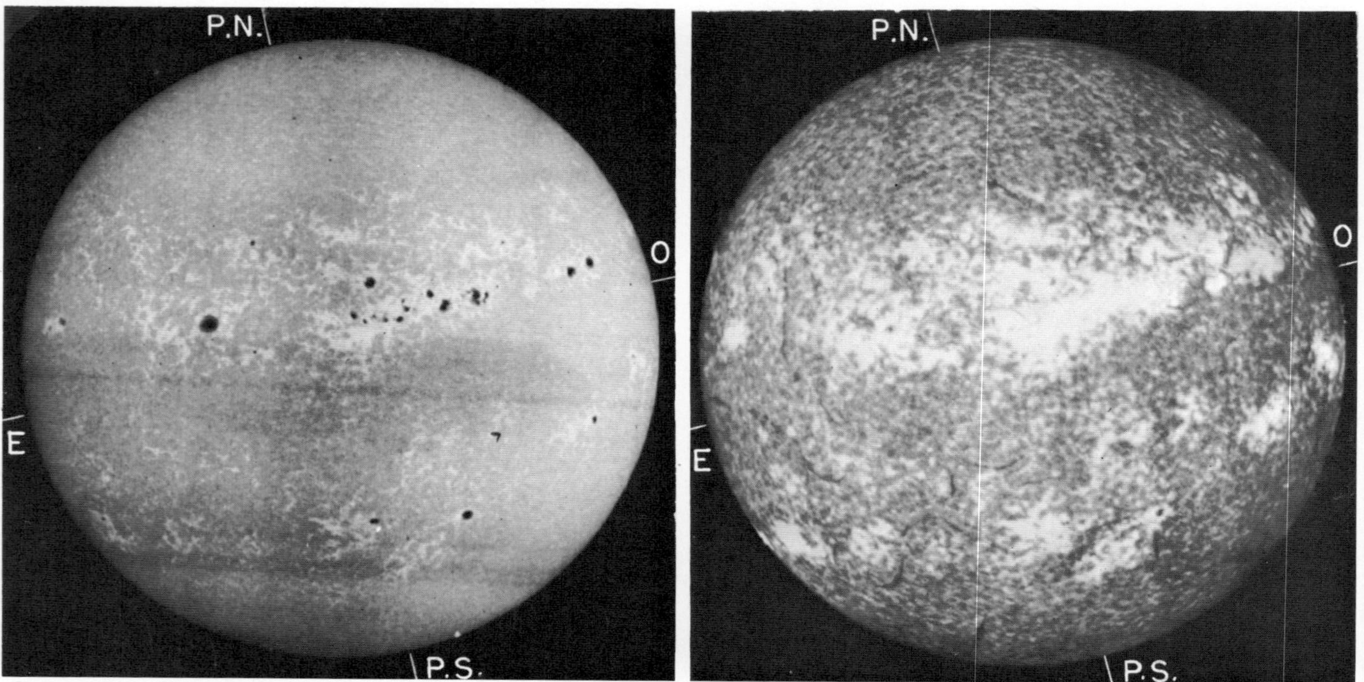

FIG. 9–15. CALCIUM SPECTROHELIOGRAMS

Left: Solar image in the violet wing of the K line (K_{1V}). *Right:* Solar image in center of K line (K_3). August 14, 1947, $7^h\ 45^m$ G.C.T. (Courtesy, L. d'Azambuja, Meudon Observatory.)

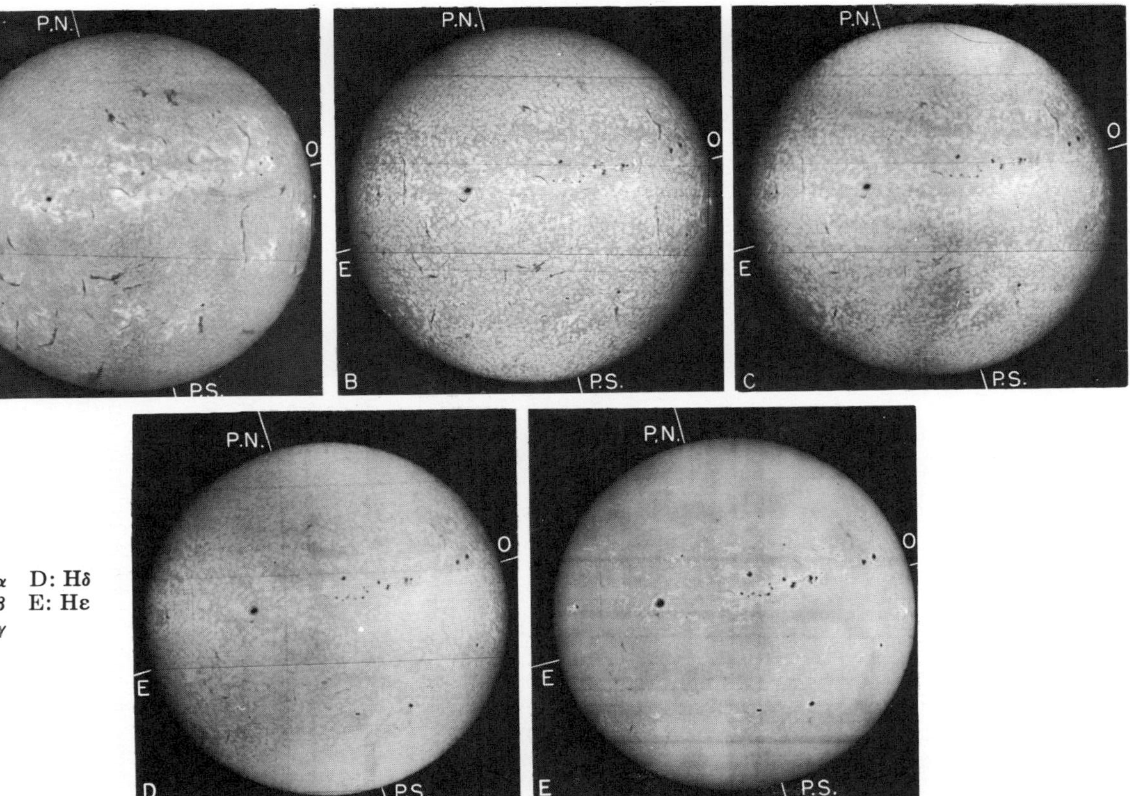

A: $H\alpha$ D: $H\delta$
B: $H\beta$ E: $H\varepsilon$
C: $H\gamma$

FIG. 9–16. HYDROGEN SPECTROHELIOGRAMS

Notice that the filaments and plages become less and less conspicuous from $H\alpha$ to $H\delta$. $H\varepsilon$ is submerged in the wing of the "*H*" (Ca II) line (compare Fig. 9–15 left for the *K* line wing, which is identical). The superposition of *H* (Ca II) and $H\varepsilon$ which absorbs slightly over the plages causes a weakening of the plages in *E*. (Courtesy, L. d'Azambuja, Meudon Observatory, August 14, 1947.)

The magnetograph clearly showed that there existed on the solar surface large areas with measurable magnetic fields. The strongest features were the *bipolar magnetic regions* (BMR's) — two adjoining areas of opposite polarity. Although they frequently contain spots, some do not and they often extend to latitudes higher than those in which spots are numerous. Furthermore, BMR's obey Hale's law of sunspot polarity. Generally they develop as small compact areas that gradually expand; wherever the field is stronger than about two gauss, plages appear and if the field becomes strong enough sunspots occur. The magnetic flux increases for a while, reaches a maximum and declines as the field weakens as the area grows. The field persists after not only the spots but also the faculae have disappeared. Striking and concurrent changes occur in chromosphere and corona. Babcock noted that delicate flocculi which constitute the fine structure of the chromosphere on Hα spectroheliograms appear to be randomly oriented in field-free regions but seemed to exhibit a noticeable alignment in fields greater than a gauss. Filaments tend to lie on the border between the two regions of opposite polarity or along the poleward side. The arches observed in the corona above plages may define the field at considerable heights above the solar surface. A one-to-one correspondence has been found between the plages and the BMR's.

The correlation between the bipolar magnetic regions and the plages was beautifully demonstrated by Leighton, whose observations although restricted to fields greater than 20 gauss had the advantage of greater angular resolution than those of Babcock. He found all regions showing a measurable longitudinal field to be identified with a prominent calcium plage area, although the latter often extended over areas larger than those corresponding to fields greater than 20 gauss. The patterns changed with time in the same way as did the plage areas. He also suggested that the fields were nearly normal to the solar surface, and that all calcium plages are produced by the action of magnetic fields. We may adopt 40 gauss as a typical value for the field of a plage.

Babcock has also identified rare *unipolar* (UM regions) which are extensive areas of only a single polarity and fields not more than about 2 gauss. Some may be remnants of bipolar magnetic regions. They seem to favor the declining phase of a sunspot cycle.

The importance of these regions lies in their terrestrially recorded effects. During the life of one particular UM region, cosmic ray counters indicated increased activity each time it crossed the central meridian of the sun, and three days later geomagnetic storms occurred. Babcock suggests that magnetic lines of force arising from a UM region define what is observable as a coronal stream which may extend out as far as the earth.

The magnetic fields observed by Babcock and also by the Crimean observers show both short-period fluctuations and gradual long-period changes. The short-period changes may be due to instabilities in convection, although an upper limit of a few gauss can be assigned to fields of individual granules. The longer-term variations are connected with large-scale hydrodynamical motions of magnetized material.

It seems quite clear that the explanation of sunspots, magnetized (BMR) regions, and probably other solar magnetic phenomena as well, is to be sought in the same mechanism. Among working hypotheses proposed, that of the toroidal magnetized ring, received much attention by Alfvén (1950), Walén (1949), Bullard (1956), Parker (1955), Ferraro and Kendall (1956), and others. The toroidal ring concept is not really a theory of sunspots, it is only a magnetohydrodynamical model that has been invoked to explain some of the most essential observations. H. W. Babcock noted that a BMR originated as though loops of a submerged toroidal field were brought to the surface, a magnetized region, a plage, and a spot appearing in that order. When the area of the BMR increases, the magnetic field weakens and these phenomena vanish in the reverse order. The BMR disappears by expansion, never by shrinking and sinking. The enhanced activity in a spot neighborhood is to be at least partly attributed to the action of the magnetic field upon the attenuated gases of the chromosphere and corona. Apparently, the lines of force are able to penetrate with ease into the coronal regions.

Magnetic phenomena are so widespread upon the sun that we would expect it to possess a *dipolar field*, i.e., a "general" field somewhat similar to that of the earth. Such a field does indeed exist, although early attempts by Hale and others to measure it were inconclusive.

9-10. The Sun's General Magnetic Field and Its Relation to Other Phenomena. Measurements by Thiessen and Kiepenheuer suggested that this polar field must be very small. Historically, the strongest argument for its existence are the polar rays observed near sunspot minimum (Gold, 1958). H. C. van de Hulst (1950) showed, however, that these structures can be explained by a field less than about a fifth of a gauss. The corresponding field at the solar surface would be higher, perhaps about a gauss. Definitive evidence for a "general" dipolar solar magnetic field is usually found only in latitudes greater than 60°. From Zeeman effect measurements the Babcocks find this mean field to be of the order of 1–2 gauss and that it has a variable, irregular, fine structure. The existence of this general field has been confirmed by von Kluber (1960) and also by Högbom's (1960) observations of the radio-frequency occultation of the Crab Nebula. The latter finds an upper limit of about 2.5 gauss for the magnetic field at the photosphere in high latitudes.

From 1953 to 1957, the solar poloidal field had opposite polarity to that of the earth's field. Then, between March and July, 1957 (near the date of maximum of the sunspot cycle) the south polar field reversed its sign, whereas the polarity of the north polar field did not change sign until November, 1958. For more than a year the sun had two magnetic poles of the same polarity (H. D. Babcock, 1959). The maximum of the sunspot cycle in the northern hemisphere came a year later than it did in the southern hemisphere. Hence, in each instance, the field reversal seemed to be connected with the spot maximum. It would appear that the sun's general field and the sunspot fields are in some way related to one another.

Recently, H. W. Babcock has indicated how all solar magnetic phenomena might be tied together in one theoretical picture. These phenomena include (1) the basically axisymmetrical solar dipolar field just described, (2) the reversal of magnetic polarity in the dipolar field, (3) the equatorial acceleration of the sun, (4) the eleven-year sunspot cycle including the change in average spot latitude during the cycles, (5) magnetic fields in spots and Hale's law of sunspot polarity, (6) the fact that preceding spots of a group tend to be larger and to last longer than following spots, (7) an apparent tilt of spot axes in the direction of rotation as indicated by an east-west asymmetry in the apparent numbers of spots, (8) orientation of spot pairs such that preceding spot group tends to fall closer to the equator, (9) tendency of sunspots to reappear at longitudes where previous spots have appeared, (10) hydrodynamical motions near larger spots, (11) occurrence of BMR's which obey Hale's polarity law, and (12) successive appearance of plages, faculae, and sunspots as field strength increases.

A basic postulate underlying Babcock's hypothesis is that the poloidal (dipolar) field of the sun and the toroidal field which is manifest in BMR's and sunspots *are both simply manifestations of the same general magnetic field*. The non-uniform rotation of the sun is responsible for the production of a toroidal field. Babcock's model may be summarized as follows: Initially, the submerged lines of force of the dipolar poloidal field lie in meridian planes in the shallow upper layers of the sun. Because of non-uniform solar rotation, the lines of force get pulled out in longitude to form toroidal magnetic fields. See Fig. 9–17.

The magnetic energy of these fields becomes increased at the expense of rotational kinetic energy. This "winding up" to form a tight spiral proceeds until the magnetic field intensity of these submerged toroidal fields reaches at some point or points a value sufficient for instability to occur. Such regions tend to be buoyed up to the surface to form BMR's. As this region expands into the chromosphere and corona, sizable loops of magnetic flux are detached. These loops may drift away from the sun carrying large numbers of charged particles.

In the initial stage, the lines of force are assumed to lie in a sheath of thickness about $0.05R$ (R = solar radius) in regions closer to the solar equator than about 30°. Nearer to the pole the sheath gradually deepens. In the vicinity of the equator the average field, H_0, is about 5 gauss. After winding of submerged lines of force has continued for about three years, a point at the equator will have gained 5.6 turns compared to one at a

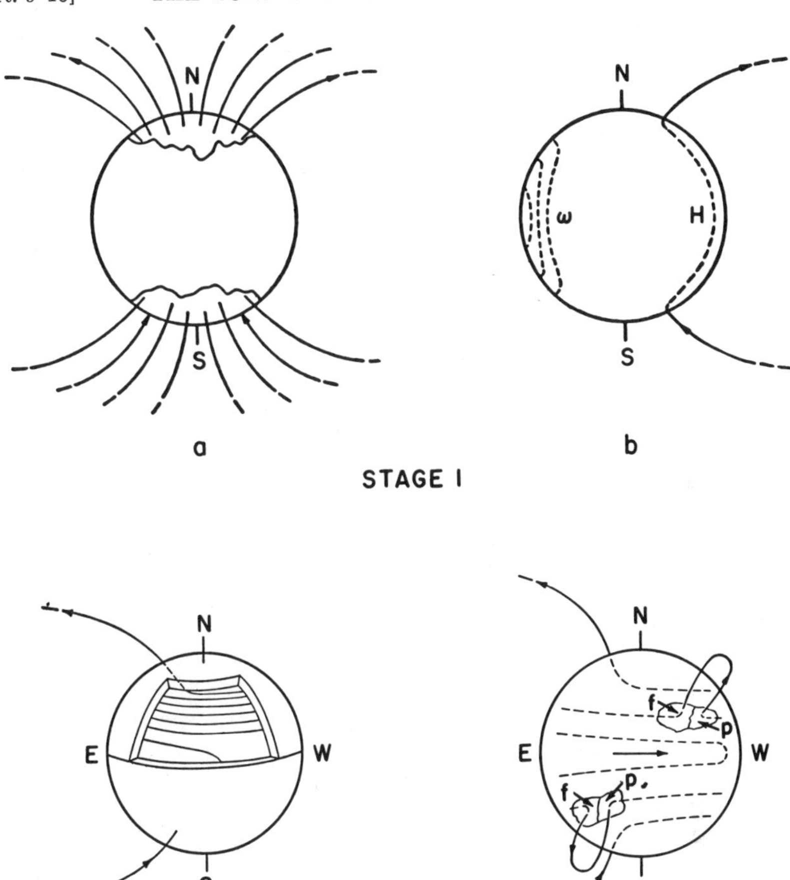

FIG. 9-17. STAGES IN THE DEVELOPMENT OF A BIPOLAR FIELD FROM A POLOIDAL FIELD

Stage 1: (a) The main dipolar field of stage 1 has a mean intensity in the polar caps of 1–2 gauss, and its total magnetic flux is 8×10^{21} maxwells. The lines of force emanate only from the polar regions. (b) The submerged part of each line of force is relatively shallow; together they lie in a layer which is thinnest between the 30° parallels of latitude. The lines labeled by ω are cross-sections of the surfaces of constant angular velocity or "isotachs."

Stage 2: The submerged lines of force have been drawn out in longitude and wrapped around the sun by the differential rotation, with a consequent amplification of field strength that depends on latitude.

Stage 3: Bipolar magnetic regions (BMR's) are formed where buoyant flux loops of the submerged toroidal field are brought to the surface. The BMR's continue to expand, and the flux loops rise higher into the corona.

(Courtesy, H. W. Babcock, *Ap. J.* **133,** 572, 1961, University of Chicago Press.)

latitude $\phi = 55°$, but since the angular velocity of solar rotation varies as $\sin^2 \phi$, the winding becomes tighter in middle latitudes than it is near the equator. Hence, instabilities set in first at medium latitudes. Babcock chooses as the epoch of this initial stage a time midway between sunspot maxima.

Since the lines of force are anchored in the ionized gas, a cylindrical volume element of plasma whose axis lies in the meridian plane will be distorted by the differential rotation in the sense of an increase in length and a decrease of cross-section. If the cross-section is decreased to a third of its initial value, the field intensity will increase three times and the magnetic energy $H^2/8\pi$ nine times. Instability occurs first at about 30° latitude and as the magnetic amplification continues, instabilities occur closer and closer to the equator. Babcock estimates that the poloidal field is amplified about 50 times before the flux is pushed through the surface and BMR's appear.

The appearance of BMR's and spots will have something of a sporadic character because the underlying toroidal flux strands will be distorted because of turbulence. Further, they may be twisted into "ropes" because of larger forward velocities of shallower layers. These flux ropes have a vortex motion that passes along the sub-photospheric regions to the point where the BMR's pass through the surface. Here they produce the cyclonic-type motion first observed by Hale in Hα spectroheliograms. Babcock's analysis indicates that during the eleven years of a solar cycle, a flux rope length of the order of 700,000,000 km may be developed. This suffices to produce the required number (2000 or 3000) of BMR's per cycle.

When the flux loops penetrate the surface, two regions of opposite polarity are produced, thus forming a BMR (Fig. 9–18). As the loops pass upward into coronal regions, extremely complicated developments may occur. As a BMR interacts with the main dipolar field, lines of force may be cut and reunited with other lines of force in such a way that detached loops that can extend to great distances may be formed. Such loops may occur not only near the equator but also near the pole. Some of the magnetic flux neutralizes the original poloidal field and replaces it with one of opposite polarity, but about 99 per cent is left over to form detached blobs that can serve as magnetic "bottles" entrapping charged particles with a substantial energy range. Alternately, much of the magnetic flux may be destroyed to produce joule heating — perhaps as much as 10^{36} ergs per solar cycle.

At the end of the sunspot cycle, the BMR fields will have disappeared into the corona and a new poloidal field of reversed polarity will have been formed. The sequence of events is then repeated.

Babcock's model represents an attempt to understand basic solar magnetic phenomena within the framework of a unified picture. It gives a semi-quantitative interpretation of many of the characteristic features of sunspots and the solar cycle, but a detailed working out of its implications will require some fundamental progress in basic magnetohydrodynamics. Magnetic energy is derived at the expense of the so-called equatorial acceleration of the sun. Hence some mechanism must be found for maintaining the non-uniform solar rotations; otherwise it would be wiped out in a few thousand years.

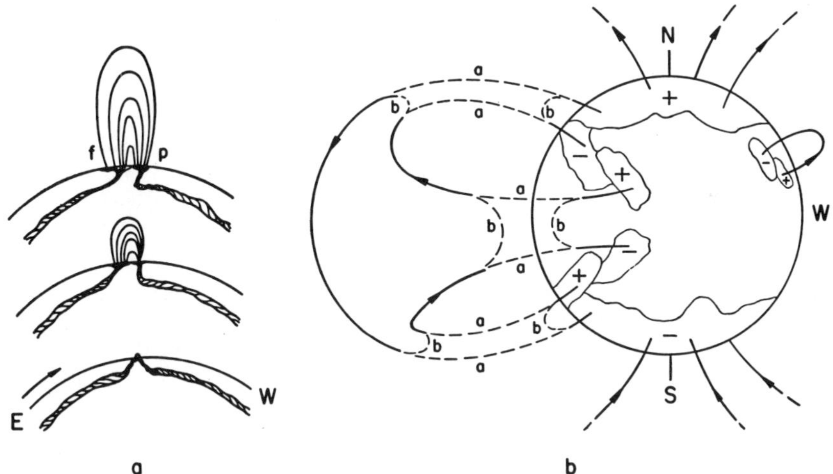

FIG. 9–18. LATER STAGES IN THE DEVELOPMENT OF BIPOLAR FIELDS AND THE REGENERATION OF A NEW DIPOLAR FIELD

(a) Formation of a BMR occurs when a constriction in a submerged flux rope is brought to the surface by magnetic buoyancy. *Lower:* The region is compact and symmetric; it may quickly produce spots and other forms of "activity" in both *p* and *f* parts. *Middle:* Some days later, all the magnetic flux arches into the atmosphere, and on the surface the lines of force have begun to spread out. Differential rotation has advanced the BMR with respect to the submerged flux rope, so that the *p* part is more compact and has a higher field strength; spots are more likely to be found within it. *Upper:* The *p* and *f* parts of the BMR continue to spread, and, as the field intensity diminishes, flocculi and other evidences of activity gradually disappear. Flow of plasma along the flux rope into the expanding BMR has resulted in new constrictions which may produce additional BMR's in the near vicinity.

(b) The expanding lines of force above older BMR's move out to approach the lines of force of the main dipole field. Severing and reconnection gradually occur, so that parts *a* are replaced by parts *b* and a portion of the main field is neutralized. Also a large flux loop of low intensity is liberated in the corona. Continuation of the process results in the formation of a new main dipolar field of reversed polarity.

(Courtesy, H. W. Babcock, *Ap. J.* **133,** 572, 1961. University of Chicago Press.)

There still remain many problems. The coolness of a spot and its sharp boundary present difficulties in any theory. One would expect radiation to leak in from the side and produce a fuzzy-edged spot in contradiction to observation. How does a spot maintain itself in mechanical equilibrium for the required lifetimes?

Plages present another set of interesting problems. They may not be in radiative equilibrium but may in fact be areas to which energy may be supplied by mechanisms such as the dissipation of magnetohydrodynamic waves. Some theories of chromospheric and coronal heating attribute much of their energy supply to this cause, which would be particularly effective in magnetized areas (see particularly Osterbrock, 1961).

Plage areas seem to be associated with domains of enhanced emission in the overlying chromosphere and corona. The best evidence is provided by radio-frequency radiation in the centimeter range. Superposed on the normal thermal radiation from the quiet sun at wavelengths shorter than 10 cm is often a slowly varying component with a period equal to the synodic period of the sun. At the November 23, 1946, eclipse, Covington noted that when a sunspot area was occulted by the moon the heretofore enhanced 10.7 cm radiation was cut down. Later work showed that the positions and shapes of the r-f sources fitted the faculae and in fact persisted with them after spots had disappeared. At 3 cm the r-f radiation emerges from facular regions and is circularly polarized because of the local plage magnetic field. At 20 cm (see, e.g., Christiansen and Matthewson, 1959) the radiation comes from a height of about 40,000 km, i.e., well in the corona, is not polarized, but still agrees well with the facular areas. The equivalent temperature, about a million degrees, suggests a coronal region of normal temperature but perhaps enhanced density.

9–11. The Nature of the Solar Chromosphere. Above the photosphere of the sun lies a thin envelope of relatively transparent gases which is called the *chromosphere*. On superficial examination it appears to be separated from the photosphere and from the tenuous corona as well (see Figs. 9–19 and 9–20). We may define the position of the base of the chromosphere as follows: Consider a ray of radiation in the continuum at $\lambda 5000$ drawn tangent to the edge of the sun, such that the total optical depth along the ray is 1.0 (see Fig. 9–21). The point where this ray intersects a radius drawn from the sun at an angle of 90° defines the bottom of the solar chromosphere. The radial optical depth, i.e., the optical depth of point C as seen along the line CD, is about 0.003.

For convenience in discussion the chromosphere is sometimes divided into three layers, the *low chromosphere*, about 4000 km, where hydrogen is sometimes becoming ionized; the *middle chromosphere*, from about 5000 to 8000 km; and the *upper chromosphere*, about 8000 to 15,000 km. Different writers use different conventions.

Under conditions of good seeing the chromosphere exhibits a distinctly "spiky" structure, indicating that above about 4000 km the fundamental structure is basically many individual prominences often called *spicules*. Athay, Billings, Evans, and Roberts observed the green coronal emission line (produced at a gas kinetic temperature of about a million degrees) at heights as low as 8000 km, i.e., below the tops of many of the spicules.

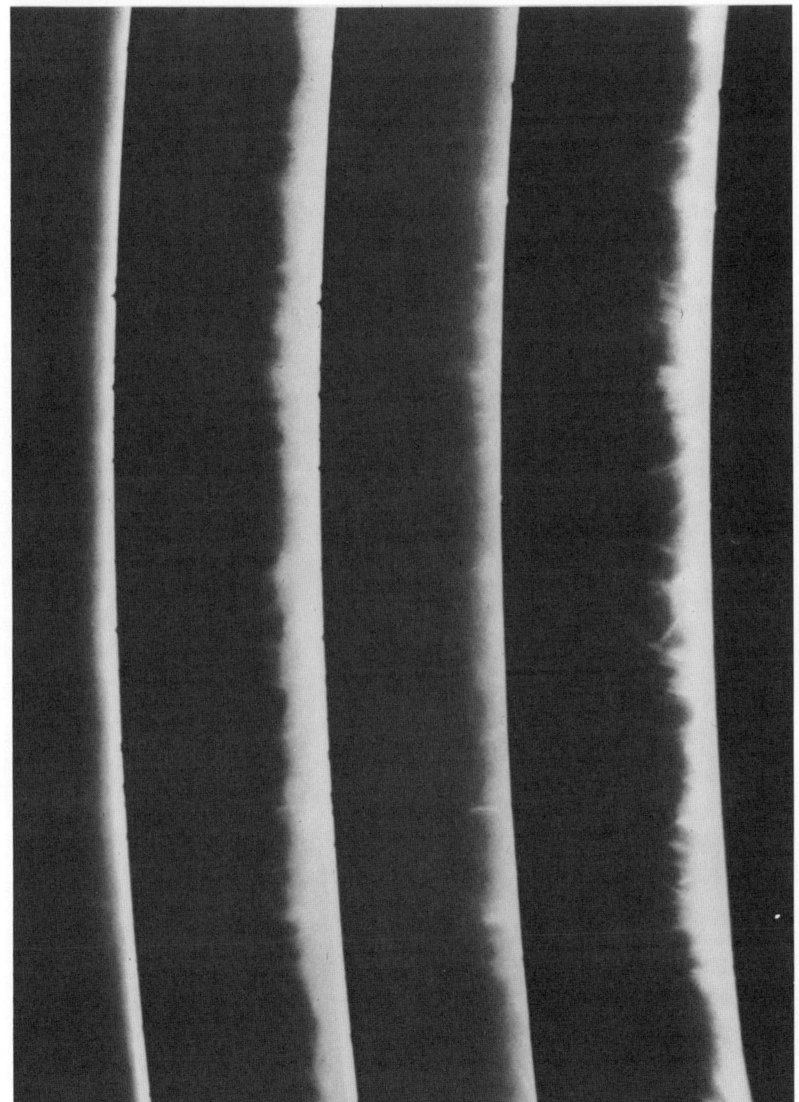

FIG. 9-19. SPICULES AT THE LIMB OF THE SUN

Photographed in Hα August 31, 1958 by R. B. Dunn.
The top strip and third from top strip are photographed away from the center of the line. (Courtesy, Sacramento Peak Observatory, Geophysics Research Directorate, AFCRC.)

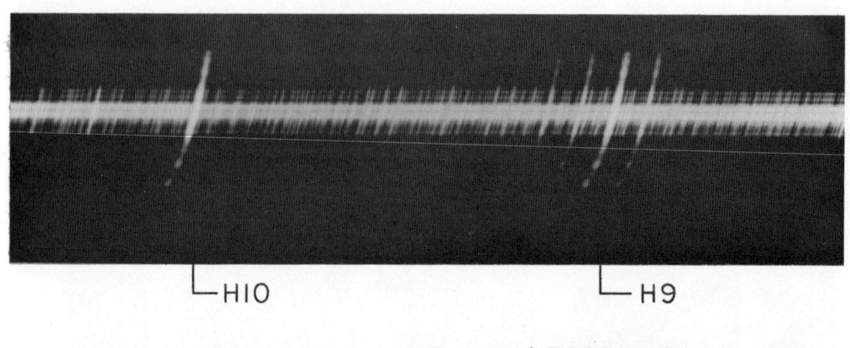

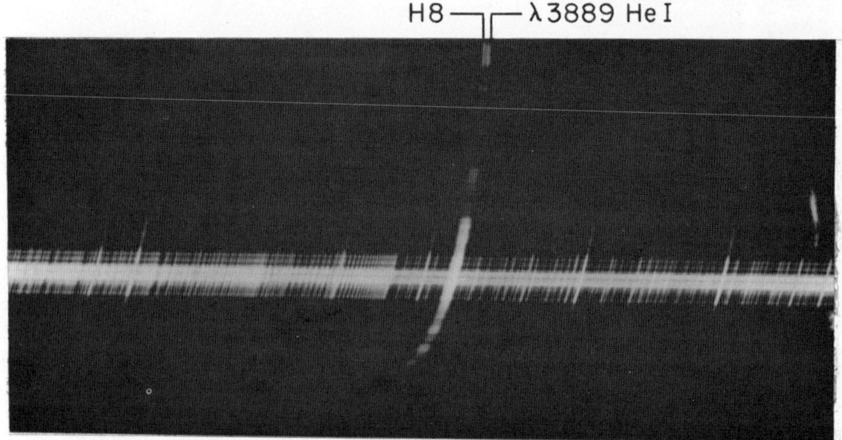

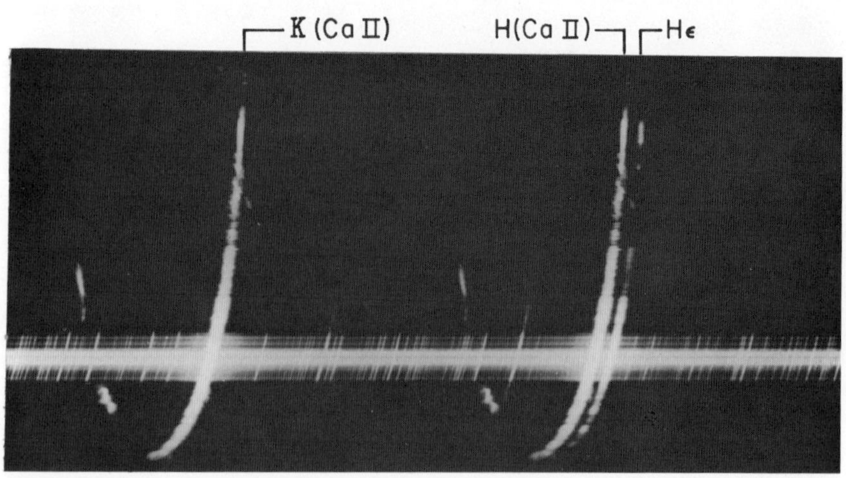

FIG. 9–20. THE FLASH SPECTRUM OF THE SUN, AS OBSERVED AT THE TOTAL ECLIPSE OF OCTOBER 12, 1958

The jagged edge of the moon produces horizontal streaks in the spectrum. The

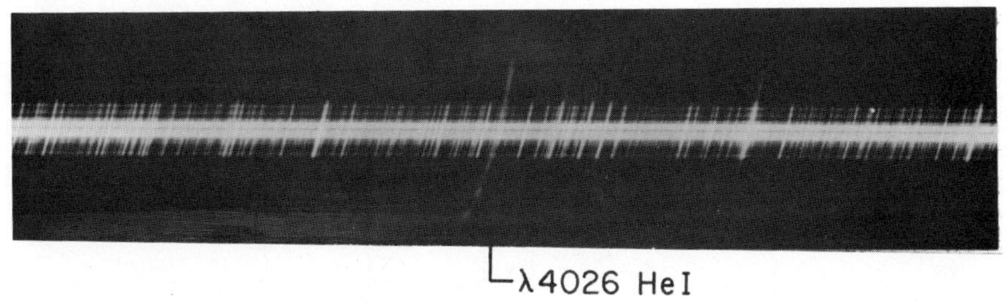

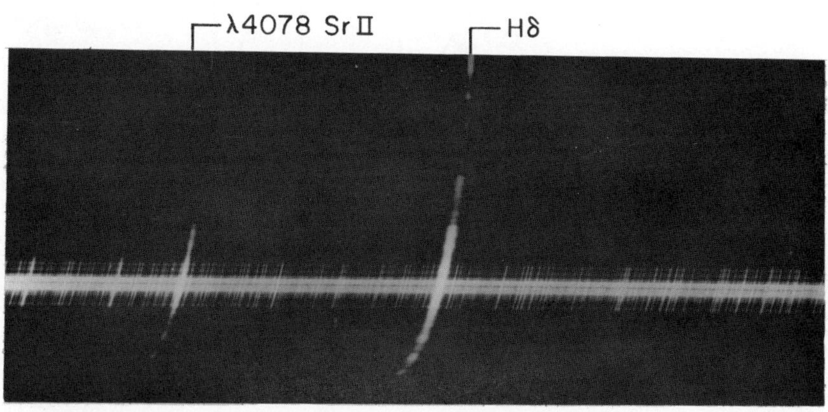

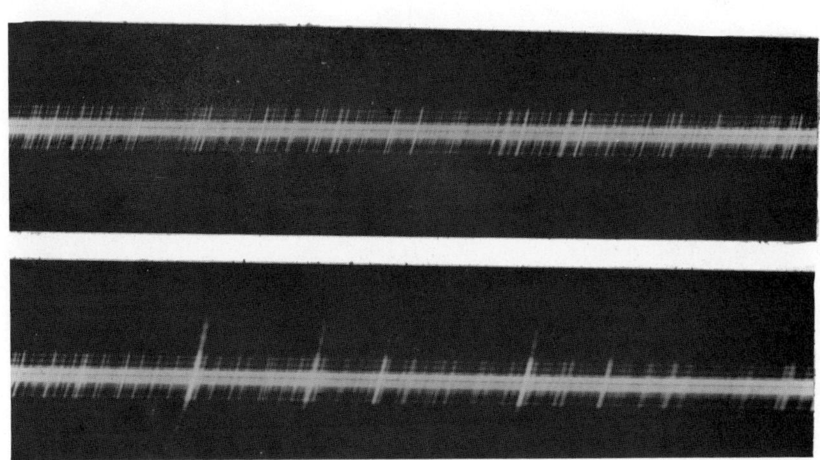

continuum appears at the lowest level. Notice the long arcs produced by the hydrogen lines Hδ, Hε, λ3889 H8, λ3835 H9, λ3797 H10, and helium λ4026, λ3889. Notice the CN band just to the left of H8. (Courtesy Z. Suemoto, University of Tokyo Observatory.)

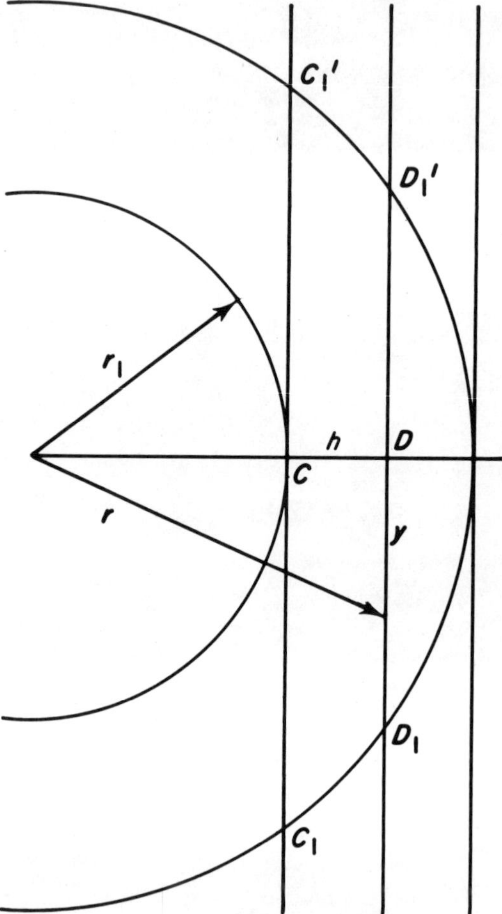

Fig. 9–21. Definitions of Some Quantities Pertaining to the Chromosphere

The plane of the paper contains the line of sight and the solar radius. At an instant of time t, the ray D_1DD_1' directed toward the observer is tangent to the edge of the moon; r is the radius vector drawn to an arbitrary radiating volume along the line of sight, and r_1 is the radius of the sun. C lies at the bottom of the chromosphere. The optical depth in the continuum along the line C_1CC_1' is 1.0.

This observation and other evidence suggests that the spicules are cool elements in a hot corona. Thus the chromosphere is placed between the rather opaque photosphere with its temperature of about 6000°K and the optically thin corona with its gas kinetic temperature of a million degrees. It is connected more closely with the photosphere, perhaps, than with the corona. Nevertheless, an important fact brought out by eclipse observations is that there exists a demarcation line between the chromosphere and the photosphere. The chromospheric density gradient at levels, say 500 km above the well-defined limb of the sun, is much less steep than the density gradient in the photospheric layers. The observed sharpness of the limb of the sun is in agreement with the hypothesis that the negative hydrogen ion is responsible for the opacity and that the photospheric layers are in hydrostatic equilibrium. Were the chromospheric density gradient to persist into the photosphere, the edge of the solar disk would be noticeably fuzzy. (See Fig. 9–19.)

The physical and mechanical state of the chromosphere is enormously complicated. The lower regions are strongly influenced by the radiation field of the photosphere, whereas the upper regions are affected by radiation and thermal conduction from the corona. Secondly, the non-uniform structure of the chromosphere as evidenced by the spicules means that sharp temperature and finally also density gradients (lateral as well as vertical) exist. As indicated in Art. 9–15 this inhomogeneity complicates the interpretation of the observations. Thirdly, the chromosphere deviates strongly from thermodynamic equilibrium so that it is not possible to specify a unique temperature that would characterize all physical processes occurring in these layers (see, e.g., Athay and House, 1962). Instead, one may define a "temperature parameter" of the chromosphere in terms of any one of the following processes:

1. The distribution of electron velocities
   a. as measured from the Balmer continuum (see Eq. 4–159)
   b. as measured from the profiles of emission lines (Eq. 3–90). The application of this method assumes $T_e = T_i$
   c. as estimated from the intensity of radio waves.
2. Distribution of atoms among excited levels (Boltzmann formula).
3. Ionization equilibrium (Saha equation).

The "temperature" of the chromospheric radiation field varies from line to line in a complicated way and one must in principle solve the transfer equations taking into account the interlocking between all levels and coupling with the continuum in an inhomogeneous atmosphere. Furthermore, energy is probably supplied to the upper corona and chromosphere by the dissipation of the energy of acoustical and magnetohydrodynamical waves. The problem has to be attacked by successive approximations with increasingly difficult mathematical complications (Thomas and Athay, 1961; Osterbrock, 1961).

9–12. Techniques for Observations of the Solar Chromosphere. Observations of the solar chromosphere involve two types of investigation. In one we are concerned with the structures and motions of individual elements, e.g., spicules observed in monochromatic radiation such as Hα; in the other we are concerned with average or integrated properties of chromospheric domains large compared with individual elements, e.g., observations of the flash spectrum or r-f measurements.

(a) *Coronagraph and Spectroscopic Method.* The fine (spicular) structure of the chromosphere can be studied outside of eclipse with the coronagraph (B. Lyot and subsequently R. B. Dunn and others). Likewise spectroscopic studies of the strongest lines can be carried out at the limb of the sun by occulting the edge of the solar disk, but a large amount of scattered light always remains.

(b) *Eclipse Method.* The best observations of the solar chromosphere so far secured (1962) are those obtained at the time of total solar eclipse. Weak as well as strong lines can be observed, and we can determine the distances to which the various radiations extend from the sun.

Many observers have employed essentially a slitless spectrograph. Just before totality, the dark-line spectrum of the photosphere vanishes, and the bright lines of the chromospheric spectrum appear. At this moment a spectrogram of the sun is taken; the narrow crescent of the sun serves as a slit and we obtain, in effect, a photograph of the chromospheric arc in each of the radiations it emits. (See Fig. 9–20.) Menzel developed a jumping-film method to secure a series of photographs at different heights above the limb. Photometric measurements of the intensities of chromospheric areas at different points enable the observer to derive density or rather emissivity gradients in the solar atmosphere. Consider the emission at the center of an arc of a flash spectrum image. The radiation received corresponds to that which is emitted along the line of sight at all points above the height h (see Fig. 9–21), i.e.,

$$I \sim \int_{x=h}^{\infty} \int_{y=-\infty}^{+\infty} S_\nu(x, y)\, e^{-\tau}\, d\tau\, dy \tag{9–22}$$

Normally I is normalized to the emission emitted in a slab 1 cm thick extending to the top of the chromosphere. If a specially symmetrical chromosphere is assumed, $S(h)$ can be determined for each observed line. Earlier observers, e.g., W. W. Campbell, S. A. Mitchell, did not calibrate their plates photometrically, but in spite of this limitation much information was obtained.

From both the observational and theoretical points of view, the fundamental paper on the chromosphere was Menzel's discussion of the eclipse observations of Campbell. He calibrated relative line intensities in terms of line strengths given by atomic theory and determined the chromospheric density distribution from the height intensity relation derived from moving-plate exposures. Menzel showed that Fraunhofer lines were formed mostly below the chromosphere in layers that produce both line and continuous absorption. From the base of the chromosphere to the bottom of the layers responsible for the Fraunhofer lines, the density increases by a factor of the order of 400. Hence the conventional use of the term "reversing layer" does not appear to be justified.

From an analysis of Mitchell's data, taking into account atmospheric scintillation and the fact that during each exposure the moon moved a finite distance, Wildt tried to get density gradients of H, He, and the metals, the electron densities and pressure of hydrogen in the solar atmosphere. Observations secured at more recent eclipses, e.g., Khartoum, 1952, have permitted a wealth of quantitative studies on density gradients and excitation conditions in the chromosphere.

(c) *Radio Method.* Radio-frequency observations of the quiet sun assist in determining the rise of temperature with optical depth in the chromosphere although they cannot fix the scale of heights. The angular resolution, of course, falls far short of separating hot and cold elements. See Fig. 9–22.

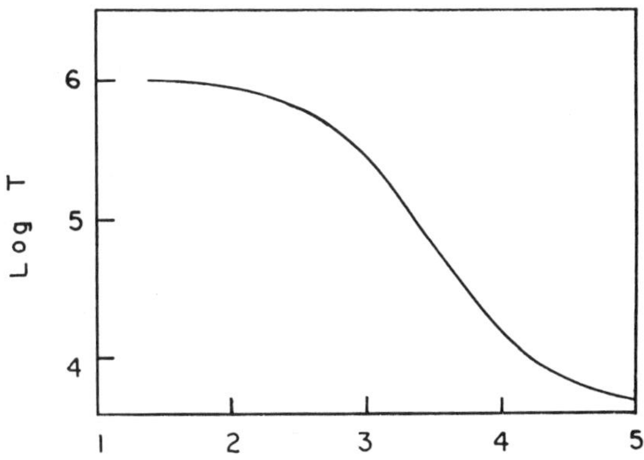

FIG. 9–22. RELATION BETWEEN RADIO-FREQUENCY RADIATION FROM THE QUIET SUN AND THE LOGARITHM OF THE FREQUENCY (IN MEGACYCLES)

The radiation of the quiet sun is expressed in terms of the logarithm of the apparent temperature. (After C. W. Allen.)

The maximum depth from which radiation of a given frequency ν emerges depends on N_ε through the plasma frequency relation Eq. 9–1. The intensity may be expressed in terms of a radiation temperature since the frequency falls in the Rayleigh-Jeans region (see Chapter 4), where $B_\nu(T)$ is proportional to T. Hence

$$T_r = \text{const} \int T_\varepsilon(h)\, e^{-\tau}\, d\tau \tag{9–23}$$

where the element of optical depth in r-f radiation (for $1 < \lambda < 10$ cm) is given roughly by

$$d\tau_\nu \cong 0.09\, \nu^{-2}\, N_\varepsilon^2\, T_\varepsilon^{-3/2}\, dh \tag{9–24}$$

Note that T_r approximately equals T_ε when $\tau \sim 1$. Since $N_\varepsilon(h)$ is known from an analysis of eclipse data, we use the observed value of $T_r(\nu)$ (see, e.g., Allen, 1956) to obtain $T_\varepsilon(h)$ or more precisely $T_\varepsilon(N_\varepsilon)$, for a chromospheric model in which T_ε and N_ε depend mostly on height. If the chromosphere consists of both hot and cold elements, radio data will supply boundary conditions for any assumed model.

(d) *Spectroheliograms.* Spectroheliograms secured in the centers of strong or even moderately strong Fraunhofer lines are monochromatic photographs of the sun in chromospheric radiation. By choosing different points in the line profile one can look to different depths in the solar chromosphere.

(e) *Ultraviolet Studies.* Ultraviolet radiation from the chromosphere may be studied with the aid of rockets, artificial satellites, and effects on the earth's ionosphere (Chapter 10).[9] Shortward of about 1500A, the dark-line Fraunhofer spectrum weakens rapidly and the solar spectrum exhibits a number of emission lines, the strongest of which are those of hydrogen and abundant light elements. In the range from 100A — 1500A most of these emissions are of chromospheric origin, and many of them must be produced in the chromosphere-corona transition zone. Although the required observational techniques are exceptionally difficult since spectrographs must be flown in rockets or placed in satellites with appropriate provision for getting data back to earth, much valuable information has been obtained. Ultraviolet spectral scans obtained by the orbiting solar observatory (OSO) have revealed a wealth of lines in the soft x-ray region (Lindsay, 1963). Many of the lines have not yet been identified (June, 1963).

9–13. Spicules. All workers on the complex problems of the chromosphere agree that this layer cannot be treated as a homogeneous medium stratified in plane parallel layers. Many data have been accumulated in support of Giovanelli's suggestion that it is made up of alternate hot and cool structures with differing dependences of N_ε and T_ε on height, h. With which observable features are those postulated hot and cold columns to be identified? There is much to suggest that the essential structures are the spicules, low, chromospheric spikelike structures with average heights of the order of 7500 km (equator) and about 9000 km (pole) but with a considerable spread in height. See Fig. 9–23. The diameters are of the order of 500–600 km.

Normally, spicules are observed at the solar limb, but de Jager (1959) identifies as spicules the fine, usually dark, mottles observed to cover about 30 per cent of the solar disk, in Hα spectroheliograms. They occur only at altitudes greater than about 2000 km. Giovanelli and Jefferies finds that these dark mottles, seen with a resolution of 5 seconds of arc are all descending. Dark mottles seen on Hα spectroheliograms with diameters of 2000–5000 km and mean lives of about 10 minutes have been identified by Bruzek (1959) as spicules. Waldmeier (1960) has called attention to structures that are larger than spicules but smaller than surges and which

[9] Ionization in the E and F layers appears to be produced mostly by radiation beyond the Lyman limit, λ912, possibly mostly x-rays, that in the D layer appears to be influenced by Lyman alpha and also x-rays. The value of the ionospheric electron densities are known from radio propagation data. If we knew the recombination rates we could assess the incident radiation required to ionize gases such as N_2 and O_2, thus using the ionosphere as a tool for chromospheric radiation.

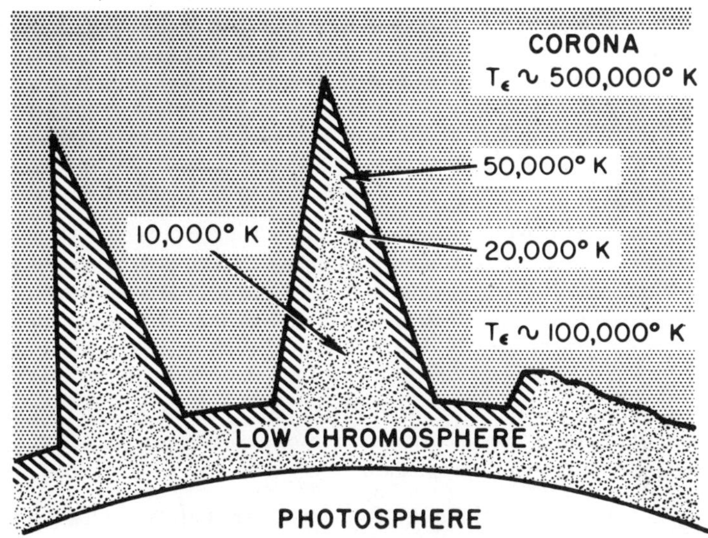

Fig. 9–23. Schematic Representation of Spicules

We have adopted the point of view that the spicules are cooler elements in a hot corona. In attempting to reconcile the complexities pointed out by Athay, Michard, Zirker, and others, we have postulated a temperature gradient between the spicule and the corona and a rise in temperature from the bottom to the top of the spicule.

appear in active regions as microflares. Athay and Thomas (1957) suggested, spicules may be masses that are ejected from the sun, travel along parabolic paths and then fall back. Their "lifetimes" expressed as the length of time a given spicule exceeds its mean height are of the order of 4 or 5 minutes.

Spicules appear to line up in the direction of the photospheric magnetic fields as indicated by the polar rays. They may show deviations as great as 30° from the normal to the solar surface as, for example, near coronal arches (Rush and Roberts, 1954). Presumably these arches define the position of the coronal magnetic fields. Hence if the spicules represent outward moving disturbances in the chromosphere they would appear to be closely affected by the magnetic field (Osterbrock, 1961).

The numbers of spicules are difficult to establish. On plates of good definition the numbers of long ones may be easily counted, but statistics for shorter ones are complicated by overlap, tilts from the normal, etc. Studies have been carried out e.g., by Lippincott, who found that for $h > 9000$ km the numbers fell off exponentially. They decreased also for smaller heights and she suggested that overlaps produced this effect. An extrapolation of the exponential law to the solar surface then gave a total number of spicules comparable with the number of granules. Athay discussed the statistics of Hα spicules observed at the solar limb for h in excess of 3000 km. At $h = 3000$ km he found about 9300 spicules on the sun covering about 0.6 per cent of the surface and concludes that the total number of spicules is much smaller than the quantity found by Lippincott.

There seems to be some evidence to suggest that spicules represent cold elements protruding into the hot corona. At the bottom of the chromosphere (or top of the photosphere) the temperature is about 4000°K and must rise towards the corona. In the spicules and similar structures the rise in temperature is relatively slow, but temperatures in excess of 20,000°K must be reached at heights greater than 5000 km.[10] At heights of about 4000 km, all line emission from neutral and singly-ionized atoms must be confined to spicules. The fact that strong coronal line emission is seen between spicules at heights below 8000 km indicates that the temperature transition from photosphere to corona must be rapid. Pronounced temperature differences between hot and cold may persist to heights as low as 1000 km (see Athay and Menzel, 1956; Athay and Roberts, 1955).

Although metallic lines clearly originate in cool elements and the coronal lines in hot ones, the exact regions of origin of hydrogen lines are not known. Possibly they come from all parts of the spicules.

Many spicule observations appear self-contradictory. Observations of these transient, delicate fine structures are severely hampered by bad seeing. It is to be hoped that our information will soon be on a more secure footing with observations obtained from stabilized balloons and (particularly for spectroscopic studies) from artificial satellites.

9–14. Significance of Spectroheliograms. Monochromatic pictures of the chromosphere in radiation of particular elements have been obtained routinely for more than sixty years. Those secured in hydrogen and Ca II radiations have supplied data of great empirical value, e.g., it has been shown that plage areas are more useful in delineating zones of solar activity than are sunspots themselves. In spite of their obvious value, spectroheliograms have resisted a quantitative interpretation in terms of an exact physical theory, although attempts have been made to give them a theoretical explanation (see, e.g., Reichel, 1954).

The appearance of spectroheliograms depends on a number of factors.

(a) For a strong line such as Ca II "K" it depends on the part of the line used, i.e., you look to different depths depending on whether you use K_1, K_2, or K_3. See Fig. 9–14.

(b) Except for the wings of strong lines, it depends critically on $\Delta\lambda$, the

[10] From a discussion of the profile of chromospheric Hα, Michard (1959) derives a gas kinetic temperature of 50,000°K for spicules (which he regards as hot elements. From theoretical considerations on the stability of the radiation field, Athay and Thomas derived electron temperatures between 24,000°K and 44,000°K. From a study of Hα and D3 in spicules, Athay (1960a) obtained a gas kinetic temperature of 50,000°K, although he later (1961) suggests that sharp temperature gradients exist between the spicule axis and the surrounding medium. On the other hand, Zirker (1962) concludes that a temperature below 20,000°K and an electron density of 10^{11} cm^{-3} would produce the observed emission and contrast, while Suemoto calls attention to the importance of self-reversal effects as far as Hα is concerned. See Fig. 9–23.

resolution of the spectroheliograph or the filter band-pass. A small $\Delta\lambda$ is required for lines of moderate intensity.

(c) Line of sight motions are particularly important, especially on the edges of lines with bell-shaped profiles. (For magnetic effects, see Leighton, 1959.)

(d) Even if lines are formed in the same atmospheric regions, the pictures are often not comparable, e.g., in ionized calcium and in hydrogen even for parts of the line corresponding to the same optical depth. If we compare Ca II K_3 and Hα, mottles that are bright in K_3 are sometimes inconspicuous or even dark in Hα. The reason for this disparate behavior is probably to be sought in the circumstance that the central intensities of Ca II "K" and Hα tend to vary in opposite directions with an increase in temperature near 6000°K. The effects of optical thickness must also be considered.

(e) Finally, we must note that we do not even have a theory of line formation that is adequate for the centers of lines. For example, the core structure of the ionized calcium "K" line with its emission and central reversal has been interpreted in terms of a model with a non-monotonic dependence of temperature on height (e.g., Goldberg and Müller) or in terms of non-coherent scattering (Miyamoto). Deviations from thermodynamic equilibrium plague all considerations pertaining to the chromosphere. Hence spectroheliograms in cores of strong lines present us with data whose exact physical interpretation is not yet possible.

It is not surprising that spectroheliograms corresponding to low levels in the chromosphere yield structures that differ considerably from those pertaining to high levels. In the centers of normal metal lines or in K_2 one observes coarse mottles about 1500–4000 miles across and about 8000 or 10,000 miles apart. These, in turn, are often arranged in a coarse large-scale structure with characteristic dimensions of the order of 30,000 miles.

Spectroheliograms made in the light of Ca I λ4227, Na "D", Fe I, and other lines show many of the same features as the Ca II images, although the plages are of relatively lower intensity. Spectroheliograms taken in weak lines resemble those of hydrogen in that dark grains appear on a bright field but no vortical structure is evident.

Of great interest is the appearance of the disk in the light of the infrared helium line λ10,830 which arises from the metastable 2^3S level. A comparison of the He I and Hα photographs taken on the same date shows that the filaments are visible in absorption on both images, but that the plages appear dark in the helium image instead of bright as on the hydrogen images.

The highest structures are those observed in the radiation of Lyman α, where Tousey finds the plage areas to be much more extensive and brilliant, although coarser, than in the K line. (See Fig. 9–24.) Helium lines correspond to very high excitation and eventually we may hope that spectroheliograms secured in λ584 (He I) and λ303.8 (He I) may tell us much more about the "hot spots" of

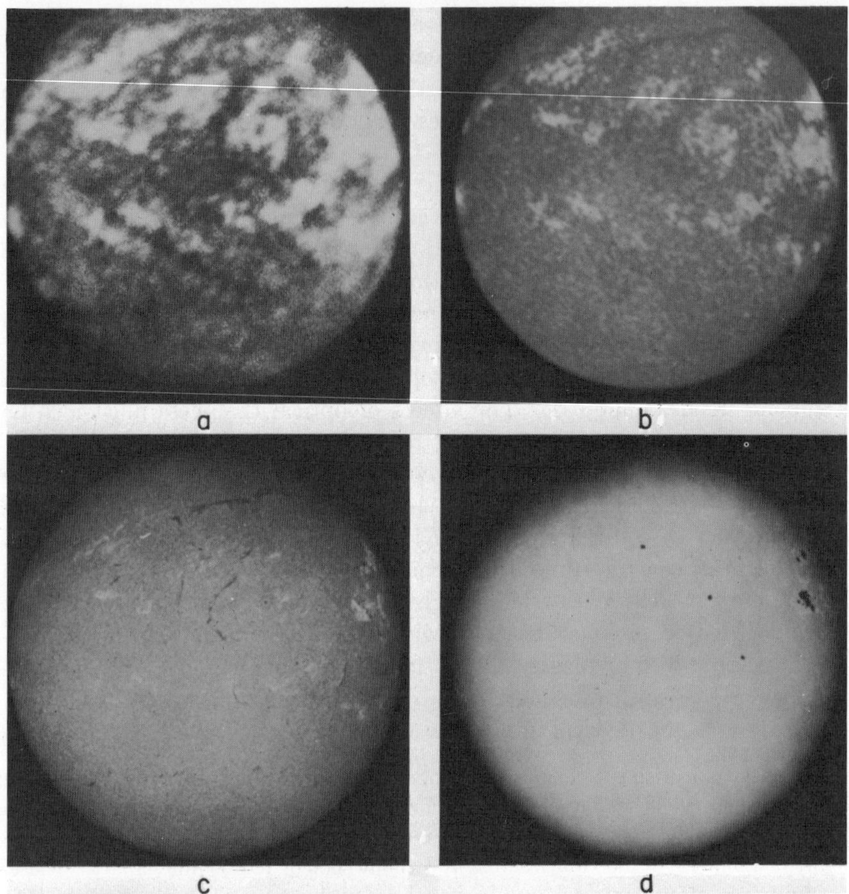

FIG. 9–24. COMPARISON OF LYMAN α, CALCIUM AND H ALPHA SPECTRO-HELIOGRAM AND WHITE LIGHT PHOTOGRAPHS

Images of the sun on March 13, 1959.

(a) Lyman-alpha of hydrogen as photographed from an Aerobee-H_1 Rocket (b) Calcium $K_{2,3,2}$ spectroheliogram (McMath-Hulbert Observatory) (c) H-alpha with 0.7A monochromatic filter (Naval Research Laboratory) (d) white light photographs (U. S. Naval Observatory. Courtesy, R. Tousey, U. S. Naval Research Laboratory, Official United States Navy Photograph.)

unusual excitation on the solar surface, and about solar activity. The resonance Mg II near λ2800 should show structures similar to those observed in Ca II "*K*." Present techniques do not yet permit us to obtain high resolution spectroheliograms of these radiations, but nevertheless it should be possible to establish correlations of highly excited chromospheric and plage areas by means of data obtained from rockets and satellites.

9–15. The Electron Density and Temperature of the Chromosphere. A determination of the electron density and temperature distribution in the chromosphere is extremely involved because of severe departures from local thermodynamical equilibrium so that the ionization temperature differs from the gas kinetic and these in turn differ from the excitation temperature, which, in fact, varies from one level to another.

The following data are available for N_ε, T_ε determinations:

1. Radio-frequency intensities from the quiet sun give T and then $T_\varepsilon(h)$, but can produce only checks on two-stream models derived from optical data. Radio data give a useful "mean" model.

2. Continuous emission in the low chromosphere is not negligible. Pagel finds that as far as the continuum is concerned, deviations from local thermodynamic equilibrium (LTE) are small right up to the very top of the photosphere. This continuous emission in the low chromosphere must be taken into account in analyzing limb darkening at the limit of the solar disk.

At the limit of the Balmer series the emission per unit volume in the frequency interval ν to $\nu + d\nu$ due to recombination of protons and electrons in the second level is

$$E\Delta\nu = 2.37 \times 10^{-33} N_i N_\varepsilon T_\varepsilon^{-3/2} \Delta\nu \tag{9–25}$$

(see, e.g., Aller, 1956, p. 147). The underlying continuum arises from the negative hydrogen ion, recombination on the third and higher levels and electron scattering. In the visual region, the continuum is produced partly by electron scattering but largely by other factors (see Thomas and Athay, 1960). Observations of the continuous radiation at times of eclipse are difficult unless a slit spectrograph is used.

3. In the very low chromosphere (below 600 km) molecular bands are useful for evaluating the excitation temperature. Bands of CO, CN, and C_2 have yielded temperatures between 4300 and 4600°K which are to be compared with Pagel's result (1960) of 3900°K $< T <$ 4700°K in optical depths between 0.01 and 0.04. Evidently, the top of the photosphere and low chromosphere represent a temperature minimum on either side of which the temperatures rise in photospheric layers or in corona.

4. The excitation of the ionosphere, properly interpreted with a non-LTE chromosphere, could put limits on models of the chromosphere.

5. The hydrogen lines extend through the entire chromosphere and may be used to measure N_ε and T_ε over a vast range. Some difficulty may be produced by the circumstance that because of its great abundance, hydrogen can contribute to emission in both hotter and cooler elements.

(a) The lower members of the Balmer series are observed to the greatest heights but are affected by self-absorption as R. N. Thomas pointed out long ago. The amount of self-reversal depends on N_2 but the b_n factor (which measures deviations from LTE) varies with n. From absolute calibration of the line profiles, one can obtain $N(n = 2)$ and ξ_T, the turbulent velocity. In LTE the emission at any point in a line profile would be $B(T)(1 - \exp(-\tau_\nu))$. Here N_2H is the number of atoms in the line of sight, α_ν is the line absorption coefficient (Chapter 7), and $\tau_\nu = \mathrm{N_2H}\alpha_\nu$. Pottasch and Thomas point out that the high opacity in the Lyman continuum does not suffice to insure LTE.

(b) For the higher members of the Balmer series (see Athay, Billings, Evans, and Roberts, 1954), b_n approaches 1, and N_ε may be computed with the aid of

$$E_{n2} = 1.78 \times 10^{-17}\, b_n\, N_i N_\varepsilon T_\varepsilon^{-3/2}\, n^{-3} \exp\left(\frac{157{,}000}{n^2 T_\varepsilon}\right) \qquad (9\text{–}26)$$

(see, e.g., Aller, 1956). Here E_{n2} is the emission per unit volume. Alternately, one may combine photometry of Hα with $N_\varepsilon T_\varepsilon^{-3/2}$ obtained from higher Balmer lines to get N_2. Further one may estimate electron density in the lower chromosphere from the principal quantum number of the last resolvable Balmer line and the Kolb-Griem theory (Eq. 7–97).

6. In the lower chromosphere one may determine excitation temperatures from relative populations of excited levels. The method may be used not only for metallic lines but also for those of helium where self-absorption effects are negligible. For example, Goldberg found helium excitation temperatures of 4300°K in the lower chromosphere; 6700°K in layers above 2300 km.

7. Ionization equilibria of metals have been employed to obtain ionization temperatures in the middle chromosphere. For example, we may use n(Fe II)/n(Fe I) in conjunction with the known Fe/H abundance ratio to solve for both N_ε and T_ε.

8. Helium lines are important for studies of the middle and high chromosphere. In neutral helium one must consider collisional as well as radiative excitations of lines. Triplet levels define an excitation temperature different from that of the singlet levels. Also, because of the metastability of the 2^3S level and electron collisions, the triplet levels are overpopulated with respect to the singlet levels (see, e.g., Athay and Menzel, 1956). Athay and Johnson conclude (1960) that helium excitation cannot be produced by back radiation from the corona but that collisional excitation at high electron temperatures is involved, $N_\varepsilon = 10^{11}$ and $T = 40{,}000$–$50{,}000$°K. For λ10,830 λ5876 to appear as emission lines against the disk (bright plages) electron densities in excess of 10^{12} cm^{-3} are needed. Lines of ionized helium are certainly produced only in regions where the temperature is very high, i.e., 25,000–40,000°K (see, e.g., Zirker, 1959; Thomas and Zirker, 1961).

9. For an atmosphere in hydrostatic equilibrium one may use the observed density distribution to get the temperature. Consider, first, the simplest situation in which all constituents of the atmosphere are evenly mixed and temperature is constant with height. The density, ρ, is then

given as a function of height, h, by direct integration of the equation of hydrostatic equilibrium

$$\frac{dP}{dh} = -g\rho \tag{9–27}$$

where

$$\rho = \left(\frac{\mu}{\mathcal{R}}\frac{1}{T}\right) P = \left(\frac{M}{kT}\right) P = \text{const } P \tag{9–28}$$

Thus

$$\rho = \rho_0 \exp\left(\frac{-Mgh}{kT}\right) = \rho_0\, e^{-ah} \tag{9–29}$$

The greater the value of kT/M, the lower the gradient. Here M is the mean atomic mass. In practical cases, M is the mass of the hydrogen atom, i.e., $M = 1.66 \times 10^{-24}$ gm, so $a = 6.60 \times 10^{-8}$ at $T = 5000°$K. Using various published data on the 1952 eclipse, Pottasch (1960) tried to derive $T(h)$ to three solar radii on the hypothesis that the corona as well as the chromosphere is in hydrostatic equilibrium, which seems unlikely.

Note that in obtaining N_ε and T_ε as a function of height in the chromosphere, one encounters two essential problems: (1) The observed quantities are gradients in line emissions; hence we have to use a non-LTE theory to go from observed brightness gradients to density gradients. (See Jefferies and Thomas, 1958, 1959; also Jefferies and White, 1960.) (2) The measurements are made in an unresolved structure, some elements of which are hot and some are cold. Metallic lines tend to arise in cool filaments, helium in hot ones, and hydrogen perhaps in both. The temperature gradients in the two regions are not known.

The problem of the chromosphere involves not only the rise to spectacular coronal temperatures but also the rise to about 6000°K at 500 km above the 4500°K boundary of the photosphere. Thomas suggested that the non-radiative energy input required to produce this temperature rise may actually exceed that supplied to the corona.

9–16. Turbulence in the Chromosphere. Microscopic motions of chromospheric gases may be studied to some extent on moving pictures of spicules, but smaller structures would escape detection. Finally, there is microturbulence which may differ in hot and cold elements. Profiles of chromospheric emission lines broadened by Doppler motions (and possibly by turbulence as well) should provide a clue to the kinetic temperature. The intensity distribution in an emission line unaffected by self-reversal will be given by Eq. 3–90

$$I(\lambda) = I_a \exp\left[-c^2(\lambda - \lambda_0)^2/\alpha^2\, \lambda_0^2\right] \tag{9–30}$$

if turbulence is not present.

Hence, measurements of the profile $I(\lambda)$, will give the most probable velocity α and hence the kinetic temperature defined by

$$\tfrac{1}{2} M\alpha^2 = kT \tag{9–31}$$

If turbulence is present and these motions can be represented by a random (Gaussian) distribution along the line of sight, α^2 will be replaced by V^2 defined by Eq. 8–107:

$$V^2 = \alpha^2 + \xi^2 \tag{9–32}$$

If the turbulent velocity, ξ, much exceeds the thermal motions, equally intense lines of all elements should have nearly the same profiles. If α greatly exceeds ξ the breadth of a line ought to decrease with increasing atomic weight.

Note that the component of turbulent velocity obtained from lines observed at the solar limb is parallel to the solar surface.

Numerous attempts have been made to measure turbulence in the solar atmosphere. Thus, Treanor (1957) developed a quantitative analysis of Fabry-Perot fringes to get chromospheric line profiles as a function of height and applied this technique to measure the *D*3 profile which he found nearly independent of height from 1500 to 4250 km. Clube (1958) extended the study to both Hα and *D*3 and found $\xi \sim 16$ km sec^{-1} (in helium-emitting regions) in agreement with Treanor's results. Studies by Michard with better time-resolution indicated that the *D*3 profile widened with height.

Gas kinetic temperatures and turbulence in the chromosphere were assessed from high resolution spectra of the February, 1952 eclipse by Redman and Suemoto (1954). From line profiles corrected for self-absorption (plus Stark effect for hydrogen), they concluded that turbulent velocity increased linearly with height, from $\xi_0 = 2.5$ km sec^{-1} at the base to 16 km sec^{-1} at a height of 2600 km, and that the temperature of the lower chromosphere did not exceed 10,000°K. In another study Unno (1959b) finds $\xi = 1.9$ and 3.5 km sec^{-1} at heights of 500 and 1000 km, respectively.

On the other hand, Athay and Thomas propose a gas kinetic temperature of 40,000°K in which case a turbulent velocity much smaller than 16 km sec^{-1} is indicated. The two concepts of high gas kinetic temperature and high turbulence are mutually exclusive since they represent alternative explanations of the same phenomena.

In theory, one can get the radial component of the turbulent velocity from profiles of centers of strong lines observed on the disk, but here we must assess the relative roles of thermal processes, deviations from local thermodynamic equilibrium, and non-coherent scattering. In summary, it appears that turbulence may increase with height, but we have yet to distinguish between spicular and interspicular regions. It is not even certain that the lines are widened by turbulence rather than a high gas kinetic temperature.

9–17. Models of the Chromosphere. Attempts have been made by numerous investigators to obtain models for the chromosphere. Some models, e.g., Piddington (1950) have been based primarily on radio data and give smooth temperature increases and electron density decreases with height. Others have been based on optical data obtained from the flash spectrum (e.g., Athay and Menzel, 1956), while yet others have attempted to use both optical and radio data. Figs. 9-25 and 9-26 show plots of log T and log N_e with height for various models. The Woolley-Allen (1950) model suggests a sharp transition from the chromosphere to the corona; Oster (1957) attempted to handle the difficult problem of the transition region between corona and chromosphere.

Hot and cold element models have been required to explain the diffi-

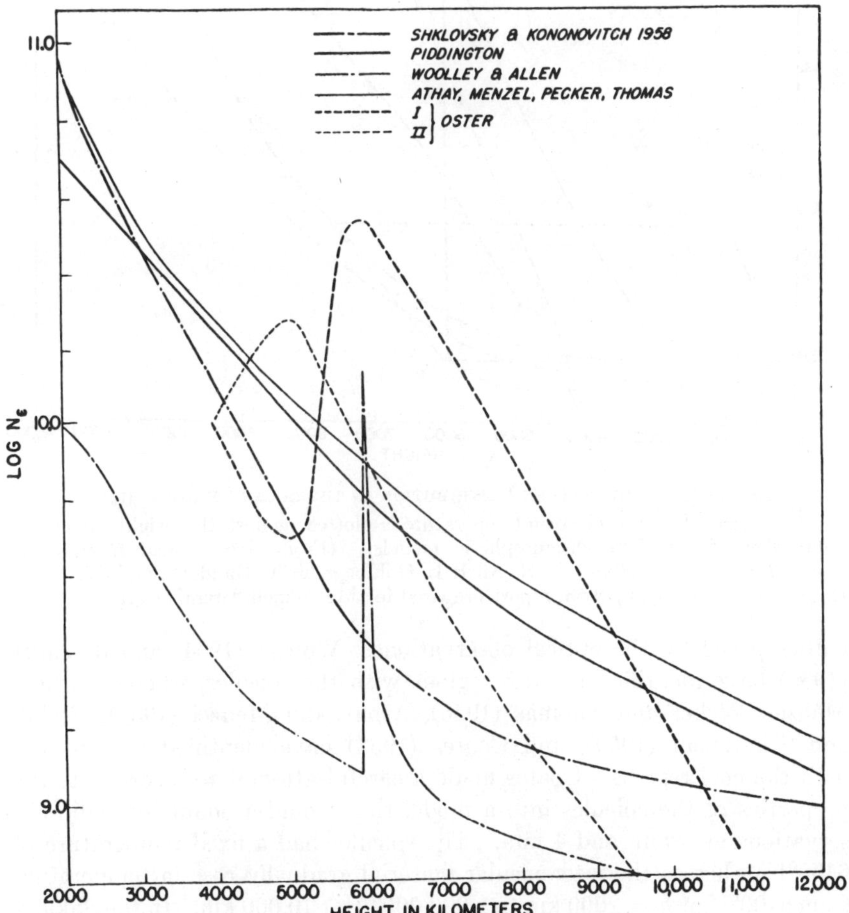

Fig. 9-25. The Electron Density in the Chromosphere

The logarithm of the electron density expressed in particles per cm³ is plotted against height in kilometers for various chromospheric models that have been proposed. (From *Astronomical Experiments proposed for earth satellites*, L. H. Aller, L. Goldberg, F. T. Haddock, and W. Liller, University of Michigan, 1958; report prepared for McDonnell Aircraft Corp.)

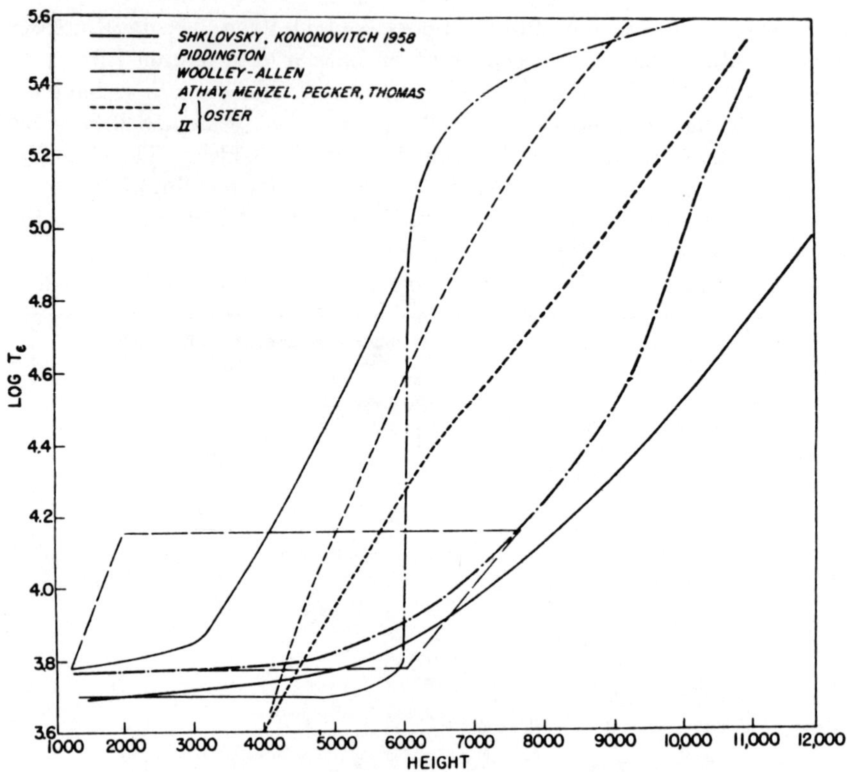

FIG. 9–26. TEMPERATURE DISTRIBUTION IN THE SOLAR CHROMOSPHERE

The logarithm of the electron temperature is plotted against the height above the photosphere for various chromospheric models. (From *Astronomical Experiments proposed for earth satellites*, L. H. Aller, L. Goldberg, F. T. Haddock, and W. Liller, University of Michigan, 1958; report prepared for McDonnell Aircraft Corp.)

culties posed by the optical observations. Woltjer (1954) and de Jager (1959) have identified the hot regions with the spicules, whereas Athay, Menzel, Pecker and Thomas (1956), Athay and Menzel (1956), Pecker and Schatzman (1959), and Coates (1958) have identified the spicules with the cool regions. Coates made a careful attempt to incorporate the properties of the spicules into a model that would account for radio observations at 8 mm and 4 mm. The spicules had a fixed temperature of 6400°K, whereas the interspicular material gradually rose in temperature from 6400°K at $h = 2000$ km to 1,000,000°K at 10,000 km. In the Shklovsky-Kononovitch (1958) model, which is based on both radio and optical data, the hot regions whose temperature is 12,000°K extend to a height of about 8000 km and cover less than 10 per cent of the solar surface. The cool regions rise from about 5800°K at 1000 km to 8000°K at 6000 km.

A promising theoretical model has been proposed recently by Athay (1961). One must remark that calculations of theoretical model chromospheres are extremely difficult because of severe departures from thermodynamic equilibrium and breakdown of the Saha equation (Pottasch and Thomas, 1960). Hence very involved calculations are required (Thomas and Athay, 1961).

The discordances between the various models clearly attest to the difficulties inherent in the problem. Table 9–2 gives a zero-order, i.e., homogeneous model derived by Boury and the writer from radio data. It serves as a basis for the suggested empirical two-element non-homogeneous model given in Table 9–3. The spicules are identified with the cool elements and the interspicular spaces with the hot ones. Other workers place the onset of inhomogeneities and the steep temperature rise near 1300 km.

TABLE 9–2

A "Zero-Order" Model for the Chromosphere Derived from Radio Data

Height (km)	log N_ε	log T	Height (km)	log N_ε	log T
1900	11.00	3.84	7100	9.41	4.70
2500	10.68	3.84	7500	9.32	4.82
3100	10.44	3.85	8100	9.19	5.02
3500	10.30	3.86	8500	9.12	5.16
4100	10.08	3.87	9100	9.03	5.37
4500	9.94	3.89	9500	8.98	5.52
5100	9.83	3.97	10100	8.91	5.72
5500	9.78	4.04	10500	8.88	5.80
6100	9.62	4.25	11100	8.84	5.86
6500	9.55	4.48	11900	8.80	5.90

TABLE 9–3

A Suggested Working Chromosphere Model

Height (km)	Area* of Cold Element	Spicule		Interspicule	
		log N_ε†	T(°K)	log N_ε†	T(°K)
1000	1.00	11.43	6000	11.43	6300
1500	0.90	11.18	6000	11.05	9000
2000	0.90	10.95	6000	10.70	10,000
2500	0.90	10.72	6100	10.35	12,000
3000	0.90	10.50	6200	10.15	15,000
4000	0.75	10.15	6500	9.88	22,500
5000	0.60	9.88	7000	9.63	35,000
6000	0.40	9.64	8000	9.43	56,000
7000	0.30	9.45	11,000	9.27	90,000
8000	0.16	9.30	20,000	9.12	140,000
10,000	0.05	9.10	53,000	8.88	500,000

* Fraction of total area covered by spicule.
† The electron density is given in electrons cm^{-3}.

The structure of the chromosphere cannot be solved without observations covering the region from λ2900 to the soft x-rays. The transition region between the upper chromosphere and the corona present peculiarly difficult problems. Temperature gradients must be particularly steep and energy must flow downwards from the corona to the phostophere by thermal conduction. One must consider photoionization, collisional ionization, and radiative recombination as well as radiative and collisional excitation of the individual lines. Most of the energy radiated by this region must fall in the rocket ultraviolet. The most ambitious attempt to compute the properties of this region was that made by Oster (1956) who proposed two models. He predicted for the two models theoretical fluxes at the earth's distances from the sun of 0.8 and 2.5 ergs cm^{-2} sec^{-1} in λ384 He I line. Hinteregger, Damon, Heroux, and Hall (1960) found a flux of 1.2 ergs cm^{-2} sec^{-1} for this line.

Particularly important, therefore, are the observations such as those secured by Friedman and Tousey and their respective groups at the Naval Research Laboratory and by Rense and his associates at the University of Colorado. They employed spectrographs and photon counters in rockets fired above the earth's atmosphere. From 2900A to 1500A the Fraunhofer spectrum is observed with steadily decreasing intensity. There is an abrupt drop near 2100A, possibly due to molecular absorption: NO, CO (Goldberg), or photoionization of aluminum (McAllister). The spectrum at shorter wavelengths arises from the corona and chromosphere and consists mostly of emission lines, although Tousey, Dewiler, and Purcell observed a solar continuum extending to 1000A. Its intensity is enhanced near Lyman α, possibly due to a charge transfer continuum of the excited hydrogen molecule. No absorption Fraunhofer lines are observed below 1325A. See Fig. 9–27.

The strongest ultraviolet emission line is that of Lyman α, which appears to have a flux of about 6 ergs cm^{-2} sec^{-1} at the earth's distance. This flux is remarkably constant and shows no great enhancement during solar flares. Profile studies have been made by Rense and by Purcell, and Tousey (1960) found the line to be broad with wings extending about 1A on either side. The profile shows a sharp core produced by cool hydrogen between the E layer of the earth's atmosphere and the sun. There is also a more diffuse absorption core produced by non-LTE processes in the solar atmosphere. The profiles differ in plage and unexcited regions.

From rockets fired during an eclipse, Chubb, Friedman, and Kreplin (1960) found that at totality the flux in Lyman α was reduced to 0.05 per cent, indicating that this emission must originate in the chromosphere rather than in coronal regions. Goldberg suggested this line was formed in a region at 4000–6000 km where the temperature was about 6000°K, although de Jager argued that the line was produced in a much hotter,

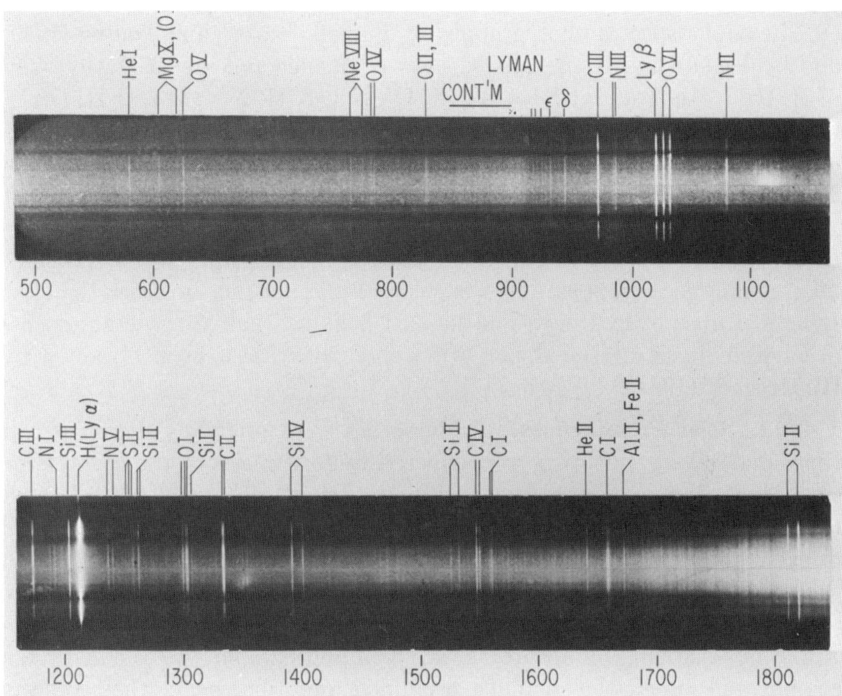

FIG. 9–27. ULTRAVIOLET SOLAR SPECTRUM

This solar spectrum was photographed from a rocket at a height of 197 km on March 13, 1959. The true solar continuum persists to about 1500A. The spectrum from λ1550 to λ920 and from λ830–λ500 is produced by scattered light. The spectrum from λ920 to λ830 is due to the Lyman continuum. (Official United States Navy Photograph; courtesy, R. Tousey, U. S. Naval Research Laboratory.)

higher level. Even if a model of the chromosphere is given, the calculation of the radiation field and the interpretation of the observed line intensities is extremely difficult. One must solve the equations of transfer and of radiative equilibrium simultaneously taking into account specific detailed mechanisms (see Athay and Thomas, 1961). For example, the region where the core of Lyman α is formed (optical depth unity) has an electron temperature of 60,000–1,000,000°K, even though the brightness temperature is about 7200°K. The core of the resonance He II line is formed in a layer whose electron temperature is near 150,000°K, even though the radiation temperature is only 20,000°K. Since the source function for a line depends on strata where the optical depth is considerably different from 1, calculation of temperature structure of the chromosphere is only the first step in the intricate process of computing the radiation field. The main resonance lines of HeI, HeII, and hydrogen are the easiest to handle.[11]

[11] Widing and Morton (1961) find that Ly α originates in regions of steep temperature gradients and suggest an electron temperature of about 70,000°K with electron densities between 3×10^9 and 3×10^{10}.

Ultraviolet spectra photographed by Purcell, Packer, and Tousey (1959) and by Violett and Rense (1959) show resonance and low excitation lines of H, He I, He II, C I, C II, C III, C IV, N I, N II, N V, O I, O II, O III, O IV, O V, O VI, Al II, Si II, Si III, Si IV, P I, S II, S VI, Fe II. In addition to these chromospheric radiations, lines of Mg X, Si XII, and Ne VIII (probably of coronal origin) and possibly Fe XVI also appear.

The Colorado work shows the resonance He II 303.8A line to be strong (flux about 0.2 to 0.8 erg cm^{-2} sec^{-1}. Rense suggests perhaps as much as 20 ergs cm^{-2} sec^{-1} appears in the solar ultraviolet flux in other than the Lyman α lines of hydrogen and ionized helium. Valuable measurements of photon fluxes in the upper atmosphere have also been obtained by Hinteregger (1961).

9–18. Solar Prominences. Prominences were originally discovered at times of total solar eclipses as bright red protuberances at the limb of the sun. Most of them do not reach heights greater than about 30,000 miles, although certain eruptive examples attain heights of several hundred thousand or even a million miles. Thus they appear entirely in coronal regions even though they exhibit spectra and level of excitation akin to spicules. They are usually observed on the limb with a spectrohelioscope, spectroheliograph, or polarizing monochromator. Still photographs fail to convey the complicated kinematics of much prominence motion; hence these objects are studied most effectively with aid of motion pictures such as those first obtained at the McMath-Hulbert Observatory (see Fig. 9–28).

One can observe prominences not only in elevation at the limb, but also in plan as dark markings on the disk. Disk observations possess certain advantages over observations at the limb. Certain types of prominences such as those associated with sunspots are low altitude objects which are visible at the limb only a day or so before solar rotation carries them from view. Hence their disk appearance, where prominences are seen in absorption against a bright chromospheric background, may give fuller and more continuous records. The greatest difficulty in interpretation of disk observations lies in the assignment of a particular disk prominence to the corresponding type observed at the limb.

Prominences are not distributed uniformly on the solar disk. They appear in two principal zones. The first group follows the sunspot zones. The second group is found in latitudes of about 45° at the start of the solar cycle, and migrates poleward. It reaches the pole near the maximum of solar activity and disappears soon thereafter. There seem to be appreciable differences between prominences associated throughout their lives with sunspots and other types.

The earliest work carried out by examination of the limb of the sun with a spectroscope and a widened slit suggested two types of prominences. Objects in which a casual inspection revealed little activity were referred

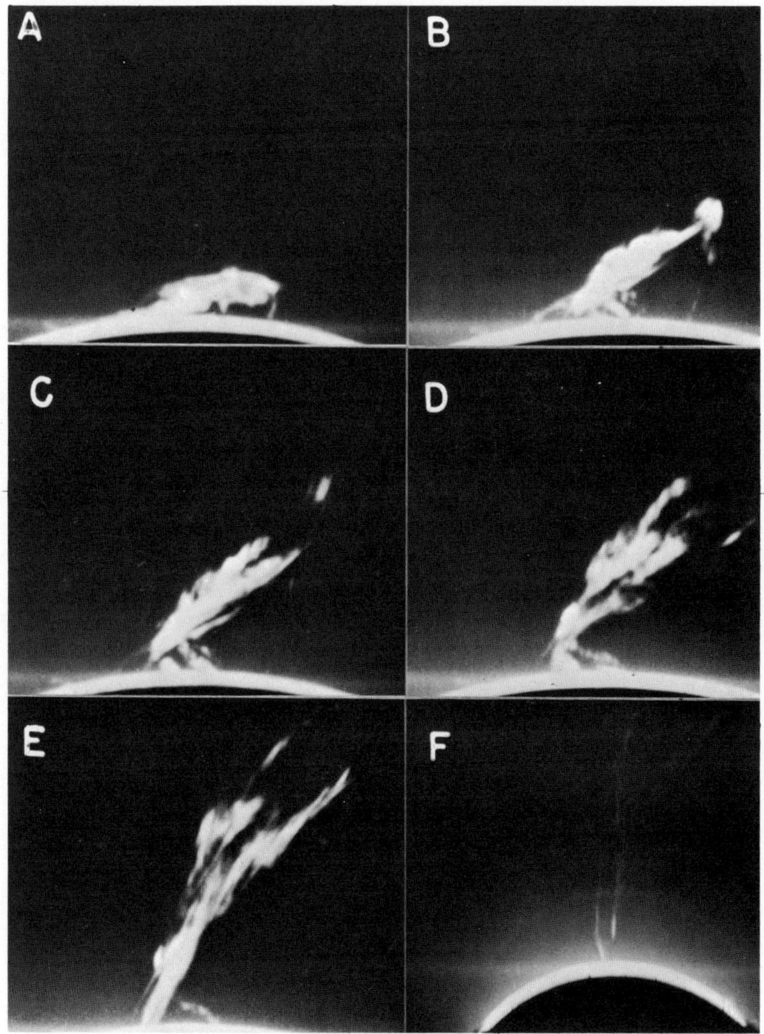

FIG. 9-28. AN ERUPTIVE TYPE PROMINENCE

Prominence of September 17, 1937, photographed in the K_2(Ca II) line. G.C.T. of exposures:

A: $14^h\ 50^m.69$ B: $14^h\ 55^m.84$ C: $15^h\ 06^m.13$
D: $15^h\ 09^m.11$ E: $15^h\ 14^m.31$ F: $16^h\ 06^m.7$

In F the prominence leaves the frame at 1,000,000 km above the sun. (McMath-Hulbert Observatory, University of Michigan.)

to as *quiescent*, those showing high velocities were called *eruptive*. More detailed studies, such as those by Pettit at Mount Wilson and by McMath and his co-workers at Michigan have shown this classification to be inadequate. Pettit employed the association of a prominence with a sunspot, its motion and structure, and whether it appeared to originate from the corona or from lower levels of the chromosphere in his classification scheme of six types: I *Active*, II *Eruptive*, III *Sunspot*, IV *Tornado*, V *Quiescent* (*Filaments*), and VI *Coronal*. There are sometimes subgroups within each class, e.g., Pettit suggests nine subdivisions of the sunspot type.

Pettit's classification is essentially descriptive; various subdivisions merge imperceptibly one into another, and there are occasional objects that seemingly do not fit into any classification scheme at all. Frequently a single prominence will evolve from one class into another. The most obvious example is the development of a quiescent into an eruptive prominence.

Alternative classification schemes have been proposed by various workers. Thus Menzel and Evans (1953) classified prominences according to whether they descended from above (coronal "rain," hedgerows, and other picturesquely named types) or whether, like spicules and surges, they rose from below. They also indicated them as belonging to sunspots or not. In classification schemes by Severny and by de Jager, motions and evolutionary histories serve as essential bases of classifications.

Quiescent prominences, Pettit's type V which often appear as dark sinuous *filaments* on the solar disk (Fig. 9–3), include the longest-lived of solar phenomena. They show a minimum of activity and motion, and sometimes present a palisaded structure with closely packed threads with diameters of the order of 450 km (Orrall and Zirker, 1961). McMath-Hulbert Observatory films show a great deal of turbulent internal motions of the order of 5–10 km sec^{-1}. See Fig. 9–29.

The thicknesses of quiescent prominences are of the order of 6000 km, their heights are about 40,000 km and their lengths are as great as 200,000 km, although objects as long as 1,000,000 km and heights of 80,000 km have been observed. They tend to attain greater heights than do prominences of other types.

The d'Azambujas (1948) observed them making an angle of about 38° with respect to meridians of solar longitude in active regions that may or may not contain spots and after that particular facular region has passed its peak of activity they move poleward; as they move they tend to lengthen and are drawn out of the meridians by solar rotation. Beyond 45° they are oriented along parallels of latitude. Equatorial filaments, seen on the limb, exhibit a maximum of structural detail while in higher latitudes such objects tend to resemble "haystacks," simply because of perspective.

As they grow older they may become active, or even eruptive, particularly if a new plage area and *center of activity* develops nearby. In some instances a filament may disappear completely only to reform in the same region with the same shape after a few days. Their behavior suggests that

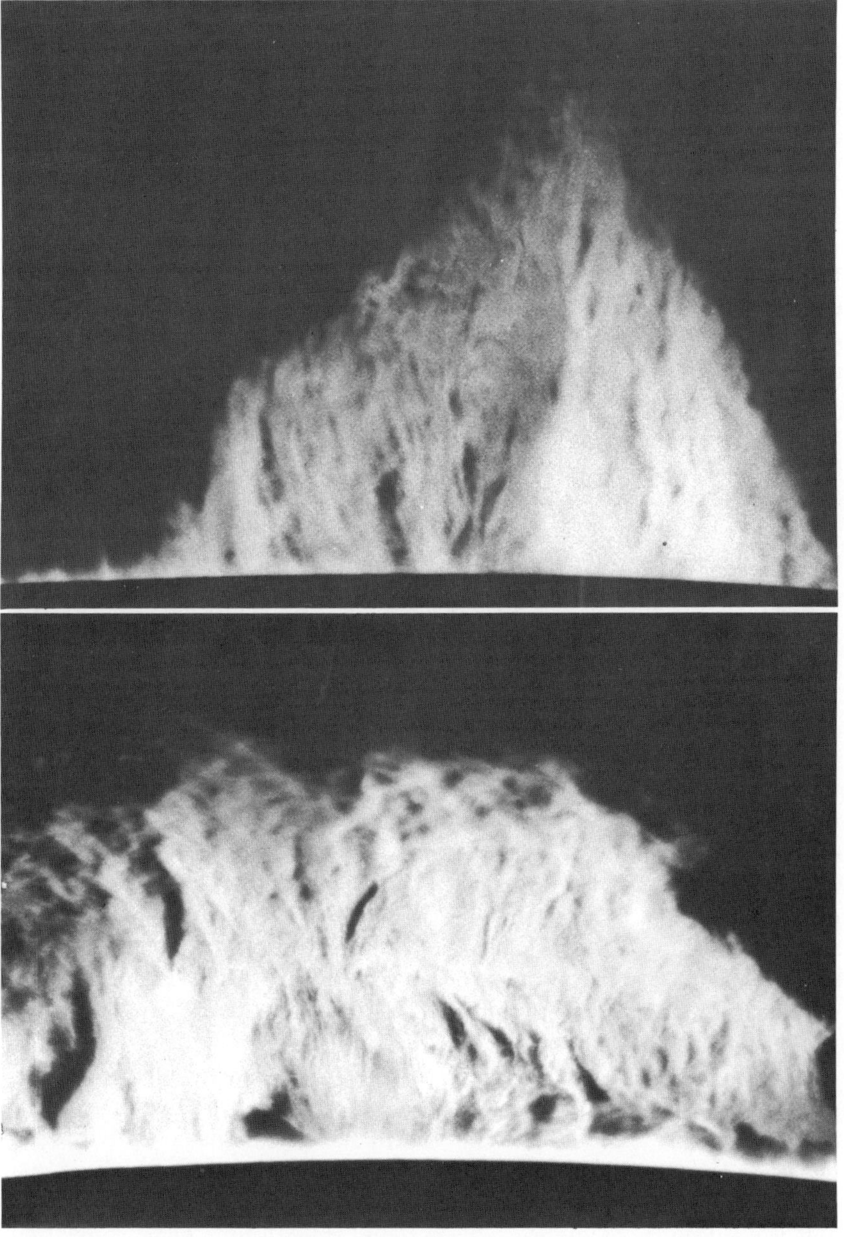

FIG. 9-29. QUIESCENT PROMINENCES

(September 12 and September 14, 1956; courtesy R. B. Dunn, Sacramento Peak Observatory, Geophysics Research Directorate, AFCRC.)

quiescent prominences represent a delicate balance between magnetic and mechanical forces which permits a condensation of coronal gases into a cooler region of a prominence.[12] When a new center of activity is formed, the fields probably suffice to produce marked instability.

An average filament lasts four solar rotations, but some have been observed to last as long as five years, possibly a consequence of the more stable magnetic configurations in the polar regions.

Type I: *Active prominences* constitute the most abundant type and occur all over the sun's disk (see Fig. 9–30). Frequently, they develop from quiescent prominences and sometimes they evolve into the spectacular eruptive type. They show a mass of filaments connected with the photosphere, and these filaments often move along a curved path into a so-called "center of attraction" at the base of the chromosphere. At times, a prominence appears to be literally pulled to shreds. Sometimes prominences standing side by side are connected by streamers in which the material actually moves in both directions; these are the so-called interactive type. Active prominences are usually found in the sunspot zones but are usually distinguished from those obviously connected with a particular sunspot group.

Type II: *Eruptive prominences* represent a special phase in the life of other types of prominences, e.g., quiescent type. They often evolve from the active type and tend to favor the sunspot zones although they usually occur away from the spots themselves. Material moves into a center of attraction at an increasing pace, and as the activity mounts, the whole prominence may rise several hundred thousand kilometers before it is pulled into the center of attraction. It may even appear to leave the sun entirely, expanding as it rises and fading while in motion (see Fig. 9–29): Velocities greater than the velocity of escape, 618 km sec^{-1}, are sometimes observed; speeds as high as 1100 km sec^{-1} have been recorded. During the whole course of development it continues to return streamers to the original centers of attraction. The record for height is held by the eruptive prominences observed by E. Pettit and J. O. Hickox on June 4, 1946, which reached a height of 1,703,000 km or 1.22 solar diameters.

Motions of prominences are best studied by moving pictures with a provision for determining the radial velocity. Eruptive prominences have been studied by various workers. Pettit concluded that eruptive prominences moved with constant speeds between sudden accelerations, but other observers, e.g., Pan Puh (1939) and Larmore (1953), have not been able to confirm these findings. They found smooth time-distance relations for prominence motions. Nevertheless, it is certain that prominences are

[12] The persistence of the fine threads without change in shape or brightness over a period of hours suggests a quasi-static equilibrium in which radiative losses neatly balance the heat supplied by conduction from the corona (Orrall and Zirker, 1961).

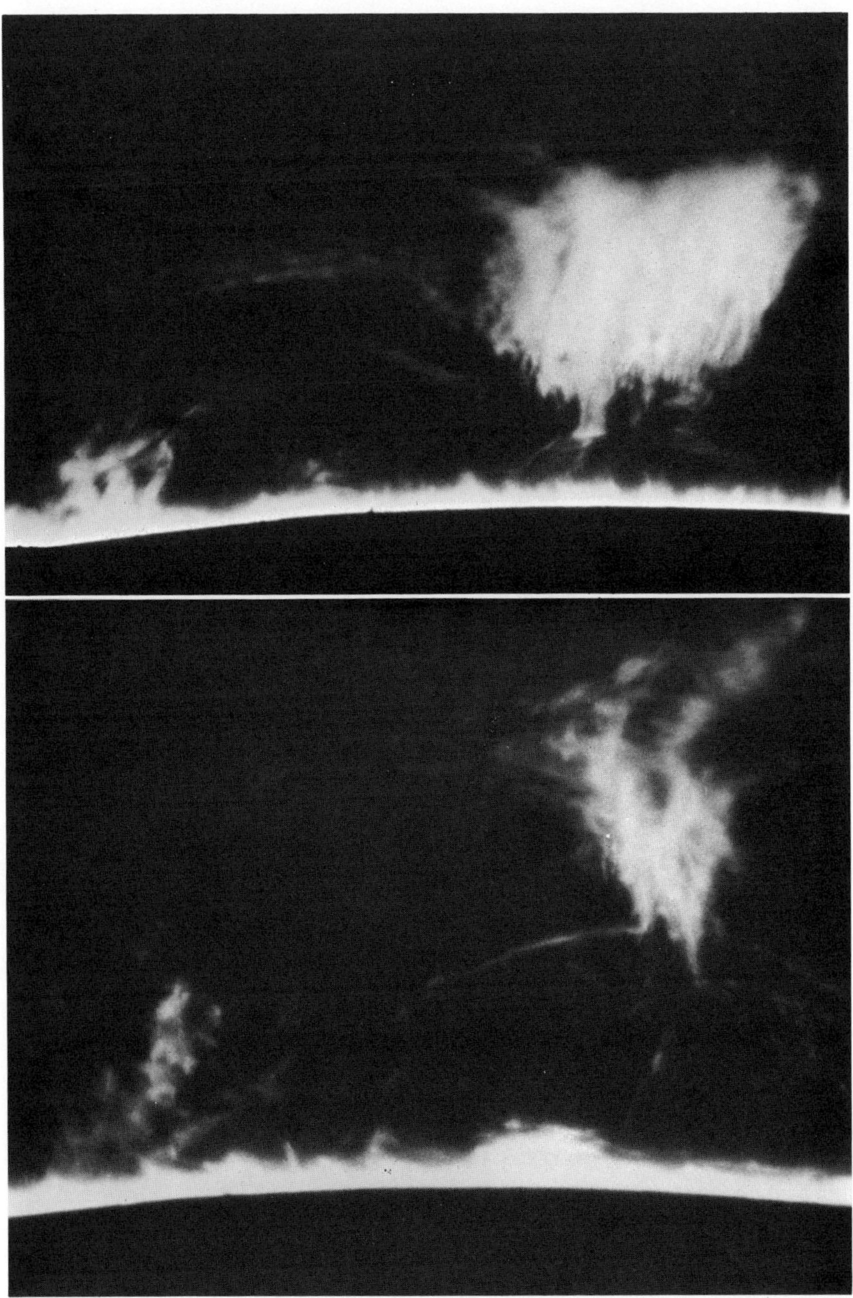

Fig. 9–30. Active Prominences

Photographed in Hα on May 5, 7, 1958. (Courtesy, R. B. Dunn, Sacramento Peak Observatory, Geophysics Research Directorate, AFCRC.)

often subjected to time-variable accelerations. Straight-line, nearly radial trajectories are favored. Pettit suggested that streamers and knots torn from active prominences also follow straight-line segments with a curved envelope, but this conclusion also is open to question. Measurements of motion in three dimensions for a large number of carefully observed prominences will be needed to solve the problem of their kinematics.

Type III: *Sunspot prominences* which occur in regions of strong magnetic fields often imitate quiescent, active, and eruptive prominences found away from spots, and in addition display several characteristic types. Cinematographic records often reveal a high level of activity in the neighborhood of spots. Long wisps may condense from coronal material in the region immediately above spots, brighten, and descend to some center of attraction in or near the sunspot group. Also there appear graceful loop structures which seem correlated with high excitation in the overlying corona, and which definitely seem to be connected with the magnetic field near a spot (Severny's electromagnetic type). See Fig. 9–31. Streamers associated with sunspots seem to move along curved trajectories, undergoing a focusing effect by the magnetic field. The type of motion does not seem to be influenced by polarity of the spot or plage area.

Type V: Most distinctive of sunspot prominences are *surges*, which were recognized as a distinct type of prominence at the McMath-Hulbert Ob-

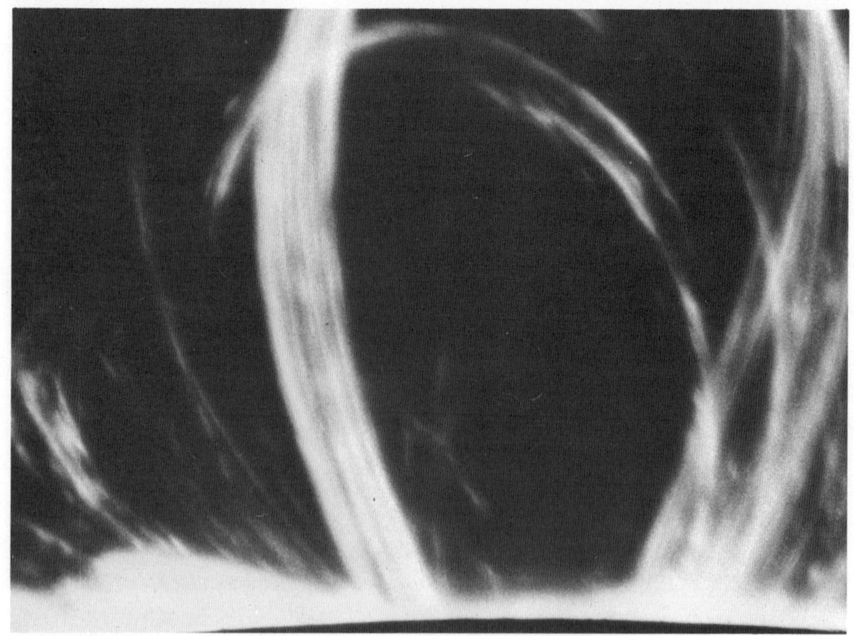

FIG. 9–31. LOOP PROMINENCES

Photographed in Hα, June 28, 1957. (Courtesy, R. B. Dunn, Sacramento Peak Observatory, Geophysics Research Directorate, AFCRC.)

servatory. McMath and Pettit described them as small jets of material rising a few thousand kilometers, or as immense tongues of gas hurled outwards to a distance of a hundred thousand kilometers, which suddenly retract or fade away. Sometimes the initial outburst breaks up and falls back in fragments. Their lifetimes range from about fifteen minutes to an hour, their velocities sometimes exceed 500 km sec^{-1}, and they protrude from the spot at all angles. They are sometimes observed to rise again and again from the same part of the spot. Upon spectroheliograms they often appear as sudden dark splotches in active spot regions. Surges bear a close relationship to solar flares. (See Art. 10–1.)

Type VI: *Coronal prominences* appear to be long, slightly curved strips at great heights above the photosphere. They descend with speeds of the order of 100 to 200 km sec^{-1} more or less haphazardly into regions of attraction, which are possibly to be identified with sunspots. Actually, they appear to be a type of sunspot prominence. See Fig. 9–32.

Fig. 9–32. Coronal Prominence Knots

Photographed September 11, 1956, in Hα. (Courtesy, R. B. Dunn, Sacramento Peak Observatory, Geophysics Research Directorate, AFCRC.)

Bright blobs and streamers often appear some 50,000–100,000 km above an active plage area. Like the surges they are connected with spot groups exhibiting flare activity. They tend to occur only in regions where the corona has an unusually high temperature and may have a very short lifetime. The delicate wisps that connect them to the underlying plage area appear to define trajectories of material moving with high velocities.

Type IV: The rare *tornado prominences* resemble cyclones or whirling waterspouts 5000 to 20,000 km in diameter and 25,000 to 100,000 km high. Pettit finds the helical motion suggested by their appearance to be confirmed by spectroscopic observations. They may evolve into eruptive or active prominences or simply disintegrate if the rotational velocity becomes sufficiently great. Some careful solar observers, however, doubt the existence of tornado prominences.

Spectra of prominences supply important data concerning their physical state. Quiescent prominences tend to show D_3(He I), H and K (Ca II), and the Balmer lines with weak lines of the metals. Other prominences, particularly those appearing over sunspots, show strong lines of Fe II, Mg II, and He I. The differences are to be attributed to varying optical thicknesses and to some degree to differences in excitation. From comparison of spectra of active and quiescent prominences observed simultaneously at the solar limb, Zirin and Tandberg-Hanssen (1960) conclude that quiescent prominences display spectra similar to that of the chromosphere at 1500 km with weak He II lines. Active prominences display broader lines and strong ionized helium lines.

Spectroscopic observations of prominences are difficult and, except for a few very strong lines, reliable intensity data can be obtained only at times of eclipses or with a coronagraph. From studies of line profiles one can estimate the gas kinetic temperature and the degree of microturbulence within the prominence. From total intensities of lines interpreted with aid of a proper curve of growth it is possible to assess the numbers of atoms in the line of sight. For example, from the Balmer lines one can obtain the numbers in the line of sight in the third, fourth, etc., hydrogenic levels, viz., N_3D and others, where D is the thickness of the prominence. Deviations from LTE make interpretation of the data difficult.

The problem of a prominence is similar to that of a cool element (spicule) in a hot gas, but its spectrum may be observed separately from that of its neighbors as is not possible usually for spicules. Furthermore, although the physical conditions in interspicular gases are not well known, the state of the corona is sufficiently well understood to fix gas density and the radiation field in the neighborhood of a prominence. Since prominences are cool objects in a hot but optically thin medium, the excitation mechanisms for individual lines are complex. From 1500A to longer wavelengths they are exposed to the normal solar photospheric radiation field, but at shorter wavelengths they are exposed to the characteristic bright lines of the solar chromosphere and x-rays from the corona. Our knowledge of this radiation field is steadily improving, thanks to data gleaned from rockets and satellites.

Within a given prominence, different spectral features may correspond to different stages in the recombination process. Thus, the He II lines in a loop prominence come from recombinations at a high temperature, while the hydrogen lines, the Ca II, *H*, and *K*, and other metallic lines arise in successively later stages of the cooling (see Zirin, 1962).

Gas kinetic temperatures in prominences have been estimated from line widths, prominence models, and hydrogen continua. (See Table 9-4.) Probably the most reliable values are those obtained from Paschen and Balmer continua, since line widths, especially in flares or active prominences, may be affected by mass motions. Temperatures as high as 30,000-50,000°K seem unlikely. The electron densities appear to be of the order of 0.5 to 3×10^{11} cm^{-3} as compared with 0.7 to 5×10^{8} cm^{-3} in the neighboring corona, where the gas kinetic temperature is about 2 million degrees. In order for a prominence to survive, it must radiate energy at least as fast as it is received by x-rays and impacts of charged particles from the surrounding corona. Hence, large, optically thick objects are favored. Prominences that appear as condensations in the corona and grow rapidly correspond to situations in which relatively dense regions of the coronal gas become thermally unstable and cool suddenly, losing energy in free-free and in bound-free emissions followed by cascade.[13] Quantitatively these events can be followed only with aid of far-ultraviolet observations that can be attained only with a systematic satellite program.

The complexities exhibited in prominence motions are well illustrated by moving pictures. A few general statements may be made, however. Different elements appear to stay well mixed in prominences despite the complexities of any motions. Lyot's visual observations with coronagraph and filters showed prominence detail to be similar in different spectral regions. Flash spectrum images of different elements but of the same arc length show similar structural details. If we compare prominence images of different intensity, we must allow for effects of self-reversal whose importance Bruck and Moss demonstrated from a comparison of Hα and D_3 (He I) intensities. It may seem strange that gases as different as H, He, Ca, and Fe move together, since many of the anticipated effects that would cause motion would also act differently on different elements. Evidently the gases tend to drag one another along in their motion. McCrea investigated this problem, employing the diffusion theory of Chapman and target areas for collisions derived from Ramsauer-type experiments. He found that if the density of a solar prominence is of the order of that of the lower chromosphere, $\sim 10^{10}$ atoms cm^{-3}, the atoms and ions would drag

[13] In the Athay-Thomas theory of the chromosphere it is proposed that the gases tend to exist in certain temperature domains to the exclusion of others. Hence coronal gases may occasionally revert to a lower temperature stable phase and appear as prominences.

TABLE 9–4

DETERMINATION OF KINETIC TEMPERATURES OF PROMINENCES

Type	T(°K)	Method	
"Cool" prominences			
quiescent	10,000–12,000	line width	Ellison and Reid (1957)
	14,000	line width	Severny (1954), Conway (1952)
	13,000	line width	P. ten Bruggencate (1953)
	12,000	Balmer and Paschen continua	Jefferies and Orrall (1961b)
	7000–10,000	(hydrogen region)	Jefferies and Orrall (1958)
	12,000	(helium region)	
"Hot" prominences			
active	7000–50,000	line width	Zirin (1956)
eruptive	30,000	line width	Zirin
limb flare	24,000	(model of prominence used to analyze continuum and line data)	Jefferies and Orrall (1961a)
chromospheric flare	12,000	Balmer and Paschen continua	Jefferies and Orrall (1961b)
flare-loop type prominence	15,000	Balmer and Paschen continua	Jefferies and Orrall (1961b)

The classification into "hot" and "cool" prominences is that proposed by Zirin and Tandberg-Hanssen (1960).

one another along even though the forces might act on those of just one of the elements. A much smaller density, however, would permit relatively large separations of different elements to occur if the forces responsible for prominence motion acted differently upon atoms of different elements. Actually, the thorough mixing of all chromospheric gases, in spite of the low density that appears to prevail in some places, suggests that forces responsible for prominence motion are non-selective as regards the atom and ion types involved. It seems unlikely that radiation pressure can play any important role in prominence motion, or equilibrium.

Hence the forces presumably are of a magnetic and hydrodynamical nature. In an electrically highly conducting medium such as the solar atmosphere the magnetic lines of force will move as though fastened to the material. Electrons are constrained to spiral in paths of small radius around the lines of force, ions may spiral in paths of larger radius but electrostatic forces prevent charge separation. Hence gases flow with ease along lines of force but perpendicular thereto only with difficulty. When magnetic forces are small compared with hydrodynamic forces, magnetic fields are dragged along with material when it moves. In certain prominences this situation may prevail, in some, particularly those near sunspots, the magnetic forces may dominate, whereas in yet others energy stored in turbulent motions $\frac{1}{2}\rho\xi^2$ may equal the magnetic energy per unit volume $H^2/8\pi$. Recently, Zirin, working with Severny in the Crimea, has measured magnetic fields of around 150 gauss in active prominences and fields of about 50 gauss in quiescent ones. In most instances, therefore, the magnetic energy is greater than the mechanical energy.

Among attempts to explain motions of a prominence we may mention those of David Evans and of Zanstra. Evans tried to account for prominence forms by the electrostatic diffusion of a charged cloud and the motions of ions in a magnetic field. By hydrodynamical experiments, Zanstra was able to reproduce certain features of solar prominences. Since prominences move, not through a vacuum but through a medium of apparently lower density, their motion should bear some resemblance to a streaming of one fluid through another (e.g., smoke through air).

Unsöld (1955) calculated the viscosity of the corona and treated a descending prominence as a freely falling body obeying Stokes' law. He derived an upper limit of about 30 km sec^{-1}, a value which could be modified by the presence of a magnetic field.

Zanstra (1955) suggested that a prominence may be nearly in mechanical equilibrium with the surrounding coronal gases, it is relatively cool and dense while the surrounding gases are hotter and more rarefied, but the gas kinetic pressures may be nearly equal. Although horizontal forces may balance, material may still flow up or down since the prominence can scarcely be in hydrostatic equilibrium.

An essential objection to all prominence theories so far proposed is that while they all give prominencelike forms, none has been able to predict the motions of any prominence from initial conditions. That is, if we evaluate the space velocity of a moving prominence knot at an early stage in its history, all theories fail completely to predict its subsequent motions. Attempts by Evans and others to derive prominence forms on the basis of various forces show that shapes of prominences can be "explained" by a host of assumptions and postulated initial conditions, but we have yet to find which, if any, of these theories will prove useful.

Although trajectories of spot prominences may be well enough defined by a magnetic field, the velocities are not explained. The general impression one obtains from movie films is of material feeding into the neighborhood of the spot, either from shredding of a nearby prominence or from some kind of condensation from the corona. The factor controlling the behavior of a prominence must range from nearly equilibrium conditions for filaments to catastrophic forces for eruptives or surges. Radiation pressure from flares has been invoked to explain prominence motion, and it does seem that surges are often connected, in time at least, with flares. If magnetic fields alone dominated their motions, prominences should not only follow curved paths but should present smooth, orderly forms. The chaotic appearance of many of them is suggestive of turbulence. The ions of ionized gas moving into a strong, inhomogeneous field would tend to follow the lines of force. Hence the gas would tend to be compressed laterally. Thus recombination may be facilitated and the prominence streamer would become visible.

Kippenhahn and Schluter (1957) have considered stabilities of prominences in magnetic fields. Shock waves and magnetohydrodynamic waves may also play important roles in chromospheric, prominence, and corona phenomena, see Art. 9–25, in addition to supplying energy. We return to some of these questions in Chapter 10.

McMath has enumerated some facts an adequate prominence theory must explain:

(1) The character of the motion and the fact that it can occur in all directions with respect to the sun. (2) No separation of gases. (3) Motions in streamers towards the chromosphere wherein sometimes several converge on one spot, and sometimes streamers from a given prominence show concurrent motion in random directions. (4) When velocity changes occur, the new velocity either up or down is greater than the preceding velocity; accelerations are high and are observed at great heights. (5) Sometimes prominences appear to leave the sun while on other occasions bright clouds form in the inner corona from which streamers descend to the sun. (6) Relative frequencies of various prominence forms. (7) Occurrence of prominence activity in all solar latitudes. (8) Prominence activity is correlated with the sunspot cycle.

Except for item 2 we do not have a satisfactory explanation of any of these facts.

Although the Doppler displacements of spectral lines indicate an actual motion of prominence material, we cannot be sure that all "motions" observed may not in part arise from a change in excitation and ionization in a previously existing medium. Much of the behavior of knots and streamers, the disintegration of prominences by shredding, and the ejection of prominences to coronal regions all suggest actual material motions. On the other hand, condensations of streamers out of material in the coronal region and fading and reappearance of filaments observed on the disk, indicate that excitation conditions must play an important role.

Observations of prominences from satellites may settle many of these questions, since excitation effects in resonance lines of hydrogen or helium are likely to differ from those in subordinate lines we observe at present.

In addition to quantitative intensity measurements on chromospheric lines in all spectral regions (particularly the far ultraviolet), data are needed on the fine structure of individual prominences, since in many instances the most delicate features appear not to have been resolved. A great advantage of prominence studies in the $\lambda 100$–$\lambda 1500$ region is that the photospheric background drops from view and disk phenomena should be easier to correlate with limb phenomena.

9–19. Optical Characteristics of the Corona. The outermost envelope of the sun, the pearly corona, is a difficult feature to observe and interpret. Yet it presents some of the most engaging problems in solar physics. Coronal radiations are observed at heights as low as perhaps 5000 km; yet it has been traced outward from the sun more than 40 solar radii where it smoothly merges into the zodiacal light. Its lower portions contain about 10^9 ions cm^{-3}; by 40 solar radii it has fallen off to about 10^3 ions or electrons per cm^3. The gas kinetic temperature, of the order of 2,000,000°K, is to be compared with chromosphere temperatures, normally between 4000 and 50,000°K and a photospheric temperature of 6000°K. Its inner structure is approximately symmetrical, its outer portions are clearly non-uniform in density.

It is important to emphasize that the solar corona has no outer boundary. It merges imperceptibly into the interplanetary medium whose properties are largely determined by the physics of the corona. In this article we discuss the corona as a solar appendage, reserving to the next chapter the relation of the corona to the outer atmosphere and radiation belts of the earth.

Because of its extremely low surface brightness, the solar corona is best observed at total eclipse. Its general appearance, as well as its detailed structure, shows a strong correlation with solar activity. At sunspot minimum there is an extension of the corona in the equatorial direction; short spikes appear in the regions of both the north and south poles. At the maximum of the solar cycle, the polar spikes are missing and the corona has a more uniform appearance. See Fig. 9–33.

Fig. 9–33. The Solar Corona August 30, 1905
(Courtesy, Lick Observatory.)

Activity in the underlying chromosphere and photosphere has a pronounced effect on the corona. Arches, fans, and fibrous filaments abound near regions of high excitation. Prominences often appear imprisoned on the interior of coronal arches, suggesting that coronal and chromospheric phenomena are closely connected. Thus, coronal, as well as chromospheric, activity is strongly affected by apparent centers of attraction and excitation of the solar surface. During early stages of development of a center of activity, rays and arches of modest dimensions appear. As time goes on, long coronal rays may eventually appear. The density of the corona above active regions is about twice that of the quiet corona at the same height.

We may regard the minimum corona as the fundamental structure, and the maximum corona is the same with rays and arches superposed. At this time the number of separate active areas is so great that arches and rays overlap to form a nearly circular structure.

Motions in the corona may be studied by motion picture techniques as developed by Lyot and more recently by Dunn at Sacramento Peak. The general impression obtained is that coronal features change slowly compared with many prominences. Flares often markedly affect both the motions and intensities of neighboring coronal regions. From a study of Doppler motions in twenty-four active areas wherein he found occasional regions of microscopic motions, Newkirk (1957) concluded that expansion or contraction of a magnetic field defining a loop or arch accounts for most of the coronal motion. (See also Karimov and Shilova, 1960.) There may occur some streaming of coronal material along magnetic lines of force as well, the coronal velocities being much smaller than velocities observed in prominences, the so-called solar breeze (Chamberlain) or solar wind (Parker). Some coronal arches are shown in Figs. 9–34, 9–47, and 9–48.

The distribution of brightness in the corona as a function of distance from the sun has been studied by many observers. From a critical discussion of all observational data some years ago, Baumbach found that the mean photographic brightness of the corona averaged over all position angles could be expressed in the form

$$I(p) = 0.0532\, p^{-2.5} + 1.425\, p^{-7} + 2.565\, p^{-17} \qquad (9\text{–}33)$$

where the intensity of the center of the solar disk is taken as 10^6 and distances are measured in units of the solar radius from the center of the sun.

At one solar radius from the limb of the sun the coronal brightness is not constant, however, and at sunspot maximum it even varies from pole to equator. Coronal isophotes are not circularly symmetrical but have an ellipticity that varies with the distance from the sun, phase of the sunspot cycle, and solar activity at time of observations. Baumbach's expression can be regarded as a useful interpolation formula.

The total light of the corona has been measured visually, photoelectrically, and by bolometric methods. Results differ from eclipse to eclipse and from observer to observer. Since the corona is much brighter in its innermost portions and varies in shape, form, and intensity with the solar cycle, and since the apparent size of the moon varies from eclipse to eclipse, some differences are to be expected. There are real variations with solar activity (see, e.g., Ney *et al.*, 1961; von Klüber and Dieter, 1961).

When corrections for variation in the apparent size of the moon are applied, it is found that at solar maximum the total light of the corona is about 1.1×10^{-6} that of the sun; at solar minimum it is about 0.6×10^{-6} that of the sun. The greater brightness and density of the maximum corona is to be attributed to the numerous rays, arches, etc.

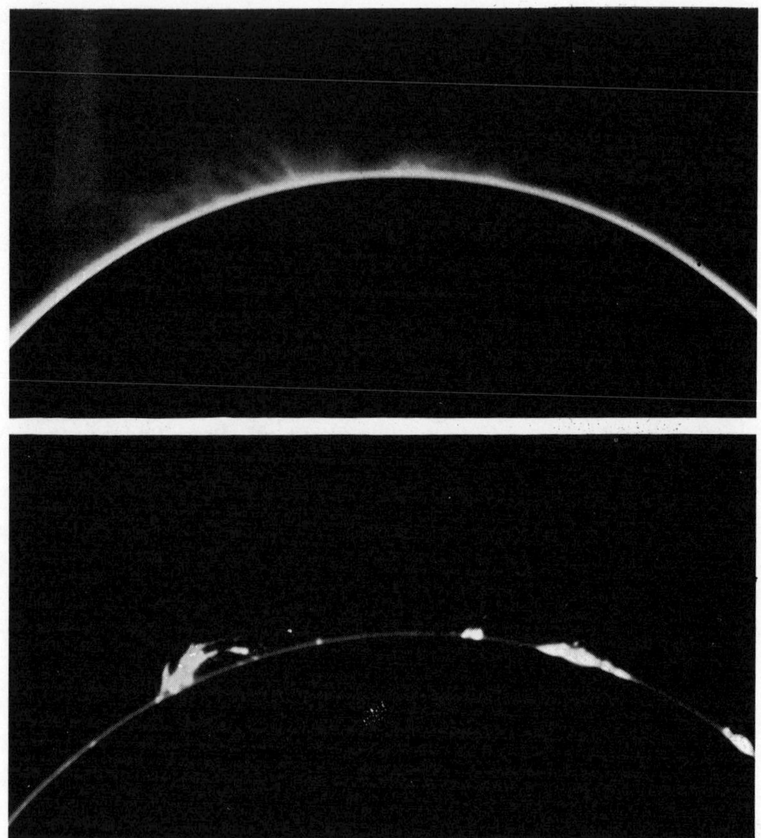

FIG. 9–34. SIMULTANEOUS PHOTOGRAPHS OF THE INNER CORONA AND THE PROMINENCES

Top: The solar corona photographed with a coronagraph and polarizing monochromatic filter isolating the λ6379 line at the center of a band 3A wide. *Bottom:* The prominences photographed simultaneously in the light of Hα with the same apparatus. (Courtesy, B. Lyot.)

Coronal optical radiation comprises three components denoted as L, K, and F. Coronal optical line radiation mostly in the inner and middle corona contributes only 1–2 per cent of the total intensity. The "white" radiation which has nearly the same energy distribution as the sun is composed of a partially polarized, continuous K component and an unpolarized or weakly polarized faint F component that shows the Fraunhofer lines. The r-f radiation is mostly of thermal origin, except in the meter range where the non-thermal component dominates.

9–20. The Optical Continuous Spectrum of the Corona and Its Interpretation. Most of the coronal light arises from a strong partially polarized continuous spectrum whose energy distribution is similar to that of the sun.

The importance of polarization measurements of the corona was first emphasized in 1879 by Schuster who pointed out that such studies might give information on the size of the scattering particles. The observations are difficult and tend to be affected by scattered light from the earth's atmosphere, particularly at large angular distance from the sun. They do show the polarization to be independent of wavelength, and that the electric vector of the light wave vibrates preferentially in the direction perpendicular to the radius drawn from the sun. Let us call the intensity of the plane-polarized component in this tangential direction I_t. The intensity of the component whose E vector vibrates along the radius drawn to the sun will be I_R. Then the degree of polarization may be defined as

$$P = (I_t - I_R)/I_0 \text{ where } I_0 = I_t + I_R \qquad (9\text{–}34)$$

If the sun were a point source and we could isolate radiation scattered from a volume element, V, it could show polarization up to 100 per cent. Since the sun fills a large solid angle as seen from a point in the corona, and the angle sun-point-observer varies over a considerable range along any given line of sight, the degree of polarization in practice never exceeds about 45 per cent. See Fig. 9–35.

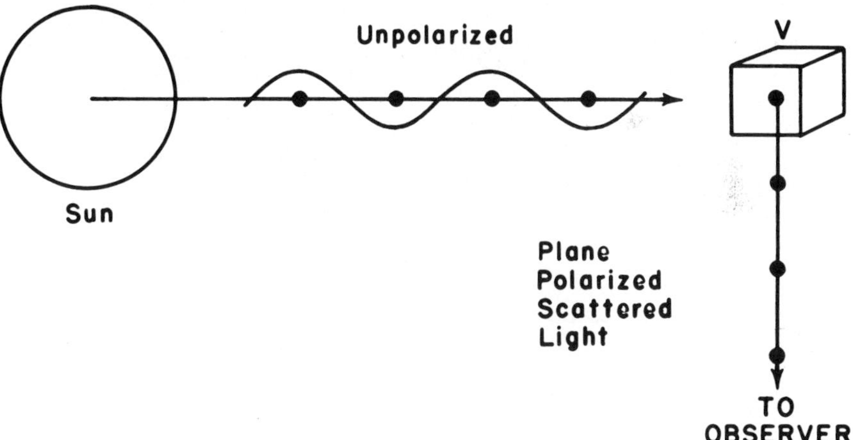

Fig. 9–35. Polarization of Light by the Corona

Unpolarized light from the sun, (regarded for the purposes of the present discussion as a point source) hits the volume element V. Electrons vibrating perpendicular to the plane of the paper scatter light with the electric vector in this plane. Light scattered by those vibrating in the plane of the paper do not reach the observer. Hence photospheric light scattered in V would be plane-polarized. The light received by an observer is only partially polarized because contributions come from volume elements along the line of sight and because the sun is not a point source. These effects can be handled quantitatively.

In Fig. 9–36 we compare the calculated and measured values of the percentage polarization as observed by Ney *et al.* (1961) at the October

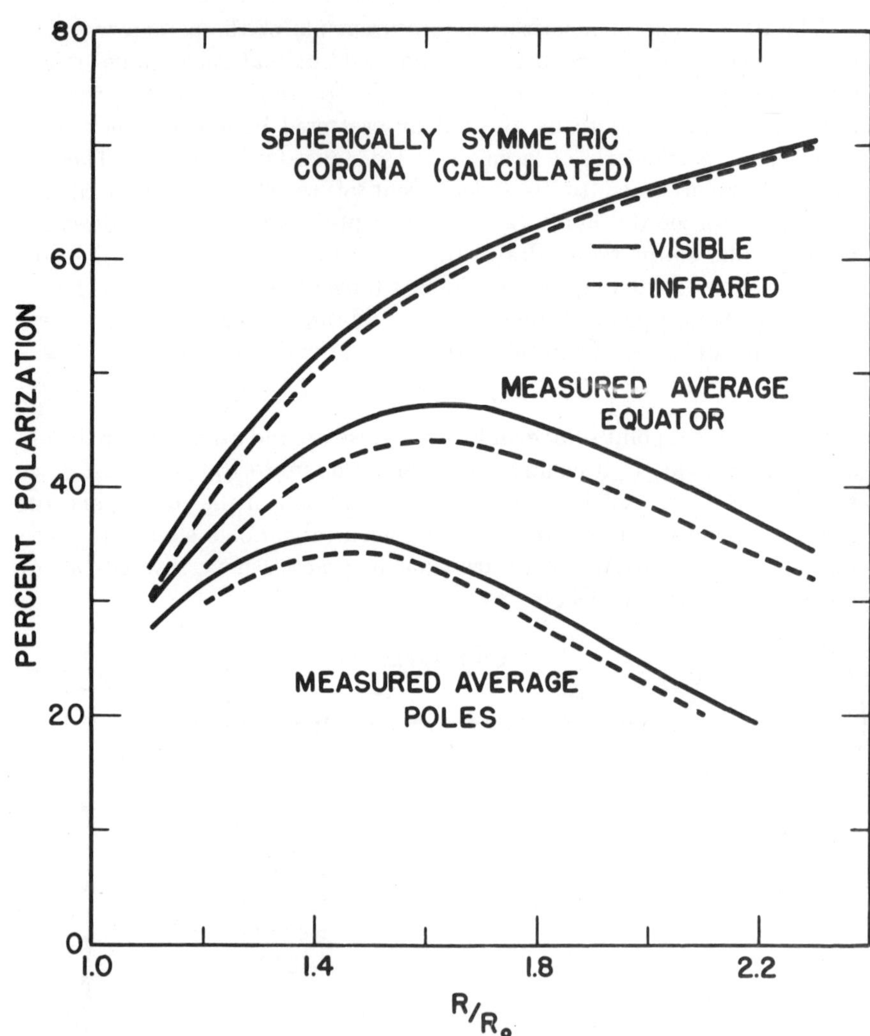

Fig. 9–36. Polarization of the Solar Corona

The observed polarization is compared with that calculated for pure electron scattering. (Courtesy, E. P. Ney, W. F. Huch, P. J. Kellogg, W. Stein, and F. Gillett, *Ap. J.* **133,** 616, 1961, University of Chicago Press.)

2, 1959, eclipse. The rapid diminution of polarization with distance shows that some source other than pure electron scattering is responsible for light from the outer corona.

If the partially polarized continuous spectrum of the inner corona arises primarily from electron scattering, as its faithful reproduction of the sun's color would indicate, very large electronic velocities would be required to obliterate completely the dark-line Fraunhofer spectrum. The H and K

lines of Ca II are almost completely washed out. The depression in the solar energy distribution caused by crowding together of many strong dark lines beyond λ3800 is reproduced, however, in the coronal spectrum. No individual lines are present; all appear to be smoothed out by the swift motions of the scattering electrons. Grotrian estimated that a thermal width for the scattering electrons of the order of 60A would be required to produce the observed effects. Then (see Chapter 3) an electron temperature of at least 300,000°K is implied.

Beyond about a solar radius the Fraunhofer absorption lines reappear. They are weaker than in the solar spectrum but of about the same width, as though a simple reflection of the solar spectrum were superposed upon a continuum. Grotrian suggested that the Fraunhofer spectrum arose from scattering by small particles related to those that cause the zodiacal light.

In order to separate the K component which has no Fraunhofer lines and evidently arises from electron scattering, from the F component which is a pure reflection of the Fraunhofer spectrum one may proceed as follows. If the Fraunhofer lines are produced only by scattering by dust and the continuum by electron scattering, the observed depth, d, of the coronal line to the photospheric line will be

$$\frac{d\text{ (coronal)}}{d\text{ (photospheric)}} = \frac{F}{F+K} = 1 - k$$

a relation which can be applied until k is about 0.2, since beyond that point the dust contribution is overwhelming.

If we assume that dust does not polarize light as it scatters it and make some assumption about the density gradient of the K corona, we can use the polarization measurements also to separate the K and F components.[14] Blackwell (1955, 1956), observing the 1954 eclipse from an airplane, measured the polarization out to 20 solar radii and the brightness to even greater distances, but at these distances a substantial fraction of the light comes from particles which are actually 50 solar radii or more away. Hence it is necessary to relate the coronal measurements to measurements of the zodiacal light as had been suggested by C. W. Allen and van de Hulst. Radio observations (see Art. 9–21) indicate that the outer corona is also inhomogeneous in structure.

If the contribution of the F corona is subtracted, van de Hulst (1950) found that the intensity distribution in the inner corona may be represented by an expression of the form

$$I(p) = \frac{1.125}{p^7} + \frac{2.565}{p^{17}} \tag{9–35}$$

[14] Actually, irregular dust particles aligned by a magnetic field or even unaligned particles could polarize the light. Careful measurements of the zodiacal light from a space station are needed.

The ratio of the F and K components varies with position angle and from eclipse to eclipse because the K corona is strongly variable and the F corona is constant. The streamers belong to the true corona, which is stronger near the solar equator than the pole, while the false corona has a marked ellipticity strongly increasing with distance from the sun (Michard, 1956).

From the coronal brightness distribution, Eq. 9–33, Baumbach calculated the emissivity $e(r)$ per unit volume per unit solid angle as a function of distance from the center of the sun. Then, on the assumption that the coronal luminosity arises from scattering of photospheric light by free electrons, he computed the electron density N_ε as a function of r. We must modify his discussion so as to take account of the light of the K component of the corona only. If

$$e(r) = A_1 r^{-a} + A_2 r^{-b} \tag{9–36}$$

we may write

$$I(p) = I_1(p) + I_2(p) = A_1 \int_{-\infty}^{+\infty} \frac{dy}{r^a} + A_2 \int_{-\infty}^{+\infty} \frac{dy}{r^b} \tag{9–37}$$

From the geometry of the problem (see Fig. 9–21, where we set $p = h + r_1$),

$$r = \sqrt{p^2 + y^2} \qquad y = p \tan \theta, \qquad r = p \sec \theta \tag{9–38}$$

Then[15]

$$I_1(p) = A_1 \int \frac{d(p \tan \theta)}{p^a \sec^a \theta} = \frac{2A_1}{p^{a-1}} \int_0^{\pi/2} \sec^{2-a} \theta \, d\theta = \sqrt{\pi} \frac{\Gamma\left(\frac{a-1}{2}\right)}{\Gamma\left(\frac{a}{2}\right)} \frac{A_1}{p^{a-1}} \tag{9–39}$$

with a similar expression for I_2. Here Γ denotes the gamma function. Thus, if $I(p)$ is given in the form of Eq. 9–35, we may compute the emissivity with the aid of Eq. 9–39. With a millionth of the sun's brightness at the center of the disk as our unit of luminosity, and the solar radius as the unit of length, the emission function becomes

$$E(r) = \frac{j(r)}{4\pi} = \frac{1.15}{r^8} + \frac{4.157}{r^{18}} \tag{9–40}$$

For isotropic radiation the scattering coefficient per electron is 0.66×10^{-24} and the corresponding scattering coefficient per unit length (solar radius) is

$$S = 0.66 \times 10^{-24}\, r_1 N_\varepsilon = 4.60 \times 10^{-14} N_\varepsilon \tag{9–41}$$

The total amount of scattered energy follows from an integration of the incident intensity over all directions:

$$j(r) = S \int I \, d\Omega = 4\pi S \bar{I} \tag{9–42}$$

[15] See, for example, B. O. Peirce, *A Short Table of Integrals* (Boston: Ginn & Co., 1929) p. 62.

where Ω is the solid angle subtended by the sun at the point r in the corona, and $\bar{I}$ is the intensity of sunlight. When $\bar{I}$ is computed, N_ε may be found from $j(r)$.

The first comprehensive discussion of the brightness, polarization, and electron density of the middle and inner corona was given by van de Hulst (1950) who took into account the anisotropy of the radiation that had been scattered and the fact that the density depends on solar cycle and latitude. More recent work by Blackwell, by Allen (1956), and by others working in the radio-frequency range has improved our knowledge of the outer corona.

9–21. The Radio-Frequency Emission of the Quiet Corona. Although the optical radiation of the corona consists mostly of scattered sunlight either by electrons (K component) or by small particles (F component) except in the metric range, the r-f radiation is largely true thermal emission produced by free-free transitions by electrons accelerated in fields of charged ions.

The rate of emission of thermal radiation per gram will be given by

$$j_\nu = 4\pi\kappa_\nu B_\nu(T) \tag{9–43}$$

See Eqs. 4–37 and 4–38, but

$$B_\nu(T) = \frac{2\nu^2 kT}{c^2} \tag{9–44}$$

an approximation that is valid because of the extremely low frequencies involved. The coefficient of absorption for free-free transitions per unit volume is (Elwert, 1948; see Eqs. 4–164, 4–165, 4–167, and 4–168)

$$(\rho\kappa_\nu) = \frac{8\pi N_i N_\varepsilon Z^2 \varepsilon^6}{3\sqrt{6\pi}\, c(mkT_\varepsilon)^{3/2}\nu^2}\, g_{\mathrm{III}} = \frac{0.0177\, N_\varepsilon^{\,2}}{\nu^2\, T_\varepsilon^{\,3/2}}\, g_{\mathrm{III}} \tag{9–45}$$

if $N_i = N_\varepsilon$. Here g_{III} is the Kramers-Gaunt correction factor. If

$$1.23 \times 10^5\, T^{1/2}\, N_\varepsilon^{\,1/3} < \nu \tag{9–46}$$

$$g_{\mathrm{III}} = \frac{\sqrt{3}}{\pi} \ln \left\{ 420 \left(\frac{T_\varepsilon}{Z}\right) N_\varepsilon^{\,-1/3} \right\} \tag{9–47}$$

alternately

$$g_{\mathrm{III}} = \frac{\sqrt{3}}{\pi} \ln \left\{ 5 \times 10^7\, \frac{T_\varepsilon^{\,3/2}}{Z\nu} \right\} \tag{9–48}$$

These equations are valid only as long as $\nu << \nu_0$ where the plasma frequency ν_0 is given by Eq. 9–1. Under these circumstances the emission of a column of length L will have an intensity

$$I_\nu = B_\nu(T)\, \overline{\rho\kappa_\nu}\, L = B_\nu\,(T)\, \tau_0 \tag{9–49}$$

if the column is optically thin. Defining the element of optical depth as

$$d\tau = \rho\kappa_\nu\, ds \tag{9–50}$$

we have

$$I_\nu(0) = \int_0^\infty B_\nu(T)\, e^{-\tau_\nu}\, d\tau_\nu \tag{9-51}$$

by the method described in Chapter 5.

If ν lies near the plasma frequency given by Eq. 9–1, complications develop. If $\nu < \nu_0$, a wave impinging on the medium from the outside cannot penetrate it at all but is simply reflected, as a broadcast wave is reflected back to the earth by the ionosphere. If it originates in the medium, it cannot escape. If ν is only slightly greater than ν_0, it will be strongly refracted since the index of refraction is given by

$$n^2 = (1 - \nu_0^2/\nu^2) \tag{9-52}$$

It will be also strongly absorbed since the absorption coefficient (Eq. 9–45) must be increased by a factor approximately $1/n$.

All low-frequency solar radio emission originates in the corona. Since the density decreases with height, the lower the frequency the higher the level at which that particular emission originates. For example, a wave of frequency 1.8 mc could escape from the corona only if it originated at a height greater than $p = 1.84$, i.e., at a height more than 585,000 km above the photosphere.

A typical example of the refraction of radio waves by the corona is illustrated in Fig. 9–37 due to Bracewell and Preston (1956), calculated for

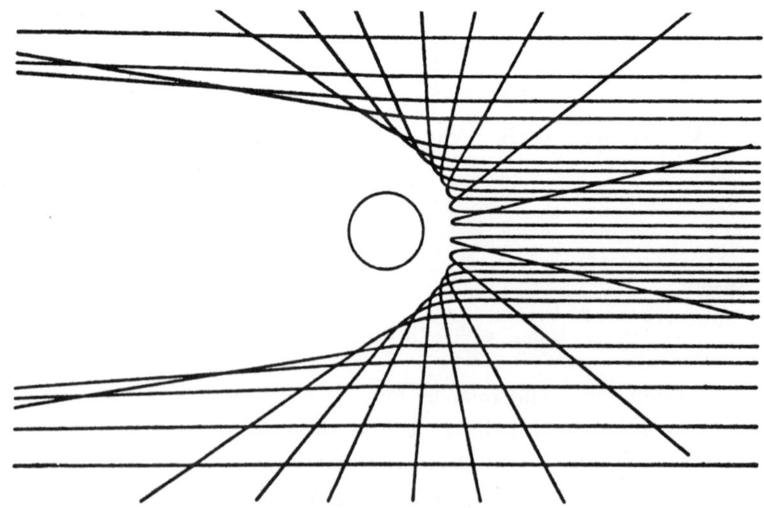

FIG. 9–37. REFRACTION OF RADIO WAVES IN THE CORONA

(From Bracewell and Preston, *Ap. J.* **123,** 14, 1956 Fig. 6, p. 19; courtesy University of Chicago Press.)

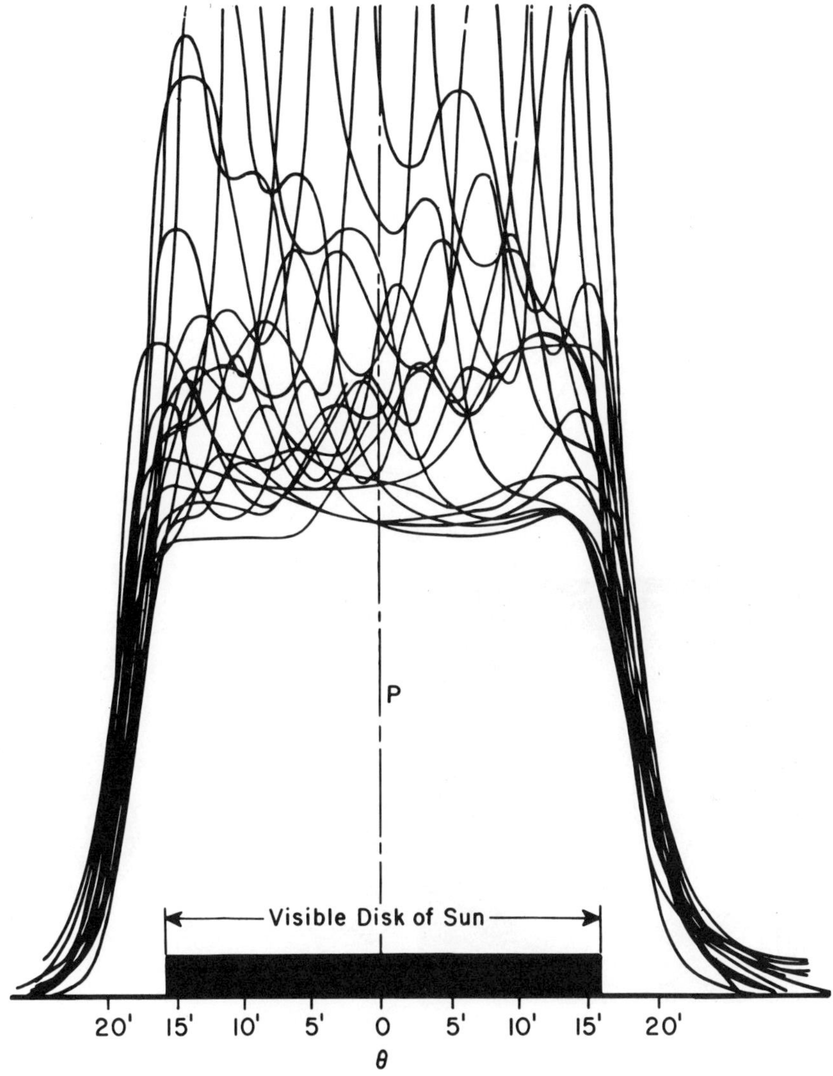

FIG. 9–38. DERIVATION OF THE QUIET SUN RADIO-FREQUENCY PROFILE FROM A SUPERPOSITION OF STRIP SCANS
(Courtesy, J. Pawsey, C. S. I. R. O. Radiophysics Laboratory.)

a symmetrical corona. These ray paths show why interpretations of solar radio data (even for the quiet sun) can become involved. In Eq. 9–50 ds is understood to be the element of length along the ray trajectory.

In order to obtain electron densities, etc., for the quiet corona one must eliminate both thermal and non-thermal "noises" produced by plage areas, flares, etc. One method (e.g., that employed at C.S.I.R.O. Radiophysics

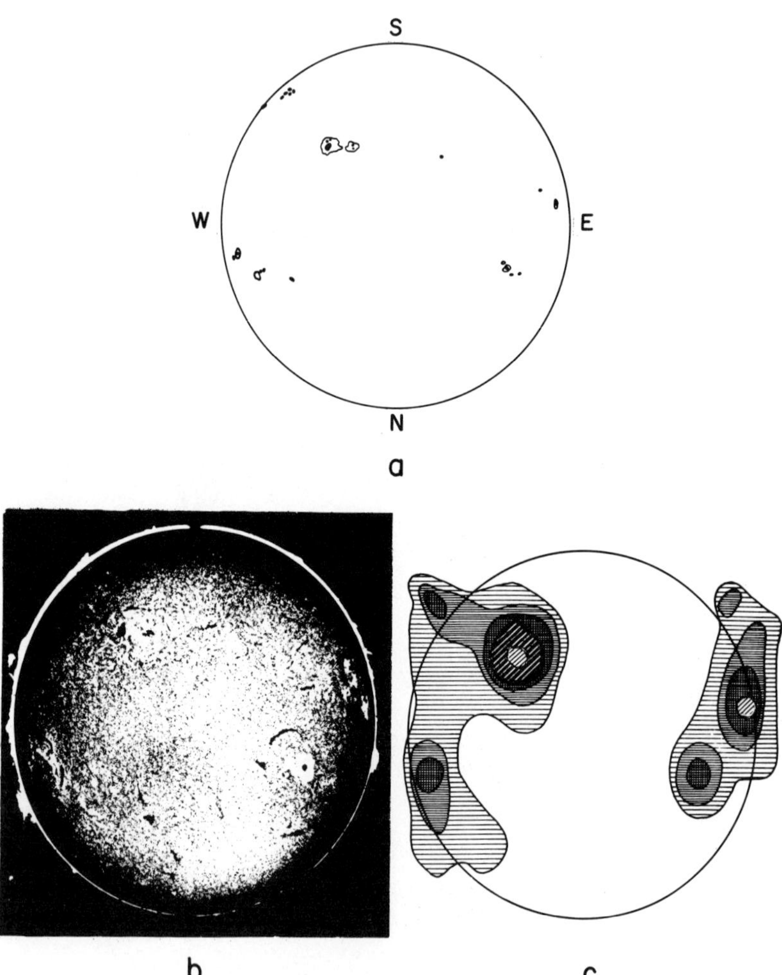

FIG. 9–39. COMPARISON OF RADIO AND OPTICAL OBSERVATIONS FOR THE ACTIVE AND QUIET SUN

The sun is shown as observed visually in white light (a) at the level of the photosphere, (b) at the chromospheric level, and (c) at coronal levels. In the coronal radio-frequency picture the quiet sun level is not shown. (After Christiansen, Mathewson, and Pawsey, *Nature* **180,** 944, 1957.)

(d) By making strip scans at various angles to the sun's axis a map of the quiet sun may be built up. Note that the radio sun not only is larger than the visible sun, but also is asymmetrical. The radio contours of the outer part show good agreement with the white light corona (e) as photographed by Waldmeier at time of the solar-minimum eclipse (June 30, 1954). (W. N. Christiansen and J. A. Warburton, *Austral. Journ. Physics*, **8,** 474, 1955.)

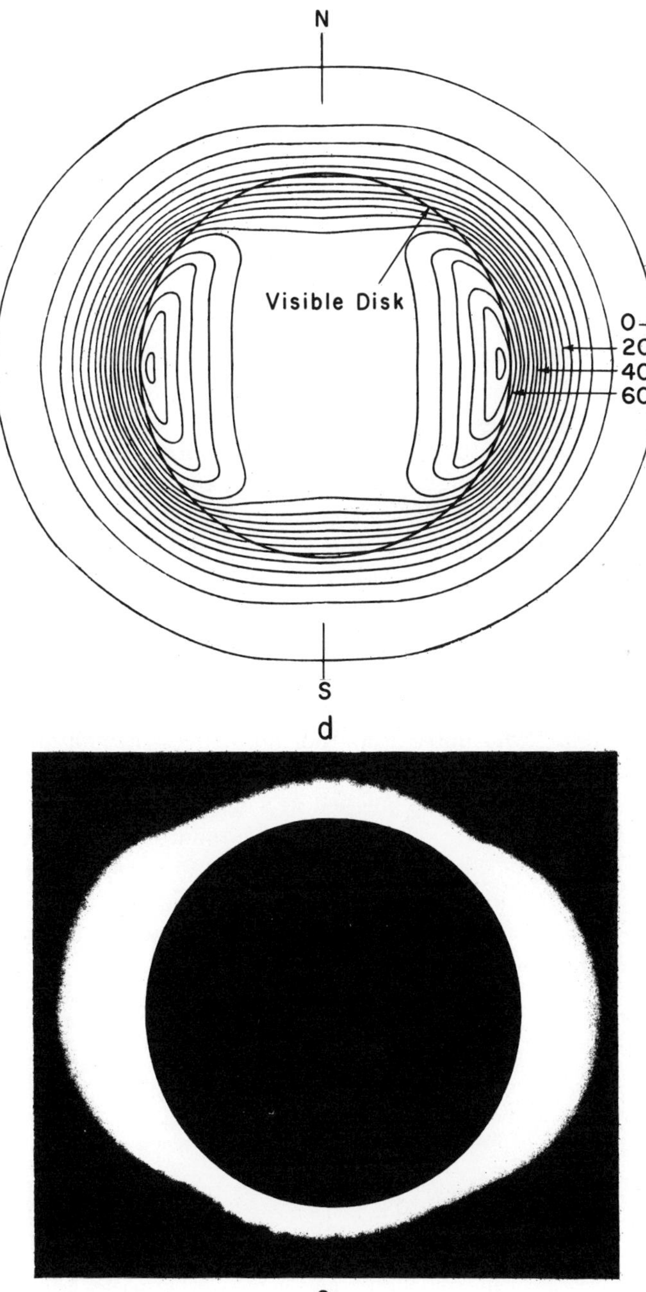
N
Visible Disk
0
20
40
60
S
d
e

Laboratory, Sydney, Australia) is to take strip scans with a multiple-element interferometer. On a particular day the scan shows the quiet sun radiation except for enhanced activity in an active region. As the sun rotates the latter shifts and by combining data from many different days a well-defined envelope of minimum activity is obtained which corresponds to the quiet sun. See Fig. 9–38. The technique can be applied in different frequencies to get the undisturbed background at different points in the corona, although at meter wavelengths it is nearly impossible to obtain the quiet corona. Eclipse observations can also be used.

By taking scans at different elevations across the solar disk one can construct an intensity contour map of the solar disk which shows an agreement in shape with the solar corona. There is a pronounced deviation from circular symmetry; the east and west limbs are definitely brighter than the polar regions (see Fig. 9–39).

The great advantage of the r-f technique is that one can observe the corona in front of the solar disk without difficulty, whereas in the optical region you can observe the corona only at the limbs. Some important applications of the r-f techniques are given in Chapter 10.

Radio measurements give data concerning the outer parts of the corona which cannot be observed effectively in the optical region. By a fortunate accident of nature the sun passes about 1°.75 from one of the strongest radio sources, the Crab Nebula. Thus the radiation from this ex-supernova can be observed as it is attenuated and refracted in the solar atmosphere out to distances of the order of 100 solar radii. Observations from 1952–1958 have been discussed by Vitkevitch and Panorkin (1959) and by Hewish (1958) who noted that the radio effects cannot be explained by simple refraction or reflection, but suggest non-isotropic scattering by irregularities that are possibly aligned in a magnetic field that is approximately radial at distances of 15 to 20 solar radii.[16] The degree of scattering depends on the solar cycle, being smallest at sunspot minimum and larger at sunspot maximum in such a way as to indicate that the scattering irregularities follow the changing form of the visible corona which is more spherical near sunspot maximum.

[16] Attempts to estimate the magnetic field strength in the vicinity of active regions in the corona have been made by methods based on measurements of the gyrofrequency. These interpretations lead to fields of the order of several hundred gauss in active regions and heights between 20,000 and 50,000 km. On the assumption that certain features of enhanced radio emission arose from synchrotron radiation, Takakura found fields of 1000–2000 gauss and 20–40 gauss in active regions of the chromosphere and corona, respectively.

We would expect much smaller fields in the quiet corona. Because of the dipolar general solar magnetic field, thermal radio emission from the northern hemisphere will contain circularly polarized components whose sense will differ from corresponding southern hemisphere radiation. Conway found the field to be smaller than 2.5 gauss in harmony with measurements by Babcock for the polar photospheric field (Art. 9–10).

Since radio waves are refracted in a complicated fashion in the corona, one cannot work out the coronal structure simply from the r-f measurements. Instead, one must construct a model, calculate its radio emission and adjust the model until the predicted and observed patterns agree. On the other hand, radio observations permit one to observe active coronal areas in front of the disk, a feat which is impossible in the optical range. Also, r-f observations have permitted analyses of non-thermal events that would hardly have been guessed from optical measurements. See Chapter 10. The density variation of the corona is shown in Fig. 9-40.

9-22. The Line Spectrum of the Solar Corona. The identification of the emission lines in the solar corona was an outstanding achievement of spectroscopy. The most conspicuous of these mysterious radiations is the green λ5303 line discovered by Harkness in 1869. Other strong lines include the red radiations at λ6375 and λ6702, three lines in the infrared at λ7892, λ10,747, and λ10,798, and one in the ultraviolet at λ3388. Between λ3328 and λ11,000 more than thirty lines have been identified while additional emissions have been found in the far ultraviolet. These include the resonance lines of Fe XV and Fe XVI.

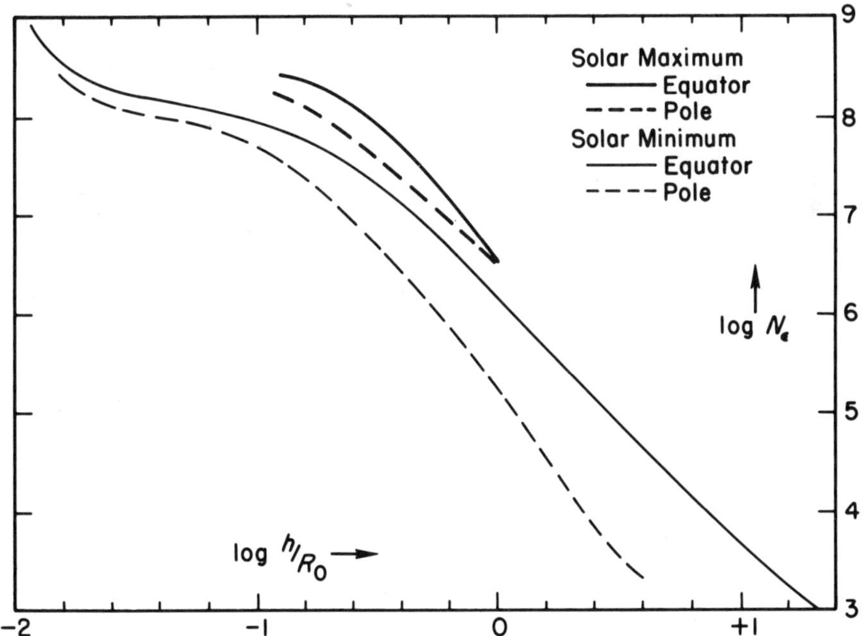

Fig. 9-40. Electron Density in Solar Corona

Logarithm of electron density is plotted against the logarithm of height above the solar surface (expressed in units of the solar radius, R_0). Data for the maximum of solar activity are taken from Newkirk; curves for solar minimum from van de Hulst and from Blackwell.

Table 9–5 lists the emission lines observed in the optical spectral region of the corona. Successive columns give the wavelength adopted from the work of S. A. Mitchell and of D. H. Menzel and W. Petrie, the intensity adopted from Grotrian, Lyot, Firor and Zirin (1962), and others, the intensity on February 15, 1946, in the repeating nova T Corona Borealis as observed at Michigan, the identifications, the transition probabilities expressed as Einstein A-values, the excitation potentials E.P. which fall mostly between 1 and 6 ev, and the ionization potentials I.P. which fall between 261 and 814 ev. The ionization potential refers to the next lower ionization stage.

TABLE 9–5

CORONAL EMISSION LINES

Wavelength (angstroms)	Intensity		Identification	A	E.P.	I.P.
	Sun	T Cor. Bor.				
λ3010	—	—	Fe XII	—	—	291
3170	—	—	Cr XI	—	—	246
3329	1.0	—	Ca XII	487	3.72	589
3388.0	16	—	Fe XIII	87	5.95	325
3454.1	2	—	—	—	—	—
3534.0	6	—	V X	—	—	206
3601.0	2.1	—	Ni XVI	193	3.44	455
3642.9	1.3	—	Ni XIII?	17	5.81	350
3685	4	—	Mn XII	—	—	288
3800.8	2.5	—	Co XII	—	—	306
3885	—	—	Mn XII	—	—	—
3987.2	0.7	6	Fe XI	9.3	4.67	261
3997	—	—	Cr XI	—	—	246
4086.4	10	—	Ca XIII	319	3.02	655
4221	3	—	Mn X	—	—	221
4232.0	2.6	—	Ni XII	237	2.93	318
4256.4	2	—	K Xi	—	—	504
4311	—	—	V X	—	—	206
4351.0	—	—	Co XV	—	—	412
4359	—	—	—	—	—	—
4412.4	—	—	A XIV	106	2.83	687
4466.9	—	—	Cr IX	—	—	185
4566.8	1.1	—	—	—	—	—
4586	—	—	—	—	—	—
4744	3	—	Ni XVII	—	—	498
5116.03	3.5	14	Ni XIII(?)	156	2.41	350
5302.86	100	100	Fe XIV	60.3	2.33	355
5446	0.5	—	Ca XV	—	4.46	820
5535	2	59	A X	126	2.24	421
5694.4	3	—	Ca XV	95	2.18	820
5774	—	—	Co XVI	—	—	444
6374.5	15	172	Fe X	69.3	1.94	233
6535.5	—	—	Mn XIII	—	—	315
6701.83	3	—	Ni XV	56.3	1.84	422
6777	—	—	K XIV	—	—	717
7059.6	15	—	Fe XV	38	31.7	390
7891.7	27	—	Fe XI	44	1.56	261
8024.21	0.5	—	Ni XV	21	3.38	422
10,746.80	55	—	Fe XIII	14	1.15	325
10,797.95	35	—	Fe XIII	9.5	2.29	325

Unlike forbidden lines in gaseous nebulae, solar coronal lines are fuzzy instead of sharp, with widths of the order of 0.8 to 1.0A, which become larger toward the red. Their radiation is unpolarized. Attempts to identify these lines proved unsuccessful for many years.

Following up a clue given by Grotrian, Edlén was able to work out identifications of the coronal lines in a consistent fashion (see Chapter 2) and show that all were forbidden lines of highly ionized atoms of abundant elements. The high level of ionization and the large Doppler line widths all speak for a gas kinetic temperature of 500,000°K or more.

Edlén showed that other ionization stages of iron and nickel are not observed because the relevant transitions fall in an inaccessible part of the spectrum or the transition probabilities are too low. Highly ionized atoms of neighboring metals might be expected to appear, but if the chemical composition of the corona is the same as that of the stellar atmospheres, their lines would be too weak to be observed against the strong background of the coronal continuum. Although calcium and argon are also present, no forbidden or permitted line of silicon, carbon, nitrogen, oxygen, or neon is possible in the optical region of the corona under the conditions that presumably exist there.

Numerous lines of these elements are visible in the rocket and satellite ultraviolet; some originate in the corona, others in the corona-chromosphere interface or hot chromospheric regions.

Lyot's and Waldmeier's coronagraphic work showed that different coronal lines have different spatial distributions. There are "red" regions of relatively low excitation where the λ6374 [Fe X] line is prominent and "green" regions where the λ5303 [Fe XIV] line and other high excitation lines show enhanced intensities. There exists a positive correlation between brighter chromospheric plages and coronal maxima, but coronal line emission tends to be strongest in spot regions. Here sometimes appears the λ5694 [Ca XV] line, which Roberts finds to be at least occasionally associated with prominences of the sunspot type. It is normally much fainter than λ5303 [Fe XIV] but Waldmeier once observed it to be three times as strong as λ5303 over an active spot group. Lyot found λ5694 to show a different behavior from that of any other coronal line. It appears only over active regions where the temperature may reach 4 or 6 million degrees, its excitation requiring an ionization energy of 814 ev (e.g., Billings and Zirin). It is enhanced in solar flares.

Although the level of excitation of the solar corona usually changes relatively slowly, the λ5303 line has been observed to change by a factor of two in an hour. Waldmeier found this green line to be especially useful as an indicator of coronal excitation. Regions of unusually strong λ5303 intensity, denoted by him as *C* regions, appear to be identical with the *M* regions postulated by geophysicists to account for the twenty-seven day interval between moderate terrestrial magnetic storms and aurorae.

The total emission in λ5303 and λ6374 closely follows the sunspot cycle, although the λ6374 maximum at the solar cycle maximum is more irregular and less pronounced than for λ5303 (Bretz and Billings, 1959). Although both electron density and temperature are higher at sunspot maximum than at minimum, the intensity of emission in both λ5303 and λ6374 is determined mostly by the electron density.

The intensities of emission lines in the inner corona are measured routinely at a number of observatories. The intensity, I, is normally defined as $I = \text{const } WP(\lambda)$, where $P(\lambda)$ is usually taken as 10^{-6} the intensity of the photospheric continuum and W is the equivalent width of the emission line in terms of the intensity of the underlying continuum. Note that W refers to a column taken along the line of sight, and hence must be interpreted with the aid of a model of the corona.

The upper levels of the coronal transitions may be excited either by collisions or by absorption by radiation from the sun. The transition probabilities are extremely low (compared with those for ordinary permitted lines) but the flux of radiation is high. In the inner corona, within about 200,000 miles from the solar surface collisional excitations prevail; in the outer corona radiative excitations must predominate.

Unless the temperature is higher than six million degrees, photoionization can be neglected in the solar corona; the number of collisional ionizations will equal simply the number of radiative recombinations. Formulas were developed by Miyamoto, Biermann, Woolley, and Allen, and others, but the most complete discussion is that given by Elwert (1954). Since we can neglect the ionization from higher levels the total number of collisional ionizations cm^{-3} sec^{-1} from the ith ionization stage may be written as

$$\mathfrak{F}_{AC} = N_A N_\varepsilon \int_{v_0}^{\infty} \sigma_{AC}\, vf\,(v)\, dv \sim 8.5 \times 10^{-6}\, N_A N_\varepsilon T_\varepsilon^{-1/2}\, \frac{\Omega e^{-\chi_i/kT}}{g_A} \tag{9–53}$$

if the target area for collisional ionization can be written in the form

$$\sigma_{AC} \sim \frac{4.17}{g_A}\, \Omega_{AC}\, v^{-2} \tag{9–54}$$

where Ω_{AC} is sometimes referred to as the *collisional strength.* It has to be calculated by quantum mechanics. Here $f(v)$ is the Maxwellian velocity distribution, N_A is the number of atoms in the lowest level of the ith ionization stage (essentially N_i), g_A is the statistical weight of the lower level, and χ_i is the ionization potential. In practice it is difficult to compute (e.g., Burgess, 1960; Schwartz and Zirin, 1959) and the cross-section formula proposed by Elwert (1953) is often used.

The number of recombinations on all bound levels per cm^3 sec^{-1} from the $(i + 1)$st to the ith ionization stage will be

$$F_C = N_{i+1} N_\varepsilon \sum_j \int_0^{\infty} \sigma_{\kappa j}(v)\, f(v)\, v\, dv = 3.26 \times 10^{-6}\, N_{i+1} N_\varepsilon Z^4 T_\varepsilon^{-3/2} \tilde{G}(T_\varepsilon) \tag{9–55}$$

by analogy with the expression for the recombination rate in hydrogen (see, e.g., Aller, 1956). Here $\sigma_{\kappa j}(v)$ is the target area for recapture of an electron moving with velocity v into a level j, and the summation is taken over all levels to which recombination can occur. Z is the charge on the nucleus and N_{i+1} is the number of atoms in the $(i + 1)$st ionization stage. The function $\tilde{G}(T_\varepsilon)$ has been calculated for hydrogenic ions, it increases monotonically with T_ε. Equating $\mathfrak{F}$ and F_c from Eqs. 9–53 and 9–55 we get

$$\frac{N_{i+1}}{N_i} = 2.6 \frac{\Omega}{g_A \, Z^4 \, \tilde{G}(T_\varepsilon)} \, T_i \, e^{-\chi/kT} \tag{9–56}$$

For any particular ion, the calculations of Ω and $\tilde{G}(T)$ are likely to be difficult. Simplified expressions have been given by Elwert (1954), and for a rough reconnaisance one may use his 1956 expression

$$\frac{N_{i+1}}{N_i} = 5 \times 10^5 \left(\frac{\chi_{\rm H}}{\chi_i}\right)^2 \frac{e^{-\chi_i/kT}}{(\chi_i/kT)} \tag{9–57}$$

where $\chi_{\rm H}$ is the ionization potential of hydrogen. Notice that the electron density cancels out, the radiation field does not enter, and if the necessary atomic parameter can be specified, the electron temperature alone controls the state of ionization.

From the observed intensity ratio of the [Fe XIV] and [Fe X] coronal lines one may calculate the coronal electron temperature if the necessary parameters are known and if these emissions both originate in the same strata. Elwert concluded that the ionization temperature of the corona was less than a million degrees (Art. 9–24).

Theoretical discussions of coronal lines closely resemble those of forbidden lines in gaseous nebulae (see, e.g., Aller, 1956) except that one must take into account radiative excitations of metastable levels also. In analogy to procedures employed in gaseous nebulae, one may derive T_ε and N_ε from simultaneous measurements of intensities of the λ6374 [Fe X] and λ5303 [Fe XIV] lines in the same layers.[17] Thus, Waldmeier (1954) found the electron density to be largest over the spot zone and to diminish towards the poles, secondary maxima being found over the polar zone of coronal activity. Likewise the electron temperature tends to be higher in these regions and at sunspot maxima. The electron density at sunspot maximum is about twice its value at sunspot minimum. Newkirk (1959, 1961) finds the electron density is about doubled over active areas.

9–23. The Coronal Spectrum in the Ultraviolet and X-ray Regions. Interpretations of the solar corona require that the line spectra be studied in the far ultraviolet and solar x-ray region, a task for satellites and

[17] Dollfus, Billings, and Rosch, however, have obtained strong evidence that emission lines of different ionization potential, e.g., λ6374 [Fe X], λ5303 [Fe XIV], and λ5602 [Ni XV] are produced in different regions of the corona. Hence comparison of their intensities cannot give local coronal temperatures.

rockets. A few coronal lines have already been identified and with improved techniques much more can be learned.

Below 1500A the proper Fraunhofer spectrum disappears and is replaced by an emission spectrum of mainly chromospheric origin. Finally, below 100A the solar spectrum is almost entirely due to the corona.

The *x-ray spectrum* due to the free-free, bound-free, and line transitions, may be calculated by procedures due to Elwert (1954). The free-free transitions or Bremsstrahlung produce a continuous spectrum with a maximum near 90A at 1,000,000°K. To these must be added the bound-free electrons by charged ions, which shifts the maximum to about 40A at 1,000,000°K. The total continuous emission is about 0.015 erg $cm^{-2}sec^{-1}$ at the earth's distance from the sun, of which the flux shortward of 20A is about 0.0001 erg $cm^{-2}sec^{-1}$ at T = 1,000,000°K (Elwert, 1960). See Fig. 9–41.

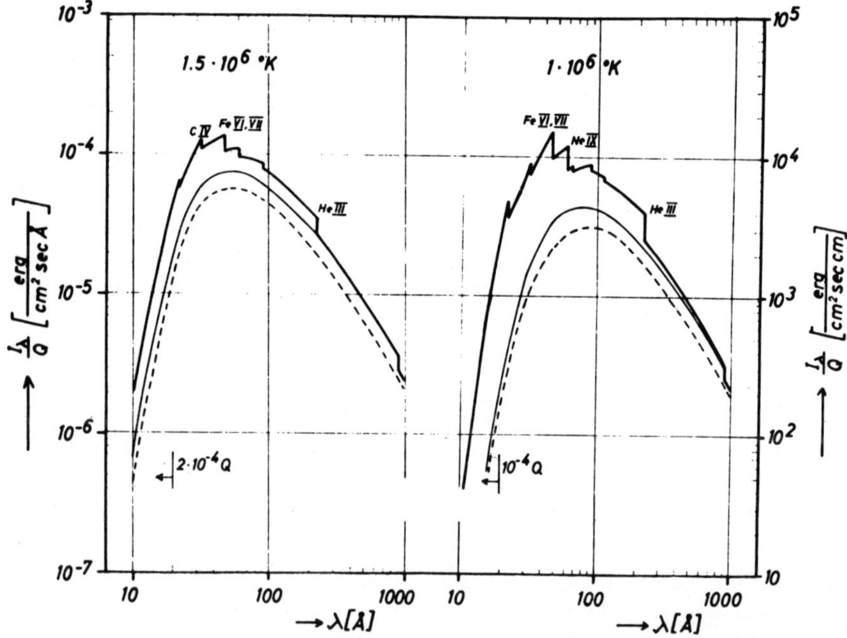

Fig. 9–41. Continuous X-ray Radiation from Solar Corona

These curves give the flux at the earth's distance, divided by an uncertainty factor Q arising from inaccuracies in the physical theory. Dashed curves give the free-free transitions; light curves H + He free-free + H free-bound; heavy curves indicate the total. (Courtesy G. Elwert, *J. Geophys. Res.* **66**, 395, Fig. 3, 1961.)

In the regions between 2 and 100A, the line spectrum, which is excited predominantly by collisions, is much more important (see Fig. 9–42). At 700,000°K, most x-rays fall between 44A and 120A. Using Elwert's method, but with revised solar abundances (Goldberg, Müller, and Aller,

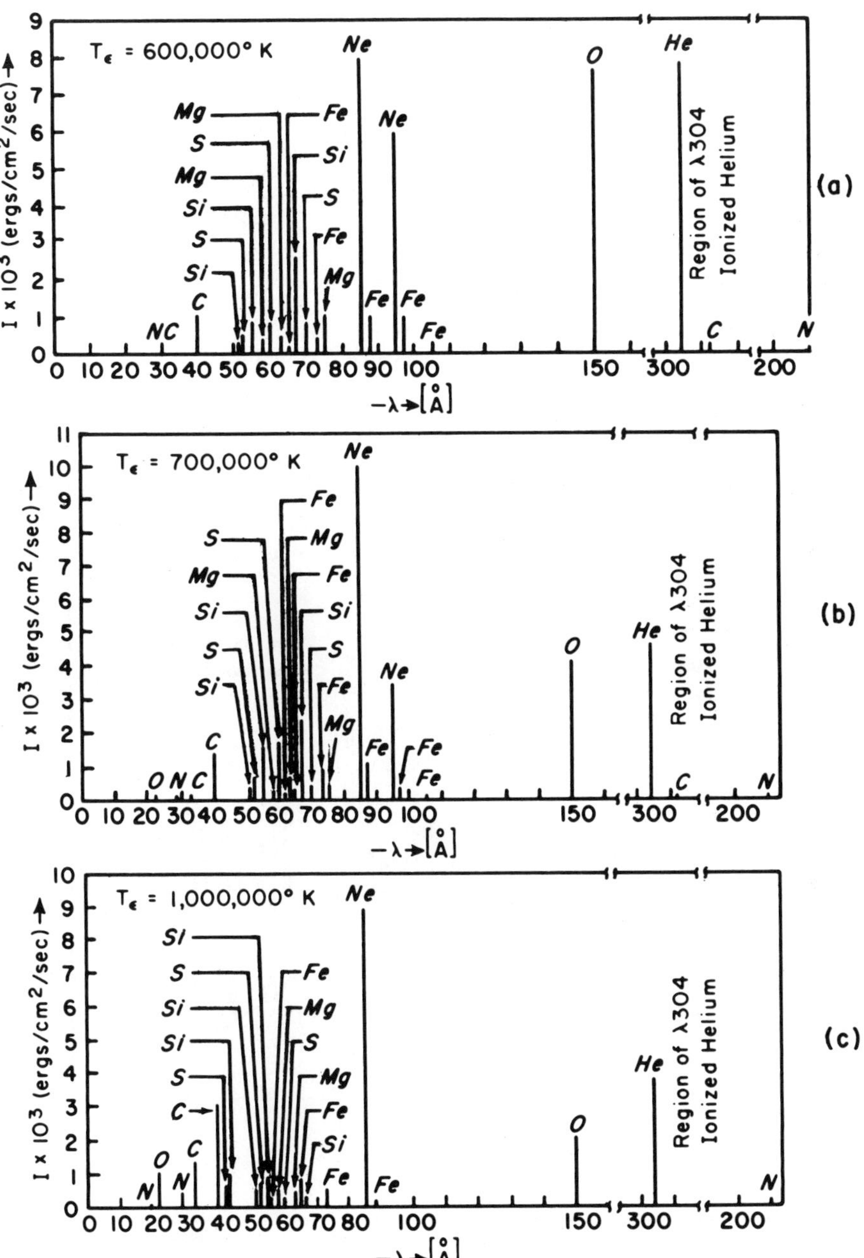

FIG. 9–42. X-RAY LINE RADIATION FROM SOLAR CORONA

The flux received at the top of the earth's atmosphere for various x-ray lines has been calculated by A. Boury using Elwert's methods. The insert gives the region λ300–λ320 which includes the resonance He II lines; contributions from chromospheric regions are not included. (From *Astronomical Experiments proposed for earth satellites*, by L. H. Aller, L. Goldberg, F. T. Haddock, and W. Liller, University of Michigan, 1958; report prepared for McDonnell Aircraft Corp.)

1960), A. Boury calculated the following values of the x-ray flux at the top of the earth's atmosphere for the solar minimum:

$T_e = 600{,}000°K$	$700{,}000°K$	$1{,}000{,}000°K$
0.040	0.036	0.029 erg $cm^{-2}sec^{-1}$

Fig. 9–42 shows that scans of the x-ray region of the coronal spectrum (such as those secured with the OSO satellite) should be able to detect relatively small changes in the excitation of the corona.

Certain line ratios could be chosen as excitation indications if the spectral details could be monitored. At solar maximum, Elwert (1960) finds that the flux may amount to 0.4 erg $cm^{-2}sec^{-1}$ for $T = 1{,}000{,}000°K$. The uncertainty in the calculations due to cross-section errors, fluctuations in electron density, etc., amount to a factor of 2 or 3. Effects of gross excitation changes are shown in Fig. 9–43 (Elwert, 1961).

So far, observations of solar x-rays have been secured from measurements

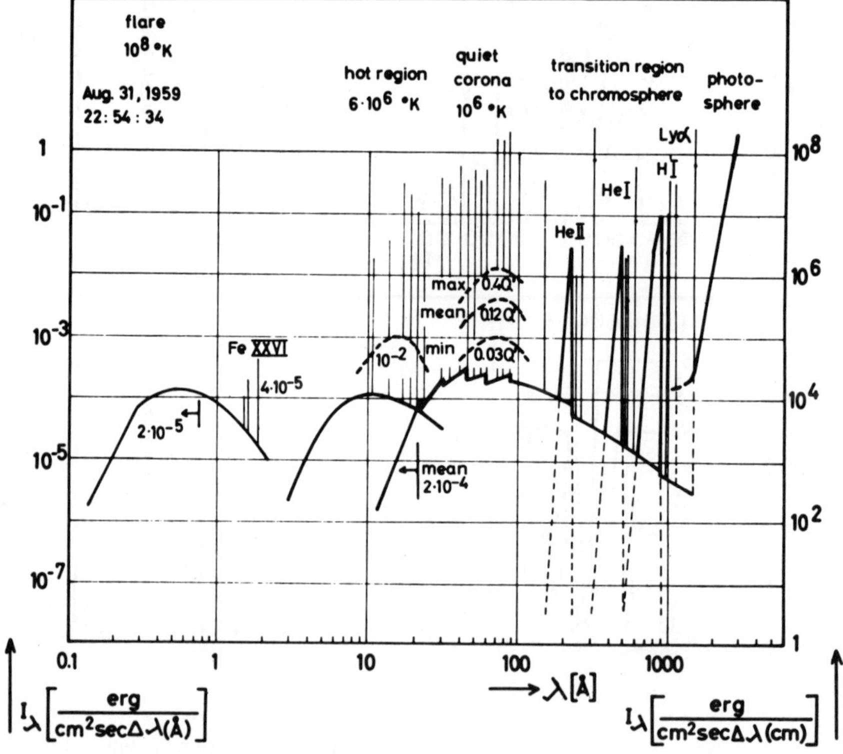

Fig. 9–43. Composite Solar X-ray Spectrum

The contributions of the transition region between the corona and chromosphere (which are taken from Oster's data) and of the quiet corona exist at all times. Those from the "hot" regions and flares can exist only under disturbed conditions. (Courtesy G. Elwert, *J. Geophys. Res.* **66,** 397, 1961.)

of the ionosphere and by rockets. The ionosphere is always "on the job," but data are difficult to interpret. It is generally believed that ionization in the *D* layer is determined under normal conditions by the Lyman-α radiation, and under unusual active sun conditions by hard x-rays (wavelengths less than 8A). The higher *E* layer is controlled by radiation in the 40–100A range by Lyman β, C III λ977 and the Lyman continuum, while the highest (*F*2) layer may be determined by radiation falling in the regions

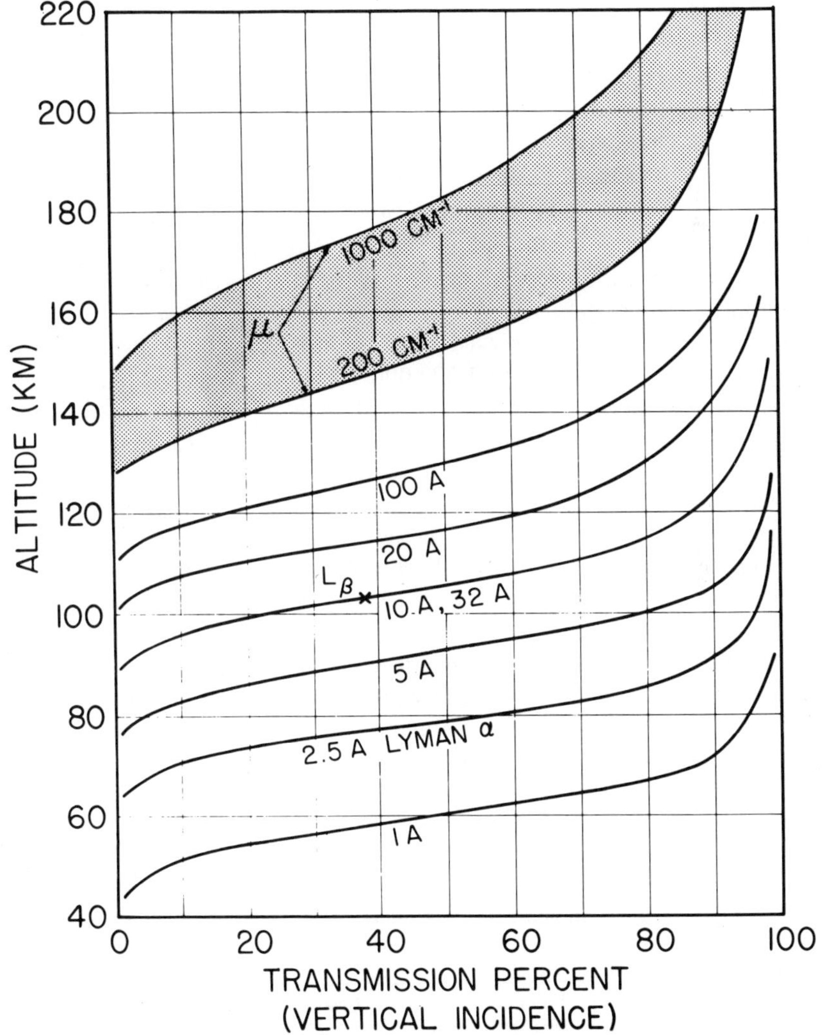

Fig. 9-44. Penetration of Solar Radiation into the Upper Atmosphere

The shaded band corresponds to the region from 100A to 1000A. Notice how hard x-rays and Lyman α can penetrate fairly deeply into the atmosphere. (Courtesy, H. Friedman.)

between 100A and Lyman α, since these wavelengths are absorbed in the highest atmospheric strata. Furthermore, since the ionosphere is sensitive to corpuscular radiation, and the mechanisms of ionization, etc., are not adequately known, it is not satisfactory as a radiation counter.

Data concerning particularly hard x-rays can be secured sometimes from balloon flights (e.g., Petersen and Winckler, 1959), but the most definitive information is that secured by Friedman, Byram, Chubb, and their associates (1954, 1955, 1956, 1957) by rockets. Unfortunately, rocket work suffers from a number of disadvantages. (a) A continuous check on the sun is not possible, (b) flights are so short we obtain sporadic snapshots frequently only after the principal phenomenon has started to subside, and (c) for certain regions of the spectrum, particularly soft x-rays, it is difficult to get above the troublesome part of the earth's atmosphere. (See Fig. 9–44.)

Until very recently (March 1962), it has not been possible to obtain actual spectra in the x-ray region. One had to compare solar radiation in one range with another for intensity and for relative variations by means of appropriate filters.[18]

For the quiet sun the observed x-radiation is probably consistent with a coronal temperature of about 1,500,000°K, but an exact value is hard to assign because of the scatter in observational values, due to disturbed conditions on the solar surface. An inspection of Fig. 9–42 shows how the predicted line spectrum changes with temperature.

Marked differences in the x-ray spectrum are noted at times of enhanced solar activity. See Fig. 9–45. Although the 50A region shows increases of several hundred per cent, the short wavelength emission is strikingly accentuated, i.e., the x-ray spectrum becomes "hardened." Elwert (1955) interpreted this phenomena by supposing that in certain active areas the temperature rose to four or six million degrees. Then there occurs a striking flux increase below about 25A, although no line radiation persists below about 6A, since (except iron) all abundant elements have K-shell x-rays only longward of 6A. Iron has a K-shell x-ray at 1.9A which might become excited at even higher temperatures. These "hot" spots in the corona are indicated by appearance of the yellow [Ca XV] line and often last only the order of an hour.

[18] For example, Friedman and associates used gas-filled counters with windows, etc., as follows:

Window	Thickness	Gas	Wavelength Range
Beryllium	0.13 mm	Neon	1–8A
Glyptal	0.0015 mm	Helium	44–100A
Lithium fluoride	2.0 mm	Nitric oxide	1180–1340A

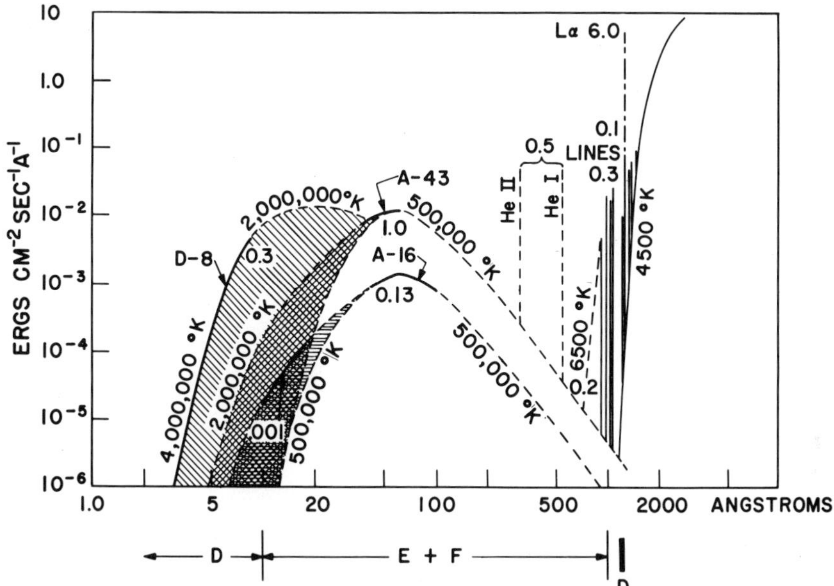

FIG. 9–45. SOLAR SPECTRUM FROM ROCKET MEASUREMENTS

Solar spectrum from rocket measurements. Solid line portions of curves are derived from measurements. Dashed curves are extrapolations. X-ray spectrum is represented as 500,000°K grey body or two different total intensities corresponding to N.R.L. Aerobees 16 and 43. Shaded areas represent short wavelength x-rays originating in hot coronal condensations. *D*-8 is data obtained by Rockoon during class 1 flare. Total flux under each curve is indicated: *D*-8, 0.3 erg cm^{-2} sec^{-1}; *A*-43, 1.0 erg cm^{-2} sec^{-1}; *A*-16, 0.13 erg cm^{-2} sec^{-1}. Measured intensity of Lyman-alpha line has varied from 0.1 to 6.0 erg cm^{-2} sec^{-1} over the past seven years. Lines in neighborhood of Lyman-alpha obtained from the spectrogram obtained by Johnson *et al.* Helium resonance lines are estimated for purposes of ionosphere calculation. Effective ionosphere regions of absorption are indicated along abscissa. (Courtesy, H. Friedman.)

On April 19, 1960, Friedman, Blake, Chubb, and Unzicker photographed the sun in the x-ray region with a pinhole camera in an Aerobee rocket at a height of 130 miles. They find that most of the x-ray emission originates in local coronal areas overlying the calcium plages. These same areas produced the slowing varying radio component and require temperatures of 0.6 to 1.5 million degrees to explain the radio data. Kawabata (1960) suggests that the brighter the plage the higher the temperatures.

The plages appear to be structures of rather fundamental import, the higher the excitation of the feature examined the more important the plages. The great importance of securing spectroheliograms of the sun in the light of the resonance lines of HI, He I, and He II is obvious.

9–24. Temperature of the Solar Corona. The coronal electron density may be obtained from electron scattering of solar photospheric light by electrons of the so-called "K" corona. No special difficulties are involved, although electron density appears to depend on the phase of the solar cycle.

A determination of the coronal temperature is a much more difficult problem. The non-appearance of the Fraunhofer lines in the K corona, due to the high speeds of free electrons that scatter light, and the absence of the recombination lines of hydrogen, indicate an electron temperature in excess of 300,000°K. If the density gradient in the corona is interpreted on the assumption it is in hydrostatic equilibrium (van de Hulst, 1950; Pottasch, 1960) a gas kinetic temperature of about 1,400,000°K is found, but there is some doubt that the corona is in hydrostatic equilibrium.

There are three methods for getting the temperature of the corona: (a) ionization equilibrium, (b) radio emission, and (c) line breadths. These methods give to some extent inconsistent results, suggesting that some of the basic assumptions are at fault.

The ionization equilibrium gives a temperature of about 800,000°K, a figure which cannot be changed very much by improvements in the cross-section (Seaton). One must compare lines originating in the same volume element.

The temperatures deduced from the r-f brightness of the sun generally exceed the ionization temperatures; they fall in the range 800,000°K–1,500,000°K.[19] From the intensities of x-rays at 10A, Elwert obtained about 1,500,000°K.

Direct measurements of widths of emission lines (see, e.g., Eq. 3–94) yield gas kinetic temperatures consistently in excess of one million degrees (see, e.g., Billings, 1959). From measurements of the profiles of the red and green coronal lines at the total solar eclipse October 12, 1958, Jarrett and von Klüber (1961) found 3.5×10^6 and 3.2×10^6°K, respectively.

Thus there is a three-fold discrepancy between ionization equilibrium, radio, and line width temperatures. The discordance cannot be blamed on the observations or atomic collision parameters. In part it may arise from inhomogeneities in the corona, in the concentration of different ions to different regions of the corona, and in a possible large-scale turbulence. Such effects as thermal convection which would tend to concentrate the more highly ionized atoms in the hotter regions must also be considered. This uncertainty in the temperature clouds attempts at quantitative theories of evaporation, solar winds, etc.

In regions of enhanced activity we can expect the temperature to rise as high as 2,000,000°K. Above active plage regions which emit λ5694 [Ca XV] yellow coronal lines, the gas may reach temperatures 4,000,000°K or even 6,000,000°K. These active regions have short lifetimes and are not to be confused with what are sometimes called "permanent" coronal con-

[19] From an interpretation of the decay of type III bursts as plasma oscillations, Warwick deduced a temperature of 1,000,000°K.

densations where the electron density is high but the temperature may be actually lower than in the surrounding corona.

The "hot spot" active regions are probably the sources of particle emissions at times of geomagnetic storms and aurorae. Rapid evaporation or even blast effects may occur at these times. Insofar as temperatures and excitation effects are concerned, observations in the visual region give only fragmentary data. Continuous monitoring by satellites can supply not only data on temperature and excitation sensitive radiations, but more particularly on areas in front of the sun's disk, where corona activity can never be observed in optical, but now only in radio frequencies.

9–25. Theories of Coronal and Chromospheric Excitation. The outer envelopes of the sun present to the theoretician a problem of gargantuan complexity. Their over-all structure and day-to-day variations are controlled by magnetohydrodynamical phenomena. From observations of the heights, intensities, breadths, and displacements of spectral lines, and from energy distributions in the continuum, we attempt to affix the temperatures and densities of the emitting gases as a function of height. Since these gases radiate under conditions deviating substantially from thermodynamic equilibrium, the observational data are to be understood only in terms of detailed mechanisms. The requirements on the theories are heavy and their critical assessment is often difficult.

A satisfactory theory of the corona, for example, must not only account for the high gas kinetic temperature and ionization observed, but also it must explain the occasional occurrences of regions of very high temperature. It must account for the shape of the corona and the form of the streamers. Finally, it must explain why the excitation tends to be greater near disturbed areas such as sunspots.

In order to answer these questions, one must compare the efficiencies of mechanisms that add energy (and mass) with the counteracting dissipative mechanisms. Most of the coronal radiation in the visible region is simply scattered sunlight and does not enter into the heat balance at all.

The dissipative agencies are easy to identify: (a) emission of radiant energy in the form of coronal forbidden lines, x-rays, and continua, (b) thermal conduction, and (c) ejection of high-speed particles, or plasma clouds at lower speeds, by evaporation or a solar "wind." The sources of energy are less easy to identify since they are most likely of a non-thermal, not necessarily visible nature.

The high excitation of the corona has been attributed to the kinetic energy of interstellar material accreted by the sun (Bondi, Hoyle, and Lyttleton), high-speed particles accelerated by fluctuating magnetic fields in the neighborhood of spots (Kiepenheuer), illumination from small intensely heated areas on the solar surface (Menzel and Goldberg), dissipation of shock waves from solar granules (Biermann, Schwarzschild, and Schatzman), and dissipation of energy from magnetohydrodynamic waves (Alfvén, Osterbrock, and others).

The heating of the chromosphere, as well as the corona, must be considered. Actually, the chromosphere dissipates more energy than does the corona. Much of the light emitted by the chromosphere (as observed in the flash spectrum for example) is simply scattered photospheric light. According to calculations by Osterbröck (1961), the most important mechanism of energy loss is emission by the negative hydrogen ion in the low chromosphere. Some emission is provided by the hydrogen lines and lines in the ultraviolet. He estimates the radiative flux of the chromosphere that is balanced by non-radiative heating to be 4×10^7 ergs cm^{-2} sec^{-1}. We estimate the total flux from the corona — line emission and thermal conduction outwards — to be 2.5×10^4 ergs cm^{-2} sec^{-1}, while perhaps ten times as much is lost by conduction back to the transition layers. Thus, as has been emphasized by Athay and Thomas, much more non-thermal energy is dissipated in the chromosphere than in the corona. The temperature of the chromosphere is lower because it emits energy much more efficiently than does the corona.

In the Schwarzschild-Biermann theory, the energy of the chromosphere and corona is supplied by a stream of acoustical waves that originate in turbulent motions of granules, transport mechanical energy through the photosphere, and dissipate it as heat in the chromosphere and corona.

As granules rise through the unstable layers, they are slowed down, and part of their mechanical energy is dissipated in the form of compressional waves. Since their formation and dissipation takes place in a random way, the energy they transmit to the overlying strata may be partly in the form of acoustical waves which are simply superposed without phase or amplitude relation with one another. The waves move with a propagation velocity V which is essentially the velocity of sound. Since w is the mean material velocity in the waves, the energy density is ρw^2, and the noise flux per unit area is of the order of $\rho w^2 V$. The exact numerical coefficient depends on the shape of the wave.

The kinetic energy of a somewhat idealized model of a granule of diameter d, density ρ, and velocity v is

$$E = \frac{1}{2} \rho \frac{\pi}{6} d^3 v^2 \tag{9–58}$$

The granular velocities v are of the order of $\frac{1}{2}$ km sec^{-1}. With $\rho = 10^{-7}$ gm cm^{-3} (corresponding to a point near the top of the convective layer), and $d = 10^8$ cm, the energy per granule will be about 6×10^{25} ergs. If, at any time, there are 10^6 granules on the surface of the sun and their lifetimes are 300 sec each, the rate of energy transport will be

$$L_{\text{granules}} = \frac{NE}{t} = 2 \times 10^{29} \text{ ergs sec}^{-1} \tag{9–59}$$

If we take the velocity of the granule, $\frac{1}{2}$ km sec^{-1}, as equivalent to the material velocity of agitation, and note that the velocity of sound in an isothermal photospheric layer is

$$V = \sqrt{\frac{k}{M}}\, T \sim 7 \text{ km sec}^{-1} \tag{9–60}$$

where M is the mass of the average atom, the upper limit to the energy transport in the wave will be

$$F_{\text{wave}} = \rho\, \overline{w^2}\, V = 1.7 \times 10^8 \text{ ergs sec}^{-1} \text{ cm}^{-2} \tag{9–61}$$

At any one instant, only a portion of the solar surface is covered by rising granules. Since the noise flux $F = \rho\overline{w^2}V$ is constant, a decrease in the density ρ will be accompanied by a rise in the velocity of agitation, w.

Not all the energy carried by the granules is converted into sound energy. If only a tenth is so transformed, the amount of energy available for heating the chromosphere and corona will be about 2×10^{28} ergs. The total amount of radiation lost by the corona, principally as free-free and bound-free transitions is 10^{23} ergs sec^{-1}, while as much as 10^{27} ergs sec^{-1} may be radiated by the chromosphere. Hence the granular noise appears capable of supplying enough energy.

After a few "wavelengths," the acoustical waves degenerate into shock waves. Unlike an electromagnetic wave, whose velocity is governed solely by the properties of the medium through which it is passing, the amplitude of a compressional wave in a fluid will have an influence on its velocity. If the amplitude is large, the top will propagate with a greater speed than the base, and the latter will tend to become distorted (see Fig. 9–46).

The number of wavelengths required for a sound wave to be transformed into a shock wave is of the order of V/w.[20]

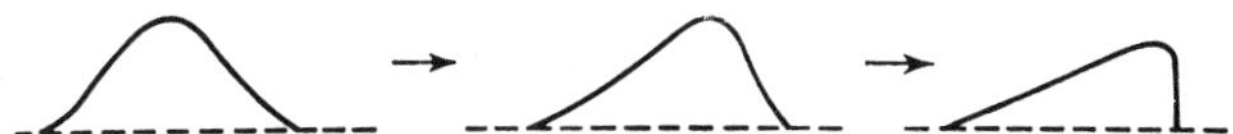

Fig. 9–46. Deformation of a Wave of Finite Amplitude into a Shock Wave

The crest of the wave moves faster than the velocity of sound, the wave becomes distorted, and finally evolves into a shock wave as the front face becomes vertical. The sketches show the velocity as a function of space coordinate for a given time.

The acoustical waves pass through the photospheric layers without appreciable dissipation of mechanical energy, since the material velocity w in an individual wave is much smaller than the sound velocity V. For

[20] Small sounds do not evolve into shock waves because they are damped out by viscosity before they travel such a distance. In an explosion, on the other hand, P is much greater than P_0 (the initial pressure) and a shock wave appears at once and energy is quickly lost. Shock waves of velocity 6000 km sec^{-1} in tubes have been observed to fall appreciably in intensity in 2 m.

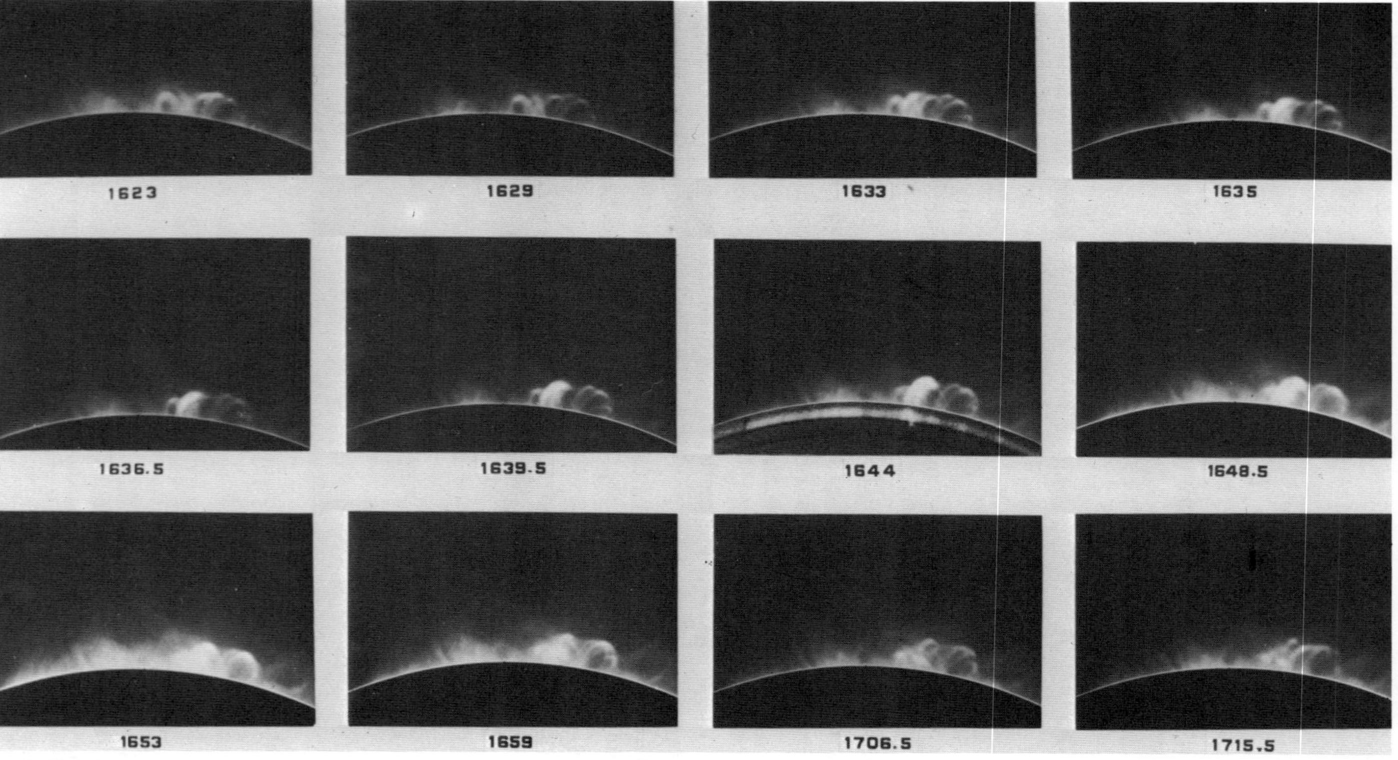

FIG. 9–47. GREEN CORONA, NOVEMBER 26, 1956
(Courtesy, R. B. Dunn, Sacramento Peak Observatory, Geophysical Research Directorate, Air Force Cambridge Research Center.)

example, in the photosphere, where $w = \frac{1}{2}$ km sec^{-1} and $V = 7$ km sec^{-1}, $w/V = 1/14$. In the chromospheric layers where the density has decreased by a factor of 100, w will have increased by a factor of 10, w/V will be 1/1.4, and the sound wave will be transformed to a shock wave.

A shock wave can be described as a surface of discontinuity moving in a fluid with a velocity U which is always greater than that of sound and is larger the greater the discontinuity in pressure and density. The material velocity w_1, the pressure p_1, and the density ρ_1 in front of the wave are related to the corresponding quantities w, p_2, ρ_2 behind the wave by means of the Hugoniot relation for the conservation of energy, and by the equation of momentum carried across the shock front.

The velocity and energy of a shock wave gradually decreases as it progresses. The discontinuity of P and p at the shock front is abrupt. As the wave passes through the gas, the latter experiences a non-reversible compression. Subsequently, it expands adiabatically, but the wave does not get back all the work done in the sudden compression; some of the latter is dissipated in heat.

We can calculate the amount of energy thus degraded if we suppose that the shape of the wake behind the front is always conserved. It is found that the greater the velocity of the material within the wave, the faster the dissipation of energy.

Schatzman has considered in some detail the transfer and dissipation of shock wave energy in the chromosphere. The compressional waves are transformed into shock waves within a few hundred kilometers. Because of the steep temperature gradients, V increases rapidly and the sound waves are refracted downward through the chromosphere and dissipated into heat.[21] Shock waves cannot move through the corona where the mean free path of an atom is much greater than the size of the wave. Energy transport in this region may be largely by conduction.

The shock wave mechanism seems to explain the observed temperature and density distributions in the chromosphere and corona. The sharp maximum of the temperature distribution in the lower corona may arise from effects of bending of sound waves by the steep gradient. If refraction were not present, the temperature maximum would be too far away from the photosphere.

Some of the mechanical energy supplied to the chromosphere and corona may persist in the form of kinetic energy of particles ejected from the sun. The kinetic energy carried by spicules in their outward motion is small compared with that embodied in the ultraviolet radiation field but

[21] Weymann (1960) has considered a chromosphere of a star whose surface gravity is much less than that of the sun. A model in which the shocks are moderate in strength but so numerous that the time lag between passage of successive shock fronts is small seems adequate for the sun but not for giants where long-period disturbances are required.

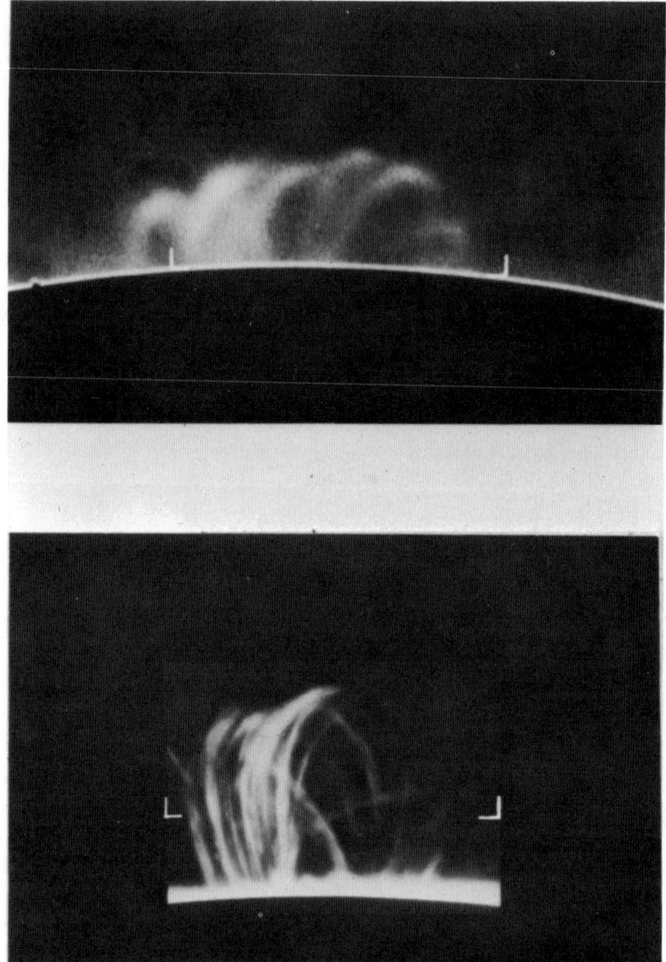

Fig. 9–48. Simultaneous Photographs of Corona (Green λ5303 Line) (*upper*) and Chromosphere (Hα Line) (*lower*)

Photographed $16^h\ 33^m$ U.T., November 22, 1956. (Courtesy, R. B. Dunn, Sacramento Peak Observatory, Geophysical Research Directorate, AFCRC.)

may be comparable with that carried out by ejected particles during the quiet sun phase (see, e.g., Athay, 1961).

Thus it appears that most of the acoustical energy is dissipated in the chromosphere, rather than in the corona. This does not mean that the chromosphere should be hotter than the corona for, as Athay and Thomas have emphasized, one must equate energy input to energy losses, and at least radiative energy losses are much more important in the chromosphere than in the corona. They conclude that most of the acoustical, shock wave, or magnetohydrodynamical energy input is dissipated in the lower layers of the chromosphere and appears as radiation in the visible spectrum. A smaller amount is dissipated in the upper chromosphere, and a still smaller amount is liberated in the corona, where ultraviolet and x-ray emission is produced. From this point of view the fundamental structure is the chromosphere; the corona is essentially a tenuous outer appendage.

The temperature structure of the chromosphere and corona is related to the composition of the sun itself. The corona has a temperature of two million degrees because hydrogen is about 10,000 times as abundant as the metals. Complex ions of metals are able to radiate much more effectively than hydrogen since their cooling capabilities (per gram) are much greater than for H or He. This low abundance in the sun permits a much higher temperature to be reached because energy dissipation is hindered. A star composed largely of iron probably would not have so hot a corona.

What can be said concerning the lower corona and the corona-chromosphere? Athay and Thomas proposed a model based on the idea that the temperature structure of the chromosphere and lower corona is determined by a non-radiative energy supply (acoustical waves, magnetohydrodynamic waves, etc.) and a radiative energy dissipation. They assumed that the energy input is less strongly dependent on the local temperature than is the radiative loss and found that for certain temperature ranges the structure would be unstable against small perturbations. These temperatures will not exist, therefore, in a steady-state model, and there will be rapid transitions from one steady-state temperature domain to another. The range in which the temperature field will be stable are those in which the energy loss is a decreasing function of the temperature. That is, if the rate of energy loss with respect to temperature is positive, the temperature configuration is stable; otherwise it is unstable. Consequently, one must identify the atomic species that produce the actual dissipation of energy by radiation. The problem is wretched because of deviations from thermodynamic equilibrium. Athay and Thomas find two domains of stable temperature: (a) the temperature is less than 12,000°K because the Balmer, Paschen, etc., lines of hydrogen dominate the energy loss; (b) the temperature lies between about 40,000°K and 70,000°K because the subordinate transitions of helium dominate. Thereafter the problem becomes one of extreme complexity, because, as the corona is approached, the number of elements contributing to the heat losses increases and the radiative transfer problem becomes difficult. At two million degrees, for example, losses by Mg X may dominate.

Hence the chromospheric model is characterized by plateaus and sharp temperature rises, a result consistent with the optical and radio data and the appearance of the chromosphere itself. The existence of stable temperature domains may also explain the origins of some prominences which appear as low-temperature regions in a hot medium. Likewise the hot and cold columns may be at least partly explained by the existence of these stable temperature domains and variations in energy input from convection cells.

Perhaps the general excitation of the corona may be explained by the hypothesis of granular noise, but its behavior in the neighborhood of spots suggests that additional causes must act. High-energy corpuscular radiation received at the earth shows that charged particles must be accelerated to high speeds by varying magnetic fields in active areas of the solar surface. Further, the coincidence of plage areas with regions of moderately strong magnetic fields indicates that magnetic effects are significant.

Thus, although more or less conventional shock waves may play an important part in heating the chromosphere and corona, it has been pointed out by Alfvén, van de Hulst, Piddington, de Jager, and Osterbrock (1961) that since the outer solar envelopes consist of partly ionized gases located in regions with finite magnetic fields, one must consider hydromagnetic effects. That is, there may occur not only longitudinal, i.e., sound waves, but also magnetohydrodynamic (*mh*), sometimes called hydromagnetic or Alfvén waves, which travel with a velocity

$$V_A = \frac{B}{\sqrt{4\pi\rho}}, \quad B = \mu H$$

The motion is transverse to the direction of propagation. These waves have been produced experimentally in incompressible fluids of high electrical conductivity (e.g., mercury) placed in a magnetic field.

In a gas which is a good electrical conductor mixtures of longitudinal and Alfvén waves are observed. There are three kinds of waves described as a "fast" mode, a "slow" mode, and the "Alfvén" mode. In a compressible gas the Alfvén mode has similar properties to those shown in an uncompressible fluid. The fast mode can propagate in any direction, moving with the velocity of sound if the magnetic field is inconsequential and with the Alfvén velocity in regions of strong magnetic field. The slow mode propagates only in directions close to the magnetic field direction. Osterbrock points out that for any reasonable assumed field strength, for the chromosphere and corona, the fast mode disturbances will propagate as sound waves in the hydrogen convection zone, and as *mh* waves in the upper chromosphere and corona.

In Osterbrock's more detailed theory, fast mode disturbances are generated in the hydrogen convection zone, pass through the photosphere, and get dissipated as shocks in the low chromosphere. He explains an increased degree of heating in plage areas by an excess generation of fast mode disturbances by turbulent magnetic fields in the convection zone that underlies the photosphere. At greater heights where the influence of the magnetic field dominates, collisions between shocks may occur with the result that energy may be fed into the slow mode and Alfvén mode. In this picture the corona is heated mostly by Alfvén waves; slow mode shocks may heat the upper chromosphere. The spicules observed at the limb are attributed to chromospheric material pushed into the corona along magnetic lines of force by shock and *mh* disturbances.

Allen finds that the accretion of the particles responsible for the zodiacal light as well as the sweeping up of interstellar material is much too small to supply the energy lost by the corona. The irregular nature of the temperature distribution as shown by the variation of $\lambda 5303$ in space and time is difficult to understand if the source of energy is external to the sun. The complicated isophotic contour of this and other lines likewise makes it extremely doubtful that the corona is in hydrostatic equilibrium. On the contrary, these irregularities strongly suggest the influence of electromagnetic forces. A magnetodynamical rather than a purely static model of the corona must be sought.

Energy losses from the corona in the form of x-rays, discrete lines, etc., can be monitored by coronagraph observations and satellites. Losses in the form of corpuscular streams can be assessed to some extent from aurorae, magnetic storms, measurements made in balloons, etc., but more directly from satellites and space probes.

Escape of corpuscular radiation may occur in various ways. For example, Chamberlain (1960) and others have suggested that high-energy particles escape by evaporation from the high-energy tail of a Maxwellian distribution, thus providing a steady small "breeze" of outward moving particles. On the other hand, Parker (1960, 1961b) argues that the interplanetary gas and solar corpuscular radiations are nothing more than the hydrodynamical continuation of an expanding solar corona. At sunspot minimum, the expansion velocity may be only about 300 km sec^{-1} with a density as low as 20 cm^3. When the sun becomes active the solar wind velocity rises to about 500 km sec^{-1} and the electron density at the earth's distance may rise to 100 cm^3 or more. The enhanced solar activity which registers in the earth's atmosphere as magnetic storms and aurorae apparently involves corpuscular radiation moving outward from the sun at 1000–2000 km sec^{-1}. Parker (1961a) describes these phenomena as hydrodynamical blast waves produced by active regions which develop suddenly in the corona and produce temperatures of the order of 4 million degrees (see Chapter 10).

No direct measurements are yet available to assess the role of thermal conduction. Energy can flow downward from the corona toward the photosphere or outward toward the earth. Chapman (1957) calculated the outflow of energy from the corona by thermal conduction, but neglected the effects of turbulence which could play a most important role in carrying away heat. Since it could increase the heat flow it could lower the electron temperature in the plasma near the earth. The calculation is difficult because of paucity of data on temperature in the outer corona. In the outer corona, losses by thermal conduction are more important than those by radiation.

The solar corona evidently reaches as far as the earth and probably some distance beyond. The inner part rotates as a solid body in consequence of the influence of solar magnetic fields, but the outer parts must "drag" behind. Interaction of these outer parts with the earth produce terrestrial effects we describe in Chapter 10.

PROBLEMS

9–1. Consider an atmosphere composed only of hydrogen and helium in hydrostatic equilibrium. Assume that the gas is fully ionized. Show that the equation of hydrostatic equilibrium is:

$$\frac{dT}{dh} + T_{\varepsilon} \frac{1}{h} \frac{dN_{\varepsilon}}{dh} = -\frac{m_{\mathrm{H}} g}{k} D$$

where $D = 0.648$ if H/He = 7 (by numbers). The acceleration of gravity is $g = 2.74 \times 10^4 R^2/R_0^2$ for the sun if R_0 is the solar radius.

9–2. Bright prominences show a continuous spectrum. If the continuous spectrum at $\lambda 5000$ is caused by electron scattering of photospheric light (the light is partially polarized) while the Balmer continuum shortward of $\lambda 3650$ is produced by recombination on the second level (see Eq. 9–25) show that by measuring the absolute intensities at $\lambda 5000$ and $\lambda 3650$ one can determine both N_ε and T_ε if the thickness and height of the prominence are known.

9–3. Let W denote the equivalent width of a coronal emission line in terms of 10^{-6} of the intensity of the continuum at the center of the disk. Show that if the line is excited by radiation from the sun, W will be independent of height for an isothermal corona. Conversely, show that if the upper level is excited by collisions W will be proportional to N_ε.

9–4. Using Milne's relation (Eq. 4–155), write out the ionization Eq. 9–56 for a coronal ion in terms of the absorption coefficients for the individual levels, the target area parameter for ionization, Ω, the ionic density, T_ε, the ionization potential, and appropriate statistical weights.

9–5. Consider the excitation of coronal lines. Let number of collisional excitation sec^{-1} cm^{-3} be

$$\mathfrak{F}_{AB} = 8.54 \times 10^{-6} N_A N_\varepsilon T_\varepsilon^{-1/2} \frac{\Omega_{AB}}{2J_A + 1} e^{-\chi_{AB}/kT_\varepsilon}$$

and let the number of radiative excitations be F_{AB}. Then, if collisional de-excitation can be neglected and we can suppose

$$\mathfrak{F}_{AB} + F_{AB} = N_B A_{BA}$$

derive an expression for emission per unit volume in the coronal lines in terms of N_A, N_ε, Ω_{AB}, A_{BA}, χ_{AB}, T_ε, and the intensity $I_{\nu AB}$.

9–6. Making use of the fact that

$$\frac{N(\text{Fe XIV})}{N(\text{Fe X})} = \frac{N(\text{Fe XIV})}{N(\text{Fe XIII})}\,\frac{N(\text{Fe XIII})}{N(\text{Fe XII})}\,\frac{N(\text{Fe XII})}{N(\text{Fe XI})}\,\frac{N(\text{Fe XI})}{N(\text{Fe X})}$$

show that by repeated applications of Eq. 9–57 one may derive T_ε from $N(\text{Fe XIV})/N(\text{Fe X})$. Making use of Eq. 9–56, what will be the error if Ω/g is consistently 20 per cent too large? 20 per cent too small?

REFERENCES

Because of the vast extent of the literature on solar phenomena, we list here only books and review articles which contain extensive bibliographies. References to more recent papers, given in the text, are found in the Bibliography.

Encyclopedia of Physics, **52**, edited by S. Flügge. Springer-Verlag, Berlin, 1959. "The Photosphere of the Sun," by L. Goldberg and A. K. Pierce; "Structure and Dynamics of the Solar Atmosphere," by C. de Jager.

NUMEROUS AUTHORS. 1953. *The Sun* (*The Solar System*, vol. I). G. P. Kuiper (ed.). University of Chicago Press, Chicago.

UNSÖLD, A. 1955. *Physik der Sternatmosphären*. Springer-Verlag, Berlin.

For the problem of the solar magnetic field, see:

BABCOCK, H. W. 1961. *Ap. J.* **133**, 572.

A general theoretical account of non-radiative energy sources in the solar chromosphere and corona is given by:

OSTERBROCK, D. 1961. *Ap. J.* **134**, 347.

LILLER, W. (ed.), 1961. *Space Astrophysics*, McGraw-Hill Book Co., Inc., New York.

CHAPTER 10

THE ACTIVE SUN: SOLAR-TERRESTRIAL RELATIONSHIPS

10–1. Introduction. The influence of the sun on terrestrial affairs has attracted attention since the time of Herschel if not before. Solar activity affects radio communications, magnetic compasses, and even power lines. An understanding of effects of solar electromagnetic and corpuscular radiation upon the earth's atmosphere will make air travel safer, will improve radio communications and ultimately may solve questions of long-range weather forecasting.

Although, at minimum, the sun affects geomagnetic activity, and of course controls the ionization in the ionosphere, we are concerned here particularly with sporadic solar activity, especially flares, radio bursts, and consequent effects on the earth. Correlations between the general level of solar activity and the state of the ionosphere, terrestrial magnetism, and frequency of occurrences of aurorae have long been well established. Radio fadeouts, disturbances of the earth's magnetic field, and bright auroral displays are commoner at times of greater solar activity (sunspot maximum), but the most striking effects are those correlated with individual flares.

The earth's upper atmosphere serves as a radiation and particle detector, capable of yielding a continuous (if not easy to interpret) record of solar activity and streams of charged particles. We mentioned (Chapter 1) that ozone which reaches a maximum concentration at a height of about 30 km, extinguishes all radiation below 2900A. Oxygen becomes photodissociated above about 100 km and exerts a pronounced effect on the physical state of the atmospheres. The far ultraviolet radiation and x-rays become extinguished in the upper atmosphere, as do most corpuscular streams of charged particles. Hence the great importance of radiation and particle detectors carried in balloons, rockets, and satellites.

Impinging ultraviolet and x-ray quanta ionize atoms and molecules to an increasingly greater degree as we ascend in the earth's atmosphere. Although the concentration of charged particles becomes easily detectable at a height of 80 km, it does not begin to influence the propagation of radio waves under normal conditions until a height of about 100 km is attained. Although rocket studies have shown that the electron density varies continuously with height, early radio reflection studies suggested that this layer of partially ionized gas, or ionosphere, was stratified into several distinguishable layers called the D, E, and F layers, situated roughly at elevations of 70 to 90, 100 to 120, and 150 to 300 km above the surface of the earth. During the day, at small zenith distances of the sun, the F-layer is broken into two sublayers, the F_1 layer about 30 km thick at 170 km elevation, and the F_2 layer at 290 km and of 50 km thickness. The E and F layers are the most important, although the behavior of the D-layer (which is not observed by rockets) is of interest because of its enhanced absorptivity at times of flare activity.

TABLE 10-1

PROPERTIES OF THE IONOSPHERE

Region	Mean Height (km)	Mean Thickness (km)	Electron Density (electrons cm^{-3})	Probable Constituent Ionized	Probable Sources of Ionization
D	80	20	Strongly variable	NO	Lyman α, x-rays
E	110	10	1.5×10^5	O, O_2	X-rays ($400 > \lambda > 100$)
F_1	170	30	2.5×10^5	O, N_2(?), A(?)	$\lambda < 210$ A
F_2	290	50	2.5×10^5	N_2	$\lambda < 912$ A

Table 10-1 summarizes some of the mean properties of the ionosphere. During solar activity the heights, thickness, and ionic concentrations vary steeply. The data in the last two columns represents current hypotheses concerning the ionosphere, which will be revised as additional data are obtained from satellites, balloons and rockets. The general run of the electron concentration with heights admits of a qualitative understanding with the aid of current theories of physical properties in the upper atmosphere, but insufficient is known about recombination and attachment coefficients, etc., to permit an unambiguous identification of the mechanisms involved.

The ionosphere is most frequently studied by radio sounding at vertical incidence. The highest frequency ν_c which is reflected from a given layer reveals the electron density in accordance with the relation [1]

$$N_e = 1.24 \times 10^{-8} \nu_c^2 \tag{10-1}$$

A variation of N_e in the E and F_2 layer with the sunspot numbers has long been known. The E, F, and F_2 layer ionization shows a close correlation with the calcium flocculi figures. Furthermore E-layer ionization is closely correlated with the 10-cm radio radiation. Stetson found the measured values of ν_c at night also depend on the sunspot numbers as though the ionization of the layers were controlled by charged particles ejected by the sun. During the IGY special radio receivers (riometers) kept a constant record of the level of opacity of the ionosphere.

[1] Interpretation of the data is often complicated by the fact that over a large range of height the electron density may differ only slightly from that in the reflection layers.

Studies of auroral displays also supply data on solar activity, although the connection may be modified to some degree by the action of the Van Allen radiation belts. Actual particle velocities can be obtained directly from Doppler displacements of auroral lines.

Rapid variations in comet tails are most likely due to corpuscular streams from the sun (Biermann), although space probes yield data that are easier to interpret.

Rockets fired during flares and detectors flown in balloons and satellites yield essential data on corpuscular streams and x-rays that do not penetrate to the lower reaches of the earth's atmosphere.

10–2. Solar Flares. The most interesting of transient events on the sun are the phenomena commonly called *flares*. With them are often associated enhanced emission in radio-frequency regions and sometimes in ultraviolet and x-ray regions as well. Flares may produce increased particle emission from the sun, sometimes blasts of ions of up to cosmic ray energies.

Flares are perhaps most aptly described as the brightenings of a previously existing plage area. They often tend to recur in the same location on the solar surface. They are usually observed only in monochromatic light, although Carrington's great flare of 1859 and a small number of other flares have been seen in white light. They are always connected with active regions on the sun (called *centers of activity*). Hence their frequency of occurrence depends on the phase of the solar cycle. At sunspot minimum months may elapse without a flare.

Flares show a vast range in size, brightness (i.e., importance) and attendant solar and terrestrial effects, from subflares to the great cosmic ray flares. There appears to be no real distinction between flares and subflares, the differences being one of degree rather than kind. The areas of subflares are often less than 0.0001 that of the solar disk but their brightness in Hα may be as great as that of a real flare. About 20,000 subflares were reported during the IGY, whereas during this same interval about 6000 ordinary flares of importance 1, 2, or 3 were reported. See Figs. 10–1 and 10–2.

Waldmeier and Bachman (1959) find flare frequency to be proportional to sunspot numbers. Of 4828 flares observed from 1945 to 1954, 79 per cent were of importance 1, 18 per cent were of importance 2, and only 3 per cent were of importance 3. The IGY data shows that "flare statistics" depend in part on the observer, it being very difficult to obtain precisely comparable data from workers with very different equipment (Dodson and Hedeman, 1960). The importance of a flare is usually assessed from its area and brightness, although the area is the really fundamental datum. A flare of area 0.0006 (in units of the solar disk) would be of importance 3, one of area 0.0012 would be 3+, whereas the largest flare observed had an area about 0.5 per cent of the area of the sun.

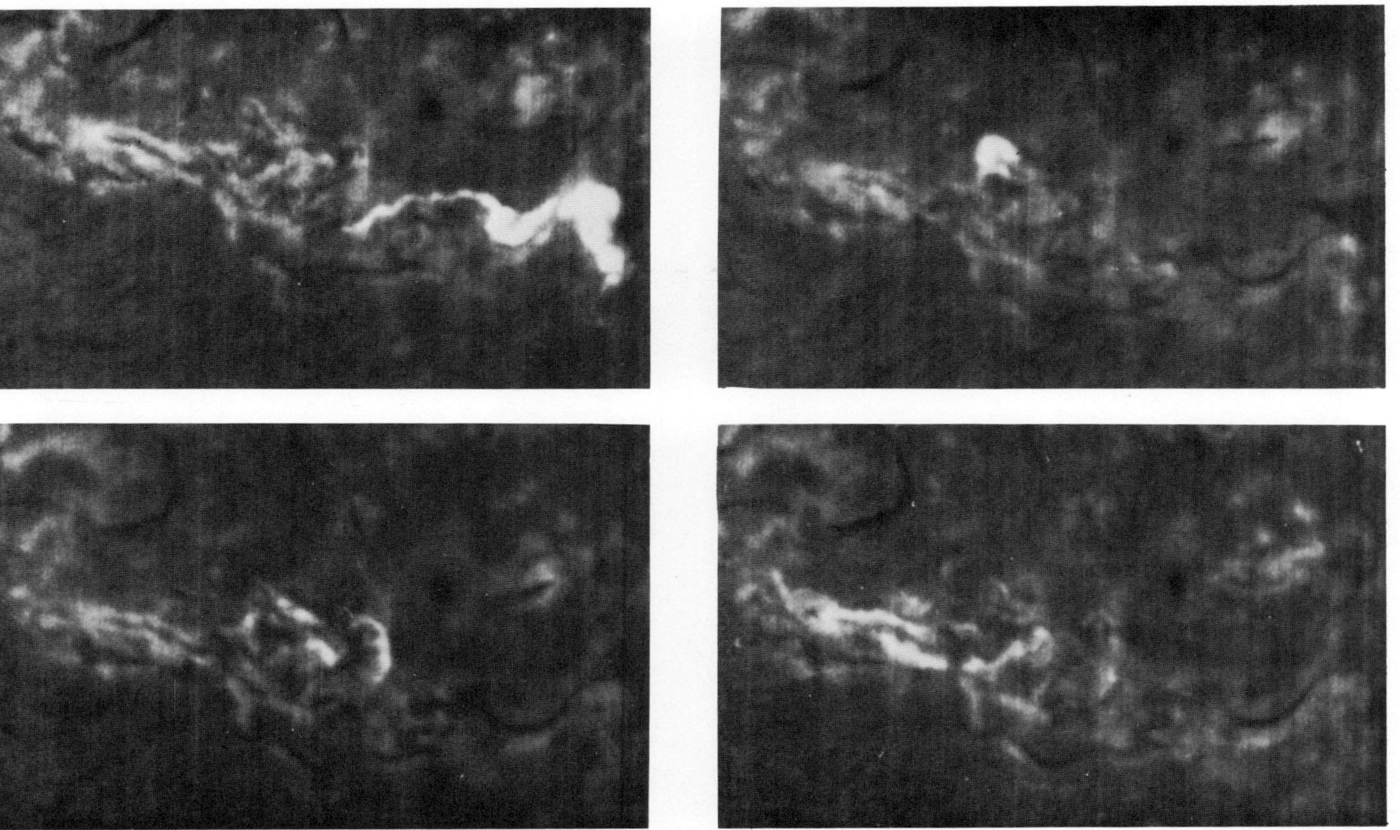

FIG. 10–1. FOUR FLARES IN A LARGE ACTIVE SOLAR AREA

August 12, 1947, Hα spectroheliogram, McMath-Hulbert Observatory.

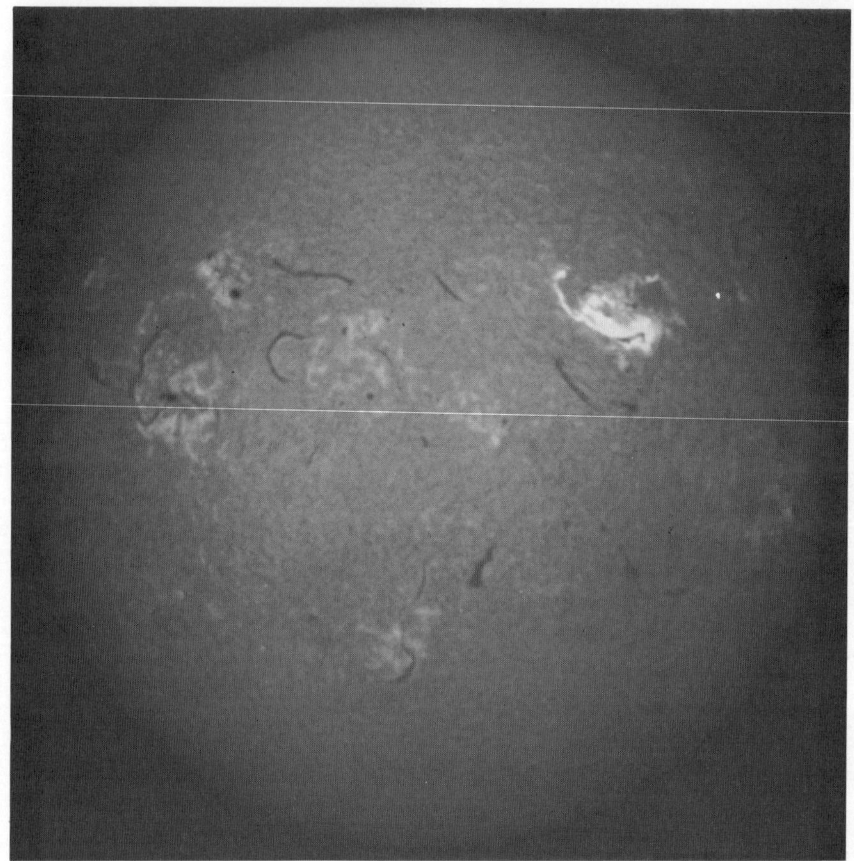

FIG. 10–2. GREAT FLARE OF 1959, JULY 16 (21^h30^m)

Hα filtroheliogram, 1959 July $16^d21^h30^m$ shows flare in northwest quadrant of solar disk. Telescope focal length 6 ft. (Courtesy, The McMath-Hulbert Observatory, The University of Michigan.)

TABLE 10–2

SOME PROPERTIES OF FLARES

Importance	1	2	3	3+
Central intensity of profile (units of local continuum) *	0.88	1.17	1.28	
Average width Δλ in A as measured on spectra *	3.0	4.5	8	15
Duration $= \bar{t}$ †	22^m	45^m	85^m	

* Dodson, Hedeman, and McMath, *Ap. J. Suppl.* no. 20 (1956).
† Waldmeier and Bachman (1959).

Furthermore, "importance" is based on Hα data, although integrated brightness, and radio frequency flux over a fixed frequency interval $\Delta\nu$, could also be used. Table 10–2 summarizes data on the central intensity of the Hα emission which falls on the central core of the line and its width as a function of the importance of the flare, although some flares that produce ionospheric disturbances may have an Hα that is not as bright as the continuum.

Flare photometry is difficult. If filter techniques are used, the band pass of the filter must be known. Also, one must allow for the fact that the neighboring regions are observed in radiation from the center of the strong red hydrogen line.

Most flares rise to maximum brightness in 3 or 4 minutes, but some rise as rapidly as 1 minute, some as slowly as 20 minutes, but the decline is always slower than the rise. A flare's duration depends more on its importance than on its intensity. Flares near the center of the disk appear to last longer than those near the limb, an effect of observational selection. (See Dodson, Hedeman, and McMath, 1956.)

Flares always occur in centers of activity. Some appear as long, bright ropes, other appear as small, bright spots whose arrangement suggests a certain chainlike nature. Although they occur most often in the region between spots or close to or over the penumbrae of spots, some especially interesting ones have been observed extending directly over spot umbrae. Flares tend to occur repeatedly, not merely in the same region, but seemingly in exactly the same small portion of the solar disk. They show a complex structure, often as though a wave of excitation ran through an existing plage area. See Fig. 10–1. Some seem to follow, at least in part, channels already established by the dark filaments (prominences seen in projection on the disk). Although filaments sometimes disappear after the outbreak of a great flare, they sometimes seem to survive bright eruptions. The brightest flares are associated with magnetically complex spot groups and tend to occur when the group is growing most rapidly. They cover a larger area (Fig. 10–2) and tend to fade away more slowly than the smaller outbursts.

According to statistical studies by J. Warwick, flares tend to occur in chromospheric regions at heights of 20,000 km. Flares studied at the limb must be distinguished from surges with which they are often associated. The great May 8, 1951, flare (see frontispiece) rose to 50,000 km in less than 90 sec and was accompanied by increase in 200 mc radiation. The great February 23, 1956, flare was seen elevated at the limb in white light. Although a flare may occur initially in the chromosphere, much of its development takes place in the corona.

One of the most fundamental observational problems concerning flares is the status of magnetic fields and magnetic field gradients in the neighborhood of flares. The plages and direct magnetic field measurements would suggest magnetic fields occur where plages appear, but the resolution of the structures of the fields is usually only 4″ or 5″ and fine details are lost. A number of workers have suggested that flares break out along neutral lines where not only the longitudinal but also the transverse fields vanish.

From a study of the great flares of June 16, 17, 18, 1959, and the 3+ flare on July 10, 14, 16, Bruzek (1960) noted that flares occur during the orderly development of a spot and its magnetic field. The greater flares and most of the minor ones originated at the edge of umbrae and near or on the "neutral" lines of zero longitudinal field strength. That is, flares originated in regions of strong magnetic fields, usually with strong transverse components but never in neutral points of zero field strength.

On the other hand, Severny (1960) found that flares occur at neutral points of crossed magnetic fields where the gradient exceeded 0.05 gauss km^{-1}. A comparison of the magnetic fields before or after a flare indicated that a considerable rearrangement, more particularly a simplification, occurs resulting in energy dissipation. Michard (1961) found from spectrographic observations that flares were associated with a considerable decrease in the total magnetic energy integrated over the active regions. Magnetic fields connected with great flares appear to be peculiar; regions of strong fields of opposite polarity are found close to one another. The rapid rearrangement of these fields seems characteristic of flare activity. Motions of the ionized gas (plasma) near the postulated neutral points of the magnetic fields are characterized by instabilities (pinch effects), which Severny suggests may produce strong Doppler broadenings. Hence the line broadening in the spectrum may depend on the direction of the line of sight.

Richardson and Minkowski, C. W. Allen and M. A. Ellison, Michard, and others, have described the *spectra* of flares. In a bright outburst, Hα develops bright generally symmetrical emission wings soon after the onset of the flare. The emission wings fade away in a few minutes although the flare may persist longer. For example, in the July 25, 1946, flare, bright Hα attained a width of 15A with a central intensity three times that of the continuum. Simultaneously, D_3 (He I) appeared in emission over the flare, although it remained in absorption over the nearby photosphere. The Si II, λ6347, λ6371, lines were reversed and the profiles of many iron and nickel lines were filled in. The continuous spectrum appeared to be brighter over a small area containing the flare. The Fe II lines are enhanced in flares, but Fe I is weak. Severny and Michard found that the metallic lines come mostly from the disturbed chromosphere. They do not appear in limb flares, but the yellow line corona is observed to increase near limb flares. An increased intensity of the Fe II lines occurs in prominences and the "hot spots" of the solar chromosphere, but the low-excitation Fe I lines are not intensified. There also occur bright helium D_3, λ10,832, and certain Paschen lines, although weaker flares show D_3 in absorption. Some

3+ flares show a strong continuum. The spectrum of a disk flare is always superposed on the Fraunhofer continuum.

The spectra of flares differ appreciably from one to another. Some show particularly strong metallic emission lines; others show mostly strong hydrogen emission lines, but the breadths of hydrogen emission lines is not a sure index of the importance of a flare, insofar as terrestrial effects are concerned. Some important flares such as the August 22, 1958, "cosmic ray" flare, showed no broad lines at all, whereas some flares with very broad hydrogen lines produced no such striking effects.

Observations of flare spectra are difficult because (a) the slit must be placed exactly on the small bright image, (b) the flares decline rapidly in brightness. Since much of the flare radiation is probably non-thermal in origin, it is desirable to secure observations over a long wavelength range, including especially the rocket ultraviolet.

At what levels do the various features of flare spectra arise? How does the excitation vary with height and time? From a study of the spectra of 60 flares, Michard (1960) concluded that the metallic line emission came mostly from deeper layers than the hydrogen lines and decreased generally faster after the maximum. The continuum is confined to the earlier phases of the flare, its spectral distribution and center-limb variation being similar to those of faculae. From the presence of emission bands of CN in the July 30, 1958, flare, Svestka (1960) concluded that it was formed in the low chromosphere.

A theoretical study of excitation conditions in flares is difficult because of probable deviations from local thermodynamic equilibrium, the assessment of which would certainly require observations in the x-ray and ultraviolet spectral regions. Various investigators have proposed specific models to account for intensities and profiles of the hydrogen lines and the general level of excitation of the spectrum. For example, Jefferies suggested models with electron densities between 5×10^{11} and 10^{13} cm^{-3}, and temperatures between 10,000 and 15,000°K and showed that they gave predictions consistent with the observed hydrogen and helium intensities.

Excitation temperatures ranging from 5500 to 45,000°K have been proposed, while electron temperatures in the range from 10,000 to 17,000°K have often been favored. The reason for the discordances is probably to be sought in deviations from thermodynamic equilibrium and the consequent non-equality of excitation and electron temperatures.

What is the cause of the great breadths of the hydrogen lines? Some investigators have attributed the broad wings to the Stark effect, but Mustel, Severny, Suemoto, and Goldberg and Müller have shown that these wings can be accounted for by radiation damping. Severny (1959) has more recently argued that the emission distribution in the far wings can not be explained by the Stark effect and must be due to Doppler motions with velocities of the order of 80–250 km sec^{-1}. Svestka suggests that the Balmer wings are formed in optically thick layers in the chromosphere-photosphere boundary where the electron density exceeds 3×10^{13} cm^{-3} and both excitation and electron temperatures are low, the thin layer being overlaid by a hotter layer where the line cores are formed.

In this connection one must mention the x-ray radiation from flares observed by Friedman from rockets at a height as low as 63 km. There occurred an enhanced *D*-layer absorption and a distinct hardening of x-rays. In 1959 they observed x-rays with energies as high as 90 kev, not emitted as brief bursts but throughout the 6 minutes the rocket was above 43 km. Energy distributions corresponding to temperatures as high as 100,000,000°K were observed. Satellite observations show that even a type I flare can produce substantial x-ray emission.[2]

Near the moment of peak intensity, high-velocity dark flocculi (darker than most filaments) are often observed. The work of M. A. Ellison, H. W. Newton, Orren C. Mohler, and Helen W. Dodson show these dark markings to be surges (see Fig. 10–3) seen in projection upon the disk.

These surges often reappear in the same location. They tend to originate (after the onset of flares) in the chromosphere and not in the corona, showing upward velocities of 100 km sec^{-1}, later downward velocities of 50 km sec^{-1}, approximately. The onset of flare activity may activate filaments. Sometimes they disappear and then reappear in the same shape as they had possessed previously.

The total energy emitted in a flare may be estimated from the Hα emission, the radio frequency emission and the speeds and densities of the emitted particles. The energies involved are of the order of 10^{30}–10^{33} ergs. In the great February 26, 1956, flare, cosmic rays alone accounted for about 10^{30} ergs, i.e., 0.1–0.2 per cent. Since all the energy was liberated in a volume about 10^{28} cm^3, the amount of energy liberated per cm^3 was 10^4–10^5 ergs. These estimates are very rough but indicate the order of magnitude of the phenomena involved.

Certainly the energy source cannot be thermal as the onset of a flare is very sudden. Possibly the requisite energy comes from sudden destruction of magnetic fields in the neighborhood of spots. Definite mechanisms have been suggested by Giovanelli and by Severny. On the basis of his observations of magnetic fields in flare regions, Severny suggested that flares begin at null points in regions of large magnetic field gradients. The flares presumably wipe out the gradients, dissipating the stored magnetic energy in a manner analogous to the "magnetic pinch" effect. Other

[2] From the available measurements for type 2 flares, Friedman and associates found a flux of 2 ergs cm^{-2} sec^{-1} for 20–100A, 0.03 ergs cm^{-2} sec^{-1} for 8–20A and 0.01 ergs cm^{-2} sec^{-1} for 2–8A. On one 2+ flare they obtained a total x-ray flux of 0.00001 ergs cm^{-2} sec^{-1} in the range 0.41–0.14A, i.e., 30–90 kev.

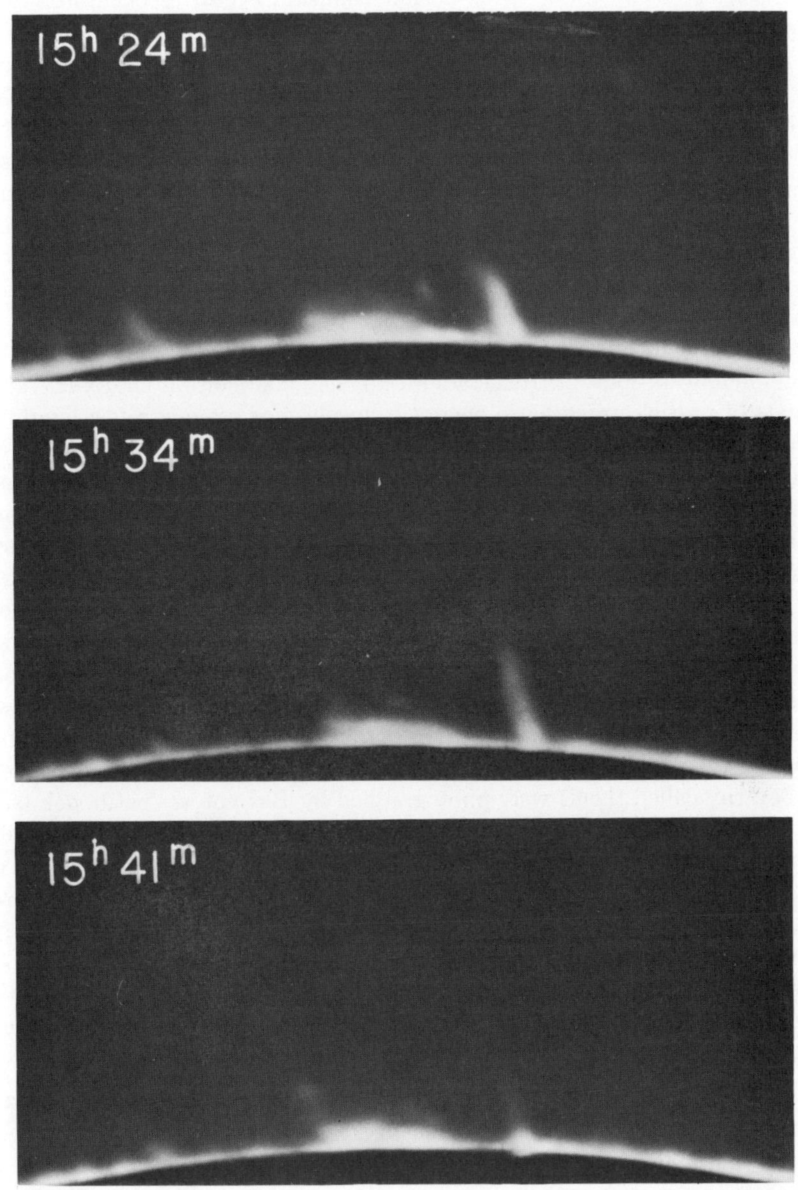

FIG. 10-3. A SURGE
Photographed August 23, 1948, at the McMath-Hulbert Observatory.

observations of magnetic fields do not confirm Severny's conclusions and the problem remains open. Flares do prefer complex magnetic groups.

Gold and Hoyle (1960) suggest that energy storage can be explained by a particular class of magnetic field where the lines of force have the general shape of twisted loops protruding above the photosphere. A sudden release of energy with dissipation as motion, in heat, acceleration of particular particles, etc., can occur when twisted loops of opposite sense and twist encounter one another and get annihilated. It has also been suggested that in the neighborhood of spots there may exist analogues of Van Allen radiation belts where electrons and ions may become trapped and build up high energies by induction processes.

The optically visible flare phenomena is apparently the least dramatic of its manifestations. Strong radio-frequency emission, intense x-ray and bursts of particles ranging in energy from plasmas moving a few hundred kilometers a second to sometimes cosmic-ray energies are among the more spectacular and probably physically more significant manifestations. Curiously, the Lyman alpha emission apparently is not enormously enhanced at times of flares. Perhaps this radiation comes from the entire chromospheric region.

A phenomenon which in some ways resembles flares but is really not connected with flares at all is Ellerman's solar "bomb." These were first observed spectroscopically as a thin band of brilliant emission 5–15A on either side of the center of Hα. Unlike flares their emission does not cross the central absorption line, i.e., the "bombs" are bright in the line wings, whereas the core is seemingly unaffected. The "bombs" (or "moustaches" as Severny called them) were photographed by B. Lyot as "petits points," i.e., small, intense, dots with diameters between $0''.4$ and $5''$ (the lower limit to their sizes is controlled by telescopic resolutions). The so-called "bombs" have lifetimes of the order of 10 minutes, and although they are likely to break out during the early or middle phases of a spot's life, they are not precursors of flares. Rather, they occur so often in the neighborhood of spots, that they can be regarded as a characteristic feature of the growth and development of a center of activity. They rarely appear in plages with no spots. (See, e.g., McMath, Mohler, and Dodson, 1960.)

10–3. Radio-Frequency Studies of Solar Flares and Related Phenomena. Important data on flares and on their influence on the corona are obtained from studies in the radio-frequency range. Transient radio frequency phenomena are often called "bursts" and have been studied in centimeter, decameter, and meter ranges.

The "bursts" observed in the cm range often show complicated polarizations and may be due to thermal plus synchrotron radiation in the flare region since they often coincide in time, height, size, and decay times with solar flares. The radio-frequency emissions in the meter range, however, are studied mainly by means of their "dynamic spectra," i.e., displays of emitted frequencies with time. In this technique, due to Paul Wild of Sydney, the output of a receiver sweeping repeatedly across a selected range of frequency is displayed on an oscilloscope screen which is photographed by a moving picture camera. Thus the changing frequency distribution of the radio-frequency "noise" from the sun is easily studied. See Fig. 10–4. The radiation whose signal is recorded is almost entirely of non-thermal origin.

It was soon recognized that various types of radiation I, II, III, IV, and V had different frequency and time dependent characteristics. Some types (II and III) show emission in discrete frequencies that are correlated with height in the solar atmosphere. These are probably due to plasma oscillations. There also exist continuum emissions that have been interpreted as *synchrotron radiation* (acceleration radiation) (see Art. 4–20) or as *Cerenkov radiation* which is produced when electrons travel through a medium at a velocity greater than the velocity of light in the medium.

The wavelength of the radiation emitted in type I events stays appreciably constant during the lifetime of the burst. They are produced high in coronal regions and show a polarization that is not correlated with spots. Type I events are not correlated with flares or centers of activity. Individual bursts of duration 0.1 to 0.2 sec occur during a noise storm but appear to originate in a fixed position (to within 80,000 miles) in a volume 30,000–200,000 miles in diameter in the corona.

The other types of bursts (II–V) all appear to have some association with great flares (sometimes all types, including the rare type IV will occur). In particular, the types II and III events are produced by rising disturbances that excite electromagnetic radiations in higher and higher levels as they arise.

First, let us discuss the type III (discrete frequency) and type V (continuous) bursts that are closely associated with one another. Type III events were once called isolated bursts. Not only do their frequencies change with time, but also the rate of change of frequency with time changes. The existence of two harmonics, approximately in the frequency ratio 2:1, suggests plasma oscillations. The ratio is not exactly 2:1 because the lower frequency falls at and is affected by the cut-off at plasma oscillation frequencies. The existence of harmonics suggests a macroscopic, non-linear oscillation of the coronal gas.

Each individual burst of a type III event lasts about 10 seconds; the whole group about one minute. Boischot, Lee, and Warwick (1960) have discussed observations of type III fast drift bursts in the frequency range 15–38 mg. These fast drift bursts last for only a few seconds at meter wavelengths, but 30 seconds or more at decameter wavelengths and show but a small frequency variation with time. These bursts account for most

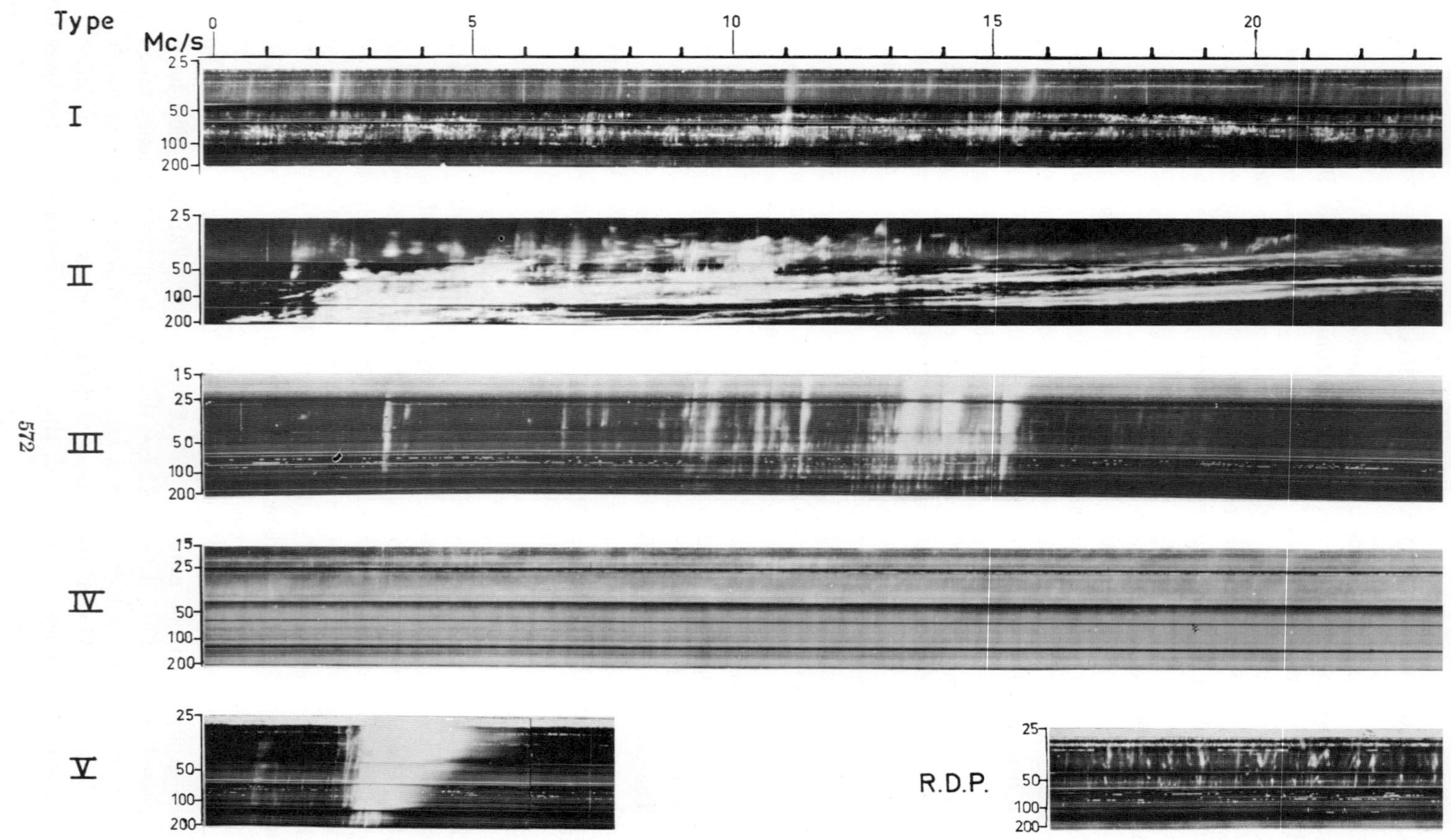
Type
Mc/s
0
5
10
15
20
I
II
III
IV
V
R.D.P.
15
25
50
100
200

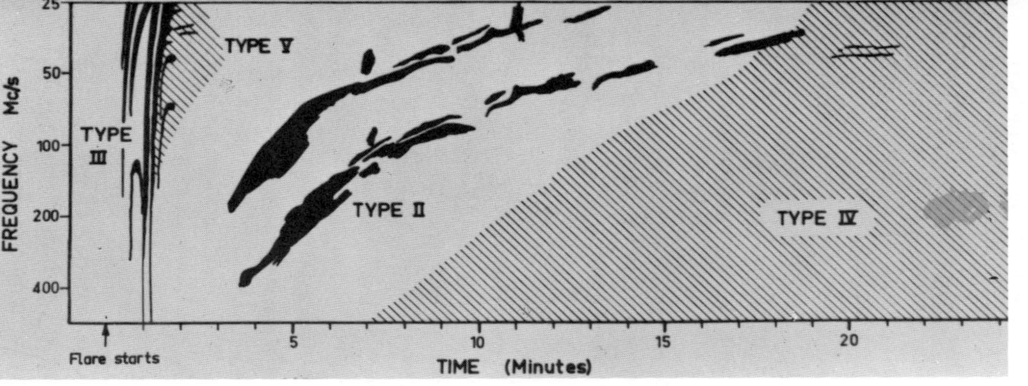

FIG. 10-4. EXAMPLES OF THE PRINCIPAL SPECTRAL TYPES OF SOLAR RADIO BURSTS AT METER WAVELENGTHS

The main figure shows records taken with the Dapto Solar Radio Spectrograph (near Sydney, Australia). The records chosen show typical examples of the six main types of events observable at meter wavelengths. Frequency is measured vertically and time horizontally, the intensity is given by the shade: intense bursts appear white, weak ones grey, and the background is black. The bright lines which are observed at a constant frequency are interference from radio transmitters.

The bottom figure shows in the same form, an idealization of the most complete sequence of events which can follow a major flare. Other details of the particular events are as follows:

Type I (22–12–59 record starts 0210 U.T.). This is a typical "noise storm" which may last for hours. This particular event occurred simultaneously with a large number of type III bursts (see below) which should not be confused with the type I.

Type II (30–11–59 record starts 0251 U.T.) This is also called a slow-drift burst. The direction of drift is always the same and the rate of drift does not vary greatly. Note, particularly near the end, the two pairs of bands, one at twice the frequency of the other.

Type III (7–10–60 record starts 0547 U.T.) These are also called fast-drift bursts. Each of the individual, almost vertical lines is a separate burst. The direction of drift is the same as for type II bursts.

Type IV (11–11–60 record starts 0408 U.T.) The example shows a form of continuum radiation (so called for its slow variation with frequency and time) which commonly follows type II bursts. It is sometimes called "type IV," but the term has been loosely applied to various forms of continuum.

Type V (4–2–60 record starts 0422 U.T.) This is a continuum event, of much shorter duration, which follows immediately after a group of type III bursts.

Drifting pairs (*D.P.*) (12–2–60 record starts 0422 U.T.) A *drifting pair* is a pair of short-duration, narrow-bandwidth bursts, very similar to one another, which occur separated in time by about two seconds. Both parts drift rapidly in frequency, either from high to low frequencies ("forward-drift pairs") or from low to high ("reverse-drift pairs"). This record shows a portion of a particularly dense storm of drifting pairs and includes drifts of both senses.

(Recorded with the Dapto Radio Spectrograph of the Division of Radiophysics C. S. I. R. O., University Grounds, Sydney, Australia. Courtesy, Paul Wild.)

of the solar emission, found in this frequency range. They may occur in regions of steep temperature gradients.

Type III bursts are interpreted as blobs of high-speed particles that move through the coronal gas and excite plasma oscillations. Each frequency corresponds to the plasma oscillation, i.e., to the density at the point of emission in accordance with Eq. 10–1, and since the frequency-time curve is very steep we conclude that the required speeds are very high — from a tenth to about three-tenths the velocity of light. Therefore an independent check on the velocities is desirable. To solve this problem, Wild used a sweep-frequency interferometer, getting the fringes as a function of time in such a way as to measure both frequency and position. He was able to verify that for a particular, fixed frequency the position of the source was constant in space. Conversely, the lower the frequency, the higher the source point above the solar limb. When he constructed the relation between electron density and height from his observations, Wild found that the quiet corona electron densities were too low, but that the density-height relation agreed with that found by Newkirk for a coronal streamer. The type III bursts may be associated with a shower of energetic particles that escape from the sun; they show a negative correlation with geomagnetic storms that, after all, involve slow-moving particles.

Type III events have been regarded as associated with flares, often small ones, but particularly those that show sudden "puffs" and dark markings that move away from flares with velocities much greater than the 500–800 km sec^{-1} often associated with flare puffs and type II events. On the other hand, Swarup, Stone, and Maxwell (1960) conclude that, although 60 per cent of type III bursts are observed to coincide with flares, the actual association amounts to only about 30 per cent — the rest are correlated by chance. They find, conversely, that 25 per cent of all flares are apparently associated with fast-drift bursts. The probability of an association increases if the flare is of greater area or height, is on the east side of the sun, or is accompanied by a surge. Few ejective solar prominences are accompanied by radio bursts, probably as a consequence of unfavorable geometry.

Type V events are associated closely with some type III events and exhibit a continuous emission over a fairly broad band, especially in the range 10 to 100 mc. They may last for several minutes after the III. Bursts observed at centimeter wavelengths (excluding major events) show better correlation with combined III, V events at meter wavelengths than with normal type III's alone.

Great flares are often characterized by type II and type IV bursts. The type II disturbances, which are often spectacular, travel outwards from the sun with velocities of 1000 to 1500 km sec^{-1}. In many instances, two distinct, harmonically related frequency bands are observed to drift from high to low frequencies at rates up to 1 mc per second. Often each drifting band is split into two (sometimes more) nearly parallel components which reproduce closely at both fundamental and harmonic frequencies. At times a herringbone structure is observed. Here the slowly drifting bands are observed to emit numerous fast-drift bursts (often in both inward and outward direction). The rare type IV bursts are sometimes preceded by a type II burst and give a continuous emission with a low frequency cut-off. The emission develops gradually, reaching a maximum after 20–40 minutes, and often lasting about 4 hours. The emitting center may have a diameter of about 10′, move upward with velocities of the order of 1000 km sec^{-1} and then stabilize. The equivalent temperature is about 10^{10} or 10^{12} °K. If a type IV continuum develops, a geomagnetic storm may occur 1.5 to 3 days later. Type II bursts do not appear to be correlated with magnetic storms unless a type IV event occurs. According to studies by Dodson and Hedeman type IV events seem to happen only when a spot umbra is covered by a flare.

Both type IV and type V events appear to be due to synchrotron radiation, whereas type II and III bursts appear to be excited by body motions of ionized gases that produce plasma oscillations. To produce intense longitudinal oscillations large charge separations are required. Plasma oscillations, however, in a box of uniform density and no magnetic field could not produce radiation since the Poynting vector would vanish. In order for radiation of frequency ν to escape at least one of the following conditions must be fulfilled: (a) the radiation is produced above the level where ν equals the plasma oscillation frequency; (b) the longitudinal oscillations are coupled with simultaneous transverse oscillations; (c) a magnetic field or extremely rapid body motions must exist; or (d) if the medium is strongly inhomogeneous, longitudinal plasma oscillations can be refracted by a density gradient, permitting radiation to escape. One or more of these conditions are likely to be fulfilled in the solar corona, but it seems likely that in most instances only a fraction of the energy stored in plasma oscillations actually escapes as radiation.

One might suppose that the splitting of the type II bursts could be caused by a coupling between gyrofrequency and plasma frequency $(f_p \pm f_m)$ in a field of 3 or 4 gauss, but there is no evidence of the polarization you would expect if this coupling existed.

A *U burst* is a fast burst that at first decreases in frequency and then increases again, after a turning point usually between 100 and 150 mc. They are usually associated with flares and attain speeds of the order of 1000 km sec^{-1}, resembling type III bursts except for inversions. Haddock suggests that U bursts probably represent harmonics of the same event. The higher-frequency burst tends to be stronger than the low-frequency one, whereas the fundamental frequency may be rarely detected. Takakura and Haddock suggested these bursts were generated by charged particle streams that do not have enough energy to overcome the magnetic field embedded within them. They reach a maximum height above the sun depending on their kinetic energy and then are pulled back down again with a velocity comparable with that with which they are ejected.

The type IV events may originate in hot coronal regions overlying flares. Takakura has considered them as synchrotron radiation due to particles with velocities between 60,000 and 270,000 km sec^{-1}. The lifetime of the radiation is determined by (1) radiation damping, (2) collisions between thermal electrons, (3) duration of supply of high velocity electrons (maybe less than one hour). If the source is in the corona, N_ε is less than 10^9 cm^{-3} and the lifetime of outbursts caused by synchrotron radiation is mostly determined by radiation damping. When the radiation source is in the chromosphere or if the magnetic field is small, the lifetime is controlled by collisions or the duration of the supply of high-velocity electrons.

Extension of the radio-frequency spectral scans to shorter and shorter wavelengths reveal further interesting characteristics of the solar bursts. Type II bursts occur only in the meter range with no concomitant activity in the cm range. On the other hand, the cm bursts seem to be correlated with production of high energy x-rays. They may consist of two components, one associated with synchrotron radiation and the other with thermal radiation.

Type IV bursts show a considerably greater complexity than previously thought (see, e.g., Kundu, 1961; Kundu and Smerd, 1962). The continuum observed at 100 mc (or at even lower frequencies) and originating at a height of 100,000 km or greater, over an area 10′ or larger in diameter, is associated with a high-frequency continuum (250–10,000 mc) that sets in near the start of the flare, is confined to altitudes less than 40,000 km, lasts an hour or less, and comes from an area 2′ or 3′. The type IV bursts are strongly associated with solar proton emission events as indicated by polar cap absorption and measurements from satellites and balloons (see Art. 10–5). See Fig. 10–5.

10–4. Centers of Activity (CA). A solar *center of activity* CA has been defined, following d'Azambuja (1953), as the sum of all optical and radio-frequency manifestations associated with the appearance, development, and decay of a magnetic region which contains a spot or spots during its life. By definition, a CA is always found in a sunspot zone, the first indication being a magnetic field and associated plages. Then spots develop usually after a few days, flares may appear, cease, and the spots then shrink and finally disappear along with associated plages. The last optical manifestation of the CA is often the formation of a stable quiescent prominence (filament) that may persist for several months unless destroyed by the

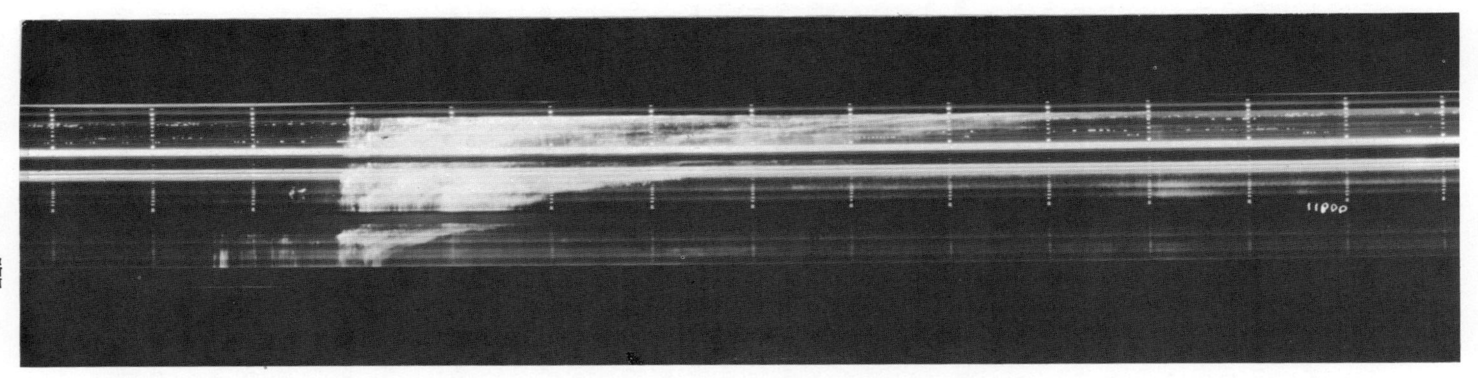

FIG. 10–5. DYNAMIC SPECTROGRAM OF A SOLAR BURST SHOWING AN INITIAL TYPE III (FAST-DRIFT) BURST, THEN A STRONG TYPE II (SLOW-DRIFT) BURST WITH CONCURRENT CONTINUUM (TYPE IV) EMISSION AT HIGH FREQUENCIES (NOVEMBER 30, 1960)

The sweep runs from 110 mc at the top to 580 mc at the bottom, and from about 17^h34^m at the left-hand side to about 18^h02^m at the right-hand side. The strong horizontal lines are interference from TV stations. (Courtesy, F. T. Haddock, The University of Michigan Radio Astronomy Observatory.)

formation of a new CA. We must recognize that terms such as "center of activity" or "active region" are descriptive only. *The basic physical agent underlying all these phenomena is the bipolar magnetic region.*

By way of illustration we summarize a description given by Dodson and Hedeman (1956) of a quickly developing (August 20–27, 1954), small, active solar region, which was studied under conditions of unusual simplicity since the sun was just emerging from minimum. There were no complications produced by overlapping CA's.

Prominences at the east limb, preceding the appearance of the spot, were the first suggestions of activity in the area where plages and spots were later to develop. A calcium plage formed about 15^h05^m U.T. on August 20, brightened (at a non-uniform rate) until August 25 and thereafter declined in area and intensity. A "green" coronal region similar in size and shape to the plage area also developed. The first sunspot appeared at 18^h00^m August 20, and the second was there by 21^h28^m. By 21^h02^m U.T. the spot group was well formed, declining after August 23, disappearing by August 26. On August 20–21 the preceding spot was larger and had a larger magnetic field, but in the period August 22–25 the reverse was true.

The first flare came only 48 hours after the initial appearance of the calcium plage, but the other four flares and enhanced radiation at meter wavelengths occurred during the period when the "following" spot and magnetic field dominated, the spots reached their maximum and declined in area. The 10-cm radiation closely followed the development of the plage and both declined more slowly than did the sunspots. The meter-wavelength excess radiation was more closely associated with flare activity and the burst-type and noise-storm events were associated with flares. On the next solar rotation virtually all evidence of activity had vanished.

This center of activity was one of the first to form after minimum in the eleven-year cycle. Its activity was very limited, no ionospheric effects were observed and geomagnetic effects were complicated. The complete lifetime of the spot group was less than six days. In more highly developed, complicated centers of activity, brilliant flares and spectacular manifestations of coronal and radio-frequency activity may occur.

10–5. Solar Activity, Flares, and Solar-Terrestrial Relationships. There is no one-to-one correspondence between solar flares and terrestrial effects. Every bright flares doesn't produce a big effect on earth. Terrestrial effects from an importance 2 flare can be more important than those from a 3+ flare. Orientation factors are important, as is also the position of the flare with respect to the spot umbra.

Events following a bright flare are: (1) sudden ionospheric disturbances, SID's, (2) magnetic "crochets," (3) cosmic-ray effects (rare), (4) polar "blackouts," (5) aurorae and effects on the magnetosphere, (6) geomagnetic storms, and (7) drag on satellites.

The first two effects are caused by enhanced ionization produced in layers below about 50 miles by direct x-ray and ultraviolet radiation from the solar flare. Their effect is almost entirely confined to the sunlit side of the earth. Short-wave 15–60 m radio waves cease to be reflected by the ionosphere (*short-wave fade-outs, SWF*). Instead they are *absorbed* in the more highly ionized D-layer. At the same time, very long wavelengths corresponding to distant thunderstorms or "parasitics" of atmospheric origin are reinforced (*sudden enhancement of atmospherics, SEA*). At yet other frequencies there occur anomalous phase differences between the ground and reflected waves, while radio noise from extra-solar-system sources may be weakened because of increased ionospheric absorption. The earth's magnetic field appears to be affected by a sudden increase in ionospheric currents. These produce so-called *magnetic crochets*, small displacements of the magnetic needle, which gradually returns to normal. These SID's usually occur soon after the onset of the flare and while it is still brightening.

The other effects (aurorae, geomagnetic storms, etc.) are produced by particles emitted by the sun. Their correlations with solar activity are well established. Consider first geomagnetic effects:

It has long been known that a compass needle shows a small-scale wavering. Part of this is due to fluctuating magnetic fields arising from motions of ions in the upper atmosphere. The small diurnal motion of the needle is due to electric currents flowing at a level of about 100 km. They are caused by dynamo action — daily varying airflow across the geomagnetic field lines induces electromotive forces, which impel electric currents in the ionosphere, whose ionization is maintained by ultraviolet light and x-rays from the sun. There are often irregular magnetic charges superposed on the diurnal motion. This magnetic disturbance often recurs at intervals of 27 days, the synodic period of the sun and then it is believed to be due to streams of charged particles emitted by certain active regions of the sun. These particular regions may correspond to areas where the $\lambda 5303$ green coronal line is strong, and may be remnants of old centers of activity. They sometimes contain long-lived quiescent prominences. Enhancement of geomagnetic activity tends to occur about three days after the passage of one of these areas across the central meridian of the sun.

At times of greatly enhanced activity, the needle shows rapid chaotic fluctuations characteristic of a magnetic storm that is often associated with radio fadeouts and bright flares. Early studies of the correlation of geomagnetic activity with sunspots show that this activity tends to be heightened to a maximum about two days after the meridian passage of a very large spot group. The smaller the spot, the less definite the correlation, unless the group happens to contain important flares. Later work showed that flares were the important element in the correlation. Strong magnetic storms are generally related to flares near the center of the disk, but sometimes storms are associated with flares near the limb. Minor storms may even be associated with coronal rather than with flare activity. Accordingly, it is likely that at times of solar flares, corpuscular streams are ejected from the sun presumably by mechanisms connected with the production of surges and types II and IV bursts. These strike the earth from several hours to three days later and produce geomagnetic storms, bright auroral and certain ionospheric disturbances that last much longer than the initial flare. The closest relationship between such ionospheric disturbances and magnetic storms and aurorae was first found at high geomagnetic latitudes. Radar studies of the ionosphere indicate that at these times streams of electrons move in the earth's upper atmosphere, in regions near the pole where the aurorae are observed. Sometimes blackouts of the background cosmic radio noise are also produced.

More recently, the study of particle emission from the sun at times of

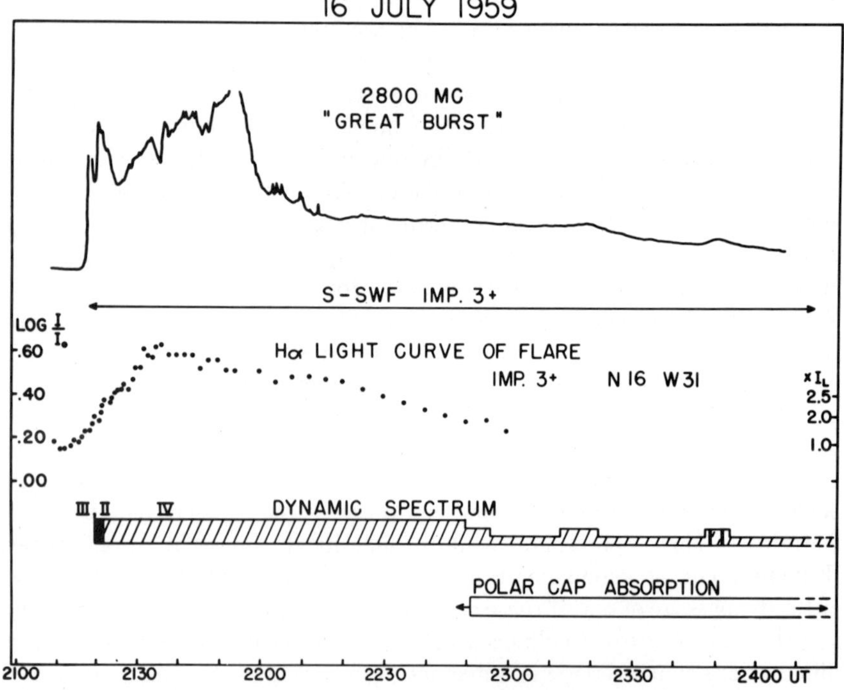

FIG. 10–6. PHENOMENA ASSOCIATED WITH THE IMPORTANCE 3+ Hα FLARE, JULY 16, 1959

Sources of data:

2800 mc burst (National Research Council of Canada)
Dynamic spectrum (Radio Observatory, University of Michigan)
Polar cap absorption (*J. Geophys. Res.* **64,** 1801, 1959)
Light curve; Hα flare (McMath-Hulbert Observatory, University of Michigan)

Studies at the Deep River Laboratory, Atomic Energy of Canada Ltd., have indicated the occurrence of an injection of cosmic rays at the surface of the earth at about 24:00 U.T. on July 16.

flares has been greatly improved by utilization of rockets and balloons to carry detectors to high altitudes, and by the use of satellites. The picture that is gradually emerging is much more complex than had been previously supposed, because of the Van Allen radiation belt or magnetosphere surrounding the earth, interplanetary magnetic fields, and the fact that flares emit particles with a range in energy, some flares emitting much more energetic particles than others. In fact, the active sun ejects particles over a range of 10^{16} in energy but over 10^7 of this range little is known.

Fig. 10–6 illustrates some of the phenomena connected with the great flare of July 16, 1959 (Fig. 10–2). Notice the great burst of 2,800 mc (10 cm) radio-frequency radiation. The SWF starts during the rising phase of the Hα light curve. Notice the type IV event and the type III event, which were followed by or associated with the ejection of the particles that produced the polar cap absorption of cosmic radio noise. (See Fig. 10–7.)

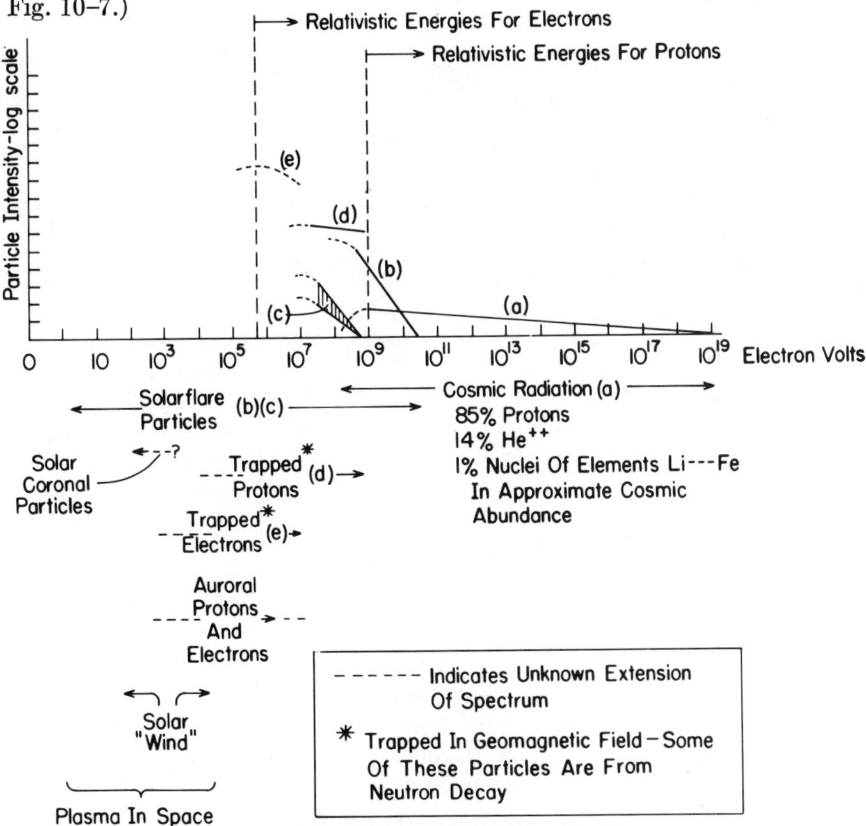

FIG. 10–7. THE SPECTRA OF ENERGETIC PARTICLES OBSERVED IN THE SOLAR SYSTEM

(Courtesy, J. A. Simpson, *Astrophysical Journal Supplement* **4,** 379, 1960, University of Chicago Press.)

10–6. Flares and Cosmic Rays. A flare may emit particles with a range of energy extending up to soft cosmic rays. The first particles may reach the earth one to five hours following a solar flare. The subsequent arrivals may continue for days. In large flares the low-energy beam may reach the earth a day following the flare, producing a geomagnetic storm, during which the particles may reach the earth at geomagnetic latitudes that the normal earth's field would exclude.

Cosmic rays are affected in two ways by flares: (a) during extremely active flares there may be an enhancement of low-energy cosmic rays, (b)

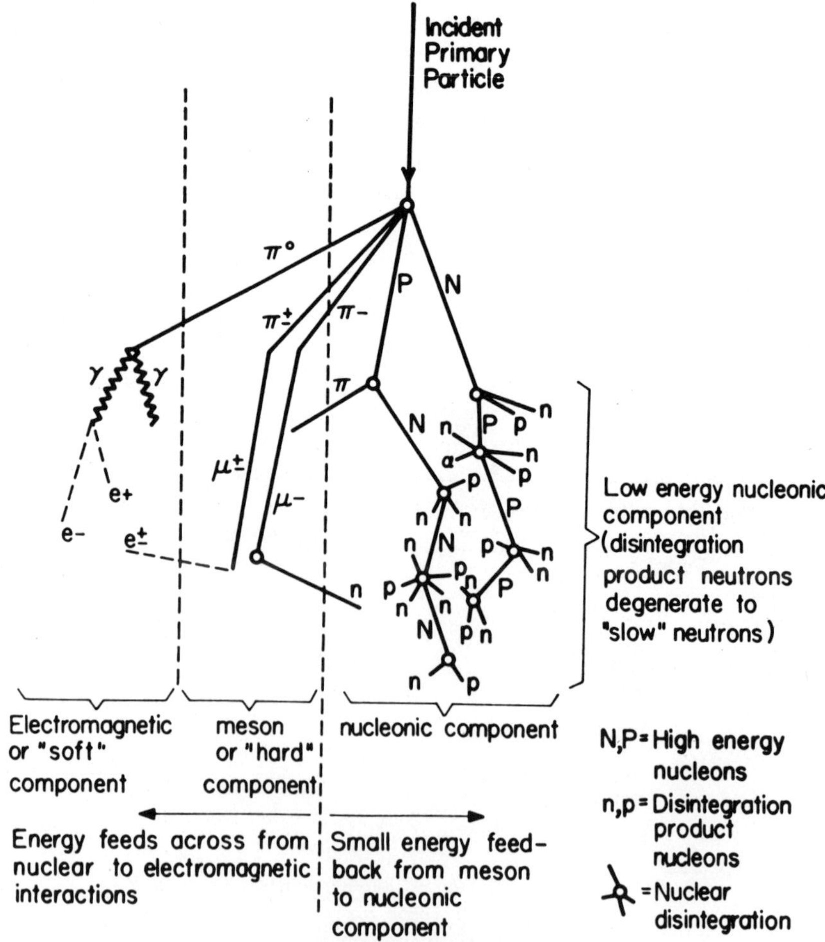

Fig. 10–8. Development of Secondary Components of Cosmic Radiation in Atmosphere from Primary Cosmic-Ray Particles

(Courtesy, J. A. Simpson, *Astrophysical Journal Supplement* **4,** 379, 1960, University of Chicago Press.)

as a result of solar activity there is a general weakening of the cosmic ray background (Forbush effect).

Fig. 10–8 depicts the formation of secondaries when a primary cosmic ray strikes an atom in the atmosphere or walls of the detector. The secondaries observed will depend on the energy of the primaries. For low-energy primaries there will be a preponderance of neutrons as compared with charged particles. For example, in a solar flare cosmic-ray event, mesons at sea level may be increased by 20 per cent but neutrons by 400 per cent.

By making observations from stations in different latitudes, it is possible to get the energy distribution of the impinging particles since the earth's magnetic field acts as an energy spectrograph. The quantity that determines whether or not a particle gets through the earth's field is not strictly the energy but the "rigidity," N, defined by $N = pc/Ze$ where p is the momentum, Ze is the charge and N is measured in units of 10^9 ev. All particles of the same rigidity will undergo the same deflection in the earth's magnetic field. At high latitudes one may use balloons to observe particles with energies as low as 1×10^9 ev (1 gev), but at the equator a particle must have more than about 20 gev to penetrate the earth's field, vertically. Detectors located in different parts of the geomagnetic field respond to different energies of solar protons provided they come from a common source. This is called the *impact zone effect.*

In Fig. 10–9 we compare neutron counts at Hobart, Tasmania (Mt. Wellington), 725 m and sea level (42° S), at Mawson (67° S) (Antarctica), and Lae (6° S) for cosmic-ray events of probable solar origin. The low-energy predominance of particles is shown by the greater enhancement at 725 m (Mt. Wellington) and the absence at Lae of neutron counts, and by the absence of any enhancement of mesons. The record is complicated by a Forbush effect.

Although indications of associations between cosmic rays and flares had been obtained earlier, the great cosmic-ray flare of February 23, 1956, one of the brightest ever observed on the sun, was the best studied. This event was of great importance because (1) the actual production of cosmic rays was observed in nature, and (2) the behavior of these cosmic rays as a function of time gave data on the then prevailing large-scale solar system magnetic field. Absorption mean free path data indicated that most of the particles were protons.

The onset of the flare occurred at 3^h31^m U.T. and an SID took place at about the same time. Strong radio-frequency outbursts at 3,000 mc and 200 mc were observed at 3^h32^m and 3^h35^m. The maximum of the optical flare occurred at 3^h42^m. Cosmic rays began to be observed at 3^h45^m at Huancayo near the equator (impact zone). The particles directly cannot reach other regions, outside the impact zone, but the cosmic rays began

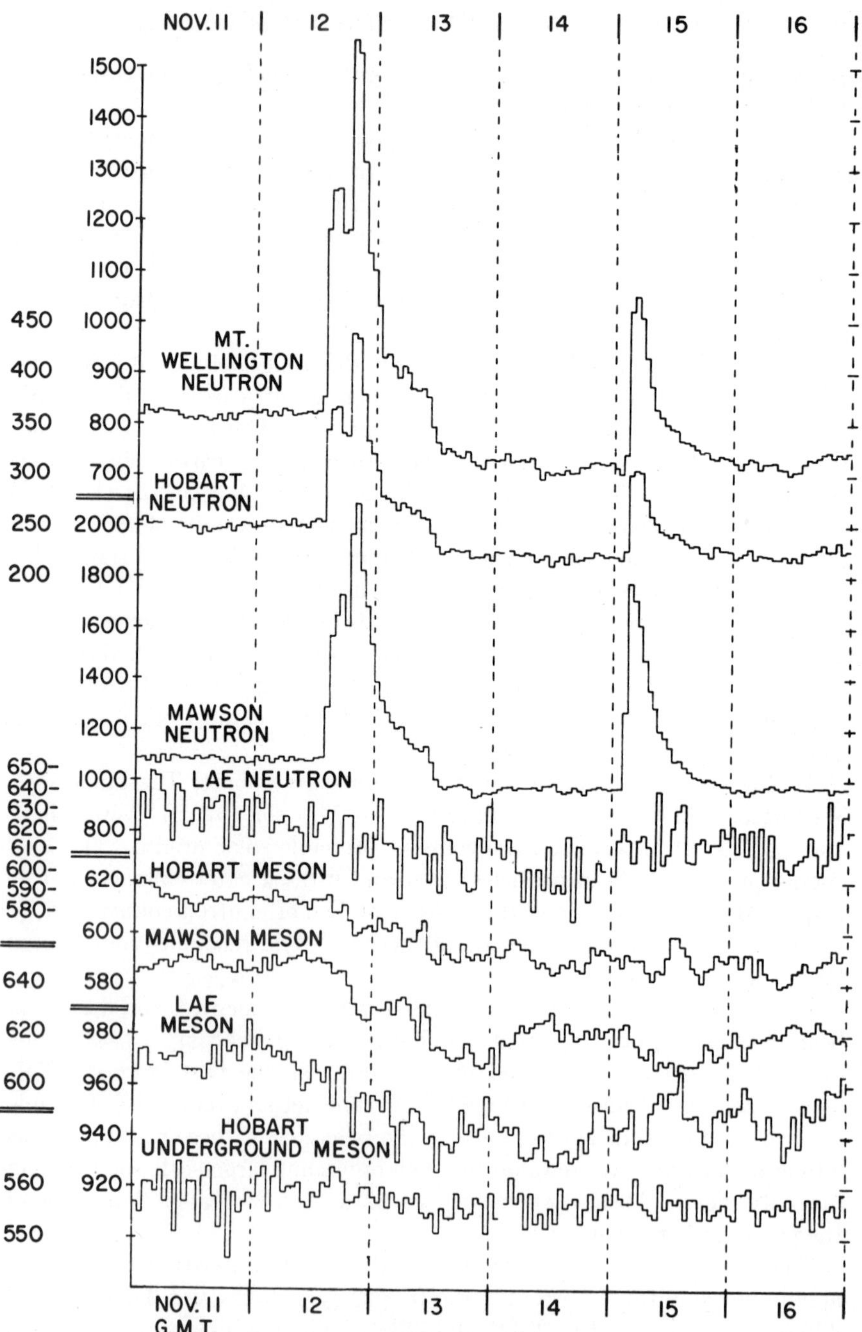

Fig. 10–9. Observations of Cosmic-Ray Events of November 11–16, 1960 at Some Australian Stations

(Courtesy, N. R. Parsons, Physics Department, University of Tasmania.)

almost simultaneously at Wellington, New Zealand (day), and Chicago (night). The maximum was reached at Chicago at 4^h00^m; thereafter a slow decline set in. From a study of such cosmic-ray records at different geomagnetic latitudes it is possible to obtain a theoretical model for the production, propagation, and escape of cosmic rays from the solar system. The intensity enhancement depended strongly on geomagnetic latitude, being 100 per cent at Chicago and 50 per cent at Huancayo. The reason for this behavior is the steep dependence of the spectrum on energy, i.e., there exists an enormous number of low-energy particles. Furthermore, the observations showed that the high-energy particles arrived ahead of those of lower energy, a spread of 10–15 minutes corresponding to an energy range of 10 gev. (See Figs. 10–10 and 10–11.)

Analysis of the data (particularly by J. A. Simpson, 1960) indicated that the first-arriving particles evidently came directly from the sun and had the highest energy in the cosmic-ray spectrum. The total time required for the release of the particles was of the order of 10^m–30^m. Furthermore, the sharp rise in cosmic-ray intensity, i.e., 8^m and 20^m to reach maxima in the impact and non-impact zones, respectively, meant that the particles must travel in simple orbits. Thus the magnetic field between the earth and the sun must have been very weak (less than 2×10^{-6} gauss) or essentially radial.

The 10-minute delay between the first arrival of 10 gev and 1 gev solar protons represents a dispersion effect in the cosmic rays and probably results from particle diffusion through relatively intense but irregular magnetic fields in the solar corona after they have been accelerated in the flare. The high-energy particles escaped first, but the source probably moved rapidly outward through coronal regions.

The decline of the cosmic ray intensity after reaching its maximum is of fundamental importance in giving information on the distribution of magnetic flelds within the solar system at the time of the outburst. We must recognize, however, that the fields may change with time and place in the solar system as a function of solar activity.

Shortly after maximum flux the directional dependence of the cosmic rays disappeared, the spatial distribution of the impinging particles was close to isotropic, even for those of highest energy. Thereafter, the particle intensity decreased with time, but the energy distribution remained the same. These results could be interpreted if we suppose that from a distance of about 1.5 astronomical units and extending out to about 5 astronomical units there exists a region of a disordered magnetic field amounting to about 2×10^{-5} gauss. The cosmic rays initially emitted by the sun are diffused and confined and scattered by these disordered fields, and the density of the cosmic-ray particles declines as $t^{-3/2}$ for all particle energies. As particles escape from the solar system through the magnetic

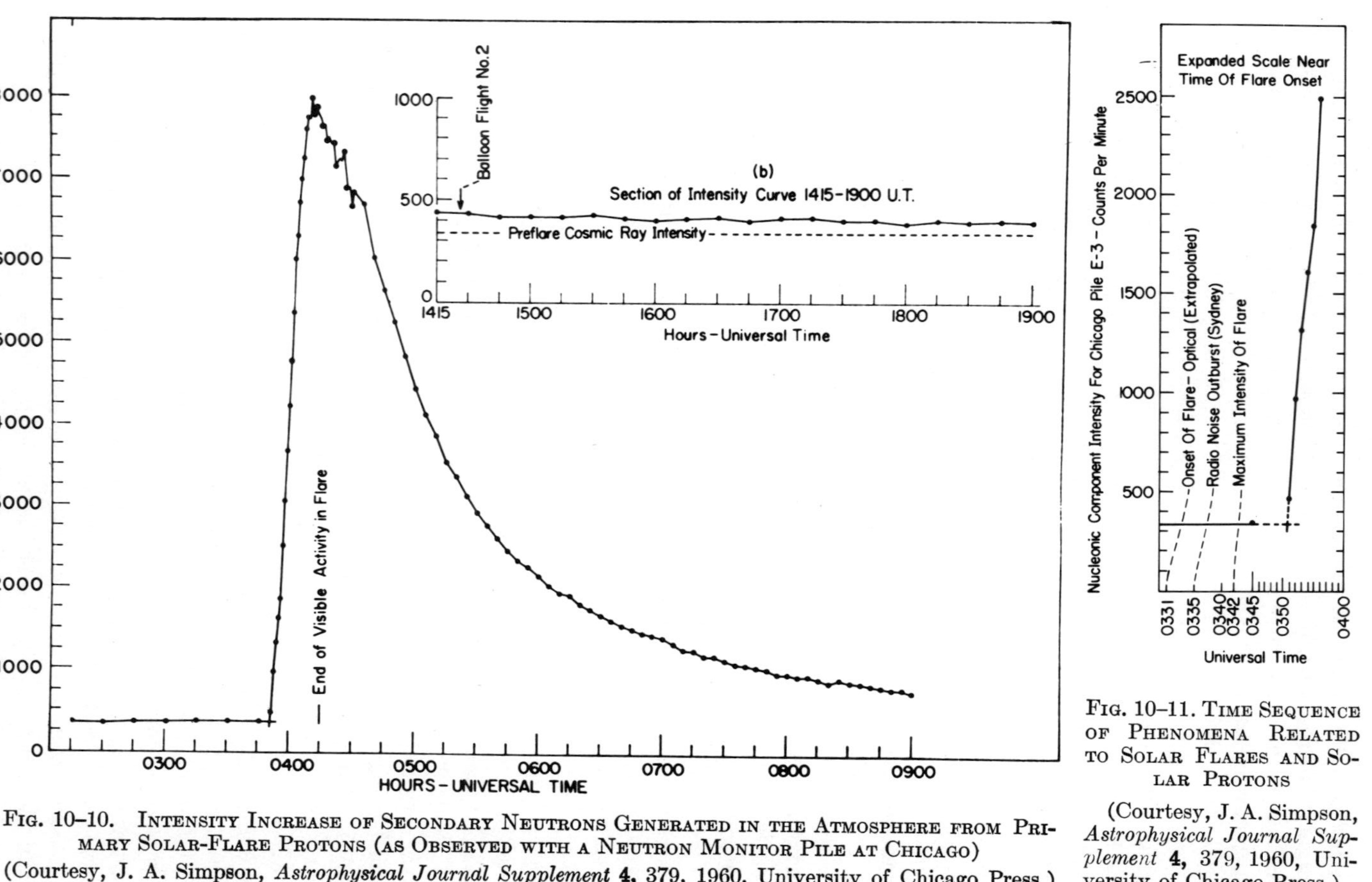

FIG. 10–10. INTENSITY INCREASE OF SECONDARY NEUTRONS GENERATED IN THE ATMOSPHERE FROM PRIMARY SOLAR-FLARE PROTONS (AS OBSERVED WITH A NEUTRON MONITOR PILE AT CHICAGO)

(Courtesy, J. A. Simpson, *Astrophysical Journal Supplement* **4,** 379, 1960, University of Chicago Press.)

FIG. 10–11. TIME SEQUENCE OF PHENOMENA RELATED TO SOLAR FLARES AND SOLAR PROTONS

(Courtesy, J. A. Simpson, *Astrophysical Journal Supplement* **4,** 379, 1960, University of Chicago Press.)

barrier, an exponential decay, e^{-at} prevails. Hence from an analysis of the flux as a function of both energy and time, it is possible to estimate both the magnitude of the field and the thickness of the diffusing region. The constancy of the spectrum with the time indicates we are within a diffusing region.

One might expect that the thickness and strength of the magnetic diffusing barrier could vary with time. The November 19, 1959, flare showed an exponential decay so the magnetic field barrier between the earth and the galaxy was thin at that time. The class 3 flare of May 4, 1960, showed sizable increases of a cosmic-ray contribution at several stations and a decay behavior clearly related to the termination of this optical flare. The early decay was exponential, the later decay quite complicated.

Cosmic-ray flares present a number of curious questions. There seem to be several types of "cosmic-ray" flares; some produce CR enhancements at sea level, others produce particles that can be detected only by detectors flown in balloons or even above the earth's atmosphere. Biswas, Fichtel, and Guss of Goddard Space Flight Center used nuclear emulsions flown in rockets fired from Ft. Churchill (where the earth's magnetic field did not cut out slow CR's) to determine chemical compositions of solar cosmic rays from the November 12, 1960, flare. They found that, in contrast to galactic cosmic rays wherein heavy nuclei are abundant, solar cosmic rays appear to have the same chemical composition as the solar photosphere. In the 1956 flare, protons seem to be more abundant than in normal cosmic rays. A similar effect appears to have been observed from CR counters in the July 15, 1959 flare.

From a study of characteristics of ten solar flares that have generated cosmic-ray increases recorded at the ground level 1942–60, Ellison, McKenna, and Reid noted that they were all associated with outstanding SID's, tended to fall west of the central meridian and north of the equator. Also, at times of maximum solar activity, high-energy particles do not appear to reach the earth. They conclude that arrival of high-energy particles capable of producing CR effects at ground level is fixed by factors other than flare importance, i.e., configuration of magnetic fields in neighborhood of spot and in interplanetary regions.

Is the flux of cosmic rays constant throughout the solar system? Attempts to answer this question have been made using meteorites and the results suggest that although there is no evidence for large-scale time changes, the CR intensity near the sun may be stronger than elsewhere, although this result is uncertain.

The second effect of the sun on cosmic rays is that of a modulation of the CR flux from outside the solar system by magnetic fields of solar origin. Thus the portion of the CR spectrum down to an energy of 1 gev increases by a factor greater than 2 between solar maximum and minimum. At lower energies even more marked effects are produced. The low energy cut-off was affected as solar activity increased.

There also occurs sharp intensity decrease in cosmic rays approximately one day after large flares (Forbush effect). See Fig. 10–12. The over-all effect is worldwide and is generally connected with phenomena such as geomagnetic storms. The intensity decrease may amount to 30% integrated over the relativistic spectrum, but for large events the energy dependence is not much affected.

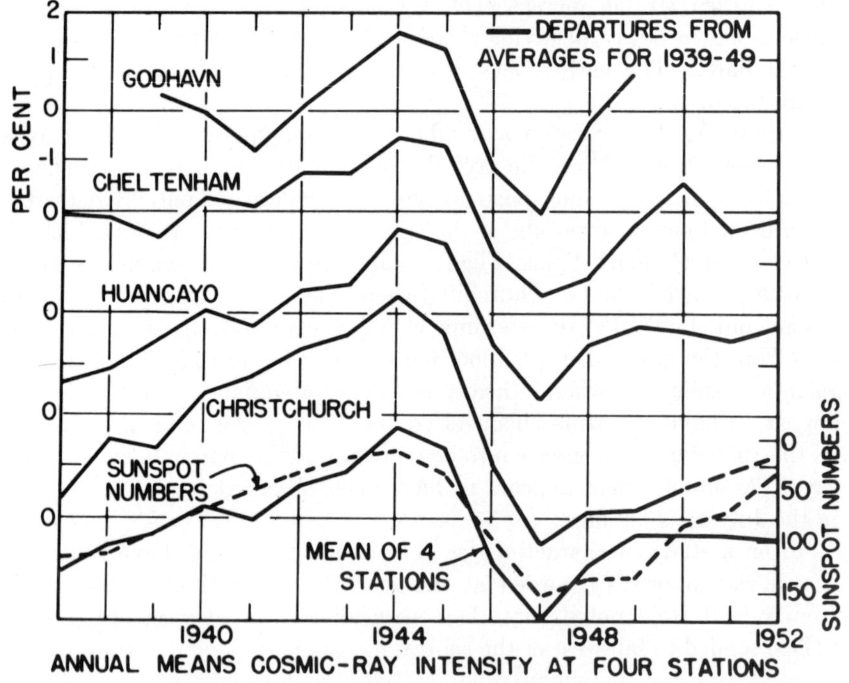

FIG. 10–12. INVERSE CORRELATION BETWEEN IONIZATION CHAMBER INTENSITY AND SOLAR ACTIVITY (FORBUSH EFFECT)

(Courtesy, J. A. Simpson, *Astrophysical Journal Supplement* **4**, 379, 1960, University of Chicago Press.)

Apparently the energy spectrum of particles with relativistic energies, as observed at solar minimum corresponds to the actual spectrum of cosmic rays from outside the solar system. See Fig. 10–13. At solar maximum we observed the resultant spectrum as modified by magnetic fields of solar origin. Just how large a volume is affected can be studied only by space probes.

Cosmic-ray flares and also the most intense outbursts of cm waves which are associated with accelerations of charged particles in the solar

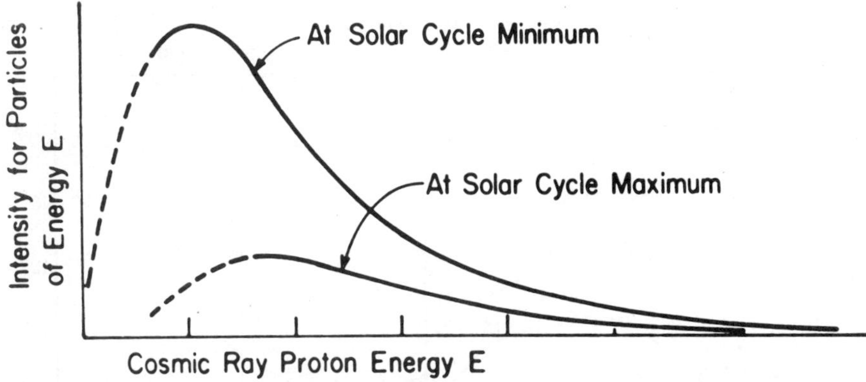

Fig. 10–13. Changes in the Cosmic-Ray Spectrum Between Solar Minimum and Solar Maximum

The changes are not drawn to scale. The behavior of the spectrum over the dashed portions of the curve is still relatively uncertain. The range of energies shown here is $0 \rightarrow 50 \times 10^9$ ev. (Courtesy, J. A. Simpson, *Astrophysical Journal Supplement* **4,** 379, 1960, University of Chicago Press.)

atmosphere tend to occur during ascending and descending phases of solar activity. Takakura and Ono (1961) conclude that these effects are of solar origin; they are not modulation effects produced by an interplanetary medium.

Although some solar flares have been identified as a source of cosmic rays, stars like the sun cannot account for the observed flux and energy distribution of cosmic rays. Insofar as the direct injection of particles with relativistic energies are concerned the mechanism fails by a factor of about a million. It is possible that non-relativistic particles may be fed into space by flares on stars and then accelerated to relativistic energies by some process such as that envisaged by Fermi.

10–7. Flares, Polar Cap Absorption (PCA), and Non-relativistic Particles. Intense solar flares can produce particles with a wide range in energy. Most such flares do not produce cosmic rays, but often produce particles capable of causing aurorae, geomagnetic storms, and often *polar cap absorption* (PCA). See Figs. 10–2 and 10–6.

Several hours after the occurrence of a brilliant flare characterized by an SID that lasts until the decline of the optical flare, and by type IV radio outbursts, there often occurs a heavy absorption of cosmic radio noise at 27 mc due to increased ionization in the 50–90 km level, presumably due to particles more energetic than those that produce magnetic storms. They are high-energy protons (10–100 Mev) that spiral in around the magnetic lines of force in geomagnetic latitudes 60–65° (on what are called *Störmer orbits* — after the Norwegian geophysicist who first systematically calculated trajectories of charged particles in the earth's magnetic fields).

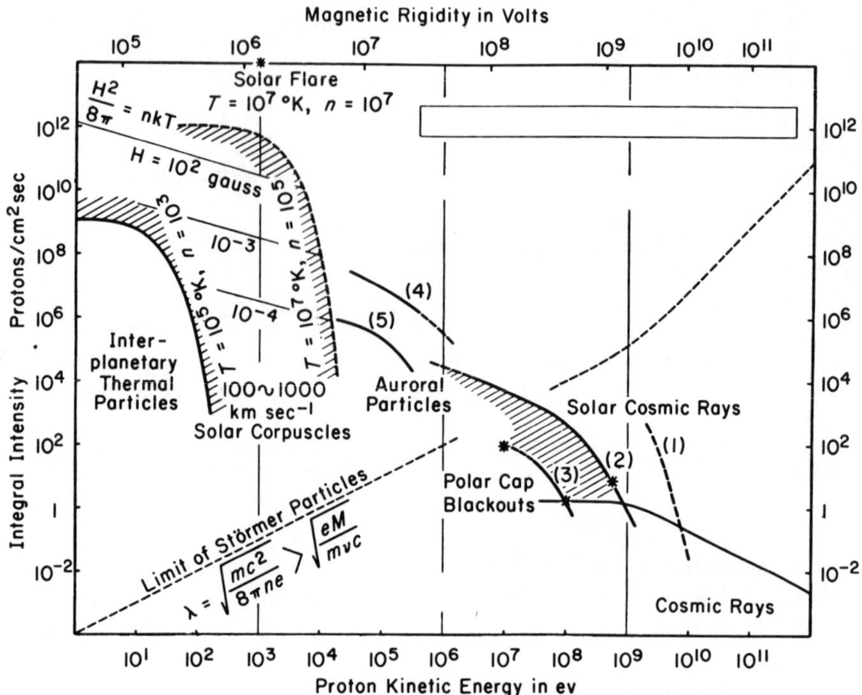

Fig. 10–14. Integral Energy Spectrum of Solar Particles Associated with Intense Solar Flares

Curve 1 indicates the spectrum of cosmic rays observed in the February 23, 1956 flare. Curves 2 and 3 define the range, 1–100 Mev, for particles responsible for polar-cap blackouts. Curves 4 and 5 define the domain of particles responsible for aurorae, while the solar plasma is indicated at the left. The dotted line (denoted as limit of Stormer particles) divides the region (lower right-hand side) where the particles follow individual orbits from the upper left-hand side region where the gas behaves as a plasma. (Courtesy, T. Obayashi and Y. Hakura, *Journal of Geophysical Research* **65,** 313, 1960.)

These particles, possibly accelerated by some "synchrotron" mechanism that produces relativistic electrons in the corona, are essentially *subcosmic* rays. They may carry their own magnetic fields with them and produce Forbush effects. See Fig. 10–14.

For ion densities of importance in the solar system, solar protons of energy above 10–100 Mev can be treated as though they were independent bodies, whose motions are not influenced by neighboring charges in the stream. Below this energy they must be treated as a conducting gas or "plasma" since the mean free path between collisions is much smaller than the dimensions of the flow. With many flares, no polar blackout is observed. There occurs only an SID, or a geomagnetic storm. Balloon flights show a sharp decline of numbers of particles with increasing energy in all solar flares.

The domain below 1 Mev includes the particles responsible for aurorae some of which appear to come directly from the sun and some of which may be accelerated in the magnetosphere. Their energies can be measured with aid of rockets fired during auroral displays or from auroral spectra.

The main body of the corpuscular cloud is a relatively high-density conducting plasma with speeds between a few hundred and 2,000 km sec^{-1}. This cloud, whose speed may depend on the importance of the flare, reaches the earth one or two days later. There, by interaction with the geomagnetic field, it produces a geomagnetic storm during which the ionization spreads along the auroral zone (producing an auroral zone blackout) which is correlated with geomagnetic and visible auroral disturbances. The actual interaction may be quite complicated. Perhaps auroral particles get trapped by the magnetic field of the corpuscular cloud, reach the earth after the arrival of the main cloud and penetrate into the polar ionosphere through the distorted outer geomagnetic field, thus producing an equatorial shift of the auroral zone during the main phase of the magnetic storm.

Chapman and Ferraro developed the idea of magnetic storms caused by plasma clouds ejected from the sun. When such a cloud encounters the earth, it acts as a diamagnetic body and pushes back the lines of force of the earth fields. Currents are induced in the thin front part of the plasma cloud at the expense of the cloud's energy. The initial compression of the earth's magnetic field causes the sudden onset of the storm, but its later development may become very complicated because (as we now know) of effects produced by the magnetosphere.

Mustel and others have suggested that weak geomagnetic activity produced by slower-moving plasma clouds might be induced not by flares at all but by particles ejected from sunspot regions or even old centers of activity where the green coronal line is strong. Such weak activity can be induced only when the source is near the solar meridian. It appears to be associated with the 27^d recurrence in geomagnetic activity.

The extreme low energy range is represented by particles that are relatively gently driven from the solar corona. These particles make up a low-energy plasma. Biermann (1957) from the acceleration, ionization and excitation of comets' tails inferred that this "wind" blows all the time. He estimated that the sun loses about 7×10^{10} gm sec^{-1} from this cause which produces a density of the order of 2 atoms cm^{-3} near the earth. Further studies have been carried out by Parker and by Chamberlain.

Effects of particle streams on motions of artificial satellites and on barometric pressure in the high atmosphere have also been observed. Thus L. G. Jacchia found that seemingly random fluctuations in the density of the upper atmosphere (as revealed by effects on satellite motion) could be correlated with changes in the 10.7 cm radiation from the sun. This radiation is a good measure for solar surface activity at many wavelengths.

Sometimes an increase in satellite drag is correlated with corpuscular streams following major solar flares, e.g., the November 12, 1960, flare produced a twofold increase in the drag on Echo I.

Solar corpuscular radiation may perhaps cause changes in the large-scale atmospheric circulation at the 300 mb level (Woodbridge, MacDonald, and Pohrte, 1959; MacDonald and Roberts, 1960). Solar particle radiation impinging on the earth near the time of maximum solar activity may, for example, affect the development of waves in westerly winds at 300 mb elevations. The meteorological effects are still uncertain.

The motions of charged particles may be strongly influenced by the interplanetary magnetic field, which may change both in time and place in response to the solar cycle. This magnetic field may be studied from its influence on ejected particles from the sun and by space probes. For example, the particles producing polar cap absorption originating west of the solar meridian reach the earth earlier and more easily than those from the east. Hence, Obayashi and Hakura (1960) suggest that the interplanetary magnetic field is formed by outward streaming solar winds carrying an imbedded solar magnetic field. This field is radial near the sun, but solar rotation produces a curvature that is convex towards the west. The space probe, Pioneer V, measured an undisturbed magnetic field of 2.5×10^{-5} gauss, whereas at times of solar activity the field amounts to 4×10^{-4} gauss.

The motions of the high-energy particles can be very complicated. Where the field is smooth and changes slowly in direction and intensity with time, the particles have orderly spiral paths. On the other hand, if the field changes rapidly in intensity so that high-field-intensity regions are interspersed with regions of low field intensity, particles may be regarded as "colliding" and being scattered by regions of strong fields — provided the scale of the strong fields is at least comparable with the Larmor radius. They will move in nearly straight lines in low-field regions, and will "diffuse" through regions of higher magnetic field. The slow-moving particles may pile up in the regions beyond the earth, producing large domains of disordered magnetic fields.

10–8. The Van Allen Radiation Belts. Probably the most important single discovery of the space research of the late fifties was the detection of a zone of geomagnetically trapped corpuscular radiation surrounding the earth. This region, originally known as the Van Allen belt, after its discoverer is now called the magnetosphere. It plays a fundamental role in the behavior of aurorae, geomagnetic storms, and related phenomena.

Experiments flown in Explorer I demonstrated the existence of particles that could penetrate matter with a thickness of 1 gm cm^{-2}, yet did not exist below an altitude of 200 miles where the air density was so low as to provide no hindrance to them. Clearly, the particles were constrained by the earth's magnetic field. All available evidence indicated that the radiation consists of protons and electrons that are trapped in the earth's field in a manner envisaged in classical theoretical studies by Poincaré, Störmer, and Alfvén. Probably, similar belts exist around other planets. Radiofrequency studies suggest that an intense magnetosphere surrounds Jupiter, while Houtgast has argued that Venus produces perturbations in the corpuscular streams from the sun, indicating that this planet also has a strong magnetic field. One of the objectives of the mariner experiment was to measure the magnetic field of Venus. The greater the magnetic moment of the planet and the less extended its atmosphere, the more favorable are the conditions for a high intensity of trapped radiation. Presumably the magnetic moment of the moon is small, so it probably has no radiation belt.

The intensity, energy spectrum, and spatial distribution of the earth's magnetosphere, all as a function of time, can be studied with the aid of probes and satellites. The lower region can be investigated with rockets. In the Argus experiment an atomic bomb was exploded in order to introduce charged particles in this zone and thus study its electrodynamical structure. Particles trapped in these zones probably spiral along the lines of force of the earth's field, are reflected as they approach the earth and gradually move eastward around the earth. Measurements made with Explorer XII revealed the terrestrial magnetosphere to be an extended volume, containing charged particles of different masses and energies. In the innermost region at about 6000 miles from the earth's center, there exist protons with energies up to more than 10 Mev. At about 12,000 miles the protons have somewhat lower average energy ranging from 100 kev to 400 kev depending on position in the magnetosphere. Their maximum intensity (which is comparable with that of the electrons present) exceeds 10^8 protons cm^{-2} sec^{-1}.

At the position of the outer Van Allen belt about 16,000 miles from the center of the earth, the penetrating components are protons of 20 Mev, electrons of 2 Mev or both. Between 23,000 and 45,000 miles there exist electrons with energies in tens of kilovolts. The intensity of these electrons with energies above 10 kev is about 10^8 particles cm^{-2} sec^{-1}.

The outer edge of the zone of trapped particles is rather abrupt. Both magnetometer and particle measurements show that beyond about 40,000 miles the geomagnetic field disappears. There is then a turbulent region about 60 miles thick between the geomagnetic field and the interplanetary magnetic field — which is of the order of a thousand times weaker than the field at the earth's surface. A low-energy proton analyzer showed that although there were no particles with energies between 100 ev and 10 kev below 40,000 miles, between 40,000 and 55,000 miles these low energy protons did appear.

The highest-energy particles seem to prefer two zones, one at about

1.5 earth radii and the other at about 4 earth radii. There is some evidence that the outer zones of the magnetosphere are more responsive to solar activity than is the inner zone. Events in the outer zone may be more intimately connected with aurorae, magnetic storms, etc., than events in the relatively stabler inner zones. The inner part of the magnetosphere is generally presumed to be produced by protons and electrons that have resulted from neutron decay. On this hypothesis, cosmic rays strike air nuclei producing (besides other particles) neutrons. The neutrons can move upward and outward through the geomagnetic field without suffering any deflections until they decay into protons and electrons. The protons carry away most of the kinetic energy of the neutrons, but the electrons may have any energy from 0 to 78.2 kev, as a result of neutron decay.

The energy spectrum of the electrons and the observed composition of the zone (protons and electrons) agree with this hypothesis. Further evidence is obtained from neutron counters flown in rockets at times of solar activity. The neutron flux in the upper atmosphere is increased at these times, due to a greater solar proton flux smashing oxygen and nitrogen nuclei. Many of the charged particles escape upwards and produce the particles in the inner magnetosphere.

The outer zone may be due to slow-speed plasma ejected from the sun that gets temporarily trapped by the earth's magnetic field. Thus, very probably the charged particles associated with auroral and geomagnetic storms may have been caught in the outer zone and dumped (perhaps quickly) into the upper atmosphere of the earth. High-energy particles from the sun such as cosmic rays and those responsible for polar cap absorption may come directly from the sun with little delay due to the action of the magnetosphere, whereas those involved in aurorae may have come from the sun in relatively low-speed plasma clouds which interact with the outer part of the magnetosphere. There, some of the particles, by interaction with magnetic fields, by influence of magnetohydrodynamic waves, etc., may build up high energies. The energy density of particles stored in the magnetic field, however, must be less than that of the magnetic field.

10–9. Aurorae. The polar aurora is one of the most beautiful sights of nature. Aurorae show a strong dependence on geomagnetic latitude and on the solar cycle. Intensive studies have resulted in a vast literature on the subject and we consider here only certain aspects connected with solar-terrestrial relationships.

Aurorae may be studied by a variety of techniques: (1) Visual observations or photographs with a high-speed camera reveal the forms, rates of change, and frequency of occurrence. From observations made at different stations, it is possible to determine heights by triangulations. (2) Photometric studies reveal brightness distributions with height. (3) Spectra help us to assign the physical mechanisms of excitation. Also, from measurements of line profiles and molecular bands one obtains the gas kinetic temperature at atmospheric heights where aurorae appear (Armstrong) and velocities and velocity distributions of incoming particles (Chamberlain, 1959). (4) Radar measurements are used to get heights of auroral arcs, etc. (5) Direct measurements of the types, energy, and fluxes of particles in and near aurorae may be made from rockets with appropriate particle detectors. For example, magnetic analyzers are used to separate protons and electrons. Electron energy analyzers, scintillation detectors, proportional counters, and Geiger counter telescopes are also used. (6) Detectors are flown in balloons to obtain *bremsstrahlung* (free-free emission) from incident electrons. (7) Satellites may be used to measure particles immediately outside the earth's atmosphere and eventually at greater distances from the earth.

The spectra of aurorae show emission bands of molecules and atomic lines of which the most prominent is the green auroral line visible also in the spectrum of the night sky. The red auroral lines are usually less conspicuous in high latitudes, although high-altitude aurorae in which the red lines are prominent have sometimes been observed there. When hydrogen lines appear in aurorae their profiles are broadened and displaced. The mean velocity of approach is about 400 km sec^{-1}, but velocities as high as 3000 km sec^{-1} have been observed. Since the stream is slowed down upon impact on the earth's atmosphere, the initial velocities must have been even higher. Thus the speeds of some particles greatly exceed the 1000 km sec^{-1} which is implied by delay times of geophysical effects following solar events.

The auroral stratum starts at 80 km and the main contribution comes from 90–120 km. Aurorae occur most frequently near geomagnetic latitude 67° and near midnight. During periods of solar activity they are seen at lower latitudes and are also observed at other times of the night. At times of solar activity, aurorae begin at the same hour in the northern and southern hemispheres, although the forms and stages of development are not the same in the two hemispheres.

From detectors flown in rockets, Davis, Berg, and Meredith (1960) found in rays and diffuse forms, the auroral emission was mostly due to electrons with energies greater than 10 kev incident on the upper atmosphere. Although protons with energies up to a few hundred kev were found on all flights, both inside and outside aurorae, no electrons were observed outside aurorae. McIlwain, however, found that in the aurorae he studied with rockets, the major fraction of the auroral light was produced by electrons with energies less than 10 kev. For example, in one aurora, three-quarters of the light was supplied by monoenergetic electrons of 6 kev. Jackson and Seddon (1960) found that aurorae produced dense cores of ionization along the lines of magnetic force.

Outstanding problems remain in the study of aurorae — problems that cannot be solved until satellites and space probes are flown outside the earth's atmosphere. For example, do the auroral particles travel from the sun as relatively low-energy protons and electrons and then become accelerated in the earth's field, or do particles with a considerable range in energy travel from the sun in a cloud which is held together by a magnetic field? In the first instance, one might expect the acceleration to occur in the outer part of the magnetosphere, which is constantly replenished by particles from the sun. In the second instance, particles flow into the atmosphere when the cloud interacts with the earth's magnetic field.

The study of solar-terrestrial relationships has made considerable progress. We can anticipate that by probes and satellites we will be able to solve some of the basic problems that have proved so obscure to date.

REFERENCES

General

Bartels, J. (ed.). 1958. *Handbuch der Physik. Geophysic III*, **49**, Springer-Verlag, Berlin.

de Jager, C. 1959. *Handbuch der Physik*, S. Flügge (ed.). Springer-Verlag, Berlin, vol. 52.

Bracewell, R. (ed.). 1959. *Paris Symposium on Radio Astronomy*. Stanford University Press, Stanford, California.

van de Hulst, H. C. (ed.). 1961. *Space Research II*. North Holland Publishing Co., Amsterdam.

Unsöld, A. 1955. *Physik der Sternatmosphären*. Springer-Verlag, Berlin.

Upper Atmosphere

A semipopular account is given by:

Massey, H. S. W., and Boyd, R. L. F. 1960. *Upper Atmosphere*. Hutchinson Press and Co., Ltd., London.

Comprehensive summaries are found in:

Ratcliffe, J. A. (ed.). 1960. *Physics of the Upper Atmosphere*. Academic Press, New York.

Solar Flares, Centers of Activity, and Related Events

Dodson, H. W., and Hedeman, E. R. 1956. *M. N.* **116**, 428.

Kiepenheuer, K. O. 1953. *Convegno Volta* **11**, 105.

D'Azambuja, M. 1953. *L'Astronomie* **67**, 430.

de Jager, C. 1959. *Op. cit.*, 183, 344.

Solar Cosmic Rays

Our discussion of the February 23, 1956, flare is taken from:

Simpson, J. A. 1960. *Ap. J. Suppl.* **4**, 379.

Aurorae

The classical reference is:

Störmer, C. 1955. *Polar Aurorae*. Oxford University Press, Fair Lawn, N. J.

For a modern treatment, see especially:

Chamberlain, J. W. 1961. *Physics of the Aurora and Airglow*. Academic Press, New York.

Advances in studies of the interplanetary medium, solar activity, and solar-terrestrial relationships are so rapid that any bibliography quickly becomes obsolete. The following journals are of particular interest:

Astrophysical Journal and *Supplements*, University of Chicago Press, Chicago.
Journal of Geophysical Research.

See also:

Bates, D. R. (ed.). 1963. *Theoretical Interpretation of Upper Atmospheric Emission.* Pergamon Press, London.

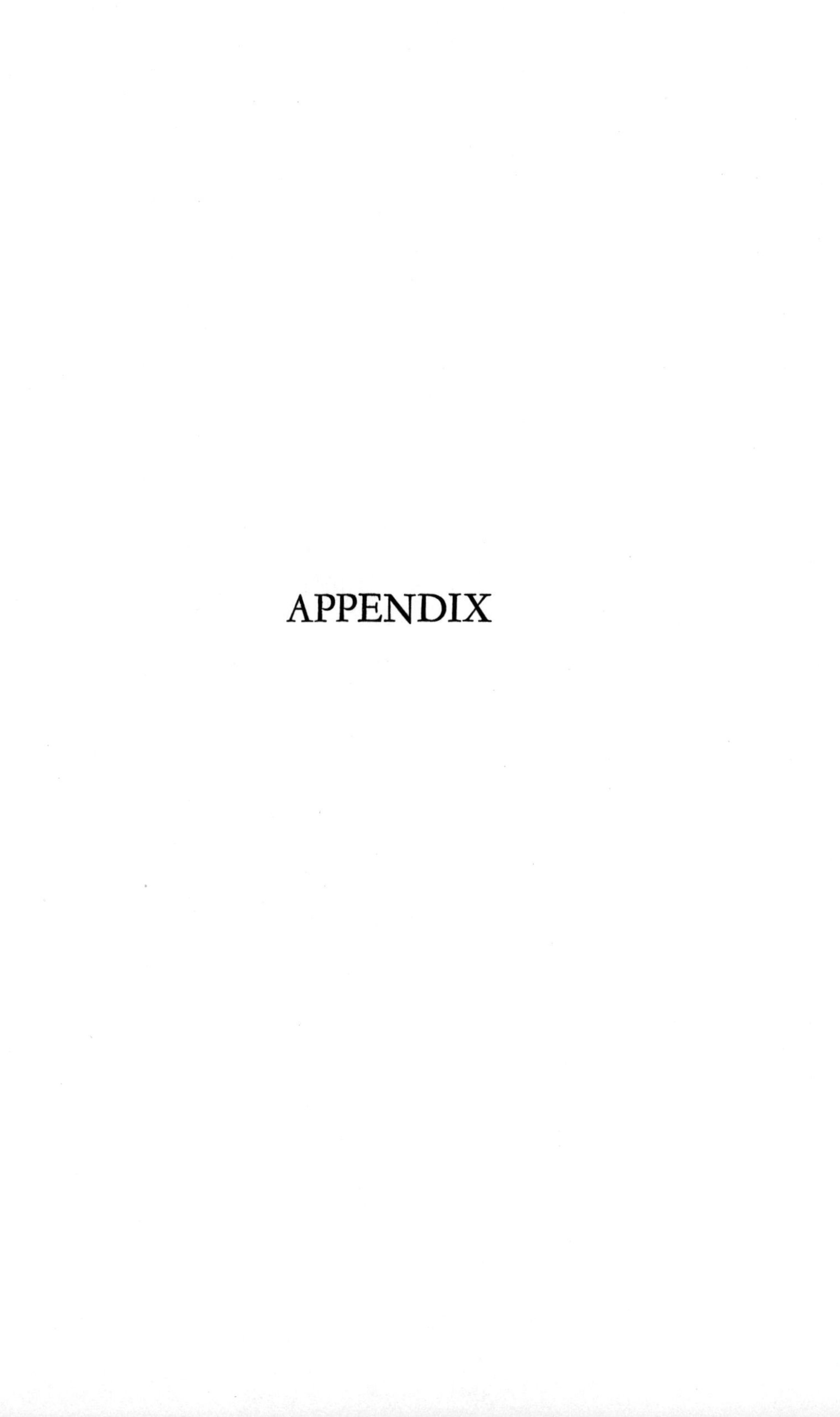

APPENDIX

APPENDIX

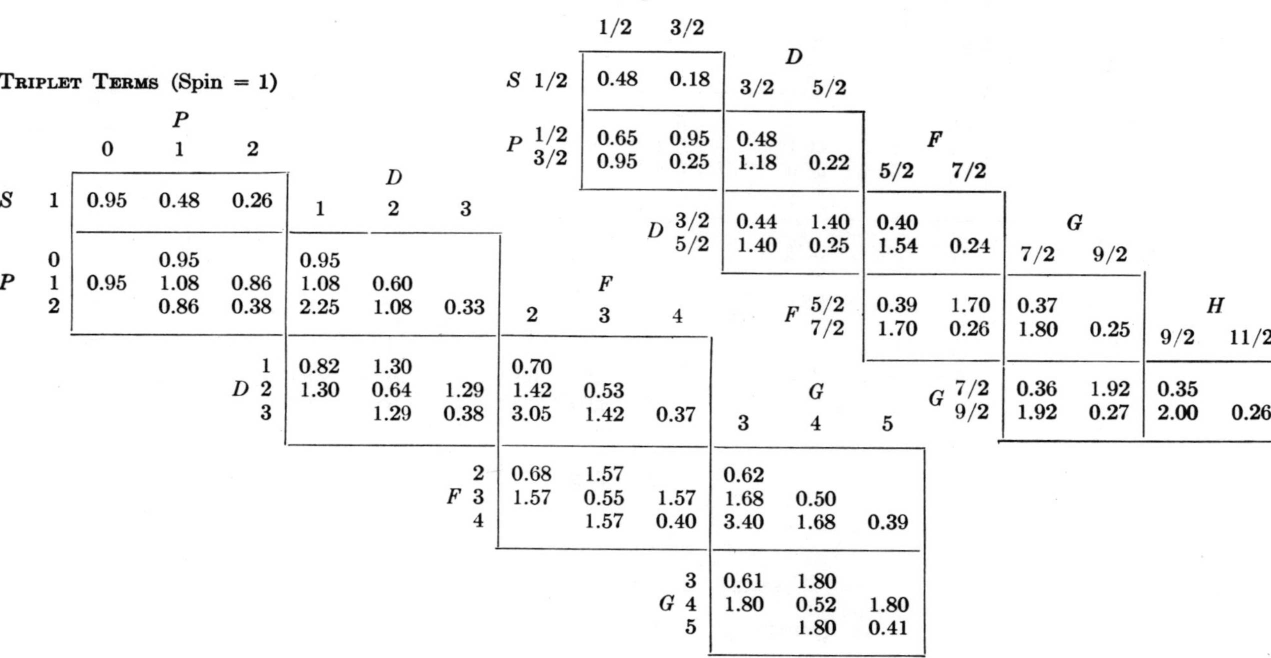

TABLE A-1

$$\log \frac{\Sigma s}{s}$$

DOUBLET TERMS (Spin = 1/2)

	P 1/2	P 3/2	D 3/2	D 5/2	F 5/2	F 7/2	G 7/2	G 9/2	H 9/2	H 11/2
S 1/2	0.48	0.18								
P 1/2	0.65	0.95	0.48							
P 3/2	0.95	0.25	1.18	0.22						
D 3/2			0.44	1.40	0.40					
D 5/2			1.40	0.25	1.54	0.24				
F 5/2					0.39	1.70	0.37			
F 7/2					1.70	0.26	1.80	0.25		
G 7/2							0.36	1.92	0.35	
G 9/2							1.92	0.27	2.00	0.26

TRIPLET TERMS (Spin = 1)

	P 0	P 1	P 2	D 1	D 2	D 3	F 2	F 3	F 4	G 3	G 4	G 5
S 1	0.95	0.48	0.26									
P 0		0.95		0.95								
P 1	0.95	1.08	0.86	1.08	0.60							
P 2		0.86	0.38	2.25	1.08	0.33						
D 1				0.82	1.30		0.70					
D 2				1.30	0.64	1.29	1.42	0.53				
D 3					1.29	0.38	3.05	1.42	0.37			
F 2							0.68	1.57		0.62		
F 3							1.57	0.55	1.57	1.68	0.50	
F 4								1.57	0.40	3.40	1.68	0.39
G 3										0.61	1.80	
G 4										1.80	0.52	1.80
G 5											1.80	0.41

TABLE A–1 (*Continued*)

QUARTET TERMS

(Spin = 3/2)

		P 1/2	P 3/2	P 5/2	D 1/2	D 3/2	D 5/2	D 7/2	F 3/2	F 5/2	F 7/2	F 9/2	G 5/2	G 7/2	G 9/2	G 11/2
P	1/2	1.55	0.86		1.08	1.08										
	3/2	0.86	1.36	0.82	1.77	0.97	0.68									
	5/2		0.82	0.46		2.00	1.05	0.40								
D	1/2				1.30	1.30			1.00							
	3/2				1.30	1.10	1.15		1.40	0.80						
	5/2					1.15	0.76	1.24	2.52	1.28	0.61					
	7/2						1.24	0.46		2.70	1.39	0.45				
F	3/2								0.94	1.55			0.84			
	5/2								1.55	0.83	1.42		1.64	0.72		
	7/2									1.42	0.66	1.52	3.00	1.52	0.59	
	9/2										1.52	.49		3.00	1.64	0.48
G	5/2												0.83	1.74		
	7/2												1.74	0.74	1.62	
	9/2													1.62	0.63	1.74
	11/2														1.74	0.50

TABLE A-1 (*Continued*)

QUINTET TERMS

(Spin = 2)

		P 1	P 2	P 3	D 0	D 1	D 2	D 3	D 4	F 1	F 2	F 3	F 4	F 5	G 2	G 3	G 4	G 5	G 6
P	1	1.28	0.82																
	2	0.82	1.55	0.81															
	3		0.81	0.51															
D	0	1.40				1.40													
	1	1.05	1.52		1.40	2.00	1.16												
	2	1.16	0.93	1.89		1.16	1.30	1.10											
	3		0.73	1.03			1.10	0.85	1.22										
	4			0.44				1.22	0.52										
F	1				1.40	1.40	2.24			1.24	1.54								
	2					1.10	1.24	2.24		1.54	1.15	1.37							
	3						0.86	1.22	2.52		1.37	0.95	1.35						
	4							0.67	1.37			1.35	0.74	1.50					
	5								0.50				1.50	0.55					
G	2									1.07	1.62	3.00			1.03	1.72			
	3										0.92	1.46	2.70		1.72	0.96	1.56		
	4											0.79	1.45	3.05		1.56	0.84	1.55	
	5												0.66	1.60			1.55	0.71	1.72
	6													0.54				1.72	0.57

TABLE A–2
Relative Multiplet Strengths in *LS* Coupling

p^2s \ p^2p	^{4}D	^{4}P	^{4}S	^{2}D	^{2}P	^{2}S	^{2}F	^{2}D	^{2}P	^{2}P
^{4}P	20	12	4							
^{2}P				10	6	2				
(^{1}D) ^{2}P							10	6	2	
(^{1}S) ^{2}S										6

p^2p \ p^2d	^{2}G	^{2}F	^{2}D	^{2}P	^{2}S	^{4}F	^{4}D	^{4}P	^{2}F	^{2}D	P	^{2}D
(^{1}D) ^{2}F	108	28	4									
^{2}D		56	35	9								
^{2}P			21	27	12							
(^{3}P) ^{4}D						168	30	2				
^{4}P							90	30				
^{4}S								40				
^{2}D									84	15	1	
^{2}P										45	15	
^{2}S											20	
(^{1}S) ^{2}P												60

TABLE A-2 (*Continued*)

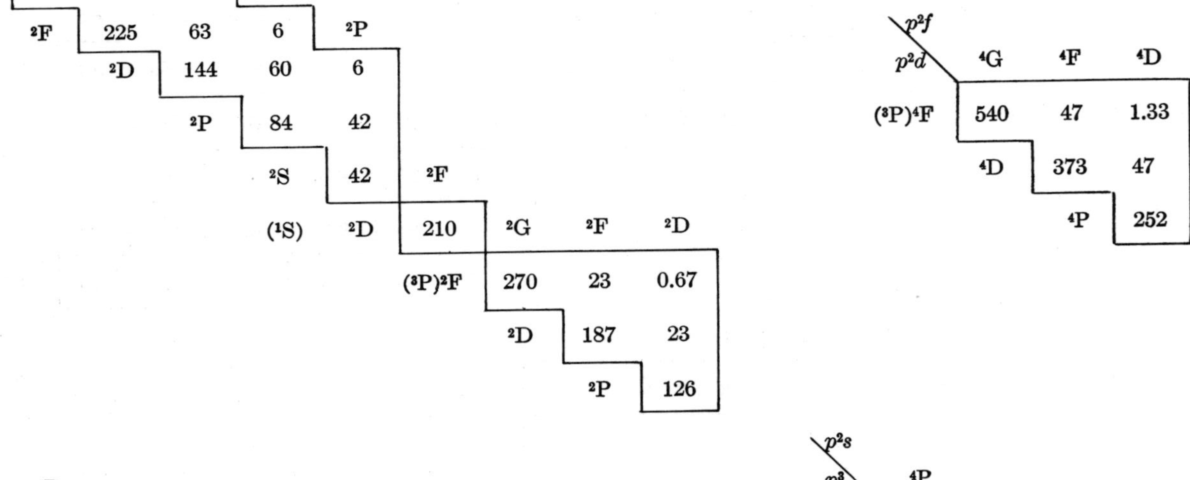

p^2d \ p^2f	²H	²G	²F	²D	²P	²F	²G	²F	²D
(¹D) ²G	330	45	3						
²F		225	63	6					
²D			144	60	6				
²P				84	42				
²S					42				
(¹S) ²D						210			
(³P)²F							270	23	0.67
²D								187	23
²P									126

p^2d \ p^2f	⁴G	⁴F	⁴D
(³P)⁴F	540	47	1.33
⁴D		373	47
⁴P			252

p^3 \ p^2d	⁴P	2²F	3²D	2²P	²S
⁴S	120				
²D		210	75	15	
²P			125	45	10

p^3 \ p^2s	⁴P	²D	²P	²S
⁴S	12			
²D		15	15	
²P		5	9	4

TABLE A–2 (*Continued*)

p^3s \ p^3p	^{3}F	^{3}D	^{3}P	^{1}F	^{1}D	^{1}P	^{3}D	^{3}P	^{3}S	^{1}D	^{1}P	^{1}S	^{5}P	^{3}P
(^{2}D)^{3}D	21	15	9											
^{1}D				7	5	3								
(^{2}P)^{3}P							15	9	3					
^{1}P										5	3	1		
(^{4}S)^{5}S													15	
^{3}S														9

p^3p \ p^3d	^{3}G	^{3}F	^{3}D	^{3}P	^{3}S	^{1}G	^{1}F	^{1}D	^{1}P	^{1}S	^{5}D	^{3}D
(^{2}D)^{3}F	162	42	6									
^{3}D		84	52	13.5								
^{3}P			31.5	40.5	18							
^{1}F						54	14	2				
^{1}D							28	17.5	4.5			
^{1}P								10.5	13.5	6		
(^{4}S)^{5}P											150	
^{3}P												90

p^3p \ p^3d	^{3}F	^{3}D	^{3}P	^{1}F	^{1}D	^{1}P
(^{2}P)^{3}D	126	22.5	1.5			
^{3}P		67.5	22.5			
^{3}S			30			
^{1}D				42	7.5	0.5
^{1}P					22.5	7.5
^{1}S						10

TABLE A–2 (*Continued*)

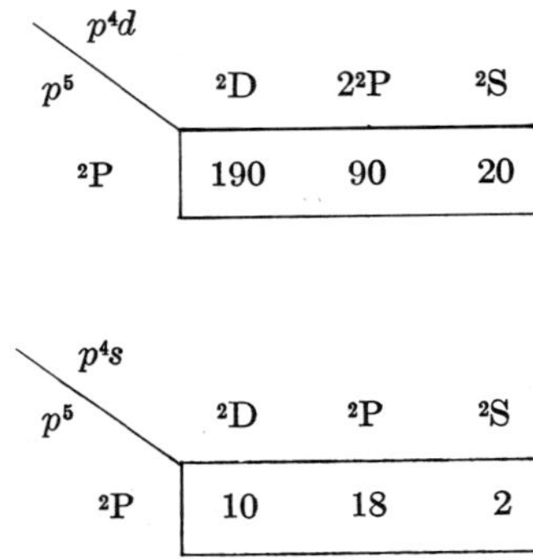

p^5 / p^4d	^2D	$2\,^2P$	^2S
^2P	190	90	20

p^5 / p^4s	^2D	^2P	^2S
^2P	10	18	2

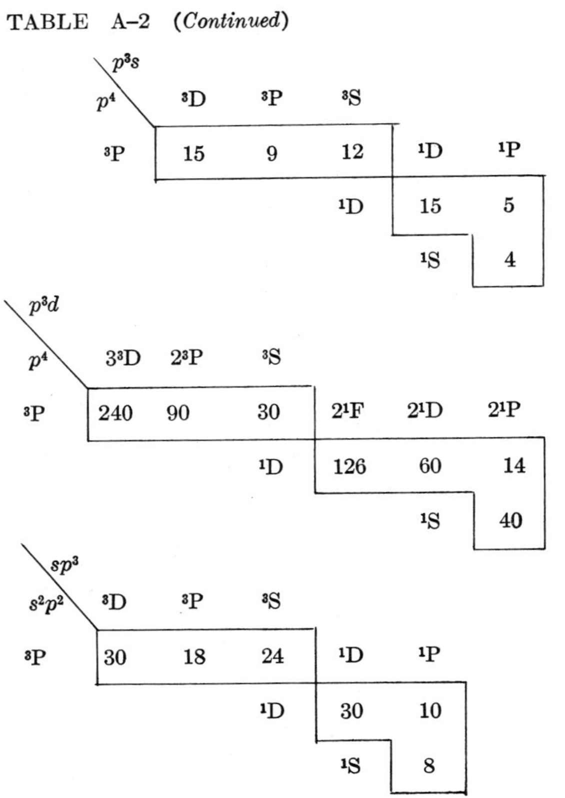

p^4 / p^3s	^3D	^3P	^3S	^1D	^1P
^3P	15	9	12		
^1D				15	5
^1S					4

p^4 / p^3d	$3\,^3D$	$2\,^3P$	^3S	$2\,^1F$	$2\,^1D$	$2\,^1P$
^3P	240	90	30			
^1D				126	60	14
^1S						40

s^2p^2 / sp^3	^3D	^3P	^3S	^1D	^1P
^3P	30	18	24		
^1D				30	10
^1S					8

TABLE A–2 (*Continued*)

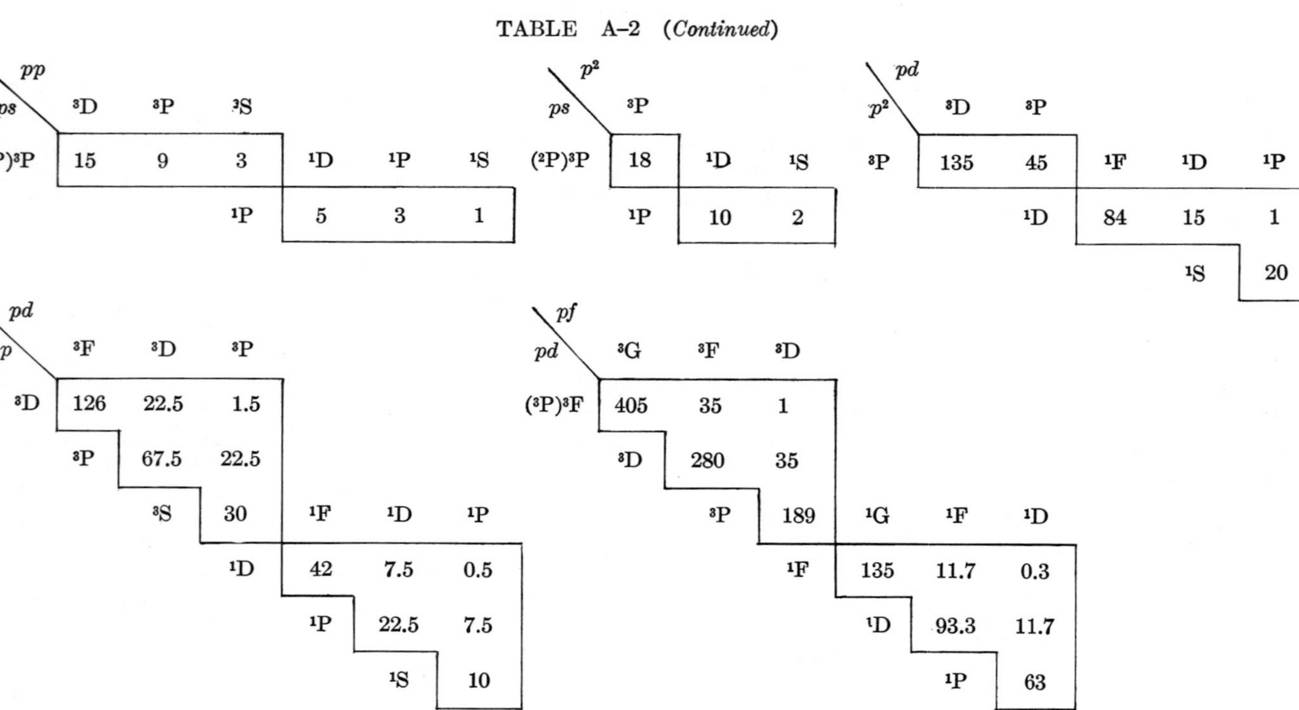

ps \ *pp*	^{3}D	^{3}P	^{3}S	^{1}D	^{1}P	^{1}S
(^{2}P)^{3}P	15	9	3			
^{1}P				5	3	1

ps \ p^2	^{3}P	^{1}D	^{1}S
(^{2}P)^{3}P	18		
^{1}P		10	2

p^2 \ *pd*	^{3}D	^{3}P	^{1}F	^{1}D	^{1}P
^{3}P	135	45			
^{1}D			84	15	1
^{1}S					20

pp \ *pd*	^{3}F	^{3}D	^{3}P	^{1}F	^{1}D	^{1}P
^{3}D	126	22.5	1.5			
^{3}P		67.5	22.5			
^{3}S			30			
^{1}D				42	7.5	0.5
^{1}P					22.5	7.5
^{1}S						10

pd \ *pf*	^{3}G	^{3}F	^{3}D	^{1}G	^{1}F	^{1}D
(^{3}P)^{3}F	405	35	1			
^{3}D		280	35			
^{3}P			189			
^{1}F				135	11.7	0.3
^{1}D					93.3	11.7
^{1}P						63

More extensive tables are given by Goldberg (*Ap. J.* **82,** 1, 1935), although the relative multiplet strengths there given must be corrected according to his data in *Ap. J.* **84,** 11, 1936.

BIBLIOGRAPHY

BIBLIOGRAPHY

References cited at the ends of chapters are not necessarily repeated here.

ABBOTT, C. G., FOWLE, F. E., and ALDRICH, L. B. 1912. *Smithsonian Astrophys. Obs. Annals* **2.**

———. 1913. *Smithsonian Astrophys. Obs. Annals* **3,** 1959.

———. 1914. *Smithsonian Astrophys. Obs. Annals.* **4,** 220.

ABOUD, A., BEHRING, W. E., and RENSE, W. A. 1956. *M. N.* **116,** 76.

———. 1959. *Ap. J.* **130,** 381.

ABT, H. 1952. *Ap. J.* **115,** 199.

———. 1957. *Ap. J.* **126,** 138, 503.

———. 1958. *Ap. J.* **127,** 658.

———. 1959. *Ap. J.* **130,** 104, 769, 824.

ADAM, M. G. 1959. *M. N.* **119,** 460.

ALDRICH, L. B., and HOOVER, W. H. 1954. *Smithsonian Astrophys. Obs. Annals* **7.**

ALFVÉN, H. 1950. *Cosmical Electrodynamics.* Clarendon Press, London.

ALLEN, C. W. 1934. *Memoirs of Commonwealth Observatory* (Canberra, Australia) **1,** No. 5; **2,** No. 6.

———. 1936. *M. N.* **96,** 842.

———. 1938. *Ap. J.* **88,** 125, 165.

———. 1940. *M. N.* **100,** 4.

———. 1949. *M. N.* **109,** 343.

———. 1950. *Observatory* **70,** 954.

———. 1955. *Astrophysical Quantities.* Athlone Press, London.

———. 1956. *M. N.* **116,** 69, 413.

———. 1960. *M. N.* **121,** 299.

———. 1961. Liége Conference, *Mem. Soc. Roy. Sci. Liége 5th Series* **4,** 241.

ALLER, L. H. 1942. *Ap. J.* **96,** 321.

———. 1946. *Ap. J.* **104,** 347.

———. 1949. *Ap. J.* **109,** 244.

———. 1956. *Gaseous Nebulae.* John Wiley and Sons, Inc., New York.

ALLER, L. H., and GREENSTEIN, J. L. 1960. *Ap. J. Suppl.* **5,** 139.

ATHAY, R. G. 1958. *Ann. d'Ap.* **21,** 98.

———. 1960a. *Ann. d'Ap.* **23,** 250.

———. 1960b. *Ap. J.* **129,** 164.

———. 1961. *J. Geophys. Res.* **66,** 385.

ATHAY, R. G., BILLINGS, D. E., EVANS, J. W., and ROBERTS, W. O. 1954. *Ap. J.* **120,** 94.

ATHAY, R. G., and HOUSE, L. L. 1962. *Ap. J.* **135,** 500.

ATHAY, R. G., and JOHNSON, H. R. 1960. *Ap. J.* **131,** 413.

ATHAY, R. G., and MENZEL, D. H. 1956. *Ap. J.* **123,** 285.

ATHAY, R. G., and ROBERTS, W. O. 1955. *Ap. J.* **121,** 231.

ATHAY, R. G., and THOMAS, R. N. 1957. *Ap. J.* **125,** 804.

BAADE, W. 1944. *Ap. J.* **100,** 137.

BABCOCK, H. D. 1945. *Ap. J.* **102,** 154.

———. 1959. *Ap. J.* **130,** 364.

BABCOCK, H. W. 1953. *Ap. J.* **118,** 387.

———. 1961. *Ap. J.* **133,** 572.

BABCOCK, H. W., and BABCOCK, H. D. 1955. *Ap. J.* **121,** 349.

BAHNG, J., and SCHWARZSCHILD, M. 1961. *Ap. J.* **134,** 312.

BARANGER, M. 1958. *Phys. Rev.* **111,** 494.

BARBIER, D. 1956. *Ann. d'Ap.* **19,** 223.

BARBIER, D., and CHALONGE, D. 1940. *Ann. d'Ap.* **3,** No. 2.

BARRETT, A. H. 1958. *Proceedings IRE* **46,** 250.

BASCHEK, B. 1959. *Zeits. f. Ap.* **48,** 95.

———. 1960. *Zeits. f. Ap.* **50,** 296.

BATES, D. R. 1951. *M. N.* **111,** 303.

———. 1952. *M. N.* **112,** 40.

BATES, D. R., and DAMGAARD, A. 1949. *Phil. Trans. R. S.* **242,** 1.

BECKER, W. 1938. *Zeits. f. Ap.* **15,** 225.

———. 1941. *Astr. Nach.* **272,** 180.

BEHRING, W. E., MCALISTER, H., and RENSE, W. A. 1958. *Ap. J.* **127,** 676.

BELL, G. DAVIS, KING, R. B., and ROUTLY, P. 1958. *Ap. J.* **127,** 775.

———. 1959. *Ap. J.* **129,** 437.

BERG, H. F., ALI, A. W., LINCKE, R., and GRIEM, H. R. 1962. *Phys. Rev.* **125,** 199.

BERGER, J. 1958. *Journ. Obs.* **41,** 108.

BHATNAGAR, P. L., KROOK, M., MENZEL, D. H., and THOMAS, R. N. 1955. *Vistas in Astronomy.* A. Beer (ed.). Pergamon Press, Ltd., Oxford, England. Vol. 1, p. 296.

BIERMANN, L. 1947. *Nach. Akad. Wiss. Gött. Math. Phys. Klasse* **3,** 12. (Mg II) (Si II).

———. 1957. *Observatory* **77,** 109.

———. 1958. *Electromagnetic Phenomena in Cosmical Physics,* I.A.U. Symposium No. 6, p. 248.

BIERMANN, L., and LÜBECK, K. 1948. *Zeits. f. Ap.* **25,** 325. (Mg II) (Si II).

———. 1949. *Zeits. f. Ap.* **26,** 145. (Mg II) (Si II).

BILLINGS, D. E. 1959. *Ap. J.* **130,** 961.

BLACKWELL, D. E. 1955. *M. N.* **115,** 429.

———. 1956. *M. N.* **116,** 56.

BLACKWELL, D. E., DEWHIRST, D. W., and DOLLFUS, A. 1959. *M. N.* **119,** 98.

BLESS, R. 1960. *Ap. J.* **132,** 532.

BÖHM, K. H. 1954. *Zeits. f. Ap.* **35,** 179.

———. 1958. *Zeits. f. Ap.* **46,** 245.

———. 1960. *Stellar Atmospheres,* vol. 6 of *Stars and Stellar Systems.* J. L. Greenstein (ed.). University of Chicago Press, Chicago.

———. 1961. *Ap. J.* **134,** 264.

BÖHM, K. H., and RICHTER, E. 1959. *Zeits. f. Ap.* **48,** 231.

———. 1960. *Zeits. f. Ap.* **50,** 79.

BÖHM-VITENSE, E. 1955. *Zeits. f. Ap.* **36,** 145.

———. 1958. *Zeits. f. Ap.* **46,** 108.

———. 1960. *Zeits. f. Ap.* **49,** 243.

BOHM, D., and ALLER, L. H. 1947. *Ap. J.* **105,** 131.

BOISCHOT, A., LEE, R. H., and WARWICK, J. 1960. *Ap. J.* **131,** 61.

BOUIGUE, R. 1954. *Ann. d'Ap.* **17,** 35, 104.

BRACEWELL, R. N., and PRESTON, G. W. 1956. *Ap. J.* **123,** 14.

BRANSCOMB, L. M., and PAGEL, B. E. J. 1958. *M. N.* **118,** 258.

BRANSCOMB, L. M., and SMITH, S. J. 1955. *Phys. Rev.* **98,** 1028.

BRAY, R. J. 1956. *M. N.* **116,** 394.

BRAY, R. J., and LOUGHHEAD, R. E. 1958. *Aust. J. Phys.* **11,** 185, 507

———. 1959. *Aust. J. Phys.* **12,** 320.

———. 1960. *Aust. J. Phys.* **13,** 139, 739.

———. 1961. *Aust. J. Phys.* **14,** 14.

BRETZ, M. 1956. *Zeits. f. Ap.* **38,** 259.

BRETZ, M. and BILLINGS, D. E. 1959. *Ap. J.* **129,** 13[illegible]

BROWN, R. H., and TWISS, R. Q. 1954. *Phil. Mag.* **45,** 663.

———. 1956. *Nature* **178,** 1046.

BROXON, J. W. 1942. *Phys. Rev.* **62,** 521.

BRUNNER, W. 1930. *Astr. Mitt. Zurich,* No. 124, p. 67.

BRUZEK, A. 1959. *Zeits. f. Ap.* **47.** 191.

———. 1960. *Zeits. f. Ap.* **50,** 1[illegible].

BULLARD, E. C. 1955. *Vistas in Astronomy.* A. Beer (ed.). Pergamon Press, Ltd., Oxford, England. Vol. 1, p. 685.

BURBIDGE, E. M., BURBIDGE, G. R., FOWLER, W. A., and HOYLE, F. 1957. *Rev. Mod. Phys.* **29**, 547.
BURBIDGE, E. M., BURBIDGE, G. R., SANDAGE, A. R., and WILDEY, R. 1960. *Liege Colloquium* **16**, 427.
BURGESS, A. 1960. *Ap. J.* **132**, 503.
BURGESS, A., VAN REGEMORTER, H., and SEATON, M. 1960. Quoted in Allen, *op. cit.*, 1961.
BUSBRIDGE, I. W., and STIBBS, D. W. N. 1954. *M. N.* **114**, 2.
BYRAM, E. T., CHUBB, T., and FRIEDMAN, H. 1954. *Phys. Rev.* **92**, 1066.
———. 1956. *J. Geophys. Res.* **61**, 251.

CAYREL, R. 1957. Thesis, Paris.
CAYREL, R., and TRAVING, G. 1960. *Zeits. f. Ap.* **50**, 239.
CAYREL-DE-STROBEL, G. 1957. *Ann. d'Ap.* **20**, 55.
CHALONGE, D. 1956. *Ann. d'Ap.* **19**, 258.
CHALONGE, D., and CANAVAGGIA, R. 1946. *Ann. d'Ap.* **9**, 143.
CHALONGE, D., CANAVAGGIA, R., EGGER-MOREAU, M., and OZOIL-PELTY, H. 1950. *Ann. d'Ap.* **13**, 355.
CHALONGE, D., and DIVAN, L. 1952. *Ann. d'Ap.* **15**, 201.
CHALONGE, D., and KOURGANOFF, V. 1946. *Ann. d'Ap.* **9**, 69.
CHAMBERLAIN, J. W. 1954. *Ap. J.* **120**, 364.
———. 1960. *Ap. J.* **131**, 47.
———. 1961. *Ap. J.* **133**, 675.
CHANDRASEKHAR, S. 1936. *M. N.* **96**, 21.
———. 1946. *Ap. J.* **104**, 444.
———. 1952. *Phil. Mag.* **43**, 501.
———. 1957. *Daedalus* **86**, 323.
———. 1958. *Ap. J.* **128**, 114, 633.
———. 1961. *Hydrodynamics and Hydromagnetic Stability*. Clarendon Press, Oxford.
CHAPMAN, S. 1957. *Smithsonian Contributions to Astrophysics* **2**, 1.
CHRISTIANSEN, W. N. 1960. *Ann. d'Ap.* **23**, 102.
CHUBB, T. A., FRIEDMAN, H., and KREPLIN, R. W. 1960. *J. Geophys. Res.* **65**, 1831.
CLIMENHAGA, J. 1961. *Publ. Dominion Astrophys. Obs.* **16**, 307.
COATES, R. J. 1958. *Ap. J.* **128**, 83.
CODE, A. 1960. *Stellar Atmospheres*, vol. 6 of *Stars and Stellar Systems*. J. L. Greenstein (ed.). University of Chicago Press, Chicago, p. 50.
CONWAY, M. T. 1952. *M. N.* **112**, 55.
COWLEY, A. 1962. *Publ. A. J. P.* **74**, 223.
COWLING, T. G. 1957. *Magnetohydrodynamics*. Interscience Press, New York.

DANIELSON, R. E. 1961. *Ap. J.* **34**, 275.
DAS, A. K., and RAMANATHAN, A. S. 1953. *Zeits. f. Ap.* **32**, 91.
DAVIES R. D., VERSCHUUR, G. L., and WILD, P. A. T. 1962. *Nature* **195**, 563.
DAVIS, D. N. 1947. *Ap. J.* **106**, 28.
DAVIS, L. R., BERG, O. E., and MEREDITH, L. H. 1960. *IGY Rocket Report No. 6*, 143.
D'AZAMBUJA, L. M. 1948. *Ann. Obs. Meudon* **6**, Fasc. 8.
DE JAGER, C. 1959. *Handbuch der Physik*. S. Flugge/Marburg (ed.). Springer-Verlag, Berlin. Vol. 52, p. 80.
DEUTSCH, A. J. 1956. *Ap. J.* **123**, 210.
———. 1956. *Pub. A. S. P.* **68**, 92.
———. 1958. *Symposium International Astronomical Union* **6**, 209.
DODSON, H. W., HEDEMAN, E. R., and MCMATH, R. R. 1956. *Ap. J. Suppl.* **2**, 241.
DUNKELMAN, L., and SCOLNIK, R. 1952. *Journ. Opt. Soc. Am.* **42**, 876.

EDDINGTON, A. S. 1929. *M. N.* **89**, 620.
EGGEN O. 1956. *A. J.* **61**, 361.
EHMANN, 1961. *J. Chem. Education* **38**, 51.
ELLISON, M. A., and REID, J. H. 1957. *Publ. Obs. Edinburgh* **2**, No. 2.

ELSASSER, W. M. 1956. *Rev. Mod. Phys.* **28,** 135.
ELSTE, G. 1955. *Zeits. f. Ap.* **37,** 201.
ELSTE, G., and JUGAKU, J. 1957. *Ap. J.* **125,** 742.
ELSTE, G., JUGAKU, J., and ALLER, L. H. 1956. Publ. *A. J. P.* **68,** 23.
ELVIUS, T. 1955. *Stockholm Observatory Annals* **18,** Nos. 7 and 9.
———. 1956. *Stockholm Observatory Annals* **19,** No. 3.
ELWERT, G. 1948. *Zeits. f. Ap.* **25,** 310.
———. 1948. *Zeits. f. Naturforsch.* **3a,** 477.
———. 1952. *Zeits. f. Naturforsch.* **7a,** 202, 432.
———. 1954. *Zeits. f. Naturforsch.* **9a,** 648.
———. 1956. *Zeits. f. Ap.* **41,** 67.
———. 1961. *J. Geophys. Res.* **61,** 391.

FAULKNER, D. J., and MUGGLESTONE, D. 1962. *M. N.* **124,** 11.
FELLGETT, P. 1959. *M. N.* **119,** 475.
FERRARO, V. C. A., and KENDALL, P. C. 1956. *Ap. J.* **124,** 443.
FIROR, J., and ZIRIN, F. 1962. *Ap. J.* **135,** 122.
FOLEY, H. M. 1946. *Phys. Rev.* **69,** 616.
FRIEDMAN, H. 1955. *Annales de Géophysique* **11,** 174.
———. 1957. Ninth Report of Commission for Study of Relation Between Solar and Terrestrial Phenomena, Paris, p. 41.
FUCHS, R. 1951. *Zeits. f. Phys.* **130,** 69.

GALT, J. A., SLATER, C. H., and SHUTER, W. L. H. 1960. *M. N.* **120,** 187.
GARSTANG, R. H. 1954. **114,** 118.
———. 1955. *Vistas in Astronomy.* A. Beer (ed.). Pergamon Press, Ltd., Oxford, England, vol. I, p. 268.
GASCOIGNE, S. C. B. 1950. *M. N.* **110,** 15.
GAUSTAD, J., and SCHWARZSCHILD, M. 1960. *M. N.* **121,** 260.
GELTMAN, S. 1956. *Phys. Rev.* **104,** 346
GOLD, T. 1958. *Electromagnetic Phenomena in Cosmical Physics.* B. Lehnert (ed.). Cambridge University Press, Cambridge, England, p. 275.
GOLD, T., and HOYLE, F. 1960. *M. N.* **120,** 89.
GOLDBERG, L. 1939. *Ap. J.* **90,** 414.
———. 1958. *Ap. J.* **127,** 308.
GOLDBERG, L., MOHLER, O. C., and MÜLLER, E. 1959. *Ap. J.* **129,** 119.
GOTTSCHALK, W. H. 1948. *Ap. J.* **108,** 326.
GRASBERGER, W. 1957. *Ap. J.* **125,** 750.
GREAVES, W. M. H. 1940. *M. N.* **100,** 189.
———. 1956. *Vistas in Astronomy.* A. Beer (ed.). Pergamon Press, Ltd., Oxford, England, vol. II, p. 1320.
GREEN, L. C. 1949. *Ap. J.* **109,** 289. (Ca I)
GREEN, L. C., RUSH, P. R., and CHANDLER, C. D. 1957. *Ap. J. Suppl.* No. 26.
GREEN, L. C. and WEBER, N. 1950. *Ap. J.* **111,** 587. (4s-p, 3d-f) (Ca II)
———. 1951. *Ap. J.* **113,** 690. (4s-p, 3d-f) (Ca II)
GRIEM, H. 1960. *Ap. J.* **132,** 883.
GRIEM, H., KOLB, A. C., and SHEN, K. Y. 1959. *Phys. Rev.* **116,** 4.
———. 1962. *Ap. J.* **135,** 272.
GRIEM, H., and SHEN, K. Y. 1961. *Phys. Rev.* **122,** 1490.
GYLDENKERNE, K. 1958. *Ann. d'Ap.* **21,** 26.

HACK, M. 1953. *Ann. d'Ap.* **16,** 417.
———. 1956. *Mem. Soc. Astr. Italy* **27,** 201, 249, 469, 547.
———. 1959. *Mem. Soc. Astr. Italy* **30,** No. 1, No. 2.
HALL, J. 1941. *Ap. J.* **94,** 71.
HARRIS, D. 1949. *Ap. J.* **109,** 53.
HART, A. B. 1954. *M. N.* **114,** 17.
———. 1956. *M. N.* **116,** 38.

HEARD, J. F. 1956. *Vistas in Astronomy.* A. Beer (ed.). Pergamon Press, Ltd., Oxford, England, vol. II, p. 1357.
HENYEY, L. G. 1946. *Ap. J.* **103,** 332.
HENYEY, L. G., and GRASBERGER, W. 1955. *Ap. J.* **122,** 498.
HERBIG, G. H., and SPALDING, J. F. 1955. *Ap. J.* **121,** 118.
HEWISH, A. 1958. *M. N.* **118,** 535.
HINDMARSH, W. R. 1955. *M. N.* **115,** 270.
———. 1959. *M. N.* **119,** 11.
HINTEREGGER, H. E. 1961. *Colloquium d'Université de Liége* **20,** 111.
HINTEREGGER, H. E., DAMON, K. R., HEROUX, L., and HALL, L. A. 1960. *International Space Science Symposium.* North Holland Publishing Co.
HITOTUYANAGI, Z., and INABA, H. 1954. *Scientific Reports Tohoku University,* Ser. I, **38,** No. 3.
HOGBOM, J. A. 1960. *M. N.* **120,** 530.
HOLSTEIN, T. 1950. *Phys. Rev.* **79,** 744.
HOSSACK, W. R. 1953. *Journ. Roy. Ast. Soc. Canada* **47,** 195.
HOUTGAST, J., and VAN SLUITERS. 1948. *Bull. Astr. Institute Netherlands* **10,** 325.
HUANG, S. S. 1948. *Ap. J.* **108,** 354. (He I)
———. 1952. *Ap. J.* **115,** 529.
HUANG, S. S., and STRUVE, O. 1952. *Ap. J.* **116,** 410.
———. 1954. *Ann. d'Ap.* **17,** 85.
———. 1955. *Ap. J.* **121,** 84.
———. 1956. *Ap. J.* **123,** 231.
HUBENET, H. 1954. *Zeits. f. Ap.* **34,** 110.
HULDT, L., and LAGERQUIST, A. 1952. *J. Opt. Soc. Amer.* **42,** 142.
HUNGER, K. 1955. *Zeits. f. Ap.* **36,** 42.
———. 1956. *Zeits. f. Ap.* **38,** 36.
———. 1960. *Zeits. f. Ap.* **49,** 131.

INABA, H. 1955. *Scientific Reports Tohoku University* **39,** 60.
INGLIS, D. R., and TELLER, E. 1939. *Ap. J.* **90,** 439.

JACKSON, J. E., and SEDDON, J. E. 1960. *IGY Rocket Report No. 6.*
JARRETT, A. H., and VON KLUBER, H. 1955. *M. N.* **115,** 343.
———. 1961. *M. N.* **122,** 223.
JASCHEK, M., and JASCHEK, C. 1959. *Zeits. f. Ap.* **47,** 29.
———. 1960. *Zeits. f. Ap.* **50,** 155.
JEFFERIES, J. T. 1960. *Ap. J.* **132,** 775.
JEFFERIES, J. T., and ORRALL, F. Q. 1958. *Ap. J.* **127,** 714.
———. 1961. *Ap. J.* **133,** 946, 963; **134,** 747.
———. 1962. *Ap. J.* **135,** 521.
JEFFERIES, J. T., and THOMAS, R. N. 1958. *Ap. J.* **127,** 667.
———. 1959. *Ap. J.* **129,** 401.
JEFFERIES, J. T., and WHITE, O. R. 1960. *Ap. J.* **132,** 767.
JOHNSON, F. S. 1954. *Journ. Meteorology* **11,** 431.
JOHNSON, H. L., and IRIARTE, B. 1958. *Lowell Observatory Bull.* **4,** 47.
JUGAKU, J. 1959. *Publ. Astr. Soc. Japan* **11,** 161.

KARIMOV, M., and SHILOVA, N. 1960. *Publ. Astr. Inst. Acad. Sci. Kazakh SSR* **9,** 10.
KARZAS, W. J., and LATTER, R. 1960. *Ap. J. Suppl.* No. 55.
KAWABATA, K. 1960. *Publ. Astron. Soc. Japan* **12,** 513.
KEENAN, P. C. 1948. *Ap. J.* **101,** 420.
———. 1954. *Ap. J.* **120,** 484.
KEENAN, P. C., and MORGAN, W. W. 1941. *Ap. J.* **94,** 501.
KELLY, P., and ARMSTRONG, B. H. 1959. *Ap. J.* **129,** 786.
KIENLE, H. 1941. *Zeits. f. Ap.* **20,** 239.
KIENLE, H., WEMPE, J., and BEILEKE, F. 1940. *Zeits. f. Ap.* **20,** 91.
KIENLE, H., WEMPE, J., and STRASSL, H. 1938. *Zeits. f. Ap.* **16,** 201.

KIEPENHEUER, K. O. 1953. *The Solar System* (*The Sun*, vol. 1). G. P. Kuiper (ed.). University of Chicago Press, Chicago, chap. 4.
———. 1957. *Zeits. f. Ap.* **42,** 209.
KINMAN, T. D. 1952. *M. N.* **112,** 425.
———. 1953. *M. N.* **113,** 613.
KIPPENHAHN, R., and SCHLUTER, A. 1957. *Zeits. f. Ap.* **43,** 36.
KIPPENHAHN, R., ST. TEMESVARY, S., and BIERMANN, L. 1958. *Zeits. f. Ap.* **46,** 257.
KOLB, A. C., and GRIEM, H. 1958. *Phys. Rev.* **111,** 514.
KOPFERMANN, H., and WESSEL, G. 1951. *Zeits. f. Phys.* **130,** 100.
KORN, J. 1940. *Astron. Nach.* **270,** 105.
KRAUS, J. D. 1950. *Antennas.* McGraw-Hill Book Co., Inc., New York.
KRISTENSON, H. 1951. *Stockholm Annals* **17,** No. 1.
———. 1955. *Stockholm Annals* **18,** No. 5.
KRON, J., and STEBBINS, J. 1957. *Ap. J.* **126,** 266.
KUIPER, G. P. 1938. *Ap. J.* **88,** 472.
KUNDU, M. R. 1959. *Ann. d'Ap.* **22,** 1.
———. 1961. *Ap. J.* **134,** 96.
KUNDU, M. R., and FIROR, J. W. 1961. *Ap. J.* **134,** 389.
KUSCH, H. J. 1958. *Zeits. f. Ap.* **45,** 1.

LABORDE, G. 1961. *Ann. d'Ap.* **24,** 89.
LABS, D. 1951. *Zeits. f. Ap.* **29,** 199.
———. 1954. *Zeits. f. Ap.* **34,** 173.
———. 1957. *Heidelberg Symposium über Probleme der Spectrophotometrie.* Springer-Verlag, Heidelberg.
LABS, D., and NECKEL, H. 1962. *Ap. J.* **135,** 969.
LARMORE, L. 1953. *Ap. J.* **118,** 436.
LEDOUX, R., SCHWARZSCHILD, M., and SPIEGEL, E. A. 1961. *Ap. J.* **133,** 184.
LEIGHTON, R. B. 1957. *Publ. Ast. Soc. Pac.* **69,** 497.
———. 1959. *Ap. J.* **130,** 366.
LEIGHTON, R. B., NOYES, R. W., and SIMON, G. W. 1962. *Ap. J.* **135,** 474.
LILLER, W., and LEWIS, E. M. 1952. *Ap. J.* **116,** 428.
LINDHOLM, E. 1946. *Ark. Mat. Astron. Fys.* **32A,** No. 17.
LINDSAY, J. 1963. Quoted by H. Newell in *Science* **139,** 469.
LIPPINCOTT, S. L. 1955. *Ann. d'Ap.* **18,** 113.
———. 1957. *Smithsonian Contributions to Astrophysics* **2,** 15.

MACDONALD, N. J., and ROBERTS, W. O. 1960. *J. Geophys. Res.* **65.**
MACRIS, C. 1953. *Ann. d'Ap.* **16,** 19.
MADLOW, W. 1962. *Zeits. f. Ap.* **55,** 29.
MAKITA, M., and MORIMOTO, M. 1960. *Publ. Astron. Soc. Japan* **12,** 63.
MALKUS, W. V. R. 1954. *Proc. Roy. Soc.* **225,** 185.
MARTINI, A., and MASANI, A. 1957. *Milano Merate Obs. Contr.*, No. 119.
MASTRUP, F., and WIESE, W. 1958. *Zeits. f. Ap.* **44,** 259.
MATSUSHIMA, S. 1955. *Ap. J. Suppl.* **1,** No. 12, 479.
MATTIG, W. 1953. *Zeits. f. Ap.* **31,** 273.
———. 1958. *Zeits. f. Ap.* **44,** 280.
MCDONALD, J. K. 1953. *Publ. Dom. Ap. Obs.* **9,** 269.
———. 1954. *Publ. Dom. Ap. Obs.* **10,** No. 4.
MCILWAIN, C. E. 1960. *IGY Rocket Report No. 6*, 157.
MCLEAN, E. A., FANEUFF, C. E., KOLB, A. C., and GRIEM, H. R. 1960. *Physics of Fluids* **3,** 843.
MCMATH, R. R., MOHLER, O. C., and DODSON, H. 1960. *Proc. Natl. Acad. Sci.* **46,** 165.
MELBOURNE, W. G. 1960. *Ap. J.* **132,** 101.
MENZEL, D. H. 1936. *Ap. J.* **84,** 462.
———. 1947. *Ap. J.* **105,** 126.
MENZEL, D. H., and EVANS, D. S., 1953. *Convegno Volta* **11,** 119.

MENZEL, D. H., and PEKERIS, C. L. 1936. *M. N.* **96,** 77.
MENZEL, D. H., and SEN, H. K. 1949. *Ap. J.* **110,** 1.
———. 1951. *Ap. J.* **113,** 482.
MERRILL, P. W. 1952. *Science* **115,** 484.
———. 1956. *Publ. Astron. Soc. Pac.* **68,** 70, 162, 356.
MICHARD, R. 1951. *Ann. d'Ap.* **14,** 101.
———. 1953. *Ann. d'Ap.* **16,** 217.
———. 1956. *Ann. d'Ap.* **19,** 229.
———. 1959. *Ann. d'Ap.* **22,** 547, 887.
MILFORD, N. 1950. *Ann. d'Ap.* **13,** 251.
MINNAERT, M., HOVEN VAN GENDEREN, E. V. D., and VAN DIGGELEN, J. 1949. *Bull. Astr. Instit. Netherlands* **11,** 55.
MITCHELL, W. 1959. *Ap. J.* **129.**
MIYAMOTO, S. 1953. *Zeits. f. Ap.* **31,** 282.
———. 1954a. *Publ. Astr. Soc. Japan* **5,** 142; **6,** 140, 150, 207.
———. 1954b. *Zeits. f. Ap.* **35,** 145.
MOLL, W. J. H., BURGER, H. C., and VAN DER BILT, J. 1925. *Bull. Astr. Instit. Netherlands* **3,** 83.
MORGAN, W., and KEENAN, P. C. 1951. *Astrophysics.* J. A. Hynek (ed). McGraw-Hill Book Co., Inc., New York.
MORIYAMA, F. 1961. *Annals Tokyo Observatory* **7,** 132.
MORTON, D. C., and WIDING, K. G. 1961. *Ap. J.* **133,** 596.
MOZER, B., and BARANGER, M. 1960. *Phys. Rev.* **118,** 626.
MULDERS, G. F. W. 1935. *Zeits. f. Ap.* **11,** 132.
MÜNCH, G. 1945. *Ap. J.* **102,** 385.
———. 1946. *Ap. J.* **104,** 87.
———. 1949. *Ap. J.* **109,** 275.
———. 1958. *Ap. J.* **127,** 644.

NECKEL, H. 1958. *Zeits f. Ap.* **44,** 153, 160.
NEWKIRK, G. 1957. *Ap. J.* **125,** 571.
———. 1959. *Paris Symposium on Radio Astronomy.* R. Bracewell (ed.). Stanford University Press, Stanford, Calif.
———. 1961. *Ap. J.* **133,** 983.
NEWTON, H. W., and NUNN, M. L. 1952. *M. N.* **111,** 413.
NEY, E. P., HUCH, W. F., KELLOGG, P. J., STEIN, W., and GILLETT, F. 1961. *Ap. J.* **133,** 616.

OBAYASHI, T., and HAKURA, Y. 1960. *J. Geophys. Res.* **65,** 3143.
OHMURA, T., and OHMURA, H. 1960. *Ap. J.* **131,** 8.
OKE, J. B. 1960. *Ap. J.* **131,** 358.
ORRALL, F. Q., and ZIRKER, J. B. 1961. *Ap. J.* **134,** 63, 72.
OSAWA, K. 1956. *Ap. J.* **123,** 513.
OSTER, L. 1957. *Zeits. f. Ap.* **42,** 228.
OSTROVSKY, YU. I., and PENKIN, N. P. 1957. *Optika I Spektr.* **3,** 193, 391.

PAGEL, B. E. J. 1959. *M. N.* **119,** 609.
———. 1960. *Ap. J.* **132,** 790.
PANNEKOEK, A. 1935. *Publ. Astr. Instit. Netherlands.*
PAN PUH. 1939. *Ann. Obs. Meudon* **8,** Fasc. 4.
PARENAGO, P. P., and MASSEVITCH, A. G. 1950. *A. J. USSR* **27,** 137.
PARKER, E. N. 1955. **121,** 491; **122,** 293.
———. 1960. *Ap. J.* **132,** 821.
———. 1961a. *Ap. J.* **133,** 1014.
———. 1961b. *Ap. J.* **134,** 20.
PARKER, J. M. 1955. *Ap. J.* **121,** 131.
PECKER, CHARLOTTE. 1953. *Ann. d'Ap.* **16,** 321.
PECKER, J. C. 1950. *Ann. d'Ap.* **13,** 294, 319, 433.

———. 1960. *Royal Society of Liége*. S. Rosseland (ed.). "Modèles d'Étoiles et Évolution Stellaire," vol. 16, 343.

Pecker, J. C., and van Regemorter, H. 1955. *Contr. Instit. Astr. (Paris) No. 197.*

Pecker, J. C., and Schatzman, E. 1959. *Astrophysique Générale*. Masson and Co., Paris.

Petersen, L. E., and Winckler, J. R. 1959. *J. Geophys. Res.* **64,** 697.

Petrie, R. M. 1950. *Publ. Dom. Ap. Obs. Victoria* **8,** 341.

———. 1953. *Publ. Dom. Ap. Obs. Victoria* **9,** 251.

———. 1956. *Publ. Dom. Ap. Obs. Victoria* **10,** 287.

———. 1956. *Vistas in Astronomy*. A. Beer (ed.). Pergamon Press, Ltd., Oxford, England, vol. II, p. 1346.

Pettit, E. 1932. *Ap. J.* **75,** 185.

———. 1940. *Ap. J.* **91,** 159.

Pettit, E., and Nicholson, S. B. 1928. *Ap. J.* **68,** 279.

Peyturaux, R. 1955. *Ann. d'Ap.* **18,** 34.

Phillips, J. G. 1952. *Ap. J.* **115,** 183.

Piddington, J. H. 1950. *Proc. Roy. Soc. London A,* **203,** 417.

Pierce, A. K. 1954. *Ap. J.* **120,** 221.

Plaskett, H. H. 1936. *M. N.* **96,** 402.

———. 1941. *M. N.* **101,** 1.

———. 1954. *M. N.* **114,** 251.

———. 1955. *M. N.* **115,** 256.

———. 1956. *Vistas in Astronomy*. A. Beer (ed.). Pergamon Press, Ltd., Oxford, England.

———. 1959. **119,** 197.

Plaut, L. 1953. *Publ. Astr. Laboratory Groningen No. 55.*

Pottasch, S. R. 1960. *Ap. J.* **131,** 68.

Pottasch, S. R., and Thomas, R. N. 1959. *Ap. J.* **130,** 941.

———. 1960. *Ap. J.* **132,** 195.

Priester, W. 1953. *Zeits. f. Ap.* **32,** 300.

———. 1955. *Zeits. f. Ap.* **36,** 230.

Proisy, P. 1958. *Ann. d'Ap.* **21,** 151.

Przybylski, A. 1953. *M. N.* **113,** 683.

Purcell, J. D., Packer, D. M., and Tousey, R. 1959. *Nature* **184,** 8.

———. 1960. *IGY Rocket Report No. 6*, p. 191.

Purcell, J. D., and Tousey, R. 1960. *J. Geophys. Res.* **65,** 187.

Raudenbusch, H. 1938. *Astr. Nach.* **226,** 301.

Rayleigh, J. W. S. 1916. *Phil. Mag. Series 6* **32,** 529.

Redman, R. O. 1943. *M. N.* **103,** 174.

Redman, R. O., and Suemoto, Z. 1954. *M. N.* **114,** 524.

Reichel, M. 1954. *Zeits. f. Ap.* **33,** 79.

Richardson, R. S. 1933. *Ap. J.* **78,** 359.

Rogerson, J. B. 1961. *Ap. J.* **134,** 331.

Rosch, J. 1959. *Ann. d'Ap.* **22,** 571, 584.

Rouse, C. A. 1961. *Ap. J.* **134,** 435.

———. 1962. *Ap. J.* **135,** 599.

Rozis-Saulgeot, A. M. 1959. *Ann. d'Ap.* **22,** 177.

———. 1960. *Ann. d'Ap.* **23,** 204.

Rudkjøbing, M. 1947. *Copenhagen Medd. No. 145.*

Rush, J. H., and Roberts, W. O. 1954. *Austral. Journ. Phys.* **7,** 230.

Russell, H. N. 1934. *Ap. J.* **79,** 317.

———. 1948. *Harvard Observatory Centennial Symposia*. Harvard University Press, Cambridge, Massachusetts.

Russell, H. N., and Moore, C. E. 1940. *The Masses of the Stars*. University of Chicago Press, Chicago.

Saito, S. 1954. *Contr. Inst. Ap. Univ. Kyoto,* No. 48.

———. 1956. *Contr. Inst. Ap. Univ. Kyoto,* No. 69.
Saito, Ueno S., and Jugaku, J. 1954. *Kyoto Ap. Contr.*, No. 43.
Sandage, A. 1957. *Ap. J.* **125,** 435.
Sandage, A. R., and Eggen, O. J. 1959. *M. N.* **119,** 278.
Savedoff, M. P. 1952. *Ap. J.* **115,** 509.
Schatzman, E. 1949. *Ann. d'Ap.* **13,** 203.
Schluter, A., and St. Temesvary, S. 1958. *Electromagnetic Phenomena in Cosmical Physics.* I. A. U. Symposium No. 6.
Schroter, E. H. 1957. *Zeits. f. Ap.* **41,** 141.
———. 1958. *Zeits. f. Ap.* **45,** 68.
Schwartz, S. B., and Zirin, H. 1959. *Ap. J.* **130,** 384.
Schwarzschild, M. 1948. *Ap. J.* **107,** 1.
———. 1958. *Structure and Evolution of Stars.* Princeton University Press, Princeton.
———. 1959. *Ap. J.* **130,** 345.
Seaton, M. J. 1958. *Rev. Mod. Phys.* **30,** 989, 992.
———. 1960. *Progress in Physics* **23,** 313. Physical Society, London.
Severny, A. B. 1954. *Isv. Crimean Ast. Obs.* **12,** 33.
———. 1959. *Isv. Crimean Ast. Obs.* **21,** 131.
———. 1960. *Isv. Crimean Ast. Obs.* **22,** 67; **23,** 311.
Shklovsky, I. S., and Kononovich, E. V. 1958. *Russ. Ast. J.* **35,** 17.
Siedentopf, H. 1932. *Astron. Nach.* **224,** 273.
———. 1933. *Astron. Nach.* **247,** 297; **249,** 53.
———. 1935. *Astron. Nach.* **255,** 157.
Skumanich, A. 1955. *Ap. J.* **121,** 404, 408.
Slettebak, A. 1956. *Ap. J.* **124,** 173.
Slettebak, A., and Howard, R. F. 1955. *Ap. J.* **121,** 102.
Sobolev, V. A. 1949. *Russ. Ast. J.* **26,** 129.
———. 1951. *Russ. Ast. J.* **28,** 355.
———. 1954. *Russ. Ast. J.* **31,** 231.
Spiegel, E. 1960. *Ap. J.* **132,** 716.
Spitzer, L. 1939. *Phys. Rev.* **55,** 699; **56,** 39.
———. 1940. *Phys. Rev.* **58,** 348.
Stair, R. 1951. *J. Res. Nat. Bur. Stand.* **46,** 353.
———. 1952. *J. Res. Nat. Bur. Stand.* **49,** 227.
———. 1954. *J. Res. Nat. Bur. Stand.* **53,** 113.
Stawikowski, A. 1959. *Torun Obs. Bull. No. 22.*
Stebbins, J., and Kron, G. 1957. *Ap. J.* **126,** 266.
Stecher, T. 1962. *Ap. J.* **136,** 686.
Stecher, T., and Milligan, J. 1962. *Ap. J.* **136,** 1.
Stepanov, V. E. 1958. *Publ. Crimean Obs.* **18,** 136.
———. 1960. *Publ. Crimean Obs.* **19,** 20.
Stone, P. H., and Gaustad, J. E. 1961. *Ap. J.* **134,** 456.
Strand, K. A. 1954. *A. J.* **59,** 61.
———. 1957. *Jour. Roy. Astr. Soc. Canada* **51,** 46.
Strömgren, B. 1937. *Handbuch der Experimental Physik.* W. Wien and F. Harms (eds.). Akademische Verlagsgesellschaft, vol. 26, p. 231.
———. 1958. *Observatory* **78.**
Strömgren, B., and Gyldenkerne, K. 1955. *Ap. J.* **128,** 38, 43.
Suemoto, Z. 1949. *Publ. Ast. Soc. Japan* **1,** 78.
———. 1957. *M. N.* **117,** 2.
Svestka, Z. 1960. *Bull. Astron. Inst. Czech.* **11,** 167.
Swarup, G., Stone, P. H., and Maxwell, A. 1960. *Ap. J.* **131,** 725.
Sykes, J. B. 1953. *M. N.* **113,** 198.

Takakura, T. 1962. *Tokyo Astr. Obs.* Reprint 224.
Takakura, T., and Ono, M. 1962. *Tokyo Astr. Obs.* Reprint 225.
Ten Bruggencate, P. 1953. *Convegno Volta* **11,** 163.
Ten Bruggencate, P., Gollnow, H., and Jager, F. W. 1950. *Zeits. f. Ap.* **27,** 223.

THIESSEN, G. 1955. *Zeits. f. Ap.* **35,** 237.
THOMAS, D. V. 1958. *M. N.* **118,** 458.
THOMAS, R. N. 1948. *Ap. J.* **108,** 142.
———. 1949. *Ap. J.* **109,** 480, 500.
———. 1957. *Ap. J.* **125,** 260.
THOMAS, R. N., and ZIRKER, J. B. 1961. *Ap. J.* **133,** 588.
TREANOR, P. J. 1957. *M. N.* **117,** 22.
———. 1960. *M. N.* **120,** 412; **121,** 503.

UBEROI, M. S. 1955. *Ap. J.* **122,** 466.
UENO, S., SAITO, S., and JUGAKU, J. 1954. *Contr. Astrophys. Inst. Kyoto No. 43.*
UNDERHILL, A. 1951. *Publ. Dom. Ap. Obs. Victoria* **8,** 337.
———. 1957. *Publ. Dom. Ap. Obs. Victoria* **10,** 357.
———. 1962. *Publ. Dom. Ap. Obs. Victoria* **11.**
———. 1962. *Quarterly Jour. Roy. Astron. Soc.* **3,** 7.
UNDERHILL, A., and WADDELL, J. H. 1959. *Natl. Bureau Standards Circular No. 603.*
UNNO, W. 1956. *Publ. Ast. Soc. Japan* **8,** 108.
———. 1957. *Ap. J.* **126,** 259.
———. 1959a. *Ann. d'Ap.* **22,** 430.
———. 1959b. *Ap. J.* **129,** 375, 388.
UNSÖLD, A. 1955. *Physik der Sternatmosphären.* Springer-Verlag, Berlin.

VAN ALLEN, J. A. 1960. *Science in Space.* L. V. Berkner (ed.). Natl. Acad. Sciences, Washington, D. C.
VAN DE HULST, H. C. 1950. *B. A. N.* **11,** 135, 410.
VAN DE KAMP, P. 1954. *A. J.* **59,** 447.
VAN REGEMORTER, H. 1959. *Ann. d'Ap.* **22,** 249, 341, 363, 681.
VARDYA, M. S. 1960. *Ap. J. Suppl.* **4,** 281.
VIOLETT, T., and RENSE, W. A. 1959. *Ap. J.* **130,** 954.
VITENSE, E. 1953. *Zeits. f. Ap.* **32,** 135.
VOIGT, H. H. 1950. *Zeits. f. Ap.* **27,** 82.
———. 1956. *Zeits. f. Ap.* **40,** 157.
———. 1959. *Zeits. f. Ap.* **47,** 144.
VON KLUBER, H. 1958. *M. N.* **118,** 201.
———. 1960. *Vistas in Astronomy.* A. Beers (ed.). Pergamon Press, Ltd., Oxford, England. Vol. 3, p. 47.

WADDELL. J. W. 1958. *Ap. J.* **127,** 284.
WALDMEIER, M. 1954. *Zeits. f. Ap.* **35,** 1, 95.
———. 1960. *Zeits. f. Ap.* **50,** 225.
WALDMEIER, M., and BACHMANN, H. *Zeits. f. Ap.* **47,** 81.
WALEN, C. 1949. *On the Vibratory Rotation of the Sun.* H. Lindstahl's Bokhandel, Stockholm.
WALRAVEN, T. and WALRAVEN, J. H. 1960. *B. A. N.* **15,** 67.
WARWICK, J. 1951. *Proc. Nat. Acad. Sci.* **37,** 196.
———. 1955. *Zeits. f. Ap.* **35,** 245.
———. 1957. *Ann. d'Ap.* **20,** 165.
WASIUTYNSKI, J. 1958. *Ann. d'Ap.* **21,** 119.
WEIDEMANN, V. 1960. *Ap. J.* **131,** 641.
WESTERLUND, B. 1951–53. *Uppsala Astr. Obs. Annales* 3, 6, 8, and 10.
WEYMANN, R. 1960. *Ap. J.* **132,** 452.
WIDING, K. G., and MORTON, D. C. 1961. *Ap. J.* **133,** 596.
WILDT, R. 1939. *Ap. J.* **90,** 619.
———. 1947. *Relations entre les Phénomènes Solaires et Géophysique.* Centre National de la Recherche Scientifique, Paris, p. 7.
———. 1956. *Ap. J.* **123,** 107.
WILLIAMS, R. 1938. *Publ. Obs. Michigan* **7,** 93.
———. 1939. *Publ. Obs. Michigan* **7,** 147, 159.

WILSON, O. C. 1959. *Ap. J.* **130,** 497.
WOLTJER, L. 1954. *B. A. N.* **12,** 165.
WOODBRIDGE, D. D., MACDONALD, N. J., and POHRTE, T. W. 1959. *J. Geophys. Res.* **64,** 331.
WOOLLEY, R. v. d. R. 1934. *M. N.* **94,** 631.
———. 1938. *M. N.* **98,** 624.
———. 1941. *M. N.* **101,** 52.
———. 1943. *M. N.* **103,** 193.
WOOLLEY, R. v. d. R., and ALLEN, C. W. 1948. *M. N.* **108,** 292.
———. 1950. *M. N.* **110,** 358.
WOOLLEY, R. v. d. R., and GASCOIGNE, S. C. B. 1948. *M. N.* **108,** 491.
WRUBEL, M. 1949. *Ap. J.* **109,** 71.
———. 1950. *Ap. J.* **111,** 157.
———. 1954. *Ap. J.* **119,** 51.
———. 1960. Quoted by L. H. Aller in *Stellar Atmospheres,* vol. 6 of *Stars and Stellar Systems.* J. L. Greenstein (ed.). University of Chicago Press, Chicago.
WYLLER, A. A. 1957. *Ap. J.* **125,** 177.
———. 1958, *Ap. J.* **127,** 506.
———. 1961. *Ap. J.* **134,** 805.

YOSS, K. 1961. *Ap. J.* **134,** 809.

ZANSTRA, H. 1941. *M. N.* **101,** 250, 273.
———. 1946. *M. N.* **106,** 225.
———. 1955. *Vistas in Astronomy.* A. Beer (ed.). Pergamon Press, Ltd., Oxford, England, vol. I, p. 256.
ZIRIN, H. 1956. *Ap. J.* **124,** 451.
———. 1962. *Ap. J.* **135,** 521.
ZIRIN, H., and TANDBERG-HANSSEN, E. 1960. *Ap. J.* **131,** 717.
ZIRKER, J. B. 1959. *Ap. J.* **129,** 424.
ZWAAN, C. 1960. *B. A. N.* **14,** 288.

INDEX OF SYMBOLS

Since an effort has been made to retain, as far as possible, the conventional notations of physical and astrophysical literature, a given letter must often be used with more than one meaning. For each symbol, where appropriate, we give the article in which it is defined, the meaning specified, and other references in the text. If a symbol is defined and used on only one page in the text, it is usually not listed here.

W_ν	7–3	Equivalent width of a line in frequency units, 301, 408
W_λ	7–3	Equivalent width of a line expressed in wavelength units, 302, 361, 363, 365
x	5–1	Geometrical depth in a stellar atmosphere, 200
X_n	4–16	hRZ^2/n^2kT, 183, 184, 185, 186
y	1–5	Distance modulus of a star, 21
y	5–4	$\log_{10}\tau(\lambda_0)$, 210, 358
Y	5–5	See Eq. 5–53, 219
Z	2–3	Atomic number, 37, 540
$Z(y)$	8–6	See Eq. 8–80, 364
α	2–13	Fine structure constant, 77
α	3–6	Most probable velocity of a molecule, 108, 109, 110, 111, 504
α	4–8	Susceptibility of medium, 163 only
α	7–8	Line asymmetry parameter for helium, 335, 336, 337 only
α_ν	4–9	Atomic absorption coefficient, 166 et seq.
$\alpha_n(\nu)$	4–16	Continuous atomic absorption coefficient for nth level, 182, 187
α_0	7–6	Fictitious absorption coefficient at line center, 323, 373
$\alpha_\lambda{}^0(H^-)$	5–3	Absorption coefficient of negative hydrogen ion, computed per hydrogen atom at an electron pressure of one dyne cm^{-2}, 207, 210, 211
$\alpha_\lambda(H)$	5–3	Absorption coefficient of atomic hydrogen per atom, 207, 210
β	4–5	Equivalent beam area of an antenna (Chapter 4 only), 154
β	7–5	Line shift due to collisions, 318
β	7–7	Ratio of electrostatic field E to mean field E_0, 320
β	9–3	Temperature gradient, 451, 453
γ	3–2	Ratio of specific heats, c_p/c_v, 96, 450, 453, 454
γ, γ_c	4–8	Classical damping constant, 161, 164, 166, 167, 168, 339
γ_6	7–5	van der Waals iteration damping constant, 321
γ_n	7–5	Damping constant for self-broadening, 322
Γ	2–7	Spin-orbit interaction energy, 53
Γ	7–5	Gamma function, 315, 316
Γ_n, Γ_{rad}	4–12	Quantum mechanical damping constant, 172, 173, 176, 178, 302, 324, 339, 355, 379
Γ_{coll}	7–5	Collisional damping constant, 317, 318, 320

INDEX OF NAMES

INDEX OF SUBJECTS